Global Perspectives on Tropical Cyclones

From Science to Mitigation

World Scientific Series on Asia-Pacific Weather and Climate

The **World Scientific Series on Meteorology of East Asia (WSS-MEA)** is being expanded into a new book series entitled **World Scientific Series on Asia-Pacific Weather and Climate (WSS-APWC)**. This expansion reflects the increasing priority being given by World Scientific Publishing Co. to the weather and climate variability of all time scales over the entire Asia and Pacific region and surrounding areas, since they are often felt globally and their impacts on society are often enormous. The APWC series will also continue the collaboration with the World Meteorological Organization Working Group on Tropical Meteorology Research, now a component of the World Weather Research Programme, and other international weather and climate programs in publication projects.

The Editorial Board of the APWC series welcomes book or program/project proposals relevant to weather and climate or related fields, including interdisciplinary topics. Interested scientists please contact either of the two Co-Chief Editors:

Professor Chih-Pei CHANG (cpchang@nps.edu) who also serves as the chair of the Editorial Board, and Professor Congbin FU (fcb@mail.tea.ac.cn).

Vol. 1: East Asia and Western Pacific Meteorology and Climate: Selected Papers of the Fourth Conference
edited by C.-P. Chang, Guoxiong Wu, Ben Jou & C. Y. Lam

Vol. 2: East Asian Monsoon
edited by C.-P. Chang

Vol. 3: Observation, Theory, and Modeling of Atmospheric Variability Selected Papers of Nanjing Institute of Meteorology Alumni in Commemoration of Professor Jijia Zhang
edited by Xun Zhu, Xiaofan Li, Ming Cai, Shuntai Zhou, Yuejian Zhu, Fei-Fei Jin, Xiaolei Zou & Minghua Zhang

Vol. 4: Global Perspectives on Tropical Cyclones: From Science to Mitigation
edited by Johnny C. L. Chan & Jeffrey D. Kepert

Forthcoming

Vol. 5: The Global Monsoon System: Research and Forecast (2nd ed.)
edited by C.-P. Chang, Yihui Ding, R. H. Johnson, G. N.-C. Lau, Bing Wang & Tetsuzo Yasunari

WORLD SCIENTIFIC SERIES ON ASIA-PACIFIC WEATHER AND CLIMATE

Vol. 4

Global Perspectives on Tropical Cyclones

From Science to Mitigation

edited by

Johnny C. L. Chan

City University of Hong Kong, Hong Kong, China

Jeffrey D. Kepert

Centre for Australian Weather and Climate Research, Australia

World Scientific

NEW JERSEY • LONDON • SINGAPORE • BEIJING • SHANGHAI • HONG KONG • TAIPEI • CHENNAI

Published by

World Scientific Publishing Co. Pte. Ltd.
5 Toh Tuck Link, Singapore 596224
USA office: 27 Warren Street, Suite 401-402, Hackensack, NJ 07601
UK office: 57 Shelton Street, Covent Garden, London WC2H 9HE

British Library Cataloguing-in-Publication Data
A catalogue record for this book is available from the British Library.

World Scientific Series on Asia-Pacific Weather and Climate — Vol. 4
GLOBAL PERSPECTIVES ON TROPICAL CYCLONES
From Science to Mitigation

ISBN-13 978-981-4293-47-1
ISBN-10 981-4293-47-4

Printed in Singapore by Mainland Press Pte Ltd.

Preface

This book is a completely rewritten, updated and expanded new edition of *Global Perspectives on Tropical Cyclones*, which in turn was a revision of *A Global View of Tropical Cyclones*. Like its predecessors, this volume arises from a quadrennial event, The International Workshop on Tropical Cyclones (IWTC) of the World Meteorological Organisation, in this case the sixth, held in San Jose, Costa Rica, during December 2006. Unlike the previous volumes, this is published by a specialist publisher, World Scientific Publishing Company, in the hope that the material will have a much wider circulation.

In his prologue to the first volume, Prof. Bill Gray lamented the growing specialisation of the field, and the difficulty of remaining an expert in all topics related to tropical cyclone observations, forecasting, research and warning. Inevitably, this trend has continued apace — in fact, accelerated — in the ensuing two decades. Enormous progress in understanding various aspects of tropical cyclones has been achieved, which results in a substantial expansion in the scope and topics in this current volume. The almost-exponential growth in satellite observations and the application of ever-increasing computer power are particularly of note, and their influence pervades the entire volume. Theoretical advances have followed from sophisticated mathematics, more powerful computations, and been informed and tested by increasingly comprehensive and accurate observations. Substantial improvements in forecast accuracy have yielded significant benefits, yet mitigation remains a challenging problem with growing human vulnerability and expectations. The increasing interest in how tropical cyclone activity might change under global warming leads to the introduction in this volume of chapters on tropical cyclones and climate, ranging from climate variability on various time scales, through seasonal forecasting, to future projections.

As with the previous volumes in the series, chapter authors were invited based on the list of topic chairs and research rapporteurs who attended the IWTC, supplemented by other experts in the particular topics. We are grateful to all the authors who have devoted considerable time and effort to extending and refining the material presented at the IWTC into a form that is more permanent and comprehensive. We sincerely thank the many reviewers who painstakingly read, commented on, and made an immense contribution to improving each chapter of the book. We also thank World Scientific Publishing Company for taking on the publishing responsibility. Finally, we thank the World Meteorological Organisation for their support of the International Workshops on Tropical Cyclones, without which this book could not exist. We believe that the book represents the most comprehensive summary of the current state of the science on tropical cyclones, and trust that it will prove useful to all with a scientific interest in tropical cyclones, be they new to the field or experienced professionals.

Johnny Chan
Jeff Kepert
7 December 2009

Contents

Preface v

Part I. Theory of Tropical Cyclones

Chapter 1
Tropical Cyclone Structure and Dynamics 3
Jeffrey D. Kepert

Chapter 2
Tropical Cyclone Formation 55
Kevin J. Tory and William M. Frank

Chapter 3
Air-Sea Interactions in Tropical Cyclones 93
Lynn K. Shay

Chapter 4
Movement of Tropical Cyclones 133
Johnny C. L. Chan

Chapter 5
The Extratropical Transition of Tropical Cyclones: Structural Characteristics, Downstream Impacts, and Forecast Challenges 149
Patrick A. Harr

Part II. Observations of Tropical Cyclones

Chapter 6
Observing and Analyzing the Near-Surface Wind Field in Tropical Cyclones 177
Mark D. Powell

Chapter 7
Satellite Observations of Tropical Cyclones 201
Christopher Velden and Jeffrey Hawkins

Chapter 8
Aircraft Observations of Tropical Cyclones 227
Sim D. Aberson, Joseph Cione, Chun-Chieh Wu, Michael M. Bell, Jeffrey Halverson, Chris Fogarty and Martin Weissmann

Part III. Climate Variations of Tropical Cyclone Activity

Chapter 9
Tropical Cyclones and Climate Change: A Review 243
Thomas Knutson, Chris Landsea and Kerry Emanuel

Part IV. Forecasting of Tropical Cyclones

Chapter 10
Track and Structure Forecasts of Tropical Cyclones 287
Julian Heming and Jim Goerss

Chapter 11
The Influence of Natural Climate Variability on Tropical Cyclones, and Seasonal Forecasts of Tropical Cyclone Activity 325
Suzana J. Camargo, Adam H. Sobel, Anthony G. Barnston and Philip J. Klotzbach

Part V. Hydrological Aspects of Tropical Cyclones

Chapter 12
Storm Surge Modeling and Applications in Coastal Areas 363
Shishir K. Dube, Tad S. Murty, Jesse C. Feyen, Reggina Cabrera, Bruce A. Harper, Jerad D. Bales and Saud Amer

Part VI. Societal Impacts of Tropical Cyclones

Chapter 13
Disaster Mitigation and Societal Impacts 409
David King, Jim Davidson and Linda Anderson-Berry

PART I

Theory of Tropical Cyclones

Chapter 1

Tropical Cyclone Structure and Dynamics

Jeffrey D. Kepert

Centre for Australian Weather and Climate Research,
Bureau of Meteorology, Melbourne Vic 3000, Australia
J.Kepert@bom.gov.au

Tropical cyclones are intense warm-cored cyclonic vortices that form over warm tropical oceans. Tropical cyclones can vary significantly from one to another, and from day to day, in intensity, size, boundary layer structure, spiral banding, eye structure and degree of symmetry. Significant progress has been made in understanding the factors that determine these differences and govern the storms' evolution. In particular, major advances have occurred in our understanding of the tropical cyclone boundary layer, of the dynamics and role of spiral bands, and of the interaction between the cyclone and its environment. This chapter reviews tropical cyclone structure and structure change, with an emphasis on these advances in knowledge since the last volume in this series.

1. Introduction

Tropical cyclones are intense atmospheric vortices that form over the warm tropical oceans. The inner part of the storm becomes nearly axisymmetric as the storm reaches maturity, and its strongest winds surround a relatively calm eye, whose diameter is typically in the range of 20 to 100 km. The pressure deficit in the eye can reach 10% of the ambient pressure, and hydrostatic calculations show that such a marked deficit requires that the cyclone's warm core extend through much of the troposphere (Haurwitz, 1935). The eye is partly or wholly surrounded by a ring of deep convective cloud that rises from just above the sea surface to the tropopause, sloping outwards as it rises, and which contains the strongest winds. In intense storms, this eyewall takes the appearance of a giant amphitheatre. The symmetric component of the cyclonic winds is nearly in gradient wind balance except in the surface frictional boundary layer and the outflow layer. The warm core aloft and gradient balance imply that the strength of the wind decreases with height above the boundary layer.

Superimposed on the dominant primary axisymmetric circulation are a weaker symmetric secondary circulation and various asymmetries. The secondary circulation consists of inflow concentrated in the boundary layer, upflow in the eyewall and spiral rainbands, and outflow in a thin layer beneath the tropopause. Surface friction and latent heat release are responsible for the secondary circulation, while interactions with surrounding weather systems also contribute at times. Air-sea exchange processes in the boundary layer weaken the storm due to momentum transfer to the sea, contribute to the secondary circulation, and provide energy through enthalpy fluxes from the sea (Ooyama, 1969). Angular momentum imported by the friction-induced component of the secondary circulation is largely or entirely destroyed by friction, but the diabatically forced component of the secondary circulation can provide a net import of angular momentum to spin up the storm. The latent heat release within the eyewall may also cause subsidence within the eye, creating the warm core and also the characteristic cloud-free circle near the middle of the central dense overcast on satellite images of intense storms. The upper tropospheric outflow layer is markedly less symmetric than the rest of the storm, and may feature anticyclonic absolute vorticity and hence inertial instability. The tendency for stronger environmental winds and potential vorticity gradients in the upper troposphere contributes to these outflow asymmetries.

Asymmetries of the inner core can consist of transient propagating features due to instabilities of the basic flow, or of longer-lived quasi-stationary features resulting from interactions between the cyclone and its environment. Such asymmetries usually have a signal in the kinematic, thermodynamic and cloud fields. The most intense storms are usually more symmetric. Outside the core region, the storm grades smoothly into its environment. As the stabilising effect of the strong rotation in the inner core recedes, the strengths of the symmetric and asymmetric flow become comparable.

The lifecycle of tropical cyclones ranges from a day or two up to several weeks. In the early stages, the primary circulation is weak and is usually slow to build strength and symmetry. This chapter mostly applies to storms with maximum surface winds exceeding 15–20 $\mathrm{m\,s^{-1}}$, while the earlier phase is considered in Chapter 2. The aim here is to describe the structure of a mature storm in the context of its dynamics and thermodynamics. This chapter has a strong bias towards recent progress on understanding mature structure, whilst endeavouring to include sufficient of the earlier material to be reasonably self-contained. In particular, there is a deliberate focus on the important recent advances in understanding the boundary layer (section 5), on the dynamics and consequences of waves and instabilities (section 7), and on the response of the storm to its environmental (section 8). The fundamentally important secondary circulation and its role in the cyclones intensification is described in section 3, leading to theories of potential intensity in section 4. Parametric wind models are updated in section 6. Predecessors to the present chapter (Holland, 1987; Willoughby, 1995) are warmly recommended to those readers wanting more detail on other matters. Similarly, readers with a particular interest in air-sea interaction, the observational diagnosis of structure and structure change, or in forecast techniques, are referred to the appropriate chapters in this volume.

2. Equations and Balance Considerations

In the nearly axisymmetric cyclone core, storm-centred cylindrical coordinates are appropriate and the equations of motion may be written:

$$\frac{du}{dt} - \left(f + \frac{v}{r}\right) v = -\frac{1}{\rho}\frac{\partial p}{\partial r} + \frac{\partial \overline{u'w'}}{\partial z} \tag{1}$$

$$\frac{dv}{dt} + \left(f + \frac{v}{r}\right) u = -\frac{1}{r\rho}\frac{\partial p}{\partial \lambda} + \frac{\partial \overline{v'w'}}{\partial z} \tag{2}$$

$$\frac{dw}{dt} = -\frac{1}{\rho}\frac{\partial p}{\partial z} - g \tag{3}$$

$$\frac{\partial \rho}{\partial t} + \frac{1}{r}\frac{\partial \rho r u}{\partial r} + \frac{1}{r}\frac{\partial \rho v}{\partial \lambda} + \frac{\partial \rho w}{\partial z} = 0 \tag{4}$$

$$\frac{d\theta}{dt} = \dot{\theta} + \frac{\partial \overline{\theta' w'}}{\partial z} \tag{5}$$

$$\frac{dq}{\partial t} = E + \frac{\partial \overline{q'w'}}{\partial z} \tag{6}$$

$$\theta = T\left(\frac{p_0}{p}\right)^{R/c_p} \tag{7}$$

$$p = \rho R_d T_v \tag{8}$$

$$T_v = T[1 + q(1/\epsilon - 1)] \approx T(1 + 0.608q) \tag{9}$$

where the meanings of all symbols are given in Table 2. In these equations, some small terms have been omitted, including the horizontal components of the turbulent flux and the vertical components of the Coriolis acceleration. While their neglect is well justified in most of the atmosphere, it is less supported but conventional to do so in the tropical cyclone core.

These equations may be applied either in a coordinate system fixed to the earth's surface at the instantaneous cyclone location, or in one moving with the storm. In the latter case, the storm movement vector should strictly be constant, since otherwise extra terms are needed in the equations to account for the acceleration of the coordinate system.

In tropical cyclones, as elsewhere in the atmosphere, many of these equations contain two or three terms that are in approximate balance, with the remaining terms being substantially smaller. For instance, friction and radial accelerations are small except in the boundary and outflow layers. Equation (1) reduces (for storm-following coordinates) to the gradient wind equation,

$$\frac{v_{gr}^2}{r} + f v_{gr} = \frac{1}{\rho}\frac{\partial p}{\partial r}. \tag{10}$$

Similarly, (3) reduces to the hydrostatic equation,

$$\frac{\partial p}{\partial z} \approx -\rho g \tag{11}$$

although this approximation is less satisfactory in the eyewall, where larger vertical accelerations and small horizontal scales prevail. Combining (10) and (11) with (8) yields a gradient-wind form of the thermal wind equation,

$$\xi \frac{\partial v}{\partial \log p} \approx -R_d \frac{\partial T_v}{\partial r} \tag{12}$$

where the inertia parameter $\xi = 2v/r + f$ replaces f in the normal geostrophic form of the equation. From (12), a warm cored vortex will have cyclonic winds decreasing with height.

The vertical component of the absolute angular momentum

$$M_a = rv + f_0 r^2/2 \tag{13}$$

where f_0 is the Coriolis parameter at the cyclone centre, is useful for developing physical understanding, as the azimuthal momentum equation (2) may be replaced by

$$\frac{dM_a}{dt} = -\frac{1}{\rho}\frac{\partial p}{\partial \lambda} + \frac{\partial \overline{M_a' w'}}{\partial z}. \tag{14}$$

Thus M_a is conserved for axisymmetric frictionless flows. Note that useful interpretation requires that the coordinate system and storm axes coincide. The absolute vorticity in cylindrical coordinates is

$$\eta = \frac{1}{r}\frac{\partial rv}{\partial r} - \frac{1}{r}\frac{\partial u}{\partial \lambda} + f = \zeta + f \tag{15}$$

and in an axisymmetric vortex is related to the angular momentum through

$$\eta = \frac{1}{r}\frac{\partial M_a}{\partial r}. \tag{16}$$

Similarly, the inertia parameter is related to the angular momentum by

$$\xi = \frac{2M_a}{r^2}. \tag{17}$$

Static stability is frequently defined through the Brunt-Väisälä frequency

$$N^2 = \frac{g}{\theta_v}\frac{\partial \theta_v}{\partial z}. \tag{18}$$

Where $N^2 > 0$, the atmosphere is statically stable and vertical parcel perturbations lead to an oscillation with frequency N. For $N^2 < 0$, the atmosphere is statically unstable and vertical perturbations grow. Another measure of the static stability is the parcel buoyancy, $-g(\rho_{\text{ref}} - \rho_{\text{parcel}})/\rho_{\text{parcel}}$. Care is needed as the parcel buoyancy depends on the reference state ρ_{ref}, and while the sum of the buoyancy and perturbation-pressure[1] forces is unique, their partitioning will depend on the choice of reference state (Smith *et al.*, 2005). Note that the buoyancy vector will have a radial component (small in a tropical cyclone), as well as a vertical component.

The inertial stability

$$I^2 = \xi\eta = \frac{1}{r^3}\frac{\partial M_a^2}{\partial r} \tag{19}$$

describes the stability of the cyclone vortex to radial perturbations in an analogous manner to

[1]i.e. relative to the reference state.

that in which N^2 describes vertical perturbations. That is, radial displacements of a parcel will be strongly resisted in a region of high inertial stability, and weakly so in a region of low I. Where $I^2 > 0$, such displacements will lead to a radial oscillation of frequency I, but where $I^2 < 0$, the vortex is inertially unstable and such perturbations will grow exponentially. The inertial stability varies widely through the storm, with high stability ($I \sim 100f$) inside of the radius of maximum winds (RMW), an abrupt decline to lower stability ($I \sim 10f$) across the RMW followed by a gradual decline to large radii. The anticyclonic outflow layer has markedly lower inertial instability and may be inertially unstable, contributing to the marked asymmetries often found there. Inertial stability is much more variable through the storm than is static stability, and contributes to marked variations in, inter alia, the response to convective heating and the boundary layer structure. Much of the storm structure can be understood in terms of inertial stability variations.

The horizontal scale of the adjustment of the storm to added heat or momentum is governed by the relative magnitudes of the Rossby radius L_R and the forcing scale L, or by their squared ratio, the rotational Froude number $F_R = (L/L_R)^2$. The Rossby radius is defined by

$$L_R = \frac{NH}{I} \tag{20}$$

where NH is the internal gravity wave speed. Note that while it is customary to refer to *the* Rossby radius, in principle *several* radii may be relevant depending on the vertical structure of the wave, but in practice the gravest mode is the most important. The Rossby adjustment problem in dynamical meteorology studies the final steady-state energy and its partitioning between kinetic and potential components as the result of a transient or steady forcing (e.g., Gill, 1982, section 7.2). Schubert *et al.* (1980) have derived analogous results for the geostrophic adjustment of vortices on an f-plane. The details are somewhat dependent on the precise form of the forcing, but are summarised in Table 1.

Table 1. Final states in the geostrophic adjustment problem for a vortex.

Scale of forcing	Forcing	Result
Small $L \ll L_R$	Wind	Kinetic energy mostly retained, mass adjusts to balance wind field
	Mass	Most energy lost as inertia-gravity waves
Large $L \gg L_R$	Wind	Most energy lost as inertia-gravity waves
	Mass	Potential energy mostly retained, wind adjusts to balance mass field

The saturation equivalent potential temperature θ_e^* is defined as the potential temperature that a saturated air parcel would have if all its moisture was condensed out and used to warm the parcel, and may be approximated by

$$\theta_e^* \approx \theta \exp(L_v q_s / c_p T). \tag{21}$$

θ_e^* is a function of T and p only which increases with rising T and falling p; thus the highest surface values in the atmosphere occur over the warm tropical oceans in the centre of tropical cyclones. The equivalent potential temperature θ_e is defined as the θ_e^* a parcel would have if it was adiabatically expanded to saturation. Emanuel (1994, Chapter 4) gives an exact formula.

2.1. *Potential Vorticity*

"Potential vorticity thinking" has become an important tool for understanding atmospheric dynamics in recent decades. The Rossby-Ertel potential vorticity Q is defined as

$$Q = \rho^{-1} \boldsymbol{\eta} \cdot \nabla \theta \tag{22}$$

and has two properties that are crucial to this approach: (i) *conservation*: that PV is conserved for adiabatic frictionless motions, and (ii) *invertibility*: that given suitable boundary conditions and a balance relationship, the full

3-dimensional mass and wind fields can be derived from the potential vorticity distribution (Hoskins *et al.*, 1985). The importance of latent heating in tropical cyclones limits the applicability of the first property, but PV thinking has nevertheless become a cornerstone of current understanding.

The PV budget equation may be written

$$\frac{dQ}{dt} = \frac{1}{\rho}(\boldsymbol{\eta} \cdot \nabla\dot{\theta} + (\nabla \times \mathbf{F}) \cdot \nabla\theta)$$
$$= \frac{1}{\rho}\nabla \cdot (\dot{\theta}\boldsymbol{\eta} + \mathbf{F} \times \nabla\theta) \qquad (23)$$

respectively, with symbols defined in Table 2. In deep convection, the maximum heating typically occurs in the middle to upper troposphere. It follows that the typical effect of convective heating in a vortex is to generate cyclonic PV in the lower levels and destroy it aloft. The second equality in (23) shows that this dipole can be written as the divergence of a flux vector, $\nabla \cdot (\dot{\theta}\boldsymbol{\eta})$, hence the effect of heating may be interpreted as a transport of PV along the vorticity vector (Raymond, 1992).

A complementary view of PV dynamics was provided by Haynes and McIntyre (1987, 1990), who showed that the local change may be written as the divergence of a flux that is directed along the isentropes,

$$\frac{\partial(\sigma Q)}{\partial t} = -\nabla_\theta \cdot \mathbf{J}, \qquad (24)$$

where σ is the mass density in isentropic coordinates and not a vertical coordinate. The flux $\mathbf{J}$ has components due to advection, diabatic heating and friction, which under the hydrostatic approximation[2] in isentropic coordinates are

$$\mathbf{J} = (u, v, 0)\sigma Q + (\dot{\theta}\partial v/\partial\theta, -\dot{\theta}\partial u/\partial\theta, 0)$$
$$+ (-F_y, F_x, 0) \qquad (25)$$

with symbols defined in Table 2. This form provides a powerful constraint on the effects of heating. Since the "generation" of PV by diabatic heating can be written as a flux divergence along the isentropes, diabatic PV "generation" must be balanced by an equal and opposite "destruction" somewhere else at the same isentropic level, and heating can be interpreted as producing a purely horizontal (in isentropic coordinates) flux of PV. Haynes and McIntyre (1987, 1990) highlight two statements: (i) There can be no net transport of Rossby-Ertel PV across any isentropic surface, and (ii) PV can neither be created nor destroyed within a layer bounded by two isentropic surfaces[3]. These statements are direct consequences of (24), and "net transport", "created" and "destroyed" should be understood in that context. Finally, note that the effect of diabatic heating $\dot{\theta}$ in (24) is twofold: it produces a horizontal flux through the second term in the right-hand side of (25), and it alters the isentropic mass density σ by causing mass to cross the isentropic surfaces. However, while mass may cross the surfaces and change σ, the PV mixing ratio σQ may not. Thus if heating is causing isentropic surfaces to move together, as happens for example in the lower troposphere with convective heating maximised in the mid-troposphere, then σ will decrease. If in addition the vertical shear is small so that the diabatic heating term of $\mathbf{J}$ in (25) is small, then Q will tend to increase to compensate and thereby keep σQ constant. The opposite applies above the heating maximum. If $\mathbf{J}$ is not small, this compensating tendency will nevertheless operate in addition to the horizontal fluxes implied by (24).

3. The Secondary Circulation

The response of axisymmetric tropical cyclones in hydrostatic and gradient-wind balance to axisymmetric sources of heat $\dot{\theta}$ and momentum

[2]See Haynes and McIntyre (1987) for the nonhydrostatic version.
[3]Although net PV can be created or destroyed where the layer terminates at the ground.

F_λ is described by the Sawyer-Eliassen equation,

$$\frac{\partial}{\partial r}\left\{\frac{1}{\rho r\theta}\left[N^2\frac{\partial\psi}{\partial r}-B\frac{\partial\psi}{\partial z}\right]\right\}+\frac{\partial}{\partial z}\left\{\frac{1}{\rho r\theta}\left[\left(I^2-\frac{CB}{g}\right)\frac{\partial\psi}{\partial z}-B\frac{\partial\psi}{\partial r}\right]\right\}=g\frac{\partial(\dot{\theta}/\theta^2)}{\partial r}+\frac{\partial(C\dot{\theta}/\theta^2)}{\partial z}+\frac{\partial(\xi F_\lambda/\theta)}{\partial z}. \quad (26)$$

Here, $C \equiv v^2/r + fv$ is the radial acceleration, $B \equiv g/\theta\, \partial\theta/\partial r = \partial C/\partial z - CN^2/g$ is the baroclinicity and the equality in the definition of B is the thermal wind equation. Equation (26) describes the toroidal circulation generated by axisymmetric forcings $\dot{\theta}$ and F_λ, and determined also by the particular boundary conditions chosen. This toroidal circulation (u, w) is described by its streamfunction ψ, where

$$u = -\frac{1}{\rho r}\frac{\partial\psi}{\partial z}, w = \frac{1}{\rho r}\frac{\partial\psi}{\partial r}. \quad (27)$$

Equation (26) is derived in pressure-based vertical coordinates by Willoughby (1979), Shapiro and Willoughby (1982), Holland and Merrill (1984) and Willoughby (1995). Here the equation is given in height coordinates for consistency with section 2 and as it makes the surface boundary condition simpler. The principal difference is that the second term on the RHS is absent in pressure-based coordinates. Derivations in height coordinates are given by Smith *et al.* (2005) and Pendergrass and Willoughby (2009).

Equation (26) is elliptic provided that the discriminant

$$D = N^2\left(I^2 - \frac{CB}{g}\right) - B^2 \quad (28)$$

is positive, a condition that Smith *et al.* (2005) show is equivalent to the potential vorticity Q being positive.

The forcings on the RHS of (26) are differentiated; thus the solution to localised forcings will consist of pairs of counter-rotating gyres. The strength and extent of these gyres will depend on the vortex structure and the boundary conditions on ψ; here the vortex structure is measured in terms of its modified inertial stability $I^2 - CB/g$, static stability N^2 and baroclinicity B. Note that (26) is linear in the forcings so that solutions obtained from individual forcings can be simply added to determine the response to the combined forcing.

The circulation induced by convective heating in the eyewall is a pair of counter-rotating gyres. The outer gyre is in an environment of weak inertial stability, so it has substantial radial extent and the motion is largely horizontal. In contrast, the inner gyre is restricted horizontally by the high inertial stability of the core and by the cylindrical geometry. Heating cannot directly change angular momentum, so the updraft between the gyres tends to be tilted to lie along a constant angular momentum surface, and hence the eyewall will tend to follow a constant M_a surface. In weak storms, the inner circulation always extends to the axis, but for $v_m \gtrsim 35\,\mathrm{m\,s^{-1}}$ the inertial stability may be sufficient to confine the circulation closer to the eyewall. These circulations produce the maximum pressure falls and tangential wind acceleration immediately inside of the eyewall, so the RMW contracts inwards as the storm intensifies. This *eyewall contraction process* has been shown to be the most common mode of cyclone intensification in the North Atlantic basin (Willoughby, 1990b). Simultaneously, the secondary circulation decelerates the flow near the storm axis, causing the radial profile of azimuthal wind to become U-shaped (Willoughby, 1978; Shapiro and Willoughby, 1982; Willoughby *et al.*, 1982). The precise details of the subsidence in the eye depend on the eye radial profile of wind: a V-shaped profile is more inertially stable inside the eye than a U-shaped, so an intense storm with a V-shaped profile has subsidence confined too close to the eyewall, while one with a U-shaped profile has subsidence extending towards the centre (Schubert *et al.*, 2007).

The surface friction-induced circulation is shallow because the static stability prevents the outflowing air from rising into the upper troposphere. Arguments that the friction-induced circulation cannot intensify the storm (e.g., Ooyama, 1969, 1982) imply that the inwards advection of angular momentum by this circulation not exceed the frictional torque that destroys angular momentum. Balanced models are somewhat limited here as the boundary layer flow is in fact not in gradient balance (Ooyama, 1969; Smith and Montgomery, 2008) but nevertheless, more sophisticated models confirm the general structure of the induced flow and that the net angular momentum import and frictional torque nearly balance (Kepert and Wang 2001; see also section 5.1.2).

Momentum forcing in the upper troposphere induces a shallow, radially extensive pair of gyres, because of the weak inertial stability there. This forcing is important to understanding the interactions between the cyclone and its environment, since upper troughs may induce an eddy flux convergence of angular momentum in the upper part of the storm.

Numerous authors have given the details of the secondary circulation in particular situations, including Willoughby (1979, 1990b), Pfeffer and Challa (1981, 1992), Shapiro and Willoughby (1982), Holland and Merrill (1984) and Pendergrass and Willoughby (2009).

3.1. *Cooperative Intensification*

These responses of the secondary circulation to the various forcings provide a framework for understanding the energetics of the tropical cyclone as a whole. The boundary layer circulation can only spin down the storm, since the inwards advection of angular momentum by the frictional inflow is countered by the surface frictional torque and by outwards advection in the return branch of the gyre. Indeed, this outflow spins down a storm in a matter of days (Eliassen and Lystad, 1977; Montgomery *et al.*, 2001; Reasor *et al.*, 2000). Nor can surface friction lift surface air to the upper tropospheric outflow channel against atmospheric stability. However, it does define a radius inside of which there is frictionally forced ascent of warm moist air out of the boundary layer. While this air is too cool to overcome the basic stability gradient, release of latent heat allows the air to ascend higher. This latent heat release in and surrounding the eyewall induces a much deeper secondary circulation, with weak inflow through much of the troposphere and outflow just below the tropopause. The low–mid tropospheric inflow induced by the heating is sufficient to cancel the outflow component of the frictionally-forced circulation in balanced models[4], at least in intensifying and steady-state storms, producing the observed deep weak inflow in the mid troposphere in real storms. The role of boundary-layer convergence in defining the regions of deep ascent is supported by observations showing that the vertical mass flux is maximised near 1-km altitude in the eyewall, rainband and stratiform rain regions (Black *et al.*, 1996).

As the primary circulation intensifies, the inertial stability of the core increases, which modifies the character of the secondary circulation. The boundary between frictionally-forced upflow and downflow migrates inwards, confining the latent heating to be closer to the centre and increasing the efficiency by which heating spins up the storm (Hack and Schubert, 1986). The heating-induced secondary circulation also contracts and strengthens the downdraft in the eye. There is thus a cooperative interaction between the primary and secondary circulations in an intensifying storm, first discussed in detail by Ooyama (1969) in the context of a 3-layer balanced model and further elicudated by Ooyama (1982). Note that this feedback implies that tropical cyclone intensification and maintenance is inherently a nonlinear process. While the Sawyer-Eliassen equation (26) is linear in the forcings, tropical cyclone intensification is nonlinear because, inter alia, the friction-induced

[4]But not the thin layer of outflow above the boundary layer due to the readjustment of supergradient flow, as discussed in section 5.

circulation defines the regions favourable for deep convection.

The intensification is eventually limited as the increase in strength of the upper warm core increases the static stability and limits the ability of the upflowing boundary layer air to ascend. Then the azimuthal-mean updraft becomes moist neutral to the ascent along the slanting angular momentum surfaces. This limit is assumed by Emanuel's potential intensity theory (section 4). We shall see later that other processes, such as the formation of secondary wind maxima and instabilities of the eyewall itself, may also act to limit the intensification.

How important is the assumption of thermal wind balance in the above arguments? For quantitative prediction, the requirement of balance is a significant limitation and the theory is only strictly valid for weak forcings and slow secondary circulations. In much of the boundary layer, the flow is significantly subgradient, with the lack of balance there recognised as a serious limitation of balanced models as far back as Ooyama (1969). However, the more sophisticated boundary layer models discussed in section 5 do qualitatively reproduce the main features of the balance model: a strong near-surface inflow layer with ascent up out of the boundary layer in the core, and subsidence into the boundary layer at larger radii. Above the boundary layer, analyses of observations confirm gradient balance to within 1 or $2\,\mathrm{m\,s^{-1}}$ (Willoughby, 1990a, 1991; Kepert, 2006a,b), although one modelling study suggested a thin annulus of supergradient flow on the inner edge of the eyewall (Zhang *et al.*, 2001). Overall, the evidence largely supports the balance assumption in this part of the storm and the cooperative intensification analysis is therefore reasonable.

From this perspective, what is the role of deep convection in a tropical cyclone[5]? Its horizontal location is largely defined by the region of frictionally-forced ascent. Thus the most obvious impact is that it determines the vertical profile of heating, thereby affecting the structure of the secondary circulation. The secondary circulation in a real tropical cyclone is not the spatially-smooth field suggested by the solution of the Sawyer-Eliassen equation, but rather a substantial proportion ($\sim$ 65%) of the total upwards transport in the eyewall and in rainbands is concentrated into buoyant updraft cores that occupy a small ($<$ 10%) of the area of each region. Most of the updraft cores are positively buoyant, and these accomplish about 40% of the total transport, but occupy $<$ 5% of the total area. These statistics are from Eastin *et al.* (2005a,b), but similar results were found from observational analyses by Jorgensen *et al.* (1985) and Black *et al.* (1996)), and from high-resolution modelling by Braun (2002), Braun *et al.* (2006) and Cram *et al.* (2007). Ooyama (1982) argues that buoyant energy release is necessary since it enables midtropospheric entrainment in the cloud bands and hence causes the deep inflow that imports angular momentum to intensify the vortex. The upwards transport from these updrafts is partially opposed by downdraft cores which, as we shall see in section 5, act to cool and dry the boundary layer. If there were no surface fluxes, these downdrafts would reduce or eliminate the ability of the air flowing up and out of the boundary layer to overcome the stability gradient and ascend to the tropopause. Over the warm tropical oceans, the thermodynamic fluxes that enable the recovery of the boundary layer from these downdrafts are thus crucial to maintenance of the storm. Following landfall, these fluxes are greatly reduced and significant outflow may be observed above the boundary layer, consistent with the collapse of the latent-heat induced part of the secondary circulation and also with the increased frictional forcing.

[5]Note that nonprecipitating clouds do not provide net heating since their evaporation consumes the latent heat released by their formation, although they do generate some vertical transport.

3.2. *Eyewall Replacement Cycles*

Convective rings other than the eyewall may also occur in tropical cyclones, particularly intense ones. Such rings will generate a secondary circulation consisting of a pair of gyres similar to that generated by the eyewall heating, although the precise details will differ because of differing inertial and static stabilities and baroclinicity at the larger radius. In intense hurricanes ($v_m \gtrsim 50\,\mathrm{m\,s^{-1}}$), the appearance of an outer eyewall or concentric ring usually marks the end of the period of intensification, and possibly a weakening of the storm. This occurs because the secondary circulation of the outer ring competes with the eyewall and its wind maximum by two mechanisms. Firstly, the outer ring's secondary circulation produces broad subsidence over the eyewall and hence surface divergence and outflow there, which acts directly to weaken the inner wind maximum (Willoughby *et al.*, 1982). Secondly, the convection in the outer ring consumes high-enthalpy air from the inflowing boundary layer and frequently replaces it with cool dry downdrafts (section 7.1), thus shutting off part of the energy supply to the eyewall. Often, these processes lead to the destruction of the inner eyewall and its wind maximum. The outer ring then becomes the main eyewall and intensification may resume as it contracts. This sequence of events is known as the eyewall replacement cycle, and intense storms may undergo several such cycles. Detailed case studies may be found in Willoughby *et al.* (1982), Willoughby (1990a) and Willoughby (1995), while a satellite view of the sequence is shown in Chapter 7.

The above mechanism describes the evolution once the convective ring has formed, but is silent on the question of where the ring comes from. Time-sequence imagery from microwave satellite sensors or radar suggest that the "wrapping-up" of a spiral rainband is a common route, but does not reveal the underlying dynamics. Experiments with barotropic models have shown that the interaction between a small intense vortex and a large weaker one may lead to the smaller vortex straining out the larger one into an annulus of elevated vorticity surrounding the smaller vortex, with a moat of low vorticity in between (Kuo *et al.*, 2004). In velocity space, this configuration corresponds to an outer wind maximum. Other outcomes are possible from this interaction, and this scenario depends on a certain range of initial vortex sizes, relative intensities and separations. Rozoff *et al.* (2006) have shown that the strong radial shear in the moat between the concentric eyewalls leads to rapid filamentation which they hypothesise acts to suppress convection there in conjunction with the subsidence forced by the respective secondary circulations. However, Wang (2008) has shown that the filamentation zone, instead of suppressing deep convection, actually provides a favourable environment for organising inner spiral rainbands, and that the low-reflectivity moat outside the eyewall is primarily controlled by subsidence from the secondary circulation and from downdrafts from anvil precipitation outside of the eyewall. In contrast, Nong and Emanuel (2003) have shown using an axisymmetric model that a WISHE-like instability (see section 4) may strengthen an existing outer local wind maximum, although this mechanism cannot explain the initial origins of the finite-amplitude instability. Terwey and Montgomery (2008) propose a hypothesis for secondary eyewall formation which includes the WISHE process, and in which the necessary wind maximum is generated by a combination of convective scale, vortex Rossby wave (section 7.2) scale and system scale processes. Further research is clearly needed to understand the formation mechanism(s) of secondary eyewalls, and is a requirement for improved intensity forecasts.

3.3. *The Outflow Layer*

The outflow layer consists of a small cyclonic core surrounded by an extensive asymmetric anticyclone. It is easy to understand why the flow should turn anticyclonic: M_a is destroyed in the frictional inflow layer but approximately conserved in the eyewall updraft and outflow, hence the outflow must have lower M_a than lower in the storm at the same radius, eventually to the extent that the flow turns anticyclonic. However, some studies of the upper troposphere surrounding a cyclone reveal areas of anticyclonic PV (e.g., Molinari *et al.*, 1998; Shapiro and Franklin, 1995; Wu and Cheng, 1999), as shown in Fig. 1. A similar feature occurs in idealised numerical simulations (e.g., Shapiro, 1992). The simple argument using M_a is deficient in that it cannot explain the sign change of η that apparently occurs. Similarly, advective rearrangement in adiabatic frictionless PV dynamics cannot create PV of the opposite sign. Shapiro (1992) shows that the upper-level negative PV was generated in his model by twisting terms that arise from horizontal gradients of the cumulus mass flux. A further clue to the dynamics can be found in the Haynes and McIntyre (1987) form of the PV budget equation, (24). The isentropic surfaces in the outflow layer bend downwards into the mid troposphere in the storm core, because the core is warm. If convective processes are leading to an increase in low- and mid-tropospheric PV at the core, there must be a compensating decrease elsewhere at the same isentropic level. It is plausible, although not yet established in this framework, to hypothesise that this decrease may be sufficient to lead to locally anticyclonic PV adjacent to the storm. The Haynes and McIntyre form of the PV budget equation shows that such behaviour should not be regarded as being anomalous but rather, as consistent with the generation of strong cyclonic PV in the core of the intensifying storm. Examination of Fig. 1 as well as the full sequence from which it came (Wu and Cheng, 1999) tends to support the idea that the upper anticyclonic PV develops in conjunction with intensification of the cyclone core.

The outflow layer is thus inertially either weakly stable or unstable. This contrasts markedly with the rest of the storm and has at least two important consequences. Firstly, it is much less symmetric than the lower part of the storm, with the outflow being often organised

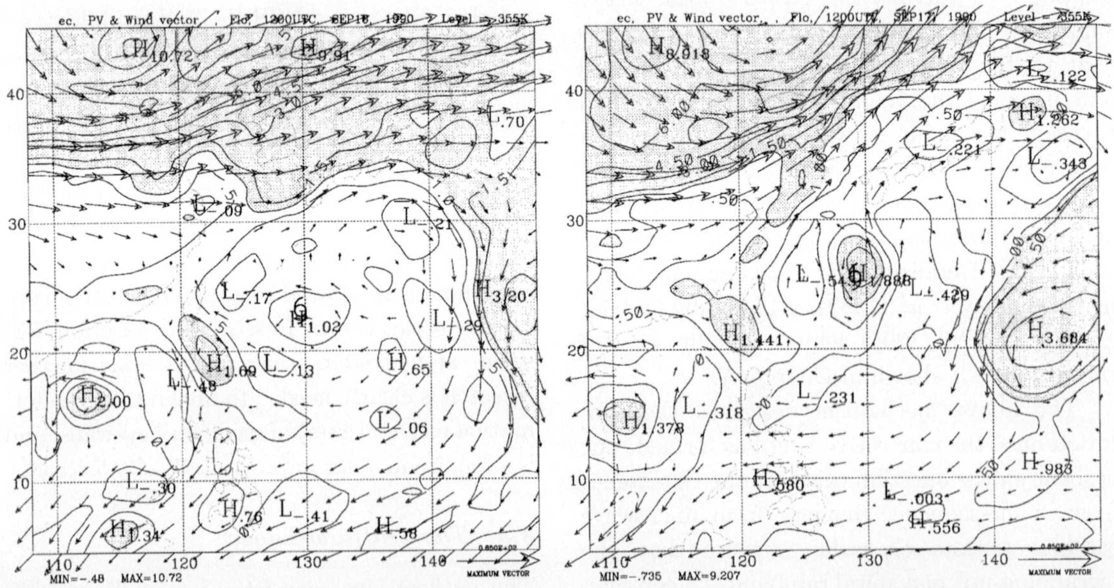

Figure 1. Wind vectors and PV on the $\theta = 355$ K surface at (left) 1200UTC 16 Sep and (right) 1200UTC 17 Sep 1990 during SuperTyphoon Flo. PV contoured every 0.5 PVU up to 1.5 PVU and every 1.5 PVU thereafter, with values greater than 1 PVU shaded. Note the development of anticyclonic PV surrounding the core, as the core intensified. From Wu and Cheng (1999).

into one or two jets. These may form through inertial instability, but in any case, low inertial stability favours the outflow of the air exhausted from the eyewall convection and hence intensification. Secondly, the low inertial instability provides a conduit through which the environment can directly influence the storm core. In the upper troposphere, the storm core is not protected by an annulus of inertially stable air as in the lower troposphere. Moreover, the low stability ensures that environmental forcing will produce a marked radial-flow response. We will see in section 8 that upper tropospheric interactions between the cyclone and its environment can have a major effect on the storm's evolution.

4. Potential Intensity Theories

Potential intensity (PI) theories attempt to answer the question: What is the maximum equilibrium intensity that a tropical cyclone can reach in a given thermodynamic environment? Here we review two branches of theory: Holland (1997) and predecessors calculate the surface pressure drop realisable by convective heating, while Emanuel's (Emanuel, 1986, 1988, 1995; Bister and Emanuel, 1998) balances the rate of extraction of enthalpy from the sea surface against frictional destruction of kinetic energy.

Emanuel's PI theory (EPI) makes an analogy with the Carnot cycle, in which the cyclone's secondary circulation contains the classical Carnot-cycle elements of isothermal expansion of the inflowing boundary layer air, adiabatic expansion as it ascends in the eyewall and flows outwards in the upper troposphere, isothermal compression at large radius and descent back to the boundary layer. Although the analogy is not exact (since the outer part of the cycle is not closed in real tropical cyclones), an important consequence is that the efficiency of the tropical cyclone as a heat engine can be calculated. Multiplying this efficiency by the surface enthalpy gain gives the amount of work the cyclone can do. Assuming that all of this work is expended in frictional dissipation leads to Emanuel's expression for the potential intensity,

$$v_m^2 = \frac{T_s - T_o}{T_o}\frac{C_E}{C_D}(k_0^* - k) \tag{29}$$

The full derivation can be found in Bister and Emanuel (1998) and a simplified version in Emanuel (2003).

The EPI model consists of two components, the boundary layer and the rest of the cyclone. The former is represented by a linearised depth-averaged model, whose role is to determine the dynamical and thermodynamical properties of the frictionally forced updraft near the RMW. The role of the surface enthalpy fluxes in making the expansion of the inflowing boundary layer air isothermal rather than adiabatic is emphasised in this part of the model through the acronym WISHE, for wind-induced surface heat exchange. The remainder of the model includes the important assumption that the cyclone core is neutral to moist slantwise convection; that is, that the θ_e^* and angular momentum surfaces coincide. This assumption amounts to a convection parameterisation because it determines how the effects of the latent heating are distributed in the rest of the cyclone, given the boundary layer updraft properties (Smith, 2000). EPI theory can thus be regarded as a steady-state version of the cooperative intensification theory discussed in section 3.1, in which the neutrality to moist convection (i) is consistent with the cyclone having reached peak intensity as it implies that there is just sufficient latent heat release to allow the boundary layer air to ascend, but not enough to intensify the storm further, and (ii) enables an analytical solution for the intensity of the storm.

Equation (29) shows that the EPI depends upon the ratio C_E/C_D of the surface enthalpy and drag coefficients. Similar results have been obtained by prognostic tropical cyclone models, with Ooyama (1969) and Rosenthal (1971) being the first to identify the importance of this ratio. Measurements generally show that C_E is nearly constant with wind speed, while C_D was long thought to increase linearly with wind speed. The resulting decrease in the ratio at high winds was an obstacle for EPI theory. This difficulty has apparently been removed by recent measurements showing that C_D ceases to increase, and in fact may decrease, once the wind speed exceeds about $35\,\mathrm{m\,s^{-1}}$ (Chapter 3).

Persing and Montgomery (2003) used an axisymmetric model to test EPI theory. They found that, when run at high resolution, the model predicted intensity averaging about $40\,\mathrm{m\,s^{-1}}$ higher than the EPI, a phenomenon they dubbed "superintensity". They argued that superintensity occurred when the model developed small-scale streamwise vortices along the inner edge of the eyewall, which efficiently mixed the high θ_e air from the eye boundary layer into the eyewall updraft, increasing the energy content of this air and hence the storm intensity. The low-level air within the eye has high values of θ_e since it has been in contact with the warm ocean surface at lower pressure than that beneath the eyewall. However, Bryan and Rotunno (2009a) presented simulations in which this mechanism was removed by artificially setting the surface moisture fluxes in the eye to zero, and showed that this change had only a negligibly small impact on intensity. Bryan and Rotunno (2009b) further demonstrated that the simulated intensity in an axisymmetric model has a strong sensitivity to the parameterisation of horizontal diffusion, and showed that the incorrect representation of this process was responsible for the results of Persing and Montgomery (2003). While entrainment of the eye reservoir of high θ_e air does not explain superintensity of the axisymmetric mean vortex, numerical (Braun, 2002; Cram *et al.*, 2007) and observational (Eastin *et al.*, 2005a,b) evidence suggests that it may be important to the development of individual buoyant updrafts on shorter timescales.

Observational comparisons tend to show that EPI is an upper bound to observed intensity (e.g., Emanuel, 2000; Tonkin *et al.*, 2000). A notable exception is an analysis of Hurricane Isabel (2002) based on dropsonde data, in which the intensity substantially exceeded the EPI (Montgomery *et al.*, 2006).

In contrast to EPI, the Holland (1997) PI theory (HPI) calculates the hydrostatic pressure fall resulting from the vertical redistribution of moist entropy obtained at the sea surface. The model defines an eyewall saturation moist adiabat whose θ_e^* is set by the surface conditions, surrounding a parameterised eye. The model thus contains an explicit feedback, since reduced hydrostatic pressure increases the surface θ_e^* (section 2) and hence energy available. Unlike EPI theory, HPI does not explicitly incorporate an energy budget, although an energy balance is used to exclude certain unrealistic occurrences. Boundary layer processes are included implicitly since the thermodynamics of the ascending air depends on the SST, prescribed boundary-layer humidity and surface pressure, but there is no explicit boundary-layer submodel. Holland (1997) assumes that the ascending air's θ_e^* is determined by the surface conditions beneath the eyewall, but discusses the possibility of higher intensity if some process acted to transfer the higher-θ_e^* air from the eye boundary layer into the eyewall updraft.

While these theories produce similar predictions for typical tropical environments and are reasonably consistent with observations, they are fundamentally different in approach and make some significantly different predictions. For example, they have opposite sensitivities to the assumed boundary layer relative humidity: Emanuel's theory predicts the strongest storms for low RH because that maximises the surface enthalpy flux, while Holland's predicts the strongest storms for large RH since that maximises the eyewall θ_e^*. This and other differences between the theories are detailed in Camp and Montgomery (2001).

The Emanuel and Holland PI theories, and indeed many numerical simulations of tropical cyclones, assume axisymmetry. Bryan and Rotunno (2009b) have examined the sensitivity of such simulations to various parameterisations in the model, including the horizontal diffusion, the hydrometeor fall velocity, and the ratio C_E/C_D. They found that intensity was strongly sensitive to each of these factors, and also that the magnitude of sensitivity of the latter two factors was quite dependent on the horizontal diffusion. In an axisymmetric model, the horizontal diffusion represents a range of asymmetric mixing processes, including eyewall mesovortices, vortex Rossby waves and boundary layer roll vortices as well as turbulence. It affects the simulated intensity by controlling the degree to which frontogenesis can collapse the eyewall to a singularity (Emanuel, 1997). Bryan and Rotunno (2009b) argue for sufficiently large horizontal diffusivity in axisymmetric models as to have a significant impact on the simulated intensity, and note that such values substantially reduce the sensitivity of intensity to the ratio C_E/C_D, from that in earlier studies.

Tropical cyclones intensify at a finite rate and thus require time to achieve their potential intensity. Factors such as the favourability of the environment determine how quickly, and whether, they get there. Indeed, the majority of storms do not reach their PI. Emanuel (2000) studied the frequency distribution of storm relative intensity; that is, peak intensity normalised by the potential intensity at the same time. He showed that the frequency distribution of relative intensity for storms whose intensity was not limited by landfall or transition over colder water could be modelled as the combination of two uniform distributions, with all normalised intensities from zero (marginal tropical cyclone) to hurricane intensity being about equally likely, then a step to another near-constant probability for all storms from hurricane intensity to a normalised intensity of one (i.e. a storm at its PI). In particular, it is relatively rare for a storm to attain its PI.

An earlier theory of hurricane intensification was known as CISK, for conditional instability of the second kind. The linear CISK theory of Charney and Eliassen (1964) differs substantially from the above theories, because it is fuelled by pre-existing atmospheric conditional instability rather than by enthalpy drawn from the sea beneath the storm. Ooyama's similar (1964) theory later matured into his nonlinear cooperative-intensification theory (Ooyama, 1969, 1982), presented in section 3.1 in the context of a continuous balanced-vortex model rather than his 3-layer model. Smith (1997) compares these models in detail, and concludes that the differences between EPI theory and Ooyama's cooperative intensification are largely restricted to technical details and degrees of emphasis on the relative importance of surface fluxes. Importantly, and in contrast to Charney and Eliassen (1964), cooperative intensification and EPI agree that both surface fluxes and convection are necessary for intensification.

5. Boundary Layer Structure

Boundary layer processes determine the supply of thermodynamic energy to the storm and the dissipation of momentum through friction, processes that we have seen are crucial to the storm's existence. In addition, it is through the boundary layer that much of the impact on humanity occurs. Recent observational and theoretical advances have led to substantial progress on understanding this part of the storm.

5.1. *Mean Structure*

5.1.1. *Observations*

Observed wind profiles in tropical cyclones frequently show a marked low level wind maximum. This maximum typically occurs around 300 to 800 m height near the eyewall, and 1 to 2 km at larger radius. It has been observed by dropsonde (Franklin *et al.*, 2003; Kepert, 2006a,b), wind profiler (e.g., Knupp *et al.*, 2000, 2005), and Doppler radar (e.g., Marks *et al.*, 1999), although it has been the recent advent of the GPS dropsonde that has emphasised the ubiquity of this jet. The broad maximum is generally more or less obscured by smaller-scale fluctuations due to turbulence in individual soundings, thus some form of averaging is needed to expose it clearly. For example, Fig. 2 shows the observed mean wind speed profile, normalised by the wind speed at the common aircraft flight level of 700 hPa or about 3 km, from the eyewall of seven hurricanes. The low level jet is clearly apparent. Below the jet, in the lowest 100–200 m, the wind speed increases nearly logarithmically with height (Franklin *et al.*, 2003; Powell *et al.*, 2003), consistent with classical theory for a neutrally-stratified surface layer.

A substantial amount of between-storm variability is apparent in Fig. 2. The strength of the normalised maximum varies from 1.12 to 1.3, and its height from 300 to 800 m, while the strength of the normalised surface wind speed is between 0.82 and 0.96.

Along with variation between storms, there is substantial variation within each storm. The decrease in height of the low-level jet with decreasing radius becomes quite marked across the eyewall, and continues to the centre of the storm (Franklin *et al.*, 2003; Kepert, 2006a,b). Figure 3 shows the observed mean storm-relative wind profile in several annular regions in Hurricane Mitch (1998); it is clear that the depth of the frictional inflow layer and the jet height decrease towards the centre. Moreover, the azimuthal-wind maximum is generally near the top of, but still within, the frictional inflow layer. Within the eye itself, both individual and mean soundings generally show little, if any, evidence of frictional retardation at the surface.

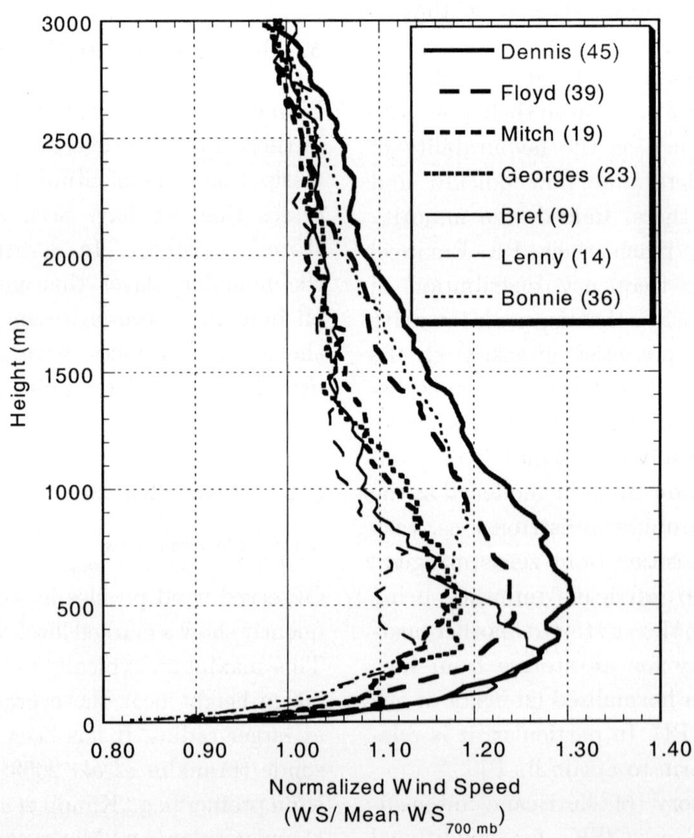

Figure 2. Mean observed eyewall wind speed profile in seven hurricanes, normalised by the wind speed at 700 hPa. From Franklin *et al.* (2003).

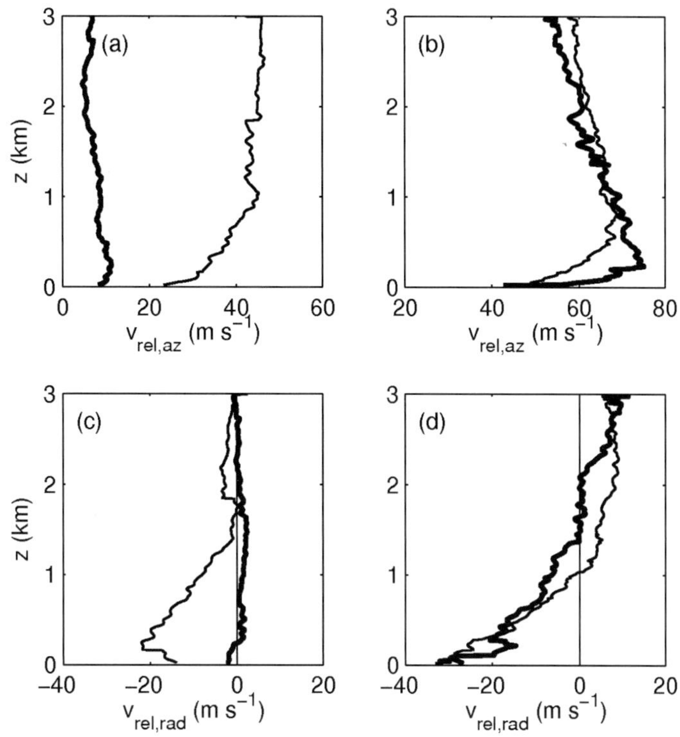

Figure 3. Observed wind profiles in Hurricane Mitch (1998). (a) Mean profiles of storm-relative azimuthal wind over radius ranges 0–15 km (heavy) and 40–100 km (light). (b) As for (a) except over radius ranges 15–25 km (heavy) and 25–40 km (light). (c, d) As for (a, b), but for the storm-relative radial wind component. From Kepert (2006b).

This radial variation in wind structure is accompanied by a variation in the surface wind factor (SWF); that is, the ratio of the near-surface wind speed to that at some higher level. Franklin *et al.* (2003) showed that the hitherto widely used value of 0.8 is appropriate for the outer vortex, but increases to 0.9 near the eyewall. They also found that the SWF varied with reference height, recommended higher values in the outer vortex near convection than in its absence, and noted higher values on the left of the storm track than on the right. These new values revised long-standing operational practice at the United States National Hurricane Center.

The SWF has been analysed in individual storms by Kepert (2006a,b) and Schwendike and Kepert (2008). The increase towards the storm centre is clearly marked, and most storms display higher values on the left of track than on the right. However other factors, including proximity to land in the case of Hurricane Mitch (1998), can also produce a significant SWF asymmetry. Surface wind data from the airborne step-frequency microwave radiometer (SFMR) usually shows an increase in SWF towards the centre, and often a left-right asymmetry as well (Powell *et al.*, 2009).

There is also a marked azimuthal variation in tropical cyclone boundary layer wind structure. Figure 4 shows the observed profiles in Hurricane Georges (1998). It is clear that a large part of the variability between profiles is due to their position within the storm (Kepert, 2006a,b); this relationship is especially striking as observations nearby in storm-relative space are not necessarily nearby in time. The height and strength of the low-level jet, and the depth and strength of the inflow layer, vary consistently around the storm. Analyses of Hurricanes Mitch (Kepert, 2006b),

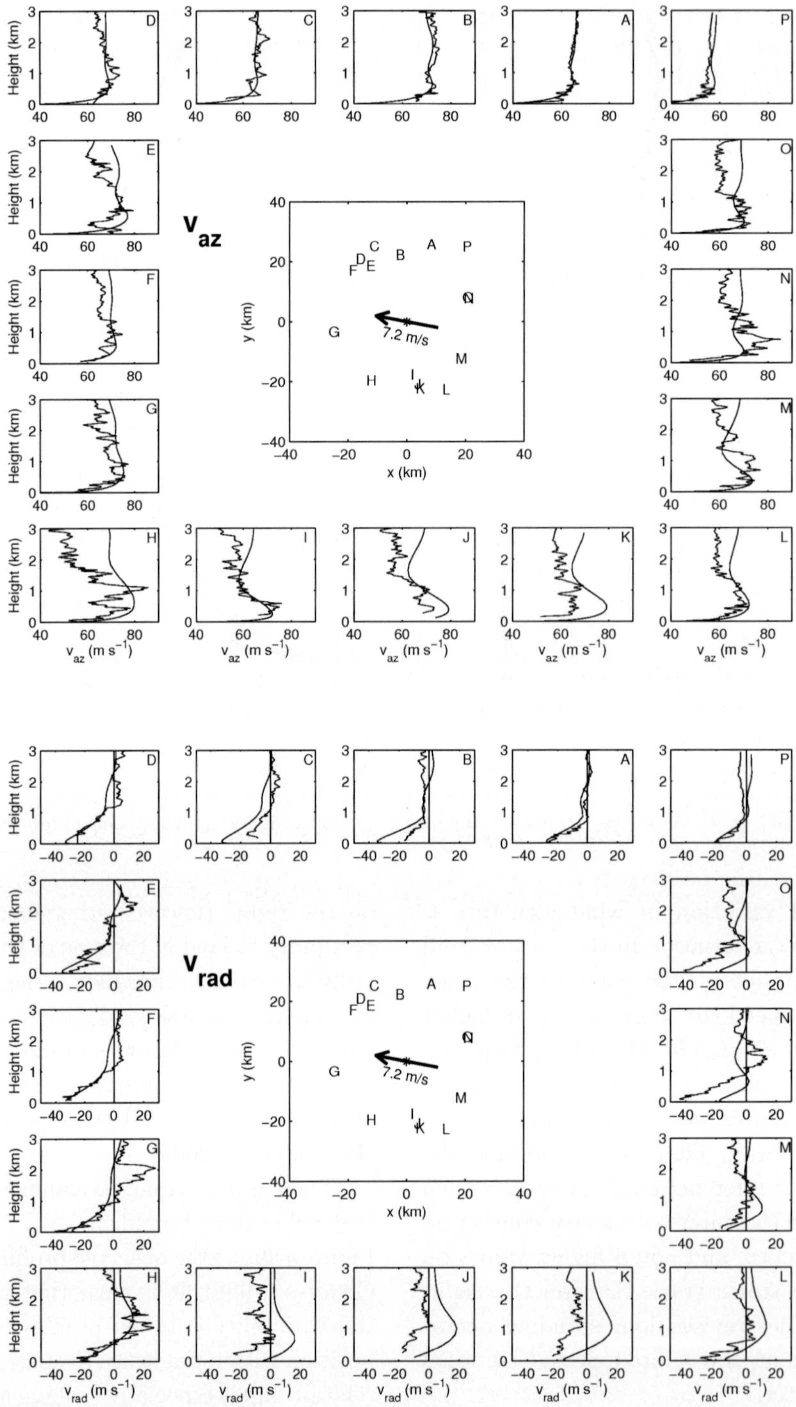

Figure 4. Profiles of the storm-relative azimuthal and radial wind components observed by dropsondes (curves with small-scale fluctuations) and represented in the model of Kepert and Wang (2001) (smooth curves) in and near the eyewall of Hurricane Georges (1998). The model values were interpolated from the model grid to the observed dropsonde trajectory. The storm-relative position of each sonde as it fell through a height of 1 km and the storm motion are shown in the central panel. From Kepert (2006a).

Danielle and Isabel (Schwendike and Kepert, 2008; Bell and Montgomery, 2008) similarly show a consistent spatial variation of wind profile shape within each storm.

The strongest surface winds are generally found in the right forward quadrant in the northern hemisphere, although exceptions to this situation are not uncommon. The rightward location can be understood to first order as the sum of the motion vector and the symmetric storm circulation. Maximum earth-relative inflow occurs to the right rear, with weaker or absent inflow to the left. The net surface wind asymmetry is thus roughly a right-rear to left-front throughflow (e.g., Powell, 1982; Black *et al.*, 1988; Powell and Houston, 1998; Kepert, 2006a).

Several of the theoretical studies reviewed below have predicted that the upper boundary layer jet is supergradient. Balance in this situation has been analysed by Kepert (2006a), Kepert (2006b), Schwendike and Kepert (2008) and Bell and Montgomery (2008), who found that Hurricanes Mitch (1998) and Isabel (2002) had azimuthal-mean jets that were ~15% supergradient, but that Georges (1998) and Danielle (1998) did not. These differences highlight the inter-storm variatiability already described, and are discussed further below.

Several studies have discussed boundary layer asymmetries due to proximity to land. Here, the higher roughness over land induces increased inflow, which is typically drier, and produces a flow asymmetry that may extend into the eyewall. Analyses of these phenomena have been presented for Hurricanes Bonnie (Schneider and Barnes, 2005), Danny (Kepert, 2002a), Floyd (Kepert, 2002b) and Mitch (Kepert, 2006b). Simulation with simplified models shows that the cause of the flow asymmetry is asymmetric friction, analogous to that produced by storm motion (Kepert, 2006b; Wong and Chan, 2007; May *et al.*, 2008).

Boundary-layer thermodynamics was analysed along inflow trajectories in Hurricane Bonnie (1998) by Wroe and Barnes (2003). They found little increase in inflow θ_e to within about 1.5 times the RMW, despite surface fluxes of over 500 W m^{-2}, since moist convection removes energy from the boundary layer at the same rate as the sea supplies it. Inwards of this, the storm's secondary circulation suppresses convection while the surface fluxes continue to increase, giving an increase in θ_e. This structure is consistent with observations showing an annulus of increased sea–air temperature contrast and reduced humidity somewhat outside of the RMW (Cione *et al.*, 2000; Barnes and Bogner, 2001). To achieve energy balance, Wroe and Barnes (2003) found it was necessary to either allow for some entrainment through the boundary layer top, or for turbulence dissipative heating. The energetics of the boundary layer are of prime importance, in light of the considerable sensitivity of storm intensity to the energy content of the boundary layer air beneath the eyewall seen in section 4. The impact of spiral band convection on boundary layer thermodynamics is discussed further in section 7.

5.1.2. *Theory and Modelling*

A satisfactory theory of the tropical cyclone boundary layer is thus challenged to explain the rich observed structure variation between storms, and with radius and azimuth within individual storms. Idealised models of the tropical cyclone boundary layer date back to the 1960s and beyond, and may be classified by their dimension:

- *1-D column:* Moss and Rosenthal (1975); Powell (1980).
- *1-D depth-averaged axisymmetric:* Smith (1968, 2003); Smith and Montgomery (2008); Smith and Vogl (2008).
- *2-D axisymmetric:* Rosenthal (1962); Carrier (1971); Eliassen (1971); Kuo (1971, 1982); Eliassen and Lystad (1977); Mallett (2000); Montgomery *et al.* (2001).
- *2-D depth-averaged:* Shapiro (1983); Vickery *et al.* (2000).
- *3-D:* Kepert (2001); Kepert and Wang (2001).

The calculated flow from several of the 2-D axisymmetric models, and from the 3-D models, display upper boundary layer wind maxima. These simulations have much in common with the observations; the azimuthal maximum is contained within the frictional inflow layer, the maximum becomes more marked towards the centre of the storm, and the depth of the boundary layer decreases towards the centre of the storm. This last property is because the boundary layer has high Rossby-number Ekman layer scaling such that the depth varies inversely with the square root of the inertial stability (Rosenthal, 1962; Eliassen and Lystad, 1977; Kepert, 2001). Indeed, these studies showed that the tropical cyclone boundary layer is a modified Ekman spiral, in which the inertial stability parameter I replaces the Coriolis parameter f, and the spiral is "stretched" in the cross-stream direction by a factor $(\xi/\eta)^{1/2}$. The upper boundary layer wind maximum is similar to that in the Ekman spiral, and a few percent supergradient in these simple models.

These simpler models ignore the influence of vertical advection; including this process gives a markedly more supergradient wind maximum. Analysis of the momentum budget equations shows that the supergradient flow is generated by inwards advection of angular momentum. The inflow is ultimately frictionally generated, but is maintained at the jet height against the outwards acceleration due to gradient imbalance by diffusive and advective transport from below; thus it is stronger in a model that contains vertical advection (Kepert and Wang, 2001). Kepert (2006b) further relates this effect to a modification of the Ekman-like solution as follows: if the vertical advection is zero or neglected, the oscillation and decay length-scales in the Ekman-like solution are equal. Introducing vertical advection makes these scales unequal; in an updraft, the oscillation scale is longer than that for decay, while in a downdraft the opposite applies. Thus the flow will exhibit larger oscillations with height near the boundary layer top in an updraft than in a downdraft. This effect is strongest where upwards motion is strongest, so the boundary layer jet is most marked beneath the eyewall and near rainbands, as is also apparent in the models of Kuo (1971, 1982).

The height-resolving 2-D and 3-D models generally show an increase in the SWF towards the centre of the storm. The depth-averaged axisymmetric and 2-D models show a similar increase in the relative strength of the boundary layer mean wind towards the centre. In both cases, this is due to advection of angular momentum by the frictional inflow maintaining relatively stronger near-surface winds than in boundary layers with straight flow, and the effect is strongest where both the inflow and radial gradient of angular momentum are strong; that is, near the eyewall and possibly also near updrafts associated with secondary wind maxima in outer convective rings. It is important to note that these models do not include any enhancement of the turbulent transport by moist convection; thus the effect is purely dynamical.

Having established that radial advection plays an important role in shaping the structure of the axisymmetric boundary layer, it is perhaps not surprising to learn that azimuthal advection also has a strong influence. Kepert (2001) showed that the motion-induced asymmetry has the horizontal structure of a wavenumber-1 inertia wave. Such waves normally propagate, with the phase speed varying rapidly with radius and are not observed. However, the effect of vertical diffusion is to retard the wave propagation: both the vertical tilt (in azimuth) of the phase lines and the decay of the wave amplitude adjust so as to bring the wave to a halt, locked in position with the asymmetric friction forcing it at the surface. There are two such waves, corresponding to the anticyclonically- and cyclonically-propagating inertia waves, but the stalled version of the anticyclonically-propagating one dominates. It rotates anticyclonically with increasing height, and has a depth scale several times that of the symmetric component. The inclusion of nonlinear processes, including vertical advection, tends to strengthen the asymmetric flow over that predicted by the linear analysis (Kepert and Wang, 2001). Analysis of dropsonde observations in Hurricanes Georges, Mitch and Danielle, but not Isabel, clearly show the predicted anticyclonic rotation of the phase of the asymmetry with height (Kepert, 2006a,b; Schwendike and Kepert, 2008).

This stalled wave structure, when combined with the symmetric component, is able to explain several well-known features of the tropical cyclone boundary layer. In the northern hemisphere, the surface earth-relative wind maximum is in the right forward quadrant, and the inflow angles are greatest on the right of track and least on the left. It also predicts the more recent observational findings, including that the boundary layer jet is more marked, more strongly supergradient, and closer to the surface on the left of track, and that the SWF is higher on the left than on the right. These higher-dimension models demonstrate an important fact: that the boundary layer of a tropical cyclone is distinct from that in much of the rest of the atmosphere, and cannot be satisfactorily understood by 1-D models that assume horizontally homogeneity. Rather, radial advection of angular momentum by the frictional inflow and asymmetric frictional forcing play a substantial role in determining the boundary layer structure and depth.

Variations in angular momentum advection play an important role in determining the spatial variation in tropical cyclone boundary layer wind profile structure. The radial gradient in angular momentum varies greatly between storms — some storms have relatively "flat" radial variation in the wind strength, or equivalently, are inertially stable and have a strong radial gradient of angular momentum throughout, while others have "peaked" profiles, and are inertially near-neutral with a relatively weak angular momentum gradient outside the RMW. Kepert and Wang (2001) presented model calculations of the resulting boundary layer structure for these extremes. The "flat" case had a weakly (5–10%) supergradient jet extending from the RMW to large radii, while the "peaked" one had a strongly (25%) supergradient jet confined to the vicinity of the RMW. The frictional inflow and eyewall updraft are also relatively stronger in the "peaked" case. Observational confirmation of these differences has been provided by Kepert (2006a,b), Bell and Montgomery (2008) and Schwendike and Kepert (2008), who found that of four storms analysed (Danielle, Georges and Mitch of 1998, and Isabel of 2002), two had markedly supergradient flow in the upper boundary layer beneath the eyewall, and two were indistinguishable from balance. The difference in the structure of these storms is as predicted by Kepert and Wang (2001): the "peaked" storms had supergradient flow, while the "flat" ones did not. Figure 4 illustrates the modelled wind profiles in Hurricane Georges corresponding to the dropsonde observations, and shows that the model is able to reproduce much of the around-storm variation in structure, in both the azimuthal and radial flow components. Thus these analyses provide strong confirmation of the theoretical predictions, being able to explain not just the general features, but also the differences among storms. It is likely that the differences in the shape of the vertical profiles of wind in Fig. 2 is due to differences in the structure of the respective storms.

5.2. *Boundary Layer Stability Effects*

Stability is known to have a profound influence on the atmospheric boundary layer. Richardson number-based arguments show this effect will be smaller in the tropical cyclone boundary layer than elsewhere, but not negligible. Powell and Black (1990) demonstrate that it is indeed important, producing a variation in the observed SWF of 0.1, of similar magnitude to the dynamical variations discussed above, with lower values corresponding to stable conditions.

Little further work has been done on this factor, but it is clearly important to complete understanding and the ability to accurately predict the tropical cyclone boundary layer flow.

5.3. *Transients and Instabilities of the Boundary Layer*

The tropical cyclone boundary layer supports a number of instabilities and transient structures. There is some overlap between this section and section 7.3 which considers general instabilities leading to spiral band structures. Here, attention is restricted to phenomena that are clearly part of the boundary layer, chiefly rolls. Section 7.3 discusses fine-scale spiral bands that occur only partly within the boundary layer.

Boundary-layer rolls are very common in the atmospheric boundary layer (e.g., Etling and Brown, 1993), and the circulations of these rolls can produce highly organized surface winds perturbations. Wurman and Winslow (1998) presented the first Doppler radar evidence for their existence in tropical cyclones, indicating intense horizontal roll vortices with an average width of 600 m roughly aligned with the mean azimuthal wind in Hurricane Fran near landfall. The amplitude of variation of wind speed in these rolls was large: bands of 40–60 m s^{-1} flow alternated with 15–35 m s^{-1}. Katsaros *et al.* (2002) examined synthetic-aperture radar images of Hurricanes Mitch and Floyd and also found periodic kilometre-scale variation. More recently, Morrison *et al.* (2005) describe features that are significantly less streaky in appearance, to the extent that it is not entirely clear that they are the same phenomenon. The different radar technology used may have contributed to the differences.

Lorsolo *et al.* (2008) analysed the observed finescale structure of the boundary layer wind field in two landfalling hurricanes from tower and Dopper radar measurements, and found numerous spatially coherent linear features in

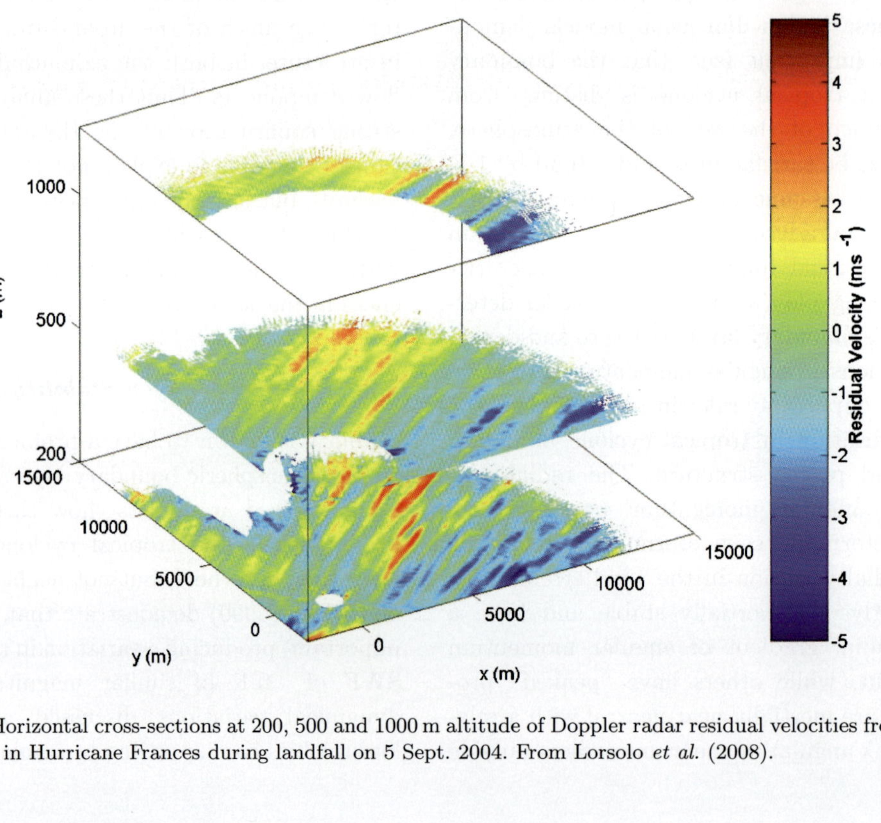

Figure 5. Horizontal cross-sections at 200, 500 and 1000 m altitude of Doppler radar residual velocities from a single radar volume in Hurricane Frances during landfall on 5 Sept. 2004. From Lorsolo *et al.* (2008).

the residual velocity field (Fig. 5). These features have an amplitude of up to $\pm 6\,\mathrm{m\,s^{-1}}$, are coherent through the depth of the boundary layer (500–1000 m), have a mean wavelength of $\sim$600 m, and are oriented on average 7° to the left of the surface flow, with a slight tendency for the orientation to rotate anticyclonically with increasing height. Radar- and anemometer-measured winds are in good agreement, demonstrating that the roll circulation extends to the surface, albeit with other scales of motion superimposed. However, they did not have sufficient data to determine whether these features did possess the full boundary layer roll circulation. Thus these features could contribute to observed damage streaks (e.g., Wakimoto and Black, 1994).

Theoretical analyses of roll development in tropical cyclones have been provided by Foster (2005) and Nolan (2005). Foster (2005) argues that the tropical cyclone boundary layer is an ideal environment for roll development. His argument extends the classical theory of roll development as an inflection-point instability of the frictionally-induced cross-isobar flow to the case of a tropical cyclone. Here, the cross-stream shear and hence instability are strong because the boundary layer is relatively shallow, and because the cross-stream component in analytical solutions is stronger than in classical Ekman-like solutions for straight flow (section 5.1.2). Nolan (2005) presents a stability analysis of a symmetric vortex, and finds both symmetric and asymmetric responses. The instabilities acquire some energy from the shear in the radial flow near the top of the boundary layer. In this regard they are similar to the rolls of Foster (2005). However, Nolan shows that the vertical shear of the azimuthal wind can also contribute energy to the instability, and that the relative importance of these mechanisms depends on both the inertial stability of the storm and the orientation of the mode.

6. Parametric Representation of Tropical Cyclones

Operationally, tropical cyclone are frequently defined by a few parameters, such as the intensity, RMW, and so forth. This supposes an ability to reconstruct the vortex based on those parameters. Such reconstructions are frequently used to force storm surge and wave models, or models of wind damage applied to an urban area, and are thus useful for applications ranging from operational forecasting and warning, to climatological risk assessment and engineering design. They also have a valuable role for initialising numerical models, both for idealised studies and for improving the initial vortex specification in numerical weather prediction models. Lastly, they may provide a framework within which to analyse observations.

A large number of such representations exist. Here, we will consider those of Holland (1980) and Willoughby *et al.* (2006). We omit one popular choice, the modified Rankine vortex, which was however discussed by Holland (1987) and Willoughby (1995).

The Holland (1980) profile model is an extension and improvement of earlier work by Schloemer (1954), in which the radial pressure profile is[6]

$$p(r) = p_c + \Delta p \exp(-(r_m/r)^b). \qquad (30)$$

Here, p_c is the central pressure, Δp is the pressure drop from the environment to the centre, r_m is the radius of maximum cyclostrophic winds, and b a shape parameter. Holland's original analysis assumed constant density with radius, but for lower tropospheric applications where the warm core is weak, it is better to take the virtual temperature T_v as constant with radius and allow ρ to vary through (8). The density at r_m is then $\rho_m = (p_c + \Delta p/e)/(R_d T_v)$ giving the maximum cyclostrophic wind v_{mc}, $v_{mc}^2 = e\Delta p/(b\rho_m)$ where e is the base of natural

[6]Noting that Holland uses $a = r_m^b$.

logarithms. The central pressure deficit is thus $\Delta p = p_c v_{mc}^2 e/(bR_d T_v - v_{mc}^2)$, whence the gradient wind equation (10) gives

$$v_{gr}(r) = -\frac{fr}{2} + \sqrt{\frac{v_m^2\,(r_m/r)^b \exp[1-(r_m/r)^b]}{1+\{\exp[1-(r_m/r)^b]-1\}\,v_m^2/(bR_d T_v)} + \left(\frac{fr}{2}\right)^2}, \tag{31}$$

a form in which the small difference between gradient and cyclostrophic balance near r_m is clear. In situations where a surface wind is required, boundary layer effects are usually crudely represented by reducing the gradient wind by an empirical factor of around 0.8, applying a cross-isobar angle of 20 to 30°, and adding on a wavenumber-1 asymmetry to represent the motion-induced asymmetry, although the use of a simplified dynamical boundary layer model would likely be an improvement.

The Holland profile has been by far the most extensively used parametric profile. Its popularity is probably due to a balance between its simplicity, with few physical parameters to choose, and its ability to give reasonably accurate representations of the majority of cyclone wind fields. Nevertheless, detailed comparison with observations reveals the following limitations: (i) the belt of strong winds near the RMW tends to be too wide, (ii) the wind decreases too rapidly with radius outside two or three times the RMW and (iii) the area of light winds inside the eye can be too large (Willoughby and Rahn, 2004). Other parametric profiles tend to share these limitations, particularly the first.

Recently, Willoughby *et al.* (2006) have presented a parametric profile which overcomes some of the limitations of other profiles, including that of Holland (1980). In particular, it allows a sharper maximum at the RMW, and allows the shape of the profile within and outside of the RMW to be independently adjusted. Unlike Holland's, which is written in terms of pressure and must be differentiated to find the equivalent gradient-wind, Willoughby *et al.*'s is written in terms of wind and must be integrated to find the pressure.

The Willoughby *et al.* (2006) wind profile may be written

$$v_1(r) = (v_{m1}+v_{m2})(r/r_m)^{n_1}, \quad \text{inside blending zone} \tag{32}$$

$$v_2(r) = v_{m1}\exp((r_m-r)/L_1) + v_{m2}\exp((r_m-r)/L_2), \quad \text{outside blending zone} \tag{33}$$

$$v(r) = (1-\alpha(r))v_1(r) + \alpha(r)v_2(r), \quad \text{within blending zone.} \tag{34}$$

The profile consists of the weighted mean of an eye profile v_1 with shape defined by n_1 ($n_1 < 2$), and an outer wind profile v_2 which is the sum of two exponentials of length scales L_1 and L_2, and amplitudes v_{m1} and v_{m2}. The maximum wind is $v_m = v_{m1} + v_{m2}$ at the RMW r_m. The weighting function $\alpha(r)$ is a 9th-order polynomial which increases monotonically from 0 to 1 across a blending zone of width $2L_b$ which contains r_m, with four continuous derivatives at each end of the blending zone. The location of the blending zone is determined by the requirement that the maximum wind occur at r_m, and found by solving $\partial v/\partial r = 0$ at r_m. Physically, the eye-profile v_1 can range through solid-body rotation ($n_1 = 1$) to something more U-shaped. The profile with only one exponential (i.e. $v_{m2} = 0$) is often adequate except for storms with a large radius of gales. It should be noted that these parameters are not independent; for example, more intense storms are usually more peaked (smaller exponential length scales), have more U-shaped wind profiles within the eye (larger n_1) and smaller RMW. The corresponding radial pressure profile can be obtained by radially integrating the gradient wind equation applied to (32), best done taking the T_v constant with radius as above. The profile

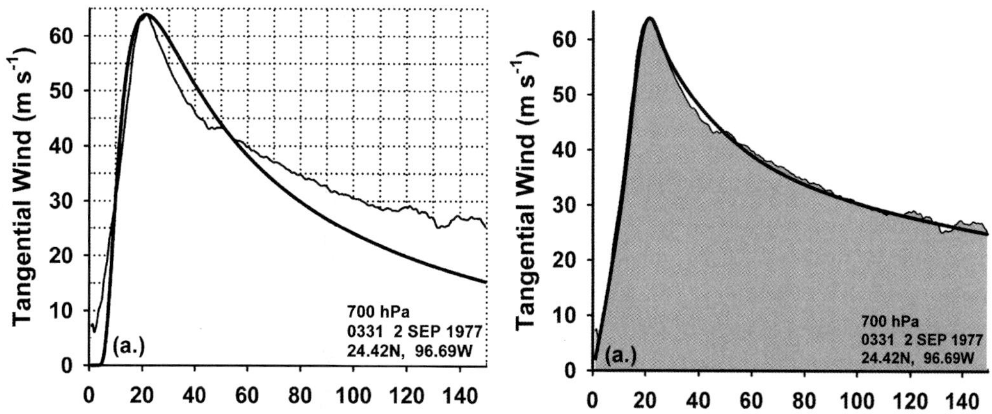

Figure 6. Observed azimuthal-mean azimuthal wind profile at 700 hPa in Hurricane Anita on 2 Sept. 1977, together with fits of equations (31) (left panel) and (32) (right panel) to the observations. From Willoughby and Rahn (2004) and Willoughby *et al.* (2006).

is not amenable to analytical integration and a numerical method is needed (Kepert, 2002c, appendix 4.A2).

The Willoughby *et al.* (2006) profile agrees better with aircraft observations than does Holland (1980), as shown for example in Fig. 6, but has so far received little use. A reason for this may be that fitting it to data involves dealing with a greater number of parameters, and that the fitting process can be poorly conditioned unless some of the parameters are held constant, as recommended by Willoughby *et al.* (2006).

Examination of aircraft data reveals that immediately outside of the RMW, the azimuthal-mean tangential wind decays relatively slowly, leading to a vorticity structure in which the very high values within the RMW are surrounded by a skirt of significant cyclonic vorticity (Mallen *et al.*, 2005). This skirt is absent in some parametric profiles, which instead have an annulus of anticyclonic relative vorticity some distance outside the RMW. While the differences in the wind field are small, the implications for vortex dynamics can be large, since (as will be seen) it is the radial vorticity structure that determines the dynamics of vortex Rossby waves and hence influences the response of the vortex to environmental vertical shear.

Freshwater flooding is a major cause of loss of life and property damage in landfalling tropical cyclones, so there is a need for parametric models of rainfall distribution as well as for wind. A simple model, R-CLIPER, described in Tuleya *et al.* (2007), is limited in that it generates an axisymmetric rain field. Recently, Lonfat *et al.* (2007) describe a rainfall model that includes representations of the spatial asymmetries caused by environmental shear as well as topographic rainfall enhancement.

7. Spiral Bands, Waves and Instabilities

Early radar observations revealed the ubiquitous presence of large-scale spiral bands of cloud and precipitation extending outwards from the eyewall (e.g., Maynard, 1945; Wexler, 1947). More recently, high-resolution radar measurements and careful analysis have shown the presence of abundant bands on much smaller scales as well (e.g., Gall *et al.*, 1998; Kusunoki and Mashiko, 2006). This section will discuss the structure, properties and dynamics of these various bands.

A substantial body of work has attempted to relate observed spiral bands to inertia-buoyancy waves. The connection has not been particularly successful, partly because of the observed scale interactions between cumulus cells and the rain band, and partly because of the difficulty of matching the observed wavelength and propagation characteristics of the bands to the theory. This body of work will not be discussed here; instead, the reviews of Willoughby (1988, 1995) are recommended.

More recent work (Montgomery and Kallenbach, 1997; McWilliams *et al.*, 2003) has elucidated the properties of vortex Rossby waves (VRWs). These are similar to the planetary Rossby wave, except that it is the cyclone's radial vorticity gradient, rather than the planetary vorticity gradient, that provides the restoring influence. They are not identical, however, and the circular geometry leads to some significant differences. The relationship with vortex merger and the resulting vorticity filamentation will be explored, as will their role in cyclone intensification. Possibly the most spectacular manifestation of VRWs is the formation of eyewall mesovortices, which will be seen to have an important influence on cyclone structure, dynamics and thermodynamics. Vortex Rossby waves possess critical and turning radii and so are confined to relatively narrow radial bands.

Spiral rainbands are sometimes called "feeder bands", a term that implies that they help the development of the storm by providing a conduit for moisture to reach the eyewall. It will be seen that they may also transfer PV inwards. On the other hand, they involve the release of latent heat away from the eye, which is less efficient for intensification in balanced vortex models (Hack and Schubert, 1986). The question of the role of rainbands in storm energetics is important, but will be seen to be unresolved so far.

7.1. *Rainband Structure*

Rainbands are trailing spiral bands containing a mixture of stratiform and convective cloud and precipitation. Several observational studies have detailed the structure of individual bands (Barnes *et al.*, 1983; Powell, 1990a,b). There can be significant differences of detail between such case studies, and it is not clear as to the factors that determine these differences[7]. The crossing angle of the band axis is typically around 20°. Because bands spiral cyclonically inwards and the flow above the boundary layer is nearly circular, the cross-band relative flow is typically from the inside of the band outwards except in the lowest few kilometres. There the flow component perpendicular to the band is in the opposite direction, from the outside in. At these levels, the band forms a partial barrier to the cross-flow, with strong inwards flow on the outer side of the band and either weaker or zero inflow on the inner. Strong low-level convergence and upwards motion are associated with this barrier. The along-band and azimuthal flow components usually show a maximum on the outer side of the band axis, which is least qualitatively consistent with the greater radial advection of angular momentum there.

Convective cells, and their associated updrafts and downdrafts, are most common on the inward side of the band axis and at the upstream end of the band, near the low-level convergence maximum. The band-relative cross-flow leads to a stratiform anvil extending mainly towards the outer side of the band, with a smaller or absent anvil on the inner side. Similarly to squall-line convection, mesoscale subsidence of cool dry air beneath the anvil leads to a mass of low θ_e air above the boundary layer. This air forms part of the inflow into the band, where it promotes the development of convective downdrafts. The convective-scale vertical motions on the inner side of the band thus play an important vertical mixing role, bringing low θ_e midtropospheric air to the surface in cool dry downdrafts and transporting high θ_e air upwards. The low-level θ_e may be as much as 20 K lower on the inner side of the band than on the outer.

[7] Care is also needed in interpretation because while most studies utilise storm-centred cylindrical coordinate system, a few use a coordinate system aligned with the band axis.

That the rainband "consumes" low-level θ_e in this manner, thereby depriving the eyewall updraft of some fraction of energy, seems to be the typical but is not the only configuration. A rainband in Hurricane Gilbert showed no such decrease of θ_e on the inside of the band, and a strong radial gradient on the outer side which was too large to be due to surface fluxes alone. The inflow boundary layer that contained this strong gradient lay beneath a layer of strong outflow of high θ_e air, and the downward flux through the top of the boundary layer was, in conjunction with the surface fluxes, sufficient to explain the strong radial θ_e gradient. The outflow layer in this case also precluded the usual inflow of low θ_e subsiding air from beneath the stratiform anvil into the outer side of the band, necessary for the development of downdrafts and the resulting wakes on the inner side of the band (Barnes and Powell, 1995).

A wide rainband in an incipient tropical cyclone (Ryan *et al.*, 1992) had a similar but less organised structure. Principal differences were that the convergence was over ~60 km and convection over ~80 km widths, rather than the 5–10 km and 10–25 km respectively of the above studies. The convergence was from both sides of the band, and was *not* concentrated into a narrow band. This band produced just a weak cool pool from shallow downdrafts, rather than a strong cold wake from marked downdrafts.

7.2. *Vortex Rossby Waves*

The similarity between planetary Rossby waves and tropical cyclone spiral bands was first noted by MacDonald (1968), with a major interest being the upgradient flux of angular momentum due to trailing waves. Montgomery and Kallenbach (1997) derived the dispersion relation for the nondivergent barotropic inviscid case:

$$\omega = n\overline{\Omega}_0 + \frac{n}{r}\frac{\overline{\zeta}_0'}{k^2 + n^2/r^2} \tag{35}$$

where the instantaneous radial wave number is

$$k(t) = k_0 - nt\overline{\Omega}_0' \tag{36}$$

and symbols are defined in Table 2. Radial and azimuthal phase and group velocities derived from (35) are given in Montgomery and Kallenbach (1997). Comparing (35) to the dispersion relation for nondivergent Rossby waves on a β-plane reveals strong similarities, justifying their nomenclature of "vortex Rossby wave". An important difference in cyclones is that the radial wavenumber k varies with time and position, because of the radial shear in the vortex basic state flow. Equation (36) thereby implies that the radial phase speed of a wave will approach 0 as time increases. Möller and Montgomery (2000, Eq. (3.3)) extend (35) to the case of waves with vertical structure.

Consider a northern hemisphere tropical cyclone with $\overline{\Omega}_0 > 0, \overline{\zeta}_0 > 0, \overline{\Omega}_0' < 0$ and $\overline{\zeta}_0' < 0$. Segments of trailing spirals can be described by $k > 0$ and $n > 0$. Then the azimuthal phase velocity ω/k is less than the flow speed $r\Omega_0$ at that radius and the waves retrogress, similar to planetary Rossby waves. The radial group velocity is initially outwards, but decreases to 0 with time due to the increase in radial wave number. Thus the increase in $|k|$ with time causes the waves to possess a stagnation radius, so they are confined to the vortex vicinity and cannot radiate to infinity. The growth of k in time also implies that inwards-propagating, leading spirals will eventually become outwards-propagating and trailing, consistent with observations.

Studies of the interaction of constant vorticity patches in, for example, nondivergent barotropic "contour dynamics" models, frequently show the formation of trailing filaments of vorticity as in, for example, Ritchie and Holland (1993) and Guinn and Schubert (1993), similar to the filaments that arise in the VRW theory. Differences between contour dynamics models and the VRW analysis above is that the vorticity filaments in the former are simply advected, as there is no continuous radial vorticity gradient in such models to provide the Rossby wave propagation mechanism. The contour dynamics models, on the other hand, demonstrate several mechanisms for the formation of the filaments, including the stretching out and shearing of one vortex patch by another and the inherent instability of sufficiently elliptical vortex patches (Melander *et al.*, 1987; Ritchie and Holland, 1993; Velasco Fuentes, 2004).

The filaments of vorticity around the main vortex, in either class of model, imply wind perturbations. Aircraft observations (Samsury and Zipser, 1995) show that most secondary wind maxima in tropical cyclones are co-located with reflectivity maxima, suggesting that vorticity bands are rainbands. In contrast, most rainbands lack a detectable wind maximum, so the converse may not be true.

Numerical simulations of sufficient resolution show spiral band features in the inner core with a high correlation between precipitation, clouds and PV in the bands, implying that these bands are coupled to VRWs (Chen and Yau, 2001; Wang, 2002a,b). Empirical normal mode analysis showed that 70–80% of the wave activity of the larger-scale asymmetries ($n = 1, 2$) could be attributed to VRWs, with gravity waves playing a much smaller role (Chen *et al.*, 2003).

Observations also provide strong support for the connection between these bands and VRWs, with Reasor *et al.* (2000) demonstrating a strong correlation between $n = 2$ bands of Doppler-derived vorticity and reflectivity in Hurricane Olivia. Corbosiero *et al.* (2006) analysed radar data in Hurricane Elena (1985). In the six hours leading up to peak intensity, the eye became elliptical in shape, with the major axis rotating at about half the wind angular velocity. The elliptical shape was due to a pair of eyewall mesovortices (section 7.4). Periodically, $n = 2$ components of reflectivity broke off the outside of the eye and propagated azimuthally at 68% of the mean flow (i.e. retrograde motion) and outwards at about $5.2\,\mathrm{m\,s^{-1}}$. Both motion components were shown to be consistent with the phase and group velocities implied by (35). Figure 7 shows the sequence of radar reflectivity corresponding to one of these instances.

While VRW theory is quite consistent with both observations and simulations in full-physics models, two significant limitations are apparent. Firstly, the agreement at wavenumbers $n = 3$ and higher is less satisfactory, possibly because on smaller scales the dynamics is more influenced by the coupling to convection and also because of the inherent noisiness of nonlinear fluid dynamics at such scales. The smaller-scale bands, and their possible mechanisms, are discussed further below. The second limitation is that the VRWs have a stagnation radius that is typically about 3 times the RMW, and also that the radial gradient of PV becomes weak at around this radius (Montgomery and Kallenbach, 1997; Wang, 2002a,b). Hence the theory of Montgomery and Kallenbach (1997) is not applicable outside of the vortex core. McWilliams *et al.* (2003) present a more comprehensive theory that includes this case, and show that the waves become nondispersive with only azimuthal propagation (see their appendix A.2), consistent with the results of the contour dynamics calculations cited above.

7.3. *Finer Scale Bands*

Aside from the large-scale spiral bands, Fig. 8 shows evidence for numerous bands on the ~10 km scale from radar observations and models (Gall *et al.*, 1998; Kusunoki and Mashiko, 2006). Unlike the larger-scale bands, these bands are simply advected by the mean flow, displaying even less propagation than VRW theory would imply. Similar fine-scale bands have been found in high-resolution numerical simulations by Yau *et al.* (2004) and Romine and Wilhelmson (2006). The modelled bands have a high (but not complete) correlation between PV, w, precipitation and low-level wind maxima, suggesting a significant role for moist dynamics. Cross-sections of the bands from the above observational and

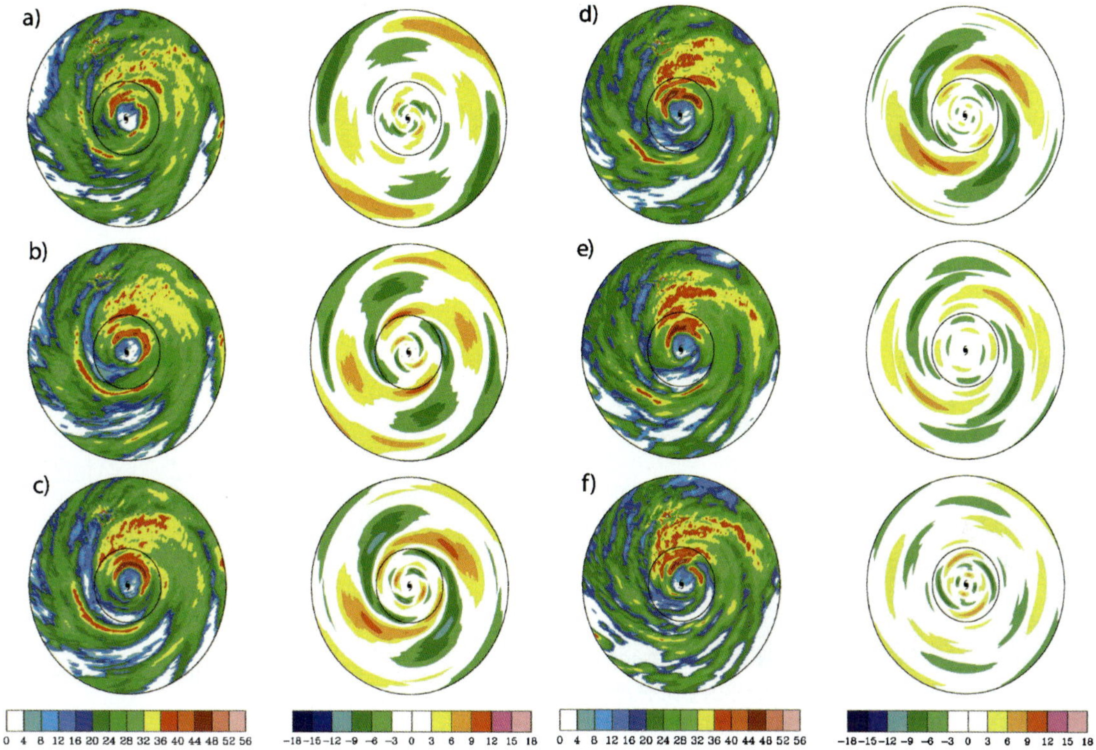

Figure 7. Radar reflectivity (left) and $n = 2$ asymmetry (right) of Elena at (a) 1600, (b) 1620, (c) 1640, (d) 1700, (e) 1720 and (f) 1740 UTC 1 Sep. The circles mark the 50- and 150-km radii. From Corbosiero *et al.* (2006).

modelling studies are compared in Fig. 8. Instability of the strong low-level shear in the radial inflow, somewhat similar to that that leads to the formation of boundary layer rolls but modified by moisture and stability variations, may also play an important role (Nolan, 2005; Foster, 2005; Romine and Wilhelmson, 2006). To date, comprehensive analyses of wave dynamics exist only for larger-scale waves, and analyses of flow stability exist only for relatively idealised situations, so the relative importance of radiating PV waves versus flow instability in the dynamics of these finer-scale bands is unclear.

7.4. *Eyewall Mesovortices*

The existence of small vortices, of ~10 km scale, abutting the inner edge of the eyewall, has been known for some time and linked to the occasional occurrence of a markedly polygonal structure of the inner edge of the eyewall on radar reflectivity images (e.g., Muramatsu, 1986). These polygonal eyewalls typically had between 3 and 6 sides, and rotated somewhat more slowly than the azimuthal flow. Aircraft observations (Marks and Black, 1990; Black and Marks, 1991) showed that in extreme cases, an eyewall mesovortex (EMV) could possess local wind and pressure perturbations of magnitude approaching that of the primary vortex core. A spectacular recent example was the six EMVs observed in Hurricane Isabel, shown in Fig. 9 (Kossin and Schubert, 2005).

Kossin and Eastin (2001) used aircraft data to show that the wind and thermodynamic structure of strong tropical cyclones evolve between two distinct regimes. In regime I, the radial profile of wind across the eye is U-shaped, with maximum angular velocity within

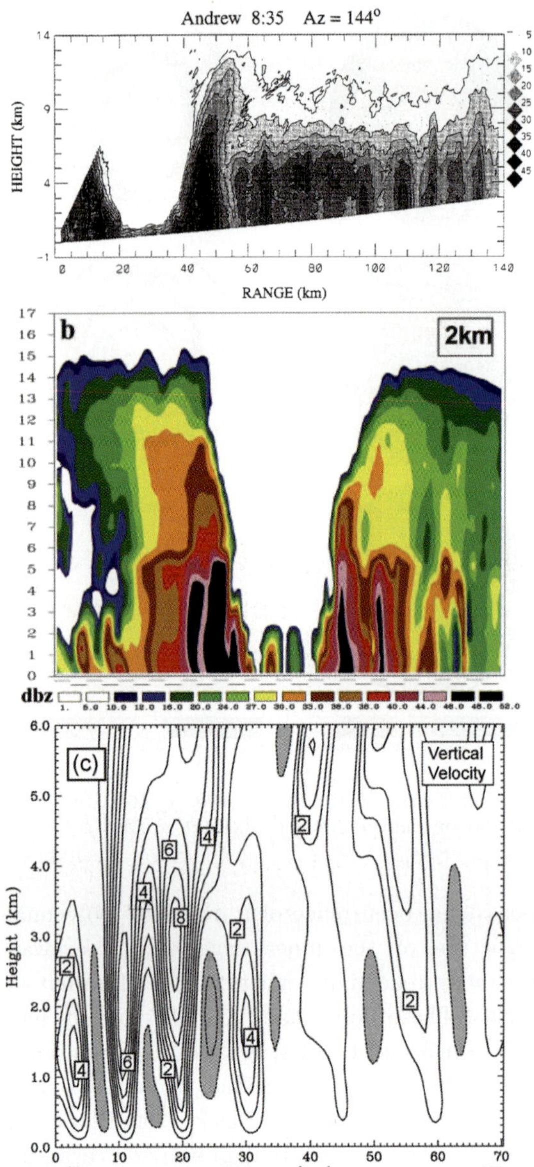

Figure 8. Radius-height sections of fine-scale bands in tropical cyclones. Top: Radar reflectivity observations in Hurricane Andrew, from Gall *et al.* (1998). Middle: Simulated reflectivity in a high-resolution simulation of Hurricane Andrew, from Yau *et al.* (2004). Bottom: Vertical velocity in a high-resolution simulation of Hurricane Opal, from Romine and Wilhelmson (2006).

Figure 9. Defense Meteorological Satellite Program visible image of Hurricane Isabel at 1315 UTC 12 Sep 2003. The six mesovortices — one at the centre and five surrounding it — cause the starfish pattern in the eye. From Kossin and Schubert (2005).

the eyewall, and an annular ring of vorticity just inside of the RMW. The eye is typically warm and dry, with elevated values of θ_e in the eyewall and lower values within the eye. Regime II, in contrast, is characterised by a V-shaped wind structure within the eye, with vorticity and angular velocity maxima near the centre of the eye. The air in the eye is relatively moist, with a θ_e maximum near the vortex centre. The transition from regime II to I occurs relatively slowly, as a consequence of the secondary circulation within the eye (section 3). In contrast, the reverse transition can be very rapid, and may occur in less than an hour. An example of the changes in the dynamical and thermodynamic variables during this transition is shown in Fig. 10. The simultaneous change in kinematic and thermodynamic variables suggests the sudden onset of intense horizontal mixing between the eyewall and the eye. This mixing appears to be caused by the onset of eyewall mesovortices (EMVs), consistent with the barotropic instability of regime I. Kossin and Schubert (2003) discuss how this mixing is distinct from a horizontal diffusion process.

This barotropic instability is similar to that in other strongly sheared flows. The regime I hurricane is approximately a hollow tower of

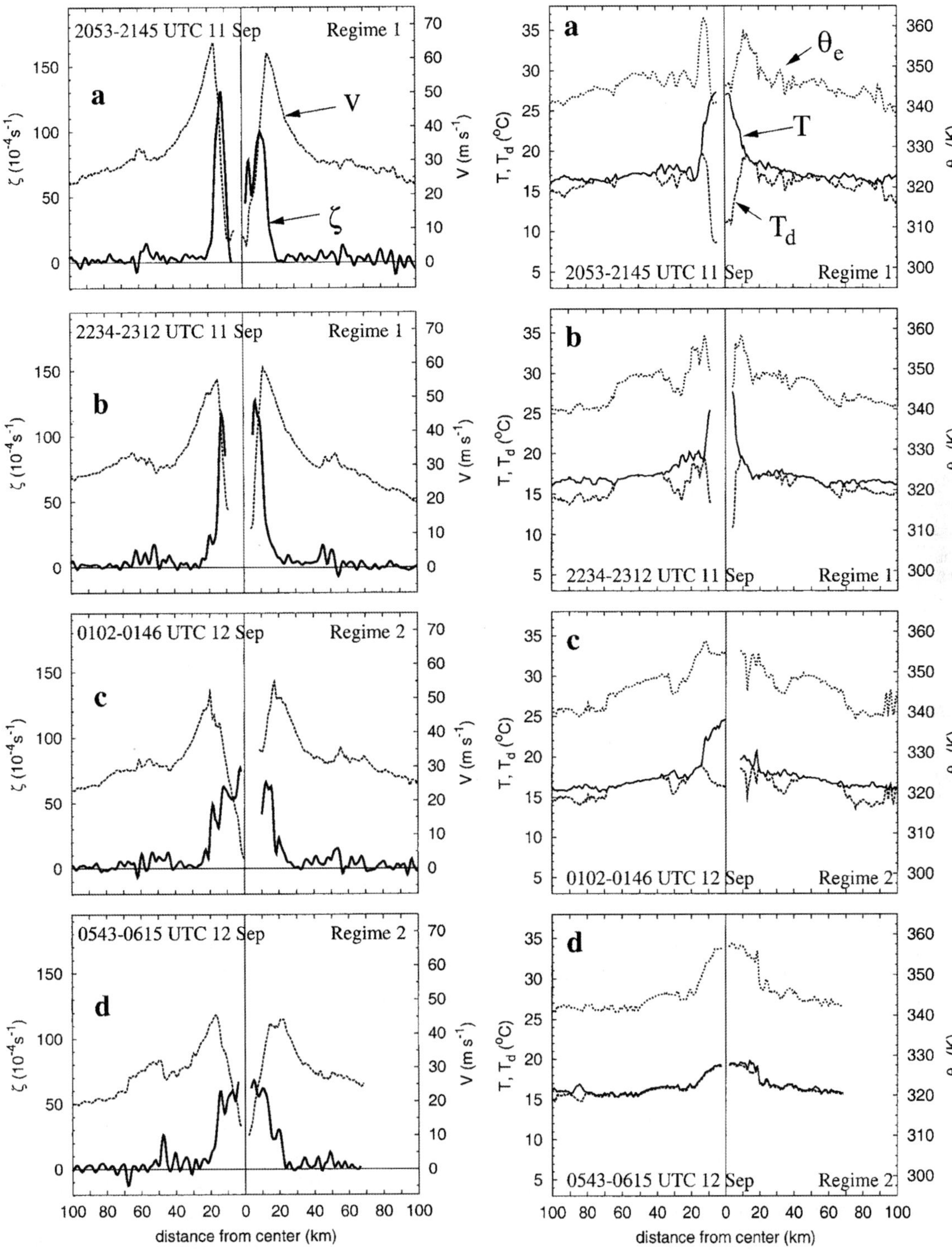

Figure 10. Time-sequence of observed flight-level dynamical (left) and thermodynamic (right) parameters at 850 hPa in Hurricane Diana (1984), showing the transition from Regime I to II. During the transition, v goes from a U-shaped to a V-shaped profile, ζ and θ_e from hollow ring to monopole structures, and the air in the eye becomes saturated. From Kossin and Eastin (2001).

vorticity (located just inside of the RMW), with less vortical flow inside and out. Vortex Rossby waves can propagate on both surfaces of the tower: on the inner face, the propagation is with (i.e. faster than) the flow, while on the outer face, it is against (slower). It is thus possible that the inner and outer waves can phase lock and mutually amplify, leading to exponential growth of the waves. This process was modelled in an unforced barotropic model by Schubert *et al.* (1999), who showed that the vortex ring could break down into several mesovortices, which subsequently merged to form a vortex monopole. Thus the wind profile went from U-shaped to V-shaped; precisely the change noted in aircraft data by Kossin and Eastin (2001). A range of final vortex structures is possible, depending on the size, width and strength of the initial vortex ring. Kossin and Schubert (2001) explore the part of the parameter space relevant to tropical cyclones. The typical response is for the vortex ring to break into a relatively large number of mesoscale vortices, which subsequently merge. These mergers may proceed to give a single discrete vortex monopole, or to a quasi-stable "vortex crystal", an asymmetric lattice of mesovortices rotating as a solid body. The flow due to such a structure consists of near-straight line segments, making up a persistent polygonal shape. Such lattices can also undergo internal rearrangements — e.g. from a pentagon with a central mesovortex to a hexagon, and back again. The former of these structures is strikingly similar to the well-known instance of mesovortices in Hurricane Isabel (Fig. 9).

Detailed observations of eyewall vorticity maxima have been presented in Hurricane Hugo by Marks *et al.* (2008) and in Hurricane Isabel by Aberson *et al.* (2006). The vortex in Hugo had a diameter of ~1 km at 450 m altitude with wind and pressure perturbations of 23 $\mathrm{m\,s^{-1}}$ and 12.5 hPa respectively, a peak updraft of 21 $\mathrm{m\,s^{-1}}$ and a peak cyclonic vorticity of $1.25 \times 10^{-1}\,\mathrm{s^{-1}}$. Hugo's eye also contained a larger-scale cyclonic circulation, about 15 km across, that orbitted the eye with a period of ~19 min, or roughly the angular velocity of the axisymmetric mean vortex, consistent with the wavenumber-1 asymmetry investigated by Nolan and Montgomery (2000) and Nolan *et al.* (2001). The vortex in Isabel produced a dropsonde horizontal wind observation of 107 $\mathrm{m\,s^{-1}}$ and an updraft of 25 $\mathrm{m\,s^{-1}}$ when the vortex mean low-level winds were around 70 to 80 $\mathrm{m\,s^{-1}}$. Radar reflectivity data suggest a horizontal scale similar to that in Hugo. If such wind anomalies extend to the ground, they likely represent a mechanism for increased damage potential that has not been widely appreciated.

These dramatic rearrangements of vorticity are accompanied (in the model) by equally spectacular pressure changes. In the change from the U-shaped (vorticity hollow tower) to the V-shaped (vortex monopole) structures, the winds inside the eye accelerate dramatically, while the maximum wind decreases. Integrating the gradient wind equation inwards, the net effect is that the central pressure falls substantially, even while the intensity as measured by the strongest winds decreases. Similar vortices have been produced in the laboratory: Montgomery *et al.* (2002) describe a water-flow apparatus which produces a curved shear layer, with primary and secondary circulations and aspect ratio similar to a hurricane. Two quasi-steady vortices, together with intermittent secondary vortices, form from shear instability of the curvilinear shear layer on the inner side of the "eyewall". The peak tangential velocity occurred within the mesovortices, and was ~50% stronger than that of the parent vortex.

An eye structure intermediate between the Kossin and Eastin (2001) regimes and possessed by tropical cyclones with an unusually high degree of axisymmetry was identified by Knaff *et al.* (2003, 2008) and named "annular hurricanes". They are also known as "truck tyres" from their appearance on IR satellite imagery. These storms apparently form from the asymmetric mixing of eye and eyewall, possibly by mesovortices, but the mixing does not proceed all the way to the monopole structure of regime II. Annular hurricanes form within certain specific and relatively rare environmental conditions, including weak westward and equatorward environmental shear and favourable thermodynamics, as measured by the PI. In addition to the large, symmetric eye, they are also unusually symmetric outside the eye, with little evidence of outer rainbands. Significantly for forecasting, they maintain intensity longer and weaken more slowly than other tropical cyclones, and are thus a significant source of intensity forecast error.

Much of the interest in EMVs has been in highly symmetric storms with a clear eye in the cirrus overcast, since this facilitates their identification from satellite imagery. However, numerical simulations (Braun *et al.*, 2006; Halverson *et al.*, 2006) have shown that EMVs also exist in sheared storms, where the storm axis is tilted. A tilted axis has enhanced low-level convergence and ascent on the downtilt side, leading to increased rainfall downtilt-left in the northern hemisphere (section 8). The opposite applies uptilt. These papers showed that EMV-like features may orbit the tilted eyewall, with their updraft and vorticity intensifying as they move into the favourable downtilt area, and weakening as they leave it. Cyclonic advection of the enhanced EMV convection led to the coldest cloud tops being on the upshear side of the storm in Hurricane Erin.

7.5. *The Effect of Spiral Bands on Storm Intensity*

Spiral bands may act to reduce intensity, since they consume some of the energy-rich boundary layer inflow and replace it with cool dry downdrafts (section 7.1). However, this energy is replenished by surface fluxes (section 5.1) so this effect may not be large. The heat released in these bands contributes little to storm intensity, since the efficiency of heat release away from the inner core is less (Hack and Schubert, 1986).

On the other hand, spiral cloud bands may form a "buffer" that helps to protect the storm from the effects of environmental shear — there is some observational evidence that large storms are better able to resist shear than small.

Samsury and Zipser (1995) have shown that at least some spiral bands accompany along-band wind maxima, and hence vorticity bands. Their results may well be an underestimate of the frequency with which rainbands contain elevated vorticity, since they had observations at only one level and reported only the incidence of local wind maxima, which would be present for only sufficiently strong vorticity maxima. To the extent that rainbands are vorticity bands, they are vorticity bands that are tilted downshear in a stable sheared flow and will support an upgradient, or inwards, flux of vorticity (Carr and Williams, 1989; Montgomery and Kallenbach, 1997; Nolan and Farrell, 1999; Möller and Montgomery, 1999, 2000; May and Holland, 1999; Chen *et al.*, 2003). Such bands will tend to intensify the storm. On the other hand, Nolan and Grasso (2003) and Nolan *et al.* (2007) have argued that the adjustment process by which the bands form will tend to weaken the storm, and that the azimuthal-mean effect of the heating may be stronger than the vorticity flux.

8. Environmental Influences

Real tropical cyclones are embedded in an environment that includes the surrounding atmosphere, ocean and land, and which has a profound effect on the formation, development, maturity and decay of the storm. In this section, we will focus on the atmospheric part of the environment, with ocean effects in Chapter 3. Important topics include the effects of vertical shear on storm structure and intensity change, and the effects of upper troughs.

Shear is long recognised as an important factor, starting with the original ventilation concept; that is, the tendency for the sheared environment to move the upper level warm core away from the surface circulation (Simpson and Riehl, 1958). Observational and modelling studies agree that there is a shear threshold for weakening of a mature storm, that weakening may not occur immediately with the onset of shear, and that the detailed response depends on storm intensity and structure. Wind shear causes a substantial asymmetry in the storm's precipitation field. Whether this translates into an asymmetry in the precipitation swath depends on the shear-relative motion, with along-shear motion resulting in an asymmetric precipitation swath. The environmental moisture around a storm can also influence its response to shear.

Trough interactions are an important effect that is difficult to diagnose, with opinion divided as to their role overall and in individual cases. Dynamically, most studies are founded on the idea that eddy fluxes of angular momentum may most easily influence the storm core in the upper levels due to the low inertial stability there (section 3). However, upper troughs often coincide with an increase in the vertical shear. Evidence seems to be that both intensifying and weakening influences are possible, but the problem of differentiating "good" from "bad" troughs is far from solved. They also form an essential component of extra-tropical transition, discussed in Chapter 5.

When a number of significant environmental influences exist, determining the net effect of the combination can be complex. For instance, the weakening effect of strong wind shear may be reduced if the storm is over a region of high ocean heat content, or increased if the environmental air is unusually dry.

8.1. *Dry Dynamics*

Imposing vertical shear on a vortex will cause it to tilt in the direction of the shear. The resulting evolution can be qualitatively understood by considering a two-level system, in which the vortex is located at different positions in the two levels. The PV inversion principal (section 2.1) shows that the upper level of the vortex will project downwards and generate a cyclonic flow displaced from the lower vortex, while the lower vortex will similarly generate an upper-level circulation. These induced circulations will cause the upper and lower parts of the vortex to orbit each other; that is, the tilted vortex axis will precess with time (Jones, 1995). The rate of precession depends on the strength of the flow induced by the displaced PV fields at the two levels. It will therefore be stronger for a larger or more intense system and for lower static stability, since these changes both act to increase the penetration depth and strength of the induced flow (Jones, 1995; Smith *et al.*, 2000). If the shear is sufficiently strong, then the upper and lower vortices will separate. The threshold for this separation will depend on the strength of the interaction. In these dry simulations, the precession of the tilt axis continues until the vortex is tilted up-shear, whereupon the shear acts to temporarily reduce the tilt. However, following further precession, the tilt is again downshear and continues to increase.

For the flow in a tilted vortex to remain balanced, a warm anomaly must develop on the up-tilt side, and a cold anomaly down-tilt, so that (through hydrostatic balance) the pressure centre can have a similar tilt to the wind centre. Thus the isentropes are perturbed upwards on the down-tilt side, and downwards uptilt. As the flow follows these isentropes, maximum ascent occurs to the right of the tilt vector, with maximum descent to the left in the northern hemisphere. The strength of the vertical motion asymmetry and the potential temperature anomalies increase with the vortex tilt, until the vortex begins to weaken (Jones, 1995).

Vortex Rossby wave dynamics provide an alternative view of the evolution of the vortex tilt (Reasor and Montgomery, 2001; Schecter *et al.*, 2002). Here, the tilted vortex is decomposed into the sum of a vertical axisymmetric vortex, and a baroclinic perturbation with azimuthal wavenumber one and vertical structure of the first internal mode. That is, it consists of a cyclonic perturbation underlying an anticyclone on one side of the storm, with the opposite on the other side. Adding this perturbation to the axisymmetric mean vortex gives a vortex with a tilted axis. The evolution of the sum can then be understood in terms of the VRW dynamics of the perturbation. These calculations assumed quasigeostrophic dynamics and so strictly apply only to weak tropical cyclones.

The response falls into one of two categories, depending on L_R (Reasor and Montgomery, 2001; Reasor *et al.*, 2004). For L_R larger than the horizontal scale of the vortex, the tilt mode projects onto an azimuthal wavenumber one quasi mode which rotates cyclonically with time and decays very slowly. The decay is essentially negligible on meteorologically relevant time-scales. This case is similar to the solutions of Jones (1995, 2000b) except that the vortex tilt decays slowly instead of growing indefinitely. At smaller L_R, the initial perturbations project onto a continuous spectrum of VRWs whose integrated perturbation energy decays algebraically to zero. This effect can be regarded as either due to constructive interference, or to the spiral wind-up of the waves. The original calculations used a quasigeostrophic model (Reasor and Montgomery, 2001), while later calculations for finite Ro (Reasor *et al.*, 2004) demonstrated a stronger resistance to tilt in this case, and showed that a vortex in shear arrives at an equilibrium with the vortex tilted to the left of the shear vector (in the northern hemisphere). The degree of tilt is greater in the outer vortex than in the core.

A distinct, but complementary view of the small L_R case is given by Schecter *et al.* (2002) and Schecter and Montgomery (2004), who show that the tilt mode decays exponentially with time due to resonant damping with a critical layer. The baroclinic mode propagates cyclonically, and so there exists a critical radius r^* at which the cyclone's azimuthal velocity matches the phase speed of the mode, and is thus the radius where the mean flow and the wave most readily interact. This interaction damps the wave, with concomitant decay of the vortex tilt, provided that the radial gradient of potential vorticity at r^* is negative. The decay is initially exponential, with rate proportional to the PV gradient at r^*. Hence for a zero gradient, the wave is neutral, while for a reversed (i.e. positive) gradient, the wave will grow and the vortex tilt will increase with time. The rate is higher for strong static stability and also depends on the internal Rossby radius of deformation L_R. The details of this latter dependence are sensitive to the structure of the axisymmetric vortex, and Schecter *et al.* (2002) present examples showing that the decay rate may either increase or decrease with increasing L_R.

The most important consequence of Schecter *et al.* (2002)'s results is that a tilted vortex will not align unless the radial gradient of PV at the critical radius is negative. For a zero gradient, the relevant modes are undamped and the vortex will not align. This situation may occur either in the limit of infinite L_R in any vortex since then $r^* \rightarrow \infty$, or in "compact" vortices in which the PV is entirely bounded by some radius. Furthermore, in vortices where the radial PV gradient is positive at r^*, the tilt mode will grow and the vortex is unstable to small tilts. Reasor *et al.* (2004) shows that many of the vortices in the simulations of Jones (1995, 2000b) possess this latter characteristic, and thus explains why many of her simulations show a general tendency for the vortex tilt to increase with time. The growth of tilt with time is therefore not a universal characteristic. Aircraft data show that the radial vorticity gradient in the lower to mid troposphere is generally negative to 3 times the RMW, a range which would normally include the critical radius r^*, and therefore real tropical cyclones are inherently stable to shear-induced tilt (Mallen *et al.*, 2005).

For initially baroclinic dry vortices, the situation is more complex. Simulations by Jones (2000b) show that the low- and mid-level centres orbit each other, while the upper level centre is usually advected away, and may shed vorticity filaments followed by axisymmetrization (e.g., Möller and Montgomery, 2000). These vortices had the same radial structure as her studies with initially barotropic vortices, so the growth of tilt is consistent with the above VRW-based analysis. Frank and Ritchie (1999) presented a calculation in which the surface-500 hPa tilt grows indefinitely, but the upper-level centre has relatively less movement. Their initial vortex was broader and weaker than that of Jones (1995), but its stability to tilt is not known.

In this section, we have seen two distinct but related groups of theory describing the evolution of a vortex in vertical sheared environment. First we considered the mutual interaction of horizontally displaced upper and lower tropospheric masses, where the interaction proceeds via the vertical projection of the lower mass into the upper troposphere and vice versa and leads to a mutual rotation about each other, and hence a precession of the tilt axis. This analysis also explains a marked wavenumber one asymmetry in the vertical motion in the storm, with the maximum updraft on the downtilt side. The second body of theory decomposes the tilted vortex into a vertical axisymmetric vortex and a first internal mode, azimuthal wavenumber one perturbation, and describes the evolution of the tilt through the vortex Rossby wave dynamics applied to the perturbation. The response falls into two regimes: the quasimode with cyclonic progression accompanied by slow decay, and the spiral wind-up or resonant damping mode where the perturbation decays rapidly and the vortex thereby strongly resists the tilting effect of the environmental shear. The latter theory importantly identifies stability criteria that decide whether the vortex is inherently stable or easily sheared apart, with many of the vortices used for calculations with the first theory falling into the unstable category. Aircraft measurements show that most real tropical cyclones have a structure that is stable to tilt. However, the basic mechanism for precession of the tilt axis in slowly-decaying tilts in the second body of theory is essentially that for precession in the first. Moreover, the PV-interaction theory predicts a strong asymmetry in vertical motion that is highly consistent with observations of rainfall and cloudiness in tropical cyclones, and is therefore important to understanding and predicting this aspect of real cyclones. A further limitation of the VRW-based theory is that it has so far considered only barotropic vortices, and while it has been most fruitful for understanding the dynamics, this assumption may omit some potentially important aspects of the problem.

8.2. *The Influence of Moisture*

Idealised "full-physics" simulations of tropical cyclones in sheared environments have been presented by (for example) Wang and Holland (1996), Frank and Ritchie (1999, 2001), Peng *et al.* (1999), Kimball and Evans (2002), Wong and Chan (2004), and Ritchie and Frank (2007). The evolution of the tilt in all of these simulations is in stark contrast to the dry simulations of Jones (1995, 2000a,b) and Frank and Ritchie (1999), in that precession of the axis does not occur and a near-equilibrium of downshear-left tilt develops, except when the shear is sufficient to destroy the storm. The natural conclusion drawn from these results was that the secondary circulation produced by the diabatic heating acts to couple the upper- and lower-level centres and resist the tilting effects of shear. A more complete explanation was provided by Schecter and Montgomery (2007), who showed that clouds in the core of a monotonic vortex can cause or accelerate the decay of discrete VRWs, thereby increasing the resilience of the vortex to tilt (section 8.1). Another factor may have been that the diabatic processes cause the initial vortex to evolve towards a structure that is stable to tilt. Unfortunately, of the full-physics simulations listed above, only Wang and Holland (1996) provide sufficient information to determine the tilt-resilience of their vortex. Their initial vortices have monotonic radial PV gradients and are therefore tilt-resilient, except in the upper troposphere, where an upper anticyclone leads to a reversed gradient at about 300 km radius (their Fig. 1).

Moist simulations that consider a range of wind shears show that the effect of shear on intensity is larger for stronger shear, with a threshold of about 8–10 $\mathrm{m\,s^{-1}}$ of shear for weakening to occur. There is often a time lag between the onset of shear and the onset of weakening, of from a few hours (very strong shear) to over 36 hours for 5 $\mathrm{m\,s^{-1}}$ shear (Frank and Ritchie, 2001; Wong and Chan, 2004). Frank and Ritchie (2001) show that the storm weakens from the top down, and identify the following sequence of events: (i) the development of strong inner core asymmetries, (ii) the asymmetries in the upper core (where the storm circulation is relatively weak) become strong enough to mix heat and PV outwards, allowing the shear to ventilate the eye; (iii) the loss of the upper warm core causes the surface pressure to rise, weakening the storm; and (iv) downshear advection of the weakened upper levels causes downshear tilt of the storm. The top-down weakening ends when the ventilated layer can descend no further, due to the greater strength and inertial stability of the vortex at lower levels. Wong and Chan (2004) found that the sensitivity of the cyclone to shear depended upon the size of the storm, with smaller storms being less resistant.

Environmental shear causes marked asymmetries in vertical motion, cloudiness and rainfall in these moist simulations. The initial cause of these asymmetries is as in the dry simulations discussed above, but as the storm develops further, latent heat release destroys the cold downshear anomaly and the maximum updraft moves to the downshear-left quadrant, together with the maximum cloud water. The rainfall maximum occurs slightly downstream on the left of the shear (Fig. 11), due to advection of the precipitation as it falls (Wang and Holland, 1996; Frank and Ritchie, 1999, 2001; Wong and Chan, 2004). The difference between simulations on f- and β-planes seems to be largely that the β-gyres (see Chapter 4) are strongest at the surface and hence induce a few $\mathrm{m\,s^{-1}}$ of equatorwards and eastwards shear over the storm core (Peng *et al.*, 1999; Ritchie and Frank, 2007). Thus the effective total shear is less for easterly environmental shear than for westerly, as the former is partly cancelled by the β-gyre-induced shear (Peng *et al.*, 1999; Ritchie and Frank, 2007).

The effect of these asymmetries on the rainfall swath depends on the relative directions of the shear and the storm motion. Where the shear is along-track, the rainfall occurs predominantly on the left of track. Conversely, for cross-track shear, this left-of-shear rain is distributed approximately symmetrically about the track (Rogers *et al.*, 2003).

8.3. *Observations of Shear and Intensity Change*

Several studies have examined the intensity change in tropical cyclones as a function of environmental shear. Statistical intensity forecast schemes (DeMaria and Kaplan, 1994, 1999; DeMaria *et al.*, 2005) have universally included environmental vertical shear is a statistically significant predictor, with the threshold between weakening and strengthening lying at about 10 $\mathrm{m\,s^{-1}}$ of 850–200 hPa shear. Gallina and Velden (2002) found a strong linear correlation between shear and intensity change, with the transition between strengthening and weakening occurring at a shear of 7–8 $\mathrm{m\,s^{-1}}$ in the North Atlantic, and 9–10 $\mathrm{m\,s^{-1}}$ in the North West Pacific. They also noted a time lag between the onset of shear, and weakening. Paterson *et al.* (2005) similarly found that shear values above about 10 $\mathrm{m\,s^{-1}}$ favoured weakening, and above 12 $\mathrm{m\,s^{-1}}$ rapid weakening, in the Australian region, with a time lag of 12 to 36 hr between the onsets of shear and weakening. While this study found that weak shear (2–4 $\mathrm{m\,s^{-1}}$) generally led to rapid intensification, a small group of storms failed to intensify even when the shear was very weak.

The Saharan Air Layer (SAL) is a layer of dry, dusty air between about 850 and 500 hPa, overlying a marked temperature inversion, with an easterly jet at about 700 hPa. When this feature engulfs a hurricane, the ventilation due

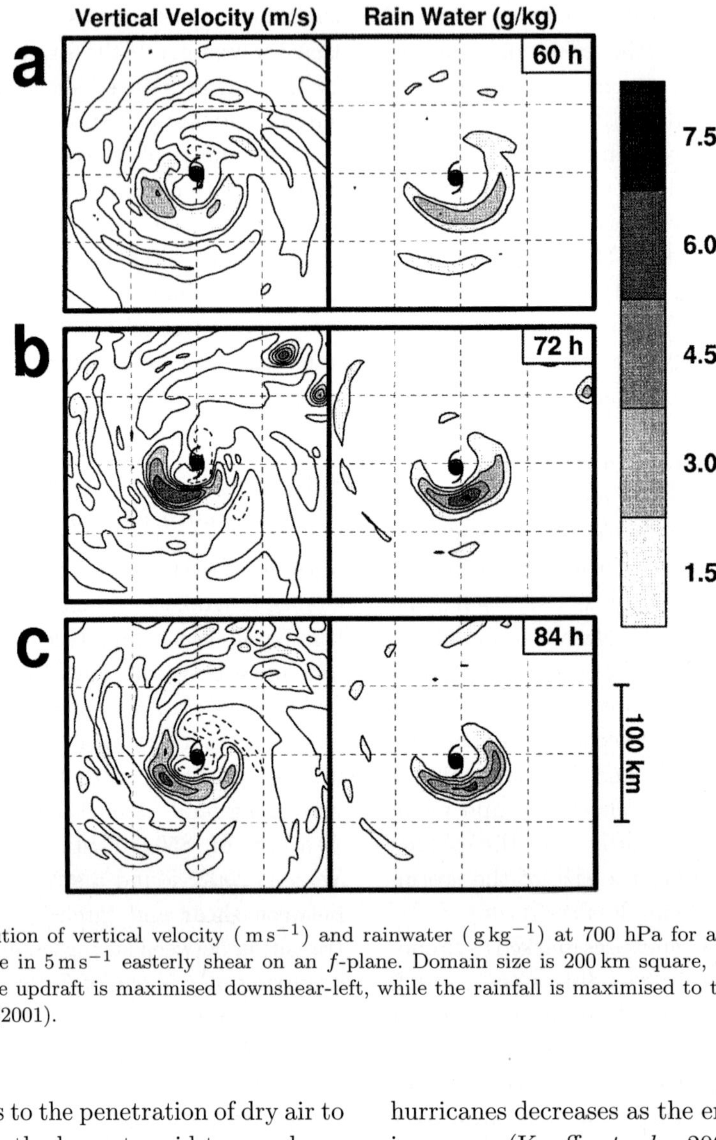

Figure 11. Evolution of vertical velocity ($\mathrm{m\,s^{-1}}$) and rainwater ($\mathrm{g\,kg^{-1}}$) at 700 hPa for an idealised simulation of a tropical cyclone in $5\,\mathrm{m\,s^{-1}}$ easterly shear on an f-plane. Domain size is 200 km square, dashed contours show negative values. The updraft is maximised downshear-left, while the rainfall is maximised to the left of shear. From Frank and Ritchie (2001).

to the shear leads to the penetration of dry air to the storm core in the lower to mid-troposphere, causing convective downdrafts that cool and dry the boundary layer. The combination of shear, dry air and the low-level inversion can lead to rapid destruction of hurricanes (Dunion and Velden, 2004). The importance of the combination of dry environmental air and shear was also recognised by Emanuel *et al.* (2004), who developed an empirical parameterisation of these effects for incorporation into an axisymmetric intensity prediction model.

Satellite remote-sensed temperature profiles show that the height of the warm core in hurricanes decreases as the environmental shear increases (Knaff *et al.*, 2004). The effect is not large, with the difference in height of the peak mean temperature anomaly between the low- ($<3.8\,\mathrm{m\,s^{-1}}$) and high-shear ($>6.8\,\mathrm{m\,s^{-1}}$) groups of about 1 km, although the coarse vertical resolution of the satellite sounder makes accurate estimation of the difference difficult. Nevertheless, the trend is consistent with the moist simulations discussed above, and particularly with Frank and Ritchie (2001)'s finding that storms in full-physics simulations weaken from the top down in environmental shear.

8.4. *Observations of the Convective Asymmetry*

Azimuthal variations in cloudiness and precipitation are a common and widely recognised symptom of the asymmetric vertical motion resulting from environmental vertical wind shear. Such asymmetries have been presented in case studies of individual storms, and are confirmed in surveys of large numbers of storms.

Detailed studies of the evolution of two hurricanes in sheared environments, Jimena (1991) and Olivia (1994) were presented by Black *et al.* (2002). At the start of the respective study periods, each storm had a small eye of 16–18 km radius, maximum winds of approximately 57 $m\,s^{-1}$ and lay over SSTs in excess of 28°C. Jimena experienced two days of 13–20 $m\,s^{-1}$ easterly shear, during which it maintained constant intensity or weakened slightly. Olivia intensified during a day of 8 $m\,s^{-1}$ easterly shear, but weakened rapidly the following day as the shear reversed and increased to greater than 15 $m\,s^{-1}$ from the west, and the storm moved over cooler water. The evolution of Olivia on the second day, over a period of about 3.5 hours during which the vertical shear over the inner core increased from 3–5 to 15 $m\,s^{-1}$ over the lowest 10 km, was studied in detail by Reasor *et al.* (2000) and Black *et al.* (2002). The azimuthal-mean circulation dramatically weakened, especially at higher levels, and the RMW expanded. The radar reflectivity became markedly asymmetric, with peak reflectivities to the left of the vertical shear vector, consistent with a strong asymmetry that developed in the eyewall vertical motion with the maximum updraft to the downshear-left. The vortex tilt over the 0.75–6 km layer increased from zero to about 3 km. The eye became elliptical and trailing spirals of vorticity with azimuthal wavenumber 2 formed.

The convective structure in both storms was strongly influenced by the environmental shear. Jimena displayed a strong tendency for convection to be to the left of the shear vector on both days. Individual convective cells formed on the downshear side of the eyewall and propagated around to the left at 60–80% of the tangential wind velocity, consistent with VRW propagation. Most dissipated near the upshear side, although one long-lived cluster was tracked for a full circuit. The observed buoyant updrafts accelerated to the left of the shear vector, rising through the freezing level on the upshear side whereupon the hydrometeors fell out or froze, resulting in the decrease in reflectivity. On the right of shear, lower tropospheric precipitation-driven downdrafts lay beneath glaciated upper tropospheric updrafts. Olivia similarly displayed a relatively weak asymmetry while the shear was weak, with eyewall cells able to be tracked right around the eyewall. Outer convection was close to circular, and may have represented the beginnings of an eyewall replacement cycle. As the shear increased and the storm moved over colder water and weakened, the reflectivity in and outside of the eyewall became strongly asymmetric and individual eyewall cells became short-lived and confined to the left-of-shear semicircle. Figure 12 shows aircraft data from the storm at this stage. The strong left-of-shear radar reflectivity signature is highly characteristic of this situation, and is accompanied by maximum radial inflow and vertical motion at the downshear end of the eyewall.

Hurricane Norbert of 1984 (Marks *et al.*, 1992) similarly had a strong radar reflectivity asymmetry, with inner core precipitation predominantly to the left of the inner-core shear vector. The eyewall vertical motion was consistent with the rainfall structure, being upwards in the right-front quadrant, increasing to strongly upwards to the left, with descent to the rear and right. The earth-relative radial and azimuthal winds were kinematically equivalent to a front-to-back crossflow at 1-km height, reversing to back-to-front above 3 km, consistent with the shear structure, although the flow up to 3-km height also contained characteristics of the motion-induced boundary-layer

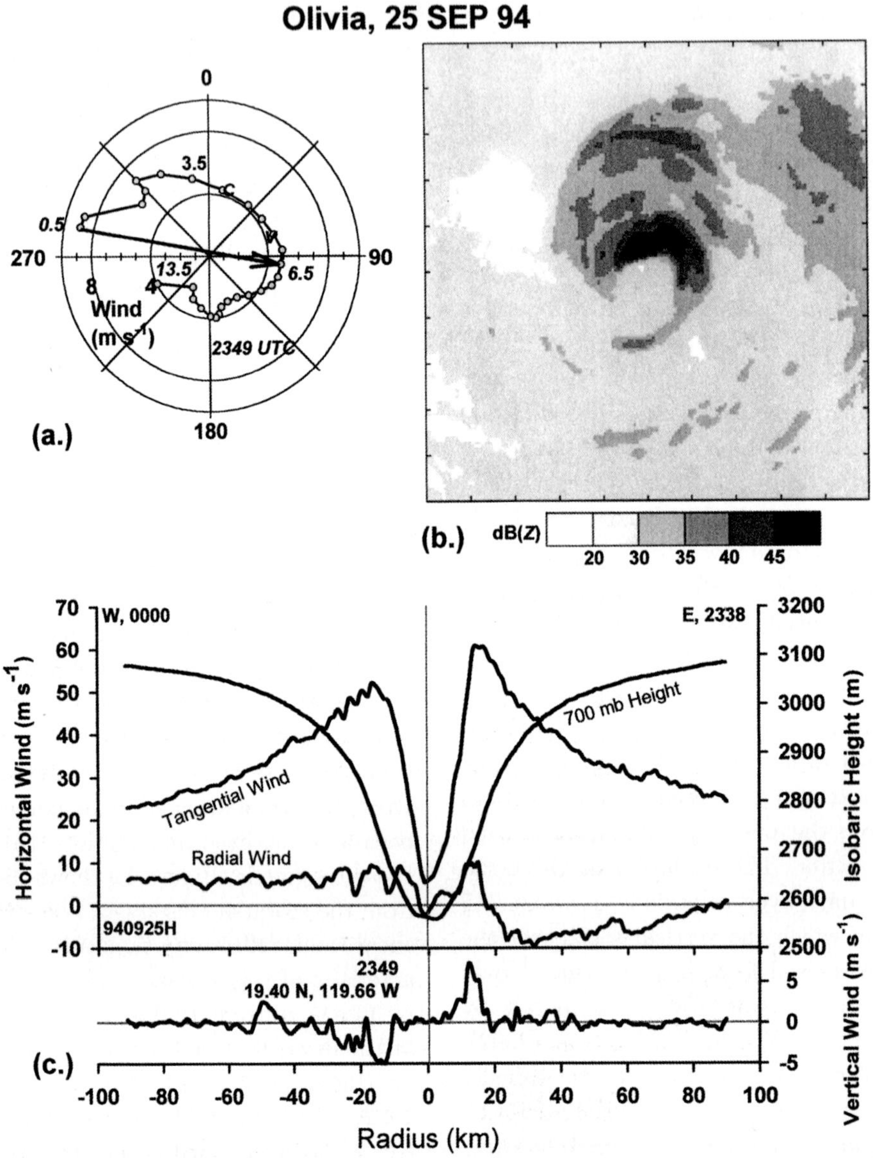

Figure 12. Hurricane Olivia late in the day on 25 Sep 1994: (a) Doppler-derived relative-wind hodograph, storm motion (C) and vertically-averaged earth-relative wind (V) at 2349 UTC. Winds are averaged over a 20–30 km storm-centred annulus. The arrow represents the largest shear. (b) Radar reflectivity PPI composite for 2338-2359 UTC, domain size is 240 k square. (c) An east-west profile of flight-level observations at 700 hPa, 2338–0000 UTC. From Black *et al.* (2002).

asymmetric flow (section 5). Hurricane Gloria of 1985 (Franklin *et al.*, 1993) similarly had the maximum updraft at 500 hPa on the down-shear side of the eyewall, and descent on the upshear side. The eyewall radar reflectivity was a maximum in a band extending cyclonically from the updraft core to the downdraft; that is, predominantly on the left of the shear vector.

A more extensive survey of convective asymmetry, using lightning flash data as a proxy for convection, was conducted by Corbosiero and Molinari (2002, 2003). Lightning is strongly

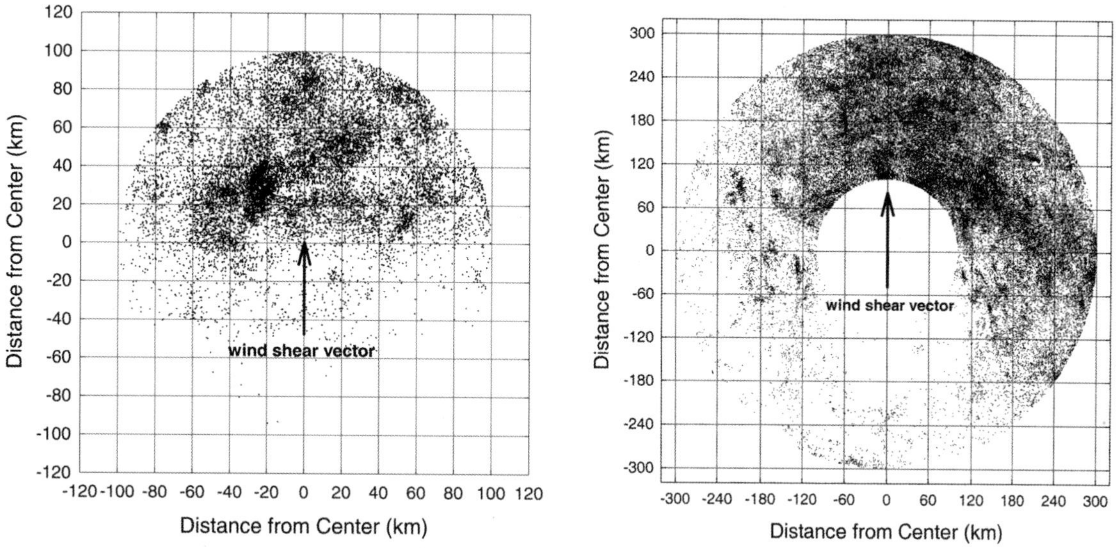

Figure 13. Locations of lightning flashes (left) within 100 km and (right) 100–300 km of the storm centres for environmental shear exceeding $10\,\mathrm{m\,s^{-1}}$. The data have been rotated so that the wind shear vector is pointing upwards on the page. From Corbosiero and Molinari (2002).

correlated with the direction of the environmental shear. Over 90% of flashes occur in the downshear half of the storm once the shear exceeds $5\,\mathrm{m\,s^{-1}}$. There is a slight preference for downshear left in the innermost 100 km, and a stronger preference for downshear right in the 100–300 km annulus, as shown in Fig. 13. The downshear-left preference in the core is more marked in more intense storms, probably because the strong rotation twists updrafts initiated downshear helically around to the left, then upshear, of the storm centre. In most of the storms, shear and the motion-induced frictional asymmetry (section 5) would place the convective maximum in similar quadrants, so it is difficult to separate the effects. In storms where the expected maxima were in different quadrants, the shear-induced asymmetry dominates that due to motion.

The rainfall asymmetry due to shear and storm motion, as measured by instruments on the TRMM satellite, was reported by Lonfat (2004) and Chen *et al.* (2006). A consistent downshear-left asymmetry is apparent (downshear-right in the southern hemisphere). It is similar to that in the lightning data of Corbosiero and Molinari (2003), except that the rain asymmetry is most marked for the weaker storms, and is downshear-left at all radii. The shear-induced asymmetry is reasonably independent of storm intensity for strong shear $>7.5\,\mathrm{m\,s^{-1}}$, but weaker shear $<7.5\,\mathrm{m\,s^{-1}}$ impacts most strongly on less intense storms. The expected right-front motion-induced convective maximum (section 5.1.2) is swamped by the effects of shear when the shear is strong, but some signal remains for weak shear outside the storm inner core. The preferred downshear-right convection in the outer core often occurs in the form of a quasistationary, persistent "principal band", accompanied by weaker mobile "secondary bands" (Willoughby *et al.*, 1984).

Annular hurricanes (section 7.4), characterised partly by their unusually high degree of symmetry on satellite images, require weak polewards and westwards environmental shear for their formation (Knaff *et al.*, 2003). The environmental shear cancels out the β-gyre-induced shear over the core. They thus represent a null case for shear-induced convective asymmetry.

8.5. *Observations of Axis Tilt*

Actual observations of axis tilt through the depth of the troposphere in tropical cyclones are rare, due to the difficulty of obtaining sufficient data to locate a well-defined centre at multiple levels. The wind centre of Hurricane Norbert, located from airborne Doppler radar, described a roughly elliptical anticyclonic spiral with height, with long axis about 4 km oriented approximately along the inner core shear of about $10\,\mathrm{m\,s^{-1}}$ over 10 km (Marks *et al.*, 1992). Franklin *et al.* (1993) found that the centre of the very intense Hurricane Gloria had only slight displacement with height, in spite of significant environmental shear. The pressure centres of Hurricane Bonnie (1998), located from aircraft data at 850, 600 and 300 hPa and at the surface from dropsonde observations, displayed a surface — 300 hPa tilt of about 45 km oriented about 45° left of the environmental 850 — 200 hPa shear of $11.9\,\mathrm{m\,s^{-1}}$ (Kepert, 2005). Most of the tilt was between the 600 and 300 hPa levels, where the shear was also strongest. Simulation of this storm by Rogers *et al.* (2003) showed a similar but smaller tilt of about 15 km over the surface — 400 hPa layer, with the simulated rainfall asymmetry in good agreement with that observed by airborne radar. The agreement between observed and modelled tilt is reasonable, with the difference being partly because the observations are 6 hours earlier than the simulation, that the shear was slowly weakening, and that the model shear was presented over a thinner layer. Finally, Zehr (2003) compared aircraft and satellite-estimated centre positions of Hurricane Bertha at three levels. Initial tilts of up to 104 km, slightly to the left of the environmental shear, were diagnosed, although later in the period the upper-tropospheric tilt reversed while the shear only slowly weakened.

8.6. *Shear Associated with Upper-level Troughs*

Environmental shear generally results in a tropical cyclone weakening (section 8.3), except when the shear is due to an approaching upper trough. This situation is complex, and a trough may cause either weakening or strengthening depending on the particular circumstances. It can also be difficult to extract the trough effect from other intensity change influences. Nevertheless, this is an important situation, since the low inertial stability of the outflow layer provides an avenue whereby the environment can directly affect the core of the storm. In this section, we restrict attention to the case where the cyclone remains tropical after the interaction. The important process of extratropical transition is considered in Chapter 5.

Hurricane Elena (1985) twice experienced marked intensity changes, each about 30 hours after periods of significant upper level eddy flux of angular momentum (Molinari and Vollaro, 1989, 1990). Major intensification occurred following the passage of a mid-latitude trough polewards of the storm, which caused a large inwards angular momentum eddy flux. An in–up–out circulation, consisting of a thin upper tropospheric outflow and a deep inflow layer connected by a band of ascent, developed inwards of the momentum source, as predicted by balanced vortex theory (section 3), and propagated inwards. When it reached the core, a secondary wind maximum developed and contracted, resulting in an extended period of rapid intensification. A PV analysis of the event (Molinari *et al.*, 1995) showed that the trough interaction was not a simple superposition of the trough's PV over Elena's. Rather, the upper-level PV anomaly that eventually approached the centre was much narrower than the original trough because the outflow anticyclone of Elena contributed to synoptic-scale wave breaking, so that interaction with a substantial layer of vertical shear that might have prevented intensification was avoided. Moreover, the enhanced convection caused by these developments eroded the PV anomaly, and prevented weakening of the cyclone as it moved to the east. A similar mechanism occurred in Hurricane Danny (Molinari *et al.*, 1998).

The rapid intensification of Supertyphoon Flo was also linked to a period of enhanced eddy flux convergence of angular momentum, which preceded the rapid intensification by about a day. Titley and Elsberry (2000) characterise this phase as a preconditioning, in which a cyclonic wind burst develops above 200 hPa, decreasing the vertical shear near this level. The warm core moves upwards, and at 200 hPa the outflow increases, with inflow immediately below. These components are consistent with the upper level forcing of a balanced symmetric cyclone as discussed in section 3. They apparently developed in response to the eddy flux convergence of angular momentum and acted as a catalyst for the subsequent rapid intensification to a supertyphoon. The rapid intensification itself was apparently an internal readjustment of the storm, probably accompanied with an eyewall replacement cycle. It occurred during a period of declining upper level outflow and angular momentum forcing. Later, Flo underwent a similarly rapid weakening during a period of modest environmental wind shear, positive eddy flux convergence, warm SSTs, and vigorous outflow. The crucial difference appears to be that the angular momentum convergence extended downward to about 500 hPa. It is the vertical derivative of the momentum forcing that drives the secondary circulation (26), so this deep eddy flux convergence of angular momentum would have resulted in a deep secondary circulation, with lower-tropospheric heat release and relatively inefficient energy conversion. The rapid intensification of Flo can be contrasted to the much more modest evolution of Typhoons Ed (Titley and Elsberry, 2000) and Gene (Wu and Cheng, 1999), a comparison which emphasises the subtle variations in environmental forcing that often controls the pace and nature of intensity change.

The rapid intensifications of Hurricanes Bertha (1996) and Erin (2001) were analysed by Shapiro and Möller (2005), using piecewise PV inversion to isolate the effects of the trough. Bertha's intensification was directly augmented by a trough, but trough interaction did not contribute to Erin's intensification. Rather, other factors sufficed.

Hanley *et al.* (2001) surveyed 10 years of North Atlantic storms, and subdivided trough interactions according to whether the upper PV mass was closer than 400 km from the storm or further away, respectively "superposition" and "distant interaction" events. Each type of interaction can precede either intensification or weakening. Provided other unfavourable indicators, such as cold water or landfall, are absent, 78% of superposition cases intensify in the North Atlantic, soon after the small-scale upper tropospheric PV mass approaches the storm centre. In these cases, as in Hurricane Elena, the PV maximum weakens and never crosses the storm to produce weakening. Distant interaction leads to intensification in 61% of cases. In these situations, the upper PV maximum remains well to the west of the storm while strong upper-level divergence and a marked jet stream develop downshear. These features probably result from cooperative interaction between the systems. The maximum divergence and the equatorwards entrance region of the jet are located near the cyclone, leading to strong upwards motion that favours intensification. The differences between this case and that of unfavourable distant interaction are subtle, being largely a matter of a slightly stronger and larger PV maximum giving about 5 m s^{-1} more vertical shear over the cyclone.

A similar delicate balance of opposing effects was noted by DeMaria *et al.* (1993), who found that upper level eddy flux convergence of angular momentum was on its own unfavourable for intensification, due to the association with shear. However, when the effects of shear were separately accounted for in the regression, the eddy flux convergence had an intensifying effect, which was strongest at a lead time of 48 hours and within 600 km of the storm centre. This parameter was used as a predictor in SHIPS, the Statistical Hurricane Intensity Prediction Scheme, up until 2002. (DeMaria *et al.*, 2005).

The much-studied rapid intensification of Hurricane Opal (1995) provides an excellent example of the forecast and diagnosis difficulties. Several careful studies (Bosart *et al.*, 2000; Persing *et al.*, 2002; Möller and Shapiro, 2002; Shapiro and Möller, 2003) have analysed the role of the approaching upper trough without arriving at a consensus. Bosart *et al.* (2000) noted that upper-level divergence near the trough and jet extended equatorwards and coincided with Opal, just before an increase in convection and the onset of rapid intensification. Persing *et al.* (2002) and Möller and Shapiro (2002) identified positive eddy vorticity fluxes in the vicinity of Opal, but a balanced vortex calculation showed that the forcing of the intensification was dominated by the mean (i.e. symmetric) terms. Shapiro and Möller (2003) used piecewise potential vorticity inversion to isolate the effects of the upper-level trough. They likewise found no indication that it was a significant contributor to Opal's lower-tropospheric intensification. The diagnosis of Hurricane Opal was complicated by the fact that the rapid intensification began as or slightly before it passed over an oceanic warm core ring (Hong *et al.*, 2000; Shay *et al.*, 2000), which would have also contributed. The subsequent rapid weakening before landfall further highlighted the forecast challenges.

9. Summary

The fundamental view of tropical cyclone intensification as a cooperative process between the primary and secondary circulations has now stood for four decades, and continues to underpin our understanding of tropical cyclones. Potential intensity theories rest on this idea. Diagnosis the secondary circulation and its effects contributed to the discovery of eyewall contraction as the means by which most storms intensify, and to the eyewall replacement cycle as a means by which intensity can vary due to internal processes within the storm.

More recently, observations, theory and numerical simulation have combined to demonstrate that small-scale features such as vortex Rossby waves and eyewall mesovortices are common in the inner core of intense tropical cyclones. The details of the primary circulation determines the nature and extent of these features, while the secondary circulation can help create the necessary conditions. Their effects can include rapid changes in the inner-core structure of the tropical cyclone, implying that intensity forecasting will continue to be difficult.

The storm's environment has long been known to influence its development. The effects of upper trough interactions include the consequences of their modulation of the secondary circulation. Understanding of the effects of environmental shear has grown to include the dynamics of a tilted vortex, from both potential vorticity and vortex Rossby wave perspectives. Fine details of the vortex's primary structure are important to the response to shear. That response to shear can include intensity change, and marked asymmetries in vertical motion, cloud and rainfall.

Much of the inflow in the secondary circulation occurs in the boundary layer. The boundary layer is far from horizontally homogeneous, with marked variation in height and structure evident from observations. Theory and simulation have successfully explained these differences. The local boundary layer is strongly influenced by the details of the vortex primary structure and by position within the storm.

In spite of the overall complexity of the dynamics, tropical cyclones can be usefully described parametrically with relatively few parameters. As with other aspects of the atmosphere, the attractor is locally of low dimension.

In summary, tropical cyclone structure and intensity are sensitive to a variety of internal processes and external factors. Integrating these influences can be difficult in practice, and so it remains unclear what the ultimate limits of intensity and structure prediction are.

Table 2. Symbols and abbreviations used in this chapter.

Symbol	Meaning
b	Shape parameter in Holland vortex profile
c_p	Heat capacity of air at constant pressure
c_{pd}	c_p for dry air
d/dt	Total derivative, $d/dt = \partial/\partial t + u\,\partial/\partial r + v/r\,\partial/\partial\lambda + w\,\partial/\partial z$
e	Base of natural logarithms
f	Coriolis parameter
f_0	Coriolis parameter at cyclone centre
g	Gravitational acceleration
k	Radial wavenumber
k	Specific enthalpy
k_0	Radial wavenumber at $t = 0$
k_0^*	Saturation specific enthalpy at sea surface
n	Azimuthal wavenumber
p	Pressure
p_c	Central pressure
q	Specific humidity
$\overline{q'w'}$	Vertical turbulent flux of q
q_s	Saturation specific humidity
r	Radial coordinate
r_m	Radius of maximum wind
r^*	Critical radius for vortex Rossby waves
t	Time
u	Radial velocity
$\overline{u'w'}$	Vertical turbulent flux of u
$\overline{v'w'}$	Vertical turbulent flux of v
v	Azimuthal velocity
v_{gr}	Gradient wind
v_m	Maximum azimuthal wind
w	Vertical velocity
z	Height coordinate
B	Baroclinicity $g/\theta\,\partial\theta/\partial r$
C	Radial acceleration $v^2/r + fv$
C_D	Drag coefficient
C_E	Enthalpy transfer coefficient
E	Net evaporation rate (i.e. evaporation — condensation)
EPI	Potential intensity theory according to Emanuel
$\mathbf{F}$	Body force per unit mass (e.g. due to friction)
F_λ	Azimuthal component of $\mathbf{F}$
F_x, F_y	Horizontal components of $\mathbf{F}$
F_R	Rotational Froude number
H	Equivalent depth of the atmosphere
HPI	Potential intensity theory according to Holland
I	Inertial stability
L_R	Rossby deformation radius
L_v	Latent heat of vapourisation
M_a	Vertical component of absolute angular momentum
$\overline{M_a'w'}$	Vertical turbulent flux of M_a
N	Brunt-Väisälä frequency
PI	Potential intensity
PV	Potential vorticity
Q	Rossby-Ertel potential vorticity
R	Gas constant
R_d	Gas constant for dry air
RMW	Radius of maximum wind
SWF	Surface wind factor
T	Temperature
T_o	Outflow temperature
T_s	Surface temperature
T_v	Virtual temperature
VRW	Vortex Rossby wave
β	Meridional gradient of f
Δp	Central pressure deficit
ϵ	Ratio of molecular weight of water to that of dry air
ζ	Relative vorticity
$\overline{\zeta}_0$	Relative vorticity of basic-state vortex
$\overline{\zeta}_0'$	Radial derivative of $\overline{\zeta}_0$
η	Vertical component of absolute vorticity
$\boldsymbol{\eta}$	Absolute vorticity vector
θ	Potential temperature
$\dot{\theta}$	Diabatic heating rate
θ_e	Equivalent potential temperature
θ_e^*	Saturation equivalent potential temperature
θ_v	Virtual potential temperature
$\overline{\theta'w'}$	Vertical turbulent flux of θ
λ	Azimuth, measured anticlockwise from the positive x-axis
ξ	Inertia parameter
ρ	Air density in height coordinates (i.e. SI units of $\mathrm{kg\,m^{-3}}$)
ρ_{parcel}	Density of air parcel
ρ_{ref}	Density of reference atmospheric profile
σ	Air density in isentropic coordinates (i.e. SI units of $\mathrm{kg\,m^{-2}K^{-1}}$)
ω	Frequency
$\overline{\Omega}_0$	Angular velocity of basic-state vortex
$\overline{\Omega}_0'$	Radial derivative of $\overline{\Omega}_0$
∇	del operator of differential vector calculus
$\nabla_\theta\cdot$	Horizontal divergence in isentropic coordinates

Acknowledgements

In preparing this chapter, I drew on the earlier chapters by Holland (1987) and Willoughby (1995), and the excellent efforts of the various topic chairs and research rapporteurs at the 4th, 5th and 6th International Workshops on Tropical Cyclones, including Kerry Emanuel,

Russell Elsberry, Greg Holland, John Knaff, Frank Marks, John Molinari, Liz Ritchie, Roger Smith, and Hugh Willoughby. The opinions expressed by the author in this chapter are scientifically based and are personal opinions. They are not an expression of Australian Government policy.

References

Aberson, S. D., M. T. Montgomery, M. Bell, and M. Black, 2006: Hurricane Isabel (2003): New insights into the physics of intense storms. Part II: Extreme localized wind. *Bull. Amer. Meteor. Soc.*, **86**, 1349–1354.

Barnes, G. M. and P. B. Bogner, 2001: Comments on "surface observations in the hurricane environment". *Mon. Wea. Rev.*, **129**, 1267–1269.

Barnes, G. M. and M. D. Powell, 1995: Evolution of the inflow boundary layer of Hurricane Gilbert (1988). *Mon. Wea. Rev.*, **123**, 2348–2368.

Barnes, G. M., E. J. Zipser, D. Jorgensen, and F. D. Marks, Jr., 1983: Mesoscale and convective structure of a hurricane rainband. *J. Atmos. Sci.*, **40**, 2125–2137.

Bell, M. M. and M. T. Montgomery, 2008: Observed structure, evolution and potential intensity of category five Hurricane Isabel (2003) from 12–14 September. *Mon. Wea. Rev.*, **136**, 2023–2046.

Bister, M. and K. A. Emanuel, 1998: Dissipative heating and hurricane intensity. *Meteorol. Atmos. Phys.*, **50**, 233–240.

Black, M. L., R. W. Burpee, and F. D. Marks, Jr., 1996: Vertical motion characteristics of tropical cyclones determined with airborne Doppler radial velocities. *J. Atmos. Sci.*, **53**, 1887–1909.

Black, M. L., J. F. Gamache, F. D. Marks, Jr., C. E. Samsury, and H. E. Willoughby, 2002: Eastern Pacific Hurricanes Jimena of 1991 and Olivia of 1994: The effect of vertical shear on structure and intensity. *Mon. Wea. Rev.*, **130**, 2291–2312.

Black, P. G., R. L. Elsberry, L. K. Shay, R. P. Partridge, and J. Hawkins, 1988: Atmospheric boundary layer and oceanic mixed layer observations in Hurricane Josephine from air-deployed drifting buoys and research aircraft. *J. Atmos. Ocean. Tech.*, **5**, 683–698.

Black, P. G. and F. D. Marks, 1991: The structure of an eyewall meso-vortex in Hurricane Hugo (1989). *Preprints, 19th Conf. on Hurricanes and Tropical Meteorology*, Miami, FL, *Amer. Meteor. Soc.*, 579–582.

Bosart, L. F., C. S. Velden, W. E. Bracken, J. Molinari, and P. G. Black, 2000: Environmental influences on the rapid intensification of Hurricane Opal (1995) over the Gulf of Mexico. *Mon. Wea. Rev.*, **128**, 322–352.

Braun, S. A., 2002: A cloud-resolving simulation of Hurricane Bob (1991): Storm structure and eyewall buoyancy. *Mon. Wea. Rev.*, **130**, 1573–1592.

Braun, S. A., M. T. Montgomery, and Z. Pu, 2006: High-resolution simulation of Hurricane Bonnie (1998). Part I: The organization of eyewall vertical motion. *J. Atmos. Sci.*, **63**, 19–42.

Bryan, G. H. and R. Rotunno, 2009a: The influence of near-surface, high-entropy air in hurricane eyes on maximum hurricane intensity. *J. Atmos. Sci.*, **66**, 148–158.

———, 2009b: The maximum intensity of tropical cyclones in axisymmetric numerical model simulations. *Mon. Wea. Rev.*, **137**, 1770–1789.

Camp, J. P. and M. T. Montgomery, 2001: Hurricane maximum intensity: Past and present. *Mon. Wea. Rev.*, **129**, 1704–1717.

Carr, L. E., III and R. T. Williams, 1989: Barotropic vortex stability to perturbations from axisymmetry. *J. Atmos. Sci.*, **46**, 3177–3191.

Carrier, G. F., 1971: Swirling flow boundary layers. *J. Fluid. Mech.*, **49**, 133–144.

Charney, J. G. and A. Eliassen, 1964: On the growth of the hurricane depression. *J. Atmos. Sci.*, **21**, 68–75.

Chen, S. S., J. A. Knaff, and F. D. Marks, Jr., 2006: Effects of vertical wind shear and storm motion on tropical cyclone rainfall asymmetries deduced from TRMM. *Mon. Wea. Rev.*, **134**, 3190–3208.

Chen, Y., G. Brunet, and M. K. Yau, 2003: Spiral bands in a simulated hurricane. Part II: Wave activity diagnostics. *J. Atmos. Sci.*, **60**, 1239–1256.

Chen, Y. and M. K. Yau, 2001: Spiral bands in a simulated hurricane. Part I: Vortex Rossby wave verification. *J. Atmos. Sci.*, **58**, 2128–2145.

Cione, J. J., P. G. Black, and S. H. Houston, 2000: Surface observations in the hurricane environment. *Mon. Wea. Rev.*, **128**, 1550–1561.

Corbosiero, K. L. and J. Molinari, 2002: The effects of vertical wind shear on the distribution of convection in tropical cyclones. *Mon. Wea. Rev.*, **130**, 2110–2123.

———, 2003: The relationship between storm motion, vertical wind shear, and convective asymmetries in tropical cyclones. *J. Atmos. Sci.*, **60**, 366–376.

Corbosiero, K. L., J. Molinari, and A. R. Aiyyer, 2006: The structure and evolution of Hurricane Elena (1985). Part II: Convective asymmetries and evidence for vortex-Rossby waves. *Mon. Wea. Rev.*, **134**, 3073–3091.

Cram, T. A., J. Persing, M. T. Montgomery, and S. A. Braun, 2007: A Lagrangian trajectory view on transport and mixing processes between the eye, eyewall, and environment using a high-resolution simulation of Hurricane Bonnie (1998). *J. Atmos. Sci.*, **64**, 1835–1856.

DeMaria, M., J.-J. Baik, and J. Kaplan, 1993: Upper-level eddy angular momentum fluxes and tropical cyclone intensity change. *J. Atmos. Sci.*, **50**, 1133–1147.

DeMaria, M. and J. Kaplan, 1994: A statistical hurricane intensity prediction scheme (SHIPS) for the Atlantic basin. *Wea. Forecasting*, **9**, 209–220.

———, 1999: An updated statistical hurricane intensity prediction scheme (SHIPS) for the Atlantic and eastern North Pacific basins. *Wea. Forecasting*, **14**, 326–337.

DeMaria, M., M. Mainelli, L. K. Shay, J. A. Knaff, and J. Kaplan, 2005: Further improvements to the Statistical Hurricane Intensity Prediction Scheme (SHIPS). *Wea. Forecasting*, **20**, 531–543.

Dunion, J. P. and C. S. Velden, 2004: The impact of the Saharan air layer on Atlantic tropical cyclone activity. *Bull. Amer. Meteor. Soc.*, **85**, 353–365.

Eastin, M. D., W. M. Gray, and P. G. Black, 2005a: Buoyancy of convective vertical motions in the inner core of intense hurricanes. Part I: General statistics. *Mon. Wea. Rev.*, **133**, 188–208.

———, 2005b: Buoyancy of convective vertical motions in the inner core of intense hurricanes. Part II: Case studies. *Mon. Wea. Rev.*, **133**, 209–227.

Eliassen, A., 1971: On the Ekman layer in a circular vortex. *J. Met. Soc. Japan*, **49**, 784–789.

Eliassen, A. and M. Lystad, 1977: The Ekman layer of a circular vortex. A numerical and theoretical study. *Geophysica Norvegica*, **7**, 1–16.

Emanuel, K. A., 1986: An air-sea interaction theory for tropical cyclones. Part I: Steady-state maintenance. *J. Atmos. Sci.*, **43**, 585–604.

———, 1988: The maximum intensity of hurricanes. *J. Atmos. Sci.*, **45**, 1143–1155.

———, 1994: *Atmospheric Convection.* Oxford University Press, 580 pp.

———, 1995: Sensitivity of tropical cyclones to surface exchange coefficients and a revised steady-state model incorporating eye dynamics. *J. Atmos. Sci.*, **52**, 3969–3976.

———, 1997: Some aspects of hurricane inner-core dynamics and energetics. *J. Atmos. Sci.*, **54**, 1014–1026.

———, 2000: A statistical analysis of tropical cyclone intensity. *Mon. Wea. Rev.*, **128**, 1139–1152.

———, 2003: Tropical cyclones. *Annu. Rev. Earth Planet. Sci.*, **31**, 75–104.

Emanuel, K. A., C. DesAutels, C. Holloway, and R. Korty, 2004: Environmental control of tropical cyclone intensity. *J. Atmos. Sci.*, **61**, 843–858.

Etling, D. and R. A. Brown, 1993: Roll vortices in the planetary boundary layer: A review. *Boundary Layer Meteorology*, **65**, 215–248.

Foster, R. C., 2005: Why rolls are prevalent in the hurricane boundary layer. *J. Atmos Sci.*, **62**, 2647–2661.

Frank, W. M. and E. A. Ritchie, 1999: Effects of environmental flow upon tropical cyclone structure. *Mon. Wea. Rev.*, **127**, 2044–2061.

———, 2001: Effects of vertical wind shear on the intensity and structure of numerically simulated hurricanes. *Mon. Wea. Rev.*, **129**, 2249–2269.

Franklin, J. L., M. L. Black, and K. Valde, 2003: GPS dropwindsonde wind profiles in hurricanes and their operational implications. *Wea. Forecasting*, **18**, 32–44.

Franklin, J. L., S. J. Lord, S. E. Feuer, and F. D. Marks, Jr., 1993: The kinematic structure of Hurricane Gloria (1985) determined from nested analyses of dropwindsonde and Doppler radar data. *Mon. Wea. Rev.*, **121**, 2433–2451.

Gall, R., J. Tuttle, and P. Hildebrand, 1998: Small-scale spiral bands observed in Hurricanes Andrew, Hugo, and Erin. *Mon. Wea. Rev.*, **126**, 1749–1766.

Gallina, G. M. and C. S. Velden, 2002: Environmental vertical wind shear and tropical cyclone intensity change utilizing enhanced satellite derived wind information. *Extended abstracts, 25th Conference on Hurricanes and Tropical Meteorology*, Amer. Meteorol. Soc., San Diego, CA, 29 April–3 May, pp. 172–173.

Gill, A. E., 1982: *Atmosphere–Ocean Dynamics.* Academic Press, 662 pp.

Guinn, T. A. and W. H. Schubert, 1993: Hurricane spiral bands. *J. Atmos. Sci.*, **50**, 3380–3403.

Hack, J. J. and W. H. Schubert, 1986: Nonlinear response of atmospheric vortices to heating by organised cumulus convection. *J. Atmos. Sci.*, **43**, 1559–1573.

Halverson, J. B., J. Simpson, G. Heymsfield, H. Pierce, T. Hock, and L. Ritchie, 2006: Warm core structure of Hurricane Erin diagnosed from high altitude dropsondes during CAMEX-4. *J. Atmos. Sci.*, **63**, 309–324.

Hanley, D., J. Molinari, and D. Keyser, 2001: A composite study of the interactions between tropical cyclones and upper-tropospheric troughs. *Mon. Wea. Rev.*, **129**, 2570–2584.

Haurwitz, B., 1935: The height of tropical cyclones and the eye of the storm. *Mon. Wea. Rev.*, **63**, 45–49.

Haynes, P. H. and M. E. McIntyre, 1987: On the evolution of vorticity and potential vorticity in the presence of diabatic heating and frictional or other forces. *J. Atmos. Sci.*, **44**, 828–841.

———, 1990: On the conservation and impermeability theorems for potential vorticity. *J. Atmos. Sci.*, **47**, 2021–2031.

Holland, G. J., 1980: An analytic model of the wind and pressue profiles in hurricanes. *Mon. Wea. Rev.*, **108**, 1212–1218.

———, 1987: Mature structure and structure change. *A Global View Of Tropical Cyclones*, Elsberry, R. L., Ed., Office of Naval Research, pp. 13–52.

———, 1997: The maximum potential intensity of tropical cyclones. *J. Atmos. Sci.*, **54**, 2519–2541.

Holland, G. J. and R. T. Merrill, 1984: On the dynamics of tropical cyclone structural changes. *Quart. J. Roy. Meteor. Soc.*, **110**, 723–745.

Hong, X., S. W. Chang, S. Raman, L. K. Shay, and R. Hodur, 2000: The interaction between Hurricane Opal (1995) and a warm core ring in the Gulf of Mexico. *Mon. Wea. Rev.*, **128**, 1347–1365.

Hoskins, B. J., M. E. McIntyre, and A. Robertson, 1985: On the use and significance of isentropic potential vorticity maps. *Quart. J. Roy. Meteor. Soc.*, **111**, 877–946.

Jones, S. C., 1995: The evolution of vortices in vertical shear. Part I: Initially barotropic vortices. *Quart. J. Roy. Meteor. Soc.*, **121**, 821–851.

———, 2000a: The evolution of vortices in vertical shear. II: Large-scale asymmetries. *Quart. J. Roy. Meteor. Soc.*, **126**, 3137–3160.

———, 2000b: The evolution of vortices in vertical shear. III: Baroclinic vortices. *Quart. J. Roy. Meteor. Soc.*, **126**, 3161–3186.

Jorgensen, D. P., E. J. Zipser, and M. A. LeMone, 1985: Vertical motions in intense hurricanes. *J. Atmos. Sci.*, **42**, 839–856.

Katsaros, K. B., P. W. Vachon, W. T. Liu, and P. G. Black, 2002: Microwave remote sensing of tropical cyclones from space. *J. Oceanogr.*, **58**, 137–151.

Kepert, J. D., 2001: The dynamics of boundary layer jets within the tropical cyclone core. Part I: Linear theory. *J. Atmos. Sci.*, **58**, 2469–2484.

———, 2002a: An analysis of some tropical cyclone boundary layer wind observations. *Extended abstracts, 25th Conference on Hurricanes and Tropical Meteorology*, Amer. Meteor. Soc., San Diego, California, 29 April–3 May, pp. 615–616.

———, 2002b: Modelling the tropical cyclone boundary layer wind-field at landfall. *Extended abstracts, 14th BMRC Modelling Workshop: Modelling and Predicting Extreme Events, Melbourne, Australia, 11-13 November*, Hollis, A. J. and P. J. Meighen, Eds., Bureau of Meteorology Research Centre, pp. 81–84.

———, 2002c: The Wind-Field Structure of the Tropical Cyclone Boundary-Layer. Ph.D. thesis, Department of Mathematics and Statistics, Monash University.

———, 2005: Objective analysis of tropical cyclone location and motion from high density observations. *Mon. Wea. Rev.*, **133**, 2406–2421.

———, 2006a: Observed boundary–layer wind structure and balance in the hurricane core. Part I: Hurricane Georges. *J. Atmos. Sci.*, **63**, 2169–2193.

———, 2006b: Observed boundary–layer wind structure and balance in the hurricane core. Part II: Hurricane Mitch. *J. Atmos. Sci.*, **63**, 2194–2211.

Kepert, J. D. and Y. Wang, 2001: The dynamics of boundary layer jets within the tropical cyclone core. Part II: Nonlinear enhancement. *J. Atmos. Sci.*, **58**, 2485–2501.

Kimball, S. K. and J. L. Evans, 2002: Idealized simulations of hurricane-trough interaction. *Mon. Wea. Rev.*, **130**, 2210–2227.

Knaff, J. A., T. A. Cram, A. B. Schumacher, J. P. Kossin, and M. DeMaria, 2008: Objective identification of annular hurricanes. *Wea. Forecasting*, **23**, 17–28.

Knaff, J. A., J. P. Kossin, and M. DeMaria, 2003: Annular hurricanes. *Wea. Forecasting*, **18**, 204–223.

Knaff, J. A., S. A. Seseske, M. DeMaria, and J. L. Demuth, 2004: On the influences of vertical wind shear on symmetric tropical cyclone structure derived from AMSU. *Mon. Wea. Rev.*, **132**, 2503–2510.

Knupp, K. R., J. Walters, and M. Biggerstaff, 2005: Doppler profiler and radar observations of boundary layer variability during the landfall of tropical storm Gabrielle. *J. Atmos. Sci.*, **63**, 234–251.

Knupp, K. R., J. Walters, and E. W. M. Jr, 2000: Doppler profiler observations of Hurricane Georges at landfall. *Geophys. Res. Lett.*, **27**, 3361–3364.

Kossin, J. P. and M. D. Eastin, 2001: Two distinct regimes in the kinematic and thermodynamic structure of the hurricane eye and eyewall. *J. Atmos. Sci.*, **58**, 1079–1090.

Kossin, J. P. and W. H. Schubert, 2001: Mesovortices, polygonal flow patterns, and rapid pressure falls in hurricane-like vortices. *J. Atmos. Sci.*, **58**, 2196–2209.

———, 2003: Diffusion versus advective rearrangement of a circular vortex sheet. *J. Atmos. Sci.*, **60**, 586–589.

———, 2005: Mesovortices in Hurricane Isabel. *Bull. Amer. Meteor. Soc.*, **85**, 151–153.

Kuo, H.-C., L.-Y. Lin, C.-P. Chang, and R. T. Williams, 2004: The formation of concentric vorticity structures in typhoons. *J. Atmos. Sci.*, **61**, 2722–2734.

Kuo, H. L., 1971: Axisymmetric flow in the boundary layer of a maintained vortex. *J. Atmos. Sci.*, **28**, 20–41.

———, 1982: Vortex boundary layer under quadratic surface stress. *Boundary-Layer Meteorol.*, **22**, 151–169.

Kusunoki, K. and W. Mashiko, 2006: Doppler radar investigations of the inner core of Typhoon Songda (2004): Polygonal/elliptical eyewalls, eye contraction, and small-scale spiral bands. *Extended Abstracts, 27th Conference on Hurricanes and Tropical Meteorology*, Amer. Meteor. Soc., Monterey, CA, 24–28 April, CD–ROM.

Lonfat, M., 2004: Tropical Cyclone Rainfall: An Observational and Numerical Study of the Structure and Governing Physical Processes. Ph.D. thesis, Rosenthiel School of Marine and Atmospheric Sciences, University of Miami.

Lonfat, M., R. Rogers, T. Marchok, and F. D. Marks, Jr., 2007: A parametric model for predicting hurricane rainfall. *Mon. Wea. Rev.*, **135**, 3086–3097.

Lorsolo, S., J. L. Schroeder, P. Dodge, and F. D. Marks, Jr., 2008: An observational study of hurricane boundary layer small-scale coherent structures. *Mon. Wea. Rev.*, **136**, 2871–2893.

MacDonald, N. J., 1968: The evidence for the existence of Rossby-like waves in the hurricane vortex. *Tellus*, **20**, 138–150.

Mallen, K. J., M. T. Montgomery, and B. Wang, 2005: Reexamining the near core radial structure of the tropical cyclone primary circulation: Implications for vortex resiliency. *J. Atmos. Sci.*, **62**, 408–425.

Mallett, W. A., 2000: The Dynamics of the Atmospheric Boundary Layer of a Mature Tropical Cyclone: A Numerical Investigation. Ph.D. thesis, James Cook University, Australia.

Marks, F. D., P. G. Black, M. T. Montgomery, and R. W. Burpee, 2008: Structure of the eye and eyewall of Hurricane Hugo (1989). *Mon. Wea. Rev.*, **136**, 1237–1259.

Marks, F. D., Jr. and P. G. Black, 1990: Close encounter with an intense mesoscale vortex within Hurricane Hugo (September 15, 1989). *Extended Abstracts, Fourth Conf. on Mesoscale Processes*, Amer. Meteor. Soc., Boulder, CO, pp. 114–115.

Marks, F. D., Jr., P. Dodge, and C. Sandin, 1999: Observations of hurricane atmospheric boundary layer structure at landfall. *Extended abstracts, 23rd Conference on Hurricanes and Tropical Meteorology*, Amer. Meteorol. Soc., Dallas, TX, Jan. 10-15, pp. 1051–1055.

Marks, F. D., Jr., R. A. Houze, and J. F. Gamache, 1992: Dual-aircraft investigation of the inner core of Hurricane Norbert. Part I: Kinematic structure. *J. Atmos. Sci.*, **49**, 919–942.

May, P. T. and G. J. Holland, 1999: The role of potential vorticity generation in tropical cyclone rainbands. *J. Atmos. Sci.*, **56**, 1224–1228.

May, P. T., J. D. Kepert, and T. D. Keenan, 2008: Polarimetric radar observations of the persistently asymmetric structure of Tropical Cyclone Ingrid. *Mon. Wea. Rev.*, **136**, 616–630.

Maynard, R. H., 1945: Radar and weather. *J. Meteor.*, **2**, 214–226.

McWilliams, J. C., L. P. Graves, and M. T. Montgomery, 2003: A formal theory for vortex Rossby waves and vortex evolution. *Geophys. Astrophys. Fluid Dyn.*, **97**, 275–309.

Melander, M. V., J. C. McWilliams, and N. J. Zabusky, 1987: Axisymmetrization and vorticity-gradient intensification of an isolated two-dimensional vortex through filamentation. *J. Fluid Mech.*, **178**, 137–159.

Molinari, J., S. Skubis, and D. Vollaro, 1995: External influences on hurricane intensity. Part III: Potential vorticity evolution. *J. Atmos. Sci.*, **52**, 3593–3606.

Molinari, J., S. Skubis, D. Vollaro, F. Alsheimer, and H. E. Willoughby, 1998: Potential vorticity analysis of tropical cyclone intensification. *J. Atmos. Sci.*, **55**, 2632–2644.

Molinari, J. and D. Vollaro, 1989: External influences on hurricane intensity. Part I: Outflow layer eddy momentum fluxes. *J. Atmos. Sci.*, **46**, 1093–1105.

———, 1990: External influences on hurricane intensity. Part II: Vertical structure and response to the hurricane vortex. *J. Atmos. Sci.*, **47**, 1902–1918.

Möller, J. D. and M. T. Montgomery, 1999: Vortex Rossby waves and hurricane intensification in a barotropic model. *J. Atmos. Sci.*, **56**, 1674–1687.

———, 2000: Tropical cyclone evolution via potential vorticity anomalies in a three-dimensional balance model. *J. Atmos. Sci.*, **57**, 3366–3387.

Möller, J. D. and L. Shapiro, 2002: Balanced contributions to the intensification of Hurricane Opal as diagnosed from a GFDL model forecast. *Mon. Wea. Rev.*, **130**, 1866–1881.

Montgomery, M. T., M. M. Bell, S. D. Aberson, and M. L. Black, 2006: Hurricane Isabel (2003): New insights into the physics of intense storms. Part I. *Bull. Amer. Meteor. Soc.*, **86**, 1335–1347.

Montgomery, M. T. and R. J. Kallenbach, 1997: A theory of vortex Rossby waves and its application to spiral bands and intensity changes in hurricanes. *Quart. J. Roy. Meteor. Soc.*, **123**, 435–465.

Montgomery, M. T., H. D. Snell, and Z. Yang, 2001: Axisymmetric spindown dynamics of hurricane-like vortices. *J. Atmos. Sci.*, **58**, 421–435.

Montgomery, M. T., V. A. Vladimirov, and P. V. Denissenko, 2002: An experimental study on hurricane mesovortices. *J. Fluid Mech.*, **471**, 1–32.

Morrison, I., S. Businger, F. D. Marks, Jr., P. Dodge, and J. Businger, 2005: An observational case for the prevalence of roll vortices in the hurricane boundary layer. *J. Atmos. Sci.*, **62**, 2662–2673.

Moss, M. S. and S. L. Rosenthal, 1975: On the estimation of planetary boundary layer variables in mature hurricanes. *Mon. Wea. Rev.*, **103**, 980–988.

Muramatsu, T., 1986: The structure of polygonal eye of a typhoon. *J. Meteor. Soc. Japan*, **64**, 913–921.

Nolan, D. S., 2005: Instabilities in hurricane-like boundary layers. *Dynamics of Atmospheres and Oceans*, **40**, 209–236.

Nolan, D. S. and B. F. Farrell, 1999: The intensification of two-dimensional swirling flows by stochastic asymmetric forcing. *J. Atmos. Sci.*, **56**, 3937–3962.

Nolan, D. S. and L. D. Grasso, 2003: Nonhydrostatic, three-dimensional perturbations to balanced, hurricane-like vortices. Part II: Symmetric response and nonlinear simulations. *J. Atmos. Sci.*, **60**, 2717–2745.

Nolan, D. S. and M. T. Montgomery, 2000: The algebraic growth of wavenumber one disturbances in hurricane-like vortices. *J. Atmos. Sci.*, **57**, 3514–3538.

Nolan, D. S., M. T. Montgomery, and L. D. Grasso, 2001: The wavenumber-one instability and trochoidal motion of hurricane-like vortices. *J. Atmos. Sci.*, **58**, 3243–3270.

Nolan, D. S., Y. Moon, and D. P. Stern, 2007: Tropical cyclone intensification from asymmetric convection: energetics and efficiency. *J. Atmos. Sci.*, **64**, 3377–3405.

Nong, S. and K. Emanuel, 2003: A numerical study of the genesis of concentric eyewalls in hurricanes. *Quart. J. Roy. Meteor. Soc.*, **129**, 3323–3338.

Ooyama, K. V., 1964: A dynamical model for the study of tropical cyclone development. *Geofis. Intern.*, **4**, 187–198.

———, 1969: Numerical simulation of the life cycle of tropical cyclones. *J. Atmos. Sci.*, **26**, 3–40.

———, 1982: Conceptual evolution of the theory and modelling of the tropical cyclone. *J. Meteor. Soc. Japan*, **60**, 369–380.

Paterson, L. A., B. N. Hanstrum, N. E. Davidson, and H. C. Weber, 2005: Influence of environmental vertical wind shear on the intensity of hurricane-strength tropical cyclones in the Australian region. *Mon. Wea. Rev.*, **133**, 3644–3660.

Pendergrass, A. G. and H. E. Willoughby, 2009: Diabatically induced secondary flows in tropical cyclones. Part I: Quasi-steady forcing. *Mon. Wea. Rev.*, **137**, 805–821.

Peng, M. S., B.-F. Jeng, and R. T. Williams, 1999: A numerical study on tropical cyclone intensification. Part I: Beta effect and mean flow effect. *J. Atmos. Sci.*, **56**, 1404–1423.

Persing, J. and M. T. Montgomery, 2003: Hurricane superintensity. *J. Atmos. Sci.*, **60**, 2349–2371.

Persing, J., M. T. Montgomery, and R. E. Tuleya, 2002: Environmental interactions in the GFDL hurricane model for Hurricane Opal. *Mon. Wea. Rev.*, **130**, 298–317.

Pfeffer, R. L. and M. Challa, 1981: A numerical study of the role of eddy fluxes of momentum in the development of Atlantic hurricanes. *J. Atmos. Sci.*, **38**, 2393–2398.

———, 1992: The role of environmental asymmetries in Atlantic hurricane formation. *J. Atmos. Sci.*, **49**, 1051–1059.

Powell, M. D., 1980: Evaluation of diagnostic marine boundary-layer models applied to hurricanes. *Mon. Wea. Rev.*, **108**, 757–765.

———, 1982: The transition of the Hurricane Frederic boundary-layer wind field from the open Gulf of Mexico to landfall. *Mon. Wea. Rev.*, **110**, 1912–1932.

———, 1990a: Boundary layer structure and dynamics in outer hurricane bands. Part I:

Mesoscale rainfall and kinematic structure. *Mon. Wea. Rev.*, **118**, 891–917.

———, 1990b: Boundary layer structure and dynamics in outer hurricane bands. Part II: Downdraft modification and mixed layer recovery. *Mon. Wea. Rev.*, **118**, 918–938.

Powell, M. D. and P. G. Black, 1990: The relationship of hurricane reconnaissance flight-level wind measurements to winds measured by NOAA's oceanic platforms. *J. Wind Eng. Ind. Aero.*, **36**, 381–392.

Powell, M. D. and S. H. Houston, 1998: Surface wind fields of 1995 Hurricane Erin, Opal, Luis, Marilyn, and Roxanne at landfall. *Wea. Forecasting*, **11**, 1259–1273.

Powell, M. D., E. W. Uhlhorn, and J. D. Kepert, 2009: Estimating maximum surface winds from hurricane reconnaissance measurements. *Wea. Forecasting*, **24**, 868–883, doi:10.1175/2008WAF2007087.1.

Powell, M. D., P. J. Vickery, and T. A. Reinhold, 2003: Reduced drag coefficient for high wind speeds in tropical cyclones. *Nature*, **422**, 279–283.

Raymond, D. J., 1992: Nonlinear balance and potential-vorticity thinking at large Rossby number. *Quart. J. Roy. Meteor. Soc.*, **118**, 987–1015.

Reasor, P. D. and M. T. Montgomery, 2001: Three-dimensional alignment and corotation of weak, TC-like vortices via linear vortex Rossby waves. *J. Atmos. Sci.*, **58**, 2306–2330.

Reasor, P. D., M. T. Montgomery, and L. D. Grasso, 2004: A new look at the problem of tropical cyclones in vertical shear flow: Vortex resiliency. *J. Atmos. Sci.*, **61**, 3–22.

Reasor, P. D., M. T. Montgomery, F. D. Marks, Jr., and J. F. Gamache, 2000: Low-wavenumber structure and evolution of the hurricane inner core observed by airborne dual-Doppler radar. *Mon. Wea. Rev.*, **128**, 1653–1680.

Ritchie, E. A. and W. M. Frank, 2007: Interactions between simulated tropical cyclones and an environment with a variable Coriolis parameter. *Mon. Wea. Rev.*, **135**, 1889–1905.

Ritchie, E. A. and G. J. Holland, 1993: On the interaction of tropical-cyclone-scale vortices. II: Discrete vortex patches. *Quart. J. Roy. Meteor. Soc.*, **119**, 1363–1379.

Rogers, R., S. Chen, J. Tenerelli, and H. Willoughby, 2003: A numerical study of the impact of vertical shear on the distribution of rainfall in Hurricane Bonnie (1998). *Mon. Wea. Rev.*, **131**, 1577–1599.

Romine, G. S. and R. B. Wilhelmson, 2006: Finescale spiral band features within a numerical simulation of Hurricane Opal (1995). *Mon. Wea. Rev.*, **134**, 1121–1139.

Rosenthal, S. L., 1962: A theoretical analysis of the field of motion in the hurricane boundary layer. National hurricane research project report no. 56, U. S. Department of Commerece.

———, 1971: The response of a tropical cyclone model to variations in boundary layer parameters, initial conditions, lateral boundary conditions, and domain size. *Mon. Wea. Rev.*, **99**, 767–777.

Rozoff, C. M., W. H. Schubert, and B. D. McNoldy, 2006: Rapid filamentation zones in intense tropical cyclones. *J. Atmos. Sci.*, **63**, 325–340.

Ryan, B. F., G. M. Barnes, and E. J. Zipser, 1992: A wide rainband in a developing tropical cyclone. *Mon. Wea. Rev.*, **120**, 431–447.

Samsury, C. E. and E. J. Zipser, 1995: Secondary wind maxima in hurricanes: Airflow and relationship to rainbands. *Mon. Wea. Rev.*, **123**, 3502–3517.

Schecter, D. A. and M. T. Montgomery, 2004: Damping and pumping of a vortex Rossby wave in a monotonic cyclone: Critical layer stirring versus inertia-buoyancy wave emission. *Phys. Fluids*, **16**, 1334–1348.

———, 2007: Waves in a cloudy vortex. *J. Atmos. Sci.*, **64**, 314–337.

Schecter, D. A., M. T. Montgomery, and P. D. Reasor, 2002: A theory for the vertical alignment of a quasigeostrophic vortex. *J. Atmos. Sci.*, **59**, 150–168.

Schloemer, R. W., 1954: Analysis and synthesis of hurricane wind patterns over Lake Okechobee, FL. Hydromet report 31, Govt. Printing Office, No. C30,70:31.

Schneider, R. S. and G. M. Barnes, 2005: Low-level thermodynamic, kinematic, and reflectivity fields of Hurricane Bonnie (1998) at landfall. *Mon. Wea. Rev.*, **133**, 3243–3259.

Schubert, W. H., J. J. Hack, P. L. Silva, and S. R. Fulton, 1980: Geostrophic adjustment in an axisymmetric vortex. *J. Atmos. Sci.*, **37**, 1464–1484.

Schubert, W. H., M. T. Montgomery, R. K. Taft, T. A. Guinn, S. R. Fulton, P. Kossin, and J. P. Edwards, 1999: Polygonal eyewalls, asymmetric eye contraction, and potential vorticity mixing in hurricanes. *J. Atmos. Sci.*, **56**, 1197–1223.

Schubert, W. H., C. M. Rozoff, J. L. Vigh, B. D. McNoldy, and J. P. Kossin, 2007: On the distribution of subsidence in the hurricane eye. *Quart. J. Roy. Meteor. Soc.*, **133**, 595–605.

Schwendike, J. and J. D. Kepert, 2008: The boundary–layer winds in Hurricanes Danielle (1998) and Isabel (2003). *Mon. Wea. Rev.*, **136**, 3168–3192.

Shapiro, L. J., 1983: The asymmetric boundary layer flow under a translating hurricane. *J. Atmos. Sci.*, **40**, 1984–1998.

———, 1992: Hurricane vortex motion and evolution in a three-layer model. *J. Atmos. Sci.*, **49**, 140–153.

Shapiro, L. J. and J. L. Franklin, 1995: Potential vorticity in Hurricane Gloria. *Mon. Wea. Rev.*, **123**, 1465–1475.

Shapiro, L. J. and J. Möller, 2003: Influence of atmospheric asymmetries on the intensification of Hurricane Opal: Piecewise PV inversion diagnosis of a GFDL model forecast. *Mon. Wea. Rev.*, **131**, 1637–1649.

———, 2005: Influence of atmospheric asymmetries on the intensification of GFDL model forecast hurricanes. *Mon. Wea. Rev.*, **133**, 2860–2875.

Shapiro, L. J. and H. E. Willoughby, 1982: The response of balanced hurricanes to local sources of heat and momentum. *J. Atmos. Sci.*, **39**, 378–394.

Shay, L. K., G. J. Goni, and P. G. Black, 2000: Effects of a warm oceanic feature on Hurricane Opal. *Mon. Wea. Rev.*, **128**, 1366–1383.

Simpson, R. H. and H. Riehl, 1958: Mid-troposphere ventilation as a constraint on hurricane development and maintenance. *Proc. Tenth Conf on Hurricanes, Miami Beach, FL*, AMS, D4.1–D4.10.

Smith, R. K., 1968: The surface boundary layer of a hurricane — Part I. *Tellus*, **20**, 473–483.

———, 1997: On the theory of CISK. *Quart. J. Roy. Meteor. Soc.*, **123**, 407–418.

———, 2000: The role of cumulus convection in hurricanes and its representation in hurricane models. *Rev. Geophys.*, **38**, 465–489.

———, 2003: A simple model of the hurricane boundary layer. *Quart. J. Roy. Meteor. Soc.*, **129**, 1007–1027.

Smith, R. K. and M. T. Montgomery, 2008: Balanced boundary layers used in hurricane models. *Quart. J. Roy. Meteor. Soc.*, **134**, 1385–1395.

Smith, R. K., M. T. Montgomery, and H. Zhu, 2005: Buoyancy in tropical cyclones and other rapidly rotating atmospheric vortices. *Dyn. Atmos. Oceans*, **40**, 189–208.

Smith, R. K., W. Ulrich, and G. Sneddon, 2000: The dynamics of hurricane-like vortices in vertical shear flows. *Quart. J. Roy. Meteor. Soc.*, **126**, 2653–2670.

Smith, R. K. and S. Vogl, 2008: A simple model of the hurricane boundary layer revisited. *Quart. J. Roy. Meteor. Soc.*, **134**, 337–351.

Terwey, W. D. and M. T. Montgomery, 2008: Secondary eyewall formation in two idealized, full-physics modeled hurricanes. *J. Geophys. Res.*, **113**, D12112, doi:10.1029/2007JD008897.

Titley, D. W. and R. L. Elsberry, 2000: Large intensity changes in tropical cyclones: A case study of Supertyphoon Flo during TCM-90. *Mon. Wea. Rev.*, **128**, 3556–3573.

Tonkin, H., G. J. Holland, N. Holbrook, and A. Henderson-Sellers, 2000: An evaluation of thermodynamic estimates of climatological maximum potential tropical cyclone intensity. *Mon. Wea. Rev.*, **128**, 2000.

Tuleya, R. E., M. DeMaria, and J. R. Kuligowski, 2007: Evaluation of GFDL and simple statistical model rainfall forecasts for U.S. landfalling tropical storms. *Wea. Forecasting*, **22**, 56–70.

Velasco Fuentes, O. U., 2004: Vortex filamentation: its onset and its role on axisymmetrisation and merger. *Dyn. Atmos. Ocean.*, **40**, 23–42.

Vickery, P. J., P. F. Skerjl, A. C. Steckley, and L. A. Twisdale, 2000: Hurricane wind field model for use in hurricane simulations. *J. Engineering Structures*, **126**, 1203–1221.

Wakimoto, R. M. and P. G. Black, 1994: Damage survey of Hurricane Andrew and its relationship to the eyewall. *Bull. Amer. Meteor. Soc.*, **75**, 189–200.

Wang, Y., 2002a: Vortex Rossby waves in a numerically simulated tropical cyclone. Part I: Overall structure, potential vorticity, and kinetic energy budgets. *J. Atmos. Sci.*, **59**, 1213–1238.

———, 2002b: Vortex Rossby waves in a numerically simulated tropical cyclone. Part II: The role in tropical cyclone structure and intensity changes. *J. Atmos. Sci.*, **59**, 1239–1262.

———, 2008: Rapid filamentation zone in a numerically simulated tropical cyclone. *J. Atmos. Sci.*, **65**, 1158–1181.

Wang, Y. and G. Holland, 1996: Tropical cyclone motion and evolution in vertical shear. *J. Atmos. Sci.*, **53**, 3313–3332.

Wexler, H., 1947: Structure of hurricanes as determined by radar. *J. N. Y. Acad. Sci.*, **48**, 821–845.

Willoughby, H. E., 1978: A possible mechanism for the formation of hurricane rainbands. *J. Atmos. Sci.*, **35**, 838–848.

———, 1979: Forced secondary circulations in hurricanes. *J. Geophys. Res.*, **84C**, 3173–3183.

———, 1988: The dynamics of the tropical cyclone core. *Aust. Meteor. Mag.*, **36**, 183–191.
———, 1990a: Gradient balance in tropical cyclones. *J. Atmos. Sci.*, **47**, 265–274.
———, 1990b: Temporal changes of the primary circulation in tropical cyclones. *J. Atmos. Sci.*, **47**, 242–264.
———, 1991: Reply. *J. Atmos. Sci.*, **48**, 1209–1212.
———, 1995: Mature structure and evolution. *Global Perspectives On Tropical Cyclones*, Elsberry, R. L., Ed., World Meteorological Organisation, pp. 21–62.
Willoughby, H. E., J. A. Clos, and M. G. Shoreibah, 1982: Concentric eyewalls, secondary wind maxima, and the evolution of the hurricane vortex. *J. Atmos. Sci.*, **39**, 395–411.
Willoughby, H. E., R. W. R. Darling, and M. E. Rahn, 2006: Parametric presentation of the primary hurricane vortex. Part II: A new family of sectionally continous profiles. *Mon. Wea. Rev.*, **134**, 1102–1120.
Willoughby, H. E., F. D. Marks, Jr., and R. J. Feinberg, 1984: Stationary and moving convective bands in hurricanes. *J. Atmos. Sci.*, **41**, 3189–3211.
Willoughby, H. E. and M. E. Rahn, 2004: Parametric presentation of the primary hurricane vortex. Part I: Observations and evaluation of the Holland (1980) model. *Mon. Wea. Rev.*, **132**, 3033–3048.
Wong, M. L. M. and J. C. L. Chan, 2004: Tropical cyclone intensity in vertical winds shear. *J. Atmos. Sci.*, **61**, 1859–1876.
———, 2007: Modeling the effects of land-sea roughness contrast on tropical cyclone winds. *J. Atmos. Sci.*, **64**, 3249–3264.
Wroe, D. R. and G. M. Barnes, 2003: Inflow layer energetics of Hurricane Bonnie (1998) near landfall. *Mon. Wea. Rev.*, **131**, 1600–1612.
Wu, C.-C. and H.-J. Cheng, 1999: An observational study of environmental influences on the intensity changes of Typhoons Flo (1990) and Gene (1990). *Mon. Wea. Rev.*, **127**, 3003–3031.
Wurman, J. and J. Winslow, 1998: Intense subkilometre-scale boundary layer rolls observed in Hurricane Fran. *Science*, **280**, 555–557.
Yau, M. K., Y. Liu, D.-L. Zhang, and Y. Chen, 2004: A multiscale numerical study of Hurricane Andrew (1992). Part VI: Small-scale inner-core structures and wind streaks. *Mon. Wea. Rev.*, **132**, 1410–1433.
Zehr, R. M., 2003: Environmental vertical wind shear with Hurricane Bertha (1996). *Wea. Forecasting*, **18**, 345–356.
Zhang, D.-L., Y. Liu, and M. K. Yau, 2001: A multiscale numerical study of Hurricane Andrew (1992). Part IV: Unbalanced flows. *Mon. Wea. Rev.*, **129**, 92–107.

Chapter 2

Tropical Cyclone Formation

Kevin J. Tory* and William M. Frank

Centre for Australian Weather and Climate Research,
Bureau of Meteorology, Melbourne Vic 3000, Australia
**K.tory@bom.gov.au*

This chapter reviews what is known about the development of environments in which tropical cyclones form, and the theories proposed to describe the construction of the tropical cyclone vortex core. Vorticity dynamics, cloud thermodynamics, and the latest observational and modelling studies are assessed, to arrive at a consolidated theory of tropical cyclone formation. The chapter focuses predominantly on advances in knowledge since the previous volume in this series.

Tropical cyclones form in environments of enhanced vorticity, humidity and deep convection, reduced wind shear, and a deep layer of warm sea surface temperatures. Transient features such as tropical waves and baroclinic disturbances can perturb this environment, providing locally enhanced regions of favourability. The intense surface-concentrated vortex core that defines the tropical cyclone is constructed when the thermodynamic environment favours the outbreak of widespread and persistent deep convection. The system-scale, troposphere-deep, in-up-out circulation that develops converges and thus enhances vorticity in the low- to middle-troposphere. Buoyancy-driven vertical accelerations in clouds generate vorticity anomalies that contribute to the core vortex construction through merger and axisymmetrization of like-signed anomalies, and the preferential segregation of cyclonic and anticyclonic anomalies. In this way cloud-scale to synoptic-scale processes contribute to tropical cyclone formation.

1. Introduction

Tropical cyclone (TC) formation involves numerous processes that occur on synoptic to convective scales. The thermodynamic and kinematic evolution during formation has often been divided into a sequence of events that represent fundamental stages of the overall formation process, i.e., event A must occur before event B can proceed, and event B must be complete before event C can proceed, and so on. These events have often been labelled with terms such as "genesis", "formation", "development" and "intensification". Not all theories, however, agree on the event sequences or event details, which has led to confusion when the above terms have been used to describe aspects of the TC formation process.

In the first volume, Frank (1987) defined "genesis" as the transition from a tropical disturbance to a tropical depression, (i.e., the formation of a rotational circulation with the scale of a few hundred km), and the terms "development" and "intensification" were used to define the transitions from a tropical depression to a tropical storm (sustained winds of 17 m s^{-1}) and a tropical storm to a hurricane (sustained winds of 33 m s^{-1}, also known as severe tropical cyclone and typhoon) respectively. In the second volume McBride (1995) introduced the terms "large-scale formation" and "core formation". The former describes the development of an environment conducive to TC formation, while the latter describes the transition to tropical storm. Like Frank, McBride used "intensification" to define system spin-up beyond tropical storm intensity. The two stage process proposed by Karyampudi and Pierce (2002) was used to define TC formation in Chapter 2 of the IWTC-VI report (Tory and Montgomery, 2006). The two stages are consistent with McBride's "large-scale" and "core" formation definitions.

This chapter is divided along these lines. The first section discusses the development of a large-scale environment favourable for TC formation (Stage 1), while the second looks at how circulations of at least tropical storm intensity form in that environment (Stage 2). We note here that while the main processes that characterize the two stages are quite distinct, they do overlap in time. A more physically based definition of TC genesis is the sequence of events that lead to the development of a warm-cored tropical vortex of sufficient strength to allow it to continue to intensify solely due to its own interactions with the warm underlying sea (Montgomery *et al.*, 2006). While in theory this definition gives a definitive measure of the completion time of genesis, in practice it is difficult to determine in both observations and model simulations.

There seems to be a broad consensus regarding the climatological conditions associated with tropical cyclogenesis. Briegel and Frank (1997) summarized seasonal studies of cyclogenesis (e.g., Gray, 1968, 1979, 1985). Collectively, they show that, in the current climate, most tropical cyclones form in regions that exhibit five characteristics:

(i) sea-surface temperatures above 26.5–27.0°C coupled with a relatively deep oceanic mixed layer (about 50 m);
(ii) a deep surface-based layer of conditional instability;
(iii) enhanced values of cyclonic low-level absolute vorticity;
(iv) organized deep convection in an area with large-scale mean ascent and high mid-level humidity;
(v) weak to moderate (preferably easterly) vertical wind shear.[1]

However, these characteristics are not fully independent, and they tend to be present over widespread areas of the tropical oceans much of the time (e.g., characteristics (i) and (ii) represent the summer-time climatology over much of the tropical oceans). Since tropical cyclogenesis is infrequent at any one location, the conditions that cause individual storms to form must differ from the above conditions in nature and/or degree.

Because the above five characteristics were determined from detailed climatologies of environments in which storms did and did not form, it is not immediately clear how much of characteristics (iii)–(v) are necessary environmental conditions set up by Stage 1 processes and how much they are consequences of development. While all three favour development, characteristics (iii) and (iv) are significantly enhanced by convection during Stage 2, which moistens the low- to middle-troposphere and provides a vertical mass flux that converges environmental vorticity.

2. Stage 1: Large-Scale Effects on TC Genesis

2.1. *Background*

In an average year between 80 and 90 TCs form over the tropical oceans (e.g., Gray, 1985; Frank and Young, 2006). The mean seasonal climatology of tropical cyclogenesis has been relatively well known since Gray (1968, 1975) and has been discussed at length in the previous volumes in this series. A brief review and update of recent work on climatological aspects of TC genesis is presented in the following section.

Less is known about the precise kinematic conditions (characteristics (iii)–(v)) present at

[1]It is not clear whether the shear direction is important for TC formation or whether easterly shear is more common in favourable TC formation environments.

the time and place of TC formation, the ranges over which these conditions can vary while still being sufficient to support genesis, and the manner in which large-scale circulations, such as tropical waves, bring about these conditions. Progress has been slower in these latter areas due largely to the fact that virtually all of the storms form over warm, tropical oceans in locations that are nearly devoid of in-situ observations. However, there has been a great deal of research into the large-scale effects on TC genesis in recent years using data from remote sensing instrumentation and the somewhat limited data from field programs, as well as global analysis fields and numerical simulations.

2.2. *Large-Scale Influences on TC Genesis*

Composite analyses (e.g., McBride and Zehr, 1981) and numerous observational case studies (e.g., Zehr, 1992) have shown that most storms form in regions that exhibit both anticyclonic upper-level relative vorticity and cyclonic low-level vorticity in the immediate vicinity of the pre-storm disturbance. Further, genesis is favored in regions of weak or easterly 850–200 hPa vertical wind shear (supported by modelling studies beginning with Kurihara and Tuleya, 1981), although there is often strong vertical shear just a few hundred km away from the incipient storm location.

Other case studies support the concept that tropical cyclogenesis may require the atmosphere to be preconditioned to high values of humidity in the lower to middle troposphere to inhibit negative effects of downdrafts (Bister and Emanuel, 1997).

Several researchers have developed TC genesis prediction schemes using large-scale statistical predictors (recent examples include Hennon and Hobgood, 2003; McDonnel and Holbrook, 2004). These schemes often show skill at detecting genesis and in some cases predicting it with lead times of up to 48h. In general, the predictors derived in these forecast schemes tend to support the importance of Gray's individual climatological genesis parameters in storm formation. However, they fall short of defining the precise range of conditions sufficient to cause TC genesis, and they do not identify the paths by which these conditions are reached. This limitation suggests that not all important genesis ingredients have been incorporated in these schemes. Indeed transient features such as tropical waves cannot be explicitly included.

Since the previous volume, the most active research topic concerning large-scale effects on TC genesis has been the role played by tropical waves. The remainder of this section will address recent work in this area.

2.3. *Tropical Waves and Tropical Cyclogenesis*

About 80–90% of all tropical cyclones form within 20° of the equator. Gray (1968) noted that except for the areas of the N. Atlantic and NW Pacific poleward of 20°, the vast majority of TCs form within or near the circulation of a monsoon trough, a configuration of the Intertropical Convergence Zone (ITCZ) in which there are westerly winds on the equatorward side of the trough axis (e.g., Briegel and Frank, 1997; Ritchie and Holland, 1999). This raises the possibility that tropical cyclogenesis events may be modulated by the family of equatorial and near-equatorial waves that propagate zonally in this band. The waves could make conditions for TC formation more favorable by organizing deep convection and/or by altering the flow in preferred areas. Indeed, a large fraction of tropical cyclones form where tropical waves enhance characteristics (iii)–(v) listed in section 1 (Frank and Roundy, 2006).

A monsoon trough provides an environment that satisfies all of the criteria for genesis mentioned above. Both observational studies (e.g., Ferreira and Schubert, 1997; Wang and Magnusdottir, 2006) and model simulations (e.g., Ferreira and Schubert, 1997; Wang and

Frank, 1999; Wang and Magnusdottir, 2005) suggest that a monsoon trough may break down spontaneously to form tropical cyclones if it remains unperturbed for a long enough period of time. However, the genesis process can be accelerated by external influences in preferred portions of the monsoon trough. It is probable that most storms form in conjunction with some sort of external perturbation of the trough and that pure ITCZ breakdown occurs only during times when such interactions are absent.

The ITCZ is perturbed by a variety of external circulations. For example, Briegel and Frank (1997) showed that typhoons in the NW Pacific have a tendency to form in the ITCZ when a midlatitude trough is north-northwest of the eastern end of the monsoon trough. However, the most common modulations of the large-scale flow in the low latitudes are due to zonally propagating equatorial and near-equatorial waves.

A recent global climatology of tropical wave activity (Roundy and Frank, 2004) analysed the contributions to the total variance in outgoing long-wave radiation (OLR) by each of the five prominent wave types with periods of two days or longer. One of these is the Madden-Julian Oscillation or MJO (Madden and Julian, 1994). Three others are classical equatorial waves that, in idealized form, are either symmetric or antisymmetric about the equator. These are equatorial Rossby waves, mixed Rossby-gravity waves, and Kelvin waves. The fifth wave type is confined to a single hemisphere, is westward propagating, and is known as the tropical-depression-type wave (also referred to as easterly waves or African easterly waves). There have been several studies of the theoretical structures of these waves (e.g., Matsuno, 1966; Zhang and Webster, 1989), and their observed structures (e.g., Reed *et al.*, 1977; Liebmann and Hendon, 1990; Lau and Lau, 1990; Hendon and Liebmann, 1991; Takayabu and Nitta, 1993; Dunkerton, 1993, Madden and Julian, 1994; Takayabu, 1994; Dunkerton and Baldwin, 1995; Wheeler and Kiladis, 1999; Wheeler *et al.*, 2000; Kiladis *et al.*, 2005).

Roundy and Frank (2004) found that the low-frequency MJO and equatorial Rossby waves combined to explain at least 15% of the total OLR variance between 20N and 20S latitude, and in some regions they explained at least 50% of the subseasonal local variance. The higher frequency waves produced smaller percentages of the total variance in the tropics, but in many areas, particularly in the Northern Hemisphere, they were major contributors to the variance. Using OLR as an indicator of rainfall in the tropics, Roundy and Frank (2004) showed that the tropical waves play a major role in modulating the tropical rainfall. The circulations of these waves modulate the large-scale wind fields also.

The idea that waves can trigger tropical cyclogenesis is not new. It has been known for decades that about half of the hurricanes in the North Atlantic form from African waves (e.g., Frank and Clark, 1980), and there have been many studies of this type of genesis (e.g., Carlson, 1969; Landsea, 1993; Thorncroft and Hodges, 2001, and others). Molinari and Vollaro (2000), Maloney and Hartmann (2000a,b, 2001), Hartmann and Maloney (2001) and others have shown effects of the MJO on North Pacific tropical cyclogenesis. Dickinson and Molinari (2002) and Molinari *et al.* (2007) analyzed events of cyclogenesis in the Northwest Pacific associated with mixed Rossby-gravity waves, and equatorial Rossby waves respectively. Bessafi and Wheeler (2006) and Hall *et al.* (2001) analyzed the relationships between the MJO and equatorial Rossby waves, mixed Rossby-gravity waves, and Kelvin waves and genesis over the southern Indian Ocean and in the Australian region, respectively.

Zonally propagating waves in the tropical atmosphere play major roles in tropical cyclogenesis by enhancing characteristics (iii)–(v) (section 1) locally in already largely favorable areas. They can enhance the potential for genesis in several ways: by increasing upward vertical motion and convection (which also tends to lead to increased deep-layer moisture), by increasing the low-level vorticity, and/or by altering the local vertical shear pattern. Some types of waves may trigger convection also in regions of converging zonal wind, where wave energy tends to accumulate (Ritchie and Holland, 1999). The larger-scale waves, such as the MJO and equatorial Rossby waves, can alter the mean zonal wind over large enough time and space scales to affect the mean flow in which the smaller and higher frequency waves propagate. Finally, some of the lower frequency waves might affect sea surface temperatures sufficiently to have some effect on genesis, but this is not explored in the current study.

Figure 1 shows the fraction of the total TCs in each basin that form within the negative OLR anomaly region of each major tropical wave type. The analysis is for all storms forming within 20° of the equator in each basin. For each filtered wave type the negative OLR anomaly (considered to be the most convectively active portion of the wave) covers about 50% of the analysis area, so a number above 50% indicates that TCs tend to occur most often in that active region.

In all six basins the storms form in the negative OLR anomalies of the MJO band between about 60%–70% of the time, and these results are significantly different from the 50% line at the 99% level (97% for the Atlantic) using a bootstrap test. Storms have a clear tendency also to form in the negative OLR anomaly regions of equatorial Rossby waves in all basins. For equatorial Rossby waves the departures from the 50% line are all significant at the 99% level.

The mixed Rossby-gravity-tropical depression-type combined wave band shows a clear tendency also for storms to form in the negative OLR anomaly regions, with percentages ranging from a low of about 65% in the South Pacific to a high of around 85% in the Atlantic. Departures are significant at better than the 99% level in all basins. The only two basins with percentages below 75% are in the Southern Hemisphere, where this wave band is least active. Interpretation of these results is complicated because a portion of the signals of the storms themselves fall into this band. Storms show a small preference for forming in the negative anomalies of the Kelvin wave band in all basins. Although the percentages of storms that do so are only 55%–65%, the departures from 50% are significant at the 97% level or above.

Figure 2 shows a sequence of composites of low-level wind and OLR anomalies, filtered for various wave types, at the time of tropical cyclogenesis. The figures shown are for the NE Pacific basin, but the results for each of the five other basins (NW Pacific, N. Atlantic, N. Indian Ocean, S. Indian Ocean, SW Pacific) show qualitatively similar results in terms of the patterns of the anomalies and the phase relationships between the genesis location and the anomalies. The large dot in the centre of each figure indicates the mean location of the forming TC at the time it was first declared to be a tropical depression. These and related results show that most tropical cyclones form within the convectively enhanced phases of the MJO, equatorial Rossby waves, mixed-Rossby-gravity waves, and/or tropical-depression-type waves. (There can be more than one wave type active at a location.) TCs tend to form within the waves where the low-level cyclonic circulations are enhanced. Kelvin waves seem to be poorly related to genesis[2] except perhaps in the NE Pacific.

The results illustrated by Fig. 2 support the general findings of the referenced case studies of relationships between tropical waves and genesis. The waves alter the vorticity and vertical shear fields, both of which are known to affect TC formation. Further, the

[2]The relatively low numbers of Kelvin wave genesis events is presumably because Kelvin waves have maximum amplitude at the equator.

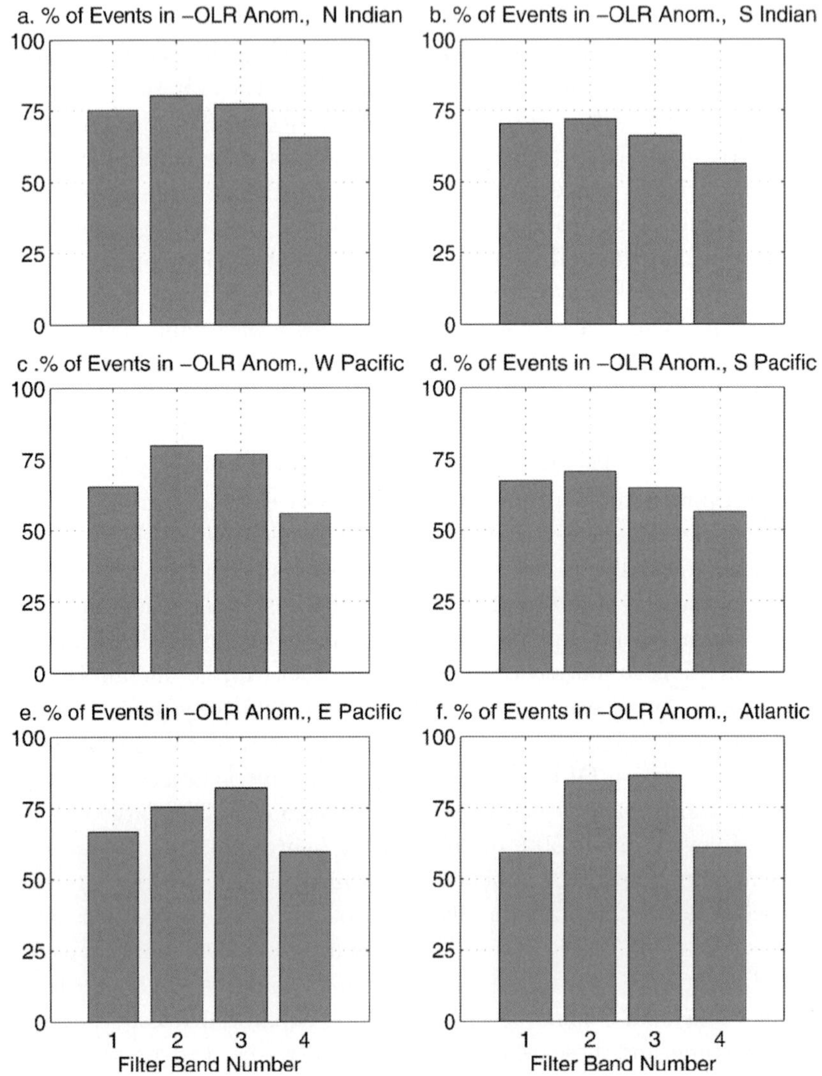

Figure 1. Each panel shows, for the indicated cyclone basin, the percentages of all TC formations that occur within the negative OLR anomaly region for each of four filter bands. These filter bands are, from left to right, the MJO, equatorial Rossby waves, a combined band spanning mixed-Rossby-gravity waves and tropical depression-type waves, and Kelvin waves. The negative OLR anomaly regions (assumed convectively active) each cover about 50% of the basin, so values above 50% indicate a preference for TCs to form in the convectively active portion of that wave type. From Frank and Roundy (2006).

waves tend to form persistent regions of wave-enhanced convection.[3] It should be noted that the anomaly wind fields shown in these figures are significantly weaker than the circulation anomalies that are caused by any single wave. This results both from the inevitable smoothing associated with compositing and the fact that not all tropical cyclones form in the most favorable region of a given wave. Tropical cyclones forming in a monsoon trough environment do not require external wave forcing to trigger genesis, but Roundy and Frank

[3]Wave-enhanced convection is deep moist convection that has been enhanced by the dynamical and/or thermodynamical fields of a large-scale propagating wave.

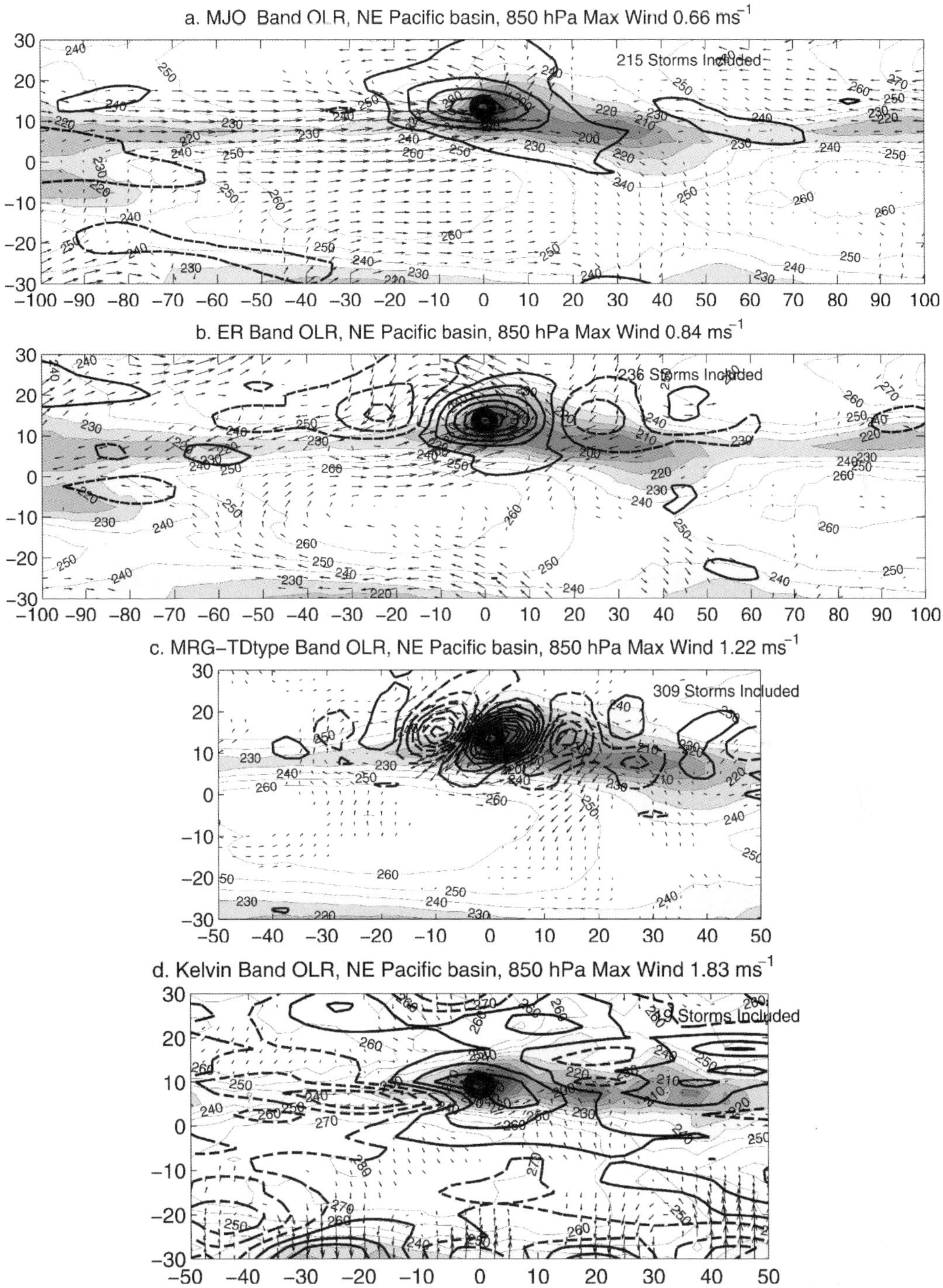

Figure 2. Each panel shows composites of the wave-band-filtered 850 hPa wind anomalies (arrows) and OLR anomalies (heavy solid and dashed lines for negative and positive anomalies respectively) for (a) Madden-Julian oscillation, (b) Equatorial Rossby waves, (c) Mixed Rossby-gravity waves and (d) Kelvin waves, at the time and place of TC genesis in the NE Pacific basin. The background shading shows the mean background OLR values (darker shading is more negative). The large black dot near the centre of each panel indicates the mean genesis location and the centre of the compositing grid. From Frank and Roundy (2006).

(2004) believe that the most common form of genesis occurs when tropical waves interact with the monsoon trough circulation to form local regions that exceed the necessary conditions for genesis.

We will see later that during formation it may take several days before a vortex develops of sufficient resilience to resist environmental shear and dry air. During this period, the TC embryo can easily be destroyed. An observational analysis found that most TCs that form in African easterly waves develop in a region of closed streamlines relative to the mean zonal flow (Dunkerton *et al.*, 2008). These authors argue there is very minimal flow exchange across the closed streamlines and thus the enclosed circulation is effectively protected from outside negative influences of vertical and horizontal shear, and dry air. They coined the analogy "marsupial pouch" to describe this protected environment in which the TC embryo develops. The "marsupial pouch" develops within the parent wave (Stage 1) within which a core vortex forms (Stage 2) of sufficient intensity and scale for it to take on its own propagation characteristics and separate from the parent wave. It is not yet clear how important this process is in other TC formation environments.

2.4. *Higher Latitude Precursors of Tropical Cyclogenesis*

While the vast majority of tropical cyclogenesis in the deep tropics has a close connection to tropical waves, a large number of tropical cyclones form poleward of 20° latitude and sometimes poleward of 30°, especially in the Atlantic basin (Bosart and Bartlo, 1991; Montgomery and Farrell, 1993; Davis and Bosart, 2003, 2004; McTaggart-Cowan *et al.*, 2006, 2007a,b). McTaggart-Cowan *et al.* (2008) documented different types of synoptic-scale precursors of Atlantic genesis cases and found that synoptic-scale upward motion induced by baroclinic disturbances of extratropical origin was important in approximately 40% of all Atlantic genesis cases. These events included upper-tropospheric troughs and also cases exhibiting weak lower tropospheric baroclinity connected with subtropical decaying frontal zones.

Like tropical waves the extratropical precursors provide anomalously favourable conditions for genesis, i.e., environments conducive to organized convection, and often enhanced lower tropospheric vorticity. In extreme cases, a surface cyclone of baroclinic origin is strong enough to self amplify and thus bypass much of the vortex core construction processes described in section 3 (Davis and Bosart, 2003). A brief discussion on the role of extratropical precursors and organized convection can be found in section 3.5.1.

To characterize the transformation of extratropical precursors into tropical depressions, Davis and Bosart used the term "tropical transition". This refers to the phenomenological change from extratropical to tropical characteristics that is often accompanied by a transition from a cold core vortex (vorticity maximized in the mid-troposphere) to a warm core vortex (vorticity maximized near the surface).[4] While such a cold-core to warm-core transition can describe the transition of all mid-tropospheric vortices (e.g., mesoscale convective vortices), the thermodynamic changes are often more pronounced in tropical transition, especially in strong tropical transition events in which the precursor disturbance is a baroclinic cyclone.

3. Stage 2: Vortex Construction

3.1. *Background*

The tropical cyclone, like an internal combustion engine, requires a crank-start before

[4]The cold and warm references to the vortex core refer to the radial temperature gradient required to provide thermal wind balance with the vertical gradient in circulation.

it becomes self-sustaining. The system is generally considered to have become self-sustaining at about tropical depression intensity, i.e., when the energy derived locally from surface sensible heat and moisture fluxes is sufficient to maintain the circulation against friction and intensify the storm (e.g., Rotunno and Emanuel, 1987). Since the previous volume there has been a surge of interest in the "genesis problem" (i.e., unravelling the details of the TC crank-start) giving rise to a number of theories.

Observations of low-level vorticity enhancement in mesoscale convective systems (MCSs) (e.g., Velasco and Fritsch, 1987; Zehr, 1992) during TC formation placed the MCS (or multiple MCSs) at the core of the theories. Conceptual models of both mid-latitude terrestrial and tropical oceanic MCSs (e.g., see Houze, 2004) usually contain a relatively narrow convective precipitation region beside an expansive stratiform precipitation region. A schematic representation of the MCS horizontal divergence/convergence profile is presented in Fig. 3. In the convective region an in-up-out circulation is accompanied by convergence in the low- to middle-troposphere with divergence above.[5] In the stratiform region there is typically mid-level convergence with divergence above and below. Vorticity is enhanced in convergent flow and weakened in divergent flow, which means low- to middle-level pre-existing vorticity should be enhanced in the MCS convective regions, and middle-level pre-existing vorticity should be enhanced in the MCS stratiform regions. However, of significance to the TC genesis problem, the divergence below the MCS stratiform region weakens the low-level vorticity, and thus opposes the intensification necessary for TC formation. This problem is referred to in later sections as the "kinematic roadblock". A "thermodynamic roadblock" has also been proposed (e.g., Bergeron, 1950; Rotunno and Emanuel, 1987), based on the concern that downdrafts suppress convection[6] by bringing low-entropy air to the surface, which stabilizes the surface layer.

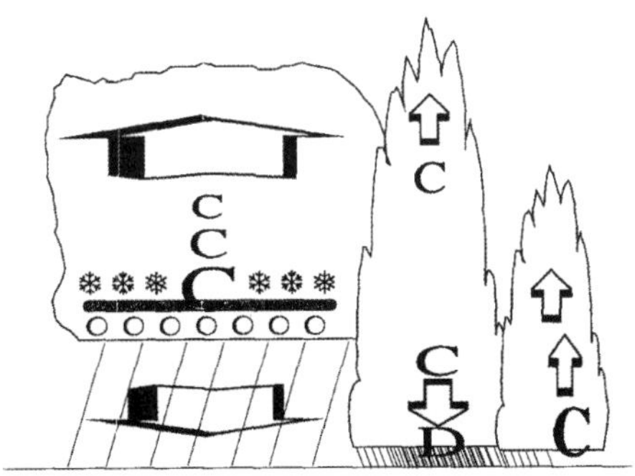

Figure 3. A schematic summary of the convergence features observed in the studies of Mapes and Houze (1995, 1993). Intense surface convergence prevails in young convective cells (right), while elevated convergence and surface divergence characterize older cells with their well-developed convective downdrafts (centre). Midlevel convergence, with its peak value atop the melting layer, is observed in stratiform rain areas (left). Arrows indicate vertical winds associated with the convergence-divergence features. (The figure and caption are reproduced from Fig. 3 of Mapes and Houze, 1995.)

Before proceeding to the TC genesis crank-start theories it is worth reviewing some general theory relevant to the TC genesis problem, including vorticity tendency, thermodynamics of convection, and the nature of Mesoscale Convective Systems (MCSs).

3.2. *General Theory*

3.2.1. *Vorticity tendency fundamentals*

In terms of the vertical component of absolute vorticity (η) the Haynes and McIntyre (1987, hereafter HM87) vorticity conservation principle states that for pressure surfaces that do not intersect a boundary there can be no *net* change in η between two pressure surfaces, and thus no *net* transport of η across pressure surfaces. (The

[5]Low-level divergence associated with downdrafts in the older convective cells reduces the net low-level convergence, but the convective region generally remains convergent on the whole.

[6]Modelling studies presented in section 3.5 suggest that downdrafts may in fact help promote further convection. However, downdrafts do reduce the net upward mass flux for a given release of conditional instability, which slows the system-scale development (section 3.3.1).

same applies to PV substance across isentropic surfaces, see section 2.1 of Chapter 1.)

In a pressure based vertical coordinate system the η tendency equation can be expressed in flux form with zero sources and sinks (HM87),

$$\frac{\partial \eta}{\partial t} + \nabla.\mathbf{J} = 0 \tag{1}$$

where η resembles a tracer-like quantity and $\mathbf{J}$ is the η flux. Expanding $\mathbf{J}$ into three components the full η tendency equation can be expressed as,

$$\begin{aligned} \frac{\partial \eta}{\partial t} &= -\nabla.[\mathbf{J}_a + \mathbf{J}_\omega + \mathbf{J}_F] \\ \mathbf{J}_a &= (u, v, 0)\eta \\ \mathbf{J}_\omega &= \omega\frac{\partial}{\partial p}(v, -u, 0) \\ \mathbf{J}_F &= (-F_y, F_x, 0) \end{aligned} \tag{2}$$

where u and v are the horizontal (parallel to pressure surfaces) wind components, ω is vertical motion, p is pressure, and F_x and F_y are the zonal and meridional components of the local frictional or other forces. (Compare Eqs. (1) and (2) with Eqs. (24) and (25) of Chapter 1.) The vertical component of all three fluxes is zero, which gives rise to the HM87 statement of zero net change in η between pressure surfaces. The divergence of $\mathbf{J}_a$ describes the tendency of a materially conserved quantity and is hereafter termed the advective flux (HM87). It describes η tendency associated with horizontal η advection, and horizontal convergence and divergence of η that "concentrates" and "dilutes" η respectively within a pressure layer[7] (see Fig. 4). It follows that the advective flux can only contribute to a change in η magnitude (not sign). In contrast the divergence of $\mathbf{J}_\omega$ and $\mathbf{J}_F$ introduces nearby η sources and sinks within the pressure layer, which can introduce a local η sign change. Mathematically $\mathbf{J}_\omega$ and $\mathbf{J}_F$ resemble η fluxes from the source to the sink, but they are not advective in nature. (Hereafter $\mathbf{J}_\omega$ and $\mathbf{J}_F$ are labelled non-advective-ω and non-advective-F fluxes respectively.) Common examples of $\mathbf{J}_\omega$ and $\mathbf{J}_F$ resulting in distinct η sources and sinks are the generation of counter-rotating vortices from respectively, the tilting of horizontal vorticity into the vertical in a thunderstorm updraft, and the generation of wake vortices by an obstacle in a fluid.

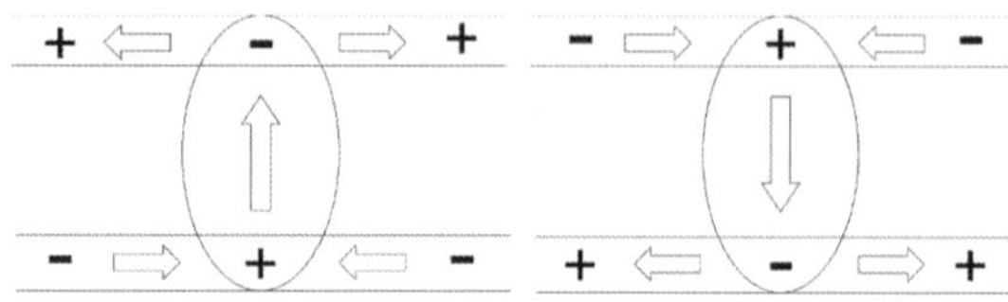

Figure 4. Schematic representation of an updraft and downdraft. The ovals represent surfaces of $\omega = 0$, and the horizontal lines denote upper and lower pressure layers. Cyclonic and anticyclonic η tendency from $\nabla.\mathbf{J}_a$ (Eq. (2)) is indicated by + and − symbols where the flow is horizontally convergent and divergent respectively (assuming cyclonic background η).

Expanding Eq. (2) by applying the divergence operator to the fluxes yields the material form of the η tendency equation,

$$\begin{aligned} \frac{\partial \eta}{\partial t} &= -\mathbf{u}_h.(\nabla_h \eta) - (\nabla.\mathbf{u}_h)\eta \\ &\quad - \left(\frac{\partial \omega}{\partial x}\frac{\partial v}{\partial p} - \frac{\partial \omega}{\partial y}\frac{\partial u}{\partial p}\right) - \omega\frac{\partial \eta}{\partial p} \\ &\quad - \left(\frac{\partial F_y}{\partial x} - \frac{\partial F_x}{\partial y}\right), \end{aligned} \tag{3}$$

where the subscript h represents horizontal components. The terms on the right hand side are respectively, horizontal advection, convergence/divergence (or stretching/squashing), tilting, vertical advection and the frictional tendency (Holton, 2004). Recognition of the relationship between Eqs. (2) and (3) provides useful physical insight into the advective and non-advective-ω fluxes. $\nabla.\mathbf{J}_a$ is equivalent to the sum of horizontal advection and convergence of η, and $\nabla.\mathbf{J}_\omega$ is equivalent to tilting and vertical advection of η. While the material tendency formulation (Eq. (3)) can assist with the physical interpretation of Eq. (2), HM87 warn that calculating the material tendency terms directly

[7]Here the pressure layer is an open ended three-dimensional volume between two pressure surfaces in which η is conserved. Conservation of η can also be considered on a pressure surface, i.e., in two dimensions.

from NWP and observational data can give misleading results because of non-trivial cancellation between terms.

It is worth considering two simple idealized examples to illustrate η tendency associated with the three **J** fluxes. The open arrows in Fig. 4 represent the flow associated with an updraft (left) and downdraft (right), and the horizontal lines are pressure surfaces that span upper and lower pressure layers. The + and − symbols indicate regions of convergent and divergent flow where η is "concentrated" and "diluted" respectively. The horizontal arrows could also represent the orientation of the $\mathbf{J}_a$ vectors. From Eq. (2) the magnitude of η will increase and decrease where these vectors converge and diverge respectively.

In Fig. 5 additional flow information is given to illustrate the changing circulation with height required for the non-advective-ω term to contribute to η tendency (Eq. (2)). The up- and downdraft centres are colocated with the centre of the background circulation, which is warm-cored above and cold-cored below. Figure 5 shows that for cyclonic circulations, an updraft in a warm cored environment and a downdraft in a cold cored environment, induce a cyclonic η source (+) towards the inner side of the circulation and a sink (−) towards the outside. The opposite orientation is induced in updrafts in a cold-cored circulation and downdrafts in

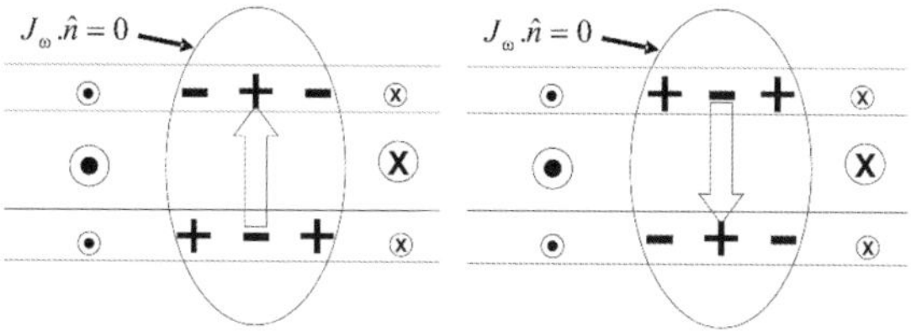

Figure 5. As in Fig. 4 except for the non-advective-ω tendency ($\nabla.\mathbf{J}_\omega$, Eq. (2)). The background cyclonic circulation is described by the $\otimes$ and $\odot$ symbols representing flow into and out of the page respectively. (Upper and lower halves are warm and cold cored respectively.) From Eq. (2) $\mathbf{J}_\omega.\hat{\mathbf{n}} = 0$ on the pressure surfaces and $\omega = 0$ surfaces (ovals).

a warm-cored circulation. These tendencies can be confirmed by considering the signs of ω and $\partial v/\partial p$, to reveal the orientation of $\mathbf{J}_\omega$ (Eq. (2)).

To demonstrate the η tendency associated with $\mathbf{J}_F$ consider the background circulation in Fig. 5 and assume friction acts in the opposite direction to the flow. Then F_y is positive to the left and negative to the right of the up- and downdrafts, and F_x is zero in the plane of the page. Thus $\mathbf{J}_F$ points to the right of the wind (i.e., outwards), and is divergent across the updraft, which is consistent with friction acting to spin down the circulation (Eq. (2)).

It is often useful to consider volume averaged η tendency. Integrating Eq. (2) over an arbitrary volume (V) and using the divergence theorem to express the terms on the RHS of Eq. (2) as a surface integral, the net tendency inside the volume becomes,

$$\iiint_V \frac{\partial \eta}{\partial t} dV = -\iint_S (\mathbf{J}_a + \mathbf{J}_\omega + \mathbf{J}_F).\hat{\mathbf{n}} dS. \tag{4}$$

Here $\hat{\mathbf{n}}$ is a unit vector normal to the surface S bounding the volume V. $\mathbf{J}_a.\hat{\mathbf{n}}$, $\mathbf{J}_\omega.\hat{\mathbf{n}}$ and $\mathbf{J}_F.\hat{\mathbf{n}}$ represent the advective, non-advective-ω and non-advective-F fluxes normal to S respectively. From Eq. (4) volume averaged changes in η must be associated with a non-zero net **J** flux through the surface S. Volume averaged η budget calculations can be simplified by choosing pressure surfaces to define the upper and lower volume boundaries in order to exploit the zero flux components normal to pressure surfaces. Then the budget is reduced to the integral of the fluxes normal to the lateral boundary of the volume.[8]

Additional insight into the advective and non-advective-ω fluxes can be gained by considering hypothetical volumes bounded above and below by pressure surfaces (that do not intersect a boundary) and laterally by surfaces of $\mathbf{u} = 0$ and $\omega = 0$ respectively. Consider first the latter example, which is illustrated in Fig. 5. The

[8]The same conclusions can be reached by performing an area integral over the x–y plane on the LHS of Eq. (2). Then the RHS is reduced to a line integral using the divergence theorem (e.g., Raymond *et al.*, 1998).

oval represents the surface of $\omega = 0$ upon which $\mathbf{J}_\omega.\hat{\mathbf{n}} = 0$ (Eq (2)). It follows that the net non-advective-ω η tendency must be zero within each pressure layer of up- and downdrafts, which from Eq. (3) means that the sum of tilting and vertical advection must be zero in up- and downdrafts (Tory and Montgomery, 2008). Non-zero local η tendencies can be present within the volume (i.e., the + and − symbols in Fig. 5 contained within the ovals). Note that while the tendency from $\nabla.\mathbf{J}_\omega$ must sum to zero across the up- or downdraft the tendency is not necessarily zero on the $\omega = 0$ surface. The former example ($\mathbf{u} = 0$) is illustrated in the left panel of Fig. 6, a schematic illustrating how the $\nabla.\mathbf{J}_a$ and $\nabla.\mathbf{J}_\omega$ tendencies work together within a pressure layer in a cyclonic η environment. Inside the surface of $\mathbf{J}_a.\hat{\mathbf{n}} = 0$ there can be no net change in η from the advective flux term. An inner convergent region (dark shading) is thus bounded by a divergent region (light shading) and the $\mathbf{J}_a$ flux vector points inwards. It follows that η is "concentrated" and "diluted" in the convergent and divergent regions respectively. (The surface of zero flow, on which $\mathbf{J}_a.\hat{\mathbf{n}} = 0$, can be relative to a moving reference frame.) In the convergent region, enlarged in the right panel, there is net upward motion in which two updrafts and a downdraft are embedded. As noted above, the net η tendency associated with the $\mathbf{J}_\omega$ flux is zero within the surfaces of $\mathbf{J}_\omega.\hat{\mathbf{n}} = 0$, but local sources and sinks may form within. η tendency dipoles form in this case because the up- and downdrafts are not centred on the background circulation (which is located somewhere between the up- and downdrafts). The vorticity anomalies outlive the vertical flows that induce them, and they may interact with time in a manner that helps contribute to the core intensification (section 3.5.3).

It follows that for volume averaged η tendency budgets only up- and downdrafts that span the volume boundary contribute to the volume averaged η tendency through the non-advective-ω term.

It can be useful to consider η tendency for the special case of axisymmetric flows using cylindrical coordinates. The tangential wind tendency equation (Eq. (2) of Chapter 1) becomes,

$$\frac{\partial v}{\partial t} = -\left(u\eta + \omega\frac{\partial v}{\partial p} - F_\lambda\right), \tag{5}$$

where u and v are the radial and tangential wind components, F_λ is the frictional force in the direction of the tangential wind, and the tangential gradient terms are zero in axisymmetric flows. In axisymmetric flows,

$$\eta = f + \frac{1}{r}\frac{\partial rv}{\partial r}. \tag{6}$$

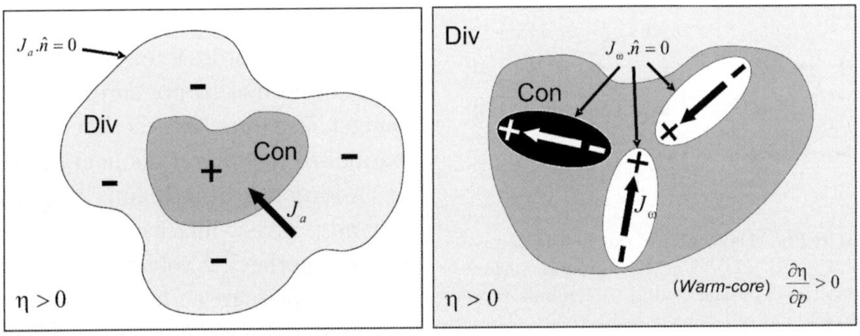

Figure 6. Horizontal schematic of advective (left) and non-advective-ω (right) η tendency. The pale and dark shading represent regions of net divergence and convergence respectively. The + and − symbols indicate positive and negative η tendency. In the left panel $(u, v) = 0$ on the outer surface ($\mathbf{J}_a.\hat{n} = 0$). The thick arrow gives the direction of $\mathbf{J}_a$. Up- (white) and downdrafts (black) are embedded in the convergent region (right panel). The thick arrows represent $\mathbf{J}_\omega$ when the convergent region is embedded in a warm-cored background cyclonic circulation (NH).

Taking the horizontal divergence of Eq. (5) gives,

$$\frac{\partial}{\partial t}\left(\frac{1}{r}\frac{\partial rv}{\partial r}\right) = -\frac{1}{r}\frac{\partial}{\partial r}\left[r\left(u\eta + \omega\frac{\partial v}{\partial p} - F_\lambda\right)\right]. \tag{7}$$

Thus, combining Eqs. (6) and (7), and remembering $\frac{\partial f}{\partial t} = 0$ in an earth-relative reference frame, we arrive at the flux form of the absolute vorticity tendency equation for axisymmetric flows in cylindrical coordinates,

$$\frac{\partial \eta}{\partial t} = -\frac{1}{r}\frac{\partial}{\partial r}\left[r\left(u\eta + \omega\frac{\partial v}{\partial p} - F_\lambda\right)\right], \tag{8}$$

where the three terms inside the curved brackets are $\mathbf{J}_a$, $\mathbf{J}_\omega$ and $\mathbf{J}_F$ respectively. Equation (8) can be expressed in the material form as,

$$\frac{\partial \eta}{\partial t} = -\frac{1}{r}\frac{\partial (ru)}{\partial r}\eta - u\frac{\partial \eta}{\partial r} - \frac{1}{r}\frac{\partial \omega}{\partial r}\frac{\partial v}{\partial p}$$
$$-\omega\frac{\partial \eta}{\partial p} + \frac{1}{r}\frac{\partial (rF_\lambda)}{\partial r}, \tag{9}$$

where the terms on the RHS are respectively, the horizontal convergence, horizontal advection, tilting, vertical advection and the frictional tendency. Equations (5) and (9) are used in section 3.2.4.

3.2.2. *Thermodynamic Considerations*

The cyclonic environment, in which the TC forms, intensifies when vorticity is converged in the lower troposphere. This convergence results when buoyancy forces, generated by diabatic heating in convective regions, drive a troposphere-deep vertical mass flux on the scale of the convective region, i.e., the system-scale secondary circulation. Ekman convergence in the boundary layer also contributes to the system-scale secondary circulation, although the associated η convergence tendency is largely cancelled by friction. In the absence of convective heating the frictionally induced secondary circulation is shallow, with inflow in the boundary layer and outflow immediately above. Convective heating warms the air so that it can ascend to the upper troposphere. Thus convection plays an important role in maintaining the troposphere-deep secondary circulation, but it is likely that only locally buoyant convective towers will contribute to storm intensification (Chapter 1, section 3). Due to the role of buoyancy forcing in driving the system-scale secondary circulation during TC development, an understanding of the thermodynamic processes in the convective region is important for understanding tropical cyclone formation.

For the in-up-out circulation to most effectively spin-up the system-scale primary circulation the thermodynamic conditions must favour convection that drives a troposphere-deep vertical mass flux. Evaporation of rain and cloud water can introduce a local downward mass-flux (e.g., convective downdrafts, stratiform subsidence), particularly in the low- to middle-troposphere. On the system-scale this opposition to the total upward mass-flux reduces the lower-level convergent inflow, and consequently the mean system-scale η concentration.

An estimate for the evaporative downdraft potential of a parcel is given by the downdraft convective available potential energy (DCAPE, Emanuel, 1994). If rain or cloud water is evaporated into the parcel it will saturate at the parcel's wet-bulb temperature. If evaporation is maximized as the parcel falls (i.e., the parcel remains saturated) it will follow a moist adiabat on a thermodynamic diagram such as the Skew-T log-P. Figure 7 illustrates a possible parcel path satisfying the above thermodynamic processes. The shaded area represents the parcel's DCAPE. The area can be divided into two sub areas, the DCAPE that would be generated if the parcel was initially saturated (labeled saturated DCAPE in Fig. 7), and the remaining area (labeled unsaturated DCAPE). This sub-division is important for illustrating processes required to reduce DCAPE. It can be seen from the figure that moistening of the convective region environment reduces the unsaturated DCAPE, and stabilization of the environment below the parcel (increasing $\frac{\partial \theta}{\partial z}$) reduces the saturated DCAPE. Clearly

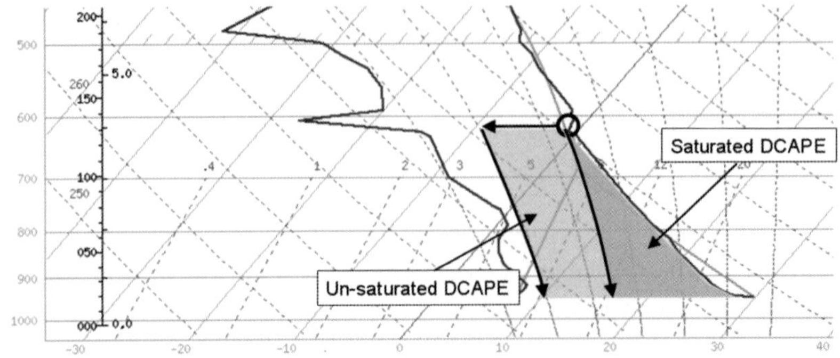

Figure 7. DCAPE (shaded) for a parcel (black circle) as depicted on an F-160 thermodynamic diagram. Dark grey lines represent temperature and dew-point temperature, Black arrows represent theoretical downdraft parcel paths for an initially saturated and an initially unsaturated parcel, which gives rise to the "saturated" and "un-saturated" DCAPE division described in the text. (This dry, deeply mixed boundary layer example was chosen to most clearly illustrate the two DCAPE subdivisions.)

moistening to saturation alone is not sufficient to inhibit downdrafts. Persistent convection both moistens and stabilizes through mixing of up- and downdraft air with the immediate environment. It follows then that for a somewhat contained region of persistent convection (i.e., protected from dry air intrusions or leakage of moist air, Dunkerton *et al.*, 2008), the lower- to middle-troposphere will both moisten and stabilize with time, and the potential for downdrafts will decrease with time. This transition should increase the efficiency of the mean in-up-out circulation and consequently the system-scale spin-up efficiency.

While stabilization of the lower troposphere reduces the potential for downdrafts it also reduces conditional instability and CAPE. Thus to reduce downdraft potential and maximise convective potential, an optimal thermodynamic profile might include a shallow layer of conditional instability near the surface, with a moist neutral, saturated layer above.

The total upward mass flux in an isolated convective system is proportional to the total diabatic heating, and the total diabatic heating (condensation minus evaporation) is proportional to the surface rainfall.[9] Thus surface rainfall provides a good indication of the temporally and spatially integrated vertical mass flux over the lifetime of the system (e.g., Raymond and Sessions, 2007). However, vertical mass flux estimates based on surface rainfall do not identify vertical variations in mass flux, which can be important when focusing on smaller time-scales and horizontal sub-regions of the convective system. The stratiform region of an MCS, for example (see section 3.2.3), may contain significant rainfall, but the low-level vertical mass-flux is downward. The competing effects of stratiform and convective processes during TC formation are likely to influence the low-level spin-up rate (e.g., Tory *et al.*, 2006, 2007).

3.2.3. *Mesoscale Convective Systems*

It is worth devoting some time to reviewing MCSs given their important role in the TC formation process. A detailed schematic of an idealized MCS typical of tropical oceanic regions is presented in Fig. 8. The kinematic and thermodynamic structure is illustrated in some detail. Distinct regions of both convective and stratiform precipitation are evident. The cool and saturated convective downdrafts bring moderately low-θ_e air (342 K, see circled numbers) to the surface. Buoyancy forces spread this denser

[9] This assumes no lateral losses of water or ice from the system, and that all precipitation reaching the surface is liquid.

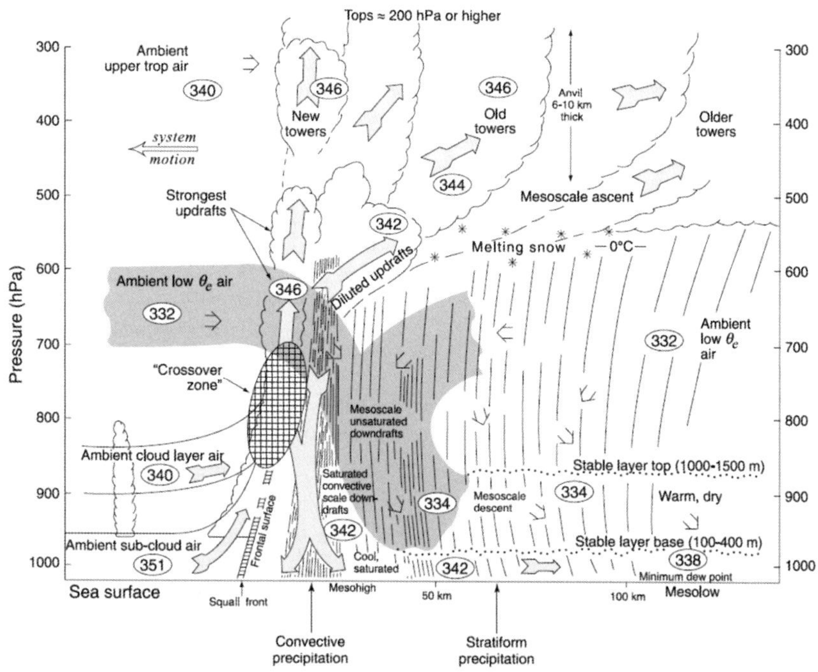

Figure 8. Idealization of a tropical oceanic MCS with leading-line/trailing stratiform structure. Parcels of subcloud boundary layer air rise and form the basic convective updrafts. Ambient cloud layer air is entrained into the updrafts. The updraft parcels rise till they lose their buoyancy by entrainment or by encountering a stable layer in the environment. Entrainment of ambient low-θ_e air weakens updrafts and forms convective-scale downdrafts, which sink to the surface in the convective precipitation zone. Note that the system has three-dimensionality such that the updraft and downdraft trajectories are not collocated, and the convective region contains a "crossover zone" where convective-scale updrafts and downdrafts coexist. From Houze (2004). (Reproduced by permission of American Physical Union.)

air outwards, causing lifting of warmer air (to the left) at a squall-line front. The relatively high-θ_e air of the sub-cloud layer feeds new convective towers when lifted at the frontal surface. Entrainment of mid-level dry ambient air (low-θ_e) dilutes updrafts, and initiates the convective-scale downdrafts through the evaporation of cloud and rain water. The constant advance of the convective downdraft outflow pushes the squall front forwards (left in Fig. 8), which triggers fresh bursts of convective towers. The old towers are carried rearward where the remnants become the stratiform cloud deck. Melting snow at the freezing level, and the evaporation of rain below, maintained by the mid-level inflow of ambient low-θ_e air (i.e., dry air), provide diabatic cooling that promotes unsaturated downdrafts on the mesoscale. The schematic suggests the mesoscale downdrafts do not in fact reach the surface. Instead they are undercut by the cooler but higher θ_e air sourced in the convective saturated downdrafts. The potential for this cold layer to suppress convection over much of the MCS (Bergeron, 1950; Rotunno and Emanuel, 1987) raised concerns of the TC genesis thermodynamic roadblock. However, this relatively uniform, low-level layered temperature distribution in the stratiform precipitation region illustrated in Fig. 8 is not consistent with the patchy distribution of both high (>350 K) and moderately low-θ_e (343–345 K) in the observed tropical oceanic MCS boundary layers reported by Kingsmill and Houze (1999). These authors suggested that either the moderately low-θ_e air was carried downwards from levels only just above the boundary layer (see also Betts, 1976), or there was considerable mixing causing the dilution of low-θ_e air (about

335 K) in downdrafts initiated in the mid-levels. Either way it suggests the thermodynamic roadblock may not be significant in many tropical oceanic MCSs. Indeed modelling studies suggest the triggering of convection by downdraft outflows may be important for maintaining convection (section 3.5).

The MCS in Fig. 8 is self-sustaining while in this favourable environment, and may persist for a number of days. The stratiform region typically grows with time as the deep convection is relatively short lived, but the remnant buoyant cloud that feeds the stratiform survives for some time. The low- to middle-level convergence in the convective updrafts favours vorticity enhancement, but it is opposed by low-level divergence in convective downdrafts and stratiform subsidence regions. Clearly the relative scales and intensity of these MCS components determine the net vertical profile of divergence over the system, and the potential of the system on the whole to enhance or weaken vorticity. The MCSs analysed by Mapes and Houze (1995) were on average convergent in the low-middle troposphere (see their Fig. 11), but there was significant variation in the vertical profiles of divergence between MCSs and at different stages of the MCS life cycles.

The pertinent question arising from the previous subsection is how does this picture change as the environmental humidity and static stability increases (DCAPE decreases)? The aforementioned points suggest that the intensity of the convective downdrafts should weaken, and the updraft towers should become more intense as the in-cloud evaporation of entrained dry air is reduced. The squall front may slow or stall as the source of cool dry air dissipates and the towers may become larger and longer lived. The reduction of stratiform evaporation should weaken the stratiform circulation. If the moisture that feeds into the stratiform deck is reduced then the areal coverage of stratiform cloud might decrease. Most of the above points imply either a reduction in low-level divergence or an increase in low-level convergence. Thus, provided the low-level conditional instability is not exhausted, moistening of the MCS environment will likely favour greater numbers of larger, longer-lived, and more intense convective towers. These towers will drive the significant troposphere-deep upward mass flux necessary for rapid system-scale intensification of the primary circulation.

This speculated connection between moistening and MCS mass-flux changes is partly supported by observations of higher mid-level θ_e surrounding deep and broad convection than that surrounding deep and narrow convection (Kingsmill and Houze, 1999), in which the higher θ_e values were mostly due to increased humidity.

The link between increased low-level convergence and broad regions of convection is consistent with MCS infrared satellite signatures, of large areas of very cold cloud tops in MCSs prior to a significant enhancement of low-level vorticity (e.g., Zehr, 1992). An example of broad deep convection and the low-level spin-up during TC formation was observed at high spatial resolution during the development of Hurricane Ophelia (Houze *et al.*, 2009). Large areas of strong low-level convergence in long-lived and very active convective regions have also been observed in non-TC-formation observational studies (Zipser and Gautier, 1978; Mapes and Houze, 1992).

The schematic of Fig. 8 shows a two-dimensional, squall-line tropical MCS, with well-defined structures that typically persist throughout the MCS life-span. However, tropical MCSs can be three-dimensional in structure, with more than one convective and stratiform region (Houze, 2004). Convective complexes can also contain multiple MCSs with attendant MCVs of largely varying intensities and sizes that influence their interaction and future development (e.g., Simpson *et al.*, 1997). Furthermore, MCVs can outlive their parent MCS, and new convection can reform some time later (e.g., Simpson *et al.*, 1997; Houze, 2004) because convection can be favoured in the remnant MCV (e.g., Raymond and Jiang, 1990). Thus a conceptual model of a well organised, two-dimensional, maturing MCS with a gradually moistening environment that facilitates a gentle transition from weak to strong low-level convergence, may be overly simplistic. A convective complex comprised of multiple MCSs, with a disorganised distribution of convective and stratiform regions, each at differing levels of maturity and humidity may evolve in fits and starts as they interact, all the while contributing to a net moistening and low-level stabilization of the system as a whole, particularly if contained in a region of closed circulation protected from horizontal and vertical shear and the intrusion of dry air.

Interestingly, the system transition from weak to strong low-level convergence ensures the mean vorticity enhancement develops at progressively lower levels, which is commonly observed during TC formation (Davidson *et al.*, 1990; Simpson *et al.*, 1997; Raymond *et al.*, 1998), and supported by modelling studies (e.g., Rogers and Fritsch, 2001; Montgomery *et al.*, 2006; Nolan, 2007). These observations inspired hypotheses of vorticity "growing downwards from the mid-levels" during TC formation (Davidson *et al.*, 1990).

While the preceding discussion makes it clear that the stratiform divergence profile acts to weaken low-level vorticity and thus opposes TC formation, the associated MCV that may develop can enhance the potential for an outbreak of deep convection (e.g., Bister and Emanuel, 1997). We suggest that if an intensifying MCV is to remain in thermal wind balance, the air below the level of maximum vorticity must cool and the air above must warm. The cooling and warming magnitude decreases with vertical distance from the level of maximum vorticity. These temperature changes increase the low-level conditional *instability* (by enhancing the temperature differential between the SST and low-levels) and increase the *stability* above. Convection that does develop in this environment will have enhanced low-level convergence that accelerates the low-level spin-up (Montgomery *et al.*, 2006; Raymond and Sessions, 2007).

While outbreaks of deep convection on the MCV scale have not been conclusively observed, numerous examples of infrared imagery during TC genesis in the literature show bursts of very cold cloud-top temperatures on horizontal scales in excess of 100 km (e.g., Zehr, 1992; Harr *et al.*, 1996a,b; Harr and Elsberry, 1996; Ritchie and Holland, 1997; Simpson *et al.*, 1997; Ritchie *et al.*, 2003; Ritchie, 2003; Tory *et al.*, 2006). Tory *et al.* (2006) found that convective rain rates determined from Tropical Rainfall Measuring Mission (TRMM) data correlated well with cloud top temperature (CTT) thresholds less than 187 K. During a particularly large convective blow-up, convective rain rates extended to 100 km in scale, which suggests the region likely contained multiple convective towers, and significant deep upward mass flux.

3.2.4. *System-Scale Intensification: Eliassen Balanced Flow Evolution*

The Eliassen balanced vortex model (Eliassen, 1951) has been used to investigate TC intensification in idealized flow configurations (e.g., Willoughby, 1979; Shapiro and Willoughby, 1982; Schubert and Hack, 1982; Hack and Schubert, 1986). These analytic and numerical studies are axisymmetric and the flow evolves in a state of gradient wind balance and hydrostatic balance. Although the genesis environment can deviate considerably from axisymmetry, the Eliassen balanced vortex model has proved useful in TC genesis studies (Hendricks *et al.*, 2004; Montgomery *et al.*, 2006). These authors extracted the heat and momentum sources from fully three-dimensional numerical simulations and applied them to the Eliassen balanced vortex model after approximating the heat and momentum sources as steady and axisymmetric. The secondary circulations diagnosed from the Eliassen balanced vortex model were found to be very similar to the azimuthally averaged system-scale secondary circulations of the three-dimensional numerical simulations. In particular it was concluded that the indirect effect of the multiple, relatively small-scaled, short-lived convective events, was to enhance the system-scale secondary circulation. Hereafter we label this system-scale secondary circulation intensification the Eliassen balanced flow evolution. In each of the above studies the Sawyer-Eliassen equation was used to diagnose the secondary circulation induced by the imposition of diabatic or frictional forcing on balanced vortices. A more detailed description of the Sawyer-Eliassen equation is given in section 3 of Chapter 1, and an adapted formulation to that used in Montgomery *et al.* (2006) is presented here.

$$\frac{\partial}{\partial r}\left(\frac{N^2}{r}\frac{\partial \psi}{\partial r} - \frac{\xi}{r}\frac{\partial v}{\partial z}\frac{\partial \psi}{\partial z}\right) + \frac{\partial}{\partial z}\left(-\frac{\xi}{r}\frac{\partial v}{\partial z}\frac{\partial \psi}{\partial r} + \frac{I^2}{r}\frac{\partial \psi}{\partial z}\right) = \frac{\partial Q}{\partial r} - \frac{\partial}{\partial z}(\xi F). \tag{10}$$

The toroidal streamfunction ψ is related to the toroidal circulation (u, w) where $u = -\frac{1}{r}\frac{\partial \psi}{\partial z}$ and $w = \frac{1}{r}\frac{\partial \psi}{\partial r}$. The primary circulation is described by the tangential wind v. Here $\xi = (f + 2v/r)$, $I^2 = \xi\eta$ is the inertial (centrifugal) stability parameter, z and r are the vertical pseudo-height and radial coordinates respectively, and Q and F are the diabatic heating and momentum forcing terms arising from azimuthally averaging the diabatic heating and eddy flux terms for Q and frictional and eddy momentum flux terms for F.

The fundamental geometry of the toroidal circulation diagnosed by Eq. (10) in response to diabatic heating is most easily visualized when the equation is approximated for the simple case where $\frac{\partial v}{\partial z}$ and F are negligible, and N^2 and I^2 are assumed to be positive with slowly-varying radial and vertical dependencies respectively. Then Eq. (10) is approximated by,

$$N^2\frac{\partial w}{\partial r} - I^2\frac{\partial u}{\partial z} \approx \frac{\partial Q}{\partial r}. \tag{11}$$

For simplicity assume an idealized TC genesis environment with the diabatic heating rate maximized at the centre of a large-scale circulation.[10] Then $\frac{\partial Q}{\partial r} < 0$, and, $\frac{\partial w}{\partial r}$ and $\frac{\partial u}{\partial z}$ are negative and positive respectively, which gives rise to a secondary circulation in the vicinity of the heating comprised of upward motion with inflow at low-levels and outflow above. Compensating subsidence will be present outside the heat source. The geometry of the secondary circulation is dependent on the ratio of N^2 to I^2 (Eq. (11)). Large or small values of this ratio lead to shallow-broad or tall-thin secondary circulations respectively. Imposing a boundary condition that confines the secondary circulation to a depth H ensures that the secondary circulation geometry varies horizontally on the scale of the Rossby radius of deformation (L_R), where,

$$L_R = \frac{NH}{I}, \tag{12}$$

and H is the troposphere depth.

While the Sawyer-Eliassen equation diagnoses only the instantaneous secondary circulation, the tendency of the primary circulation can be determined by solving an elliptic geopotential tendency equation (Shapiro and Montgomery, 1993). For simplicity, we seek here only a qualitative understanding of the relationship between the system vortex structure, and the tangential wind and absolute vorticity tendencies, using the assumptions of gradient wind and hydrostatic balance. Assuming axisymmetry, the tangential wind and η tendency equations are given by Eqs. (5) and (9) respectively. The first terms on the RHS of Eqs. (5) and (9) show that inflow $(-u)$ enhances v and convergence $(-\frac{1}{r}\frac{\partial(ru)}{\partial r})$ enhances η, and the enhancement rate increases with increasing η.

Axisymmetric circulations with weak radial winds and weak friction are approximately in

[10]Convection typically develops off centre during TC genesis.

gradient wind balance,

$$\frac{1}{\rho}\frac{\partial p}{\partial r} \approx fv + \frac{v^2}{r}, \qquad (13)$$

where ρ is air density. For the system to intensify in gradient wind balance, any increase in v must be matched by an increase in the radial pressure gradient. The diabatic heating, and adiabatic temperature changes associated with the heating induced secondary circulation, contribute to this changing pressure gradient. In the developing TC environment tropospheric warming is accompanied by a net outward evacuation of mass and a lowering of surface pressure, such that tropospheric warming or cooling can essentially be used to diagnose pressure changes. Inside the heating region a slight excess of diabatic heating over adiabatic cooling due to ascent provides net warming and a lowering of pressure, while outside the heating region subsidence warming also contributes to lowering pressure. It follows that a subsidence gradient can contribute to the changing pressure gradient necessary to balance any acceleration of v that occurs outside the heating region.

The dependence of the secondary circulation scale on L_R becomes intuitive when continuity is considered. For a given upward mass flux there must be a balancing downward mass flux in the surrounding subsidence region. The smaller the size of the subsidence area, the greater the magnitude of the subsidence heating. For a relatively weak system (small v and η, thus small I) the subsidence must be spread over a relatively large area so that only a weak subsidence induced pressure gradient is introduced to balance the relatively small accelerations of v. In a relatively intense circulation (large v and η, thus large I) the radial inflow will induce a greater increase in v that must be balanced by a stronger change in pressure gradient, which requires a greater magnitude of subsidence locally.

In essence the secondary circulation diagnosis (Sawyer-Eliassen solutions) and the instantaneous primary circulation prognosis, describes qualitatively an iterative solution to the tangential wind tendency. This prognosis describes a contraction with time of the secondary circulation and a system spin-up that becomes more focused and concentrated towards the heat source. The prognosis also shows that the primary circulation intensification rate increases with increasing intensity (Shapiro and Willoughby, 1982; Schubert and Hack, 1982). Thus, a non-linear intensification rate[11] might be expected from a hypothetical persistent (and time invariant) heat source (Hack and Schubert, 1986).

The Sawyer-Eliassen solutions, which describe the gradual evolution of the mean vortex, are valid for a steady heat source. In the real atmosphere, however, the convective heating is temporally and spatially sporadic. Non-steady, symmetric and asymmetric heating in axisymmetric and three-dimensional vortex studies show that some fraction of the heat energy released in convection is lost to the system through the propagation of inertia-gravity waves, and some fraction is retained by the system in the form of an enhanced circulation, i.e., increased kinetic and available potential energy (e.g., Nolan and Montgomery, 2002; Nolan and Grasso, 2003; Willoughby, 2009). Willoughby (2009) experimented with a diabatic heating source that varied sinusoidally with time at a range of frequencies. For steady heating the solutions were consistent with the Sawyer-Eliassen equation, but for heating frequencies greater than the inertial frequency (I) inertia-gravity waves were generated and much of the perturbation energy supplied by the heat source was lost to the system. Intermediate heating frequencies could yield a Sawyer-Eliassen-type response at some inner-core region where the inertial

[11]This non-linear intensification mechanism is independent of any explicit feedback between moisture flux and wind speed (e.g., wind induced surface heat exchange, WISHE, Yano and Emanuel, 1991; Emanuel *et al.*, 1994).

frequency was large, with energy lost to inertia-gravity waves at larger radii where the inertial frequency was small. It follows that heating will need to be longer-lived in developing tropical disturbances than the mature tropical cyclone, due to the smaller inertial frequency of tropical disturbances.

Sporadic, multiple convective bursts in nature could be considered to be made up of some mean state of steady heating plus temporal and spatial perturbations. The atmosphere is likely to respond to the mean heating in some manner akin to the Sawyer-Eliassen response to heating (i.e., the Eliassen balanced flow evolution), while the higher frequency energy is lost to the system as inertia-gravity waves (Willoughby, 2009). It stands to reason that the more active a convective region is, the greater the pseudo-steady heating will be and the more efficient the system-scale intensification will be through the Eliassen balanced flow evolution. It is not yet clear on a quantitative level how the heat energy released in vortical hot towers (VHTs, section 3.5.3) will be partitioned into energy retained and energy lost by the system. The high vorticity of a VHT is likely to influence this energy partition.

3.3. *TC-core Formation Theories*

There are three "crank-start" genesis theories that appeared in the literature in the late 1990s, two have since been labelled "top-down" and one "bottom-up" by Hendricks *et al.* (2004) and Montgomery *et al.* (2006). These labels refer to the way in which the theories describe TC-scale vorticity enhancement. The top-down theories focus on the mesoscale convective vortex (MCV) that forms in the stratiform convergence region of the MCS (Fig. 3), and propose mechanisms for low-level vorticity enhancement associated with the MCV above. The bottom-up theory focuses on the MCS convective region, where low- to mid-level convergence directly enhances vorticity (Fig. 3). This mechanism itself is independent of the mid-level vortex structure, although the MCV can provide a thermodynamic environment that favours deep convection (section 3.2.3). While all three theories describe low-level vorticity enhancement in MCSs, the top-down theories had to overcome the kinematic genesis roadblock, i.e., how to enhance vorticity in the divergent flow beneath the MCS.

We note here that the following descriptions focus on the central themes of each theory; the core aspects that were tested by the founding authors in idealized models. We note also that the kinematic and thermodynamic roadblocks were perceived issues related to the conceptual understanding of tropical MCSs at the time. More recent modelling and observational studies (section 3.5) offer new insight that suggest the roadblocks may not be as problematic as first thought.

3.3.1. *Top-down Merger*

The top-down merger proponents (Ritchie and Holland, 1997; Simpson *et al.*, 1997) developed a theory for the surface intensification of a cyclonic vortex based on observations of MCS behaviour in TC genesis environments. The movement of MCS cloud features strongly resembled vortex interactions as represented in two-dimensional vorticity dynamics (e.g., McWilliams, 1984). The vortical behaviour of the MCS cloud masses evident in visual and infrared satellite imagery was said to be due to the interaction of the MCVs (Ritchie and Holland, 1997; Simpson *et al.*, 1997). Idealized models have been used to demonstrate that the merger of mid-level vortices in a cyclonic environment results in an intensification of the radially-averaged mid-level vorticity with increased vertical scale (Ritchie and Holland, 1997; Simpson *et al.*, 1997). An example from Ritchie and Holland (1997) is illustrated in Fig. 9. They used a five-level model on an f-plane (latitude of 15°N) with moist processes excluded to study the dry-adiabatic vortex dynamics. The model was initialized with two mid-level vortices, which subsequently merged to form one

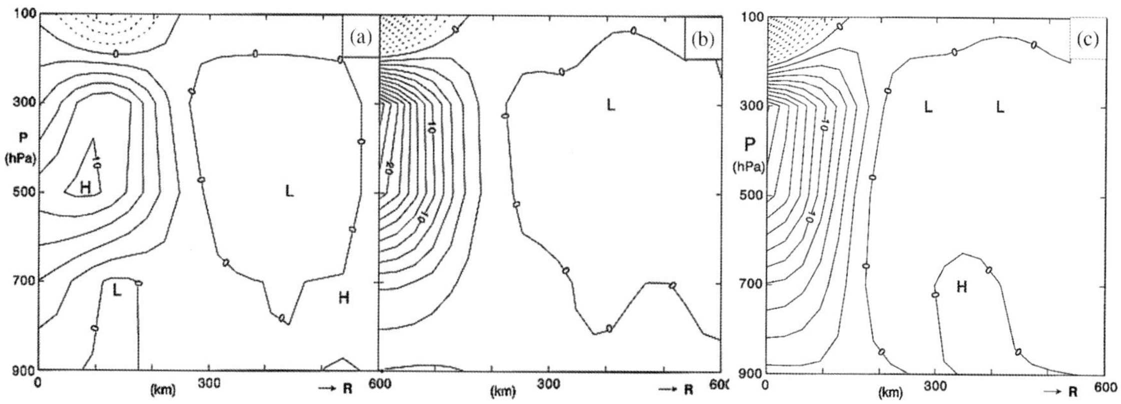

Figure 9. Azimuthal average of vorticity for merger of midlevel vortices in a baroclinic model with no diabatic heating. In (a) time = 72 h and (b) time = 120 h there is no background relative vorticity. In (c) time = 120 hours, and a monsoon trough-like background vorticity field has been created by multiplying the planetary vorticity by a factor of three below the middle troposphere, and decreasing to 1 above (contours: $2.0 \times 10^{-5}\,\mathrm{s}^{-1}$). From Ritchie and Holland (1997).

axisymmetric vortex. For the top-down merger hypothesis to provide the TC crank-start two or more sufficiently intense MCVs must interact before strong cyclonic vorticity and winds develop at the surface.

The dynamics of the vorticity enhancement below the merged vortices is well described on page 1391 of Ritchie and Holland (1997). They note that a PV anomaly imposed in a rotating environment induces an enhancement of vorticity above and below described by the Rossby penetration depth (D). Following Hoskins *et al.* (1985),

$$D = \frac{L\sqrt{f_{loc}\eta}}{N}, \tag{14}$$

L is the horizontal scale of the perturbation, $f_{loc} = f + \frac{2v}{r}$ is twice the local absolute angular velocity (f, v, and r are the Coriolis parameter, background tangential wind and radius of the background tangential wind respectively), N is the Brunt-Väisälä frequency. Ritchie and Holland noted that the merger of three-dimensional vortices will likely result in a single vortex of larger size (L), and from Eq. (14) of greater depth also. Consistent with Eq. (14) they show that the depth to which the vortex intensifies is proportional to the background rotation (Fig. 9), which suggests the efficiency of this crank-start mechanism, will increase with increasing background rotation.

Ritchie and Holland described the increased depth below the vortex centre as a downward growth of vorticity. It is important to recognise that this vorticity growth (and the matching growth above the vortex centre) is not a vertical transport or advection of vorticity. To illustrate this point, consider the thermal structure of a vortex maximized at mid-levels in thermal wind balance. Isentropes contract vertically towards the level of maximum tangential circulation, i.e., dip down from above and rise up from below, to balance the varying tangential circulation with height. If we assume, as Ritchie and Holland did, that the merger introduces an increase in horizontal size of the radially averaged vortex, then the expanding vortex will be accompanied by an expansion of the vertically contracted isentropes as the system adjusts towards thermal wind balance. For adiabatic vortex interactions, these changes in thermal structure are brought about by subsidence above and ascent below, which is opposite to that required for vorticity advection to provide the increased vortex depth. A simple assessment of this adjustment flow and the tangential circulation using the insights gained in section 3.2.1 provide a plausible explanation for the increased vortex depth. For brevity consider only the vortex enhancement below the MCV. The rising motion that lifts the isentropes will be accompanied by converging flow below and diverging flow at the level of maximum vorticity (similar to the schematic in the left panel of Fig. 4), and this converging flow provides a plausible mechanism for lower-level vortex intensification through η concentration (i.e., vortex stretching). Additionally, an examination of Fig. 5 (lower half of the left panel) suggests the non-advective-ω contribution may act to broaden the low-level vortex.

3.3.2. *Top-down Showerhead*

The top-down showerhead theory advanced the hypothesis that the crank-start can be achieved in a single MCS and its associated MCV. Emanuel (1993) and Bister and Emanuel (1997) proposed that sustained precipitation in the stratiform cloud deck would gradually saturate the relatively dry layer below the cloud, while the evaporatively-driven subsidence would advect cyclonic η or PV to the surface. The tangential wind from their control simulation (the axisymmetric, non-hydrostatic convection-resolving model of Rotunno and Emanuel, 1987) is illustrated in Fig. 10. It shows an enhancement of low-level tangential wind between 6 and 14 hours, but they do not provide a vorticity tendency analysis to confirm the hypothesis that vorticity was advected downwards to the surface. The discussion in section 3.2.1 suggests vertical advection of η should always be considered in conjunction with tilting (Eq. (3)) and tendency calculations should employ the flux formulation (Eq. (2)) to avoid cancellation between the two terms. A schematic illustrating the combined vertical η advection and tilting appropriate to the top-down showerhead scenario is depicted in the lower half of the right panel of Fig. 5. Note the positive anomaly is surrounded by negative anomalies, and all anomalies are contained within the downdraft. Any associated divergence (e.g., lower half of the right panel in Fig. 4) should dilute the anomalies but at the same time preferentially advect the anticyclonic anomalies away from the system centre.

A schematic summarizing Bister and Emanuel's full theory is presented in Fig. 11. They proposed evaporation of stratiform rain gradually humidified and cooled the lower troposphere below. The cooling was balanced dynamically by the intensifying mid-level vortex with its associated downward extension of the cyclonic circulation. Bister and Emanuel proposed that the intensification of this cold-cored circulation favoured redevelopment of convection by (i) enhancing fluxes through increased surface wind speeds, and (ii) the lower troposphere cooling "reduces the value of boundary layer θ_e that is needed for convection to occur". When the surface-layer θ_e increases sufficiently, convection

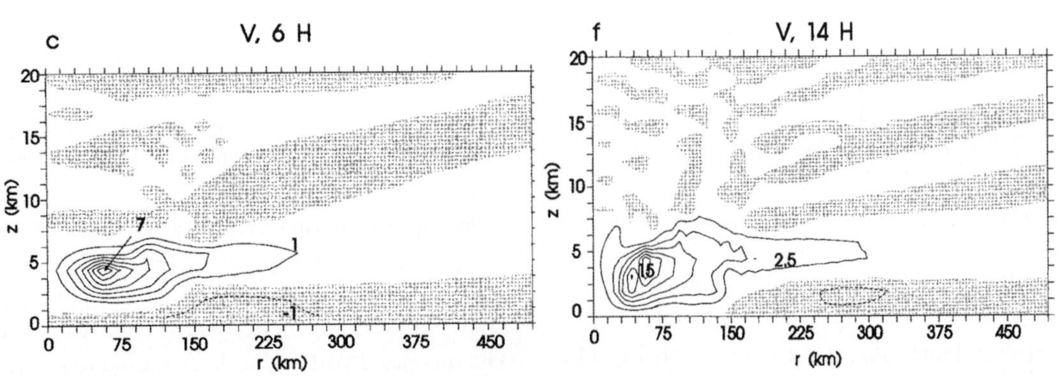

Figure 10. Average tangential velocity (m s^{-1}) for the control simulation of Bister and Emanuel (1997) between (left) 4 and 8 hours, and (right) 12 and 16 hours. Negative values are stippled.

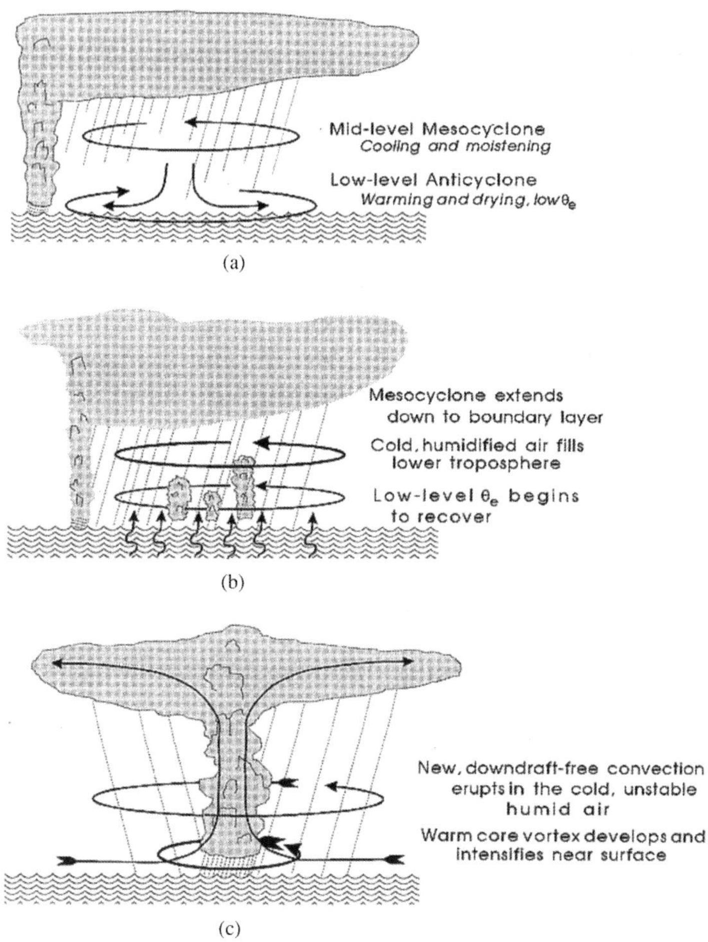

Figure 11. The Bister and Emanuel conceptual model of tropical cyclogenesis from a preexisting MCS. (a) Evaporation of stratiform precipitation cools and moistens the upper part of the lower troposphere; forced subsidence leads to warming and drying of the lower part. (b) After several hours there is a cold and relatively moist anomaly in the whole lower troposphere. (c) After some recovery of the boundary layer convection redevelops. (From Bister and Emanuel, 1997.)

with reduced downdraft potential develops in the humidified environment. The lower panel of Fig. 11 indicates schematically that downdraft free convection in a low-level cyclonic environment allows mean lower tropospheric convergence to spin up the low-level vortex as illustrated in the left panel of Fig. 4.

3.3.3. *Bottom-up*

Inspired by observations reported in Zehr (1992) of low-level vortex intensification following bursts of convection in tropical oceanic MCSs, Montgomery and Enagonio (1998) and Enagonio and Montgomery (2001) proposed convergence of ambient absolute vorticity into the convective updrafts produced strong low-level cyclonic vorticity anomalies that interacted with the MCV in such a way as to enhance the cyclonic circulation below and align it with the MCV above. In idealized numerical simulations, Montgomery and Enagonio (1998) and Enagonio and Montgomery (2001) added potential vorticity (PV) anomalies, that might be generated by deep convective cores, to a mid-level vortex resembling an MCV.

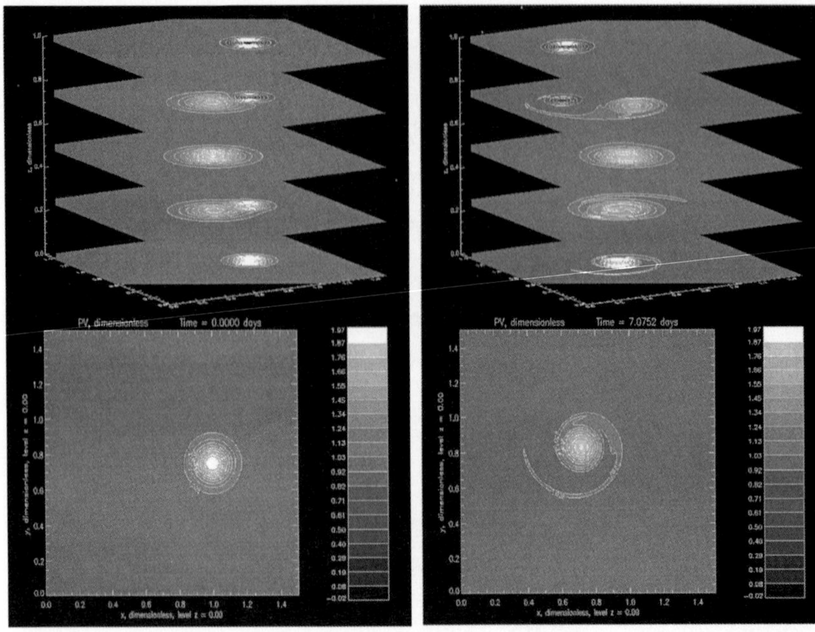

Figure 12. Contours of PV versus x and y on a number of height surfaces between and including the upper and lower boundaries, as well as a plan view of contours of PV versus x and y on the lower boundary, from the midlevel vortex with single-cluster convection experiment of Montgomery and Enagonio (1998) at (left) time $t = 0$, and (right) $t = 7$ days. In the left panel the MCV is evident in the domain centre, with a PV anomaly to its right representing the effects of low-level convergence and upper-level divergence that might be expected to develop following an episode of deep cumulonimbus convection. In the right panel the resulting vortex structure shows a near upright low- to mid-level cyclonic core with maximum intensity at low-levels. (From Montgomery and Enagonio, 1998.)

The deep convective vortices interacted with the MCV to produce a near-symmetric vortex with strong low-level vorticity, on plausible development time-scales (Fig. 12). In Fig. 12 the deep convective PV anomaly is initially off-centre relative to the MCV. With time it becomes vertically aligned with the MCV, and the surface PV immediately below the MCV is significantly enhanced. This new state of balance is achieved through an isentropic redistribution of PV from the MCV, the convective-like PV anomalies and the ambient environment (Haynes and McIntyre, 1987). Vertical motions are minimal during this PV redistribution, and it follows that advective processes dominate (first term on the RHS of Eq. (25) of Chapter 1).

A similar vorticity alignment process leading to the spin-up of the low-level circulation below an MCV was observed by Davis and Galarneau (2009).

3.4. *Theoretical Synthesis*

It is quite possible that all three TC core formation theories (section 3.3) contribute to TC genesis. The top-down proponents recognized vorticity concentration in deep convection (bottom-up) was essential for TC intensification, but believed it would not occur on a broad enough scale until the low-level cyclonic circulation was sufficiently enhanced (top-down merger, top-down showerhead), and the lower troposphere was sufficiently humidified (top-down showerhead). The bottom-up experiments, on the other hand, suggested that low-level vortex enhancement below the MCV was possible without the enhanced low-level MCV circulation and humidification. However, the preceding discussions on thermodynamic considerations (section 3.2.2), MCSs (section 3.2.3) and the System Scale Intensification (section 3.2.4) suggest low-level vorticity intensification should be more efficient in these environments of enhanced low-level circulation and reduced downdraft potential. Modelling studies (section 3.5) suggest that the bottom-up process is indeed always present, but further analysis of these studies would be required to determine the importance of the two top-down mechanisms.

All theories agree that significant intensification occurs once deep convection breaks out on a large scale. Figure 6 illustrates an intensification scenario that could equally apply to an individual convective cloud, a single MCS, or an entire convective complex. To maximise the intensification rate of the convergent region downdrafts must be minimized,[12] through moistening and stabilization of the low- to middle-troposphere (section 3.2.2). Turbulent mixing associated with the convective processes can provide a gradual change towards this desired thermodynamic state. But the convective region must remain protected from the outside environment for long enough for the deep intense upward mass-flux to develop on scales of hundreds of km (Dunkerton *et al.*, 2008).

The Eliassen balanced flow evolution (i.e., the enhancement of the system-scale secondary circulation in response to the net heating of multiple deep convective bursts, section 3.2.4) concentrates vorticity on the convective complex scale, which increases the efficiency of the top-down merger and bottom-up TC-core formation processes by enhancing the environmental vorticity where these processes take place.

The preceding paragraphs begin to provide evidence of a chain of cause and effect. The links are as follows:

(i) in a protected region of persistent convection the in-up-out circulation becomes more efficient with time,

(ii) the in-up-out circulation accelerates the primary circulation (tangential flow),

(iii) the increased circulation, increases the efficiency of the TC core formation processes.

If the TC core formation processes acted to sustain the convection, a positive feedback loop for TC formation would exist.[13] In the absence of a positive feedback some other form of convective forcing is required.

3.5. *TC Development in Observations and Numerical Models*

3.5.1. *Mesoscale Convective Forcing — Sustained Convection*

Evidence of convective forcing that initiates convection on scales of hundreds of km can be found in both observational and modelling studies. Both large-scale external forcing (e.g., tropical waves, baroclinic features) and forcing from internal processes have been identified. The precise mechanism for triggering convective outbreaks in the tropical wave troughs is not well understood. For TCs with baroclinic origins, however, systematic lifting in regions of isentropic ascent favours convection by moistening the lower and middle troposphere and releasing conditional instability (Davis and Bosart, 2003, 2004, 2006). In some cases relatively strong frontal lifting can destabilize the atmosphere and lead to a burst of convection (e.g., the genesis of Hurricane Michael, 2001). In other cases the convective forcing can be more subtle. During the development of Hurricane Humberto (2001) for example, the lifting provided a moistened environment, and enhanced vertical wind shear on the periphery of an upper-tropospheric trough provided the organizing agent for a large MCS (Davis and Bosart, 2006).

[12] In modelling studies presented in section 3.5, downdrafts are never eliminated (Montgomery *et al.*, 2006; Nolan, 2007)

[13] In the mature TC the intense core winds drive the entropy flux from the ocean surface, which feeds the eye-wall convection (e.g., Rotunno and Emanuel, 1987). Convective diabatic heating drives the Eliassen balanced flow evolution that intensifies the core (e.g., Shapiro and Willoughby, 1982).

The role of vertical wind shear in organizing convection can be important. Numerical simulations have shown that in some cases reduced background shear results in slower TC formation rates (e.g., Davis and Bosart, 2006; Musgrave *et al.*, 2008), a result that runs counter to the widely-accepted negative relationship between shear and TC development. The distinction is that shear is an organizing agent of convection but a decrease of vertical shear is helpful for subsequent intensification of the depression thus spawned. Davis and Bosart (2004) described a particularly favourable configuration of convection relative to the baroclinic cyclone wherein the convection occurs on the upshear side of the cyclone. This results in anticyclonic vorticity aloft over the cyclone that weakens the vertical shear through the troposphere, and thus reduces the negative impact of shear on the subsequent vortex intensification. Convection is known to also be stimulated in the down-tilt direction of a vortex in shear near the radius of maximum tangential velocity (Raymond and Jiang, 1990; Ritchie and Holland, 1997; Trier *et al.*, 2000; Reasor *et al.*, 2000). This includes tilted large-scale cyclonic circulations such as monsoon lows (e.g., Tory *et al.*, 2006).

Idealized modelling studies without external convective forcing (Montgomery *et al.*, 2006; Nolan, 2007) have successfully simulated TC formation. In the real world, however, where the atmosphere is generally hostile to TC formation, external convective forcing will greatly increase the likelihood of development. In both studies the convection persisted throughout the simulations even in the presence of significant downdrafts (c.f. the thermodynamic roadblock of section 3.2.3). In fact the spreading downdraft gust front boundaries acted as internal convective forcing by triggering further convection (Montgomery *et al.*, 2006; Davis and Bosart, 2006). This internal convective forcing is consistent with the self-sustaining convection mechanism Yamasaki (1984) proposed to be essential for maintaining convection on times scales great enough for TC formation to take place.

Lower tropospheric monsoonal or trade-wind surges have also been proposed as mesoscale to synoptic-scale convection triggers (Zehr, 1992; Gray, 1998; Ritchie *et al.*, 2003). Zehr (1992) identified an association between wind surges and significant vortex amplification in a high proportion of NW Pacific convective bursts during the 1983 and 1984 typhoon season. The surge often resulted from cross-equator flow accelerating into the monsoon trough, and produced a strengthening of the monsoon westerlies.

Thus it would appear that in a suitably favourable environment a combination of both external and internal convective forcing can sustain convection on sufficiently large time and space scales for TC formation to be successful.

3.5.2. *Moistening of the Middle Troposphere*

The main goal of the Tropical Experiment in Mexico (TEXMEX) was to investigate the hypothesis that "the elevation of θ_e in the middle troposphere just above a near-surface vorticity maximum is a necessary and perhaps sufficient condition for tropical cyclogenesis" (Bister and Emanuel, 1997). Essentially this hypothesis was proposed as a possible way around the thermodynamic roadblock mentioned in section 3.2.3. While a correlation between higher relative humidity and the weakening of downdraft entropy fluxes was found (Raymond *et al.*, 1998), the system-averaged low- to mid-levels were convergent throughout, suggesting the convection was never critically suppressed. Raymond and Sessions (2007) argue that while the TEXMEX observations did show an increase in the middle level θ_e, it was the stabilization at low-levels that was most important for the reduction of downdrafts (i.e., the saturation DCAPE, section 3.2.2, Fig. 7).

While the exact nature and effect of downdrafts and precipitation in moist rotating convective regions is yet to be investigated in TC genesis observations and simulations, the importance of a moist environment for TC genesis was the primary subject of Nolan (2007) in a series of idealized three-dimensional cloud-resolving TC genesis simulations. In his simulations a warm-cored intense surface vortex did not form until the relative humidity of much of the troposphere exceeded 90%. While this result is consistent with the axisymmetric genesis simulations of Bister (2001), the value of 90% is moister than the three-dimensional cloud resolving simulations of Hidalgo (2008) and Montgomery *et al.* (2006),[14] who found genesis occurred with low- to middle-troposphere relative humidity of about 80% (and even drier above 600 hPa). The differences in these results may reflect the different model configurations, including cloud microphysics and radiation schemes. Hidalgo's study was of a real event (Hurricane Diana, 1984), unlike Nolan's study which was idealised and contained no background shear and no radiation scheme. The size of the averaging domains could impact the results. Too large an area would include dry air outside the convective region and thus bias the average downwards. In a series of short-range ensemble forecasts Sippel and Zhang (2008) found deep moisture and high CAPE were the two most important factors in the initial conditions for TC formation. It is not clear yet to us whether relative humidity is useful for diagnosing imminent formation.

3.5.3. *Vortex Interactions — Vortex Upscale Cascade*

As discussed in section 3.3.1 vortices of any scale will interact when in close enough proximity, giving rise to the well recognised patterns of vortex orbiting, merger, shearing and axisymmetrization of like-signed anomalies (e.g., McWilliams, 1984; Dritschel and Waugh, 1992; Ritchie and Holland, 1993), and the repelling of opposite signed anomalies (Montgomery and Enagonio, 1998). Intense anomalies are not easily axisymmetrized by the parent vortex, instead they migrate across the ambient vorticity gradient, cyclonic anomalies up-gradient (towards increasing cyclonic vorticity) and anticyclonic anomalies down-gradient (McWilliams and Flierl, 1979; Smith and Ulrich, 1990; Schechter and Dubin, 1999). Together these processes comprise what we refer to as the vortex upscale cascade.

Early vortex interaction studies related to TC formation focused on MCV scale interactions (top-down merger). The associated MCS cloud features served as proxies for the MCVs, and the interaction of these cloud features, evident in satellite imagery, offered important clues into the vortical nature of the disturbances (Ritchie and Holland, 1997; Simpson *et al.*, 1997; Ritchie *et al.*, 2003). The interaction of vortices of MCV and cloud-scale in the manner described by the bottom-up theory has been observed also (e.g., Molinari *et al.*, 2004). In recent times much smaller scale interactions contributing to TC formation have been observed. The Doppler radar coverage of the intensification of TS Allison (2001) provided evidence of vortex interactions on a variety of scales (Sippel *et al.*, 2006), and multiple vortices as small as 1–2 km in diameter were identified. These smaller scale vortices are generated by vorticity concentration in intense convective updrafts, and have been coined "vortical hot towers" (VHTs, Hendricks *et al.*, 2004; Montgomery *et al.*, 2006).

The observed vortex interactions have been reproduced by realistic and idealised TC formation simulations, investigating the direct role of vortex interactions in the construction of the TC core. These studies include cloud resolving at 1 km grid spacing (Hidalgo, 2008), 2 km grid spacing (Montgomery *et al.*, 2006), 3 km grid spacing (Hendricks *et al.*, 2004; Montgomery *et al.*, 2006; Hidalgo, 2008), and 0.15° grid spacing with convective parameterization (Tory *et al.*, 2006). In each study the grid resolution limits the scale of the smallest vortices.

The horizontal grid spacing issue was investigated by Hidalgo (2008) with a comparison of simulations at 3 and 1 km grid spacing. He

[14]The system averaged relative humidity was not reported in Montgomery *et al.* (2006).

found about a ten times increase in the number of vorticity anomalies in the higher resolution study, with a similar decrease in scale (see also Nguyen *et al.*, 2008). The increased resolution also favoured more intense updrafts. The higher resolution study exhibited a greater proportion of anticyclonic anomalies, which is consistent with the arguments provided in section 3.2.1, that anticyclonic anomalies can only be generated by the non-advective η flux tendencies (Eq. (2), Figs. 5 and 6). The greater the number of up- and downdrafts in the central convergence region, the greater the number of vorticity dipoles generated. Also, the greater the vertical velocities the greater the magnitude of the vorticity dipoles. From a vorticity anomaly frequency distribution Davis and Bosart (2006) concluded that the anticyclonic anomalies in their study tended to be small in scale, and that they were largely cancelled, in a vorticity averaging sense, by cyclonic anomalies of similar size. This cancellation is consistent with the arguments summarized in Fig. 6.

A comparison of vortex interactions between the coarser and finer resolution studies mentioned above can be made by comparing Figs. 13 and 14. The former from Hidalgo (2008) shows many hundreds of anomalies interacting, whereas the latter from Tory *et al.* (2006) shows only two interacting anomalies. Like-signed anomalies tend to merge or to be ingested by larger like-signed anomalies. Anticyclonic anomalies in the vicinity of a stronger cyclonic core tend to migrate outwards away from the

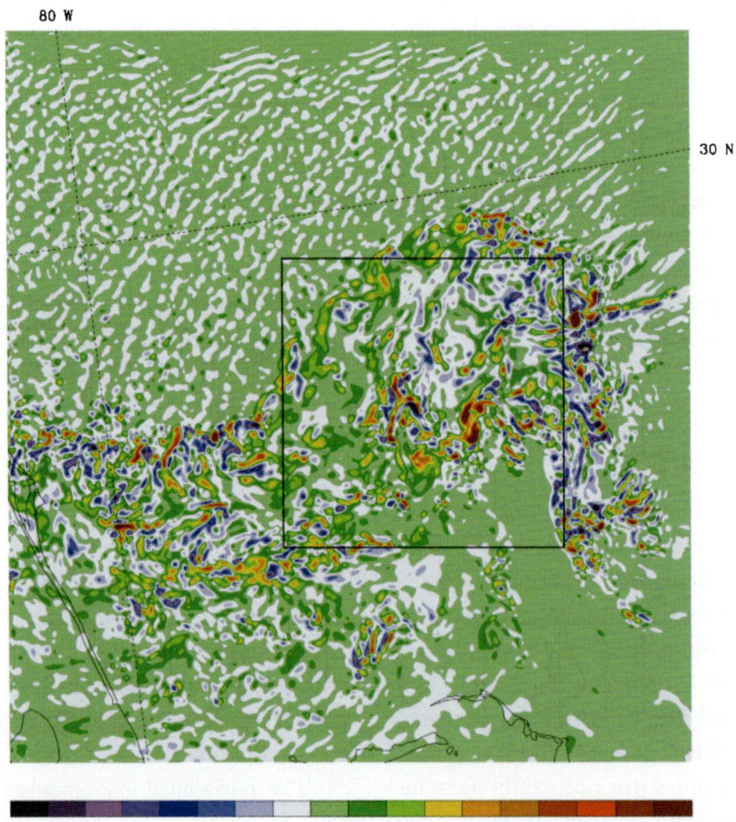

Figure 13. η on a 925 hPa surface from a cloud resolving, non-hydrostatic simulation of Hurricane Diana (1984) with grid-spacing of 1 km (Hidalgo, 2008). The warm (cyclonic) and cool (anticyclonic) colours illustrate both the local convective scale η anomaly distributions and larger-scale cyclonic-anticyclonic distributions.

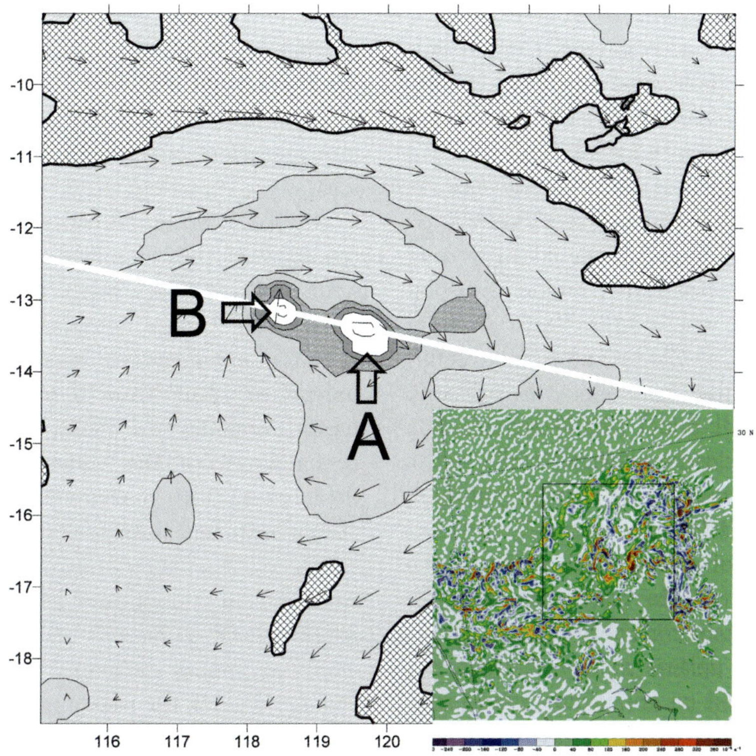

Figure 14. PV on the $\sigma = 0.85$ surface from a hydrostatic operational forecast model with convective parameterization and 0.15° grid-spacing (Tory *et al.*, 2006), with Fig. 13 super-imposed in the lower-right corner for scale comparison. Hatched areas represent anticyclonic PV. At this relatively coarse resolution only two anomalies A and B are interacting, and no anticyclonic anomalies are present in the developing core. (From Tory *et al.*, 2006 and Hidalgo, 2008.)

core centre (see also Montgomery and Enagonio, 1998; Nguyen *et al.*, 2008). This behaviour is evident in all modelling studies mentioned above, but it is increasingly more frequent in the higher resolution studies, due to the larger number of anomalies. Despite the significant differences in scale, the fundamental behaviour appears to be the same. The ensemble of convective updrafts converges η on the scale of the convective region. Convective updrafts generate η anomalies through local η convergence into the updrafts, and the tilting effects illustrated in Fig. 5. Vortex interactions act continuously to rearrange the anomalous vorticity clutter towards a single, upright, symmetric monolithic vortex core. Doppler radar observations (e.g., Reasor *et al.*, 2005; Sippel *et al.*, 2006; Houze *et al.*, 2009) confirm the vortex upscale cascade process is present in real developing TCs, and that grid-spacing of 1 km is able to resolve at least the finer scales evident in the radar data (see Fig. 17 of Houze *et al.*, 2009).

The cloud-scale processes that generate the anomalous vorticity clutter are stochastic in nature (e.g., Zhang and Sippel, 2009) and the finer details of simulated storm evolution deviate from those observed, with consequences on TC formation predictability. Tory *et al.* (2007) argue that the details of the η anomaly interactions are in general qualitatively unimportant, because the larger-scale kinematic and thermodynamic environment largely determine whether TC formation will be successful. They argued that while the large-scale environment remains favourable and the model convection is sustained, the vortex upscale cascade and system-scale intensification will ensure formation regardless of the details of the η anomaly evolution and interaction. In contrast, using both coarse and cloud-resolving simulations, Zhang and Sippel (2009) concluded that chaotic cloud-scale processes associated with small and unobservable differences in initial conditions can determine whether or not a TC will develop. It is possible that the TC genesis events Tory *et al.* (2007) investigated were supported by strong large-scale forcing (Australian region monsoon low environments), whereas the North Atlantic events investigated by Zhang and Sippel (2009) were more borderline developers. Either way these contrasting results suggest an ensemble prediction system for TC formation could be a valuable forecasting tool (Zhang and Sippel, 2009).

3.5.4. *System-scale Intensification*

The system-scale intensification can be defined by the net change in vorticity of the system. If the system is contained in some arbitrary volume then the system-scale intensification can be determined from the vorticity fluxes on the surface of the volume (HM87, section 3.2.1). Recalling that the non-advective-ω flux, $\mathbf{J}_\omega$, is negligible away from strong vertical motions (Eq. (2)), a volume can be chosen immediately surrounding an active convective region (e.g., the convergent region in Fig. 6) in which the η flux at the lateral boundaries is mostly advective. In the boundary layer the frictional flux $\mathbf{J}_F$ can be very significant, although it is largely opposed by the advective flux $\mathbf{J}_a$ associated with Ekman convergence. However, because there is no net η flux across pressure surfaces the boundary layer frictional influence on η should not affect the pressure layers above.[15] Then the change in η within the volume that represents the "system" can be determined largely from the product of η and the flow perpendicular to the chosen lateral surface. A similar relationship was used by Raymond *et al.* (1998) in their budget study of vorticity tendency during the development of Hurricane Guillermo (1991) in the eastern Pacific (although they limited their analysis and interpretations only to convergence of planetary vorticity).

If (i) the frictional contribution to vorticity tendency $\mathbf{J}_F$ only significantly affects the developing TC boundary layer, and (ii) $\mathbf{J}_\omega \approx 0$ on the chosen volume boundary, then the majority of η enhancement inside the chosen volume must result from η transport across the lateral boundary (cyclonic into and/or anticyclonic out of the volume). In the system depicted in Fig. 6 there is mean inflow and the environment outside is everywhere cyclonic, which contributes to a mean system-scale intensification in this layer. In reality there are likely to be numerous cyclonic and anticyclonic anomalies embedded in the region, and as mentioned in the previous section, sufficiently intense anomalies will self-segregate, cyclonic inwards and anticyclonic outwards. These eddy processes may proceed against the mean inflow and outflow (Nguyen *et al.*, 2008). Together the mean and eddy η contributions represent an amplification of environmental vorticity inside the system, which in turn increases the efficiency of the TC-core formation processes within.

Care is needed in interpreting these arguments because they ignore rearrangements of η within the control volume (e.g., intensification of a central vortex core). In this framework intensity is equivalent to the circulation around the control volume (circulation theory, e.g., Raymond *et al.*, 1998).

If we define the system-scale intensification as the net change in cyclonic vorticity inside some volume, then the system-scale intensification can also be monitored by measuring the change with time of volume averaged vorticity.

[15] Raymond *et al.* (1998) considered the possibility of non-trivial contribution to $\mathbf{J}_F$ above the boundary layer from cumulus friction. However, in this thought experiment, convection and the associated cumulus friction is contained within the volume, which means $\mathbf{J}_F$ from cumulus friction will be negligible on the volume boundary and will thus have negligible contribution to the volume integrated η tendency.

Horizontal averages of vertical vorticity are often displayed as time-height contour plots for both observations (Davidson *et al.*, 1990; Raymond *et al.*, 1998) and numerical model data (e.g., Rogers and Fritsch, 2001; Nolan, 2007; Hidalgo, 2008). These offer useful insight to the system-scale development as a function of height and time.

As noted in section 3.2.4 the system-scale intensification in three-dimensional modelling studies has been likened to the Eliassen balanced flow evolution (e.g., Montgomery *et al.*, 2006), which predicts a collapse in scale of the system as it intensifies and an increasing rate of intensification with time. The inward contracting and intensifying ring of subsidence, may play a role in the observed scale contraction of intensifying systems during TC formation (Zehr, 1992), as the convection suppressing subsidence impacts on the outer edges of the convective region.

3.5.5. *Summary*

In section 3.4 a plausible scenario is proposed that unites early TC genesis theories. It is hypothesized that sustained convection produces a thermodynamic adjustment that favours widespread deep convection, which facilitate system-scale intensification, and the development of a central upright vortex core. The modelling and observational studies discussed in this subsection support the notion that (i) both external and internal processes can sustain convection over sufficient time and space scales to support TC formation, and (ii) sustained convection gradually moistens the low- to middle-troposphere and stabilizes the lower troposphere. Furthermore, the observed increase with time of the system-scale low-level convergence is consistent with reduced DCAPE and downdraft mass flux (at least relative to the low-level upward mass flux). Model studies and observations show that vortex interactions are an integral part of the TC core construction, and that the ensemble of heating in the convective hot towers does invoke a secondary circulation consistent with the Eliassen balanced flow evolution (Montgomery *et al.*, 2006). Thus, multiple small-scale, relatively short-lived vortical heat sources have a combined effect on the system similar to a quasi-steady heat source (e.g., section 3.2.4).

The TC vortex core construction could be summarized by the following processes that necessarily occur in a somewhat protected environment:

(a) Sustained convection adjusts the thermodynamic profile to reduce the potential for downdrafts.
(b) Reduced downdrafts favours increased vertical mass flux efficiency, and enhanced lower tropospheric convergence. (Reduced downdrafts may favour more widespread convective outbreaks.)
(c) Deep convection provides strong diabatic heating, and low- to mid-level vorticity concentration on the system-scale, and drives the system-scale intensification.
(d) Individual convective elements provide cloud-scale vorticity anomalies through convergence and tilting effects, which feed the vortex upscale cascade.

4. Developing versus Non-Developing Systems

Idealized TC genesis studies (e.g., Montgomery *et al.*, 2006; Nolan, 2007) show that TC formation can be achieved in environments satisfying Gray's necessary conditions (section 1), without contributions from transient features such as tropical waves and baroclinic influences (section 2). Indeed, idealized simulations show that TC formation can occur in environments that are initially quite dry, with relatively weak background absolute vorticity, provided the convection remains active Nolan (2007). The convective processes slowly moisten the development region, and concentrate absolute vorticity (section 3.2), which gradually primes the region for TC formation. However, the real world is not likely to support such gradual development. Tropical disturbances are regularly subjected to horizontal deformation and vertical shear that acts to tilt and disperse developing core regions of enhanced moisture and vorticity, and/or contaminate them with intrusions of dry air. It follows that closed or otherwise protected environments, and environments with a "genesis head-start" (i.e., enhanced vorticity and relative humidity) are more likely to spawn a tropical cyclone. The genesis head-start can be provided by tropical waves or baroclinic features, which favour organized convection. Additionally the waves can provide protective closed circulations relative to the mean flow that may persist for a number of days (Dunkerton *et al.*, 2008).

The above argument would suggest that every tropical disturbance is a potential developer (provided the sea surface temperature and conditional instability is adequate to maintain convection). Development is then largely determined by how much of a head-start the system has, and how long it remains in a protected environment. Development may be modulated by the superposition of other waves propagating through the region, with phases that favour and suppress convection (e.g., Molinari *et al.*, 1998; Hanley *et al.*, 2001; Molinari *et al.*, 2007, section 2.3).

Acknowledgements

This chapter has benefited greatly from the important and insightful contributions of the following people: Lance Bosart, Noel Davidson, Tim Dunkerton, Eric Hendricks, Kevin Mallen, John Molinari, Dave Raymond, Liz Ritchie, Rob Rogers and Matthew Wheeler. Special thanks to Chris Davis for his contribution to the high-latitude TC formation section, and to Roger Smith and Jeff Kepert for their highly valuable reviews. Lastly we would like to acknowledge the very considerable contribution of Mike Montgomery to this chapter, especially the vortex construction section.

References

Bergeron, T., 1950: Über der mechanismus der ausgiebigan niederschläge. *Ber. Dtsch. Wetterd.*, **12**, 225–232.

Bessafi, M. and M. C. Wheeler, 2006: Modulation of south Indian Ocean tropical cyclones by the Madden-Julian oscillation and convectively-coupled equatorial waves. *Mon. Wea. Rev.*, **134**, 638–656.

Betts, A. K., 1976: The thermodynamic transformation of the tropical subcloud layer by precipitation and downdrafts. *J. Atmos. Sci.*, **33**, 1008–1020.

Bister, M., 2001: Effect of peripheral convection on tropical cyclone formation. *J. Atmos. Sci.*, **58**, 3463–3476.

Bister, M. and K. A. Emanuel, 1997: The genesis of Hurricane Guillermo: TEXMEX analyses and a modeling study. *Mon. Wea. Rev.*, **125**, 2662–2682.

Bosart, L. F. and J. A. Bartlo, 1991: Tropical storm formation in a baroclinic environment. *Mon. Wea. Rev.*, **119**, 1979–2013.

Briegel, L. M. and W. M. Frank, 1997: Large-scale influences on tropical cyclogenesis in the western North Pacific. *Mon. Wea. Rev.*, **125**, 1397–1413.

Carlson, T., 1969: Synoptic histories of three African disturbances that developed into Atlantic hurricanes. *Mon. Wea. Rev.*, **97**, 256–276.

Davidson, N. E., G. J. Holland, J. L. McBride, and T. D. Keenan, 1990: On the formation of AMEX tropical cyclones Irma and Jason. *Mon. Wea. Rev.*, **118**, 1981–2000.

Davis, C. A. and L. F. Bosart, 2003: Baroclinically induced tropical cyclogenesis. *Mon. Wea. Rev.*, **131**, 2730–2747.

———, 2004: The TT Problem: Forecasting the tropical transition of cyclones. *Bull. Amer. Meteor. Soc. (Map Room).*, **85**, 1657–1662.

———, 2006: The formation of Hurricane Humberto (2001): The importance of extra-tropical precursors. *Quart. J. Roy. Meteor. Soc.*, **132**, 2055–2085.

Davis, C. A. and T. J. Galarneau, 2009: The vertical structure of mesoscale convective vortices. *J. Atmos. Sci.*, **66**, 686–704.

Dickinson, M. and J. Molinari, 2002: Mixed Rossby-gravity waves and Western Pacific tropical cyclogenesis. Part I: Synoptic evolution. *J. Atmos. Sci.*, **59**, 2183–2196.

Dritschel, D. G. and D. W. Waugh, 1992: Quantification of the inelastic interaction of unequal vortices in two-dimensional vortex dynamics. *Phys. Fluids A*, **4**, 1737–1744.

Dunkerton, T. J., 1993: Observation of 3–6-day meridional wind oscillations over the tropical Pacific, 1973–1992: Vertical structure and interannual variability. *J. Atmos. Sci.*, **50**, 3292–3307.

Dunkerton, T. J. and M. P. Baldwin, 1995: Observation of 3–6-day meridional wind oscillations over the tropical Pacific, 1973–1992: Horizontal structure and propagation. *J. Atmos. Sci.*, **52**, 1585–1601.

Dunkerton, T. J., M. T. Montgomery, and Z. Wang, 2008: Tropical cyclogenesis in a tropical wave critical layer: Easterly waves. *Atmos. Chem. Phys. Disc.*, **8**, 11 149–11 292.

Eliassen, A., 1951: Slow thermally or frictionally controlled meridional circulation in a circular vortex. *Astrophys. Norv.*, **5**, 19–60.

Emanuel, K. A., 1993: The physics of tropical cyclogenesis over the eastern Pacific. Tropical cyclone disasters. *Proceedings of ICSU/WMO International Symposium on Tropical Cyclone Disasters*, Lighthill, J., Z. Zhemin, G. Holland, and K.Emanuel, Eds., Peking University Press, pp. 136–142.

———, 1994: *Atmospheric Convection.* Oxford University Press, 580 pp.

Emanuel, K. A., J. D. Neelin, and C. S. Bretherton, 1994: On large-scale circulations in convecting atmospheres. *Quart. J. Roy. Meteor. Soc.*, **120**, 1111–1430.

Enagonio, J. and M. T. Montgomery, 2001: Tropical cyclogenesis via convectively forced vortex Rossby waves in a shallow water primitive equation model. *J. Atmos. Sci.*, **58**, 685–706.

Ferreira, R. N. and W. H. Schubert, 1997: Barotropic aspects of ITCZ breakdown. *J. Atmos. Sci.*, **54**, 261–285.

Frank, N. L. and G. Clark, 1980: Atlantic tropical systems of 1979. *Mon. Wea. Rev.*, **108**, 966–972.

Frank, W. M., 1987: Tropical cyclone formation. *A Global View of Tropical Cyclones*, Elsberry, R. L., Ed., Office of Naval Research, pp. 53–90.

Frank, W. M. and P. E. Roundy, 2006: The role of tropical waves in tropical cyclogenesis. *Mon. Wea. Rev.*, **134**, 2397–2417.

Frank, W. M. and G. S. Young, 2006: The interannual variability of tropical cyclones. *Mon. Wea. Rev.*, **135**, 3587–3598.

Gray, W. M., 1968: Global view of the origin of tropical disturbances and storms. *Mon. Wea. Rev.*, **96**, 669–700.

———, 1975: Tropical cyclone genesis. *Dept. of Atmos. Sci. Paper No. 232, Colorado State University, Ft. Collins, CO*, 121.

———, 1979: Hurricanes: Their formation, structure and likely role in the tropical circulation. *Meteorology Over the Tropical Oceans.*, Shaw, D., Ed., RMS, James Glaisher House, pp. 155–218.

———, 1985: Tropical cyclone global climatology. WMO Technical Document WMO/TD-No. 72, Vol. I, WMO, Geneva, Switzerland, 3–19.

———, 1998: The formation of tropical cyclones. *Meteor. Atmos. Phys.*, **67**, 37–69.

Hack, J. J. and W. H. Schubert, 1986: Nonlinear response of atmospheric vortices to heating by organised cumulus convection. *J. Atmos. Sci.*, **43**, 1559–1573.

Hall, J. D., A. J. Matthews, and D. J. Karoly, 2001: The modulation of tropical cyclone activity in the Australian region by the Madden-Julian oscillation. *Mon. Wea. Rev.*, **129**, 2970–2982.

Hanley, D., J. Molinari, and D. Keyser, 2001: A composite study of the interactions between tropical cyclones and upper-tropospheric troughs. *Mon. Wea. Rev.*, **129**, 2570–2584.

Harr, P. A. and R. L. Elsberry, 1996: Structure of a mesoscale convective system embedded in Typhoon Robyn during TCM-93. *Mon. Wea. Rev.*, **124**, 634–652.

Harr, P. A., R. L. Elsberry, and J. C. L. Chan, 1996a: Transformation of a large monsoon depression to a tropical storm during TCM-93. *Mon. Wea. Rev.*, **124**, 2625–2643.

Harr, P. A., M. S. Kalafsky, and R. L. Elsberry, 1996b: Environmental conditions prior to formation of a midget tropical cyclone during TCM-93. *Mon. Wea. Rev.*, **124**, 1693–1710.

Hartmann, D. L. and E. D. Maloney, 2001: The Madden-Julian oscillation, barotropic dynamics, and north Pacific tropical cyclone formation. Part II: Stochastic barotropic modeling. *J. Atmos. Sci.*, **58**, 2559–2570.

Haynes, P. H. and M. E. McIntyre, 1987: On the evolution of vorticity and potential vorticity in the presence of diabatic heating and frictional or other forces. *J. Atmos. Sci.*, **44**, 828–841.

Hendon, H. H. and B. Liebmann, 1991: The structure and annual variation of antisymmetric fluctuations of tropical convection and their association with Rossby-gravity waves. *J. Atmos. Sci.*, **48**, 2127–2140.

Hendricks, E. A., M. T. Montgomery, and C. A. Davis, 2004: On the role of "vortical" hot towers in hurricane formation. *J. Atmos. Sci.*, **61**, 1209–1232.

Hennon, C. C. and J. S. Hobgood, 2003: Forecasting tropical cyclogenesis over the Atlantic basin using large-scale data. *Mon. Wea. Rev.*, **131**, 2927–2940.

Hidalgo, J. M., 2008: Vortical hot towers, their aggregate effects and their resolution dependence in the formation of Hurricane Diana (1984). PhD thesis, Colorado State University.

Holton, J. R., 2004: *An Introduction to Dynamic Meteorology.* Elsevier Academic Press, Burlington, MA, 535 pp.

Hoskins, B. J., M. E. McIntyre, and A. Robertson, 1985: On the use and significance of isentropic potential vorticity maps. *Quart. J. Roy. Meteor. Soc.*, **111**, 877–946.

Houze, R. A., 2004: Mesoscale convective systems. *Reviews of Geophysics*, **42, RG4003**, 1–43.

Houze, R. A., W.-C. Lee, and M. Bell, 2009: Convective contribution to the genesis of Hurricane Ophelia (2005). *Mon. Wea. Rev.*, in press.

Karyampudi, V. M. and H. F. Pierce, 2002: Synoptic-scale influence of the Saharan air layer on tropical cyclogenesis over the eastern Atlantic. *Mon. Wea. Rev.*, **130**, 3100–3128.

Kiladis, G. N., K. H. Straub, and P. T. Haertel, 2005: Zonal and vertical structure of the Madden-Julian oscillation. *J. Atmos. Sci.*, **62**, 2790–2809.

Kingsmill, D. E. and R. A. Houze, 1999: Thermodynamic characteristics of air flowing into and out of precipitating convection over the west Pacific warm pool. *Quart. J. Roy. Meteor. Soc.*, **125**, 1209–1229.

Kurihara, Y. and R. E. Tuleya, 1981: A numerical simulation study on the genesis of a tropical storm. *Mon. Wea. Rev.*, **109**, 1629–1653.

Landsea, C., 1993: A climatology of intense (or major) Atlantic hurricanes. *Mon. Wea. Rev.*, **121**, 1703–1713.

Lau, K.-H. and N.-C. Lau, 1990: Observed structure and propagation characteristics of tropical summertime synoptic scale disturbances. *Mon. Wea. Rev.*, **118**, 1888–1913.

Liebmann, B. and H. H. Hendon, 1990: Synoptic-scale disturbances near the equator. *J. Atmos. Sci*, **47**, 1463–1479.

Madden, R. A. and P. R. Julian, 1994: Observations of the 40–50-day tropical oscillation-A review. *Mon. Wea. Rev.*, **122**, 814–837.

Maloney, E. D. and D. L. Hartmann, 2000a: Modulation of eastern north Pacific hurricanes by the Madden-Julian oscillation. *J. Clim.*, **13**, 1451–1460.

———, 2000b: Modulation of hurricane activity in the Gulf of Mexico by the Madden-Julian oscillation. *Science*, **287**, 2002–2004.

———, 2001: The Madden-Julian oscillation, barotropic dynamics, and North Pacific tropical cyclone formation. Part I: Observations. *J. Atmos. Sci.*, **58**, 2545–2558.

Mapes, B. E. and R. A. Houze, 1992: An integrated view of the 1987 Australian monsoon and its mesoscale convective systems. Part I: Horizontal structure. *Quart. J. Roy. Meteor. Soc.*, **118**, 927–963.

———, 1993: An integrated view of the Australian monsoon and its mesoscale convective systems. Part II: Vertical structure. *Quart. J. Roy. Meteor. Soc.*, **119**, 733–754.

———, 1995: Diabatic divergent profiles in Western Pacific Mesoscale Convective Systems. *J. Atmos. Sci.*, **52**, 1807–1828.

Matsuno, T., 1966: Quasi-geostrophic motions in the equatorial area. *J. Meteor. Soc.*, **44**, 25–43.

McBride, J. L., 1995: Tropical cyclone formation. *Global Perspectives on Tropical Cyclones*, Elsberry, R. L., Ed., World Meteorological Organisation, pp. 63–105.

McBride, J. L. and R. Zehr, 1981: Observational analysis of tropical cyclone formation. Part II: Comparison of non-developing versus developing Systems. *J. Atmos. Sci.*, **38**, 1132–1151.

McDonnel, K. A. and N. J. Holbrook, 2004: A Poisson regression model of tropical cyclogenesis for the Australian-Southwest Pacific ocean region. *Wea. Forecasting*, **19**, 440–455.

McTaggart-Cowan, R., L. F. Bosart, C. A. Davis, E. H. Atallah, J. R. Gyakum, and K. A. Emanuel, 2006: Analysis of Hurricane Catarina (2004). *Mon. Wea. Rev.*, **134**, 3029–3055.

McTaggart-Cowan, R., L. F. Bosart, J. R. Gyakum, and E. H. Atallah, 2007a: Hurricane Katrina (2005). Part I: Complex life cycle of an intense tropical cyclone. *Mon. Wea. Rev.*, **135**, 3905–3926.

———, 2007b: Hurricane Katrina (2005). Part II: Evolution and hemispheric impacts of a diabatically generated warm pool. *Mon. Wea. Rev.*, **135**, 3927–3949.

McTaggart-Cowan, R., G. D. Deane, L. F. Bosart, C. A. Davis, and T. J. Galarneau, 2008: Climatology of tropical cyclogenesis in the North Atlantic (1948–2004). *Mon. Wea. Rev.*, **136**, 1284–1304.

McWilliams, J. C., 1984: The emergence of isolated coherent vortices in turbulent flow. *J. Fluid Mech.*, **146**, 21–43.

McWilliams, J. C. and G. R. Flierl, 1979: On the evolution of isolated nonlinear vortices. *J. Phys. Oceanogr.*, **9**, 1155–1182.

Molinari, J., K. Lombardo, and D. Vollaro, 2007: Tropical cyclogenesis within an equatorial Rossby wave packet. *J. Atmos. Sci.*, **64**, 1301–1317.

Molinari, J., S. Skubis, D. Vollaro, F. Alsheimer, and H. E. Willoughby, 1998: Potential vorticity analysis of tropical cyclone intensification. *J. Atmos. Sci.*, **55**, 2632–2644.

Molinari, J. and D. Vollaro, 2000: Planetary- and synoptic-scale influences on Eastern Pacific tropical cyclogenesis. *Mon. Wea. Rev.*, **128**, 3296–3307.

Molinari, J., D. Vollaro, and K. L. Corbosiero, 2004: Tropical cyclone formation in a sheared environment: A case study. *J. Atmos. Sci.*, **61**, 2493–2509.

Montgomery, M. T. and J. Enagonio, 1998: Tropical cyclogenesis via convectively forced vortex Rossby waves in a three-dimensional quasi-geostrophic model. *J. Atmos. Sci.*, **55**, 3176–3207.

Montgomery, M. T. and B. F. Farrell, 1993: Tropical cyclone formation. *J. Atmos. Sci.*, **50**, 285–310.

Montgomery, M. T., M. E. Nicholls, T. A. Cram, and A. Saunders, 2006: A vortical hot tower route to tropical cyclogenesis. *J. Atmos. Sci.*, **63**, 355–386.

Musgrave, K. D., C. A. Davis, and M. Montgomery, 2008: Numerical simulations of the formation of Hurricane Gabrielle (2001). *Mon. Wea. Rev.*, **136**, 3151–3167.

Nguyen, V. S., R. K. Smith, and M. T. Montgomery, 2008: Tropical-cyclone intensification and predictability in three dimensions. *Quart. J. Roy. Meteor. Soc.*, **134**, 563–582.

Nolan, D. S., 2007: What is the trigger for tropical cyclogenesis? *Aus. Met. Mag.*, **56**, 241–266.

Nolan, D. S. and L. D. Grasso, 2003: Nonhydrostatic, three-dimensional perturbations to balanced, hurricane-like vortices. Part II: Symmetric response and nonlinear simulations. *J. Atmos. Sci.*, **60**, 2717–2745.

Nolan, D. S. and M. T. Montgomery, 2002: Nonhydrostatic, three-dimensional perturbations to balanced, hurricane-like vortices. Part I: Linearized formulation, stability, and evolution. *J. Atmos. Sci.*, **59**, 2989–3020.

Raymond, D. J. and H. Jiang, 1990: A theory for long-lived convective systems. *J. Atmos. Sci.*, **47**, 3067–3077.

Raymond, D. J., C. López-Carrillo, and L. L. Cavazos, 1998: Case-studies of developing east Pacific easterly waves. *Quart. J. Roy. Meteor. Soc.*, **124**, 2005–2034.

Raymond, D. J. and S. L. Sessions, 2007: Evolution of convection during tropical cyclogenesis. *Geophys. Res. Lett.*, **34**, L06 811.

Reasor, P. D., M. T. Montgomery, and L. F. Bosart, 2005: Mesoscale observations of the genesis of Hurricane Dolly (1996). *J. Atmos. Sci.*, **62**, 3151–3171.

Reasor, P. D., M. T. Montgomery, F. D. Marks, Jr., and J. F. Gamache, 2000: Low-wavenumber structure and evolution of the hurricane inner core observed by airborne dual-Doppler radar. *Mon. Wea. Rev.*, **128**, 1653–1680.

Reed, R. J., D. Norquist, and E. E. Recker, 1977: The structure and properties of African wave disturbances as observed during Phase III of GATE. *Mon. Wea. Rev.*, **105**, 317–333.

Ritchie, E. A., 2003: Some aspects of mid-level vortex interaction in tropical cyclogenesis. *Chapter 12, Cloud systems, hurricanes and the Tropical Rainfall Measuring Mission (TRMM): A tribute to Dr. Joanne Simpson*, Tau, W.-K. and R. Adler, Eds., Amer. Met. Soc.

Ritchie, E. A. and G. J. Holland, 1993: On the interaction of tropical-cyclone-scale vortices. II: Discrete vortex patches. *Quart. J. Roy. Meteor. Soc.*, **119**, 1363–1379.

———, 1997: Scale interactions during the formation of Typhoon Irving. *Mon. Wea. Rev.*, **125**, 1377–1396.

———, 1999: Large-scale patterns associated with tropical cyclogenesis in the Western Pacific. *Mon. Wea. Rev.*, **127**, 2027–2043.

Ritchie, E. A., J. Simpson, W. T. Liu, J. Halverson, C. S. Velden, K. F. Brueske, and H. Pierce, 2003: Present day satellite technology for hurricane research. *Chapter 12, Hurricane: Coping with disaster*, Simpson, R., Ed., Amer. Geophys. Union, 360.

Rogers, R. F. and J. M. Fritsch, 2001: Surface cyclogenesis from convectively driven amplification of midlevel mesoscale convective vortices. *Mon. Wea. Rev.*, **129**, 605–637.

Rotunno, R. and K. A. Emanuel, 1987: An air-sea interaction theory for tropical cyclones. Part II: Evolutionary study using a nonhydrostatic axisymmetric numerical model. *J. Atmos. Sci.*, **44**, 542–561.

Roundy, P. E. and W. M. Frank, 2004: A climatology of waves in the equatorial region. *J. Atmos. Sci.*, **61**, 2105–2132.

Schechter, D. A. and D. H. E. Dubin, 1999: Vortex motion driven by a background vorticity gradient. *Phys. Rev. Lett.*, **83**, 2191–2194.

Schubert, W. H. and J. J. Hack, 1982: Inertial stability and tropical cyclone development. *J. Atmos. Sci.*, **39**, 1687–1697.

Shapiro, L. J. and H. E. Willoughby, 1982: The response of balanced hurricanes to local sources of heat and momentum. *J. Atmos. Sci.*, **39**, 378–394.

Simpson, J., E. A. Ritchie, G. J. Holland, J. Halverson, and S. Stewart, 1997: Mesoscale interactions in tropical cyclone genesis. *Mon. Wea. Rev.*, **125**, 2643–2661.

Sippel, J. A., J. W. Nielsen-Gammon, and S. E. Allen, 2006: The multiple-vortex nature of tropical cyclogenesis. *Mon. Wea. Rev.*, **134**, 1796–1814.

Sippel, J. A. and F. Zhang, 2008: A probabilistic analysis of the dynamics and predictability of tropical cyclogenesis. *J. Atmos. Sci.*, **65**, 3440–3459.

Smith, R. K. and W. Ulrich, 1990: An analytical theory of tropical cyclone motion using a barotropic model. *J. Atmos. Sci.*, **47**, 1973–1986.

Takayabu, Y. N., 1994: Large-scale cloud disturbances associated with equatorial waves. Part I: Spectral features of the cloud disturbances. *J. Meteor. Soc. Japan*, **72**, 433–448.

Takayabu, Y. N. and T. Nitta, 1993: 3–5 day disturbances coupled with convection over the tropical Pacific Ocean. *J. Meteor. Soc. Japan*, **71**, 221–245.

Thorncroft, C. and K. Hodges, 2001: African easterly wave variability and its relationship to Atlantic tropical cyclone activity. *J. Clim.*, **14**, 1166–1179.

Tory, K. J., N. E. Davidson, and M. T. Montgomery, 2007: Prediction and diagnosis of tropical cyclone formation in an NWP system. Part III: Developing and non-developing storms. *J. Atmos. Sci.*, **64**, 3195–3213.

Tory, K. J. and M. T. Montgomery, 2006: Topic 2.2: Internal influences on tropical cyclone formation. *World Meteorological Organisation. Report on the Sixth International Workshop on Tropical Cyclones.* San Jose, Costa Rica, November 2006.

———, 2008: Tropical cyclone formation: A synopsis of the internal dynamics. *Proceedings of the AMS 28th Conference on Hurricanes and Tropical Meteorology*, Orlando, Florida.

Tory, K. J., M. T. Montgomery, N. E. Davidson, and J. D. Kepert, 2006: Prediction and diagnosis of tropical cyclone formation in an NWP system. Part II: A diagnosis of tropical cyclone Chris formation. *J. Atmos. Sci.*, **63**, 3091–3113.

Trier, S. B., C. A. Davis, and J. D. Tuttle, 2000: Long-lived mesoconvective vortices and their environment. Part I: Observations from the central United States during the 1998 warm season. *Mon. Wea. Rev.*, **128**, 3376–3395.

Velasco, I. and J. Fritsch, 1987: Mesoscale Convective Complexes in the Americas. *J. Geophys. Res.*, **92(D8)**, 9591–9613.

Wang, C.-C. and G. Magnusdottir, 2005: ITCZ breakdown in three-dimensional flows. *J. Atmos. Sci.*, **62**, 1497–1512.

———, 2006: The ITCZ in the central and eastern Pacific on synoptic time scales. *Mon. Wea. Rev.*, **134**, 1405–1421.

Wang, H. and W. M. Frank, 1999: Two modes of tropical cyclogenesis: An idealized simulation. *Proceedings of the AMS 23rd Conference on Hurricanes and Tropical Meteorology*, Dallas, TX, 10–15 Jan., 923–924.

Wheeler, M. C. and G. N. Kiladis, 1999: Convectively coupled equatorial waves: analysis of clouds and temperature in the wavenumber-frequency domain. *J. Atmos. Sci.*, **56**, 374–399.

Wheeler, M. C., G. N. Kiladis, and P. J. Webster, 2000: Large-scale dynamical fields associated with convectively coupled equatorial waves. *J. Atmos. Sci.*, **57**, 613–640.

Willoughby, H. E., 1979: Forced secondary circulations in hurricanes. *J. Geophys. Res.*, **84C**, 3173–3183.

———, 2009: Diabatically induced secondary flows in tropical cyclones. Part II: Periodic forcing. *Mon. Wea. Rev.*, **137**, 822–835.

Yamasaki, M., 1984: Dynamics of convective clouds and "CISK" in vertical shear flow-with its application to easterly waves and squall-line systems. *J. Meteor. Soc. Japan*, **62**, 833–863.

Yano, J.-I. and K. A. Emanuel, 1991: An improved model of the equatorial troposphere and its coupling with the stratosphere. *J. Atmos. Sci.*, **48**, 377–389.

Zehr, R., 1992: Tropical cyclogenesis in the western North Pacific. NOAA Tech. Rep. NESDIS 61.

Zhang, C. and P. J. Webster, 1989: Effects of zonal flows on equatorially trapped waves. *J. Atmos. Sci.*, **46**, 3632–3652.

Zhang, F. and J. A. Sippel, 2009: Effects of moist convection on hurricane predictability. *J. Atmos. Sci.*, **66**, 1944–1961.

Zipser, E. J. and C. Gautier, 1978: Mesoscale events within a GATE tropical depression. *Mon. Wea. Rev.*, **106**, 789–805.

Chapter 3

Air-Sea Interactions in Tropical Cyclones

Lynn K. Shay

Division of Meteorology and Physical Oceanography,
Rosestiel School of Marine and Atmospheric Science,
University of Miami, Miami, FL 33157, USA
nshay@rsmas.miami.edu

Significant progress has been made in the area of upper ocean responses and air-sea interactions during the passage of tropical cyclones (TC) since the first edition published in 1992. In terms of the upper ocean impacts on intensity, considerable attention has focused on the three-dimensional cold wake structure and the negative feedback due to ocean mixed layer cooling by shear-induced vertical mixing in quiescent oceans. By contrast, in the western parts of the oceanic basins where the ocean state is not at rest, strong currents (e.g., Kuroshio, Gulf Stream) transport warm water poleward as part of the gyre circulation. In these regimes, the upper ocean current shears do not necessarily develop as in regions with shallow ocean mixed layers where significant sea surface cooling often occurs. Transports by these energetic western boundary currents tend to be resistive to shear-induced ocean mixing events since the mixed layer is already deep and the thermal response (or sea surface temperature cooling) tends to be minimized. The implication is since the oceanic mixed layers do not significantly cool during TC passage, there is a more sustained heat flux to the atmospheric boundary layer, thus representing an important mechanism for observed deepening of recently observed severe TCs.

In terms of the air-sea interactions, surface waves and sea spray impact the surface drag coefficient. Recent studies have shown the surface drag coefficient to level off between 28 to 33 m s^{-1} at values from 2.5 to 3.5 $\times$ 10^{-3}. While the measurement uncertainties increase for higher wind speeds, it is clear that the surface drag cannot continue to increase with wind speeds. Moreover, for intense TCs, the ratio of the enthalpy and surface drag coefficient exceeds unity (typically 1.2 to 1.5). When this ratio is less than unity, theoretical studies suggest that TCs cannot reach their maximum potential intensity.

1. Background

1.1. *Early Studies*

It has long been recognized that tropical cyclones (TCs) derive their energy primarily from the release of latent heat upon condensation of water vapor in the atmospheric boundary layer (Ooyama, 1969). That is, it is necessary to have a moisture source, such as the ocean where surface temperatures have to be sufficiently warm (often taken as $\geq$26°C) to maintain a moisture flux from the ocean to the atmosphere (Palmen, 1948). Perlroth (1969) emphasized the importance of oceanic temperatures of 26°C ($\sim$78°F) on TC intensity via this heat and moisture transfer between the two fluids. He related ocean thermal structure difference (ΔT) between the sea surface temperature (SST) and subsurface temperatures at 200-feet using historical analyses in the Atlantic Ocean basin and TC activity. Approximately 90% of the tropical storms reached TC status when $\Delta T < 4$°C (weakly stratified) while only 4% of the tropical storms reached TC status when $\Delta T > 8$°C (strongly stratified). Leipper and Volgenau (1972) subsequently demonstrated that the 26°C isotherm depth and the upper ocean's thermal structure, known as oceanic heat content (OHC), has important implications on TC intensity variations:

$$\mathrm{OHC} = c_p \int_0^{D26} \rho[T(z) - 26]dz,$$

where c_p is specific heat at constant pressure, $D26$ is the 26°C isotherm depth, and OHC is zero when SST decrease to less than 26°C. While their measurements were not acquired under directly forced TC conditions, OHC differences between pre and post TC in the Gulf of Mexico demonstrated the relative importance of this measure of thermal energy. Even these early studies pointed to the importance of subsurface stratification (upper ocean thermal structure) and warm water advection by currents on TC intensity. Obviously, there are linkages between OHC variability and air-sea fluxes on equivalent potential temperatures and TC intensity change.

1.2. *Oceanic Feedback*

The most apparent effect of TC passage is the marked SST cooling, and by proxy the OML temperature response, of typically 1 to 6°C occurring to the right (left) of the storm track by 1–2 radii of maximum winds (Rmax) in the northern (southern) hemispheres. Although warm SSTs are required to maintain the TC, these *biases* of the maximum SST decreases and OML depth increases of 20 to 40 m are due to entrainment mixing of the cooler thermocline water with the warmer OML (see Fig. 1). Entrainment mixing usually accounts for most of the observed cooling of the SSTs and OML deepening either due to vertical shear of the horizontal currents or surface-generated turbulence by wind stress or surface friction velocity (u_*) distributed across the OML. Ocean mixing and cooling are principally a function of wind-forced currents and their associated shears ($\partial v/\partial z = S$). Wind-forced motions are often associated with near-inertial flows that have a frequency shifted slightly above the local inertial period (f) by 1 to 20% depending on the TC characteristics and the strength of the underlying ocean stratification. Forced near-inertial motions represent the low-frequency cutoff of the internal wave spectra which range from the buoyancy frequency (N) to f (Gill, 1984). While the forced current structures have large vertical scales, near-inertial shears across the OML base tend to be associated with shorter vertical wavelengths (e.g., higher order baroclinic modes) that reduce the Richardson numbers (defined as the ratio of buoyancy frequency (N^2) and (S^2)) to below criticality values (Price, 1981; Sanford *et al.*, 1987; Shay *et al.*, 1992, 1998; Price *et al.*, 1994; Jacob *et al.*, 2000).

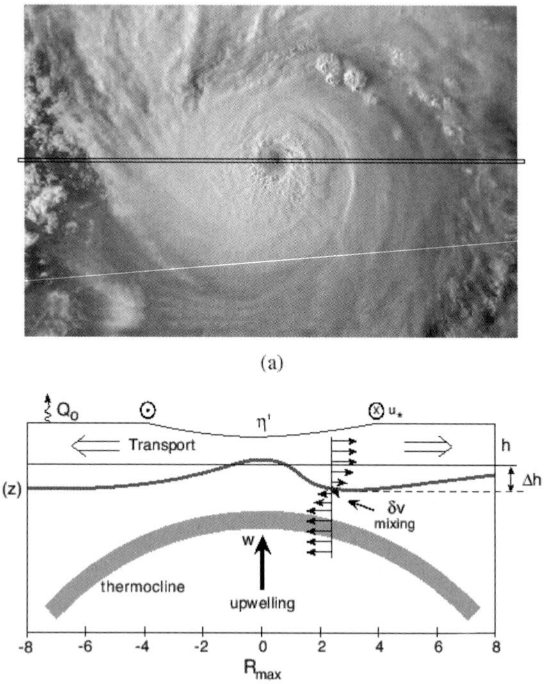

(a)

(b)

Figure 1. (a) TC image and (b) a conceptual depiction of physical processes forced by hurricane winds such as shear-induced mixing and OML deepening, upwelling due to transport away from the center, and surface heat fluxes from the ocean to the atmosphere, all of which may contribute to ocean cooling during TC passage. (Reproduced from Shay, 2009 by permission of Elsevier Press International.)

These physical processes are associated with the OML heat budget through shear-driven entrainment mixing, surface enthalpy (a combination of sensible and latent fluxes: Q_o), and horizontal advection by ocean currents under relatively (and often assumed) quiescent initial ocean conditions where the initial state is at rest. Upwelling of the cooler isotherms from the subsurface layers increases the buoyancy frequency associated with more stratified thermocline water that tends to keep the Richardson number above criticality along the track. This upwelling process, forced by wind-driven current divergence in the OML away from the TC center, increases the stratification that decreases shear-induced mixing via the Richardson number. Thus, both shear-induced mixing and divergent OML currents (i.e., momentum response) impact upper ocean temperature structure and OML depth variations.

Chang and Anthes (1978) modeled the mutual response between the upper ocean and the TC. Consistent with Fig. 1(b), they point out this SST cooling occurs behind the eye of the TC as part of the cold wake. As the SSTs continue to decrease towards the 26°C threshold, heat and moisture fluxes (enthalpy fluxes) decrease accordingly and cause *negative feedback* to the TC. The more cooling and OML deepening, the less heat and moisture transfer to the atmosphere, which tends to decrease TC intensity. These OML cooling levels have been observed (Shay *et al.*, 1992; Cione and Uhlhorn, 2003; Black *et al.*, 2007) and modeled (Price, 1981; Bender and Ginis, 2000). To simplify the thermodynamics, models have used idealized temperature profile to initialize the domain, leading to the idea of coupling of 1-dimensional (1-D) OML to atmospheric models to improve TC intensity across the cold wake (Schade and Emanuel, 1999). When TCs encounter oceanic regimes with shallow OML, the ocean thermal response to TC forcing is consistent with this *negative feedback* concept where the stratification at the OML base is weak (i.e., Gulf of Mexico common water). The upper ocean thermal structure, and hence the OML depths are not spatially uniform over most of the global ocean basins and cannot be simply characterized as idealized thermal structure in the 1-D approach (Yablonsky and Ginis, 2009; Halliwell *et al.*, 2010).

An illustration of this structural variability in water masses is shown in Fig. 2. The Eastern Pacific Ocean has thin OML, but has strong stratification at the OML base due to vertical temperature and salinity gradients. For the ocean to mix (and significantly cool in this regime), considerably more wind-driven ocean current shear needs to be generated by surface winds since the buoyancy frequency is about 20 cycles per hour (cph). At the lower latitudes (within 10 to 20° of the equator), local inertial periods (IP) are long compared to the advective time scales of the TCs. Although TC's can get intense over the warm pool, there is little evidence of strong cold wakes south of 18°N in this regime. It is only when a TC moves over the region with similar OML depths with weaker stratification (12 cph) such as Juliette in September 2001. By contrast, the vertical structure in the Loop Current water mass is weakly stratified (N ranges between 4 to 6 cph). However below that nearly isothermal layer, warm thermal structure extends to depths of more than 100 m where the 26°C isotherm depth lies at 150 m. This thermal structure behavior increases the OHC where deep, warm layers contain more heat for TC's to tap promoting intensification (Shay *et al.*, 2000; Lin *et al.*, 2005; Mainelli *et al.*, 2008). This is an example of *less negative feedback* than in cases where cooling levels of several degrees decreases the air-sea fluxes (Cione *et al.*, 2000).

1.3. *Ocean-Atmosphere Transfers*

The oceanic modulation of TC intensity has drawn more attention on gaining a better understanding of the physical interaction between the atmosphere and ocean during TC passage. There are unfortunately few observational data sets in high-wind conditions with both concurrent atmosphere and ocean measurements. Parameterizations of momentum, heat and moisture transfer are further complicated by

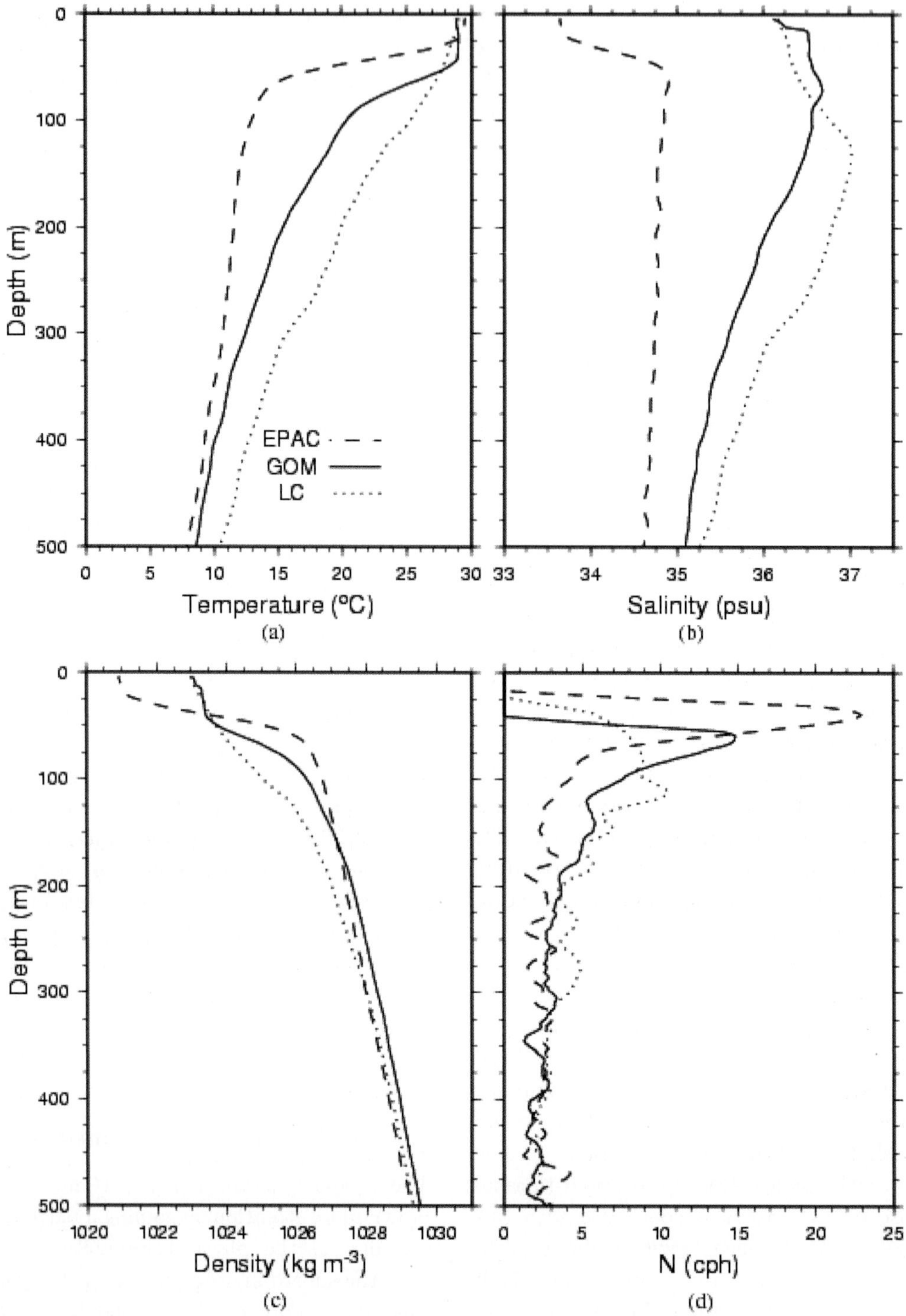

Figure 2. (a) Temperature (°C), (b) salinity (practical salinity units), (c) density (kg m^{-3}) and (d) buoyancy frequency (N: cycles per hour) profiles from the Eastern Pacific Ocean (dash), the Gulf of Mexico Common Water (solid) and the Loop Current (dot) as measured from airborne expendable ocean profilers. (Reproduced from Shay and Brewster, 2010 by permission of the American Meteorological Society.)

sea state, sea spray and the complexities of the upper ocean. Under certain conditions there should be a level of mutual dependence of the air-sea transfer processes of heat, moisture and momentum as suggested in idealized model simulations (Emanuel, 1995) that intensity is sensitive to the ratio of enthalpy transfer coefficient to drag coefficient ($c_k\ c_d^{-1}$, where c_k is the bulk enthalpy coefficient and c_d is the surface drag coefficient). The conclusion that this quantity probably lies within a rather limited range ($\leq$1.5) is commensurate with the observation that most TCs do not typically reach their maximum potential intensity (MPI).

This chapter attempts to summarize significant progress has been made in this broad area of upper ocean processes and air-sea interaction during TC passages since Ginis (1995). The intended approach is to synthesize empirical, observational, analytical and numerical results from a global perspective. In this context, this chapter begins by describing progress in upper-ocean response processes that include the SST, OML and thermocline responses. Of particular interest is to demonstrate that *in situ* and remotely sensed measurements are now being used to initialize the oceanic models to obtain an accurate basic state in section 2. Section 3 focuses on the observed and modeled responses to TC forcing including the observed and simulated ocean response during TC passages. Section 4 discusses the interactions that occur across the interface including surface winds and waves, and the communication to the atmospheric boundary layer through the momentum and enthalpy fluxes across the interface. These findings are summarized within a global context in section 5. A Glossary of Terms is given at the end of the chapter followed by a fairly complete list of references.

2. Oceanic States

Coupled models to predict hurricane intensity and structure change are being used to issue forecasts to the public who increasingly rely on the most advanced weather forecasting systems to prepare for landfall (Marks and Shay, 1998). For such models, it has become increasingly clear over the past decade that ocean models will have to include realistic initial conditions to simulate not only the oceanic response to hurricane forcing (Sanford *et al.*, 1987, 2007; Shay *et al.*, 1992, 1998; Price *et al.*, 1994; D'Asaro, 2003; Jacob and Shay, 2003; Black *et al.*, 2007; Shay and Uhlhorn, 2008; Jaimes and Shay, 2009), but also to simulate the atmospheric response to oceanic forcing (Emanuel, 2003; Bender and Ginis, 2000; Bao *et al.*, 2000; Shay *et al.*, 2000; Hong *et al.*, 2000; Chan *et al.*, 2001; Walker *et al.*, 2005; Lin *et al.*, 2005; Wu *et al.*, 2007; Ali *et al.*, 2007; Manielli *et al.*, 2008). The atmospheric response is related to the level of feedback from the oceanic responses.

2.1. *Circulation*

From a basin-scale perspective, the wind-driven upper ocean circulation transports heat from the tropics to the mid-latitudes as part of the annual cycle. Embedded within this ocean circulation pattern is a broad range of smaller-scale variability in which there is an energetic scale of $\approx$100 km known as the mesoscale. For example, the western boundary currents (e.g., Gulf Stream in the North Atlantic Ocean and Kuroshio in the Pacific Ocean basins) are characterized as mesoscale ocean features (Bane, 1994). Within these western boundary current regimes, warm core rings (WCR) and cold core rings (CCR) are formed and propagate westward. In the ocean's interior, surface Ekman transport induced by the winds is directed northward (southward) along the southern (northern) periphery of the high pressure ridge during the summer months in the northern hemisphere. That is, cooler water from the north is advected southward, and warmer water from the south moves northward. These water masses converge to form a subtropical ocean front (Weller, 1991). The thermohaline circulation arises as a result of the density changes induced by surface heating and cooling and through mixing processes between differing water masses. In this framework, deep western boundary currents exemplify this process and generally flow in a direction opposite to the wind-driven surface flow, representing an important component to the global oceanic circulation. A large part of the resultant circulation is governed by horizontal density and pressure gradients in geostrophic balance, resulting in a circulation that forms ridges and valleys on the sea-surface with surface elevation changes of about 1 m over horizontal scales of 100 km across western boundary currents.

One of the better studied oceanic circulation system is the North Atlantic Ocean gyre including the Florida Current and the Gulf Stream (Hogg and Johns, 1995). Based on an annual average, the transport from the Caribbean Sea into the Gulf of Mexico through the Yucatan Straits (Leipper, 1967) is approximately 24 Sv (Sverdrup $= 10^6\,\mathrm{m}^3\ \mathrm{s}^{-1}$), which forces the Loop Current (LC). The maximum flows through the Straits of 1–1.2 m s^{-1} transports warm subtropical water with a different temperature and salinity relationship than the background water mass between 18–26°C in the Gulf of Mexico (GOM). The anticyclonic circulation around the LC flows through the Florida Straits between United States and Cuba forming the Florida Current. The Antilles Current flows north of the Caribbean Islands and joins the Florida Current increasing its transport to about 40 Sv (Schott *et al.*, 1988). As this current turns northward, it follows the continental slope in the Middle Atlantic Bight, to eventually form the Gulf Stream core (Stommel, 1948). Once it passes Cape Hatteras, the Gulf Stream flows into the North Atlantic Ocean with increased transport due to inflow from recirculations bordering along its northern and southern flanks (Johns *et al.*, 1995). Of practical importance to TCs, warm ribbons of higher OHC values with deeper OML surrounds the coastlines in the basins. In this context, such oceanic patterns occur in all the basins and have global consequences for landfalling TCs.

2.2. *Loop Current*

The time-dependent LC can penetrate ~500 km northward of the Yucatan Strait. Recurring WCR shedding events with peak periods from 6 to 11 months (Sturges and Leben, 2000, Leben, 2005) occur when CCRs are located on the LC periphery prior to separation. These WCRs, diameters of ~200 km, propagate west to southwest at phase speeds of 5 approximately km d^{-1} (Elliot, 1982), and can remain in the Gulf of Mexico for several months. In this LC regime, OHC values relative to the 26°C isotherm depth exceeds a value of 100 kJ cm^{-2} (Leipper and Volgenau, 1972). Such OHC levels are resistive to storm-induced cooling by wind-driven current shears across a deeper OML base. Across the Yucatan Straits, background oceanic flows that are set up by large horizontal pressure gradients due to temperature and salinity variability play a significant role in altering the development of strong wind-driven current shears within the LC and WCR complex. As suggested in Fig. 3, pre- and post-Isidore conditions across the Yucatan Straits (lies between Cuba and the Yucatan Peninsula) indicate strong density and pressure gradients that are associated with the northward-flowing LC at speeds of up to 1 m s^{-1}. Horizontal gradients were sharpened after passage since Isidore cooled the Yucatan shelf waters by 4.5°C compared to less than 1°C along the western part of the Yucatan Strait and over the shelf. Strong advection of the thermal and salinity gradients through this regime impacted the oceanic response within the LC.

2.3. *Data Assimilation and Initial Conditions*

Data assimilative ocean modeling efforts and feature based model initialization are effective methods for providing initial boundary conditions to the oceanic and coupled TC prediction models (Falkovich *et al.*, 2005; Halliwell *et al.*, 2008; 2010). The ocean model must be initialized

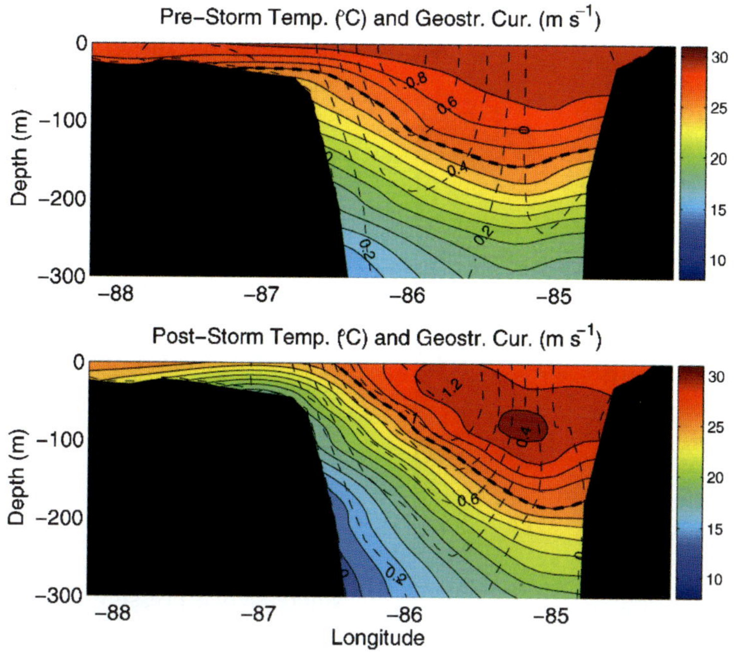

Figure 3. Pre-(upper) and post-Isidore (lower) thermal (°C: color) and northward (into the page) geostrophic velocity (m s^{-1}: dashed) cross-section from expendable profilers deployed on 19 and 23 September 2002 across the Yucatan Straits. Heavy dashed line represents the 26°C isotherm depth. (Reproduced from Shay and Uhlhorn, 2008 by permission of the American Meteorological Society.)

so that surface height anomaly (SHA) features are in the correct locations with realistic temperature and salinity profiles (and the OHC values). For example, ocean forecast systems based on Hybrid Coordinate Ocean Model (HYCOM: Chassignet *et al.*, 2003) were evaluated in the northwest Caribbean Sea and Gulf of Mexico for September 2002 prior to TCs Isidore and Lili, and for September 2004 prior to TC Ivan (Halliwell *et al.*, 2008). In the initial analysis of the pre-Isidore state from an experimental HYCOM forecast system at 0.08° resolution (Cummings, 2003), this model product assimilates both satellite altimetry SHAs (Cooper and Haines, 1996) and SSTs. Comparison of OHC maps hindcast by HYCOM to OHC maps from aircraft measurements and derived from satellite altimetry observations (left upper panels of Fig. 4) demonstrate that the analysis (labeled HYCOM OI in the figure) reproduces the LC orientation but underestimates OHC. In the NW Caribbean Sea, the thermal structure hindcast tends to follow the September climatology but does not reproduce the larger OHC values. In the hindcast, the upper ocean is less saline than both climatology and observations above 250 m (not shown) and less saline than the observations between 250 and 500-m depths. The HYCOM structure was relaxed to the three-dimensional (3-D) Modular Ocean Data Assimilation Scheme (MODAS) (Fox *et al.*, 2002) and temperature and salinity analyses generated from available *in-situ* observations. Biases in this HYCOM MODAS product and observations were reduced in both horizontal maps and vertical profiles. The OHC distribution associated with the complex boundary currents and eddies must be initialized in the ocean model to capture the ocean response to the TC forcing. Careful evaluations of these products against *in situ* observations are needed to insure that ocean features are correctly located with realistic structure as shown in Halliwell *et al.* (2010).

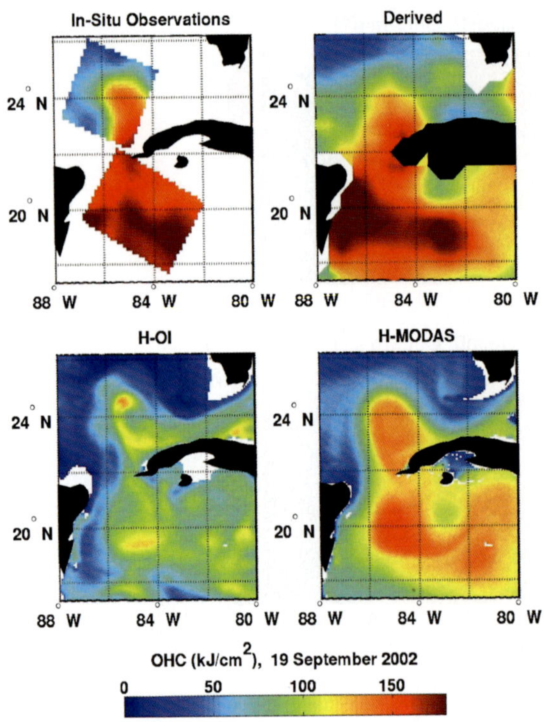

Figure 4. OHC (kJ cm^{-2}) in the northwest Caribbean Sea and southeast GOM from an objective analysis of in situ observations, altimetry, HYCOM OI, and HYCOM-MODAS on 19 September 2002 prior to hurricane Isidore. (Reproduced from Halliwell *et al.*, 2008 by permission of the American Meteorological Society.)

2.4. *Vertical Mixing Parameterizations*

The OML heat budget and ensuing air-sea fluxes are influenced by the choice of entrainment mixing parameterizations across the OML base as suggested by Fig. 1. Contrasting viewpoints on this entrainment mixing topic have been focused on documenting the differences between 1-D and 3-D responses to TC passage (Jacob *et al.*, 2000; Jacob and Shay, 2003; Yablonsky and Ginis, 2009; Halliwell *et al.*, 2008, 2010). Away from strong oceanic fronts, the 1-D approach is valid in that advective tendencies by the background currents are considered to be nonessential to first order assumed in modeling efforts (Price, 1981; Schade and Emanuel, 1999). Vertical mixing processes, predicated on lowering a bulk Richardson number to below critical values (<1) through enhanced vertical shear causing the OML to cool and deepen, dominate the oceanic response. The modified 1-D column approach follows from this general approach except that the domain is initialized with differing thermal structure. In the horizontal plane, temperature (and hence density) gradients have to geostrophically adjust prior to turning on the TC forcing. If the ocean is not in geostrophic balance, the resulting simulations (i.e., levels of SST cooling) will be unrealistic due to a mass field imbalance. The net effect is to induce excessive cooling in regions where oceanic features may be present. Indeed, the adjusted, steady-state ocean currents are small compared to a translation speed of most TCs (e.g., this ratio is similar to a Froude number). In regimes where strong pressure gradients exist, the bulk Richardson number instability, which is at the core of the "1-D assumption", remains at critical levels only for a short period (Uhlhorn, 2008) that limits SST cooling and OML deepening. In other words, pre-existing, energetic ocean currents dampen the near-inertial current response which is primarily responsible for the shear-induced mixing across the OML base.

In a 3-D ocean model, sensitivity tests have been conducted for five mixing schemes: K-Profile Parameterization (KPP: Large *et al.*, 1994); Goddard Institute for Space Studies Level-2 closure (GISS: Canuto *et al.*, 2002); Level-2.5 turbulence closure scheme (MY: Mellor and Yamada, 1982); quasi-slab dynamical instability model (PWP: Price *et al.*, 1986); and the turbulent balance model of Gaspar (KT: 1988) that is a modified version of Kraus and Turner (1967). The results indicate a broad range of OML variability and heat fluxes for quiescent ocean initial conditions (Jacob and Koblinsky, 2007). As shown in Fig. 5, the range of fluxes exceeded 500 W m^{-2} during TC Gilbert where the range of OML temperature changes ranged from 2 to 4°C except for PWP where it was almost 6°C cooling. For realistic ocean conditions for the Gilbert case, Jacob and Shay (2003) simulated OML temperatures and compared with observed

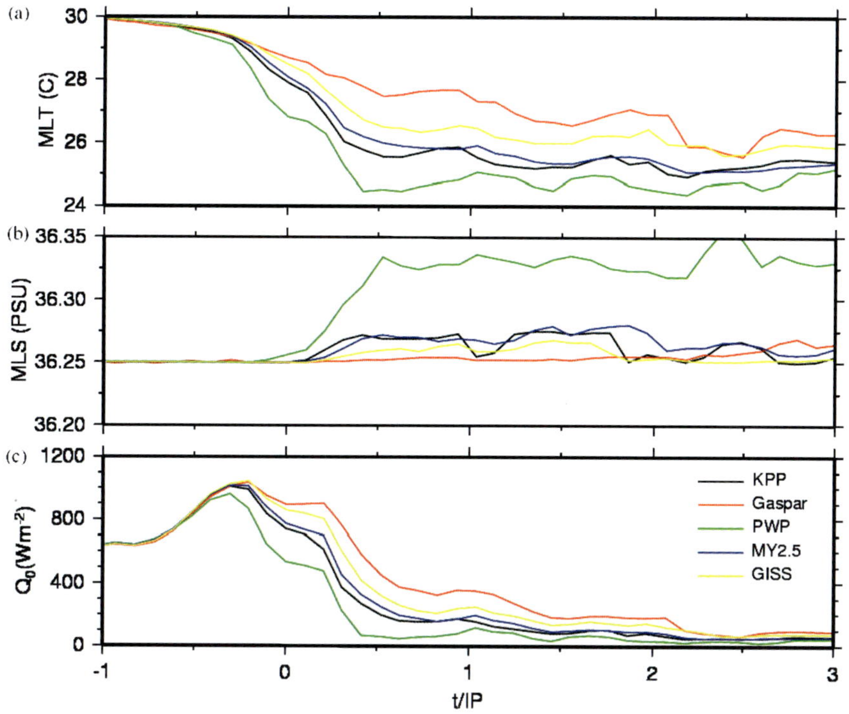

Figure 5. OML evolution at 2 Rmax right of the storm track for quiescent ocean initial conditions for the hurricane Gilbert case: (a) OML temperature (MLT: °C), (b) OML salinity (MLS: Practical Salinity Units), and (c) surface heat fluxes (Qo: W m^{-2}). (Reproduced from Jacob and Koblinsky, 2007 by permission of the American Meteorological Society.) Note that time is normalized by the local inertial period (IP), which was about 30 hours.

profile data and found that the three higher-order turbulent mixing schemes (KPP, MY, GISS) lead to a more accurate ocean response simulation with a WCR present.

The OML salinity evolution with and without precipitation forcing highlights the effect of precipitation on the upper ocean salt budget. For the no precipitation case, Jacob and Koblinsky (2007) found that the salinity variability was similar ahead of the storm center for all schemes. In the PWP scheme, salinity increases significantly in the right-rear quadrant over the first half of an IP following passage due to enhanced mixing. However, in the other mixing cases, the simulated salinity evolution is similar with minimal changes in the KT model due to less intense mixing (no shear instability). By including precipitation forcing, the OML salinity began to decrease about 0.5 IP before the TC closest approach with maximum freshening of the OML observed in the KT case. This freshening process due to enhanced precipitation increases static stability in the OML, leading to a simulated salinity balance for PWP case that is more consistent with the other schemes. The OML temperature and salinity evolution in cases with and without precipitation for PWP scheme indicates mean temperature and salinity differences of 0.5°C and 0.2 psu in the OML (Fig. 5(b)).

An average freshening of 0.2 psu in the OML is seen in the wake of the TC in all the five simulated cases when precipitation forcing was used in the model. Based on the Tropical Rain Measuring Mission (TRMM), the freshwater input (precipitation minus evaporation: P-E) was estimated for TC's Isidore and Lili by integrating the data in the along-track direction (Shay and Uhlhorn, 2008). The P-E value in Isidore was a factor of two to three times larger than in TC Lili between +/−4 Rmax. Excessive

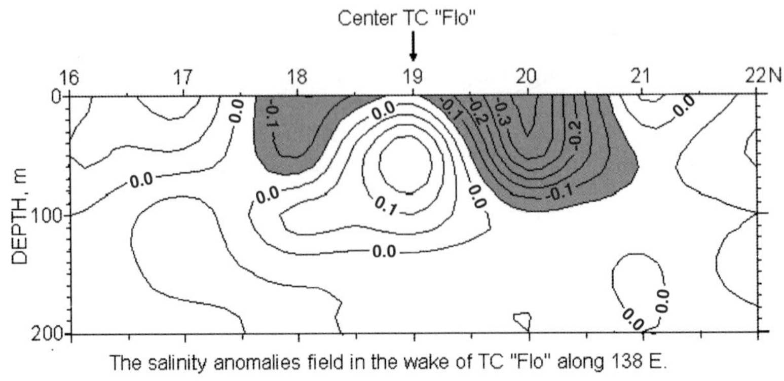

Figure 6. Salinity anomaly response (gray) to Typhoon Flo in the western North Pacific Ocean relative to the storm center along 138°E where anomalies are contoured at 0.05 practical salinity units (psu) based measurements acquired during Spectrum 90 (from Pudov and Petrichenko, 2000).

precipitation of 30 cm within the hurricane's core (~12 inches of rain) impacted the OML salinity budget during Isidore's passage. By contrast, the precipitation rate in Lili's core was about 11 cm (~5 inches) presumably due to its faster translation speed. Such rain events cause changes in the OML salinity balance of 0.2 to 0.4 psu suggested by Fig. 6 (Pudov and Petrichenko 2000). Thus, rain-rates significantly impact the OML salinity balance through the P-E term and the resulting surface buoyancy flux that must be accounted for in OML modeling efforts of the response to TC passage (Price *et al.*, 1986; Jacob and Koblinsky, 2007).

3. Ocean Processes

3.1. *Thermal Response*

Thermal structure changes observed during TCs have been well documented in the literature. In general, the measurements have emphasized the cooling of the SST and deepening of the OML. Based on over a decade of thermal structure measurements, Black (1983) documented the oceanic thermal response to TC passages. One of the key findings was the 1–6°C cooling in the cold wake where OML depths increased by 20 to 40 m depending on the strength of the TC as well as the initial stratification. Recent studies have documented the levels of oceanic cooling using satellite-derived SSTs (Monaldo *et al.*, 1997). During high winds, the SST response represents a proxy of the OML temperature changes as well mixed conditions prevail as suggested by Fig. 1.

One of the more well-documented studies using both expendable profilers and satellite-derived measurements was from TC Gilbert (1988). The sampling strategy was designed to measure the momentum and thermal structural variations from pre, during and two cold-wake experiments in a quiescent area of the western GOM (Shay *et al.*, 1992). Objectively-analyzed temperatures and OML depths revealed that Gilbert induced an SST decrease of 3.5°C to the right of the storm track associated with deeper OML of up to 70 m. The spatial evolution of the response indicated a near-inertial, wave-like pattern (Fig. 7), consistent with the conceptual model of Black (1983). Temperature gradients between the pool of cooler water and the WCR were 3°C over a 50 km distance, indicating a potential existed for strong along-track advection by wind-driven currents.

Jacob *et al.* (2000) used the conservation of heat to assess the various contributions to the observed OML heat and mass budgets during and subsequent to Gilbert's passage. Advection of temperatures by the background currents accounted for 5% of the heat budget near the track and up to 15% of the budget in the WCR, whereas the wind-forced current advection was much weaker. It should be noted that the WCR

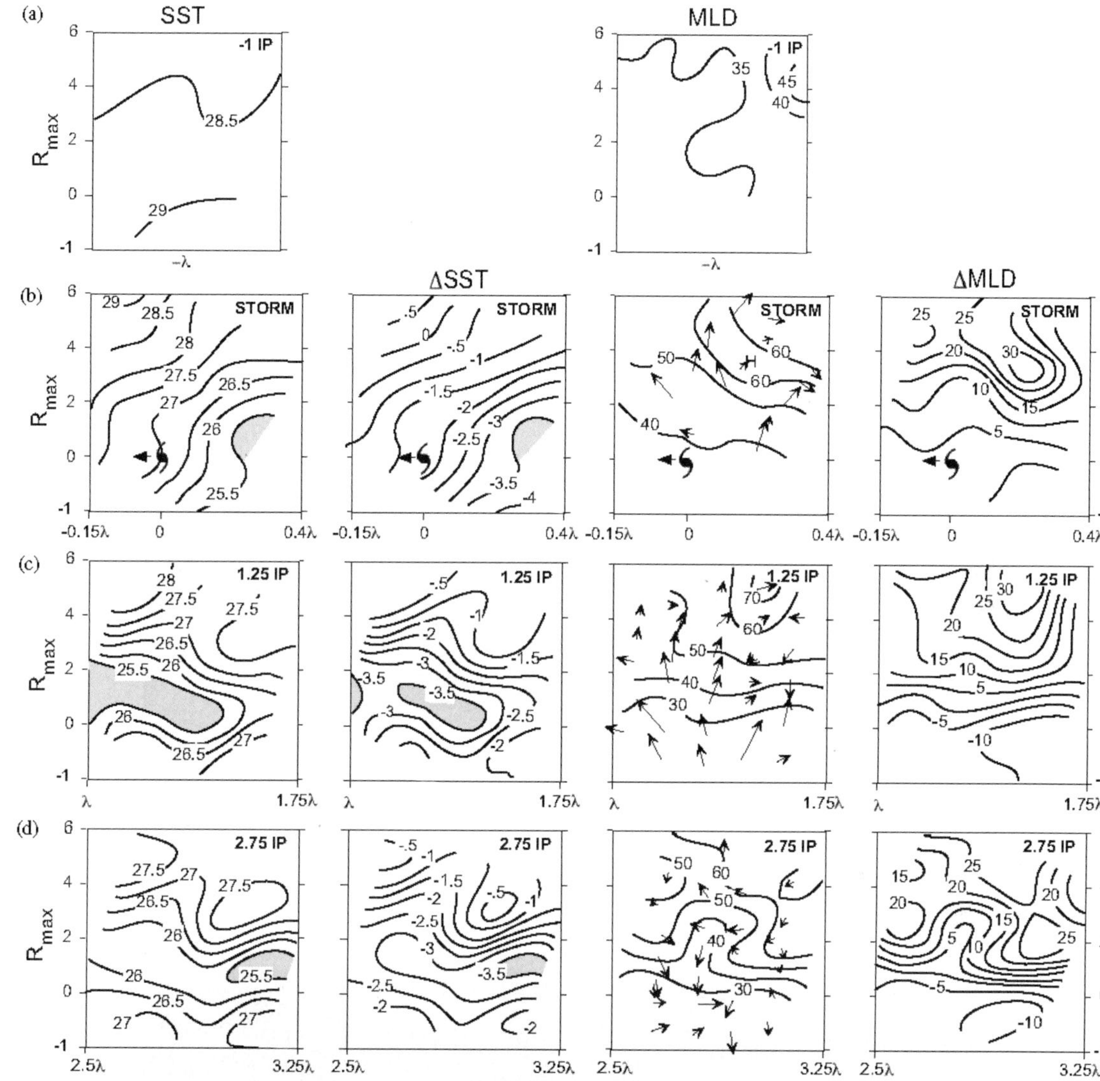

Figure 7. A sequence of analyses observed during TC Gilbert (1988) for SST (°C) and MLD (m) for (a) Pre (-IP: 15 Sept), (b) During (TC: 16 Sept), (c) Wake 1 (1.25 IP: 17 Sept), and (d) Wake 2 (2.75IP: 19 Sept) in a TC coordinate system where cross-track (Rmax = 60 km) and along-track is scaled in inertial wavelengths ($\Lambda = \text{IP} \times U_h$: 580 km). In the SST images, the maximum cooling is shaded in gray and the mean mixed layer currents (arrows) and MLD are contoured at 10 m intervals and ΔMLD is contoured at 5 m intervals. (Reproduced from Shay *et al.*, 1992 by permission of the American Geophysical Union.)

cooled by less than 1°C. Estimated surface fluxes contributed 10% to the heat balance between the track and 3 Rmax. In this quiescent ocean, cooling in the wake was dominated by entrainment heat flux induced by the current shear at the OML base by forced near-inertial currents (70 to 75%) starting just behind the eye. In front of the storm, the wind stress accounted for a similar fraction of cooling for the asymmetric, near-surface winds based on Powell and Houston (1996) approach. Shear and stress-induced mixing mechanisms accounted for a large fraction of the observed OML cooling but in different parts of experimental domain.

More recently, Walker *et al.* (2005) showed that the SST response was modulated by both WCR and CCR in the Gulf of Mexico during Ivan's passage (Fig. 8). Up to 7°C

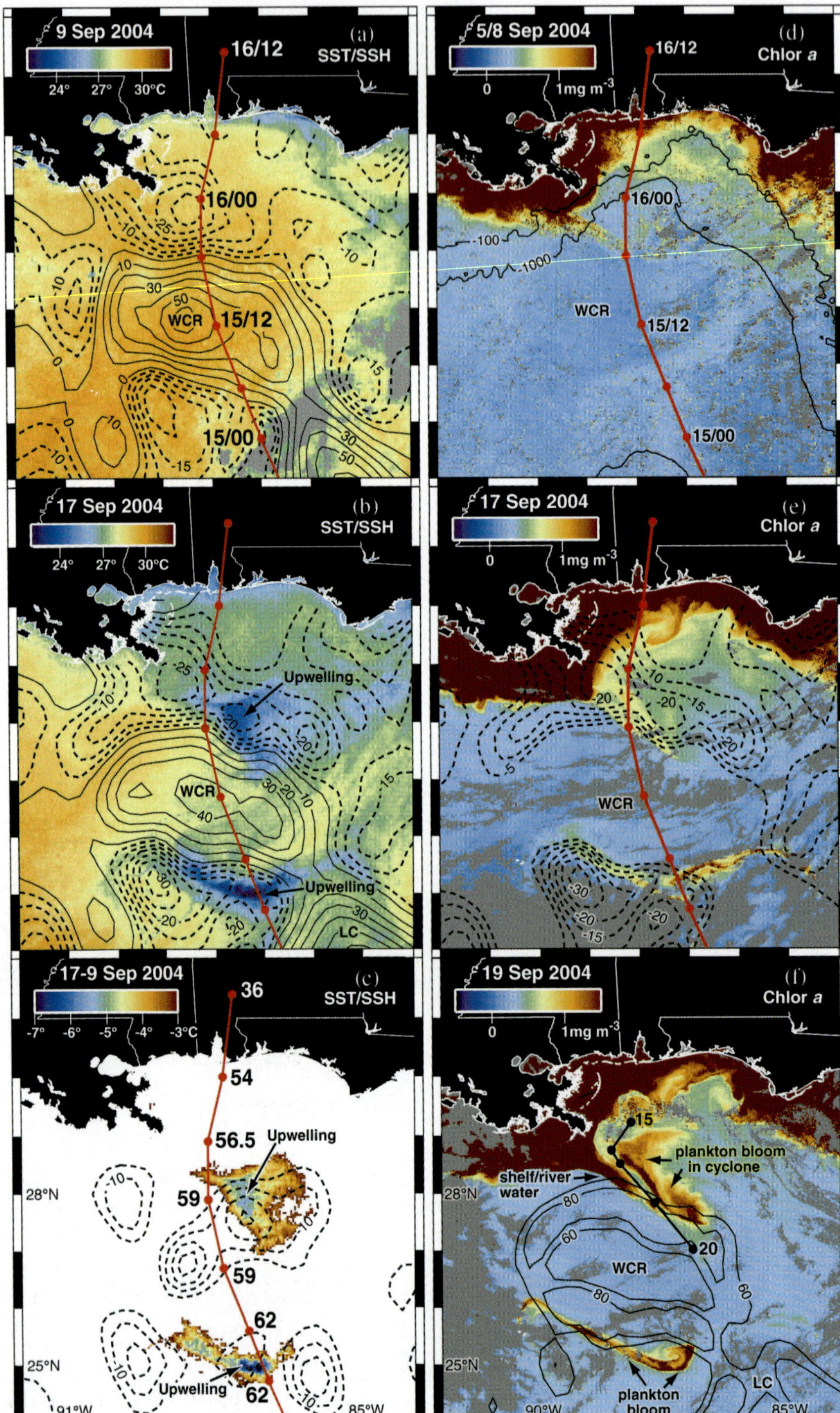

Figure 8. SSTs (color: °C) and SHA's (contours: cm) (a) pre (9 Sept) and (b) post Ivan (17 Sept.) and (c) SST and SHA differences where wind speeds from Ivan are shown in m s^{-1} and Chlorophyll a (mg m^{-3}: right panels) corresponding to similar days for (d) pre, (e) post and (f) 19 Sept. for corresponding isotachs derived from the SHA gradients. (Reproduced by combining Figs. 2 and 3 from Walker *et al.*, 2005 by permission of the American Geophysical Union.)

cooling was observed in the CCR due in part to upwelling associated with TC Ivan's wind stress curl. Upward isotherm displacements of up to 65 m were documented from altimeter-derived SHA and wind stress curl. This result was corroborated using a reduced-gravity model where for each 10 cm surface height change, the isotherm displacement is approximately 33 m. The observed 20 cm decrease in the SHA field may have been associated with the barotropic (sea-surface) trough in the cooling wake (Geisler 1970; Shay and Chang, 1997). While this depressed sea surface and the SST cooling provided *negative feedback* to Ivan, the associated phytoplankton blooms in the wake had peak concentrations of chlorophyll A (Chl-a) exceeding 1 mg m^{-3} distributed over 13,000 km^2 area detected from SeaWIFS imagery compared to background concentrations of less than 0.5 mg m^{-3}. This relative maximum occurred about 3 to 4 days following passage in the vicinity of the CCR. By contrast, the SSTs in the WCR only decreased by 1–2°C during Ivan, consistent with TC Opal, Isidore and Lili and Katrina and Rita in the GOM.

In the wake of TC Ketsana (Oct 2003) in the western Pacific ocean basin, SST imagery by the Advanced Microwave Sounding Radiometer (AMSR) and chlorophyll concentrations from the Moderate Resolution Imaging Spectroradiometer (MODIS) were combined to assess the oceanic response (Yin *et al.*, 2007). Based on post-storm measurements, Chl-a concentrations increased to 1.2 mg m^{-3} about 7 days after passage. This equates to more than an order of magnitude change induced by upwelling of nutrients from the cooler thermocline water. The maximum Chl-a level found was more than 3 mg m^{-3} in Ketsana's cold wake as this response persisted for more than two weeks following TC passage. Simulations using a Regional Ocean Modeling System with a Nitrogen-Phytoplanton-Zooplankton-Detritus (NPZD) biochemical model revealed levels of Chl-a of 0.4 to 1.7 mg m^{-3}, consistent with satellite estimates. The typhoon's wind stress curl induced significant upwelling into the OML and forced this large phytoplankton bloom. These results are in agreement with Walker *et al.* (2005) and Hanshaw *et al.* (2008) studies.

Park *et al.* (2005) combined ARGO float temperature and salinity profiles with satellite-derived SSTs to document the thermal response to typhoons in the western Pacific Ocean (WPAC) over ten-day periods (Fig. 9). They found a high correlation of OML temperature changes and SSTs, but there was a strong dependence on latitude and initial OML depth. A key result was a negative correlation between OML depth change and its initial depth. That is, the initial OML depths exceeding 50 m are considered a proxy for the Kuroshio's location and its WCR field. In the Kuroshio Current, the OML depth changes are less where the SST cooled by 1–2°C. This result has consequences for both the genesis and the development of the typhoon into a mature system over this regime. This result underscores the inherent oceanic variability in the vicinity of western boundary currents and their background eddy fields. That is, strong gradients in the OML depths and temperatures play a pivotal role in TC intensity changes discussed below.

3.2. *Ocean Current and Shear Response*

The ocean's momentum response is classified into two regimes: the directly-forced or near-field; and the evolving 3-dimensional wake or far-field (Shay, 2001). In the near-field, the cyclonically-rotating wind stress field of a TC causes OML currents of about 1 to 1.5 m s^{-1} to diverge from the storm track starting within one-quarter of an inertial wavelength (Λ) which is the product of the storm translation speed (U_h) and the local inertial period (IP) as noted above. This current divergence causes the upwelling of cooler water underneath the storm track, thereby decreasing the OML depth. Over the next half of the inertial cycle, OML currents converge towards the storm

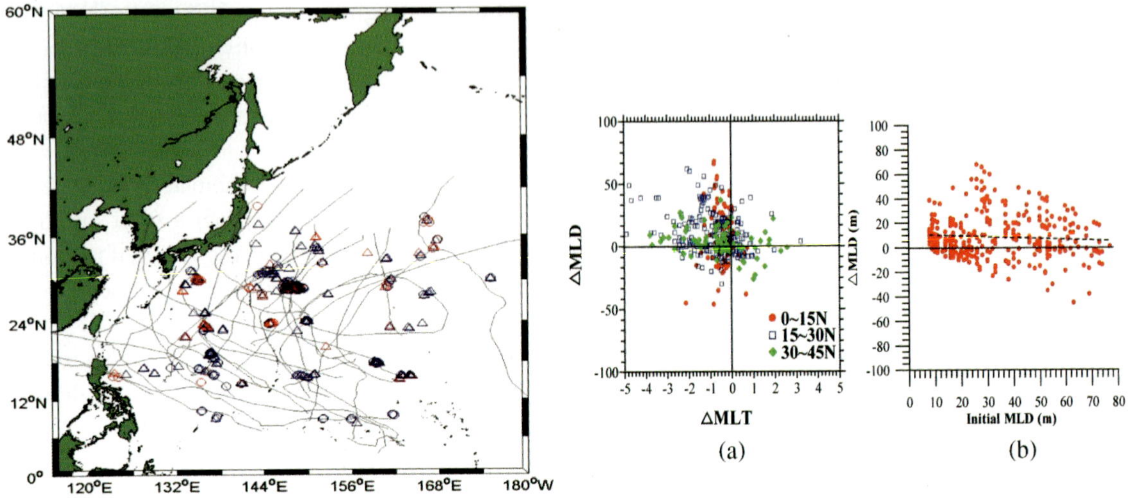

Figure 9. Distribution of typhoon tracks (left panel) relative to the ARGO float pairs in the Western Pacific Ocean from 1999–2003 and scatter (right panels) for (a) OML temperature change (ΔMLT: °C) and depth change (ΔMLD: m) binned by bands (red: 0–15°N; blue 15–30°N; green: 30–45°N) and (b) initial depth (MLD:m) and depth change (ΔMLD:m). (Reproduced from Park *et al.*, 2005 by permission of the IEEE.)

track, causing an increase in the OML depth as warmer water is downwelled into the thermocline. This alternating cycle of upwelling and downwelling of the isotherms (and isopycnals) occurs over distances of Λ and establishes horizontal pressure gradients that couple the OML to the thermocline as part of a spreading 3-dimensional wake (Price, 1983; Shay *et al.*, 1989, 1992; Zedler *et al.*, 2001; D'Asaro, 2003; Sanford *et al.*, 2007). In this context, the near-inertial response has been observed and modeled during TC passage. Within this framework, near-inertial motions are characterized by large horizontal and vertical wavelengths (Geisler, 1970) due to oscillations in the vertical current structure by low-order baroclinic modes (Gill, 1984; Shay *et al.*, 1989). This is often called the low-mode dominance. An important aspect of these forced motions is the frequency shift above the local f by about 1–20% (Price, 1983). This frequency shift is necessary for the near-inertial energy to vertically propagate from the surface-intensified OML into the thermocline. These near-inertial responses have been observed for several IP following storm passage (Price, 1981; Shay and Elsberry, 1987; Sanford *et al.*, 2007; Teague *et al.*, 2007) and have suggested a slight warming of the thermocline through entrainment mixing.

Observations of the ocean current response to TC passage have been generally sparse over the global oceans as the community has had to rely on fortuitous encounters with buoys and moorings deployed in support of other experiments or ships crossing TC wakes (Shay and Elsberry, 1987; Church *et al.*, 1986; Teague *et al.*, 2007, Zedler *et al.*, 2001). Using a network of high frequency radar stations (Yoshikawa *et al.* 2006), a strong wind-driven current response was observed during typhoon Mari's passage in the Tsushima Strait (Fig. 10). Surface currents have also been shown to impact the direction of the surface wind stress (Drennan and Shay, 2006). Notice the divergent surface velocity response exceeded 50 cm s^{-1}, which is indicative of a 3-D response (Yablonsky and Ginis, 2009; Halliwell *et al.*, 2010). As suggested in Fig. 1, the wind-forced currents directly affect the thermal, haline and mass field response to TC passage. Recent dedicated airborne field experiments to measure the momentum and thermal response during TC passage are listed in Table 1.

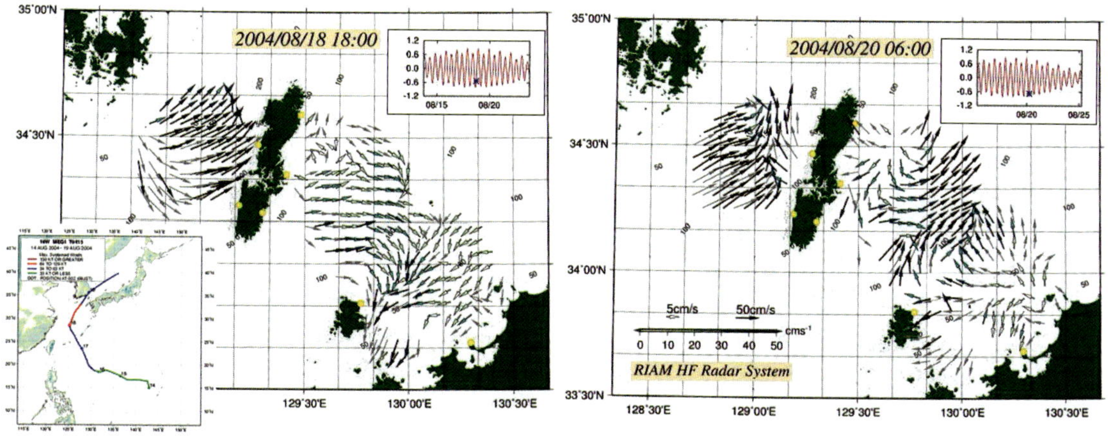

Figure 10. Surface currents observed before (left panel) and after (right panel) severe typhoon Mari (inset) in 2004 in the western Pacific Ocean relative to the tidal currents (upper right inset). (Image courtesy of Dr. Y. Yoshikawa.)

Table 1. Summary of aircraft field programs where currents and shears were observed during TC passage where EPAC-Eastern Pacific Ocean, WATL-Western Atlantic Ocean, GOM-Gulf of Mexico, AXCP-airborne expendable current profiler, AXCTD- airborne expendable conductivity temperature and depth profiler, EMAPEX-ElectroMagnetic Autonomous Profiling Experimental Floats.

TC	Basin	Year	Instrument(s)	References
Norbert	EPAC	1984	AXCP	Sanford *et al.* (1987)
Josephine	WATL	1984	AXCP	Sanford *et al.* (1987)
Gloria	WATL	1985	AXCP Drifters	Black *et al.* (1988); Price *et al.* (1994)
Gilbert	GOM	1988	AXCP	Shay *et al.* (1992)
Dennis	WATL	1998	Floats	D'Asaro (2003)
Isidore/Lili	GOM	2002	AXCP AXCTD	Shay and Uhlhorn (2008)
Frances	WATL	2004	EMAPEX floats	Sanford *et al.* (2007)
Katrina/Ike	GOM	2005	AXCTD AXCP	Jaimes and Shay (2009)
Gustav	GOM	2008	EMAPEX floats	

One clear example of a fortuitous encounter occurred during TC Ivan (2004) over an array of moored Acoustic Doppler Current Profilers (ADCP) as shown in Fig. 11. Fourteen moorings were deployed about 180 km south of Mobile, Alabama as part of the Slope to Shelf Energetics and Exchange Dynamics (SEED) project from early May through early Nov 2004 (Fig. 11). During the period of deployment TC Ivan passed directly over these ADCP moorings (Teague *et al.*, 2006, 2007), which provided the temporal evolution of the 3-D current (and shear) structure at 2–4-m vertical resolution. The response at Ivan moorings (along track and at 1.5 Rmax) is shown in Fig. 12 for the near-inertial wave band (band-pass filter of the detided current signals) where the local IP is ∼24 h at 30°N. Profiler data started at a depth of 50 m and extended to 500 m at 4-m intervals. These data were filtered between 22–28 hours then vertically averaged to estimate the depth-averaged current response. Shay (2009) found amplitudes of 4–6 $\mathrm{cm\,s^{-1}}$ in agreement with those determined from limited vertical sampling from the Hurricane Frederic current meter arrays

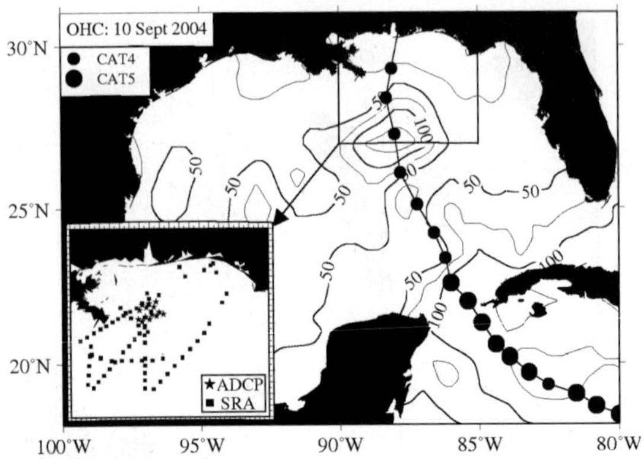

Figure 11. OHC map (contours are kJ cm^{-2}) and inset showing NRL mooring locations (stars) and scanning radar altimeter directional wave measurements (Wright *et al.*, 2001: squares) relative to Ivan's track and intensity. The OHC pattern shows the WCR encountered by Ivan prior to landfall. The cooler shelf water (OHC < 20 KJ cm^{-2}) resulted from TC Frances two weeks earlier.

(Shay and Chang, 1997). These barotropic current signals have large phase speeds and are associated with Poincaré wave motions (e.g., long waves with rotation).

Near-inertial baroclinic flows and the levels of maximum vertical shears in the momentum response to TC Ivan were energetic (Teague *et al.*, 2007). Based on profiles at 1 to 2 Rmax to the right of Ivan's track, the current profiles show clear downward propagation of the baroclinic energy ($\mathbf{V}_g$) from the wind-forced OML into the thermocline, consistent with an upward propagating phase velocity ($\mathbf{V}_p$). Baroclinic motions, which are the most energetic part of the ocean's momentum response to a moving TC, have a characteristic time scale for the phase of each mode to separate from the OML current structure when the wind stress scale ($2R_{max}$) exceeds the deformation radius of the first baroclinic mode (~30–40 km). This time scale increases with mode number due to decreasing phase speeds (Shay, 2001). This vertical propagation is primarily associated with the predominance of the anticyclonic-rotating (clockwise in northern hemisphere) energy (Leaman, 1976; Shay and Jacob, 2006). The corresponding vertical velocity signal

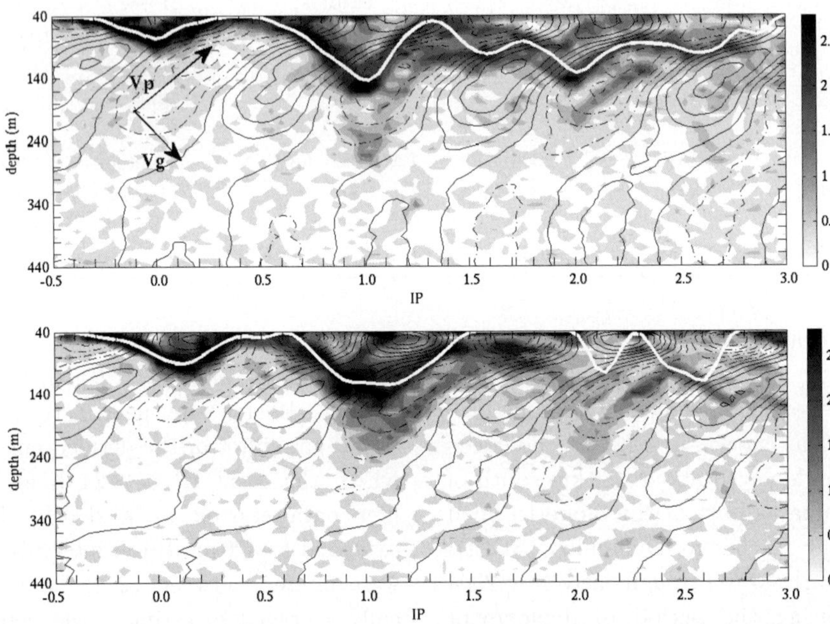

Figure 12. Band-pass filtered cross-track current structure along the track (upper panel) and at $1.5R_{max}$ (lower panel) contoured at 5 cm s^{-1} with $t = 0$ the time of TC Ivan's passage and the time axis normalized by the inertial period (~1 day). Heavy solid line represents the depth of the magnitude of the maximum shear (gray shade × 10^2 s^{-1}). Arrows represent upward phase ($\mathbf{V}_p$) with downward group velocity ($\mathbf{V}_g$) indicative of downward propagating waves into the thermocline. Time interval spans 15–19 Sept (data courtesy of Naval Research Laboratory).

in the upper ocean induced by Ivan was approximately 1 cm s^{-1}.

For forced near-inertial motions, the currents rotate anticyclonic with depth in the northern hemisphere in time as energy propagates downward into the thermocline while the phase propagates upward (Leaman, 1976). This has been observed in current profilers in Gilbert (Shay *et al.*, 1998), moored ADCP measurements in Ivan (Teague *et al.*, 2007), and the new and innovative EM/APEX floats deployed in Frances during CBLAST (Sanford *et al.*, 2005, 2007). From the Gilbert current profiles, 63 profilers revealed a predominance of the anticyclonic-rotating energy where the average ratio of the anticyclonic to cyclonic energies was 3.6, indicative of a preference for downward energy propagation from the OML. The corresponding vertical energy flux, determined by integrating the product of the excess anticyclonic energy and the vertical group velocity, indicated an average value of ~2 ergs cm^{-2} s^{-1} (Shay and Jacob, 2006). This energy flux is similar to that estimated from near-inertial current profilers deployed in a frontal regime by Kunze and Sanford (1985). This result suggests that TCs are as effective as fronts in transferring wind-forced energy into the thermocline. From a global perspective, Nilsson (1995) estimated the near-inertial wave power for TC's translating at 7 m s^{-1} to be $\sim 2.6 \times 10^{10}$ W. If this rate of 2 ergs cm^{-2} s^{-1} is used, the near-inertial wave power is estimated to be about 7.4×10^{11} W, or roughly 30 times this theoretical estimate. Similar estimates of wave power have been found in the conversion of barotropic to internal tides of 10^{12} W (Bell, 1975; Sjoberg and Stigebrandt, 1992). Thus, the estimated energy flux from TC force winds represents a realistic value based on these comparisons (Shay and Jacob, 2006).

As noted in section 2, near-inertial current shears are central to the oceanic cooling by lowering the bulk (gradient) Richardson numbers to <1 (0.25) and forcing the OML to deepen through instabilities. Uhlhorn (2008) documented the magnitude of the observed current shears range from several experiments listed in Table 1. A value of $1.4 \times 10^{-2}\,s^{-1}$ was found from TC Lili measurements in the LC to as high as $3.5 \times 10^{-2}\,s^{-1}$ in the GCW during TC Gilbert. In the regime influenced by TC Ivan with relatively shallow OML, the shear is nearly three times larger than that observed in the LC system (Shay and Uhlhorn, 2008). Higher vertical shear levels were observed from the EM/APEX floats deployed in TC Frances (Sanford *et al.*, 2005; 2007). As shown in Fig. 13, these higher shear levels forced mixing events during Frances passage by essentially lowering the gradient Richardson number to below 0.25 ($S^2 - 4N^2 > 0$) across the base of the OML at 1–2Rmax to the right of the track. During this period of intense forcing, strong near-inertial currents were excited by the TC of up to 1.5 m s^{-1} which created the vigorous shears (Sanford *et al.*, 2007). Cooling level in TC Frances was ≈2.3°C based on the combination of floats and drifters deployed in the CBLAST experiment (Black *et al.*, 2007).

One of the more interesting aspects found in the velocity shear measurements such as TC Gilbert data was that the current shears remained large over the approximate three IPs following passage in the cold wake (Shay *et al.*, 1992). However, after the first 1.5 IPs, the Richardson numbers remained above critical levels. This suggests the OML reached an equilibrium point and vertical mixing ceased (Jacob *et al.*, 2000). By contrast, vertical mixing ceased after about 0.5 IP following TC Lili, suggesting that only the near-surface isothermal layer cooled by about 1°C as the Richardson numbers were subcritical for only a short period. This result is consistent with idealized simulations with a cross-track initial flow resembling a LC where background currents were about 0.8 m s^{-1} (Uhlhorn, 2008). This is the rationale of why ocean models need realistic data assimilation to get the background oceanic conditions as accurately as possible prior to coupling an ocean to a TC model as shown above.

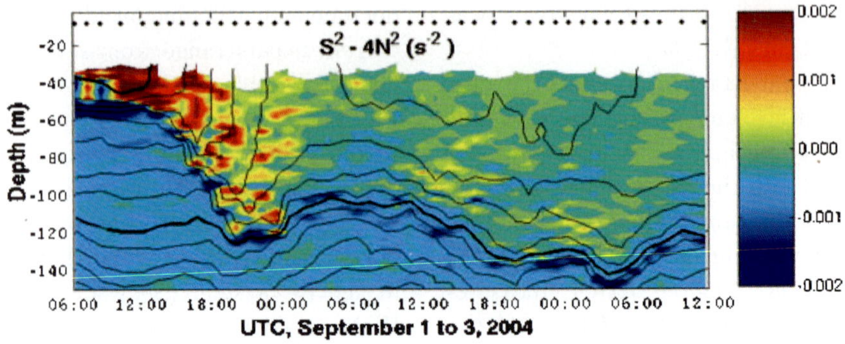

Figure 13. Reduced shear (color: s^{-2}) based on the gradient Richardson (Ri $= N^2\ S^{-2}$ where N is the buoyancy frequency and S is the shear) and the temperature (contours: °C) contoured at 0.5°C intervals number where the heavy black contours represent 29 and 25°C. The shear instability criterion is simply rewritten as $S^2 - 4N^2 > 0$ to represent mixing events (values greater than 0) in yellow and red. Blue colors represent stable conditions. Analysis is based on data from one EM/APEX float (Sanford *et al.*, 2005; 2007). The time of closest approach of TC Frances at 1700 UTC 1 Sept 2004 and black dots at the top of the panel represent individual profiles from the float. (Reproduced from Sanford *et al.*, 2007 by permission of the American Geophysical Union.)

3.3. *Integrated Thermal Structure and Layer Depths*

The thermal structure is clearly affected by the wind-forced momentum response during TC passage. However, the level of ocean cooling depends crucially on the initial OML and the 26°C isotherm depth and the strength of the stratification (N) across the base of the OML. In many basins, the 26°C isotherm depth is located near the OML base (see Fig. 2). The integrated thermal energy reflects the vertical distribution of the thermal energy in the OHC estimation. Since the amount of work done on the OML scales as the surface friction velocity (u_*^3) (Kraus and Turner, 1967), the deeper the layer (e.g., 26°C isotherm), the higher the stress-induced turbulent mixing is needed to redistribute OML properties through shear-induced instabilities. In this framework, high OHC ($100\,\mathrm{kJ\,cm^{-2}}$) values, strong buoyancy frequencies ($N \sim 20$ cph), and lower latitudes (10 to 20°) affect the vertical distribution of the thermal structure and that decrease the negative feedback during TC passage (Chang and Anthes, 1978).

It is difficult to monitor the global oceans for isotherm depths, thermal structure and OHC given the relative paucity of *in situ* profiler measurements with perhaps the exception of the global ARGO float network. Thus, satellite remote sensing offers the optimal approach to infer isotherm depths and OHC variations. Measurements from radar altimeter missions (Fig. 14) of the SHA field from NASA TOPEX, Jason-1 and 2, U.S. Navy Geosat Follow-On-Mission (GFO), Envisat and ERS-2 (Cheney *et al.*, 1994; Scharroo *et al.*, 2005) and SSTs are used in a reduced gravity model (e.g., Goni

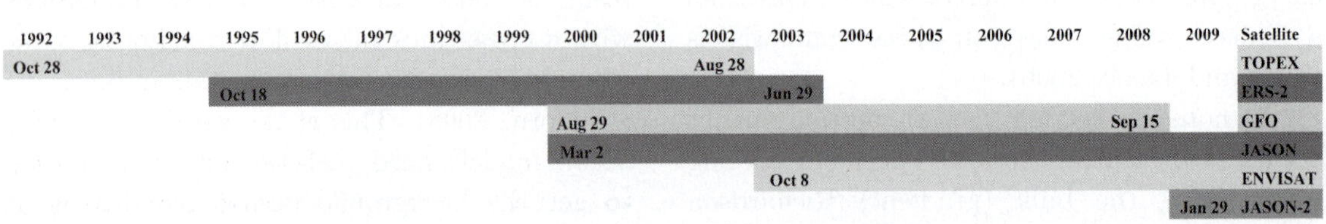

Figure 14. Bar chart showing time line of available SHA field from various altimetry platforms.

et al., 1996) with hurricane season climatology (Mainelli-Huber, 2000). Since mesoscale ocean features move only a few km d^{-1}, altimeter-derived SHA locates warm (cold) features that are usually identified as positive (negative) values as observed during TC's Opal (Shay *et al.*, 2000), Ivan (Walker *et al.*, 2005; Halliwell *et al.*, 2008), Katrina and Rita in the Gulf of Mexico (Mainelli *et al.*, 2008; Shay, 2009), Maemi (Lin *et al.*, 2005, Wu *et al.*, 2007), Chaba and Songda (Wada and Usui, 2007) in the western Pacific Ocean basin, and cyclones in the Bay of Bengal (Jena *et al.* 2006; Ali *et al.*, 2007).

In the western Pacific Ocean, Lin *et al.* (2005) analyzed remote sensing imagery prior and subsequent to the passage of TC Maemi in 2003 (Fig. 15). The intensity of Maemi increased by 36 m s^{-1} over an eddy-rich oceanic regime where SSTs exceeded 29.5°C. Based on results from the Coupled Hurricane Intensity Prediction System (Emanuel, 2003), the WCRs acted as an insulator between TCs and the deeper thermocline water (Wu *et al.*, 2007). The SST response was significantly less as the OML is much deeper in these regimes similar to findings in other basins. Without initializing this model with a WCR ocean feature, the simulated intensity was a category below the observed intensity, in accord with coupled

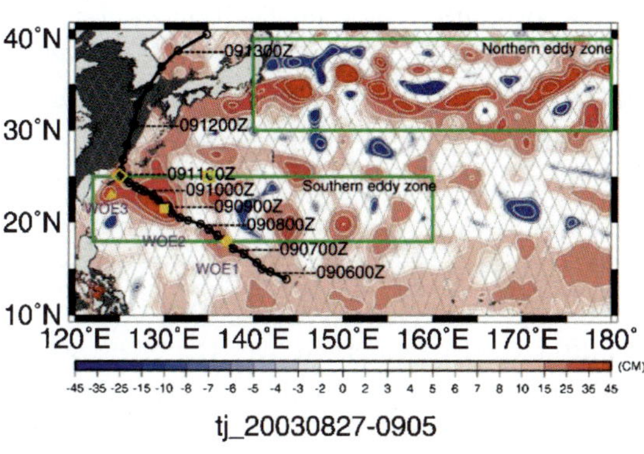

Figure 15. TOPEX-derived SHA field (cm) between 27 Aug and 5 Sept 2003 prior to TC Maemi on 6 Sept relative to the southern and northern eddy zones in the WPAC. (Reproduced from Lin *et al.*, 2005 by permission of the American Meteorological Society.)

results of Hong *et al.* (2000). Lin *et al.* (2005) point out the importance of the eddy-rich regime associated with the western boundary current or Kuroshio during TC passages. In the WPAC, the Kuroshio is a western boundary current (Stommel, 1948) that plays the same role as the Gulf Stream in the WATL by transporting heat poleward as part of an annual cycle in the gyre.

More recently, Lin *et al.* (2009) argue that there needs to be translational speed dependence on the available OHC and the 26°C isotherm depth for TCs to reach category-5 status. Based on a decade of temperature structure data from ARGO floats (Lin *et al.*, 2008), specific isotherm depths ranges and OHC values were determined for differing TC speeds. However, the resultant empirical values were based on storm speeds alone. That is, higher OHC values are needed for slower moving TCs to offset the upwelling of the isotherms compared to faster moving TCs where OHC values could be about half the values for the slower moving TC cases. One aspect of the problem that was missing in this approach was the *time available for mixing* which is related to a storm Rossby number. In addition to the storm translational speed, this revised approach would have also included latitude to estimate the local Coriolis parameter (f) and storm scale associated with the wind stress curl. Greatbatch (1983) showed the importance of these parameters on the ocean response.

In the Arabian Sea (e.g, Northern Indian Ocean) beginning in May, a cyclonically rotating atmospheric vortex forms with similar characteristics to the TC in the Atlantic Ocean and the Gulf of Mexico basins (Rao, 1984). This low-pressure center with cyclonically-rotating winds moves towards the northwest due to the migration of the InterTropical Convergence Zone (ITCZ) (Findlater, 1969). The timing of this northward movement of the ITCZ is modulated by the Madden Julian Oscillation causing a few week lag in Southwest Monsoon onset. A low-level jet is formed (e.g., Findlater Jet) due to the convergence of the ITCZ and the persistent

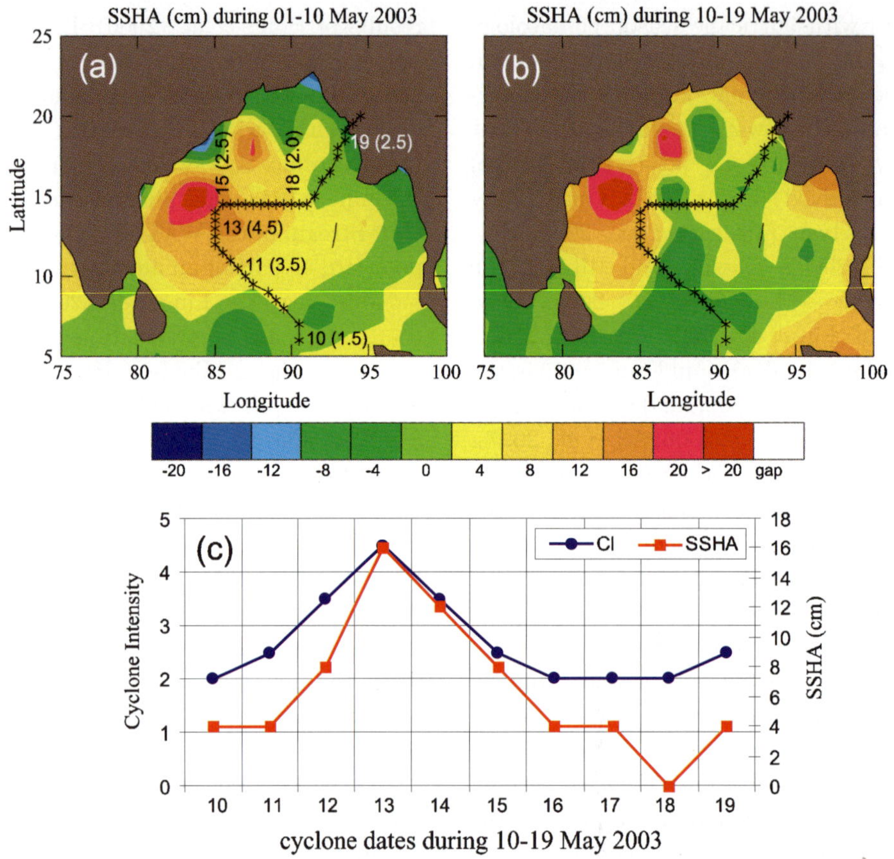

Figure 16. Composite topography from several altimetry-derived sea-SHA field (cm) from (a) 1–10 May and (b) 10–19 May 2003 relative to the TC track and (c) the along-track time series variation of the SSHA and TC index (CI) for the Bay of Bengal cyclone. (Reproduced from Ali *et al.*, 2007 by permission of the American Geophysical Union.)

northerlies over the northern portion of the Arabian Sea. In May 2003, Ali *et al.* (2007) found that this onset atmospheric vortex had characteristics similar to a TC (Fig. 16). The maximum intensity (Category-4 index) of the TC occurred over an oceanic regime where the SHA exceeded 20 cm (deeper OML) where the corresponding SST exceeded 30.5°C. The SHA field associated with this WCR is less compared to those in the GOM where SHA are 40–50 cm in surface deflection relative to mean sea surface level (Leben, 2005).

An illustration of this impact was observed during the passages of Katrina and Rita as they both deepened to a Cat-5 TC over a bulging, lobe-like structure along the LC's western flank (Scharroo *et al.*, 2005; Shay, 2009; Jaimes and Shay, 2009). As shown in Fig. 17, there is a one-to-one correlation between TC intensity (surface pressure decreases) and OHC values exceeding 100 kJ cm^{-2} in the LC. By contrast, background SSTs of more than 30°C were nearly uniformly distributed in this regime, and did not reveal the complex LC and bulging WCR structure in the central Gulf of Mexico (Sun *et al.*, 2006; Shay, 2009). Surface fluxes increased over the LC after Katrina emerged over the GOM where OHC values relative to 26°C water exceeded 100 kJ cm^{-2}. This OHC level is more than five times the threshold of 17 kJ cm^{-2} d^{-1} integrated over the TC scale (Leipper and Volgenau, 1972). Normalized OHC values vary inversely to pressure changes (surface pressure decreases, OHC increases), whereas the normalized SST

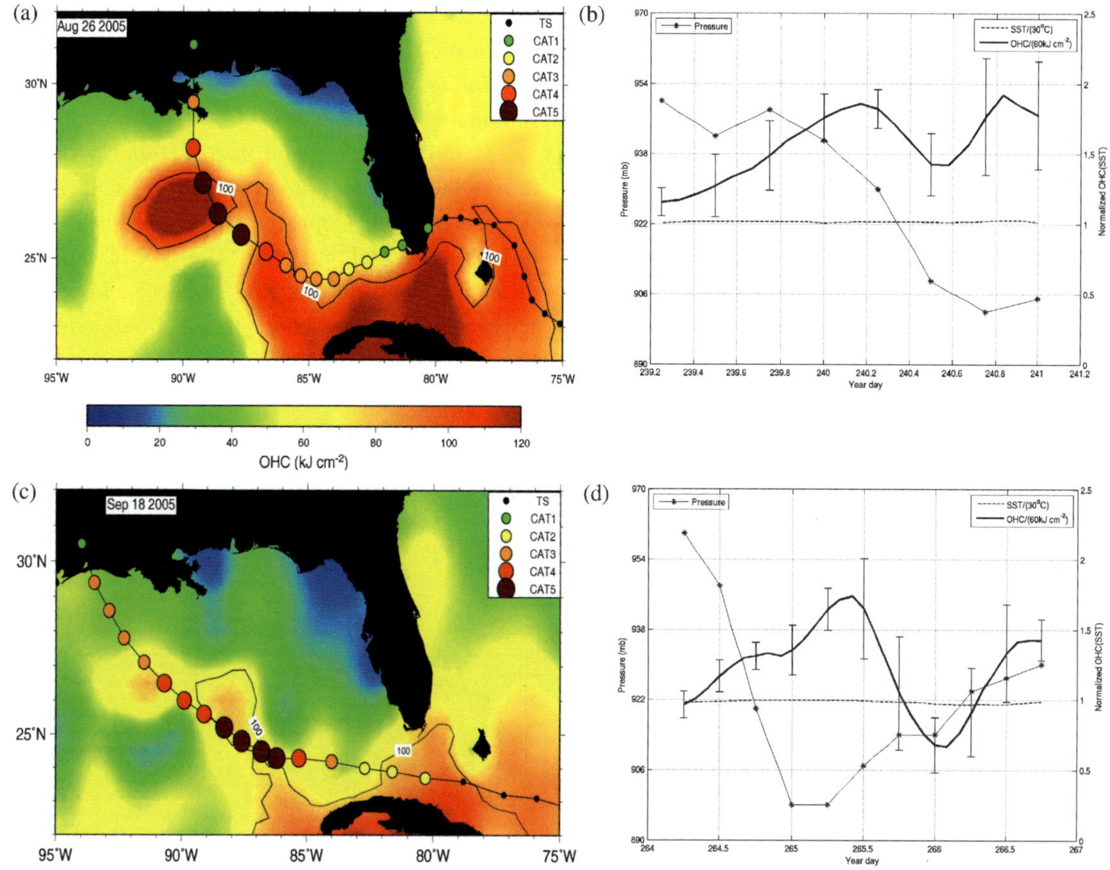

Figure 17. Left panels (a,c) represent pre-storm OHC (kJ cm^{-2}:color) and 26°C isotherm depth (m: black contour) based on a hurricane season climatology, Reynolds SSTs, Jason-1 and GFO radar altimetry measurements relative to the track and intensity of TC's (a) Katrina and (c) Rita. Right panels (b,d) represent time series of surface pressure (thin black) versus along-track SST (dashed) and OHC (thick black) variations normalized by 30°C and 60 kJ cm^{-2}, respectively. OHC uncertainties are based on 6-hourly averages in the cross-track direction. (Reproduced from Shay, 2009 by permission of Elsevier Press International.)

values are flat during both TC's. Mainelli *et al.* (2008) input OHC into the SHIPS (DeMaria *et al.*, 2005) and found that OHC is a better indicator of TC strength than just SST alone which is similar to when TC Opal (1995) encountered a WCR and deepened to a category-4 TC during favorable atmospheric conditions (Bosart *et al.*, 2000; Shay *et al.*, 2000; Hong *et al.*, 2000). Clearly, In the WPAC, Indian Ocean and GOM ocean basins, these cases are all examples of *less negative feedback* since the upper ocean cooled by less 2°C to ∼28°C in regimes of deep warm layers. Thus, the depth of the 26°C isotherm, usually located close to the depth of the OML, and the resultant OHC is relevant to TC intensity changes.

3.4. *Evaluation of OHC Estimates*

Given the availability of the satellite-derived values of OHC derived from radar altimetry, it is very important to assess them relative to *in situ* data profiles. On 15 (Post-Katrina) and 26 Sept 2005 (Post Rita), oceanic current, temperature and salinity measurements were acquired from expendables centered on the LC/WCR during IFEX (Rogers *et al.*, 2006; Jaimes and Shay, 2009). The post-Katrina experiment was conducted in an area where the TC rapidly intensified over the LC/WCR complex. To obtain the evolution of the thermal structure and OHC within the same WCR, two thermistor chain drifters were deployed from the USAF WC-130 aircraft in the path of TC Rita on 21 Sept 2005 following a similar approach used in the CBLAST experiment (Black *et al.*, 2007). Upper ocean temperature time series (Fig. 18) were obtained from two drifters closest to Rita's center (distance in upper panels) while circulating around the WCR. The two drifters were about 80 to 100 km from the WCR center where OHC values were 120 kJ cm^{-2} as the 26°C isotherm depths ranged between 90–110 m. These values were consistent with altimeter-derived OHC values (Jaimes and Shay, 2009). Rita clipped the northeastern part of the WCR as the storm was weakening prior to landfall on the Texas-Louisiana border as a Cat-3 hurricane.

Satellite-inferred and *in situ isotherm* depths and OHC values were correlated at levels of 0.9. Regression slope for the OHC is 0.9 with a bias of 1.3 kJ cm^{-2} in the WCR. For the 26°C isotherm depths, the slope was about 1.1 with a 9.3 m bias where the altimeter-derived value was larger than that from the profiler data. This larger bias was due to the advection of the CCR between the LC and WCR from the post-Rita data set (Jaimes and Shay, 2009). These estimates were also consistent with those derived

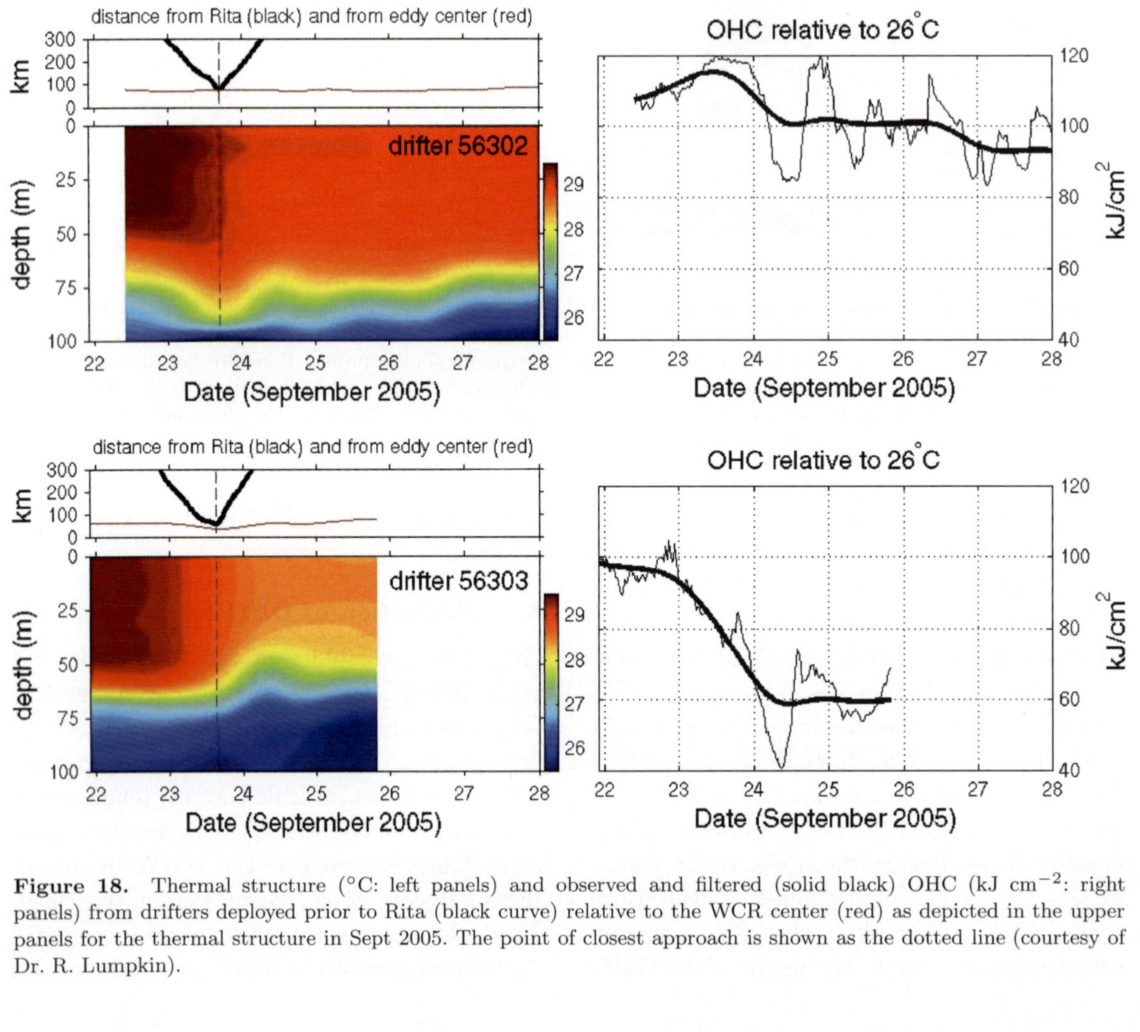

Figure 18. Thermal structure (°C: left panels) and observed and filtered (solid black) OHC (kJ cm^{-2}: right panels) from drifters deployed prior to Rita (black curve) relative to the WCR center (red) as depicted in the upper panels for the thermal structure in Sept 2005. The point of closest approach is shown as the dotted line (courtesy of Dr. R. Lumpkin).

from drifter-based measurements. While the bias in the depth is large, the result suggests this is roughly a 10–15% uncertainty in the signals where isotherm depths ranged from 90–105 m in the WCR. Several sets of profiler measurements have suggested that the OHC scales as ~1 kJ cm^{-2} m^{-1} in the LC and WCR structures.

To further illustrate the validity of this OHC approach from altimetry under non-forcing conditions, satellite-derived values were compared to *in situ* measurements from XBT transects in the eastern Pacific Ocean basin (Shay and Brewster, 2010). Six years of data along a repeated transect is used to determine OHC and compare it to the five-year average from satellite-inferred values along the same transect as well as the closest moorings (Fig. 19). There is marked agreement between the XBT, mooring and satellite-derived OHC values. In addition, there are no significant differences between the two *in situ* and the remotely sensed values at 95% confidence. Over a broader spatial scale from 2000 to 2008 during the EPAC basin hurricane season, OHC value statistics from 6,420 *in-situ* data points revealed RMS OHC differences of 13 to 20 kJ cm^{-2} or up to 15% of the maximum values. The slope of regression line for OHC values is 0.9 with an RMS difference of 17 kJ cm^{-2} where the dynamic range lies between 113 to 190 kJ cm^{-2} (not shown). This latter value is a maximum rather than an average. Thus, estimating the 26°C isotherm depth using satellite altimetry in a reduced gravity model (Goni *et al.*, 1996) allows one to determine OHC for use with SHIPS (DeMaria *et al.*, 2005; Mainelli *et al.*, 2008).

4. Air-Sea Interface

The effect of the underlying ocean has drawn attention towards gaining a better understanding

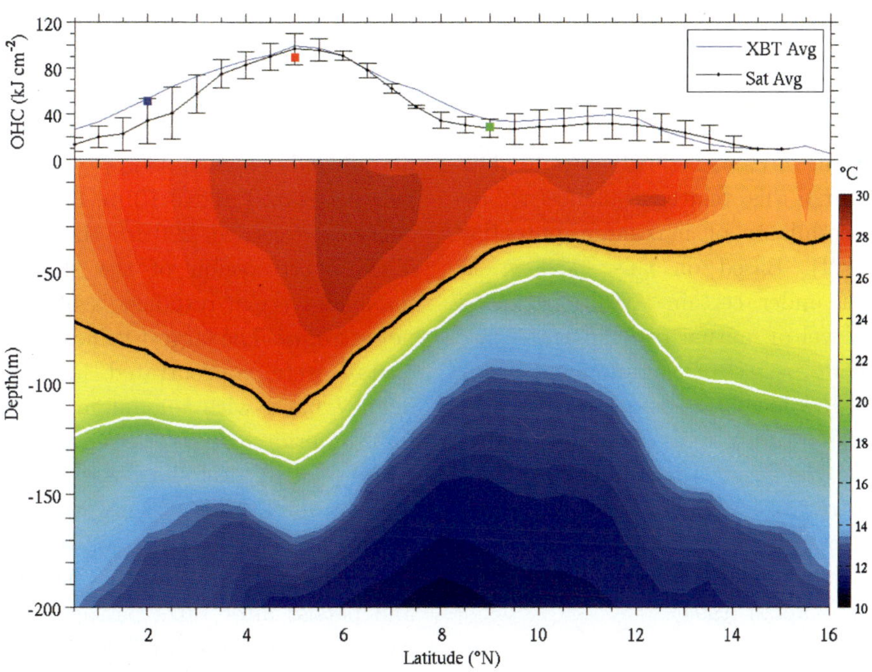

Figure 19. Average OHC (upper panel: kJ cm^{-2}) from repeat XBT transects in the eastern Pacific Ocean basin (blue line), TAO moorings at 2, 5 and 8°N (140°W: colored boxes) and the corresponding satellite-derived values (black line) with $\pm 2\sigma$ and the corresponding vertical temperature structure from the XBTs (lower panel) where the depth of the 20°C isotherm (white line) and the 26°C (black line). Data are averaged from the months of July from 2000 to 2005. (Reproduced from Shay and Brewster, 2010 by permission of the American Meteorological Society.)

of the physical interaction between the atmosphere and ocean during these events. Due to limited observations at the air-sea interface in high-wind conditions, the understanding has not progressed nearly enough to significantly improve the parameterization of momentum and energy transfers between the two fluids. The relationships of the transfer processes of small-scale roughness (Charnock, 1955) and surface-layer stability (Monin-Obukhov similarity theory) are understood under low- to moderate-wind conditions (Large and Pond, 1981), but additional phenomena not typically observed such as the sea state maturity (Donelan, 1990; Moon *et al.*, 2004a,b) and sea spray (Wang *et al.*, 2001; Andreas and Emanuel, 2001) have been shown to modulate the heat and momentum exchange. These effects under TC winds have been primarily studied in controlled laboratory experiments (Donelan *et al.*, 2004). In a TC environment, both young and mature waves are present and impact air-sea fluxes and OML and atmospheric boundary layer processes.

Intensity is maintained in part by the balance between the heat gained by the atmospheric boundary layer and the energy lost due to friction. Emanuel (1986) proposed a theory requiring this relative balance to be a *primary* modulator of maximum potential intensity (MPI). Based on this view, it is assumed that under certain conditions there should be a level of mutual dependence of the air-sea transfer processes of heat and momentum as suggested in idealized model simulations (Emanuel, 1995; Braun and Tao, 2000). This has led to the conclusion that intensity is sensitive to the ratio of enthalpy to drag coefficient ($c_k c_d^{-1}$), which lies within a limited range. However, most TCs do not reach their MPI (DeMaria and Kaplan, 1994).

4.1. *Surface Wave Field*

In 1993, TC Emily approached NOAA buoy 44014 from the southeast with sustained winds of $28\,\mathrm{m\,s^{-1}}$, the significant wave heights (H_s) reached 8 m (Fig. 20) where maximum wave spectral energies were contained in the swell portion of the spectrum (i.e., ~13 s wave) that decayed rapidly after TC Emily's passage. These wave spectral energies contained in the lower-frequency intervals associated with the swell began several hours in advance of storm passage, peaking at the point of closest approach, and decayed over 1 to 2 IPs. In the wake of a TC, the barotropic trough in the surface height field contains IP fluctuations of 4 to 7 cm due to the vertically integrated mean mass divergence field (Shay and Chang, 1997). By contrast, in baroclinic models (e.g., rigid lid), these IP oscillations do not exist as the vertical velocity is zero at the free surface as per the kinematical boundary condition. In frequency bands between 0.2–0.4 Hz, IP undulations in the surface height affect wave spectral energies for an extended period of time (Faber *et al.*, 1997) as suggested by the peaks in both wave spectral energies and significant wave heights. More recently, TC Lili approached buoy 42001 from a south-southwest direction just after reaching maximum intensity (Cat-4). As Lili passed within ~Rmax to the west of the buoy, winds reached $48\,\mathrm{m\,s^{-1}}$, significant wave heights peaked above 10 m, and wave spectral energies exceeded $220\,\mathrm{m^2\ Hz^{-1}}$ (Fig. 20(b)). This is the region of the storm where the maximum ocean response is often observed as noted above. The largest values of wave spectral energy were concentrated in the lower frequency band with maximum values at 0.4 IP (~10 h) prior to passage and persist for about 1.5 IP (~40 h) after the closest approach of Lili. Smaller amplitudes of the wave spectral energy in the higher frequency (0.2–0.4 Hz) intervals are evident 1.7 IP (~48 h) prior to passage, and persist after Lili's passage over the buoy. The phase of these oscillations, which is most pronounced between the frequency intervals 0.2–0.4 Hz, is near the local IP (~27 h).

Wang *et al.* (2005) documented the wave response to Ivan over the NRL SEED moorings (Fig. 21). Wave heights increased with peak

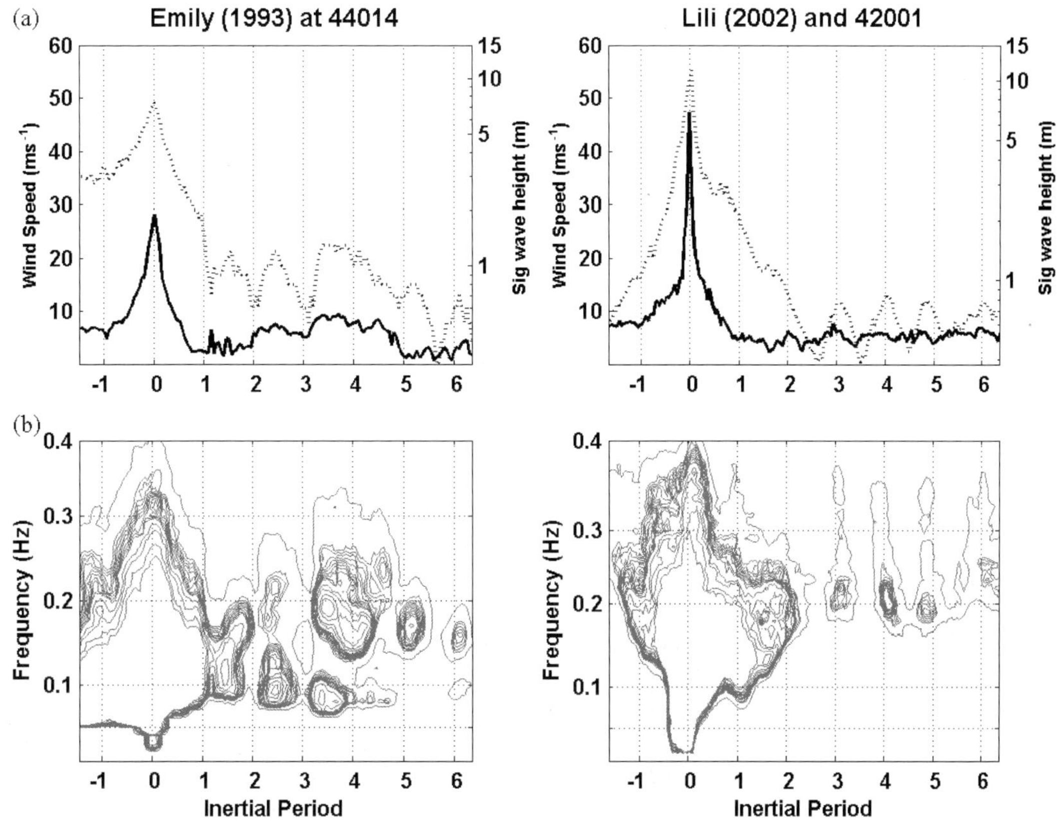

Figure 20. Time series of (a) wind speed (solid) and significant wave height (H_s: dotted), and (b) wave spectral energy (0.1 to 1 m^2 Hz^{-1}) from buoys 44014 and 42001 during and after Emily (left) and Lili (right). Time is scaled by local IP ~20 hrs for NOAA buoys 44014 and ~27 hrs for 42001.

values when the radial distance between the mooring and storm center was ~75 km (Fig. 21(c)). Hs reached maximum values of 16 m to 18 m on inshore moorings and were larger than those detected at the NDBC buoy (15.9 m). The maximum wave height was recorded to be 27.7 m at mooring 3, and wave height variations were consistent with the radial variations in the surface wind of Ivan. At Rmax, the model predicted a maximum wave height of ~21 m. Previous studies have suggested in a TC wave field that the maximum wave height approaches 1.9 × Hs, which would be consistent with these measurements. However, the moored measurements sampled only a small part of the domain influenced by Ivan's broad wind field with large changes in the bottom topography.

4.2. *Surface Winds*

Surface winds in hurricanes have been estimated remotely using the Stepped-Frequency Microwave Radiometer (SFMR) from aircraft (Uhlhorn *et al.*, 2007). With six frequencies, the SFMR measures radiative emissions, expressed in terms of brightness temperatures (T_b), from the ocean and the atmosphere. The percentage of foam coverage on the sea surface is known to increase monotonically with wind speed. At microwave frequencies, foam acts as a blackbody emitter. As foam increases, the ocean emits microwave energy more readily, and assuming a constant SST, the T_b increases. Given an accurate physical model that relates ocean surface wind speed and rain-rate to measurements of T_b at each frequency, a set of equations

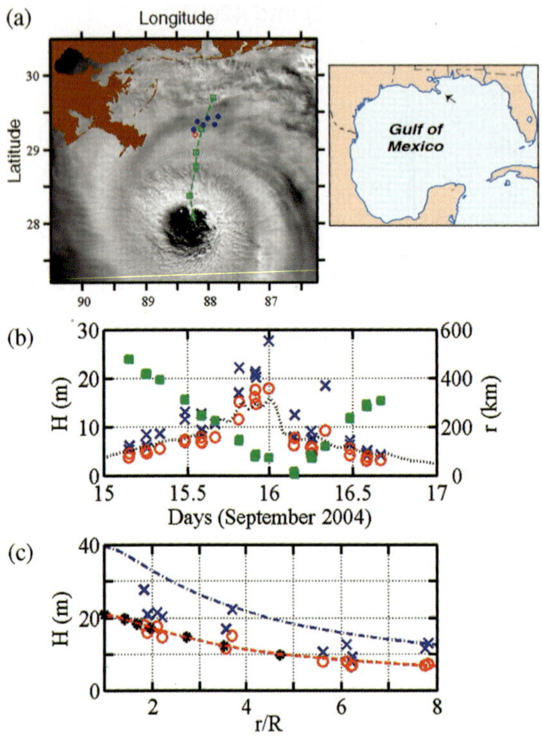

Figure 21. (a) TC Ivan image at 1850 UTC 15 Sept 2004 with the green line representing the track of Ivan at 3-h intervals moving over the SEED moorings (blue). Panel (b) represents the time evolution of Hs (circle) and Hmax (cross) at differing distances. Hs is from NDBC buoy 42040 (dotted curve) and its radial distance to Ivan's center is shown by the green squares. Panel (c) represents Hs and Hmax as a function of normalized distance from the center compared to the exponential distance: digitized values of a segment 15° CW from the forward direction of a numerically simulated wave field (black asterisks). Blue curve depicts the line of Hmax = 1.9 Hs where circles and crosses are as in panel (b). (Reproduced from Wang *et al.*, 2005 by permission of the American Association for the Advancement of Science.)

are inverted to estimate surface winds. Based on measurements during the 2005 Atlantic hurricane season, Uhlhorn *et al.* (2007) developed a new emissivity and wind speed model function based on comparisons with direct measurements of surface winds in hurricanes by GPS dropwindsondes (Fig. 22). This function eliminates a previously-documented high bias in moderate SFMR-measured wind speeds (10–50 $\mathrm{m\,s^{-1}}$), and additionally corrects an extreme wind speed (>60 $\mathrm{m\,s^{-1}}$) systematic underestimate in the past cases. The model function behaves differently below and above the hurricane wind speed threshold (32 $\mathrm{m\,s^{-1}}$), which may have implications for air-sea momentum and kinetic energy exchange.

4.3. *Surface Drag Coefficient*

Knowledge of the heat and moisture fluxes across the interface and into the atmospheric boundary layer are critical elements (Emanuel, 1995). Momentum transfer between the two fluids is characterized by the variations of wind with height and a c_d that is a function of wind speed and surface roughness. For more than a decade, GPS sondes (Hock and Franklin, 1999) have been deployed from aircraft to measure Lagrangian wind profiles in the TC boundary layer. By analyzing these profiles, Powell *et al.* (2003) found a logarithmic variation of mean wind speed in the lowest 200 m, a maximum speed at 500 m, and a gradual weakening with height to 3 km. From these estimates, the surface stress, roughness length, and neutral stability drag coefficient determined by the profile method suggest a leveling of the surface momentum flux as winds increase above hurricane-force and a slight decrease of the drag coefficient with increasing winds.

Donelan *et al.* (2004) described a series of wind-wave tank experiments that included stress measurements from hot-film anemometry and digital particle image velocimetry, and laser/line scan cameras for measuring the water surface elevation. Measurements of the drag coefficient from this laboratory experiment, referenced to the 10-m wind speed, are shown in Fig. 23. Wind speed was measured at 30-cm height in the tank and extrapolated to the 10-m using the logarithmic dependence on height. Wind speeds were verified between crest height and 30 cm for all but the two highest wind speeds. Wind changes both direction and speed over relatively short distances in a TC compared to those required to approach full wave development. The largest waves in the wind-sea move slowly compared to the wind and

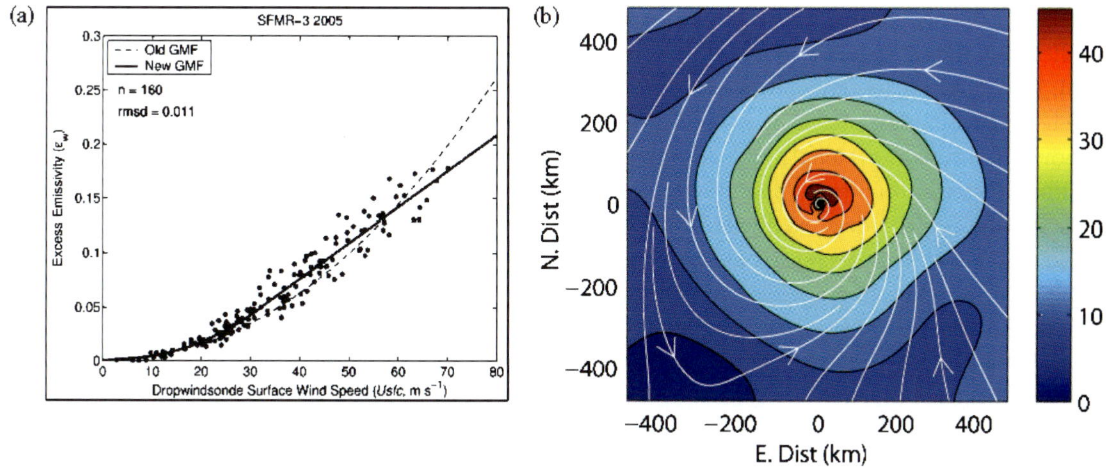

Figure 22. (a) Excess emissivity from SFMR compared to 10-m surface winds measured from GPS dropsondes. The total number of samples is 160 and the RMS difference between the SFMR model function was 0.011 (left panel) and (b) example of an HWIND wind field (Powell and Houston, 1996) and streamlines relative to the TC center (0,0) when SFMR data are included into the analysis from TC Frances (2004) where the color bar is in m s^{-1}. (Reproduced from Uhlhorn *et al.*, 2007 by permission of the American Meteorological Society.)

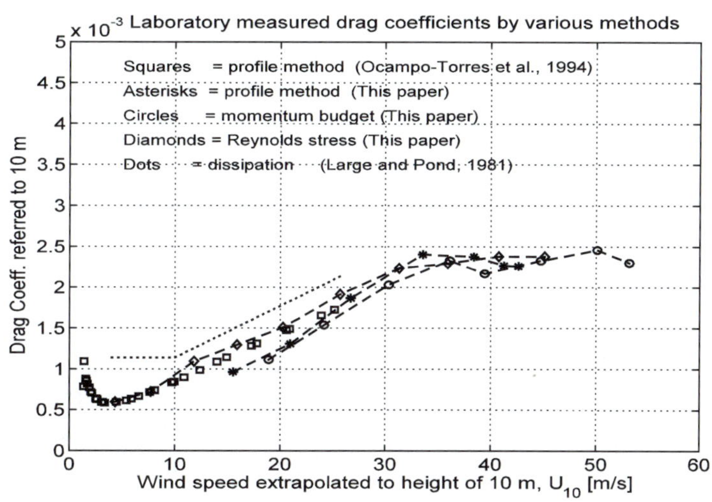

Figure 23. Laboratory measurements of the neutral stability drag coefficient ($\times 10^{-3}$) by profile, eddy correlation ("Reynolds"), and momentum budget methods. The drag coefficient refers to the wind speed measured at the standard anemometer height of 10 m. The drag coefficient formula of Large and Pond (1981) is also shown along with values from Ocampo-Torres *et al.* (1994) derived from field measurements. (Reproduced from Donelan *et al.*, 2004 by permission of the American Geophysical Union.)

often travel in directions differing from the wind. Under such circumstances, these long waves contribute to the aerodynamic roughness of the sea as hypothesized by Kitaigorodskii (1968) and demonstrated by Donelan (1990). Measurements at sea (e.g., Large and Pond, 1981) and in laboratories demonstrate the increasing aerodynamic roughness with increasing wind speed. A "saturation" of the drag coefficient does appear once the wind speed exceeds 33 m s^{-1} (Fig. 23). Beyond this speed, the surface does not become any rougher. At the highest wind speed, significant height and peak wave frequency in the laboratory were 9 cm and 1.4 Hz, respectively. From

10–26 m s^{-1} wind speeds, these measurements parallel the ocean measurements (Large and Pond, 1981), but are less energetic.

The saturation level for c_d is ~0.0025, corresponding to a roughness length of 3.35 mm from the laboratory results. Powell *et al.* (2003) found a "saturation" of the drag coefficient at 0.0026 at about 35 m s^{-1} that decreased at higher wind speeds using wind profiles normalized by a mean boundary layer wind. Shay and Jacob (2006) found a "saturation" wind drag coefficient at 30 m s^{-1} of 0.0034 derived by equating internal wave ocean fluxes to the surface winds at 10-m level. Note it was assumed that all of the wind drives a current and no separation was attempted here to partition the energy between current and wave fields. Beyond 30 m s^{-1}, the c_d began to leveled off at surface wind speeds up to 38 m s^{-1} observed in TC Gilbert. The implication here is that ocean current profilers deployed in TC can be used to trace the momentum flux, and the c_d behavior.

The momentum flux was also estimated using the current profiles recorded in water depths of less than 100 m during TC Ivan (Teague *et al.*, 2007), and discussed in terms of the neutral drag coefficient (Jarosz *et al.*, 2007). The eye of Ivan generated both strongly forced and relaxation stage responses to the surface wind field (Wang *et al.*, 2005; Teague *et al.*, 2006, 2007). Over the shelf, the response consisted of a barotropic component, and a much weaker baroclinic response. Scale analysis of recorded current velocity showed that this forced response was described by the linearized, depth-integrated along-shelf horizontal momentum balance with bottom friction. The drag coefficient was estimated based on these observed depth-averaged currents at an ADCP mooring for wind speeds between 20 and 48 m s^{-1}. In this case, surface drag increased to a peak value of 0.0026 at 32 m s^{-1} before decreasing. This inflection point is coincident with above results. Differences may also be attributed to uncertainties in the wind measurements and the use of simplified ocean dynamics.

Recently, Sanford *et al.* (2007) estimated the volume transport per unit of width based on velocity profiles in hurricane Frances using a saturated surface drag coefficient of Powell *et al.* (2003) and Large and Pond (1981) formulations. Numerical simulations from the PWP mixing model embedded in the 3-D ocean model (Price *et al.*, 1994) indicated consistent results for the volume transport and SST cooling values at two of the three floats to the right of the storm track. Along the track, however, differences differed by about 15–20%. As Sanford *et al.* (2007) point out, perhaps an azimuthal dependence in the surface drag coefficient due to surface waves (Wright *et al.*, 2001) must be included in the models as well as wind-field asymmetries common to most TCs. In this context, additional case studies must be performed to understand how the waves, winds and currents are coupled under high-wind conditions (Bender and Ginis, 2000; Bao *et al.*, 2000; Rogers *et al.*, 2006; Drennan and Shay, 2006; Black *et al.*, 2007; Chen *et al.*, 2007)

4.4. *Wind-Wave Coupling*

The momentum flux is parameterized with a non-dimensional surface roughness (or Charnock's equation, $z_{ch} = z_0 g/u_*^2$, where z_0 is the roughness length, u_* is the surface friction velocity and g is the gravitational acceleration) and the stability correction (not shown) is based on the Monin-Obukhov similarity theory, regardless of the wind speed or the sea state in the GFDL model. Moon *et al.* (2004) investigated the Charnock coefficient under TC conditions using a coupled wind-wave (CWW) model (Fig. 24). In the CWW model, the surface wave directional frequency spectrum near the spectral peak is calculated using the WAVEWATCH III (Tolman, 2002) model and the high frequency part of the spectrum was parameterized using the theoretical model of Hara and Belcher (2002). The wave spectrum is then introduced to the wave boundary layer model of Hara and Belcher (2004) to estimate the Charnock

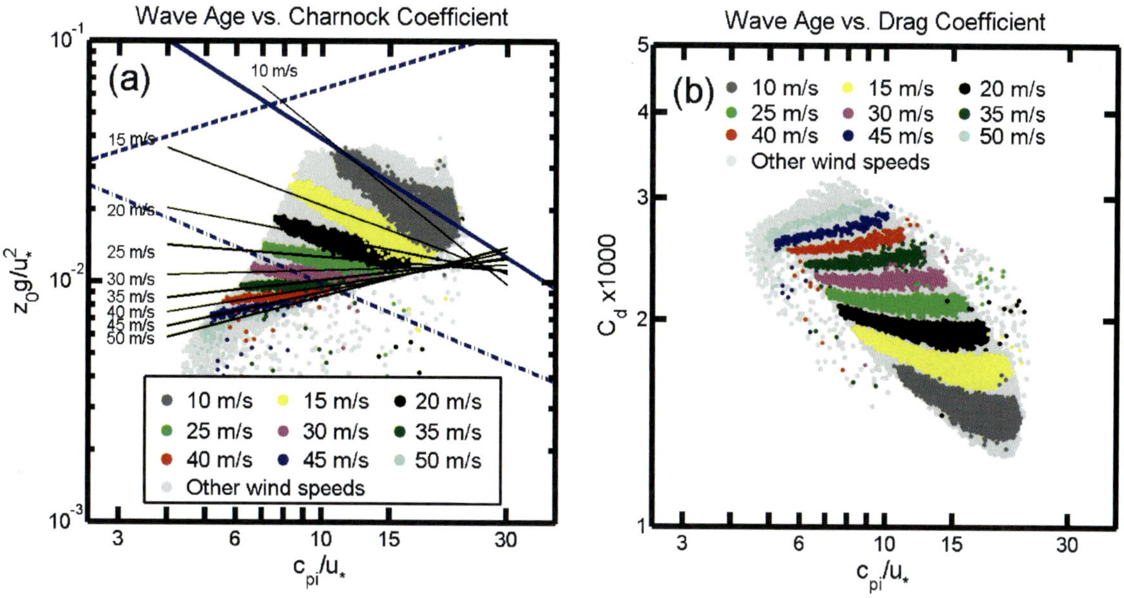

Figure 24. Scatterplots of the (a) Charnok Coefficient (z_{ch}) and (b) drag coefficients (c_d) as a function of the wave age (cp/u^*) for several TCs in the Atlantic Ocean. Differing colors represent 5-m s^{-1} intervals for surface winds for z_{ch} and c_d are the best fits for each wind speed group. In panel a, blue solid line and dash-dot represent empirical estimates for ocean and laboratory experiments (Donelan, 1990). Dotted line is the formula of Toba *et al.* (1990). (Reproduced from Moon *et al.*, 2004a by permission of the American Meteorological Society.)

coefficient at differing wave evolution stages. The regression lines between the wave age and the Charnock coefficient have a negative slope at low wind speeds but have a positive slope at higher wind speeds. For example, in Fig. 24(a), this change in slope occurs between 25–35 m s^{-1}. This behavior of the Charnock coefficient provides one explanation why c_d under a TC, where seas tend to be "young," is reduced in high wind speeds as the wave field is not fully developed.

Consistent with this change in the Charnock coefficient, c_d levels off at high wind speeds and begins to decrease at approximately 35 m s^{-1} (Fig. 25). The result also is in accord with these other studies in that the drag coefficient levels off between surface winds of 28 to 33 m s^{-1} and begins to decrease with wind speeds (Powell *et al.*, 2003; Donelan *et al.*, 2004). Notwithstanding, perhaps the most important finding of the Moon *et al.* (2004b) study is that a hurricane model must have a surface wave model to gain a more accurate prediction of surface enthalpy fluxes under the high-wind conditions.

4.5. *Enthalpy Fluxes*

Direct turbulent flux measurements were carried out in the hurricane boundary layers using a research aircraft instrumented with fast-response turbulence sensors (Black *et al.*, 2007). As shown in Fig. 26, the wind speed range for momentum and enthalpy fluxes and exchange coefficients has been extended by over 50% compared to that in previous studies. The drag coefficient (c_d) increases linearly with 10-m wind speed up to 22 m s^{-1} then level off at higher wind speed (French *et al.*, 2007). The Dalton number (c_E) is nearly constant with 10-m wind speed up to 30 m s^{-1} (Drennan *et al.*, 2007). Combining the sensible and latent heat flux measurements, Zhang (2007) derived the enthalpy flux and the exchange coefficient for enthalpy transfer (c_k) showing that there is no evidence of an increase of c_k with wind speed, in good agreement with the Humidity Exchange over the Sea (HEXOS) result (DeCosmo *et al.*, 1996). The ratio of c_k/c_d versus wind speed is relatively constant beyond

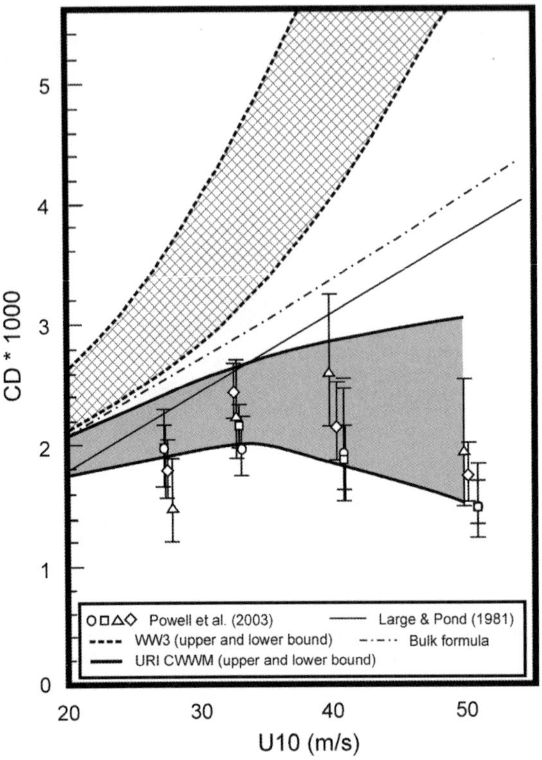

Figure 25. Drag coefficients (c_d) from various observation-based values, empirical formulas, and model outputs as a function of U_{10}. Symbols represent observations from GPS sonde wind profiles (Powell *et al.*, 2003). Vertical bars represent 95% confidence limits. Solid line is an extrapolation of the Large and Pond (1981) formula. Dash-dot line is the bulk formula used in GFDL hurricane model. Shaded and hatched areas represent ranges between upper and lower bound of C_d obtained by the URI coupled wave-wind model and an internal estimation of WAVEWATCH III, respectively. (Reproduced from Moon *et al.*, 2004b by permission of the American Meteorological Society.)

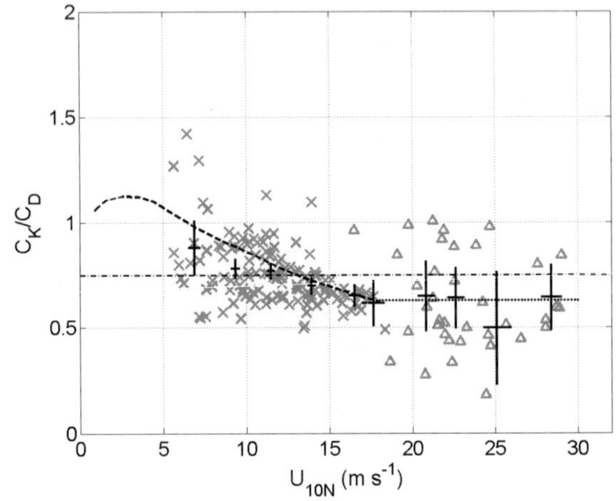

Figure 26. The ratio of c_k/c_d as a function of 10-m neutral wind speed. Data from CBLAST (Δ), and HEXOS (x) are shown. Solid black lines show the mean and 95% confidence intervals of the combined HEXOS and CBLAST field data after binning average by wind speed. The dotted black line shows the mean of the CBLAST data. The ratio based on COARE 3.0 bulk flux algorithm is shown as the dashed line. The threshold value of 0.75 suggested by Emanuel is also shown as the dash-dotted line (from Zhang, 2007).

$20\,\mathrm{m\,s^{-1}}$ for the flux runs with both momentum and enthalpy flux measurements. The average of the c_k/c_d values is 0.63, well below the 0.75 threshold for TC development suggested by Emanuel (1995).

Fluxes of heat and moisture are central to the TC intensity question and are usually determined from bulk aerodynamic formulae that utilize near-surface atmospheric observations and upper-ocean temperature data measured by ocean profiles. Atmospheric variables are estimated from GPS sondes (Hock and Franklin, 1998) deployed within the storm from both NOAA aircraft as well as Air Force Reserve reconnaissance flights. Surface winds are routinely measured by the SFMR (Uhlhorn *et al.*, 2007) and GPS sondes measure profiles of temperature, pressure and humidity from flight level to the surface. For each measurement, 10-m values of these quantities are optimally interpolated to a storm-relative grid aligned with the direction of storm motion. The SFMR wind observations are objectively analyzed (Powell and Houston, 1996) and interpolated to the same grid as the GPS thermodynamic data. Finally, the SSTs observed by expendable ocean probes on in-storm flights are interpolated to the same grid as the GPS measurements as in Isidore and Lili (Shay and Uhlhorn, 2008). Key terms in the estimates for heat and moisture fluxes are temperature and specific humidity differences and the bulk transfer coefficient (i.e., $c_k c_d^{-1}$). As most of the recent studies have indicated a leveling off of the surface drag coefficient at $30\,\mathrm{m\,s^{-1}}$, Emanuel (2003) suggests that this ratio becomes independent of wind speed and that the ratio should be O(1), since below this value intense hurricanes cannot be simulated in the numerical model. Heat exchange coefficients have been set equal to c_d (Cione *et al.*, 2000). An additional ocean forcing mechanism is the surface precipitation flux (rain rate). As noted above, freshwater input by rain alters the ocean response both by direct cooling due to rain that has a lower temperature than the SST, and by stabilizing the OML by decreasing the salinity and reducing the mixing rate (Price *et al.*, 1986; Jacob and Koblinsky, 2007).

Estimates of enthalpy fluxes during TC's Isidore and Lili were sensitive to the storm translation speed. In Isidore, peak enthalpy flux $\sim 1.7\,\mathrm{kW\,m^{-2}}$ is in the right-rear quadrant of the storm due to the high SSTs ($\sim 30°\mathrm{C}$) as there was a negligible decrease from pre-storm SST conditions, especially over the warm LC where ocean cooling was minimal (Shay and Uhlhorn, 2008). Although the maximum momentum flux (7 Pa) is in the right-front quadrant, TC Isidore's wind stress field was symmetric as it moved at only $4\,\mathrm{m\,s^{-1}}$. Estimated maximum surface enthalpy fluxes in Lili were about $1.4\,\mathrm{kW\,m^{-2}}$ due in part to the marked asymmetry associated with the faster storm translation speed ($7\,\mathrm{m\,s^{-1}}$) and smaller SSTs by about 1°C. This result highlights how modest SST differences alter the surface heat fluxes in extreme winds (Cione and Uhlhorn, 2003).

Enthalpy fluxes were integrated along the track to obtain the cross-track (radial) distributions of net sea surface heat loss (Fig. 27). The estimated surface heat loss in Isidore ($\sim 9\,\mathrm{kJ\,cm^{-2}}$) is almost a factor of two larger than in Lili ($\sim 4.5\,\mathrm{kJ\,cm^{-2}}$) due to the enhanced enthalpy fluxes, slower storm speed, and larger horizontal SST gradients along the western side of the Yucatan Strait. These inferred surface heat losses fluxes reflect the lack of the oceanic response of the LC observed during both TC's. For example, Cione and Uhlhorn (2003) argue that it is only inner-core SSTs that the storm underneath a storm if SSTs are changing by

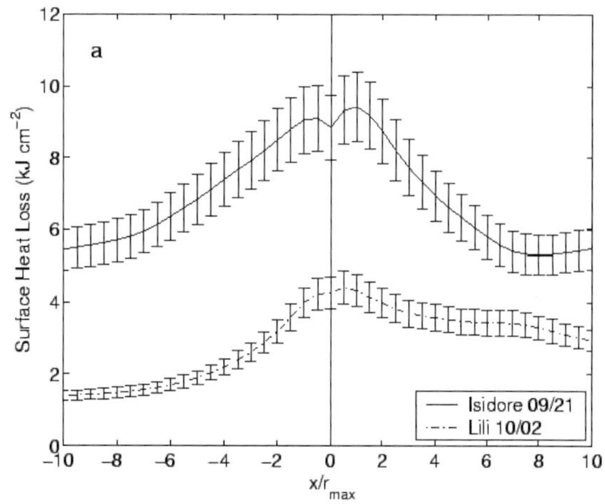

Figure 27. Integrated along-track variations in the normalized cross-track direction of the surface heat loss (upper curve) observed during TC's Isidore (solid) and Lili (dashed) with the uncertainties based on observed ocean and atmospheric data. (Reproduced from Shay and Uhlhorn, 2008 by permission of the American Meteorological Society.)

1° to 2°C. Since SSTs represent the boundary condition for OHC estimates from the satellite data, any SST decrease has to be reflected in the estimation procedure.

5. Summary and Concluding Remarks

Significant progress has been made in understanding the basic oceanic and atmospheric processes that occur during TC passage since 1995 (Ginis, 1995). There is a continuing need to isolate fundamental physical processes involved in the coupled interactions through detailed process studies using experimental, empirical, theoretical, and numerical approaches with data assimilation methods. As suggested from new observations, these approaches are needed to improve predictions of tropical cyclone intensity and structure.

Considerable ocean-atmosphere variability occurs over the storm scales that include fundamental length scales such as the radius of maximum winds and radius to gale-force winds. Here, the fundamental science questions are how the two fluids are coupled through oceanic and atmospheric boundary layer processes, and what are the primary time scales of this coupling? These questions are not easily answered as the interactions occur over a broad range of time and space scales. One school of thought is that the only important process with respect to the ocean is underneath the eyewall where ocean cooling has occurred. While it is at the eyewall where the maximum momentum and enthalpy fluxes occur, the broad surface circulation over the warm OML also has non-zero fluxes that contribute heat and moisture to the TC. The deeper the OML (and 26°C isotherm depth), more heat (OHC) is available to the storm through the enthalpy fluxes. It is not just the magnitude of the OHC, since the depth of the warm water is important to sustaining surface enthalpy fluxes. Process studies need to begin to examine these multiple scale aspects associated with the atmospheric response to ocean forcing.

To understand the coupling with the upper ocean, measurements must be made prior to the TC arrival over an oceanographic area to resolve background flows and the associated thermodynamics. Such data are needed not only to initialize ocean models (e.g., Loop Current, Kuroshio, Gulf Stream), but these data are required during the TC to examine vertical mixing processes on the upper ocean momentum and thermal response. In addition to aircraft-based sampling by AXCPs and AXCTDs and new profiling floats such as the EM/APEX and drifters, efforts along the southeastern United States are underway to deploy high frequency radars to map the surface currents to 200 km from the coast as part of an integrated ocean observing system. Such measurements would not only be invaluable to map the wind-driven surface currents during high winds, but also to map the directional wave spectra over the domain. These measurements could then be used to not only examine air-sea interactions and evaluate the coupled models, but also assess the relative importance of surface wave-current interactions in surge models.

Surface drag coefficient variations has received attention over the last five years largely through highly specialized experiments. Several treatments have come to the conclusion that there is a leveling off or saturation values of $\approx 30\,\mathrm{m\,s^{-1}} \pm 3\,\mathrm{m\,s^{-1}}$. The ratio of the enthalpy coefficient and the drag coefficient is central to air-sea fluxes impacting the TC boundary layer. In this context, the relationship between the coupled processes such as wave breaking and the generation of sea spray and how this is linked to air-sea fluxes remains a fertile research area. A key element of this topic is how the atmosphere responds to the oceanic forcing where there seems to be contrasting viewpoints. One argument is that the air-sea interactions are occurring over surface wave (wind-wave) time and space scales and induce intensity changes of more than a category. Yet empirical studies suggest the drag coefficients range between 1 and 4×10^{-3} depending on the quadrant where the surface waves change direction relative to the 10-m winds (M. Powell, personal communication, 2007). In the recent coupled model studies, the partitioning of the wind energy into the currents and waves remains an important research question. There should be more in depth experimental and empirical studies where surface waves are measured along with upper ocean currents and surface winds. Surface waves are essentially submesoscale phenomena that affect the enthalpy fluxes in differing quadrants, however, first-order balances are primarily between the atmospheric and oceanic boundary layers that constrain and impact the sea surface processes.

Acknowledgments

The author gratefully acknowledges support from the NSF through grants *ATM-01-08218, 04-44525*, NASA Hurricane Science Program and NOAA Joint Hurricane Testbed program. I continue to be grateful to the NOAA Aircraft Operation Center (Dr. Jim McFadden) who make it possible to acquire high quality data during hurricanes. Mr. Bill Teague (NRL-Stennis) and Drs. Rick Lumpkin, Peter Black, Eric Uhlhorn, Jun Zhang, Mark Powell (NOAA-AOML), Mark Donelan, Brian Haus, Will Drennan, George Halliwell (all at RSMAS), S. Daniel Jacob (University of Maryland-Baltimore County), Tom Sanford (Applied Physics Laboratory, University of Washington), Nan Walker (Louisiana State University) and Isaac Ginis (University of Rhode Island), I.-I. Lin (National Taiwan University), Y. Yoshikawa (Kyushu University) and I.-J. Moon (Cheju National University) shared material used herein. Dr. Russell Elsberry made editorial comments on the original version of this chapter for the IWTC-VI workshop edited by Dr. Hugh Willoughby. I thank Dr. Jeff Kepert for his insightful review and of course the patience of the Editor Dr. Johnny Chan. Without the cooperation of these scientists this chapter would not be possible. Ms. Jodi Brewster, Mr. Benjamin Jaimes and Mr. Pat Myers provided material and assisted with some of the graphics.

Glossary of Terms

1-D: One-Dimensional
3-D: Three-Dimensional
ADCP: Acoustic Doppler Current Profiler
AMSR: Advanced Microwave Sounding Radiometer
AXBT: Airborne Expendable Bathythermograph
AXCP: Airborne Expendable Current Profiler
AXCTD: Airborne Expendable Conductivity Temperature Depth profiler
CBLAST: Coupled Boundary Layer Air-Sea Transfer Experiment
CCR: Cold Core Rings
CHIPS: Coupled Hurricane Intensity Prediction Scheme
COARE: Coupled Ocean and Atmospheric Response Experiment
CWW: Coupled Wind Wave model
EM-APEX: Electromagnetic Autonomous Profiling Explorer Floats
EPAC: Eastern Pacific Ocean Basin
EPIC: Eastern Pacific Investigation of Climate
FC: Florida Current
GCW: Gulf of Mexico Common water
GFDL: Geophysical Fluid Dynamics Laboratory
GISS: Goddard Institute for Space Studies Level-2 closure
GOM: Gulf of Mexico
GPS: Global Positioning System
HEXOS: Humidity Exchange Over the Sea
HYCOM: Hybrid Coordinate Ocean Model
IFEX: Intensity Fluctuation Experiment
IP: Inertial Period
KPP: K-Profile Parameterization scheme
KT: Kraus, Turner turbulence balance model
MY: Mellor-Yamada 2.5 level Closure Scheme
LC: Loop Current
OML: Ocean Mixed Layer
ONR: Office of Naval Reserach
OHC: Oceanic Heat Content
OI: Optimal Interpolation
MODAS: Modular Ocean Data Assimilation Scheme
MODIS: Moderate Resolution Microwave Sounding Unit
MLD: Mixed Layer Depth
MLT: Mixed Layer Temperature
MPI: Maximum Potential Intensity
NASA: National Aeronautics and Space Administration (US)
NDBC: NOAA's National Data Buoy Center
NRL: US Navy Research Laboratories
NSF: National Science Foundation (US)
NOAA: National Oceanic and Atmospheric Administration (US)

PWP: Price, Weller and Pinkel dynamical instability model
PSU: Practical Salinity Unit (psu)
SHA: Sea-Surface Height Anomaly (SSHA)
SFMR: Stepped Frequency Microwave Radiometer
SRA: Scanning Radar Altimeter
SHIPS: Statistical Hurricane Intensity Prediction Scheme
SST: Sea Surface Temperature
TAO: Tropical Ocean Atmosphere Mooring Array as a result of TOGA
TC: Tropical Cyclone
TOPEX: The Topographic Experiment (Radar Altimeter)
TMI: Tropical Rain Measuring Mission (TRMM) Microwave Imager
TOGA: Tropical Ocean Global Atmosphere Program
WATL: Western Atlantic Ocean
WCR: warm core rings
WPAC: Western Pacific Ocean Basin
XBT: Expendable Bathythermographs

References

Ali, M. M., P. S. V. Jagadeesh, and S. Jain, 2007: Effects of eddies on Bay of Bengal cyclone intensity, *EOS*, **88**, 93,95.

Andreas, E. and K. A. Emanuel, 2001: Effects of sea spray on tropical cyclone intensity. *J. Atmos. Sci.*, **58**, 3741–3751.

Bane, J. M., Jr., 1994: The Gulf Stream System: an observational perspective. *The Oceans : Physical-Chemical Dynamics and Resources*, Eds. S. K. Majumdar, E. W. Miller, G. Forbes, R. Schmalz and A. Pannah, Pennsylvania Academy of Sciences, pp. 88–108.

Bao, J.-W., J. M. Wilczak, J. K. Choi, and L. H. Kantha, 2000: Numerical simulations of air-sea interaction under high wind conditions using a coupled model: A study of hurricane development. *Mon. Wea. Rev.*, **128**, 2190–2210.

Bell, T. H. Jr., 1975: Topographically generated internal waves in the open ocean. *J. Geophys. Res.*, **80**, 320–327.

Bender, M. and I. Ginis, 2000: Real-time simulation of hurricane-ocean interaction. *Mon. Wea. Rev.*, **128**, 917–946.

Black, P. G., 1983: Ocean temperature changes induced by tropical cyclones. PhD Dissertation, Pennsylvania State University, State College, PA, 278 pp.

Black, P. G., E. A. D'Asaro, W. Drennan, J. R. French, P. P. Niiler, T. B. Sanford, E. J. Terrill, E. J. Walsh, and J. Zhang, 2007: Air-sea exchange in hurricanes: Synthesis of observations from the Coupled Boundary Layer Air-Sea Transfer experiment. *BAMS*, **88**, 357–384.

Bosart, L., C. S. Veldon, W. E. Bracken, J. Molinari, and P. G. Black, 2000: Environmental influences on the rapid intensification of hurricane Opal (1995) over the Gulf of Mexico. *Mon. Wea. Rev.*, **128**, 322–352.

Braun, S. A. and W.-K. Tao, 2000: Sensitivity of high resolution simulations of hurricane Bob (1991) to planetary boundary layer paramterizations. *Mon. Wea. Rev.*, **128**, 3941–3961.

Chan, J. C. L., Y. Duan and L. K. Shay. 2001: Tropical cyclone intensity change from a simple ocean-atmospheric coupled model. *J. Atmos. Sci.*, **58**, 154–172.

Chang, S., and R. Anthes, 1978: Numerical simulations of the ocean's nonlinear baroclinic response to translating hurricanes. *J. Phys. Oceanogr.*, **8**, 468–480.

Charnock, H., 1955: Wind stress on a water surface. *Quart. J. Royal Meteor. Soc.*, 639–640.

Chassignet, E. P., L. Smith, G. R. Halliwell, and R. Bleck, 2003: North Atlantic simulations with the hybrid coordinate ocean model (HYCOM): Impact of the vertical coordinate choice and resolution, reference density, and thermobaricity. *J. Phys. Oceanogr.*, **33**, 2504–2526.

Chen, S., J. F. Price, W. Zhao, M. Donelan, and E. J. Walsh, 2007: The CBLAST Hurricane program and the next generation fully coupled atmosphere-wave-ocean models for hurricane research and prediction., *BAMS*, **88**, 311–317.

Cheney, R., L. Miller, R. Agreen, N. Doyle, and J. Lillibridge, 1994:, TOPEX/Poseidon: The 2-cm solution. *J. Geophys. Res.*, **99**, 24,555–24,563.

Church, J. A., T. M. Joyce, and J. F. Price, 1989: Current and density observation across the wake of Hurricane Gay, *J. Phys. Oceanogr.*, **19**, 259–265.

Cione, J. J., P. G. Black, S. H. Houston, 2000: Surface observations in the hurricane environment. *Mon. Wea. Rev.* **128**, 1550–1561.

Cione, J. J., and E. W. Uhlhorn, 2003: Sea surface temperature variability in hurricanes: Implications with respect to intensity change. *Mon. Wea. Rev.*, **131**, 1783–1796.

Cooper, M., and K. Haines, 1996: Altimetric assimilation with water conservation properties. *J. Geophys. Res.*, **101**, 1059–1077.

Cummings, J. A, 2003: Ocean data assimilation, *In* COAMPS: Version 3 Model Description, NRL Publication *NRL/PU/7500-03-448*. Naval Research Laboratory, Washington, DC, 20375, pp. 21–28.

D'Asaro, E. A., 2003: The ocean boundary layer under hurricane Dennis. *J. Phys. Oceanogr.*, **33**, 561–579.

DeCosmo, J. B., K. B. Katsaros, S. D. Smith, R. J. Anderson, W. A. Oost, K. Bumke, and H. Chadwick, 1996: Air-sea exchange of water vapor and sensible heat: The Humidity Exchange over the Sea (HEXOS) results. *J. Geophys. Res.*, **101**, 12,001–12,016.

DeMaria, M., and J. Kaplan, 1994: Sea surface temperatures and the maximum intensity of Atlantic tropical cyclones. *J. Climate*, **7**, 1324–1334.

DeMaria, M., M. Mainelli, L. K. Shay, J. Knaff, and J. Kaplan, 2005: Further improvements to the statistical hurricane intensity prediction scheme. *Wea. and Forecasting*, **20**, 531–543.

Donelan, M. A., 1990: Air-sea interaction. The Sea, LeMehaute, B. and D. Hanes, Eds., *Ocean Engineering Science*, **9**, John Wiley and Sons, pp. 239–292.

Donelan, M. A., B. K. Haus, N. Reul, W. J. Plant, M. Stiassine, H. Graber, O. Brown, and E. Saltzman, 2004: On the limiting aerodynamic roughness of the ocean in very strong winds. *Geophys. Res. Letters*, 31L18306.

Drennan, W. M. and L. K. Shay. 2006: On the variability of the fluxes of momentum and sensible heat. *Bound. Layer Meteor.*, **119(1)**, 81–107.

Drennan, W. M., J. A. Zhang, J. R. French, C. McCormick, and P. G. Black, 2007: Turbulent fluxes in the hurricane boundary layer, Part II: Latent heat flux. *J. Atmos. Sci.*, **64**, 1103–1115.

Elliot, B. A., 1982: Anticyclonic rings in the Gulf of Mexico, *J. Phys. Oceanogr.*, **12**, 1292–1309.

Emanuel, K. A., 1986: An air-sea interaction theory for tropical cyclones Part 1: Steady-State maintenance. *J. Atmos. Sci.*, **43**, 585–605.

Emanuel, K. A., 1995: Sensitivity of tropical cyclones to surface exchange and a revised steady-state model incorporating eye dynamics. *J. Atmos. Sci.*, **52**, 3969–3976.

Emanuel, K. A., 2003: A similarity hypothesis for air-sea exchange at extreme wind speeds. *J. Atmos. Sci.*, **60**, 1420–1428.

Faber, T., L. K. Shay, S. D. Jacob, S. H. Houston, and P. G. Black, 1997: Observed air-sea interactions during hurricane Emily. Preprints, *22nd Conf. on Hurricanes and Tropical Meteorology*, Fort Collins, CO, *Amer. Meteor. Soc.*, pp. 433–434.

Falkovich, A., I. Ginis, and S. Lord, 2005: Implementation of data assimilation and ocean initialization for the coupled GFDL/URI hurricane prediction system. *J. Atmos. Oceangr. Tech.*, **22**, 1918–1932.

Findlater, J., 1969: A major low-level air current near the Indian Ocean during the northern summer. *Quart. J. R. Met. Soc.*, **95**, 362–380.

Fox, D. N., W. J. Teague, C. N. Barron, M. R. Carnes, and C. M. Lee, 2002: The Modular Ocean Data Assimilation System (MODAS). *J. Atmos. Oceanogr. Tech.*, **19**, 240–252.

French, J. R., W. M. Drennan, J. A. Zhang, and P. G. Black, 2007: Turbulent fluxes in the hurricane boundary layer. Part 1: Momentum flux. *J. Atmos. Sci.*, **64**, 1089–1102.

Gaspar, P., 1988: Modeling the seasonal cycle of the upper ocean. *J. Phys. Oceanogr.*, **18**, 161–180.

Geisler, J. E., 1970: Linear theory on the response of a two layer ocean to a moving hurricane. *Geophys. Fluid Dyn.*, **1**, 249–272.

Gill, A. E., 1984: On the behavior of internal waves in the wakes of storms. *J. Phys. Oceanogr.*, **14**, 1129–1151.

Ginis, I., 1995: Interaction of tropical cyclones with the ocean. *in Global Perspective of Tropical Cyclones*, Chapter 5, Ed. R. L. Elsberry, Tech. Document WMO/TD 693, World Meteorological Organization, Geneva, Switzerland, pp. 198–260.

Goni, G. J., S. Kamholz, S. L. Garzoli, and D. B. Olson, 1996: Dynamics of the Brazil/Malvinas confluence based on inverted echo sounders and altimetry. *J. Geophys. Res.*, **95**, 22103–22120.

Greatbatch, R.J., 1983: On the response of the ocean to a moving storm: nonlinear dynamics. *J. Phys. Oceanogr.*, **13**, 357–367.

Halliwell, G., L. K. Shay, S. D. Jacob, O. Smedstad, and E. Uhlhorn, 2008: Improving ocean model initialization for coupled tropical cyclone forecast models using GODAE nowcasts. *Mon. Wea. Rev.*, **136**(7), 2576–2591.

Halliwell, G., L. K. Shay, J. Brewster, and W. Teague, 2010: Evaluation and sensitivity analysis of an ocean model to hurricane Ivan in the northern Gulf of Mexico. *Mon. Wea. Rev.* (accepted, revised and resubmitted).

Hanshaw, M. N., M. S. Lozier, and J. B. Palter, 2008: Integrated impact of tropical cyclones on sea surface cholophyll in the North Atlantic. *Geophys. Res. Lett.*, **35**, L01601 doi:10.1029/3007 GL031862.

Hara, T., and S. E. Belcher, 2004: Wind profile and drag coefficient over mature ocean surface wave spectra. *J. Phys. Oceanogr.*, **34**, 2345–2358.

Hogg, N. G. and W. E. Johns, 1995: Western boundary currents. *Rev. of Geophys.* (suppl), U.S. National Report to the International Union of Geodesy and Geophysics 1991–1994, pp. 1311–1334.

Hock, T. J., and J. L. Franklin, 1999: The NCAR GPS dropwindsonde. *BAMS*, **80**, 407–420.

Hogg, N. G. and W. E. Johns, 1995: Western boundary currents. *Rev. of Geophys.* (suppl), U.S. National Report to the International Union of Geodesy and Geophysics 1991–1994, pp. 1311–1334.

Hong, X., S. W. Chang, S. Raman, L. K. Shay, and R. Hodur, 2000: The interaction of hurricane Opal (1995) and a warm core ring in the Gulf of Mexico. *Mon. Wea. Rev.*, **128**, 1347–1365.

Jacob, D. S., L. K. Shay, A. J. Mariano, and P. G. Black, 2000: The three-dimensional mixed layer heat balance during hurricane Gilbert. *J. Phys. Oceanogr.*, **30**, 1407–1429.

Jacob, S. D., and L. K. Shay, 2003: The role of oceanic mesoscale features on the tropical cyclone- induced mixed layer response. *J. Phys. Oceanogr.*, **33**, 649–676.

Jacob, S. D., and C. Koblinsky, 2007: Effects of precipitation on the upper ocean response to a hurricane. *Mon Wea. Rev.*, **135**, 2207–2225.

Jaimes, B. and L. K. Shay, 2009: Mixed layer cooling in mesoscale eddies during Katrina and Rita. *Mon. Wea. Rev.*, **137**(12), 4188–4207.

Jarosz, E., D.A. Mitchell, D.W. Wang, and W.J. Teague, 2007: Bottom-up determination of air-sea momentum exchange under a major tropical cyclone. *Science*, **315**, 1707.

Jena, B., M. V. Rao and B. K. Sahu, 2006: TRMM-derived sea surface temperature in the wake of a cyclonic storm over the central Bay of Bengal. *Inter. J. Rem. Sens.*, **27(14)**, 3065–3072.

Johns, W. E., T. J. Shay, J. M. Bane, and D. R. Watts, 1995: Gulf Stream structure, transport and recirculation near 68°W. *J. Geophys. Res.*, **100**, 817–838.

Kitaigorodskii, S. A., 1968: On the calculation of the aerodynamic roughness of the sea surface. Izvestia. Acad. Sci. USSR, *Atmos. Oceanic Phys.*, **4**, 498–502.

Kraus, E. B. and J. S. Turner, 1967: A one-dimensional model of the seasonal thermocline. II: The general theory and its consequences. *Tellus*, **1**, 98–105.

Large, W. G., and S. Pond, 1981: Open ocean momentum flux measurements in moderate to strong winds. *J. Phys. Oceanogr.*, **11**, 324–336.

Large, W. S., J. C. McWilliams, and S. C. Doney, 1994: Oceanic vertical mixing: A review and a model with non-local boundary layer parameterization. *Rev. Geophys.*, **32**, 363–403.

Leaman, K. D., 1976: Observations of vertical polarization and energy flux of near-inertial waves. *J. Phys. Oceanogr.*, **6**, 894–908.

Leben, R. R., 2005: Altimeter derived Loop Current metrics. In *Circulation in the Gulf of Mexico: observations and models, Geophys. Monogr.*, **161**, 181–201.

Leipper, D. F., 1970: A sequence of current patterns in the Gulf of Mexico. *J. Geophys. Res.*, **75**, 637–657.

Leipper, D., and D. Volgenau, 1972: Hurricane heat potential of the Gulf of Mexico. *J. Phys. Oceanogr.*, **2**, 218–224.

Lin, I.-I., C.-C. Wu, K. A. Emanuel, I.-H. Lee, C.-R. Wu, and I.-F. Pun, 2005: The interaction of super typhoon Maemi (2003) with a warm ocean eddy. *Mon. Wea. Rev.*, **133**, 2635–2649.

Lin, I-I., I.-F. Pun and C.-C. Wu, 2008: Upper ocean thermal structure and the western north Pacific Category 5 typhoons. Part I: Ocean features and category 5 typhoons' intensification. *Mon. Wea. Rev.*, **136**, 3288–3306.

Lin, I-I., I.-F. Pun and C.-C. Wu, 2009: Upper ocean thermal structure and the western north Pacific Category 5 typhoons. Part II: Dependence on translation speed. *Mon. Wea. Rev.*, **137**, in press.

Mainelli-Huber, M., 2000: On the role of the upper ocean in tropical cyclone intensity change. M.S. Thesis, RSMAS, University of Miami, Florida, 73 pp.

Mainelli, M., M. DeMaria, L. K. Shay and G. Goni, 2008: Application of oceanic heat content estimation to operational forecasting of recent category 5 hurricanes. *Wea. Forecasting*, **23**, 3–16.

Marks, F., and L.K. Shay, 1998: Landfalling tropical cyclones: Forecast problems and associated research opportunities: Report of the 5th Prospectus Development Team to the

U.S. Weather Research Program, *BAMS*, **79**, 305–323.

Mellor, G. L., and T. Yamada, 1982: Development of a turbulence closure model for geophysical fluid problems. *Rev. Geophys. Space Phys.*, **20**, 851–875.

Monaldo, F. M, T. Sikora, S. Babin, and R. E. Sterner, 1997: Satellite imagery of sea surface cooling in the wake of hurricane Edouard. *Mon. Wea. Rev.*, **125**, 2716–2721.

Moon, I., I. Ginis, and T. Hara, 2004a: Effect of surface waves on Charnock coefficient under tropical cyclones. *Geophys. Res. Lett.*, **31**, L20302.

Moon I.-J., I. Ginis, and T. Hara, 2004b: Effect of surface waves on air–sea momentum exchange. II: Behavior of drag coefficient under tropical cyclones. *J. Atmos. Sci.*, **61**, 2334–2348.

Nilsson, J., 1995: Energy flux from traveling hurricanes to the oceanic internal wave field. *J. Phys. Oceanogr.*, **25**, 558–573.

Ocampo-Torres, F.J., M.A. Donelan, N. Merzi, and F. Jia, 1994: Laboratory measurements of mass transfer of carbon dioxide and water vapour for flow conditions. *Tellus*, **46B**, 16–32.

Ooyama, K., 1969: Numerical simulation of the life cycle of tropical cyclones. *J. Atmos. Sci.*, **26**, 3–40.

Palmen, E., 1948: On the formation and structure of tropical cyclones. *Geophyisca*, **3**, 26–38.

Park, J., K.-A. Park, K. Kim, and Y.-H. Youn, 2005: Statistical analysis of upper ocean temperature response to typhoons from ARGO floats and satellite data. IEEE, pp. 2564–2567.

Perlroth, I, 1969: Effects of oceanographic media on equatorial Atlantic hurricanes. *Tellus*, **21**, 230–244.

Powell, M. D., and S. Houston, 1996: Hurricane Andrew's landfall in South Florida. Part II: surface wind fields. *Wea. and Forecasting*, **11**, 329–349.

Powell, M. D., P. J. Vickery, and T. A. Reinhold, 2003: Reduced drag coefficient for high wind speeds in tropical cyclones. *Nature*, **422**, 279–283.

Price, J. F., 1981: Upper ocean response to a hurricane. *J. Phys. Oceanogr.*, **11**, 153–175.

Price, J. F., 1983: Internal wave wake of a moving storm. Part I: Scales, energy budget and observations. *J. Phys. Oceanogr.*, **13**, 949–965.

Price, J. F., R. A. Weller, and R. Pinkel, 1986: Diurnal cycling: Observations and models of the upper ocean response to diurnal heating, cooling and wind-mixing. *J. Geophys. Res.*, **7**, 8411–8427.

Price, J. F., T. B. Sanford, and G.Z. Forristall, 1994: Observations and simulations of the forced response to moving hurricanes. *J. Phys. Oceanogr.*, **24**, 233–260.

Pudov, V., and S. Petrichenko, 2000: Trail of a typhoon in the salinity field of the ocean upper layer. *Atmospheric and Ocean Physics* (*Izvestiya Acad. Sci.*), **36**, 700–706.

Rao, R. R., 1984: A case study on the influence of the summer monsoonal vortex on the thermal structure of upper central Arabian Sea during the phase of the MONEX-79. *Deep-Sea Res.*, **31**, 1511–1521.

Rogers, R., S. Aberson, M. Black, P. Black, J. Cione, P. Dodge, J. Dunion, J. Gamache, J. Kaplan, M. Powell, L. K. Shay, N. Surgi, and E. Uhlhorn, 2006: The intensity forecasting experiment (IFEX), a NOAA multiple year field program for improving intensity forecasts. *BAMS*, **87**(11), 1523–1537.

Sanford, T. B., P. G. Black, J. Haustein, J. W. Feeney, G. Z. Forristall, and J. F. Price, 1987: Ocean response to a hurricane. Part I: Observations. *J. Phys. Oceanogr.*, **17**, 2065–2083.

Sanford, T. B., J. H. Dunlap, J. A. Carlson, D. C. Webb and J. B. Girton, 2005: Autonomous velocity and density profiler: EM-APEX. Proceedings of the IEEE/OES 8th Working Conference on Current Measurement Technology, IEEE, pp. 152–156.

Sanford, T. B., J. F. Price, J. Girton, and D. C. Webb, 2007: Highly resolved observations and simulations of the oceanic response to a hurricane. *Geophys. Res. Lett.*, **34**, L13604, 5 pp.

Schade, L., and K. Emanuel, 1999: The ocean's effect on the intensity of tropical cyclones: Results from a simple ocean-atmosphere model. *J. Atmos. Sci.*, **56**, 642–651.

Scharroo, R., W. H. Smith, and J.L. Lillibridge, 2005: Satellite altimetry and the intensification of Hurricane Katrina. *EOS*, **86**, 366–367.

Shay, L. K. 2001: Upper ocean structure: Response to strong forcing events. In *Encyclopedia of Ocean Sciences*, Eds. R. A. Weller, S. A. Thorpe, J. Steele, Academic Press International, London, UK, pp. 3100–3114.

Shay, L. K. 2009: Upper ocean structure: A Revisit of the response to strong forcing events. In *Encyclopedia of Ocean Sciences*, Eds. R.A. Weller, S.A. Thorpe, J. Steele, Elsevier Press International, Oxford, UK, pp. 4619–4637.

Shay, L.K. and R.L. Elsberry, 1987: Near-inertial ocean current response to hurricane Frederic. *J. Phys. Oceanogr.*, **17**, 1249–1269.

Shay, L.K., P.G. Black, A.J. Mariano, J.D. Hawkins and R.L. Elsberry, 1992: Upper ocean response to hurricane Gilbert. *J. Geophys. Res.*, **97**, 20,227–20,248.

Shay, L. K., and S. W. Chang, 1997: Free surface effects on the near-inertial ocean current response to a hurricane: A revisit. *J. Phys. Oceanogr.*, **27**, 23–39.

Shay, L. K., A. J. Mariano, S. D. Jacob, and E. H. Ryan, 1998: Mean and near-inertial ocean current response to hurricane Gilbert. *J. Phys. Oceanogr.*, **28**, 858–889.

Shay, L. K., G. J. Goni, and P. G. Black, 2000: Effects of a warm oceanic feature on hurricane Opal. *Mon. Wea. Rev.*, **128**, 1366–1383.

Shay, L. K, and S. D. Jacob, 2006: Relationship between oceanic energy fluxes and surface winds during tropical cyclone passage (Chapter 5). Atmosphere-Ocean Interactions II, *Advances in Fluid Mechanics*. Ed. W. Perrie, WIT Press, Southampton, UK, pp. 115–142.

Shay, L. K., and E. W. Uhlhorn, 2008: Loop Current response to Hurricanes Isidore and Lili. *Mon. Wea. Rev.*, **136**, 3248–3274.

Shay, L. K., and J. Brewster. 2010: Eastern Pacific oceanic heat content estimation for hurricane forecasting. *Mon. Wea. Rev.*, (accepted, revised and resubmitted).

Schott, F., T. N. Lee, and R. Zantopp, 1988: Variability of the structure and transport of the Florida Current. *J. Phys. Oceanogr.*, **18**, 1209–1230.

Sjoberg, B., and A. Stigebrandt, 1992: Computations of the geographical distribution of the energy flux to mixing processes via internal tides and the associated vertical circulation in the ocean. *Deep-Sea Res.*, **39**, 269–291.

Stommel, H., 1948: The westward intensification of wind driven ocean currents. *Trans. Amer. Geophys. Union*, **29**(2), 202–206.

Sun, D. R. Gautam, G. Cervone, Z. Boyeyi, and M. Kaptos, 2006: Comment on Satellite altimetry and the intensification of Hurricane Katrina. *EOS*, **87**(8), 89.

Sturges, W., and R. Leben, 2000: Frequency of ring separations from the Loop Current in the Gulf of Mexico: A revised estimate. *J. Phys. Oceanogr.*, **30**, 1814–1819.

Teague, W. J., E. Jarosz, M. R. Carnes, D. A. Mitchell, and P. J. Hogan, 2006: Low-frequency current variability observed at the shelfbreak in the northeastern Gulf of Mexico: May–October 2004. *Cont. Shelf Res.*, doi:10.1016/j.csr.2006.08.002.

Teague, W. J., E. Jarosz, D. W. Wang, and D. A. Mitchell, 2007: Observed oceanic response over the upper continental slope and outer shelf during Hurricane Ivan. *J. Phys. Oceanogr.*, **37**, 2181–2206.

Toba, Y., N. Iida, H. Kawamura *et al.*, 1990: The wave dependence of sea surface wind stress. *J. Phys. Oceanogr.*, **20**, 705–721.

Tolman, H. L., 2002: User manual and system documentation of WAVEWATCH-III version 2.22. NOAA/NWS/NCEP/OMB Tech. Note 222, 133 pp.

Uhlhorn, E. W., P. G. Black, J. L. Franklin, M. Goodberlet, J. Carswell and A. S. Goldstein, 2007: Hurricane surface wind measurements from an operational stepped frequency microwave radiometer. *Mon. Wea. Rev.*, **135**, 3070–3085.

Uhlhorn, E. W., 2008: Gulf of Mexico Loop Current mechanical energy and vorticity response to a tropical cyclone. PhD Dissertation. RSMAS, University of Miami, Florida, 148 pp.

Wada, A., and N. Usui, 2007: Importance of tropical cyclone heat potential for tropical cyclone intensity and intensification in the western north Pacific. *J. Oceanography*, **63**, 427–447.

Walker, N., R. R. Leben, and S. Balasubramanian 2005: Hurricane forced upwelling and chlorophyll a enhancement within cold core cyclones in the Gulf of Mexico. *Geophys. Res. Letter*, **32**, L18610.

Wang, D. W., D. A. Mitchell, W. J. Teague, E. Jarosz, and M. S. Hulbert, 2005: Extreme waves under Hurricane Ivan. *Science*, **309**, 896.

Wang, Y., J. D. Kepert, and G. J. Holland, 2001: The effect of sea spray evaporation on tropical cyclone boundary layer structure and intensity. *Mon. Wea. Rev.*, **129**, 2481–2500.

Weller, R. A., 1991: Overview of the Frontal Air-Sea Interaction Experiment (FASINEX): a study of air-sea interaction in a region of strong gradients. *J. Geophys. Res.*, **96**, 8501–8516.

Wright, C. W., E. J. Walsh, D. Vandemark, W. B. Krabill, A. W. Garcia, S. Houston, M. Powell, P. Black, and F. D. Marks, 2001: Hurricane directional wave spectrum spatial variations in the open ocean. *J. Phys. Oceanogr.*, **31**, 2472–2488.

Wu, C.-C., C.-Y. Lee, and I-I Lin, 2007: The effect of the ocean eddy on tropical cyclone intensity. *J. Atmos. Sci.*, **64(10)**, 3562–3578.

Yablonsky, R. M, and I. Ginis, 2009: Limitation of one-dimensional ocean models for coupled hurricane-ocean forecasts. *Mon. Wea. Rev.*, doi 10.1175/2009MWR2863.

Yin, X., Z. Wang, Y. Liu, and Y. Xu., 2007: Ocean response to typhoon Ketsana traveling over the northwest Pacific and a numerical model approach. *Geophys. Res. Lett.*, **34**, doi:10.1029/3007GL031477.

Yoshikawa, Y., A. Masuda, K. Marbayashi, M. Ishibashi, and A. Okuno, 2006: On the accuracy of HF radar measurement in the Tsushima Strait, *J. Geophys. Res.*, **111**, C04009.

Zedler, S., T. Dickey, S. Doney, J. Price, X. Yu, and G. Mellor, 2001: Analyses and simulations of the upper ocean's response to hurricane Felix at the Bermuda Testbed Mooring site: 13–23 August 1995. *J. Geophys. Res.*, **107**, 3232.

Zhang, J. A., 2007: An airborne investigation of the atmospheric boundary layer structure in the hurricane force wind regime. Ph.D. Dissertation, RSMAS, University of Miami, Florida, 60pp.

Chapter 4

Movement of Tropical Cyclones

Johnny C. L. Chan
School of Energy and Environment, City University of Hong Kong
Tat Chee Avenue, Kowloon, Hong Kong, China
johnny.chan@cityu.edu.hk

This chapter reviews the physical processes governing the movement of tropical cyclones. Because of the extensive review in the last edition, only a brief summary of the understanding of such processes up to 1995 is presented to provide the main concepts, which include steering and the beta effect, both of which are primarily discussed from a barotropic framework. Since the last edition, more research on the baroclinic processes have been carried out and these are described mostly from the potential vorticity perspective. In addition, a new concept involving the coupling of the strong vertical motion near the eyewall and the Coriolis force (the "gamma effect") that is likely to lead to another "inherent" motion process is also discussed. More recently, some studies have been made on possible changes in the track of a tropical cyclone when it is close to a landmass. The possible effects of topography and variations in roughness and moisture availability over land in modifying the track are also presented.

1. Introduction

Among all elements of tropical cyclone (TC) forecasting, track prediction has always been the top priority for obvious reasons. If the prediction of the future location of a TC is wrong, all other predictions such as intensity, rainfall and storm surge become meaningless. A substantial amount of TC research during the last half a century has therefore been devoted to improving our understanding of the physical processes responsible for TC movement. By the mid 1990s, results from such research have led to a rather complete description of the various possible causes of TC movement, although mostly from a barotropic perspective. Elsberry (1995) has presented a thorough summary of these results in the last version of *Global Perspectives on Tropical Cyclones*. Since then, research has continued on furthering our understanding of the motion processes when baroclinic effects are involved. In addition, the effect of strong vertical motion within the eyewall on TC movement has been examined. Changes in the tracks of TCs near landfall have also been studied in more detail.

Although this chapter focuses on the results of the studies after 1995, some basic concepts of the motion process are briefly described first. Detailed discussions of such concepts can be found in Elsberry (1995). Applications of the concepts described in this chapter in actual TC forecasting are discussed in Chapter 11.

2. A Brief Review of Barotropic Processes

2.1. *Basic Concept of Steering*

The basic idea of TC movement is that the TC vortex is "steered" by its surrounding flow, which is therefore referred to as the "steering flow", analogous to an object floating in a river being pushed downstream by the river flow. Dynamically, steering is the advection of the relative vorticity (ζ) of the TC by the surrounding horizontal flow ($\mathbf{V}$), which would lead to an increase of ζ downstream of $\mathbf{V}$, that is,

$$\frac{\partial \zeta}{\partial t} = -\mathbf{V} \cdot \nabla \zeta \quad (1)$$

where ∇ is the two-dimensional (horizontal) del operator. As a TC tends to move towards the location where $\partial\zeta/\partial t$ is a maximum (Chan, 1984), this advection effect will then cause the TC to move downstream along the direction of $\mathbf{V}$, which is referred to as the "steering flow". Although this concept of steering is very simple, it has been used extensively to explain and predict TC movement with relatively good success especially in short-term ($\leqslant$ 24 h) forecasts (see Chapter 11).

In the tropics, the atmosphere can be considered, to a first approximation, to be largely barotropic. Under this assumption, the vorticity equation can be written as

$$\frac{\partial \zeta}{\partial t} = \underbrace{-\mathbf{V} \cdot \nabla \zeta}_{\text{A}} \underbrace{-\beta v}_{\text{B}} \quad (2)$$

where β is the variation of the Coriolis parameter and v the meridional wind. Comparing with Eq. (1), the only difference is the addition of Term B, which is referred to as the beta effect (see section 2.2). In general, Term A in Eq. (2), which is the steering term, dominates except when the steering flow is weak. Thus, if $\mathbf{V}$ can be calculated or predicted, it is possible to predict the future movement of a TC to a large extent. While Term B is relatively small relative to Term A, it does cause the TC movement to deviate from that specified by $\mathbf{V}$. More discussion of this point is given in Section 2.2. Such a concept has been demonstrated by Chan (1984) and many others.

Although the steering concept has been proven to be largely valid, the practical problem is in the calculation of $\mathbf{V}$. In other words, how should the "surrounding flow" be estimated? Because the TC structure is not barotropic, through what depth of the troposphere should the flow be averaged? Many studies have been devoted to addressing these issues such as the use of geopotential height gradients (Neumann, 1979; Keenan, 1982; Dong and Neumann, 1986; Dong, 1988; Velden and Leslie, 1991) or radial-band averaging of the winds (George and Gray, 1976; Chan and Gray, 1982; Chan, 1985; Franklin, 1990; Kaplan and Franklin, 1991; Roux and Marks, 1991; Marks *et al.*, 1992). One important result from these studies is that a consistent deviation exists between the TC motion vector and the estimated steering flow vector such that this difference vector has a magnitude of around 1–2.5 m s^{-1} and a direction generally pointing westward and poleward for TCs in both hemispheres (Fig. 1). In the barotropic context of Eq. (2), such a deviation must be related to Term B, which is discussed in section 2.2.

2.2. *The Beta Effect*

Chan (1982) and Holland (1983) suggested viewing the effect of the Coriolis force from the advection of the planetary vorticity [i.e., Term B in Eq. (2)] so that advection of planetary vorticity and a subsequent adjustment to a change in relative vorticity would lead to a westward and poleward motion, as portrayed in Fig. 2. By superposing the advection of the vortex due to steering and the beta effect, Chan (1982) showed that the difference in direction and/or speed can vary depending on the direction of the environmental flow.

Chan and Williams (1987) demonstrated in a simulation that if only Term B in Eq. (2) is retained (i.e., only the beta effect without any feedback to the vortex flow, which is referred to as the linear effect), the vortex would only be stretched to the west without any significant displacement (Fig. 3a), which is due to Rossby wave dispersion. Only through the inclusion of the nonlinear term can the vortex move towards the northwest (Fig. 3b). The translation speed of around 2 m s^{-1} and the direction are very similar to those observed (see Fig. 1).

Fiorino and Elsberry (1989) provided an explanation of the nonlinear effect from another

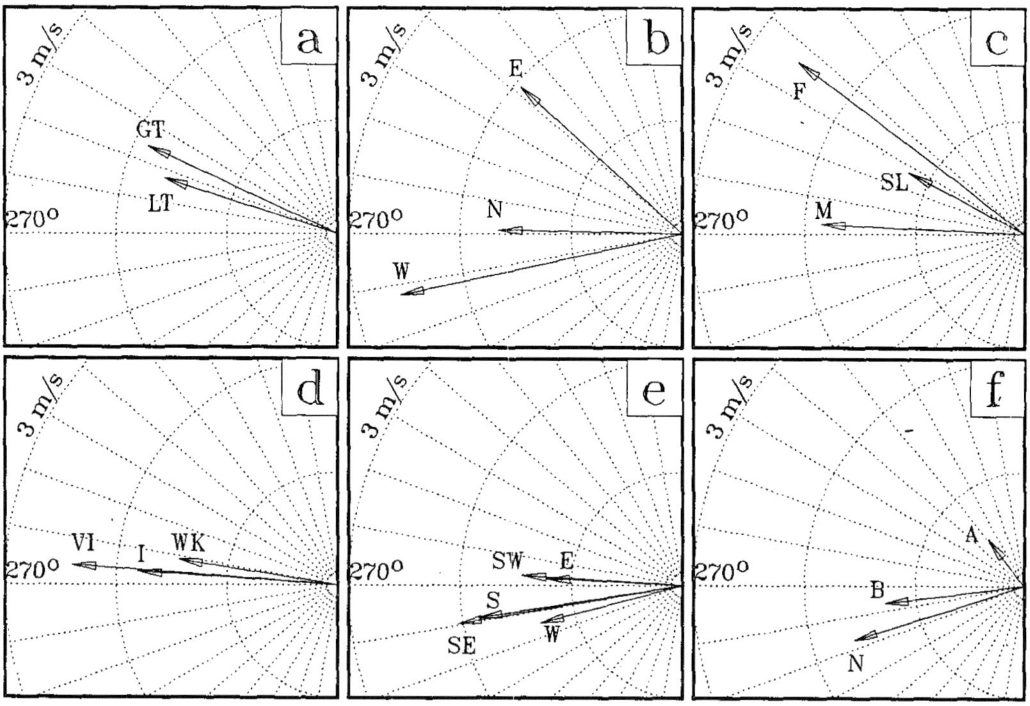

Figure 1. Difference vectors of TC motion minus the composite steering flow for various stratifications: (a) latitude relative to 20°N, (b) direction of TC movement, (c) speed of TC movement, (d) intensity, all from Chan and Gray (1982) for the western North Pacific, and (e) direction of TC movement, and (f) recurvature, both from Holland (1984). (from Carr and Elsberry, 1990).

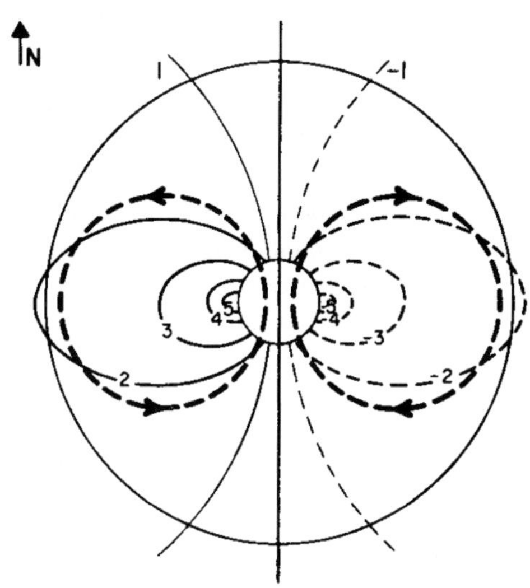

Figure 2. Field of $\partial\zeta/\partial t$ in arbitrary units, centred on a symmetric, non-divergent Northern Hemisphere cyclone due to the beta effect with no environmental flow. Heavy dashed lines show the induced secondary circulation as a result of the changes in ζ (from Holland, 1983).

perspective. They decomposed the flow and ζ into an axially-symmetric and an asymmetric component, i.e.,

$$\begin{aligned} \mathbf{V} &= \mathbf{V}_s + \mathbf{V}_a \\ \zeta &= \zeta_s + \zeta_a. \end{aligned} \tag{3}$$

where the subscripts s and a refer to the axially-symmetric and asymmetric components respectively. Equation (2) can then be written as (with the environmental flow being zero)

$$\frac{\partial \zeta_a}{\partial t} = \underbrace{-\mathbf{V}_a \cdot \nabla \zeta_s}_{\text{C}} \underbrace{-\mathbf{V}_s \cdot \nabla \zeta_a}_{\text{D}} \underbrace{-\beta v}_{\text{B}}. \tag{4}$$

Term A in Eq. (2) is thus decomposed into Terms C and D in Eq. (4), which are the asymmetric advection of symmetric vorticity (AASV) and symmetric advection of asymmetric vorticity (SAAV). The beta effect (Term B) first causes the development of ζ_a, the field of which is then rotated by the vortex flow

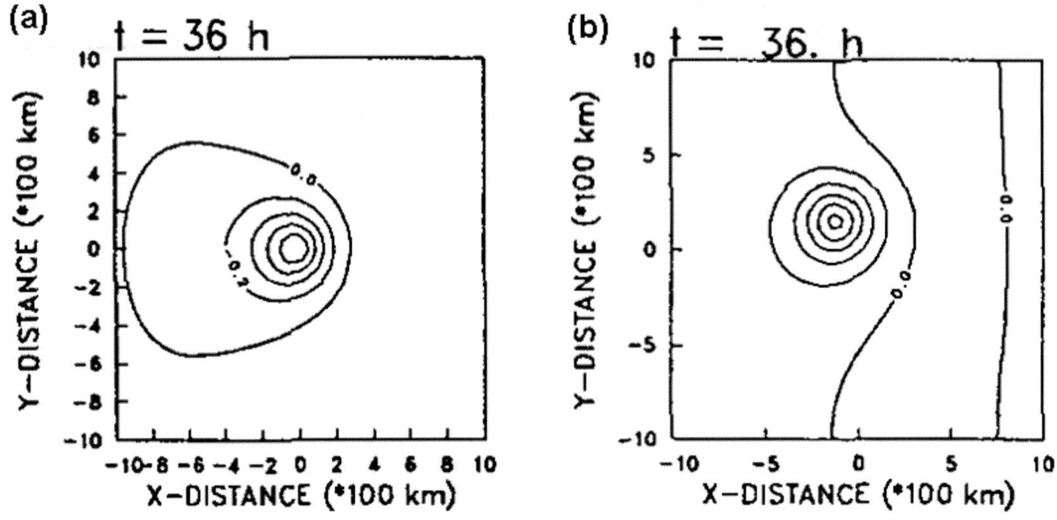

Figure 3. Streamfunction fields after integrating Eq. (2) for 36 h when (a) only the linear term (Term B) is and (b) both terms in Eq. (2) are included. Contour interval in (a) is $0.2 \times 10^5\ \mathrm{m^2\,s^{-1}}$, and in (b) $0.1 \times 10^6\ \mathrm{m^2\,s^{-1}}$ (from Chan and Williams, 1987).

(the SAAV term). Fiorino and Elsberry (1989) showed that these two terms are nearly balanced after an initial adjustment such that the asymmetric streamfunction (and the ζ_a) field has a southwest-northeast orientation (Fig. 4). The gradient of the streamfunction is $\mathbf{V}_a$, which then

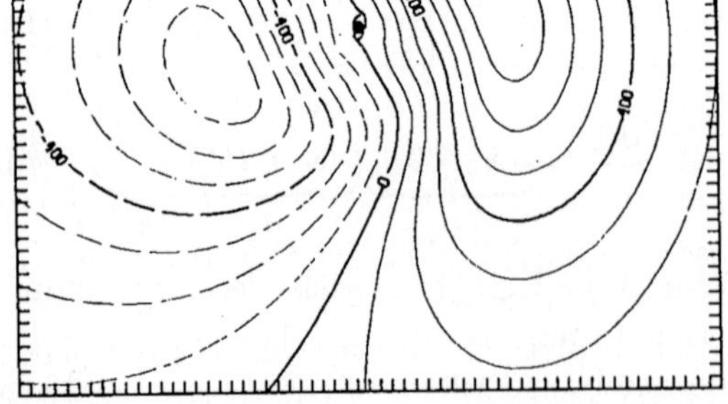

Figure 4. Asymmetric streamfunction (unit: $10^4\ \mathrm{m^2\,s^{-1}}$, contour interval: $1 \times 10^5\ \mathrm{m^2\,s^{-1}}$) after integrating Eq. (2) for 24 h when both terms in Eq. (2) are included. Solid and dashed lines indicate anticyclonic and cyclonic streamfunctions respectively (from Fiorino and Elsberry, 1989).

advects the vortex relative vorticity towards the northwest through the dominant AASV term. The asymmetric flow $\mathbf{V}_a$ near the vortex centre has been referred to as the ventilation flow and the paired circulations associated with the asymmetric streamfunction (see Fig. 4) have been known as the beta gyres.

Thus, a barotropic vortex on a beta plane in the absence of any background flow will move westward and poleward due to the nonlinear beta effect. The instantaneous effect described in Fig. 2 cannot provide an adequate explanation of this movement. Many researchers have since provided further descriptions of the beta gyres (e.g. Peng and Williams, 1990; Carr and Williams, 1989; Smith *et al.*, 1990; Willoughby, 1988, 1990; Shapiro and Ooyama, 1990).

The beta effect described above assumes a certain vortex structure. Changes in such a structure may therefore modify the beta effect. In addition, if a relative vorticity gradient exists in the surrounding flow, Terms C and D will be modified, and hence the beta effect will also be modified. These effects have been studied by many researchers (DeMaria, 1985; Evans *et al.*, 1991; Ulrich and Smith, 1991; Smith, 1991; Li and Wang, 1994; Williams and Chan, 1994; Wang and Li, 1995) and summarized in Elsberry (1995).

To summarize, the rotation of the earth introduces an inherent westward and poleward movement of a vortex that is known as the beta effect through the nonlinear interaction between the advection of the earth's vorticity and the resulting changes in relative vorticity. Because this beta effect exists for all cyclones on the earth, the movement of the TC vortex can be considered as due to the resultant of the steering flow plus the beta effect. If the surrounding flow is not uniform but possesses relative vorticity or a relative vorticity gradient, the beta effect could be modified.

Chan (1997) summarized the different processes associated with TC motion in a barotropic framework (Fig. 5). The interaction between the environmental flow and the TC vortex leads to the generation of azimuthal wavenumber-1 (WN-1, e.g. the beta gyres) and wavenumber-2 (WN-2) circulations (due to the interactions between the vortex flow and the vorticity associated with the environmental flow), as well as the advection of vortex vorticity (i.e., steering). Chan and Cheung (1998) showed that these three components can interact among themselves such that WN-1 or WN-2 can draw energy from each other or from the environmental flow so that each could grow or decay, which could then lead to different vorticity advection patterns. The sum of this resultant advection and that of steering then determines TC motion.

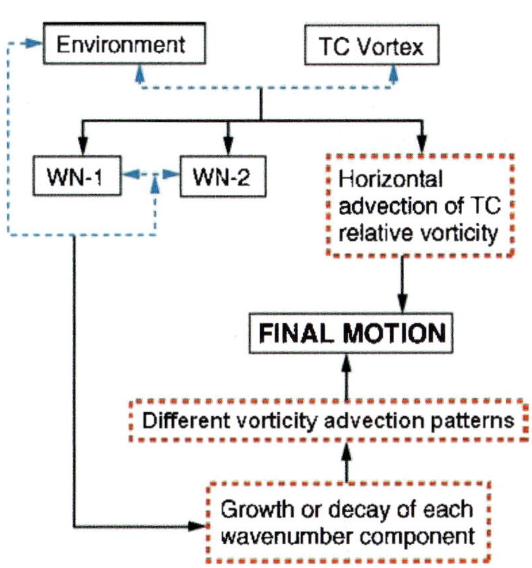

Figure 5. Schematic showing the various processes and interactions (dashed arrows) involved in TC motion in a barotropic framework. Solid boxes indicate the different components present and the dot-dashed boxes are the processes (from Chan, 1997).

3. Binary Interaction

Occasionally, two TCs are in close proximity such that their circulations interact and produce complicated tracks, with sometimes a merging of the vortices (e.g. Chan and Lam, 1989; Holland and Lander, 1993; Lander, 1994, 1995; Lander and Holland, 1994; Chan, 1995). Fujiwhara (1923) was the first to identify such possible binary vortex interactions, which therefore have been known as the Fujiwhara effect. Brand (1970) studied such interactions for a large number of binary TCs in the western North Pacific and found that the two TCs generally rotated cyclonically around each other. The rotation speed increases as the distance between the two TCs decreases. Neumann (1993) showed that the frequency of occurrence of binary systems is the highest in the western North Pacific (about once per year on the average), with the eastern North Pacific and the Australian/Southwest Pacific region having the next highest frequencies.

Lander and Holland (1994) proposed a general model that describes the interaction between two vortices, which begins with an approach of the two vortices towards each other, followed by a "capture" process, then either a final merger or an escape from the interaction (Fig. 6). They showed that merging does not always take place. Instead, the smaller cyclone is generally swept into the horizontally and vertically sheared flow of the larger cyclone so that the former dissipates before merger can occur.

To understand more of the physical mechanisms, Chang (1983) performed a numerical study of such an interaction and suggested that the convergence/divergence of the flow was responsible for the mutual attraction of the vortices. In a non-divergent atmosphere, the vortices would simply rotate around each other

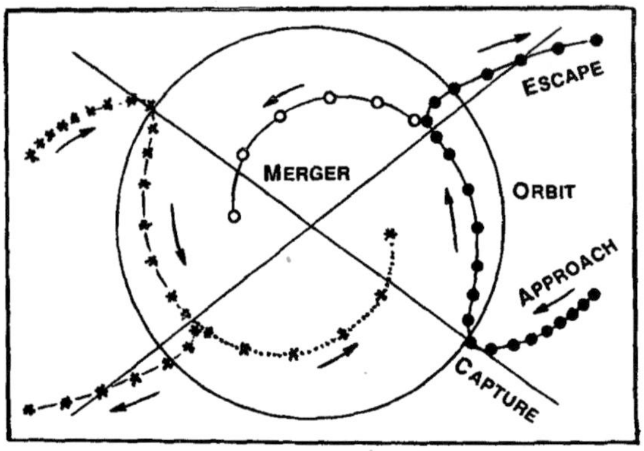

Figure 6. A schematic of binary interaction of two tropical cyclones (from Lander and Holland, 1994).

without merging. However, DeMaria and Chan (1984) showed that whether two adjacent vortices would repel or attract depends on the structure of their vorticity profiles. Further studies on binary interaction have been performed by others (e.g., Smith, 1990; Ritchie and Holland, 1994; Chan and Law, 1995). Chan and Law (1995) also showed that if the two vortices are embedded in a non-uniform environment (e.g., one that possesses a horizontal shear), the binary system can interact with the vorticity field of the environment to produce different movements of the vortices, which have also been found by Dong and Neumann (1983) based on observations. Details can be found in the summary in Elsberry (1995).

In summary, when two TCs are in close proximity, their interaction involves not only their own circulations, as suggested in Fig. 6, but also the interactions among these circulations and that of the environment through the processes described in section 2.

4. The Gamma Effect

Equation (2) was derived from the equations of motion in which the Coriolis force is only present when the Coriolis parameter is multiplied by the horizontal wind components (e.g., Holton, 1992). This assumption is usually justified using a scale analysis that the other components of the Coriolis force ($2\Omega w\cos\phi$ in the x-momentum equation and $2\Omega u\cos\phi$ in the vertical momentum equation) are small for synoptic-scale phenomena. Within the TC eyewall, vertical motion can be $\sim$1–10 m s^{-1} so that the first term is comparable in magnitude to the term $2\Omega v\sin\phi$ and should not be neglected. Similarly, vertical accelerations ($\partial w/\partial t$) within the TC eyewall are large so that in a non-hydrostatic model, the $2\Omega u\cos\phi$ term has a magnitude comparable to $\partial w/\partial t$ and therefore should also be included.

Liang and Chan (2005) found that the inclusion of the term $2\Omega w\cos\phi$ in the x-momentum equation resulted in a southwestward displacement of a TC with a speed of $\sim$1 km h^{-1} even on an f plane (Fig. 7a). On a beta plane, inclusion of this term gives a vortex track that is generally west to southwest

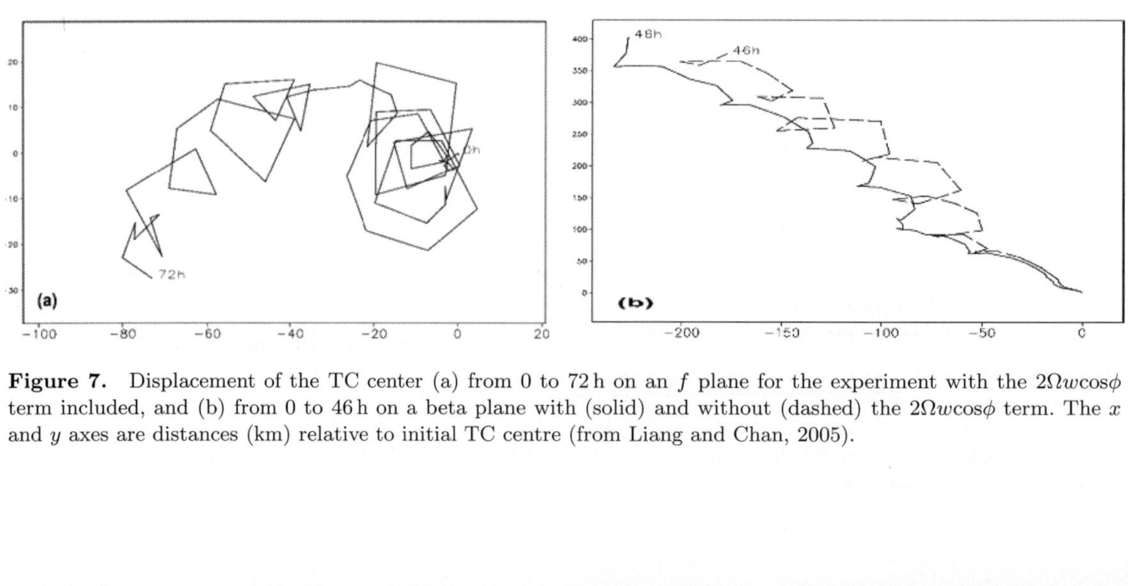

Figure 7. Displacement of the TC center (a) from 0 to 72 h on an f plane for the experiment with the $2\Omega w\cos\phi$ term included, and (b) from 0 to 46 h on a beta plane with (solid) and without (dashed) the $2\Omega w\cos\phi$ term. The x and y axes are distances (km) relative to initial TC centre (from Liang and Chan, 2005).

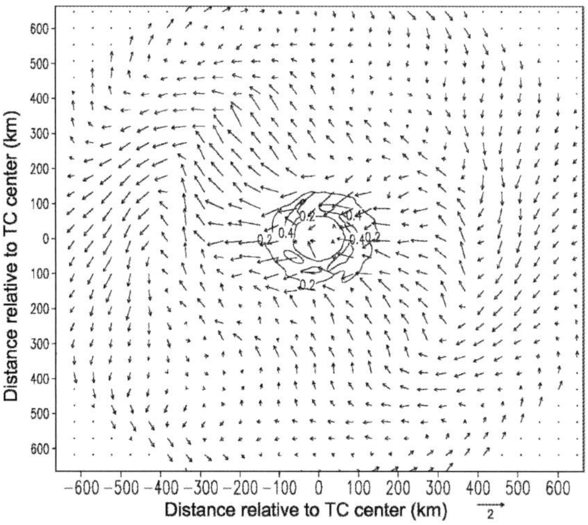

Figure 8. Asymmetric winds (vectors) and upward motion (contours, $\mathrm{m\,s^{-1}}$) in the experiment with the gamma effect included (from Liang and Chan, 2005).

of the inherent northwestward track due to the beta effect (Fig. 7b). This term is found to generate an asymmetric wind structure with a generally easterly flow near the center (Fig. 8), which therefore causes the vortex to be displaced toward the southwest.

The easterly flow can be explained by considering the vorticity equation (Liang and Chan, 2005). If the vortex is symmetric, the only term that contributes to a relative vorticity tendency on an f plane is due to the meridional gradient of w, $2\Omega\cos\phi(\partial w/\partial y)$. This term produces a negative tendency north of the eyewall to the north of the TC, but positive south of the eyewall to the south of the TC. The net adjustment of the wind to this vorticity tendency distribution is to set up a pair of gyres (the gamma gyres), with anticyclonic flow to the north and cyclonic flow to the south, which therefore produces a generally easterly flow across the TC center, as shown in Fig. 8. On a beta plane, this effect is then superposed on the inherent northwestward movement due to the beta effect.

Although this effect is not very large, it should be taken into consideration in understanding and prediction of TC motion. The effect of the term $2\Omega u\cos\phi$ has yet to be studied.

5. Baroclinic Effects

Ngan and Chan (1995) showed that in about 30% of the 66 cases they studied, the TC motion cannot be adequately explained from barotropic considerations alone. In other words, contributions from baroclinic processes appear to be important in many cases, which should be expected as two baroclinic effects exist in TCs, viz. vertical wind shear and diabatic heating. In addition, because the vortex structure varies in the vertical, the beta effect will also be different. Differential vorticity advection and vertical wind shear will therefore lead to a motion vector that cannot be explained by barotropic processes. These possible processes have been discussed in Wang and Li (1992), Shapiro (1992), Wu and Emanuel (1993), Flatau *et al.* (1994), Wang and Holland (1995), and summarized in Elsberry (1995).

In this section, these effects (diabatic heating and vertical wind shear) are combined based on the potential vorticity (PV) concept. In a baroclinic atmosphere, PV is conserved under adiabatic conditions and can only be modified by diabatic and frictional processes. Thus, the study of the physics of TC motion under baroclinic conditions can be achieved by evaluating the various terms in the PV equation in the free atmosphere (so that only the diabatic process needs to be considered), which is analogous to the absolute vorticity conservation in a barotropic atmosphere that was discussed in section 2.2.

Wu and Wang (2000) solved the PV tendency (PVT) equation in a numerical model and found that a cyclonic vortex tends to move towards the area of maximum $(\partial q_1/\partial t)$, where q_1 is the azimuthal wavenumber-one (WN-1) component of PV (q), which is an extension of the result of Chan (1984) who found from observations that a TC tends to move towards the area of maximum $(\partial\zeta/\partial t)$ (see section 2.1 and Fig. 1).

Symbolically, the time tendency of q_1 can be written as

$$\frac{\partial q_1}{\partial t} = \Lambda_1\left[\begin{Bmatrix}\text{Horizontal}\\ \text{advection}\end{Bmatrix} + \begin{Bmatrix}\text{Vertical}\\ \text{advection}\end{Bmatrix} + \begin{Bmatrix}\text{Diabatic}\\ \text{heating}\end{Bmatrix} + \{\text{Friction}\}\right] \tag{6}$$

where Λ_1 is an operator that extracts the WN-1 component of the terms. Compared with the barotropic vorticity equation (Eq. (4)), the two extra terms (other than friction which is also present in the barotropic equation) are vertical advection and diabatic heating. In a TC, vertical velocity is strongest in the eyewall, and in the tropical atmosphere, the largest vertical gradient of q is near the tropopause. In areas away from the TC eyewall, the tropopause, and the boundary layer (where friction is important), the diabatic heating associated with convection becomes the only additional term. In the Wu and Wang (2000) simulations, the vertical advection term was generally much smaller in the low to mid troposphere (Fig. 9). The sum of the vectors that point towards the area of maximum horizontal advection and diabatic heating agrees very well with the vortex motion vector.

Analogous to Eq. (4), Chan *et al.* (2002) decomposed $\mathbf{V}$ and q into their symmetric and WN-1 components so that the horizontal advection (HA) term becomes

$$\Lambda_1\{\text{HA}\} = \underbrace{-\mathbf{V}_1 \cdot \nabla q_s}_{\text{E}} \underbrace{-\mathbf{V}_s \cdot \nabla q_1}_{\text{F}} \tag{7}$$

where the subscripts s and 1 indicate the symmetric and WN-1 components. Note that the symmetric advection of symmetric PV is zero and advection of WN-1 PV by the WN-1 flow generally does not yield a WN-1 pattern.

Term E in Eq. (7) is the asymmetric advection of symmetric PV (AASPV), which may seem to be the contribution from the steering flow. However, $\mathbf{V}_1$ actually consists of the environmental flow, the "ventilation flow" associated with the β effect and any other asymmetric flow that may develop (e.g. due to asymmetric convection). In addition, while q_s is primarily equal to the PV of the vortex, any symmetric component of the environmental PV as well as symmetric convection will also have a contribution. In other words, the beta effect becomes implicit in Term E so that understanding of the vortex motion no longer requires an explicit isolation of this effect. This PV approach has the advantage that the sometimes-opposing contributions from vertical wind shear (which is also associated with a horizontal PV gradient) and the beta effect (Shapiro, 1992) do not have to be treated separately.

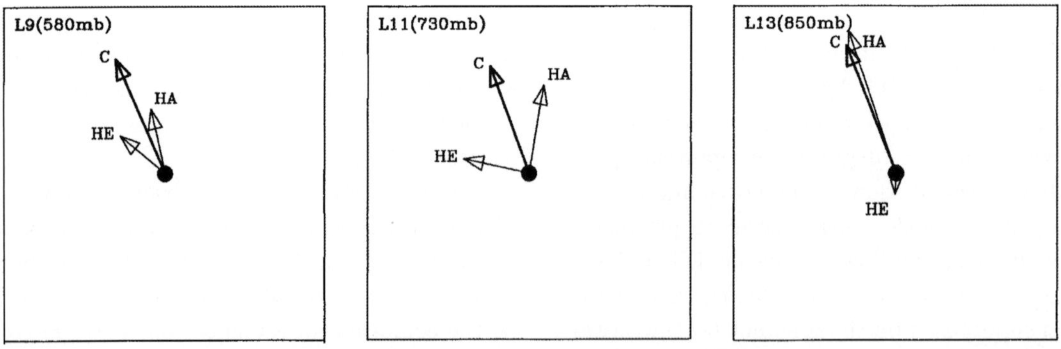

Figure 9. Vectors representing the contributions of the horizontal advection (HA) and diabatic heating (HE) terms at three levels of the atmosphere after 36 h of integration in the Wu and Wang (2000) diabatic experiment on a beta plane with a resting environment. The vector C represents the resultant of the three vectors. The vertical advection term is too small to be seen. The scale of the arrows is the same in each figure.

Similarly, Term F in Eq. (7) represents the symmetric advection of asymmetric (i.e. WN-1) PV (SAAPV). The WN-1 PV consists of the beta gyres, other asymmetric features in the environmental flow such as vertical shear, and contributions from asymmetric convection. Note that the symmetric flow contains both the vortex circulation and the symmetric component of the environmental flow.

Wu and Wang (2000) expanded the diabatic heating term DH as

$$\mathrm{DH} = g\left[-(f+\zeta)\frac{\partial Q}{\partial p} - \frac{\partial u}{\partial p}\frac{\partial Q}{\partial y} + \frac{\partial v}{\partial p}\frac{\partial Q}{\partial x}\right] \quad (8)$$

where g is the gravity and Q the convective heating rate. Thus, the DH term depends on the vertical and horizontal gradients of convective heating, and vertical wind shear. In other words, the contribution of the DH term to the WN-1 PVT is not simply through convective heating. When compared with the full vorticity equation (e.g., Holton, 1992), the first term is similar to the divergence term while the second and the third terms are analogous to the twisting term. The physical interpretations are also similar.

Chan *et al.* (2002) further noted that if the convection is rather symmetric, the magnitude of the WN-1 DH term should be small so that the HA term would dominate. A steady-moving TC (i.e., one without much change in direction or speed) is likely to be under the influence of an appreciable and less-variable environmental flow. In this case, the HA process, and especially the AASPV term, is probably more important than the DH process. If the track is not smooth, it is more likely that the latter term is having a larger contribution.

Based on the above discussion, Chan *et al.* (2002) proposed the following general conceptual framework to describe TC motion in a baroclinic atmosphere. A TC will move towards an area of maximum WN-1 PVT, which is mainly contributed by the HA and DH processes. When the TC motion is steady, the HA process dominates. In this case, although the contribution to the overall PVT comes largely from the AASPV process, that from the SAAPV process could also be appreciable. A significant contribution of the DH process can lead to a non-smooth TC track. Turning of the TC may be partly related to a rotation of the asymmetric DH distribution. Because asymmetric convection contributes to the WN-1 PV, the SAAPV process also increases in importance relative to the AASPV process. Changes in the distribution of one or more of these processes then lead to subsequent changes in the direction of TC motion.

The validity of this framework can be illustrated with the numerical simulations of Wu and Wang (2000) who studied the motion of an idealized vortex on a beta plane with no mean flow. Earlier simulations by Willoughby (1988, 1992) also supported the idea that rotation of convective heating can lead to a turning of a TC. Chan *et al.* (2002) examined over 500 cases of TCs over the western North Pacific and found that this proposed framework indeed can explain the TC motion in most cases. Thus, by considering the changes in PV due to the processes shown in Eqs. (7) and (8), the movement of TCs can be quite well understood.

To summarize, the processes governing TC motion in a baroclinic atmosphere are much more complicated. By considering the temporal changes in PV, the different processes can be much better diagnosed and understood. While the steering concept (PV advection rather than vorticity advection) may still be the dominant term in many cases, the contributions of vertical wind shear and diabatic heating must be taken into consideration. For example, the presence of vertical wind shear introduces a horizontal PV gradient that will change the PV advection. Further, a vertical wind shear can lead to an azimuthally-asymmetric distribution of convection (see Chapter 1), and hence the diabatic heating distribution. From a numerical prediction perspective, these results highlight the importance of a correct prediction of the diabatic heating, which is related to the vortex structure, convective parameterization, and moisture distribution in the model.

6. Track Changes Near Landfall

As a TC is about to make landfall, its circulation will interact with the land surface that has a different roughness and much less moisture supply. Most of the studies on the problem of landfall have focused either on the rainfall or the winds (see Chapter 6). However, recent studies have shown that such changes in roughness and moisture may modify the tracks as well. Topography has also been found to modify the TC track.

In barotropic simulations on an f plane, Kuo *et al.* (2001) showed that under certain conditions, a TC-like vortex can be "trapped" by topography due to the topographic beta effect such that the vortex has a clockwise path around an area of elevated topography. An example of this effect is shown in Fig. 10. Such a path depends on the "effective" topography, the width of the topography, and the parameters associated with the vortex such as radius and intensity. Such "looping" motion is similar to some of the tracks of TCs making landfall over the Island of Taiwan (see Kuo *et al.* 2001).

The Wong and Chan (2006) numerical simulations indicated that a TC-like vortex on an f plane will drift towards land that has a substantially large roughness. The average drift speed is about $1\,\mathrm{m\,s^{-1}}$, and the track contains oscillations (Fig. 11). Such a drift is present whether the "land" in the model has the same amount of moisture as the sea or has only 5% of the moisture of the sea. If only the land moisture is reduced but the roughness is the same as that of the sea, the vortex does not exhibit any significant motion deviations (see Fig. 1 of their paper). These results suggest that it is the discontinuity in roughness that causes the vortex to drift towards the land.

Wong and Chan (2006) explained this drift in terms of the modification of the near-surface winds by the rough land surface. Reduction of winds over land leads to a

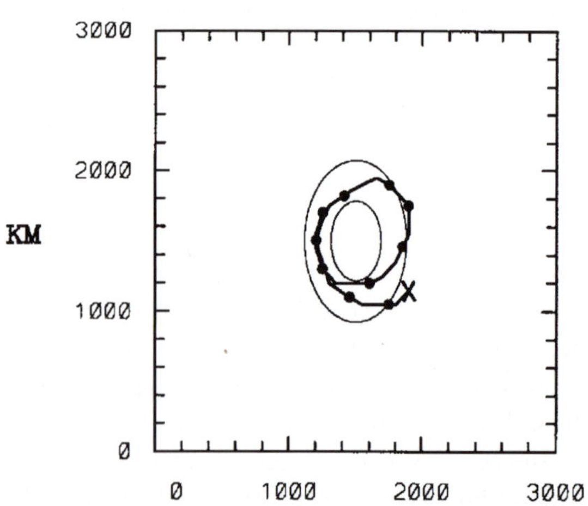

Figure 10. An example (from Kuo *et al.*, 2001) of a vortex track around an elevated topography on an f plane. The topography contours (thin solid ellipses) start at 1000 m and increment by 1000 m. The time interval between adjacent black dots is 12 h and the total integration time is 120 h. The vortex initial position is indicated by the "×" symbol.

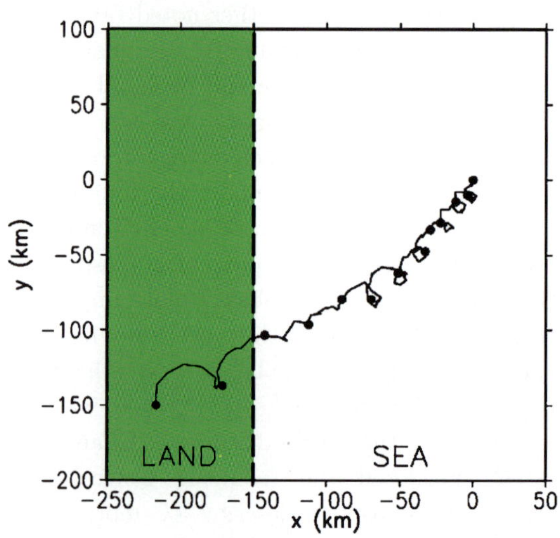

Figure 11. Track of the TC surface center on an f plane with the land (gray shaded) having a much larger roughness than the sea. The dots denote 12-hourly TC positions. The origin is the location of the domain center (re-drawn from Wong and Chan, 2006).

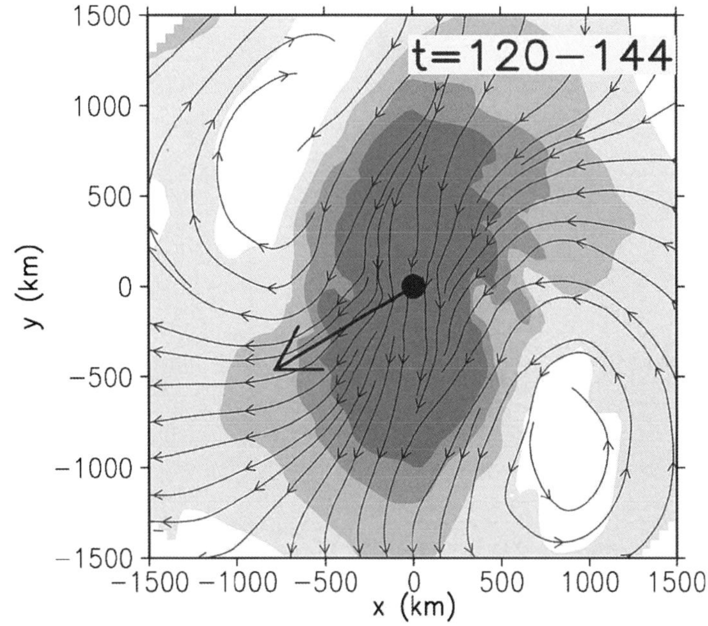

Figure 12. Time composite between 120 and 144 h of the asymmetric component of the lower tropospheric (about 900 to 550 hPa) flow of the experiment shown in Fig. 11. The TC center is marked by a dot at the origin and the big arrow, magnified by a factor of 10 such that a length equal to 1000 km on the horizontal axis represents an actual distance of 100 km, indicates the overall movement of the center during this period. The shading indicates the wind speed with the grades shown in the bar below (re-drawn from Wong and Chan, 2006).

large-scale asymmetric convergence/divergence along the coast. The large-scale boundary-layer asymmetric divergence/convergence induces an asymmetry in vertical motion. An asymmetric flow also develops in the lower troposphere because of the creation/reduction of asymmetric vorticity by the divergence term in the vorticity equation. Adjustment of the atmosphere to these changes in vorticity then generates a pair of "gyres" that steer the TC towards the coast (Fig. 12). Such changes also lead to changes in convection, and hence to the diabatic contributions to the PV tendency (see section 5). A combination of the horizontal advection, vertical advection, and the diabatic heating terms then results in the drift of the vortex towards the coast, which is evident in the PV budget calculations (Fig. 13).

These results suggest the existence of another "inherent" motion of a TC when it encounters a discontinuity in roughness such that a TC is "attracted" towards the area of higher roughness. To study whether such an inherent effect is still present in a more realistic situation, Szeto and Chan (2009) performed similar experiments on a beta plane. Under such conditions, a pair of counter-rotating gyres will form and the vortex experiences an inherent northwestward movement (in the Northern Hemisphere) (see section 2.2). Because the axis of this pair of gyres is southwest-northeast, the flow associated with these gyres is likely to interact differently with the land surface for different orientations of the coastline. Two experiments are therefore carried out, one with a north-south (NS) oriented coastline and another east-west (EW) oriented. While the dominant motion of the vortex is still towards the northwest, the discontinuity in roughness between the land and the sea results in the vortex deviating from this predominant motion, with the vortex deviating to the right of the

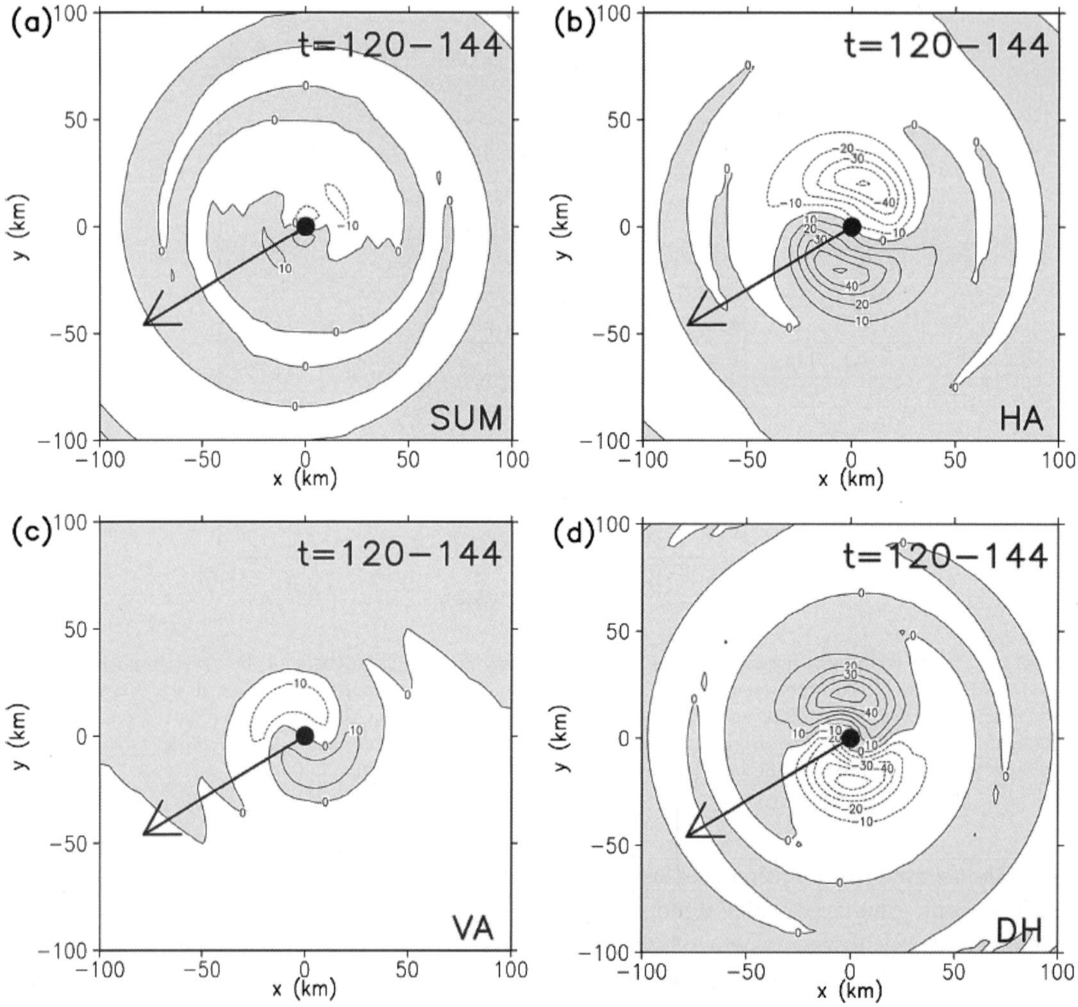

Figure 13. Wavenumber-1 component of the time composite between 120 and 144 h of the terms in the Ertel potential vorticity equation of the lower tropospheric (about 900 to 550 hPa) flow of the experiment shown in Fig. 11: (a) sum of the terms (friction term neglected), (b) horizontal advection term, (c) vertical advection term, and (d) diabatic heating term. Positive values are shaded. The big arrow indicates the overall movement of the center during that period. (Unit is 10^4 PVU s^{-1}) (re-drawn from Wong and Chan, 2006).

beta-induced track (Fig. 14a) in the NS experiment by about 30 km and to the left in the EW experiment by about 15 km (Fig. 14b) at landfall. Because the grid resolution of the simulations is 5 km, such deviations are likely to be real.

Szeto and Chan (2009) explained this deviation by examining the differences in the flow patterns with and without the land. Because of the different orientations of the coastline, the superposition of the two inherent effects (the beta gyre circulation and the land-induced circulation) leads to different "net" flows (Fig. 15), and hence different convergence/divergence patterns near the land surface in the NS and EW experiments, which leads to different advection and convection patterns. Applying arguments similar to those in Wong and Chan (2006), they explained why such differences in the deviations occur.

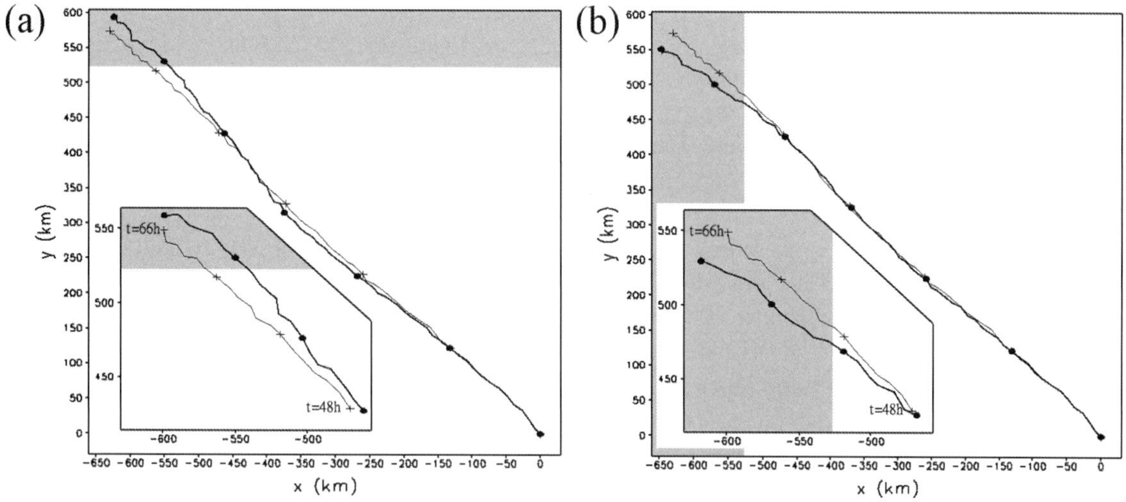

Figure 14. Track (thick solid) and 12-hourly positions (dots) of the TC surface center in the experiment with (a) an east-west oriented and (b) a north-south oriented coastline on a beta plane. The thin line in each figure denotes the track and 12-hourly positions of the TC in a control experiment with no land. Land surface is shaded and the origin is the position of TC at the initial time ($t = 0$ h). The inset on the lower-left of each figure shows the enlarged area near landfall (from Szeto and Chan, 2009).

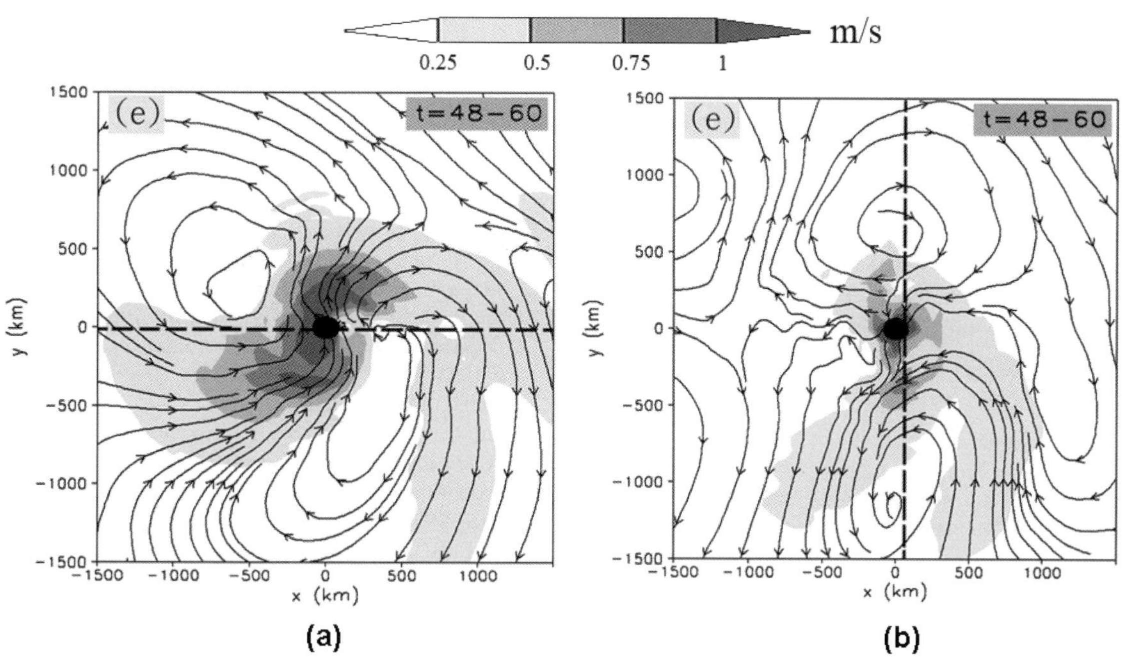

Figure 15. Time composite during the 48–60 h period of the lower tropospheric (about 900–550 hPa) flow in the beta plane experiments with (a) an east-west oriented and (b) a north-south oriented coastline relative to the control experiment (without land). The TC center is denoted by a dot at the origin. Coastline is denoted by thick dashed line, with the land being to the north in (a) and west in (b) of the line. Shading indicates speed of the flow with the scale on the top of the figure (re-drawn from Szeto and Chan, 2009).

7. Summary and Discussion

The study of TC motion has come a long way since the first study by Rossby (1948). As pointed out in the Introduction, the most important element in TC forecasting is its movement. Therefore, a majority of TC research has been focused on motion. Research activities in the 1980s and early 1990s have led to the establishment of the theory that in a barotropic framework, TC motion is mostly due to the steering and the beta effect and their nonlinear interaction. With the barotropic processes well understood, research moved to the study of the baroclinic processes in which the TC movement is explained by the PV tendency. It is noted that since the Chan *et al.* (2002) paper, research on the physics of general TC motion has been almost non-existent, which suggests that most scientists are quite content with the current theories of TC motion.

Recent research has focused on the effects of land on TC tracks, which include topography, changes in the roughness, and moisture availability. All these elements will modify the tendencies of PV due to the development of a secondary circulation that changes the advection and convection patterns, and hence TC motion.

Although we largely understand the physical processes involved in TC motion, how these processes are represented in numerical prediction models becomes important, which is discussed in Chapter 11.

References

Bell, G. J., 1980: Some aspects of tropical cyclone movement and forecasting. *Symposium on Typhoons*, World Meteorological Organization, Shanghai, China, Oct. 6–11, pp. 104–109.

Brand, S., 1970: Interaction of binary tropical cyclones of the western North Pacific. *J. Appl. Meteor.*, **9**, 433–441.

Carr, L. E., III and R. T. Williams, 1989: Barotropic vortex stability to perturbations from axisymmetry. *J. Atmos. Sci.*, **46**, 3177–3196.

Carr, L. E., III and R. L. Elsberry, 1990: Observational evidence for prediction of tropical cyclone propagation relative to environmental steering. *J. Atmos. Sci.*, **47**, 542–46.

Chan, J. C. L., 1982: On the physical processes responsible for tropical cyclone motion. *Atmospheric Science Paper No. 358*, Colorado State University, Ft. Collins, Colorado, USA, 200 pp.

Chan, J. C. L., 1984: An observational study of the physical processes responsible for tropical cyclone motion. *J. Atmos. Sci.*, **41**, 1036–1048.

Chan, J. C. L., 1985: Identification of the steering flow for tropical cyclone motion using objectively-analyzed wind fields. *Mon. Wea. Rev.*, **113**, 106–116.

Chan, J. C. L., 1995: Performance of global and regional NWP models in their prediction of Typhoon Nat (1991). *Wea. Forecasting*, **10**, 400–410.

Chan, J. C. L., 1997: Barotropic dynamics of tropical cyclone motion as inferred from the TCM-90 Final Analysis dataset. *Preprints, 22nd Conf. on Hurricanes and Trop. Meteor.*, Amer. Meteor. Soc., Ft. Collins, Colorado, USA, May 19–23, pp. 625–626.

Chan, J. C. L. and W. M. Gray, 1982: Tropical cyclone movement and surrounding flow relationships. *Mon. Wea. Rev.*, **110**, 1354–1374.

Chan, J. C. L. and R. T. Williams, 1987: Analytical and numerical studies of the beta-effect in tropical cyclone motion. Part I: Zero mean flow. *J. Atmos. Sci.*, **44**, 1257–1265.

Chan, J. C. L. and H. Lam, 1989: Performance of the ECMWF model in predicting the movement of Typhoon Wayne (1986). *Wea. Forecasting*, **4**, 234–245.

Chan, J. C. L. and A. C. K. Law, 1995: The interaction of binary vortices in a barotropic model. *Meteor. Atmos. Phys.*, **56**, 135–155.

Chan, J. C. L. and K. K. W. Cheung, 1998: Characteristics of the asymmetric circulation associated with tropical cyclone motion. *Meteor. Atmos. Phys.*, **65**, 183–196.

Chan, J. C. L., F. M. F. Ko and Y. M. Lei, 2002: Relationship between potential vorticity tendency and tropical cyclone motion. *J. Atmos. Sci.*, **59**, 1317–1336.

Chang, S. W., 1983: A numerical study of the interactions between two tropical cyclones. *Mon. Wea. Rev.*, **111**, 1806–1817.

DeMaria, M. and J. C. L. Chan, 1984: Comments on "A numerical study of the interactions between two tropical cyclones". *Mon. Wea. Rev.*, **112**, 1643–1645.

DeMaria, M., 1985: Tropical cyclone motion in a nondivergent barotropic model. *Mon. Wea. Rev.*, **113**, 1199–1210.

Dong, K., 1988: On the relationship between tropical cyclone motion and intensity. *Mon. Wea. Rev.*, **116**, 964–968.

Dong, K. and C. J. Neumann, 1983: On the relative motion of binary tropical cyclones. *Mon. Wea. Rev.*, **111**, 945–953.

Dong, K. and C. J. Neumann, 1986: The relationship between tropical cyclone motion and environmental geostrophic flows. *Mon. Wea. Rev.*, **114**, 115–122.

Elsberry, R. L., 1990: International experiments to study tropical cyclones in the western North Pacific. *Bull. Amer. Meteor. Soc.*, **71**, 1305–1316.

Elsberry, R. L. (Ed.), 1995: *Global Perspectives on Tropical Cyclones.* WMO-/TD-No. 693, World Meteor. Organization, Geneva, 289 pp.

Evans, J. L., G. J. Holland and R. L. Elsberry, 1991: Interaction between a barotropic vortex and an idealized subtropical ridge. I. Vortex motion. *J. Atmos. Sci.*, **48**, 301–314.

Fiorino, M. and R. L. Elsberry, 1989: Some aspects of vortex structure related to tropical cyclone motion. *J. Atmos. Sci.*, **46**, 975–990.

Franklin, J. L., 1990: Dropwinsonde observations of the environmental flow of hurricane Josephine (1984): Relationships to vortex motion. *Mon. Wea. Rev.*, **118**, 2732–2744.

Flatau, M., W. H. Schubert and D. E. Stevens, 1994: The role of baroclinic processes in tropical cyclone motion: The influence of vertical tilt. *J. Atmos. Sci.*, **51**, 2589–2601.

Fujiwhara, S., 1923: On the growth and decay of vertical systems. *Quart. J. Roy. Meteor. Soc.*, **49**, 75–104.

George, J. E. and W. M. Gray, 1976: Tropical cyclone recurvature and nonrecurvature as related to surrounding wind-height fields. *J. Appl. Meteor.*, **16**, 34–42.

Gross, J. M., 1991: The effect of shallow, medium and deep layer mean wind in the beta and advection model. *Preprints, 19th Conf. Hurr. Trop. Meteor.*, Amer. Meteor. Soc., Miami, Florida, USA, May 6–10, pp. 104–106.

Holland, G. J., 1983: Tropical cyclone motion: Environmental interaction plus a beta effect. *J. Atmos. Sci.*, **40**, 328–342.

Holland, G. J., 1984: Tropical cyclone motion: A comparison of theory and observation. *J. Atmos. Sci.*, **41**, 68–75.

Holland, G. J. and M. Lander, 1993: On the meandering nature of tropical cyclones. *J. Atmos. Sci.*, **50**, 1254–1266.

Holton, J. R., 1992: *An Introduction to Dynamic Meteorology. 3rd Edition*, Academic Press, 511 pp.

Hoskins, B. J., M. E. McIntyre and A. W. Robertson, 1985: On the use and significance of isentropic potential vorticity maps. *Quart. J. Roy. Meteor. Soc.*, **111**, 877–946.

Kaplan, J. and J. L. Franklin, 1991: The relationship between the motion of tropical storm Florence (1988) and its environmental flow. *Preprints, 19th Conf. Hurr. Trop. Meteor.*, Amer. Meteor. Soc., Miami, Florida, USA, May 6–10, pp. 93–97.

Keenan, T. D., 1982: A diagnostic study of tropical cyclone forecasting in Australia. *Australian Meteor. Mag.*, **30**, 69–80.

Kuo, H.-C., R. T. Williams, J.-H. Chen and Y.-L. Chen, 2001: Topographic effects on barotropic vortex motion: No mean flow. *J. Atmos. Sci.*, **58**, 1310–1327.

Lander, M. A., 1994: Description of a monsoon gyre and its effects on the tropical cyclones in the western North Pacific during August 1991. *Wea. Forecasting*, **9**, 640–654.

Lander, M. A., 1995: The merger of two tropical cyclones. *Mon. Wea. Rev.*, **123**, 2260–2265.

Lander, M. A. and G. J. Holland, 1994: On the interaction of tropical-cyclone scale vortices: Part I: Observations. *Quart. J. Roy. Meteor. Soc.*, **119**, 1347–1361.

Li, X.-F. and B. Wang, 1994: Barotropic dynamics of the beta gyres and beta drift. *J. Atmos. Sci.*, **51**, 746–756.

Liang, X. and J. C. L. Chan, 2005: The effects of the full Coriolis force on the structure and motion of a tropical cyclone. Part I: Effects due to vertical motion. *J. Atmos. Sci.*, **62**, 3825–3830.

Marks, F. D., Jr., R. A. Houze, Jr. and J. F. Gamache, 1992: Dual-aircraft investigation of the inner core of Hurricane Norbert. Part I: Kinematic structure. *J. Atmos. Sci.*, **49**, 919–942.

Masina, S. and N. Pinardi, 1993: The halting effect of baroclinicity in vortex merging. *J. Phys. Oceanogr.*, **23**, 1618–1637.

Melander, M. V., N. J. Zabusky and C. J. McWilliams, 1988: Symmetric vortex merger in two dimensions: Causes and conditions. *J. Fluid Mech.*, **195**, 303–340.

Neumann, C. J., 1979: On the use of deep-layer mean geopotential height fields in statistical prediction of tropical cyclone motion. *Preprints, 6th Conf. Prob. Statistics. Atmos. Sci.*, Amer. Meteor. Soc., Boston, MA, USA, pp. 32–38.

Neumann, C. J., 1993: Global overview. *Global Guide to Tropical Cyclone Forecasting.* World

Meteor. Organization, Geneva, Switzerland, pp. 1.1–1.56.

Ngan, K. W. and J. C. L. Chan, 1993: The physics of tropical cyclone motion as revealed by the observations from the TCM-90 Experiment. *Preprints, 20th Tech. Conf. Hurricanes and Trop. Meteor.*, San Antonio, Texas, USA, May 10–14, pp. 35–37.

Ngan, K. W. and J. C. L. Chan, 1995: Tropical cyclone motion — steering vs. propagation. *Preprints, 21st Tech. Conf. Hurricanes and Trop. Meteor.*, Amer. Meteor. Soc., Miami, Florida, USA, Apr. 24–28, pp. 23–25.

Peng, M. S. and R. T. Williams, 1990: Dynamics of vortex asymmetries and their influence on vortex motion on a beta plane. *J. Atmos. Sci.*, **47**, 1987–2003.

Rao, G. V., 1970: An analytical study of the differential frictional effect on vortex movement. *Mon. Wea. Rev.*, **98**, 132–135.

Ritchie, E. A. and G. J. Holland, 1994: On the interaction of tropical-cyclone scale vortices II: Discrete vortex patches. *Quart. J. Roy. Meteor. Soc.*, **119**, 1363–1379.

Rossby, C. G., 1939: Relation between variations in the intensity of the zonal circulation of the atmosphere and the displacements of the semi-permanent centers of action. *J. Marine Research*, **2**, 38–55.

Roux, F. and F. D. Marks, Jr., 1991: Eyewall evolution in Hurricane Hugo deduced from successive airborne Doppler observations. *Preprints, 19th Conf. Hurr. Trop. Meteor.*, Amer. Meteor. Soc., Miami, Florida, USA, May 6–10, pp. 558–563.

Shapiro, L. J., 1992: Hurricane vortex motion and evolution in a three-layer model. *J. Atmos. Sci.*, **49**, 140–53.

Shapiro, L. J. and K. V. Ooyama, 1990: Barotropic vortex evolution on a beta plane. *J. Atmos. Sci.*, **47**, 170–187.

Smith, R. K., 1991: An analytic theory of tropical cyclone motion in a barotropic shear flow. *Quart. J. Roy. Meteor. Soc.*, **117**, 685–714.

Smith, R. K., W. Ulrich and G. Dietachmayer, 1990: A numerical study of tropical cyclone motion using a barotropic model. Part I. The role of vortex asymmetries. *Quart. J. Roy. Meteor. Soc.*, **116**, 337–362.

Szeto, K. C. and J. C. L. Chan, 2009: Structure and track changes of tropical cyclones during landfall: Beta-plane simulations. *Adv. Atmos. Sci.*, in press.

Ulrich, W. and R. K. Smith, 1991: A numerical study of tropical cyclone motion using a barotropic model. II. Motion in spatially-varying large-scale flows. *Quart. J. Roy. Meteor. Soc.*, **117**, 107–124.

Velden, C. S. and L. M. Leslie, 1991: The basic relationship between tropical cyclone intensity and the depth of the environmental steering layer in the Australian region. *Weather & Forecasting*, **6**, 244–253.

Wang, B. and X.-F. Li, 1992: The beta drift of three-dimensional vortices: A numerical study. *Mon. Wea. Rev.*, **120**, 579–593.

Wang, B. and X.-F. Li, 1995: Propagation of a tropical cyclone in a meridionally varying zonal flow: An energetics analysis. *J. Atmos. Sci.*, **52**, 1421–1433.

Wang, Y. and G. J. Holland, 1995: Tropical cyclone motion and evolution in vertical shear. *J. Atmos. Sci.*, **53**, 3313–3332.

Williams, R. T. and J. C. L. Chan, 1994: Numerical and analytical studies of the beta-effect in tropical cyclone motion. Part II: Zonal mean flows. *J. Atmos. Sci.*, **51**, 1065–1076.

Willoughby, H. E., 1988: Linear motion of a shallow-water, barotropic vortex. *J. Atmos. Sci.*, **45**, 1906–1928.

Willoughby, H. E., 1990: Linear normal modes of a moving, shallow-water barotropic vortex. *J. Atmos. Sci.*, **47**, 2141–2148.

Willoughby, H. E., 1992: Linear motion of a shallow-water barotropic vortex as an initial-value problem. *J. Atmos. Sci.*, **49**, 2015–2031.

Wong, M. L. M. and J. C. L. Chan, 2006: Tropical cyclone motion in response to land surface friction. *J. Atmos. Sci.*, **63**, 1324–1337.

Wu, C. C. and K. A. Emanuel, 1993: Interaction of a baroclinic vortex with background shear: Application to hurricane movement. *J. Atmos. Sci.*, **50**, 62–76.

Wu, L. G. and B. Wang, 2000: A potential vorticity tendency diagnostic approach for tropical cyclone motion. *Mon. Wea. Rev.*, **128**, 1899–911.

Chapter 5

The Extratropical Transition of Tropical Cyclones: Structural Characteristics, Downstream Impacts, and Forecast Challenges

Patrick A. Harr

Department of Meteorology, Naval Postgraduate School
Monterey, CA 93943, USA
paharr@nps.edu

The process of extratropical transition (ET) occurs when a tropical cyclone moves poleward into the midlatitudes. During ET a decaying tropical cyclone frequently re-intensifies as an intense extratropical cyclone, which may produce intense rainfall, strong winds and hazardous seas. Changes in the structure of decaying tropical cyclone to an extratropical cyclone involve interactions among a number of complex physical processes. Furthermore variability among the interactions contributes to decreased predictability of the ET process. The variability in ET processes is related to the interaction between the decaying tropical cyclone and the midlatitude circulation into which it is moving. One of the important impacts of the ET of a tropical cyclone has been recently recognized as the influence on the multi-scale dynamics associated with the midlatitude circulation far downstream of the ET event. The downstream impacts may be related to a variety of physical processes associated with the decaying TC and the midlatitude characteristics in which the ET is occurring. In this review, the classification of ET types, important physical characteristics, interactions with the midlatitude circulation, downstream impacts, and predictability are discussed.

1. Introduction

The poleward movement and extratropical transition (ET) of a tropical cyclone (TC) initiates complex interactions with the midlatitude environment that often results in a high-impact midlatitude weather system with strong winds, high seas, and large amounts of precipitation. A TC undergoing ET may pose a serious threat by extending tropical-cyclone like conditions over a larger area, bringing the strong winds typical of major winter storms to midlatitude regions during summer or autumn. Although these extreme conditions severely impact the region of the ET, there are often significant impacts that occur downstream of an ET event. These impacts may be due to the excitation of large-scale propagating Rossby wave-like disturbances that can substantially modify the midlatitude flow downstream.

1.1. *Structural Characteristics*

The challenges involved in understanding ET contain elements of both tropical and extratropical meteorology. The acceleration of a TC into midlatitude regions leads to an expansion and increased asymmetry in the wind field, which contributes to rapid growth in ocean-surface wave heights. Due to the interaction between the decaying TC and the midlatitude westerlies, increased translation speed during ET results in a decreased warning time for hazardous weather. Furthermore, if the timing is misjudged, large errors in the forecast track may result.

Typically, the movement of a TC into the midlatitudes is accompanied by a decrease in intensity. However, the interaction with an extratropical system during ET may result in rapid re-intensification of the decaying TC as an extratropical cyclone. This major forecast challenge is complicated by the fact that the interpretation of satellite imagery appropriate for a TC would suggest that the system is weakening. Although the absolute intensity (defined by the maximum sustained wind speed anywhere within the storm circulation) of a TC undergoing ET often decreases, the expanding wind field increases the area over which a storm may impact (Evans and Hart, 2008). During ET, the forcing of ascent by the large-scale midlatitude flow influences the expansion of the region of heavy precipitation away from the center of the decaying TC to regions that typically extend poleward and downstream of the circulation center. Thus, heavy precipitation extends over a larger area such that heavy precipitation can occur over land without the TC making landfall. Because large amounts of precipitation may fall over an area ahead of the decaying TC center, the ground may be saturated prior to the arrival of the precipitation surrounding the remnant TC center. Due to the change in the mechanisms responsible for the development of precipitation, current operational techniques for quantitative precipitation forecasting are often inadequate during ET. Furthermore, the increased asymmetry in precipitation and wind fields can lead to the strongest winds being associated with dry air. Over locations such as southwest Australia, this enhances the hazard from bush fires.

1.2. *Downstream Impacts*

It is also apparent that significant downstream impacts often involve multiple ocean basins and continental areas in association with ET events. Both the local and downstream impacts of ET are related to the complex interactions among varying spatial-scales (from the convective-scale inner-core of a TC to the synoptic-to planetary-scale upper-level midlatitude troughs) and physical processes. For example, important mechanisms related to downstream impacts include forcing of a Rossby wave response by advection of vorticity due to the divergent wind from the TC core. Also, diabatically-generated Rossby waves/vortices may result from upward motion along sloping isentropic surfaces that exist at the TC midlatitude interface.

1.3. *Forecast Challenges*

Extratropical transition poses a significant challenge to the forecaster. Numerical forecasts of ET are not yet reliable, which introduces large uncertainties in the predictions of the timing of the increased translation speed, the potential for rapid re-intensification as an extratropical cyclone, the severity of the weather, and the impact on the weather downstream. Often, there is a decrease in forecast skill across the Northern Hemisphere (Fig. 1) during an ET event as occurred during Typhoon (TY) Nabi and TY Saola over the western North Pacific and Hurricane Maria over the North Atlantic in September 2005. Such decreases in forecast skill associated with ET are common to all numerical weather prediction models.

The numerical prediction of ET is hindered by the difficulty in representing both the TC and the midlatitude circulation into which it is moving. Difficulty in accurate initialization of both the TC structure and the midlatitude flow contribute to numerical forecast errors. While current operational global models may adequately represent spatial scales sufficient to forecast the evolution of the midlatitude flow, they are generally unable to represent the inner-core of a TC. Many studies indicate that fine-scale structures associated with the inner-core of a TC influence the ET process. Furthermore, physical processes (convection, boundary-layer processes, air-sea interaction) play a crucial role in ET and may not be represented adequately in current operational models. While regional and mesoscale models may provide better representation of the TC structure, representation of the large-scale midlatitude circulation into which the decaying tropical circulation is moving often suffer from boundary problems.

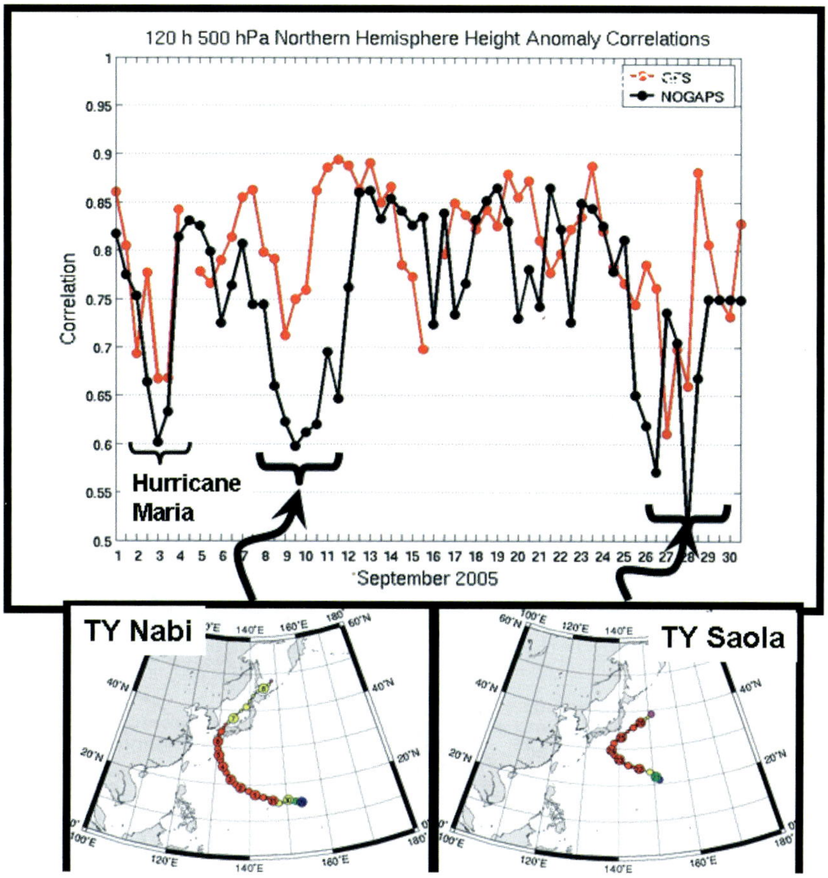

Figure 1. Anomaly correlation for 120-h forecasts of 500 hPa height over the Northern Hemisphere for the National Centers for Environmental Prediction Global Forecast System (GFS) and the Navy Operational Global Atmospheric Prediction System (NOGAPS). Track plots are from http://agora.ex.nii.ac.jp/digital-typhoon/index.html.en.

An important contributor to the lack of understanding and representation of the major processes during ET is that the majority of events occur over the ocean with a very limited number of *in-situ* observations. To increase predictability associated with ET and its downstream impacts, the importance of key *in-situ* observations and advanced assimilation of remotely-sensed satellite data must be established. This may result from targeted or adaptive observation strategies, which includes advancing data assimilation and ensemble-based techniques. Because of the varied physical characteristics associated with ET, a mix of data types and platforms may be required to increase predictability of ET-related weather.

1.4. *Overview*

Scientific issues associated with ET, downstream impacts due to ET, and forecast difficulty during ET events may be placed in a framework of mechanisms, predictability, and strategies for increasing predictability. The ET process may be characterized by complex physical interaction within three interrelated regions: the TC core, the TC-middle latitude interface, and the midlatitude impact region (Fig. 2a). For example, to understand the impact of ET on significant downstream weather events, mechanisms responsible for the generation, intensification, and propagation of the Rossby wave-like disturbances need to be identified. All three regions of the ET process

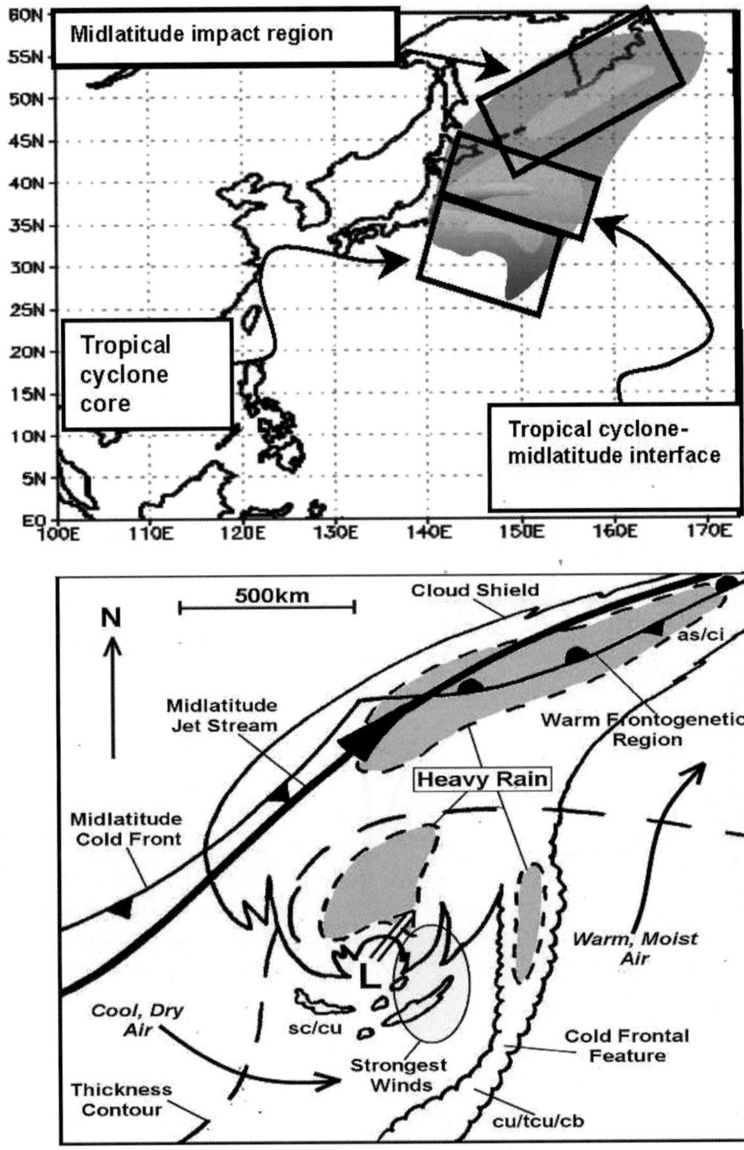

Figure 2. (a) Schematic of three regions associated with the ET of a decaying tropical cyclone over the western North Pacific. The light gray shaded region represents overall cloud patterns. Shaded regions within the light gray areas indicate regions of concentrated cloud amounts defined by convection in the tropical cyclone core region, large-scale precipitation in the tropical cyclone midlatitude interface region, and cirrus in the midlatitude impact region. (b) Schematic of the detailed structure and weather elements associated with the tropical cyclone core and tropical cyclone-midlatitude interface regions defined in (a). (Figure 2(b) kindly provided by Dr. Chris Fogarty, Canadian Hurricane Center, Halifax Nova Scotia.)

likely play important roles in defining important weather elements (Fig. 2b) and in the definition of mechanisms responsible for downstream impacts due to ET. A Rossby wave response may be forced by advection of vorticity due to the divergent wind (Sardeshmukh and Hoskins, 1988), which may result from the TC core. A similar mechanism may be associated with diabatic Rossby waves (Parker and Thorpe, 1995; Moore and Montgomery, 2004,

2005) due to upward motion along sloping isentropic surfaces that exist at the TC midlatitude interface (Harr and Elsberry, 2000). This region is often associated with predecessor rainfall events as outlined by Cote *et al.* (2007). Finally, the middle latitude impact region provides the avenue by which the wave energy impacts the middle latitude circulation into which the decaying TC is moving. Furthermore, the downstream response to ET events exhibits large spatial and temporal fluctuations, which may be related to specific characteristics of each of the three ET region.

In this chapter, the current understanding of the structural evolution of ET is summarized in section 2. Section 3 defines the critical factors associated with downstream impacts of ET, and section 4 is devoted to forecast issues. A summary of outstanding questions and future directions is provided in section 5.

2. Structural Characteristics

Several important physical characteristics have been identified to represent various stages of the ET process. These have been defined through a combination of analyses methods using a variety of diagnostic and statistical techniques. The physical mechanisms associated with each stage of ET are often related to the interaction between the decaying TC and the midlatitude circulation into which it is moving. In this section, the basic classifications of ET are defined. Then, the physical mechanisms associated with the primary features related to each stage of ET are discussed.

2.1. *Classification*

Early case studies of ET used surface analyses to classify ET as *complex* when the TC interacted with a surface baroclinic zone and *compound* when it interacted with a surface low pressure system (Sekioka, 1956, 1970, 1972a,b; Matano and Sekioka, 1971a,b; Brand and Guard, 1979). A third classification is defined when the TC remnants dissipate while moving into the midlatitude environment. This three-tier classification is used operationally by the Japan Meteorological Agency (JMA) (Kitabatake, 2002). For ET events over the southeast Indian Ocean, Foley and Hanstrum (1994) defined an ET event as cradled when large-scale easterly flow existed to the south of the decaying TC cyclone during ET. An ET was defined by Foley and Hanstrum (1994) to be captured when the cyclone became embedded in westerly flow ahead of a cold front.

In an examination of nearly 30 cases of ET over the western North Pacific, Klein *et al.* (2000) found that nearly all cases appeared to transform from a warm-core vortex into a baroclinic, extratropical cyclone in a similar manner based on infrared satellite imagery. They labeled this stage of ET the transformation stage. During transformation (Fig. 3), a recurving TC interacts with a pre-existing lower-tropospheric baroclinic zone and the associated vertical wind shear. The interactions between the TC and the midlatitude baroclinic zone involve lower-tropospheric temperature advection, establishment of a vertical motion dipole, dispersal of the TC warm core, lower-tropospheric frontogenesis, and ascent/descent along the tilted isentropic surfaces of the baroclinic zone (Harr and Elsberry, 2000; Harr *et al.*, 2000). The structure of the TC changes such that the cloud and precipitation fields associated with the remnant TC core (Figs. 2 and 4) become quite asymmetric. The convection may become asymmetric in response to vertical shear, moisture variations, SST gradients, or orography. Shimazu (1998) defined the characteristic pattern of cloud and precipitation that results from the asymmetries in the TC core as a "delta rain" region. An area of warm frontogenesis (Harr and Elsberry, 2000; Sinclair, 1993, 2002), which is labeled the TC-midlatitude interface region in Fig. 2a, frequently becomes evident just poleward of the asymmetric remnant TC core. Often, an upper-level jet streak may be found just poleward of the midlatitude interface region. In the case of favorable interactions between the decaying

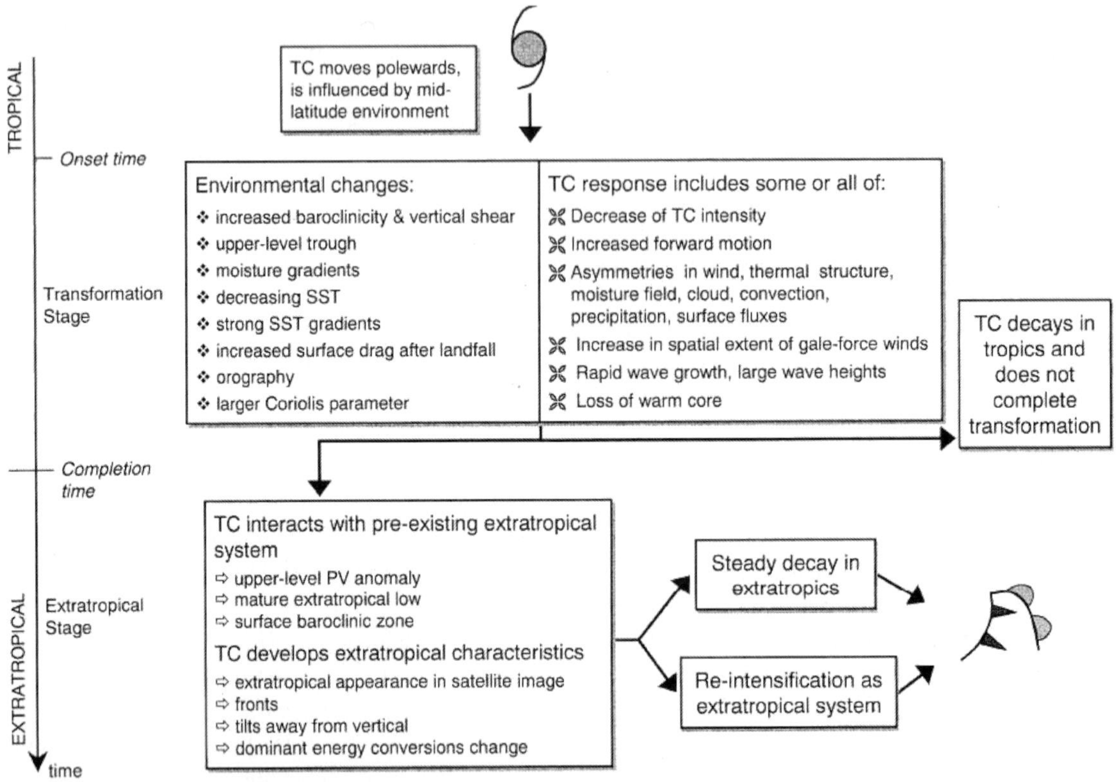

Figure 3. A two-stage classification of extratropical transition based on the classification of Klein *et al.* (2000). The onset and completion times correspond to the definitions of Evans and Hart (2003). The "tropical" and "extratropical" labels indicate approximately how the system would be regarded by an operational forecast center (after Jones *et al.*, 2003).

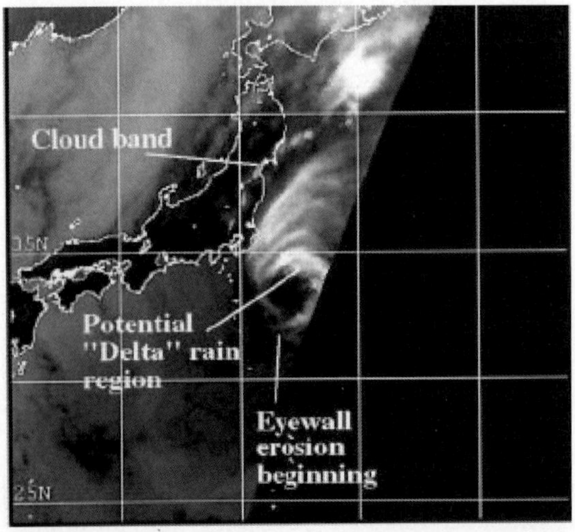

Figure 4. The transformation of TY David (1997) as defined in an 85-GHz SSM/I image at 2216 UTC 18 Sep 1997 (adapted from Klein *et al.*, 2000).

TC and the midlatitude environment into which it is moving, the decaying TC often becomes located under the right entrance region of the jet streak, which enhances the development of the extratropical cyclone (Klein *et al.*, 2002; Evans and Prater-Mayes, 2004; Ritchie and Elsberry, 2007). Similar to Klein *et al.* (2000), Evans and Hart (2003) also recognized the increasing asymmetry of the remnant TC core region in a climatological study of ET events over the North Atlantic Ocean basin.

Klein *et al.* (2000) defined a re-intensification stage of ET that begins when the remnants of the decaying TC begin to deepen as an extratropical cyclone and ends when a minimum sea-level pressure is attained. To reflect the potential for dissipation of the remnant TC into the midlatitudes, Jones *et al.* (2003) re-defined the re-intensification stage as the extratropical stage (Fig. 3). As the extratropical stage begins, the interaction between the decaying TC and low-level baroclinity begins to modify the gradient of lower-tropospheric equivalent potential temperature (θ_e). The modification of the θ_e gradient results in warm frontogenesis that begins during the transformation stage due to the deformation patterns associated with the TC interaction with the midlatitude circulations (Harr and Elsberry, 2000). Development of a well-defined cold front is often suppressed due to a direct thermal circulation that includes the descent of cold air from upstream of the re-intensifying cyclone (Palmen, 1958; Harr and Elsberry, 2000). Either re-intensification or decay can occur as the TC interacts with a preexisting extratropical system. In the absence of a preexisting extratropical system and strong baroclinic zone, the ex-TC would be expected to continue to decay.

Hart (2003) defined the physical characteristics of cyclones in a phase space spanned by lower-to upper-tropospheric thermal wind magnitudes and a representation of storm thermal symmetry. Based on these parameters, the phase space (Fig. 5) is useful for representing extratropical and TC characteristics (Evans and Hart, 2003; Arnott *et al.*, 2004). The transformation stage may be defined to begin when the cyclone trajectory passes from the symmetric warm-core region (lower right in Fig. 5) of the phase space to the asymmetric warm-core region (upper right in Fig. 5). The extratropical stage is considered to have begun when the cyclone trajectory passes into the cold-core asymmetric region (Fig. 5).

Currently, no accepted operational definition of ET exists. As can be seen in Figs. 3 and 5, a variety of factors can be used to specify the various stages of ET. Some researchers will refer to any TC that enters mid-latitudes as having

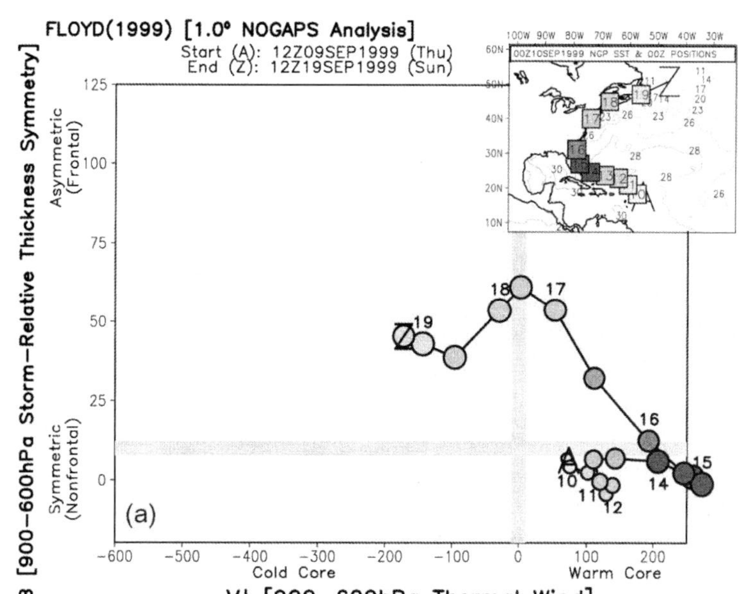

Figure 5. Cyclone phase space diagrams for the extratropical transition of Hurricane Floyd in 1999. The abscissa indicates whether the cyclone is warm or cold core in the lower troposphere (strong cold core on left; strong warm core on right). The ordinate indicates the degree of thermal asymmetry in the lower troposphere (symmetric at bottom; asymmetric at top). Here, A indicates the first time, and Z the final time plotted. Markers are plotted every 12 h with the size of the marker indicating the mean radius of the 925-hPa gale force winds and the shading indicating the minimum mean sea level pressure (>1010 hPa white; <970 hPa black). The inset gives the track of the cyclone superposed on the model analysis sea-surface temperature field (8C) [(adapted from Hart (2003)].

undergone ET. A definition of ET should satisfy the needs of the operational and research communities plus convey important information to the general public. Since ET typically occurs after the operational TC centers have discontinued official forecasts, it is important to have a definition of ET that reflects the severe weather elements that still occur during and following ET. Improved understanding of the processes through which ET occurs will aid in the development of a globally-accepted definition of ET.

2.2. *Interaction with the Midlatitude Circulation: Vertical Wind Shear and Trough Interaction*

As the decaying TC moves poleward, the impact of increased vertical wind shear contributes to the main structural changes identified during the transformation stage of ET. These changes are related to the development of a vertical tilt in the dynamic structure of the decaying TC core and development of asymmetries in the temperature, vertical motion, precipitation, and cloud fields (Jones, 1995, 2000a,b; Frank and Ritchie, 1999, 2001). The magnitude of these changes depends on the strength of the shear and the strength and size of the decaying TC (Jones, 1995, 2000a,b; Davis *et al.*, 2008).

Studies by Thorncroft and Jones (2000), Klein *et al.* (2000), Frank and Ritchie (2001), and Evans and Prater-Mayes (2004) indicate that often the upper-level potential vorticity (PV) of the decaying TC core is eroded due to the strong vertical wind shear as the TC enters the midlatitudes. Results from Jones (2000b), indicate that the erosion of the upper-level PV may be due to differential advection by the vertical shear or due to filaments of PV breaking off from the main PV anomaly at upper levels. Eventually, a westward tilt with height is necessary for re-intensification of the decaying TC as an extratropical cyclone. This occurs with the approach of an upper-level PV anomaly from the west (Palmen, 1958; Chien and Smith, 1977; Sinclair, 1993; Bosart and Lackmann, 1995; Thorncroft and Jones, 2000).

Harr and Elsberry (2000) characterized the interaction between the poleward-moving TC and the midlatitude circulation into which it is moving based on whether the primary midlatitude circulation was northwest or northeast of the decaying TC. Klein *et al.* (2000) found a statistically significant relationship such that ET during the northwest pattern resulted in lower sea-level pressures during the re-intensification stage of ET. Often, the approach of an upper-level trough from the west will result in the presence of an upper-level jet streak just poleward of the decaying TC. If the timing between the poleward-moving TC and the upper-level trough is favorable (Klein *et al.*, 2002; Atallah and Bosart, 2003; Ritchie and Elsberry, 2007; Riemer *et al.*, 2008), then the outflow from the TC may interact with the upper-level midlatitude jet (Fig. 6). In the case of the ET of Tropical Storm (TS) Banyan (Vancas, 2006), which occurred during July when the midlatitude circulation does not typically contain strong jet systems, the outflow from TS Banyan merged with the jet and resulted in an increased favorable interaction as the decaying TS moved into the right entrance region of the jet streak. At 0000 UTC 26 July (Fig. 6(a)), the outflow from TS Banyan is represented by a secondary isotach maximum south of the primary midlatitude jet. By 1200 UTC 26 July (Fig. 6(b)), the outflow merged with the midlatitude jet, which resulted in one isotach maximum that was higher than when the outflow and midlatitude jet were separated.

Davis *et al.* (2008) found that the total vertical mass flux in the middle troposphere was essentially proportional to the vertical shear, and was maximized during ET rather than during the mature hurricane phase. When the vertical shear exceeded roughly $20\,\mathrm{m\,s^{-1}}$, the mass flux decreased. A shear of approximately $15\,\mathrm{m\,s^{-1}}$ was deemed optimal for maximizing the total mass flux, which suggests that the precipitation and upper-tropospheric divergence

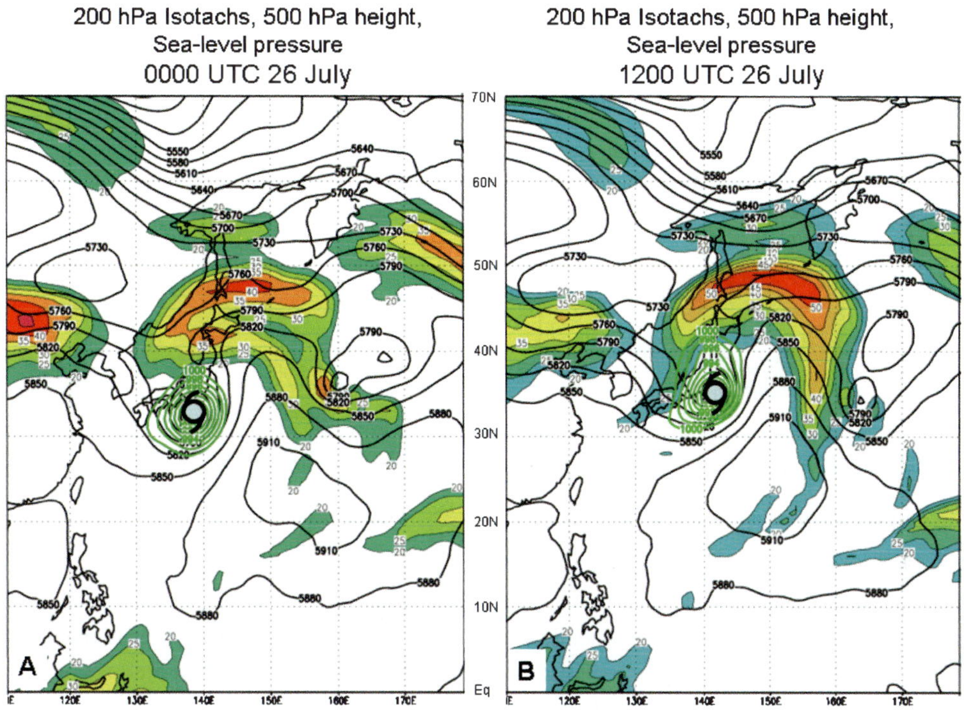

Figure 6. Isotachs at 200 hPa (shaded $\mathrm{m\,s^{-1}}$), 500 hPa height (20 m interval), and sea-level pressure contoured for values less than 1000 hPa in 2 hPa intervals for (a) 0000 UTC 26 July, (b) 1200 UTC 26 July. (Adapted from Vancas, 2006.)

will generally increase until the cyclones are well into the midlatitudes.

The approach of the upper-level trough contributes to a linkage between the trough and decaying TC such that the final westward tilt with height is achieved to promote re-intensification of the decaying TC and an extra-tropical cyclone. Because downstream development associated with ET events often occurs over data-sparse regions, satellite data are critical for examining basic structural characteristics during the ET process. Vancas (2006) utilized temperature retrievals (DeMuth *et al.* 2004) from the Advanced Microwave Sounding Unit (AMSU-A) to define the thermal characteristics of the atmosphere in the vicinity of the ET of TS Banyan (Fig. 7). At low-levels (Fig. 7(a)), the cold air that is sinking to the west and north and warm air that is rising to the east and south of the decaying center of Banyan are clearly identified. Harr *et al.* (2000) associated this thermal pattern with the baroclinic energy conversion necessary for the re-intensification of the decaying TC as an extratropical cyclone. In the vertical (Fig. 7(b)), the mid-to upper-level warm core of the decaying TC has been deformed such that it extends to the east of the center at low levels in association with the warm air advection in that region and extends west of the center at upper levels as it begins to link with the warm anomaly above the tropopause associated with the upstream midlatitude trough. The results of Vancas (2006) indicate that the AMSU-A retrievals are valuable for identifying thermal characteristics of the ET process.

In an examination of ET cases over the northwest North Pacific, Harr *et al.* (2000) found that the evolution of a TC from a warm-core vortex to a cold-core vortex was dominated by the flux of eddy angular momentum from an upstream trough rather than eddy heat flux associated with the movement of the decaying

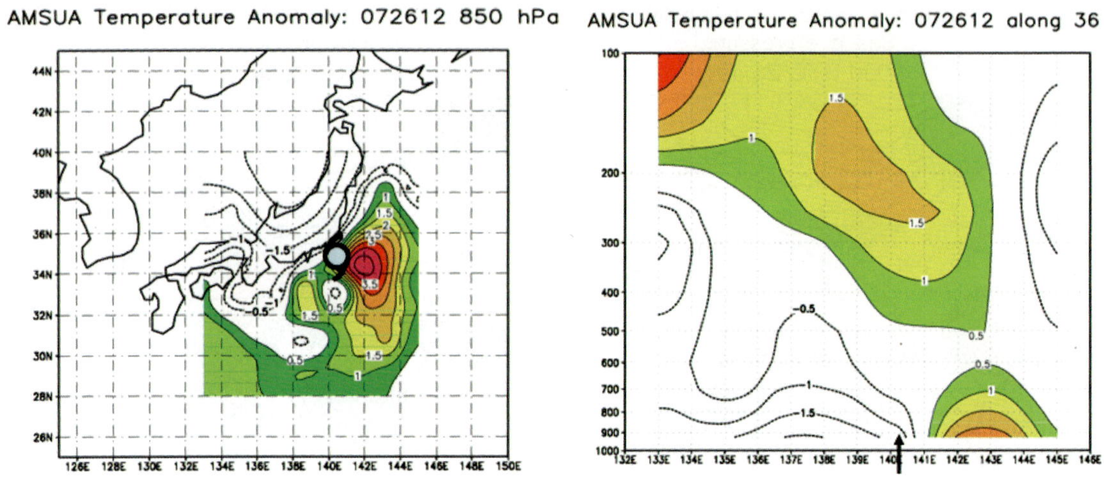

Figure 7. Temperature anomaly (K) from AMSU-A retrievals at 1200 UTC 26 July for (left) 850 hPa and (right) vertical cross section along 36° N. The tropical cyclone symbol in the left panel and the arrow in the right panel mark the location of the center of TY Banyan. (Adapted from Vancas, 2006.)

TC into the midlatitude baroclinic zone. Hart *et al.* (2006) found a similar relationship in their examination of ET cases over the North Atlantic. Petterssen–Smebye type-B cyclogenesis is defined when extratropical development occurs due the movement of an upper-level trough over a region of low-level warm advection (Petterssen and Smebye, 1971). Therefore, the development of an extratropical cyclone during ET has been described as type-B cyclogenesis modified by the lower-level remnants of a TC, which also may be enhanced due to the reduced stability associated with the presence of moist tropical air (DiMego and Bosart, 1982a,b; Bosart and Dean, 1991; Sinclair, 1993).

Klein *et al.* (2002) and Ritchie and Elsberry (2007) examined the sensitivity of the ET process to the phasing of the poleward-moving TC with an approaching midlatitude trough by altering the movement of the TC in numerical simulations. In these studies, re-intensification of the ex-TC remnants is described as being similar to Type-B extratropical cyclogenesis.

Klein *et al.* assessed the degree to which the midlatitudes were favorable by removing the ex-TC remnants and determining if: a moderate or strong cyclogenesis event would have occurred anyway; only a weak extratropical cyclone would occur; or no extratropical cyclone would develop. The technique of vortex removal and relocation was used by Klein *et al.* (2002) to provide examples of displaced-vortex simulations in which actual cases were simulated by altering the phasing of the decaying TC with the evolving midlatitude circulation. The results of the Klein *et al.* (2002) study are summarized in Fig. 8 as the re-intensification of the decaying TC is categorized relative to TC and midlatitude contributions. The outcome of the re-intensification depends on: (i) whether the transforming TC (which primarily defines the low-level thermodynamic support) is properly located relative to the dynamically-induced region in the midlatitude circulation that is favorable for cyclone development; (ii) magnitudes of the factors that define the region favorable for cyclone development; and (iii) how the transformed TC interacts with this region.

Many of the physical characteristics associated with the primary classifications of the ET process are related to interactions between the decaying TC and the midlatitude circulation into which it is moving. This results in increased vertical wind shear, modification of the thermodynamic and dynamic signatures of the mature

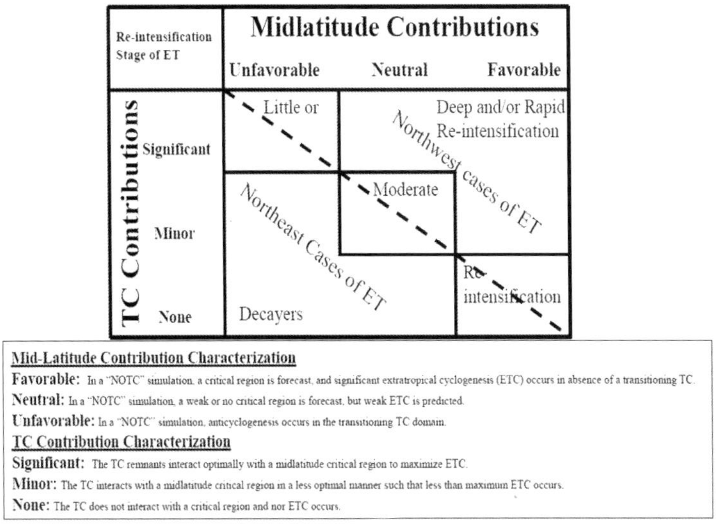

Figure 8. Characterization of the midlatitude circulation and tropical cyclone contributions to the re-intensification stage of ET depicted as a matrix, where the combinations of these contributions may produce three outcomes: (i) deep and/or rapid re-intensification (upper right; SLP at the end of ET < 980 mb); (ii) little to moderate re-intensification (along diagonal; SLP at the end of ET > 980 mb); or (iii) decay of the tropical cyclone without completing ET (lower left). Northwest and northeast synoptic patterns of ET (Harr *et al.*, 2000) are separated by the diagonal dashed line (after Klein *et al.*, 2002).

TC, vertical tilt indicative of a developing extratropical cyclone, and interaction between the TC outflow and the midlatitude jet.

2.3. *Frontogenesis*

The generation of frontal characteristics is a primary structural feature during the ET process. Harr and Elsberry (2000) examined the scalar and rotational frontogenesis patterns associated with a variety of ET cases over the western North Pacific Ocean. The typical pattern of maximum scalar frontogenesis during ET occurs north and east of the typhoon remnants (Fig. 9(a)). The primary contribution to the scalar frontogenesis is due to horizontal deformation (Fig. 9(b)) as defined by the axes of dilatation, which are nearly parallel to the isentropes. This orientation is consistent with frontogenesis in the warm frontal region when a cyclonic vortex is superimposed on a baroclinic zone (Keyser *et al.*, 1988). Upstream from the TC center, the deformation field and the sinking motion associated with cold advection contribute to a direct circulation and scalar frontolysis.

Positive rotational frontogenesis acts to rotate the gradient of potential temperature in a cyclonic direction, whereas negative rotational frontogenesis acts to rotate the potential temperature gradient in an anticyclonic direction (Keyser *et al.*, 1988). For the case of Typhoon David examined by Harr and Elsberry (2000), there was a large region of positive rotational frontogenesis along the eastern boundary of the thermal trough with regions of negative rotational frontogenesis north and east of the typhoon center. Thus the pattern of rotational frontogenesis implies cyclonic rotation of the potential temperature gradient west of the typhoon center and anticyclonic rotation to the north and east, leading to an amplification of the thermal wave. The combination of rotational frontogenesis acting to amplify the thermal wave with scalar frontogenesis acting to enhance the horizontal temperature gradients indicates that re-intensification as an extratropical cyclone is likely to occur (Harr and Elsberry, 2000).

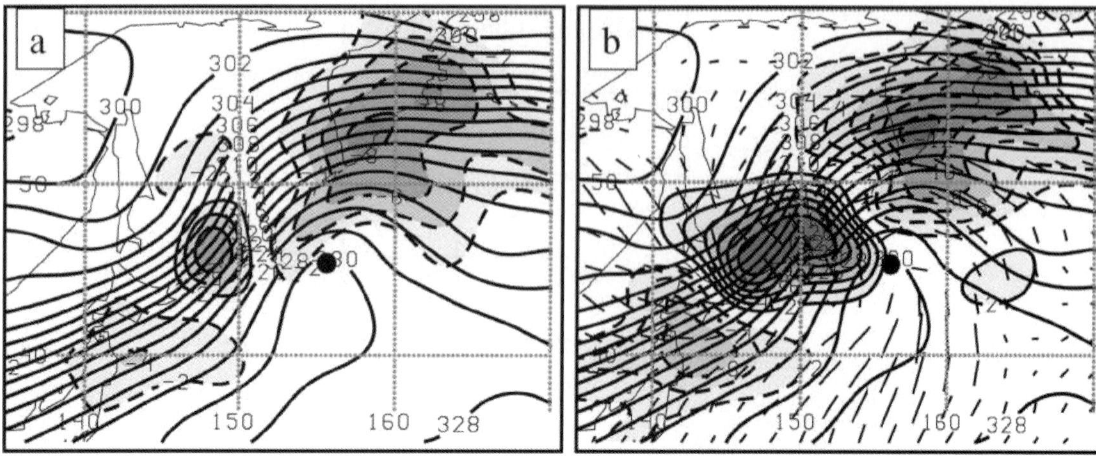

Figure 9. Potential temperature at 500 hPa (contoured at 2-K intervals) for Typhoon David at 0000 UTC 20 Sep 1997 and (a) shaded contours of scalar frontogenesis, (b) shaded contours of the contribution to scalar frontogenesis by horizontal deformation (negative values correspond to frontogenesis; positive values frontolysis), Shaded contours are at intervals of $2 \times 10^{-10\mathrm{K}}\,\mathrm{m}^{-1}\,\mathrm{s}^{-1}$ with negative contours dashed. Axes of dilatation are included in (b). The black dot marks the location of the low-level center of the typhoon. [Adapted from Harr and Elsberry (2000).]

2.4. *Precipitation*

During an ET event, the precipitation expands poleward of the center in relation to the warm frontogenesis (Fig. 2(b)) and interaction with the midlatitude jet (Harr and Elsberry, 2000; Atallah and Bosart, 2003). Furthermore, the precipitation distribution is typically maximum to the left (right) of the track in the Northern (Southern) Hemisphere. Atallah and Bosart (2003) investigated the impact of diabatic modification of the upper-level PV enables a TC to remain beneath an equatorward jet entrance region.

It is important to consider contributions to the organization of precipitation over the region of warm frontogenesis. One aspect to be investigated is the release of conditional symmetric instability (CSI; Bennetts and Hoskins, 1979; Schultz and Schumacher, 1999) through the displacement of saturated air parcels along a slanted path (Emanuel, 1983). The observation that the largest rain rates during ET tend to occur in bands that are oriented parallel to the thermal wind suggest that CSI may be an important factor during ET. In addition, a number of studies have shown that the combination of frontogenetical forcing and small symmetric stability leads to enhanced vertical motion that in turn acts to strengthen the frontogenesis (Emanuel, 1985; Thorpe and Emanuel, 1985).

2.5. *Ocean Waves*

Extratropical Transition events are often associated with extremely large surface wave fields that result from continued high winds speeds and increased translation speed of the decaying TC. MacAfee and Bowyer (2000a,b) suggest that the high forward translation speeds of ET events can result in a resonance between the ocean waves being generated and the wind system generating them. The decaying TC moves with the fetch and traps the waves within the wind system. This trapped fetch allows the waves to grow much larger than would be possible if the storm were stationary or slowly moving. The enhancement of the wave field tends to occur where the direction of storm motion is in the same sense as the wind, that is, on the right-(left-) hand side of the storm track in the Northern (Southern) Hemisphere. Because of this phenomenon, the waves associated with a decaying TC in the process of ET can pose a greater threat than those associated with a stronger TC within the tropics.

3. Downstream Impacts of Extratropical Transition

Often, the movement of the TC into the midlatitudes results in a significant perturbation to the midlatitude flow that may rapidly extend downstream of the ET event. There is often significant variability in the downstream response to an ET event. This implies that there are important sensitivities to a variety of physical mechanisms associated with the forcing on the midlatitude flow. These mechanisms may include basic baroclinic energy conversions, forcing of diabatically-forced Rossby wave-like circulations, and enhancements to downstream jet streaks. All of these factors would exhibit sensitivity to a variety of interactions with a decaying TC. Finally, use of ensemble prediction systems indicate that the downstream forcing due to an ET is typically associated with reduced predictability.

As discussed in section 2, all three regions in Fig. 2 likely play important roles in the transformation and re-intensification stages of ET. Furthermore, the relative structural changes in the stages of ET are hypothesized to be responsible for the variability in downstream development that is often associated with an ET event.

3.1. *Ridge Building*

Downstream impacts from an ET event imply that the decaying TC and ET process modify the midlatitude circulation into which the TC is moving. A key factor in the impact of ET on the midlatitude circulation is the development of a downstream ridge connected with the outflow of the TC (Atallah and Bosart, 2003). However, there is a large amount of variability associated with the development, timing, and amplitude of the downstream ridge. The dynamics of downstream ridge development has a fundamental link to diabatic heating, but also depends upon the properties of the midlatitude circulation and of the decaying TC (Riemer *et al.*, 2008; Davis *et al.*, 2008).

Vancas (2006) examined the development of a strong ridge immediately downstream of the ET of Tropical Storm (TS) Banyan. One purpose of this study of a July case was to show that under favorable phasing between a decaying TC and a midlatitude trough, downstream development was not limited to the later portion of the typhoon season over the western North Pacific, which is typically from June to October. The ET of TS Banyan initiated a high-amplitude Rossby-wave like response (Fig. 10) across the North Pacific. The placement of an upper-level ridge over western North America contributed to a period of well-above normal temperatures over this region.

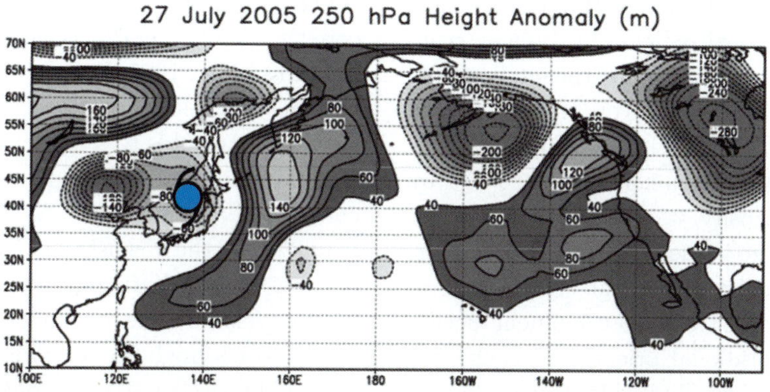

Figure 10. Average 250 hPa height anomalies for 0000 UTC and 1200 UTC 27 July 2005. The tropical cyclone symbol marks the location of TS Banyan at 1200 UTC 27 July. (Adapted from Vancas, 2006.)

Vancas (2006) identified the development of the anomalous ridge to the east of the decaying Banyan as critical to the development of the downstream circulations across the North Pacific. Vancas (2006) pointed to the relative roles of the low-level advection of warm, moist air east of the center (Fig. 7) and the export of low potential vorticity air into the midlatitudes in the outflow layer of the TC (not shown) as being important for determining the characteristic growth of the downstream ridge in the ET process.

Riemer *et al.* (2008) examined the interaction of a poleward-moving TC with a straight midlatitude jet streak using a full-physics numerical model with idealized initial conditions. When the TC interacts with the jet, a ridge-trough couplet forms in association with a distinct jet streak in the upper-level circulation. Extratropical cyclogenesis occurs immediately downstream of the ET event and downstream baroclinic development (Orlanski and Sheldon, 1995) follows. During the interaction of a TC with a straight jet defined to be at 42°N with a magnitude of $40\,\mathrm{m\,s^{-1}}$ at 175 hPa, Riemer *et al.* (2008) found that a Rossby-wave response was generated approximately 4 days into their simulation (Fig. 11). The Rossby wave packet propagated faster downstream than the ET event, which is near the estimate of Simmons and Hoskins (1975) of the relation between the group speed of a perturbation and the speed of the mean flow. It is also interesting to note that the study of the idealized case by Riemer *et al.* (2008) reveal perturbations in the midlatitude flow upstream of the TC.

Riemer *et al.* (2008) investigated the sensitivity of the downstream response to ET by varying the intensity of the midlatitude jet and examining the influence of moist processes. During interaction with a weak jet (WJET in Fig. 12), the downstream development occurs with a more meridional orientation such that the anticyclonic baroclinic life cycle (Thorncroft *et al.*, 1993) is favored. Interaction with

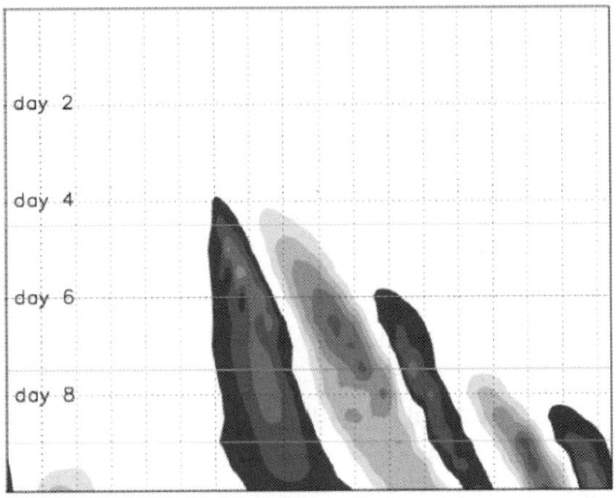

Figure 11. Hovmoller plot of 200 hPa meridional winds speed ($\mathrm{m\,s^{-1}}$) averaged over 14°N to 65°N for the straight jet experiment of Riemer *et al.* (2008). Dark to light shading defines southerly winds and light to dark shading defines northerly winds. The dotted grid lines are every 1000 km and the time axis is labeled (after Riemer *et al.*, 2008).

a strong jet (SJET in Fig. 12) the downstream impact is represented by a longer wavelength and higher group velocity with a more zonally-oriented development. The impact of moist processes is to increase the amplitude of the downstream wave pattern with minimal change to the wavelength or group speed of the downstream response.

The building of a ridge downstream of an ET event can be also examined in relation to the export of low PV air in the TC outflow and by the fact that in regions of convection and associated latent heat release, such as in the inner core of a TC, the PV is no longer conserved (Hoskins *et al.*, 1985; Raymond, 1992). The diabatic modification of PV in regions of latent heat release contributes to a reduction in upper-level PV. The enhanced ridging results in a steepening of the tropopause. As found by Riemer *et al.* (2008), the downstream ridge amplification may result in a baroclinic life cycle dominated by anticyclonic wave breaking (Thorncroft *et al.*, 1993), which might be amplified downstream of the TC by the low-PV TC outflow (Thorncroft and Jones, 2000). In the case of Hurricane Floyd

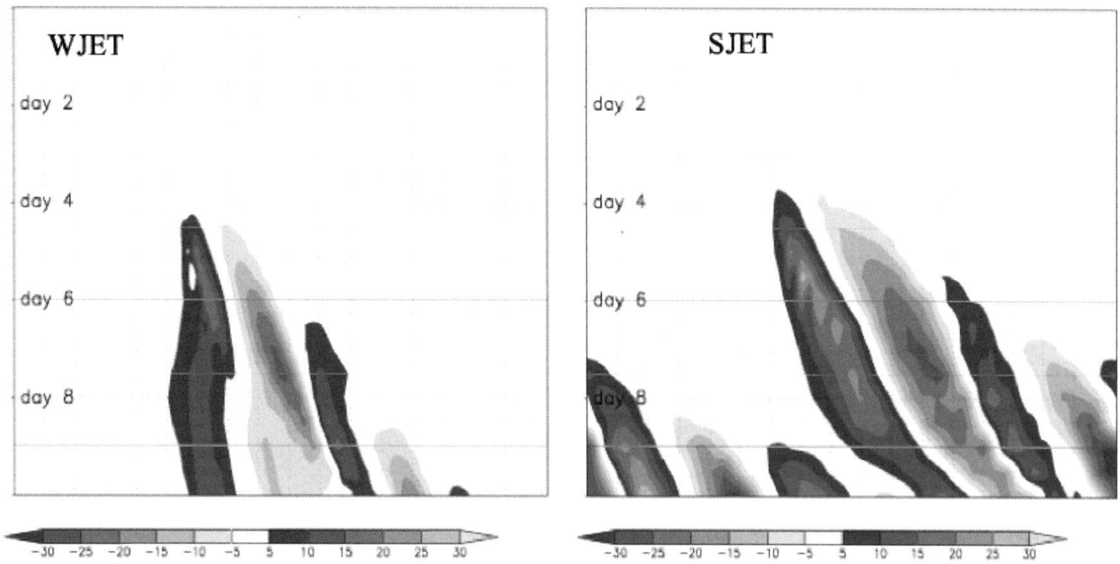

Figure 12. As in Fig. 11, except for the weak jet (WJET) and strong jet (SJET) simulations of Riemer *et al.* (2008) (adapted from Riemer *et al.*, 2008).

(1999) the diabatic outflow of the TC assisted in the formation of a tropopause fold (Atallah and Bosart, 2003). Bosart and Lackmann (1995) examined the ET of Hurricane David (1979) and found that no significant upper-level PV anomaly existed until the outflow from the hurricane modified the tropopause structure to the west (upstream) of the TC. The diabatically-induced tropopause anomaly subsequently interacted with David resulting in significant development as an extratropical system.

3.2. *Downstream Development*

Harr and Dea (2009) present an example of the influence of an ET event over the western North Pacific on the downstream circulation across the North Pacific. In this case, the midlatitude circulation across the North Pacific had a rather zonal orientation as TY Nabi recurved (Fig. 13(a)). As TY Nabi began to undergo ET (Figs. 13(b),(c)) the ridge immediately downstream of the decaying TC began to build. The importance of this process during ET was discussed above. A trough develops immediately downstream of the ridge as TY Nabi begins to complete ET (Figs. 13(c),(d)). As the ex-TY Nabi begins to re-intensify as an extratropical cyclone (Figs. 13(e),(f)), a ridge over the eastern North Pacific amplifies downstream of the central North Pacific trough. Also a cyclone over the west coast at 1200 UTC 8 September (Fig. 13(d)) that was cut-off from the main midlatitude flow (Fig. 13(f)) deepened (Figs. 13(e),(f)) as the ridge over the eastern North Pacific amplified and a shortwave trough moved down the coast of western North America.

In this example, it is clear that downstream development occurred in conjunction with the ET of TY Nabi as the ridge amplified immediately downstream of the decaying TC. Harr and Dea (2009) examined this downstream development in terms of the distribution of eddy kinetic energy (EKE). During the ET process, EKE is generated from available potential energy via baroclinic energy conversion. Ageostrophic geopotential fluxes (Orlanski and Sheldon, 1995) contribute to the transport of the EKE downstream (Fig. 14).

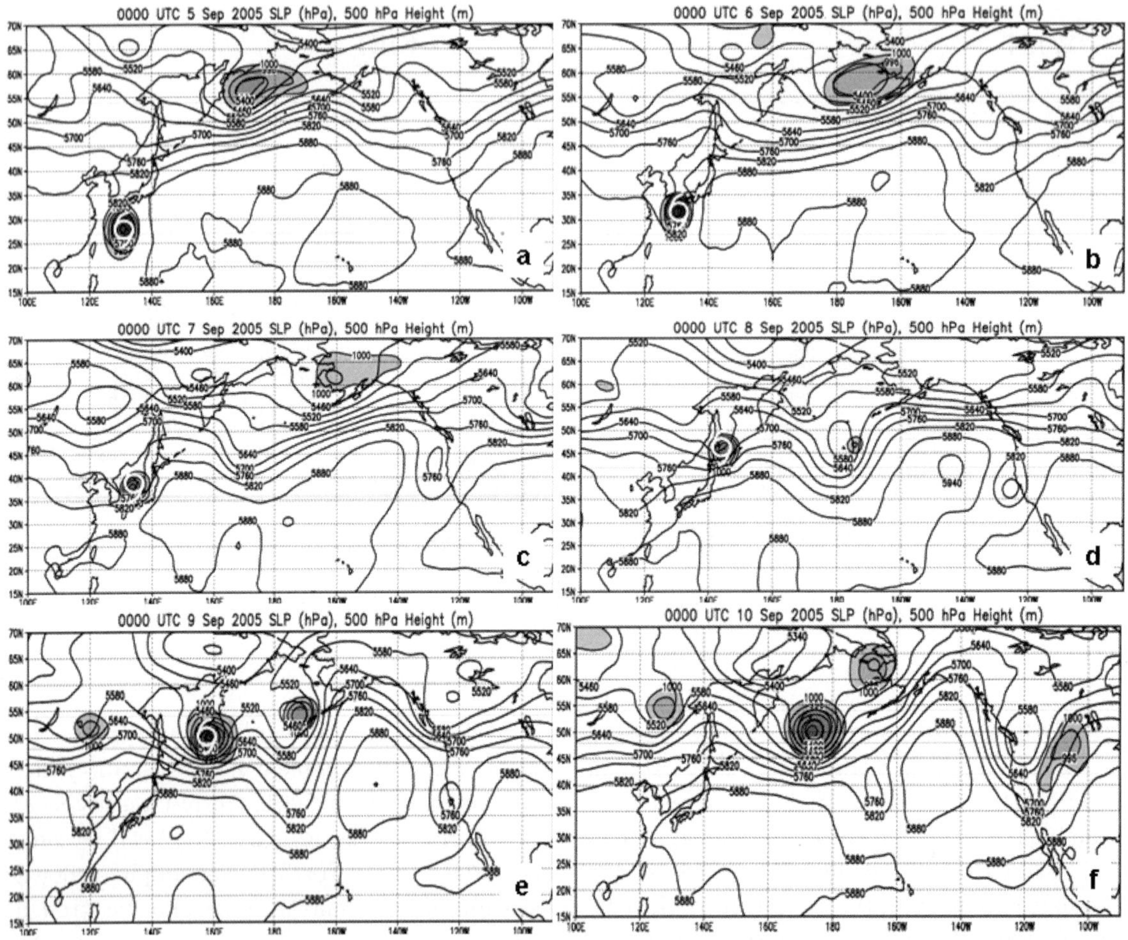

Figure 13. Sea-level pressure (hPa, shaded below 1000 hPa) and 500 hPa heights (m) for 1200 UTC 5–10 September 2005. The locations of TY Nabi as it proceeded through the ET process is marked by the tropical cyclone symbol. (After Harr and Dea, 2009.)

As TY Nabi approached southern Japan (Fig. 14(a)), there was no significant linkage between the EKE associated with Nabi and the midlatitude circulation. Twenty-four hours later (Fig. 14(b)), EKE was being transported from Nabi into the downstream trough. As the trough developed (Fig. 14(c)), EKE continued to be dispersed from Nabi and an EKE center developed on the downstream side of the central North Pacific trough. This EKE center became a source of EKE for the trough over western North America (Fig. 14(d)).

Harr and Dea (2009) examined the downstream evolution of EKE associated with a variety of ET cases over the western North Pacific (Fig. 15). The variability in downstream development is very sensitive to the processes associated with the development of the downstream ridge, which are fundamental aspects of the ET process (Klein *et al.*, 2000; Harr and Elsberry, 2000; Harr *et al.*, 2000). However, in each ET case, which is marked by the TC symbol in Fig. 15, downstream development occurred as the magnitude of eddy kinetic energy increased

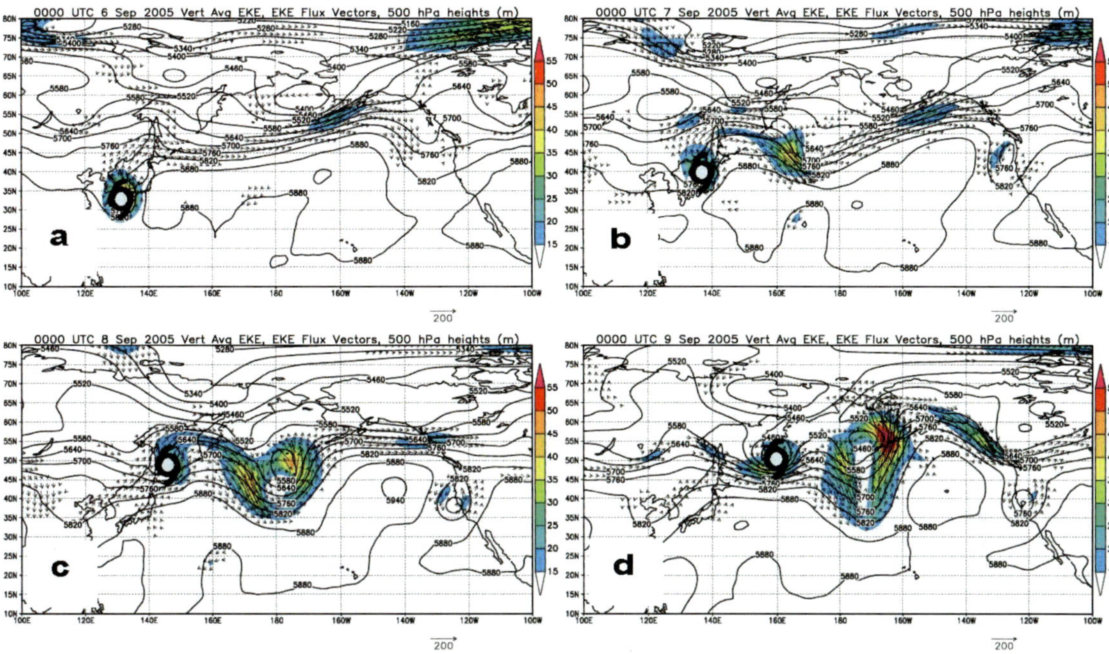

Figure 14. Vertically averaged eddy kinetic energy (shaded, MJ m^{-2}), 500 hPa heights (contours, m), and eddy kinetic energy total flux vectors (MW m^{-2}. reference vector at lower right of each panel) during the ET of TY Nabi during 0000 UTC 6–9 September 2005. (Adapted from Harr and Dea, 2009.)

successively across the western, central, then eastern North Pacific.

3.3. *Predictability*

It has been documented that many operational global numerical weather prediction models do not accurately simulate the structural characteristics associated with ET (Evans *et al.*, 2006). The ET process involves a complicated interaction among mechanisms that act over a variety of spatial and temporal scales. Based on the examples provided above, it is clear that identification of the impact of an ET event on the large-scale midlatitude circulation is based on understanding the various mechanisms associated with the ET process. Even when a numerical forecast properly represents the favorable phasing between a decaying TC and the midlatitude circulation, variability in the downstream response often exists. This indicates that additional aspects of the ET process must be understood and resolved for predictability to increase or variability among EPS members to decrease.

Harr *et al.* (2008) and Anwender *et al.* (2008) examined the predictability of the downstream impacts during ET in the context of increased spread among members of several operational global ensemble prediction systems. During the ET of Nabi (Fig. 16(a)), a Rossby wave-like response (Fig. 16(b)) was generated across the North Pacific. In association with this pattern, the variability, which was measured by the standard deviation among ensemble members from the National Centers for Environmental Prediction (NCEP) Global Forecast System (GFS), increased downstream from the location of the ET (Fig. 17).

Harr *et al.* (2008) and Anwender *et al.* (2008) used a fuzzy cluster analysis of ensemble members to show that the number of possible downstream scenarios decreased as the lead time to the ET event decreased. The decrease in

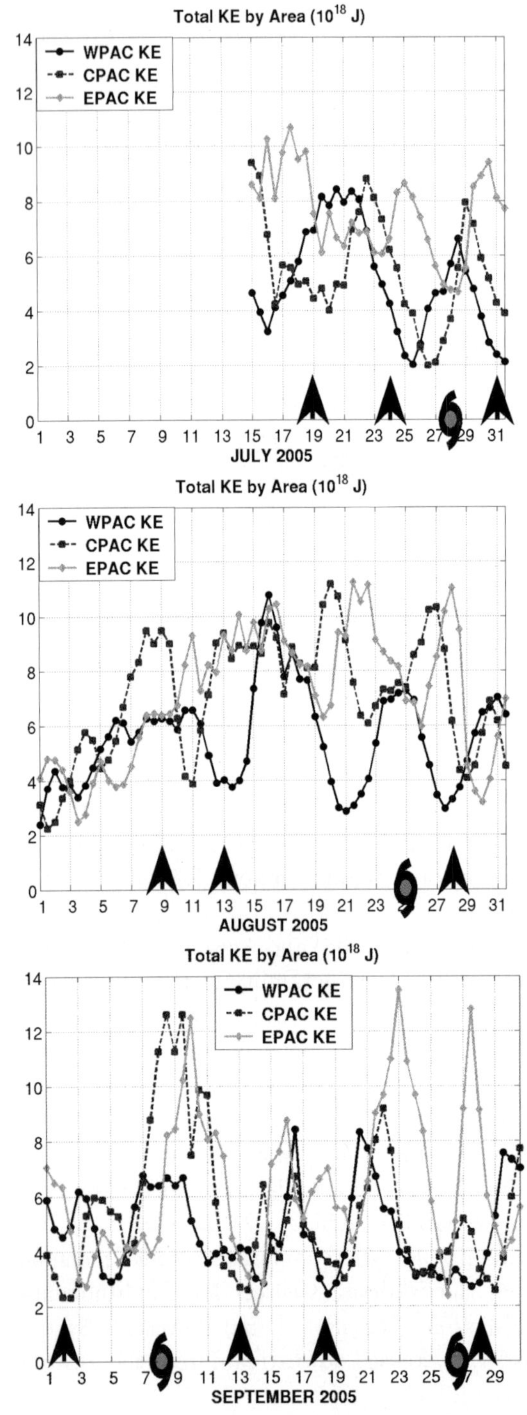

Figure 15. Volume integrated eddy kinctic energy (10^{18} J) over the North Pacific between 15 July–30 September 2005. The tropical cyclone symbol marks the dates that tropical cyclones moved poleword of 40°N. The arrows mark the last date associated with a tropical cyclone that stayed south of 40°N. (After Harr and Dea, 2009.)

the number of clusters in which the ensemble members could be partitioned was related to an increase in predictability. Therefore, at longer forecast intervals the predictability was reduced. Anwender *et al.* (2008) found this trait to be consistent in the ensemble prediction system of the European Center for Medium Range Forecasts and for a variety of ET cases.

Finally, it is not always necessary for a decaying TC to complete ET as a significant extratropical cyclone to impact the downstream midlatitude circulation. McTaggart-Cowan *et al.* (2007a, b) investigated the downstream impacts due to the poleward movement of Hurricane Katrina (2005). Although Hurricane Katrina did not re-intensify and essentially decayed during the ET process, significant downstream flow perturbations resulted from the advection of a large upper-level warm anomaly into to the midlatitudes. This anomaly resulted in a strong blocking event over the North Atlantic and reduced predictability. The upper-level warm anomaly was traced over hemispheric scales for a period of weeks.

4. Forecast Issues

The ET of a TC poses a significant challenge to the forecaster. As defined above, the process of ET represents a complex interaction among a variety of physical mechanisms associated with the TC and the midlatitude circulation into which it is moving. Furthermore, there is a degree of uncertainty in the prediction of the timing of such factors such as the increased translation speed of the decaying TC, the potential for rapid re-intensification as an extratropical cyclone, and the severity of the weather and ocean conditions during ET.

The review article by Jones *et al.* (2003) presented an extremely comprehensive study of the forecast issues associated with ET. This remains a valid summary as understanding of the many of the components of ET as defined above is lacking. In particular, the variability among the complex physical interactions during ET reduces the predictability associated with an

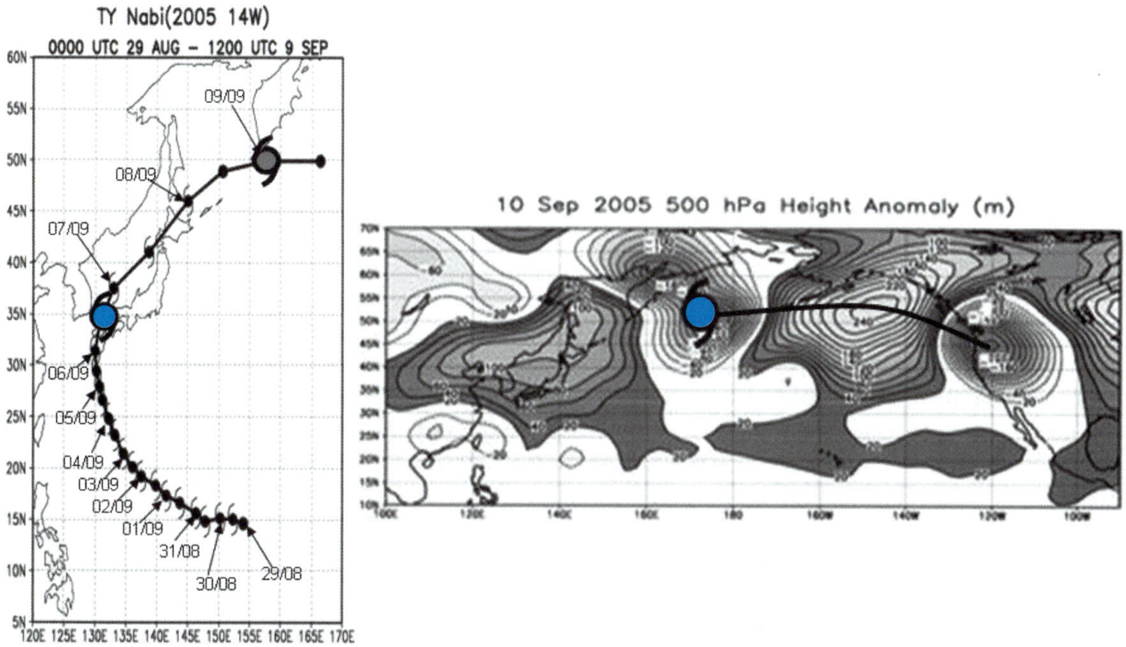

Figure 16. (a) The track of TY Nabi during 2006. The 0000 UTC positions for each date are labeled. The lightly shaded TC symbol defines the time that the transformation stage of ET began and the darkly shaded TC symbol defines the time that ET was completed. (b) The 500 hPa height anomalies valid for 0000 UTC 10 September 2005. The TC symbol marks the location of the extratropical cyclone that resulted from the ET of TY Nabi. The curved line defines the wave pattern that extended to western North America. (adapted from Harr *et al.*, 2008).

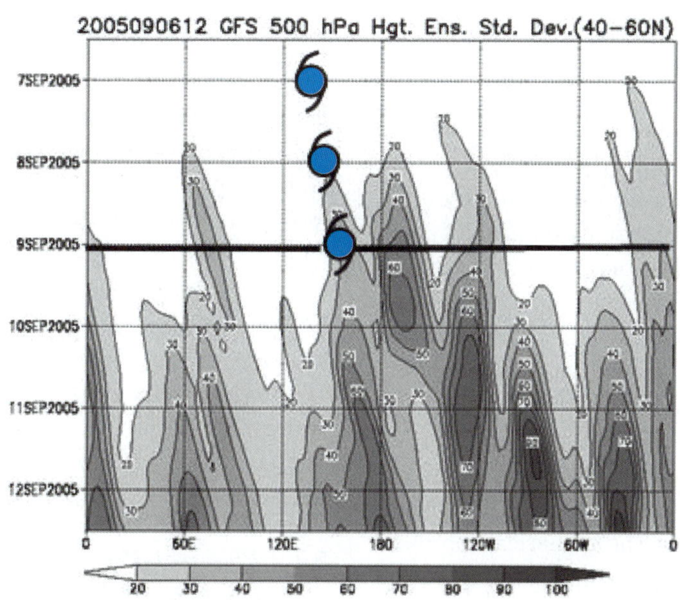

Figure 17. A Hovmoller diagram of the standard deviation (m) in 500 h Pa height among the ensemble members of the NCEP GFS ensemble prediction system. The standard deviation is averaged between 40°N and 60°N. The tropical cyclone symbols mark the longitude of TY Nabi and the black line marks the time and location associated with the completion of the ET of TY Nabi (adapted from Harr *et al.*, 2008).

ET event. The difficulty associated with forecasting of ET events is compounded by the lack of direct observations during ET. Only recent programs have resulted in *in situ* observations of ET events (Abraham *et al.*, 2004). The forecast problem associated with ET may be classified into three categories: (i) the evolution of significant weather and ocean conditions; (ii) the possible rapid reintensification of the decaying TC as an extratropical cyclone; and (iii) the downstream propagation of Rossby waves generated by perturbing the midlatitude flow.

Recent development of an operational modeling tool (MacAfee and Bowyer, 2005, 2006) has aided in forecasting of wave trajectories and significant wave heights for TCs undergoing the rapid forward translation during an ET event. Wave forecasts from this model are available within less than one minute after the forecaster produces or changes the forecast track.

The relative sensitivity of the ET process to the TC structure versus the midlatitude environment has been approached from various perspectives (Klein *et al.*, 2002; Agustí-Panareda *et al.*, 2004; Evans and Prater-Mayes, 2004; McTaggart-Cowan *et al.*, 2004). In several cases, the relative impacts of the decaying TC and midlatitude circulation has been examined via modifying initial conditions prior to numerical simulation. Although this has identified relative roles of TC- and midlatitude-related factors for individual cases, the case-to-case variability prevents broad generalizations regarding the relative importance of various processes. These types of investigations provide promise in the development of conceptual models and forecasting tools that will be of significant utility to forecasters in the near-future.

The numerical modeling studies of Evans and Prater-Mayes (2004) and Ma *et al.* (2006) highlight the sensitivity of the numerical representation of ET to details of the model configuration and initialization. Further documentation for the impact of model initial conditions on the simulation of the structure of an evolving ET is provided in Evans *et al.* (2006), who compare the structural evolution of the evolving storm as represented in the cyclone phase space of Hart (2003). Rather than requiring exact replication of analyzed and forecast structure, they follow Arnott *et al.* (2004) and cluster the identified structures into seven groups: three TC types of increasing intensity; two hybrid forms (including the transitioning phase); and two extratropical. Fogarty *et al.* (2006, 2007) highlighted the impact of using synthetic vortex insertion in a pre-ET initialization of the forecasts of Hurricane Michael (2000) and Juan (2003).

5. Summary and Future Directions

Predictability on hemispheric scales exhibits a large amount of variability in association with ET events. Perturbations to the midlatitude flow characteristics during ET result in the development of intense baroclinic development within approximately one Rossby wavelength of the ET event can result in coherent anomalies that propagate around the hemisphere over the period of weeks. Although recent research has identified important physical processes that define the ET process, there has been little progress on characterizing the influence of specific mechanisms on the variability and predictability of downstream impacts. Although there is promise for increased observations that may be obtained from in situ or remote platforms, there is a great deal of uncertainty as to what impact new or additional observations will have on forecasts sequences of the ET and related downstream flow. Therefore, key issues regarding mechanisms, observations, and impacts remain to be addressed to increase understanding of the role(s) of ET on impacts of predictability over a variety of time and space scales.

5.1. *Physical Processes and Downstream Impacts of Extratropical Transition*

Physical processes associated with the ET involve the transport of anomalous amounts of moisture and heat into the midlatitudes in conjunction with the presence of a positive potential vorticity (PV) anomaly. Therefore, diabatic processes often play a critical role in altering the distribution of PV and may modulate the impact of the PV anomaly associated with the decaying TC on the pre-existing midlatitude PV distribution. Additionally, the transport of heat and moisture has important implications on boundary layer contributions to the re-intensification of the decaying TC as an extratropical cyclone.

Often the movement of the TC into the midlatitudes imposes significant perturbation to the midlatitude flow that may rapidly extend downstream and to a lesser degree upstream of the location of the ET event. The large amount of variability in the downstream response to an ET suggests that there are important sensitivities to a variety of physical mechanisms associated with the forcing on the midlatitude flow. These mechanisms may include basic baroclinic energy conversions, forcing of diabatically-forced Rossby wave-like circulations, and enhancements to downstream jet streaks. All of these factors would exhibit sensitivity to a variety of interactions with a decaying TC. Finally, use of ensemble prediction systems indicate that the downstream forcing due to an ET is typically associated with reduced predictability.

Not only is ET associated with severe local weather conditions, the impact of an ET on the midlatitude flow has been shown to be directly linked to high-impact weather events far downstream of the original ET event. Therefore, there is a strong requirement to assess the extent of the sensitivity to various physical mechanisms during ET. These mechanisms include the impact of sensible and latent heat fluxes, precipitation distribution, frontogenesis, and baroclinic energy conversions. Finally, the character of the downstream response to ET should be placed in the framework of the mean environmental conditions (i.e., baroclinic wave guides) across the entire ocean basin in which the ET is occurring.

As indicated above, reduced predictability downstream of an ET event is often represented by increased variability among members of an ensemble prediction system (EPS). Furthermore, the initial perturbations used to initialize the EPS often define the character of the downstream variability among the members. Browning *et al.* (2000) and Klein *et al.* (2002) found that small changes to the position of the TC or the midlatitude trough that influencing the movement of the TC into the midlatitudes can have significant impacts on the forecast ET and the downstream variability. Torn (2006) employed a 90 member Weather Research and Forecasting (WRF) ensemble Kalman filter (EnKf) over a period of seven days to examine the sensitivities to small perturbations in the phasing between a TC and midlatitude trough. While the studies of Harr *et al.* (2008) and Anwender *et al.* (2008) highlighted the role(s) of perturbations associated with the TC in determining clusters of similar forecast scenarios, the EnKF method highlighted sensitivities to upstream upper-level troughs that interacted with the decaying TCs. Therefore, sensitivity of the phasing between the TC and midlatitude circulation was dependent on the location and intensity of the midlatitude troughs upstream of the poleward-moving TC.

Objective procedures that may be based on adjoint or ensemble techniques help identify regions where there is increased chance for fast-growing initial condition errors that may be detected by available measurements. The fast-growing initial-condition errors may be reduced by taking additional observations and assimilating them into the model initial conditions. Aberson (2003) documented that targeted dropsonde observations in areas with large ensemble forecast variance improved short-range forecasts of some TC tracks by as much as 25%. During the poleward movement of a TC, areas of fast-growing initial condition errors are located near the TC and over regions upstream of the TC that are associated with a midlatitude trough (Langland, 2005; Torn, 2006; Peng and Reynolds, 2006). Majumdar *et al.* (2006) noted that selection of regions of initial condition error growth varied by TC characteristics and by method (i.e., EnKf, total-energy singular vectors).

Because of the variety of physical mechanisms that impact the ET of a TC over a range of spatial and temporal scales, it follows that a wide spectrum of observations will be critical to reduce subsequent forecast errors. As the objective procedures have identified, the observations may need to be obtained in the vicinity of the decaying TC and/or over remote regions associated with important midlatitude circulations. Furthermore, observations of low-level and surface parameters are as necessary as observations of upper-level interactions between TC outflow characteristics and the midlatitude flow.

An increase in predictability associated with downstream impacts from ET may be realized by increased or improved observations of critical processes associated with ET. If reliable guidance is available such that the locations of areas that would result in maximum reduction of initial error growth could be identified, then there must be procedures available to assimilate the additional observations into the initial conditions. It is noted that targeted or adaptive observations improve average forecast skill but not every forecast will benefit from increased observations (Langland, 2005). Morss and Emanuel (2002) discuss this issue in the context of the statistical aspects of data assimilation such that true observation and background errors can only be estimated. Unfortunately, many of the assumptions used to estimate observation and background error do not strictly apply in regions of strong gradients and dynamical processes as exist in the environment of an ET event. There is a critical need to assess exactly how observations in the region of a decaying TC, the TC-midlatitude interface region, or the midlatitude impacts region will be assimilated and how they will influence the evolution of the downstream flow patterns throughout the forecast sequence.

5.2. *Observing and Forecasting of Extratropical Transition*

The difficulty in operational forecasting the ET of TCs has been highlighted relative to diagnostic analysis of observations and numerical prediction. Often, ET is associated with maintenance of TC force winds, precipitation, and ocean waves far into the midlatitudes. Since official forecasts of the TC may have been terminated (except certain regions such as in eastern Canada where the Canadian Hurricane Center issues specific forecasts during ET), it is often difficult to convey the continued threat of the damaging weather elements to the public. Therefore, a requirement for a precise definition of ET is required such that the continued threat may be conveyed to the general public. Furthermore, the definition should encompass the needs of operational and research communities for a common framework on which to examine the physical processes that occur during ET.

In recent years, several cases of ET have been observed with aircraft such that new data sets have been available for analysis of the complex physical processes that occur during ET. While these studies have provided new insights into changes of the remnant TC structure there are still requirements for increased forecast skill associated with many of the impacts of ET such as the expansion of the surface wind distribution, the distribution of heavy precipitation, occurrence of tornadoes, and the generation of extreme ocean wave conditions.

One particular forecast issue associated with ET has been identification of the timing of the ET process and the likely structural characteristics associated with the extratropical cyclones that result from the ET process. Operational forecasting of these factors has improved with the use of the cyclone phase space for summary and display of analyzed and forecast structural characteristics of a TC as it proceeds through the ET process. Furthermore, detailed numerical simulations of the ET process have been instrumental in identifying sensitivities to

TC characteristics and the midlatitude circulation into which the decaying TC is moving. In conjunction with numerical simulations, diagnostic analyses of ET events have led to specification of factors important for increasing forecast skill of many attributes of an ET event. While some diagnostic analyses have been performed on conventional atmospheric fields, others have been based on the cyclone phase space, which provides a compact framework for assessing the dynamic and thermodynamic characteristics of the ET process.

Specific recommendations for improvements in forecasting of ET events revolve around the timing of the evolutionary processes and the sensitivity of the process to various physical characteristics. These improvements are ultimately connected to observations of critical processes such that appropriate diagnosis of the ET may be conducted and the processes may be accurately depicted in the initial conditions of operational numerical forecast models. Critical to these improvements is the increased utility of satellite data and in particular microwave imagery from polar-orbiting platforms that will identify thermodynamic and moisture distributions during the ET event. Ultimately, the development of a satellite-based ET diagnostic should be developed.

In conclusion, it is noted that a universal definition of ET has not been adopted. The evolutionary nature, the complex physical interactions, and the variability associated with the outcome of ET makes it difficult to objectively identify various components of ET. From a forecasting and climatology viewpoints, an operational definition that objectively classified the start and end of the evolutionary ET process would be desirable. Incidentally, the U.S. National Hurricane Center and the Canadian Hurricane Center regularly use model-derived cyclone phase space trajectories in their operational forecast discussions. Because the societal impacts of an ET event can be devastating due to extreme conditions related to precipitation, wind speeds, and wave heights, a continuing need exists for observations and research on improved understanding of the evolution and prediction of the ET process.

Acknowledgments

This review is based in part on material from the Sixth International Workshop on Tropical Cyclones. Professors Jenni L. Evans and John R. Gyakum gratefully served as rapporteurs for the ET topic. Comments on an earlier version of this manuscript by Dr. Chris Fogarty are sincerely appreciated. Support for the preparation of this review came from the National Science Foundation, Climate and Large-Scale Dynamics Program and from the Office of Naval Research, Marine Meteorology Program.

References

Aberson, S. D., 2003: Targeted observations to improve operational tropical cyclone track forecast guidance. *Mon. Wea. Rev.*, **131**, 1613–1628.

Abraham, J., W. Strapp, C. Fogarty, and M. Wolde, 2004: Extratropical transition of Hurricane Michael: An aircraft investigation. *Bull. Amer. Meteor. Soc.*, **85**, 1323–1339.

Agusti-Panareda, A., C. D. Thorncroft, G. C. Craig, and S. L. Gray, 2004: The extratropical transition of Hurricane Irene (1999): A potential-vorticity perspective. *Quart. J. Roy. Meteor. Soc.*, **130**, 1047–1074.

Anwender, D., P. A. Harr, and S. C. Jones, 2008: Predictability associated with the downstream impacts of the extratropical transition of tropical cyclones: Case studies, *Mon. Wea. Rev.*, **136**, 3226–3247.

Arnott, J. M., J. L. Evans, and F. Chiaromonte, 2004: Characterization of extratropical transition using cluster analysis. *Mon. Wea. Rev.*, **132**, 2916–2937.

Atallah, E. H., and L. F. Bosart, 2003: The extratropical transition and precipitation distribution of Hurricane Floyd (1999). *Mon. Wea. Rev.*, **131**, 1063–1081.

Bennetts, D. A., and B. J. Hoskins, 1979: Conditional symmetric instability — A possible explanation for frontal rainbands. *Quart. J. Roy. Meteor. Soc.*, **105**, 595–602.

Bosart, L. F., and D. B. Dean, 1991: The Agnes rainstorm of June 1972: Surface feature evolution culminating in inland storm redevelopment. *Wea. Forecasting*, **6**, 515–537.

Bosart, L. F., and G. M. Lackmann, 1995: Postlandfall tropical cyclone reintensification in a weakly baroclinic environment: A case study of Hurricane David (September 1979). *Mon. Wea. Rev.*, **123**, 3268–3291.

Brand, S., and C. P. Guard, 1979: An observational study of extratropical storms that evolved from tropical cyclones in the western North Pacific. *J. Meteor. Soc. Japan*, **57**, 479–482.

Browning, K. A., A. J. Thorpe, A. Montani, D. Parsons, M. Griffiths, P. Panagi, and E. M. Dicks, 2000: Interactions of tropopause depressions with an ex-tropical cyclone and sensitivity of forecasts to analysis errors. *Mon. Wea. Rev.*, **128**, 2734–2755.

Chien, H. H., and P. J. Smith, 1977: Synoptic and kinetic energy analysis of Hurricane Camille (1969) during transit across the southeastern United States. *Mon. Wea. Rev.*, **105**, 67–77.

Cote, M. L., L. F. Bosart, D. Keyser, and M. L. Jurewicz, 2007: Heavy rainfall events preceding the arrival of tropical cyclones. *Prepints, 22nd Conference on Weather Analysis and Forecasting*, Amer. Meteor. Soc., Boston, MA.

Davis, C. A., S. C. Jones, and M. Riemer, 2008: Hurricane vortex dynamics during Atlantic extratropical transition. *J. Atmos. Sci.*, **65**, 714–736.

Demuth, J. L., M. DeMaria, and J. A. Knaff, 2004: Evaluation of Advanced Microwave Sounding Unit tropical-cyclone intensity and size estimation algorithms. *J. Appl. Meteor.*, **43**, 282–296.

DiMego, G. J., and L. F. Bosart, 1982a: The transformation of Tropical Storm Agnes into an extratropical cyclone. Part I: The observed fields and vertical motion computations. *Mon. Wea. Rev.*, **110**, 385–411.

DiMego, G. J., and L. F. Bosart, 1982b: The transformation of Tropical Storm Agnes into an extratropical cyclone. Part II: Moisture, vorticity and kinetic energy budgets. *Mon. Wea. Rev.*, **110**, 412–433.

Emanuel, K. A., 1983: The Lagrangian parcel dynamics of moist symmetric instability. *J. Atmos. Sci.*, **40**, 2368–2376.

Emanuel, K. A., 1985: Frontal circulations in the presence of small moist symmetric stability. *J. Atmos. Sci.*, **42**, 1062–1071.

Evans, C., and R. E. Hart, 2008: Analysis of the wind field evolution associated with the extratropical transition of Bonnie (1998). *Mon. Wea. Rev.*, 2047–3065.

Evans, J. L., and R. E. Hart, 2003: Objective indicators of the life cycle evolution of extratropical transition for Atlantic tropical cyclones. *Mon. Wea. Rev.*, **131**, 909–925.

Evans, J. L., and B. E. Prater-Mayes, 2004: Factors affecting the post-transition intensification of Hurricane Irene (1999). *Mon. Wea. Rev.*, **132**, 1355–1368.

Evans, J. L., J. M. Arnott, and F. Chiaromonte, 2006: Evaluation of operational model cyclone structure forecasts during extratropical transition. *Mon. Wea. Rev.*, **134**, 3054–3072.

Foley, G. R., and B. N. Hanstrum, 1994: The capture of tropical cyclones by cold fronts off the west coast of Australia. *Wea. Forecasting*, **9**, 577–592.

Fogarty, C. T., R. J. Greatbatch, and H. Ritchie, 2006: The role of anomalously warm sea surface temperatures on the intensity of Hurricane Juan (2003) during it approach to Nova Scotia. *Mon. Wea. Rev.*, **134**, 1484–1504.

Fogarty, C. T., R. J. Greatbatch, and H. Ritchie, 2007: The use of a vortex insertion technique to simulate the extratropical transition of Hurricane Michael (2000), *Wea. Forecasting*, **22**, 480–500.

Frank, W. M., and E. A. Ritchie, 1999: Effects of environmental flow upon tropical cyclone structure. *Mon. Wea. Rev.*, **127**, 2044–2061.

Frank, W. M., and E. A. Ritchie, 2001: Effects of vertical wind shear on the intensity and structure of numerically simulated hurricanes. *Mon. Wea. Rev.*, **129**, 2249–2269.

Harr, P. A., and R. L. Elsberry, 2000: Extratropical transition of tropical cyclones over the western North Pacific. Part I: Evolution of structural characteristics during the transition process. *Mon. Wea. Rev.*, **128**, 2613–2633.

Harr, P. A., R. L. Elsberry, and T. F. Hogan, 2000: Extratropical transition of tropical cyclones over the western North Pacific. Part II: The impact of midlatitude circulation characteristics. *Mon. Wea. Rev.*, **128**, 2634–2653.

Harr, P. A., D. Anwender, and S. C. Jones, 2008: Predictability associated with the downstream impacts of the extratropical transition of tropical cyclones: Methodology and a case study of Typhoon Nabi (2005). *Mon. Wea. Rev.*, **136**, 3205–3225.

Harr, P. A., and J. Dea, 2009: Downstream development associated with the extratropical transition of tropical cyclones over the western North Pacific. *Mon. Wea. Rev.*, **137**, 1295–1319.

Hart, R. E., 2003: A cyclone phase space derived from thermal wind and thermal asymmetry. *Mon. Wea. Rev.*, **131**, 585–616.

Hart, R. E., J. L. Evans, and C. Evans, 2006b: Synoptic composites of the extratropical transition lifecycle of North Atlantic tropical cyclones: Factors determining post-transition evolution. *Mon. Wea. Rev.*, **134**, 553–578.

Hoskins, B. J., M. E. McIntyre, and A. W. Robertson, 1985: On the use and significance of isentropic potential vorticity maps. *Quart. J. Roy. Meteor. Soc.*, **111**, 877–946.

Jones, S. C., 1995: The evolution of vortices in vertical shear. I: Initially barotropic vortices. *Quart. J. Roy. Meteor. Soc.*, **121**, 821–851.

Jones, S. C., 2000a: The evolution of vortices in vertical shear. II: Large-scale asymmetries. *Quart. J. Roy. Meteor. Soc.*, **126**, 3137–3160.

Jones, S. C., 2000b: The evolution of vortices in vertical shear. III: Baroclinic vortices. *Quart. J. Roy. Meteor. Soc.*, **126**, 3161–3186.

Jones, S. C., and Co-authors, 2003: The extratropical transition of tropical cyclones: Forecast challenges, current understanding and future directions. *Wea. Forecasting*, **18**, 1052–1092.

Keyser, D., M. J. Reeder, and R. J. Reed, 1988: A generalization of Petterssen's frontogenesis function and its relation to the forcing of vertical motion. *Mon. Wea. Rev.*, **116**, 762–780.

Kitabatake, N., 2002: Extratropical transformation of Typhoon Vicki (9807): Structural changes and the role of upper-tropospheric disturbances. *J. Meteor. Soc. Japan*, **80**, 229–247.

Klein, P. M., P. A. Harr, and R. L. Elsberry, 2000: Extratropical transition of western North Pacific tropical cyclones: An overview and conceptual model of the transformation stage. *Wea. Forecasting*, **15**, 373–396.

Klein, P. M., P. A. Harr, and R. L. Elsberry, 2002: Extratropical transition of western North Pacific tropical cyclones: Midlatitude and tropical cyclone contributions to reintensification. *Mon. Wea. Rev.*, **130**, 2240–2259.

Langland, R. H., 2005: Issues in targeted observations. *Quart. J. Roy. Meteor. Soc.*, **131**, 3406–3425.

Ma, S., H. Ritchie, J. R. Gyakum, J. Abraham, C. T. Fogarty, and R. McTaggart-Cowan, 2003: A study of the extratropical reintensification of former Hurricane Earl using Canadian Meteorological Centre regional analyses and ensemble forecasts. *Mon. Wea. Rev.*, **131**, 1342–1359.

MacAfee, A. W., and P. J. Bowyer, 2000a: Trapped-fetch waves in a transitioning tropical cyclone (Part I — The need and the theory). *Preprints, 24th Conf. on Hurricanes and Tropical Meteorology*, Fort Lauderdale, FL, Amer. Meteor. Soc., pp. 292–293.

MacAfee, A. W., and P. J. Bowyer, 2000b: Trapped-fetch waves in a transitioning tropical cyclone (Part II — Analytical and predictive model). *Preprints, 24th Conf. on Hurricanes and Tropical Meteorology*, Fort Lauderdale, FL, Amer. Meteor. Soc., pp. 165–166.

MacAfee, A. W., and P. J. Bowyer, 2005: The modeling of trapped-fetch waves with tropical cyclones — A desktop operational model. *Wea. Forecasting*, **20**, 245–263.

MacAfee, A. W., and P. J. Bowyer, 2006: Corregium. *Wea. Forecasting*, **21**, 429.

Majumdar, S. J., S. D. Aberson, C. H. Bishop, R. Buizza, M. S. Peng, and C. A. Reynolds, 2006: A comparison of adaptive observing guidance for Atlantic tropical cyclones. *Mon. Wea. Rev.*, **134**, 2354–2372.

Matano, H., and M. Sekioka, 1971a: On the synoptic structure of Typhoon Cora, 1969, as the compound system of tropical and extratropical cyclones. *J. Meteor. Soc. Japan*, **49**, 282–295.

Matano, H., and M. Sekioka, 1971b: Some aspects of extratropical transformation of a tropical cyclone. *J. Meteor. Soc. Japan*, **49**, 736–743.

McTaggart-Cowan, R., J. R. Gyakum, and M. K. Yau, 2004: The impact of tropical remnants on extratropical cyclogenesis: Case study of Hurricanes Danielle and Earl (1998). *Mon. Wea. Rev.*, **132**, 1933–1951.

McTaggart-Cowan, R., L. Bosart, G. Gyakum, and E. Atallah, 2007: Hurricane Katrina (2005), Part I: Complex lifecycle of an intense tropical cyclone. *Mon. Wea. Rev.*, **135**, 2905–3926.

McTaggart-Cowan, R., L Bosart, G.Gyakum, and E. Atallah, 2007: Hurricane Katrina (2005), Prt II: Evolution and hemispheric impacts of a diabatically-generated warm pool. *Mon. Wea. Rev.*, **135**, 3927–3949.

Moore, R. W., and M. T. Montgomery, 2004: Reexamining the dynamics of short-scale, diabatic Rossby waves and their role in midlatitude moist cyclogenesis. *J. Atmos. Sci.*, **61**, 754–768.

Moore, R. W., and M. T. Montgomery, 2005: Analysis of an idealized three-dimensional diabatic Rossby vortex: A coherent structure of the moist baroclinic atmosphere. *J. Atmos. Sci.*, **62**, 2703–2725.

Morss, R. E., and K. A. Emanuel, 2002: Influence of added observations on analysis and forecast errors: Results from idealized systems. *Quart. J. Roy. Meteor. Soc.*, **128**, 285–321.

Orlanski, I., and J. P. Sheldon, 1995: Stages in the energetics of baroclinic systems. *Tellus*, **47A**, 605–628.

Palmen, E., 1958: Vertical circulation and release of kinetic energy during the development of Hurricane Hazel into an extratropical storm. *Tellus*, **10**, 1–23.

Peng, M. S., and C. A. Reynolds, 2006: Sensitivity of tropical cyclone forecasts as revealed by singular vectors. *J. Atmos. Sci.*, **63**, 2508–2528.

Palmén, E., 1958: Vertical circulation and release of kinetic energy during the development of hurricane Hazel into an extratropical storm. *Tellus*, **10**, 1–23.

Parker, D. J., and A. J. Thorpe, 1995: Condition convective heating in a baroclinic atmosphere: A model of convective frontogenesis. *J. Atmos, Sci.*, **52**, 1699–1711.

Petterssen, S., and S. J. Smebye, 1971: On the development of extratropical storms. *Quart. J. Roy. Meteor. Soc.*, **97**, 457–482.

Raymond, D. J., 1992: Nonlinear balance and potential-vorticity thinking at large Rossby number. *Quart. J. Roy. Meteor. Soc.*, **118**, 987–1015.

Riemer, M., S. C. Jones, and C. A. Davis, 2008: The impact of extratropical transition on the downstream flow: An idealised modelling study with a straight jet. *Quart. J. Roy. Meteor. Soc.*, **134**, 69–91.

Ritchie, E. A., and R. L. Elsberry, 2007: Simulations of the extratropical transition of tropical cyclones: Phasing between the upper-level trough and tropical cyclone. *Mon. Wea. Rev.*, **135**, 862–876.

Sardeshmukh, P. D., and B. J. Hoskins, 1988: The generation of global rotational flow by steady idealized tropical divergence. *J. Atmos. Sci.*, **45**, 1228–1251

Schultz, D. M., and P. N. Schumacher, 1999: The use and misuse of conditional symmetric instability. *Mon. Wea. Rev.*, **127**, 2709–2732.

Sekioka, M., 1956: A hypothesis on complex of tropical and extratropical cyclones for typhoon in middle latitudes, I. Synoptic structure of typhoon Marie over the Japan Sea. *J. Meteor. Soc. Japan*, **34**, 42–53.

Sekioka, M., 1970: On the behaviour of cloud patterns as seen on satellite photographs in the transformation of a typhoon into a tropical cyclone. *J. Meteor. Soc. Japan*, **48**, 224–232.

Sekioka, M., 1972a: A kinematical consideration on behaviour of a front within a typhoon area. *Arch. Meteor. Geophys. Bioklimatol.*, **A21**, 1–12.

Sekioka, M., 1972b: Note on the extratropical transformation of a typhoon in relation with cold outbreaks. *Arch. Meteor. Geophys. Bioklimatol.*, **A21**, 413–418.

Shimazu, Y., 1998: Classification of precipitation systems in mature and early weakening stages of typhoons around Japan. *J. Meteor. Soc. Japan*, **76**, 437–445.

Simmons, A. J., and B. J. Hoskins, 1979: Downstream and upstream development of unstable baroclinic waves. *J. Atmos. Sci.*, **36**, 1239–1254.

Sinclair, M. R., 1993: Synoptic-scale diagnosis of the extratropical transition of a southwest Pacific extratropical cyclone. *Mon. Wea. Rev.*, **121**, 941–960.

Sinclair, M. R., 2002: Extratropical transition of southwest Pacific tropical cyclones. Part I: Climatology and mean structure changes. *Mon. Wea. Rev.*, **130**, 590–609.

Thorpe, A. J., and K. A. Emanuel, 1985: Frontogenesis in the presence of small stability to slantwise convection. *J. Atmos. Sci.*, **42**, 1809–1824.

Thorncroft, C. D., and S. C. Jones, 2000: The extratropical transitions of Hurricanes Felix and Iris in 1995. *Mon. Wea. Rev.*, **128**, 947–972.

Thorncroft, C. D., S. C. Jones, and M. E. McIntyre, 1993: Two paradigms of baroclinic-wave lifecycle behaviour. *Quart. J. Roy. Meteor. Soc.*, **119**, 17–55.

Torn, R. D., 2006: Initial condition sensitivities for Western Pacific extratropical transition events. *Preprints, 27th Conference on Hurricanes and Tropical Meteorology*, American Meteorological Society, Boston, MA.

Vancas, M., 2006: The extratropical transition of tropical storm Banyan. M. S. Thesis, Naval Postgraduate School. 72 pp [available at http://theses.nps.navy.mil/07Sep_Vancas.pdf].

PART II

Observations of Tropical Cyclones

Chapter 6

Observing and Analyzing the Near-Surface Wind Field in Tropical Cyclones

Mark D. Powell

National Oceanic and Atmospheric Administration
Atlantic Oceanographic and Meteorological Laboratory
Hurricane Research Division, Miami, FL, USA
Mark.Powell@noaa.gov

This paper describes the current state of the art in measuring and analyzing surface winds in tropical cyclones. Observing platforms and strategies will be reviewed, along with their advantages and limitations.

1. Introduction

Since 1997, developments in wind sensing instrumentation and sampling strategies have dramatically improved our ability to measure the intensity of tropical cyclones as well as the extent of damaging winds. In the Atlantic tropical cyclone basin, west of 60 west longitude, research and reconnaissance aircraft have been responsible for the greatest improvements, using radiometric and GPS-based remote sensing techniques. Outside the Atlantic basin, the primary observing platform remains the satellite, which is gradually improving capabilities for remotely sensing surface winds in extreme environments. The in-situ surface observing network, which should represent the ground truth for these new systems, has a mixed record; the official networks of many countries are not reliable in extreme winds and even the coastal and buoy platforms have suffered from costly failures during critical situations. In general however, over the past few years we have experienced an unprecedented increase in the quantity and quality of observations available to document the pre-landfall wind structure of tropical cyclones. The challenges facing us now include:

1) Can we improve the current coastal and inland observing network to allow high resolution documentation of extreme winds during landfall and post-landfall?
2) Do we have sufficient data coverage to resolve wind maxima and the radial and azimuthal variation in the wind field at landfall?
3) With many platforms now measuring the same patch of ocean, how do we combine them to construct our best estimate of the wind field at that location and time?
4) How do we present the information so it can be used to support a wide variety of forecast, warning, emergency response, disaster assessment, and design risk applications?
5) Will new information allow us to reexamine past tropical cyclones in an attempt to improve the quality of the historical record enough to assess climate change?

This paper examines the current state of hurricane wind measurement and analysis. We describe many of the available wind observing platforms, discuss their advantages, limitations, and how analysis of the information may be used to document and respond to a hurricane disaster. Our discussion shows that there is considerable uncertainty in assessing storm intensity in tropical cyclones, but the situation is improving.

We assign uncertainty based on specified instrument accuracy, estimates of the error in processing measurements to a standard framework, and experience using a variety of observing platforms to assess the wind field of hurricanes. When reporting uncertainty, we assume errors are Gaussian and use the estimated standard deviation expressed as a wind speed or percentage. For example, an estimate of a 50 m/s wind speed with a 10% uncertainty would mean that there is ~32% probability that the true wind could be <45 m/s or >55 m/s. Uncertainty should be an integral part of any wind product so the user has some idea of the variability in the estimate. However, assessing uncertainty is difficult due the dependence of measurements on multiple variables. For instance, if a particular observing platform has instrument bias that varies with rainfall or other environmental conditions, it may require additional quality control. Measurements at non standard heights, exposures, and averaging times require processing to be adjusted to a standard height and averaging time and exposure, all of which have their own errors that are not well known and also depend on environmental conditions. The observing platform should have a measurement strategy consistent with the field to be analyzed, and the collective distribution of the platforms should provide sufficient spatial coverage that an analysis algorithm will minimize errors of representativeness, which in turn is related to the analysis parameters, and the temporal and spatial scales of the atmospheric features.

2. Land Observing Platforms

The primary surface observing networks worldwide consist of anemometer towers placed at airports. These may be automated (e.g. Automated Surface Observing System (ASOS), Fig. 1) or subject to viewing by trained observers. The most reliable land-based anemometers include propeller vane and cup instruments. Cup inertia can cause cup anemometers to underestimate gusts and overestimate lulls. Newer instrument technologies have fewer moving parts and include acoustic and hot film instruments. The acoustic instruments appear to be able to perform in high winds but performance in heavy precipitation is not yet defined. Hot film instruments can measure erroneous peak gusts during heavy rain conditions.

Figure 1. Left: ASOS station wind exposure documentation photo looking NE (Powell *et al.*, 2004). Right: Florida Coastal Monitoring Program tower set up at Frisco NC, the day before Hurricane Isabel made landfall in September 2003 (FCMP).

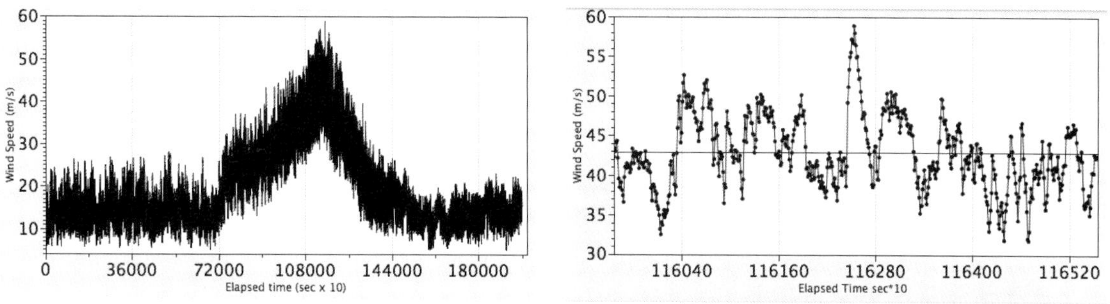

Figure 2. Time series from FCMP tower T0 during Hurricane Dennis landfall in northwest Florida (a) full record, (b) zoomed 1 min record surrounding peak gust (Ref: F. Masters and K. Gurley, Department of Civil and Coastal Engineering, University of Florida).

Designed primarily for aviation use, fundamental flaws for use in extreme events include the anemometers rated to <125 kt, lack of emergency back up power, and outdated communication systems (Powell, 1993). Many of the ASOS stations in hurricane prone areas have photographic exposure documentation available on the web site of NOAA's National Climatic Data Center (Powell *et al.*, 2004) and now internet tools such as Google Earth can help diagnose upstream terrain that can influence wind instrument exposure. Partial records of hurricane eye walls have been measured in Hurricane Marilyn (1995) in St. Thomas USVI, and at Punta Gorda in Hurricane Charley of 2004. In each case power failures prevented complete documentation of the eyewall passage across the site. Accuracy of ASOS anemometers is on the order of 1 m/s or 5% for speed and ±5 degrees before taking into account exposure.

To help fill the gap in these situations, the Texas Tech University Hurricane Research Team (TTHRT) (see www.atmo.ttu.edu/WEMITE/wemite.html) and Florida Coastal Monitoring Programs (www.ce.ufl.edu/~fcmp/), (Fig. 1) have developed portable wind measurement towers that may be *deployed* in advance of projected landfall locations. These programs have sampled numerous storms over the past few years and FCMP has pioneered real-time data transmission.

However the logistics of these deployments are arduous and to maintain safety precautions, towers have to be set up well in advance of landfall, leading to the possibility that last minute turns in the storm could cause to station to sample the storm's periphery instead of the eyewall core. Combined, the TTU and FCMP programs have sampled over 20 hurricanes since 1998. These systems performed admirably during the busy 2004–2005 hurricane seasons, helping fill the power outage void in the network of official observing stations. Hurricane Dennis of 2005 (Fig. 2) provided the highest winds sampled by the FCMP group with peak 10 min, 1 min, and 3 sec wind speeds of 42.1, 42.9, and 49.8 m/s, respectively. The TTU group sampled peak 1 min and 3 sec winds of 42 and 52 m/s in Hurricane Rita of 2005 (J. Schroeder, personal communication).

As reported by Cao *et al.*, (2007), in 2003, Typhoon Maemi passed through a dense network of sonic and vane anemometers on the Island of Miyakojima, Japan. Instantaneous winds exceeded the maximum measurable range of the anemometers in the front eyewall but post-storm calibration indicated unreliable vane anemometer readings for winds >90 m/s. Analysis of the rear side eyewall passage records indicated peak 10 min and 3 s winds of ~60 and 80 m/s, respectively (Cao *et al.*, 2007). Uncertainty of tower anemometer measurements is <5% but anemometer wind

Figure 3. Portable (SMART) Doppler radars with NEXRAD radar in background (NOAA/NSSL).

measurements are only valid within their calibrated range.

In the early 1990s the National Weather Service Next Generation radars (Fig. 3, NEXRAD www.roc.noaa.gov/) began replacing the 1950s era precipitation radars around the U.S. and additional coastal Doppler radar networks are now common in many countries. In the past several years, portable Doppler radars have been developed that can sample tropical cyclones during landfall. Doppler radars measure the velocity component of rain droplets in the direction of the transmitted beam. The main limitations of the Doppler radar are that two or more radars are generally needed to make vector wind measurements, and ground clutter usually makes it difficult to measure winds at heights $<100\,\text{m}$. NEXRAD radars measure winds at $0.5\,\text{m/s}$ resolution and use spectral width as a measure of noise. The Ground Based Velocity Track Display (GBVTD) technique uses storm motion assumptions to estimate a two dimensional wind fields from one ground based radar with an accuracy of $5\,\text{m/s}$ (Lee *et al.*, 1999). Portable Doppler radars (SMART radars and Doppler on wheels, http://www.nssl.noaa.gov/smartradars/) have added to the mix of portable networks being set up to monitor hurricane landfalls, documenting internal boundary layer development as onshore flow moves over a barrier island (Lorsolo and Schroeder, 2004) and well as relatively fine scale linear coherent features in the velocity and reflectivity fields (Wurman and Winslow, 1998, Foster *et al.*, 2004, Lorsolo *et al.*, 2008). The linear features have been attributed to roll vortices and even speculated to be responsible for damage but as yet there has not been enough evidence to document the process behind these features (see Kepert, 2009, (Chapter 1 of this volume) for more information).

Observing networks are becoming more prevalent to support a variety of activities including agriculture, transportation network monitoring, water resource, and ecosystem management. Attempts are ongoing to form networks of networks known as mesonets, with a primary example being the state of Oklahoma mesonet in the U.S. (http://okmesonet.ocs.ou.edu/). The problems facing mesonets include robustness for extreme winds, redundancy, and adherence to standards for measurement and exposure. Since the mesonets represent a variety of applications, calibrations, exposures and sampling methods vary widely so they must be documented carefully and used with caution. The ASTM standard method for characterizing surface wind using a rotating anemometer (ASTM 2005) serves as a resource for helping to standardize and document the exposure of the stations. NOAA's Forecast Systems Lab has established a nationwide mesonet (Barth *et al.*, 2002) with observations available through the internet.

3. Marine Observations

NOAA's National Data Buoy Center (NDBC) manages the network of oceanic moored buoys and coastal marine automated network (C-MAN) stations and buoys (Fig. 4). All stations have redundant sensors that are quality controlled and communicate observations via Geostationary satellites. The moored buoys consist of 5-m or 10-m anemometer heights and report on the hour (8.5 min average) as well as "continuous mode" in which consecutive 10 min mean winds are measured as well as the peak

Figure 4. Left: NOAA moored buoy (6m Nomad ship hull, NOAA NDBC), Right: Fowey Rocks C_MAN showing failed anemometer mast (extending horizontally to right, NDBC).

wind gust during each hour. The C-MAN stations also have continuous wind capability and the NDBC web site maintains exposure photos for stations closest to the coast. The NDBC platforms have a reputation as the "hurricane picket line" with the most hardened wind observing stations in the U.S., having documented eye walls in Hurricanes Eloise, Frederic, Kate, Lili, Opal, and Ivan. Unfortunately in recent years several notable failures have occurred. The anemometer mast aboard the Fowey Rocks lighthouse folded during the eyewall of Hurricane Andrew, shortly after measuring a peak wind of 147 kts at the 39 m level. Additional failures have included buoys breaking apart from their moorings during extreme waves in Hurricane Isabel (2003), Ivan (2004), Katrina (2005) and Ike (2008) as well as ship collisions, and erosion. Accuracies of NDBC moored buoys and C-MAN stations are on order of 1 /s or 10% and 10 degrees for wind direction (Gilhousen, 1987). There have been some indications that buoy measurements are affected by wave sheltering (Large, *et al.*, 1995; Hervey, 1999) Recent work by Howden *et al.*, 2008 compared sonic and mechanical anemometers aboard a 3-m discus buoy that withstood Hurricane Katrina in Mississippi Sound. They documented buoy heeling angles >20 deg for several hours caused by a combination of wind and current in waves with significant heights of 6–8 m. They found that traditional mechanical type anemometers used for marine observations suffer from a low bias caused by lack of compensation for buoy tilt. Errors caused by higher frequency buoy motion associated with waves is assumed to be reduced by averaging over periods of 10 min. In comparison the sonic anemometer (with tilt correction) was found to be robust, however, the possible effects on the measurement of precipitation and spray are unknown.

Ships of opportunity transmit wind observations to the global telecommunications system. Ship wind measurements tend to represent higher anemometer levels near 20 m and require correction for ship motion and flow distortion (Bourassa *et al.*, 2003, Cardone *et al.*, 1990). Ships without anemometers may report wind observations based on the descriptive characteristics of the sea state as described in the Beaufort scale. It is important to quality control ship observations by examining how a ship has compared to operational analyses and also checking a series of measurements from the ship for consistency relative to neighboring observations. Ships with anemometers and reporting Beaufort scale are believed to have uncertainties of 15% and 20%, respectively.

Drifting buoys (drifters) represent a network supporting oceanographic research. Drifters measure wind speeds with acoustic sensors using the Wind Observation through Ambient Noise (WOTAN) method (Vagle *et al.*, 1990) and direction with a vane and internal compass. Mean winds are computed each hour and 4 h of data are transmitted to overflying satellites every 6 h. Drifter sensor accuracy is specified at ± 1 m/s for winds <20 m/s and ± 10 degrees for wind direction. Nine drifters sampled winds in 2004 Hurricanes Frances and Jeanne. When hourly drifter data were compared to co-located Quikscat wind speeds, preliminary results show measurements were comparable (mean difference of 0.4 m/s with 1 m/s standard deviation) in winds <15 m/s but the acoustic method overestimated higher wind speeds, probably due to increased acoustic signal from wave breaking (Morzel http://www.cora.nwra.com/~morzel/drifters.frances.html). Drifting buoys can provide valuable data in remote areas but need to be checked carefully relative to nearby conventional platforms. Uncertainty is estimated at 20% but may be reduced if they compare well with higher-quality platforms.

4. Satellites

The most important space-based platform for observing ocean surface vector winds is the SeaWinds instrument aboard NASA's polar orbiting QuikScat satellite (Fig. 5, Yueh *et al.*, 2003), available since 1999. Sea Winds uses a Ku-band (13.4 GHz) scatterometer with two

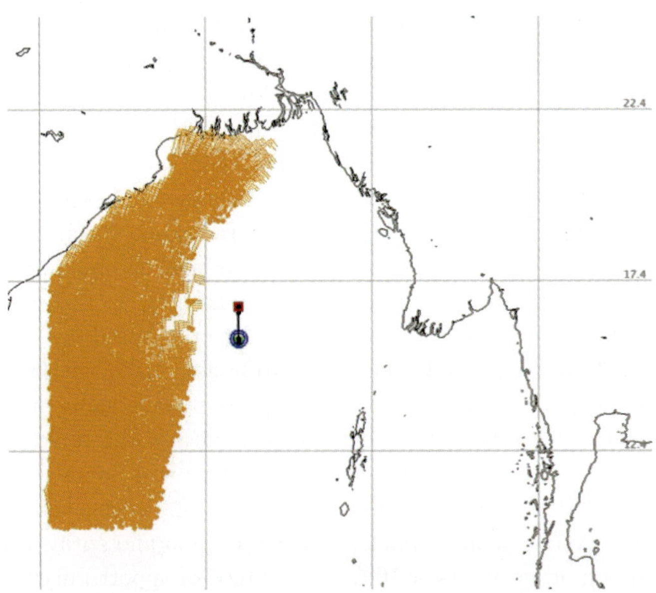

Figure 5. Data coverage plot showing distribution of QuikScat HIRES winds at 1255 UTC. Storm relative data coverage for Bay of Bengal Cyclone Sidr 14 Nov. 2007.

emitters, each with two different incident angles which respond to microwave scattering from capillary waves. For tropical cyclones QuikScat typically provides 1–2 looks per day with a 1800 km wide swath of measurements at either 25 or 12.5 km resolution more than 30 km offshore. An ambiguity removal algorithm selects one of up to six wind vectors as the most likely. QuikScat is especially helpful in identifying incipient tropical cyclones and the extent of tropical storm force (17 m/s) wind speeds. The instrument measurements are sensitive to contamination from rain but many of the poor data can be identified by an algorithm that sets rain "flags". Limitations include occasional poor data on the edges of swaths and occasional poor wind directions (caused by poor choices of the available directional ambiguities), and a tendency for the signal to saturate at wind speeds above 50 m/s due to absorption of Ku band microwave radiation. However, usually the measurement suffers from rainfall contamination well before such speeds can be measured since high winds are usually correlated with rain producing storms. Mission accuracy requirements for Quikscat are 2 m/s for wind speed and ±20 degrees for wind direction. Uncertainty is 10% in winds up to 35 m/s in areas without rain. Quikscat has surpassed all expectations and has been a critical part of NOAA's hurricane forecast and warning program such that forecasters at the National Hurricane Center have stated: "QuikSCAT has become an important tool, especially for estimating the track, intensity and size of tropical and other strong marine storms." Unfortunately Quikscat has exceeded its design life. It could fail at any time but is expected to have sufficient fuel until 2011.

Replacement strategies currently in discussion include the Hurricane Imaging Radiometer (HIRAD) and the "XOVWM" system. The HIRAD is a multifrequency radiometer similar to the Stepped Frequency Microwave Radiometer (SFMR) being flown on NOAA and U.S. Air Force Reserves hurricane reconnaissance aircraft. The instrument is capable of a 22 km wide swath with 1-3 km resolution when flown aboard a jet aircraft similar to the NOAA G4. HIRAD will be tested as a concept for future spaceborne application. The XOVWM concept is for a larger antenna than QuikSCAT with improved spatial resolution, Ku- and C- band scatterometers, and an X-band radiometer. The C-band scatterometer is less susceptible to rain contamination than Ku band, and the radiometer will allow for atmospheric attenuation correction

The European Space Agency MetOp-A satellite was launched in October of 2006 and carries the Advanced Scatterometer (A-SCAT) and Advanced Microwave Sounding Unit (AMSU) instrument. ASCAT (Fig. 6) operates at frequencies in the C-band (5 GHz), which receive less rain contamination but are more susceptible to signal saturation at wind speeds >30 m/s. ASCAT operates with two sets of antennae to provide two 550 km wide data swaths inclined 36 deg from the satellite ground track, resulting in a 650 km gap between swaths. Wind vector measurements comprise 50 km or 25 km resolution cells more than 80 km offshore and wind measurement accuracy is estimated at $2\,\mathrm{m\,s^{-1}}$ or 10% for winds of 2–$24\,\mathrm{m\,s^{-1}}$ and 20 deg for wind direction. ASCAT provides about half the resolution of QuikSCAT and 60% of the coverage of QuikSCAT, but should prove critical for estimating tropical storm force wind radii, especially in the interim period before the launch of a QuikSCAT replacement.

The Advanced Microwave Sounding Unit (AMSU, DeMuth *et al.*, 2004) can measure upper tropospheric warm temperature anomalies in the hurricane's inner core. These measurements can be used to diagnose the pressure field and, through a non linear balance equation, the winds at the 3 km level may be estimated and then adjusted to the surface. This method is relatively new and coarse in resolution (~45 km horizontal) but may be useful for diagnosing intensity when no other platforms are available or in combination with other satellite

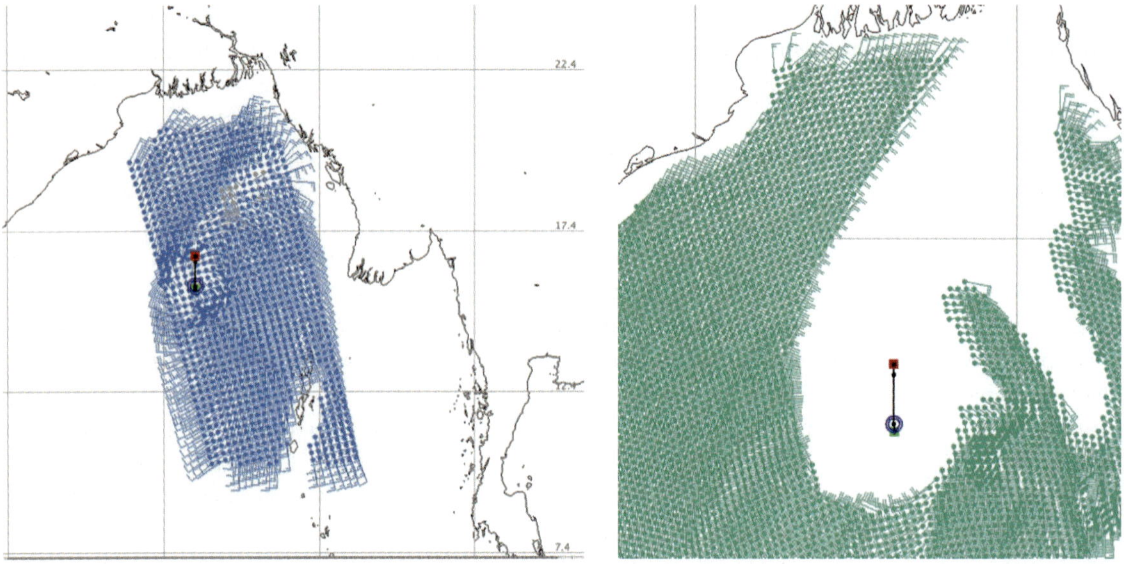

Figure 6. Data coverage plot showing distribution of ASCAT measurements (left) at 1452 UTC and WINDSAT measurements (right) at 1145 UTC in Cyclone Sidr On 14 November, 2007. Data are plotted storm relative for the center position at 1200 UTC.

platforms. GOES infrared brightness temperatures may also be used to estimate intensity through a statistical approach and are currently under development at University of Wisconsin (Kossin and Velden, 2003). The uncertainty of these methods is not yet established.

Geostationary (GOES) satellites have the ability to report visible images of clouds at intervals of 15–30 min. These measurements allow daytime tracking of low level cloud features, assignment of the motion vector to a particular height, and empirical adjustment of the observations to the surface (Dunion *et al.*, 2002). These observations have proven extremely important to fill in gaps in Atlantic basin observation coverage. Typically Atlantic basin reconnaissance aircraft fly limited radial legs in set "alpha" or Fig. 4 shaped patterns. In most cases the aircraft leg ends before establishing the extent of tropical storm force wind speeds. The GOES cloud drift winds help fill in this gap. Limitations include the lack of data when the storm is outside GOES range or has an extensive upper level cloud layer obscuring the surface (central dense overcast). Also, due to the experimental nature of the product, it is still not operationally generated during all tropical cyclone situations. Recently this technique has been applied to GOES 3.9 micron near infrared channel cloud images to allow cloud tracking at nighttime. Wind speed uncertainty is 20% due to uncertainty in cloud height and surface adjustment.

The Special Sensor Microwave Imager (SSM/I, Goodberlet *et al.*, 1989) aboard three polar orbiting satellites measures emissions in seven channels at frequencies ranging from 19–85 GHz over a 1400 km swath with a 25 km footprint and spacing. Measurements are contaminated by both cloud and rain and therefore are typically only useful on the periphery of tropical cyclones. SSMI provides wind speed measurements that are sensitive to cloud and rain contamination but occasionally help to define the extent of tropical storm force winds. Wind speed accuracy is 2 m/s for clear skies in winds <20 m/s. The Tropical Rainfall Measurement Mission (TRMM) satellite, was launched in 1997 (Connor and Chang, 2004) Like SSM/I, the TRMM Microwave Imager suffers from similar wind measurement problems problems but has an additional channel that is less susceptible to cloud contamination. Wind speed accuracy is on order of 2 m/s in rain free areas <20 m/s. The TRMM mission has far surpassed its initial 3-year term and is now expected to end in 2010.

Newer microwave sensors include the NASA Aqua satellite (AMSR-E), RADARSAT, ENVISAT (http:// envisat.esa.int/) Horstman *et al.*, 2004) and the Navy's WINDSAT. Each of these platforms will suffer from some rain contamination affects but the extent of the problem for hurricanes is not yet known. The Canadian Space Agency's RADARSAT has measured 5-cm scale surface roughness features over the sea with C-band Synthetic Aperture Radar (SAR). Swath widths are 500 km and spatial resolution is 100 m. Wind speed may be estimated using methods similar to those used by scatterometers (algorithms based on the normalized radar cross section). An advantage of RADARSAT is that the C-band microwave energy can penetrate clouds and rain, however, the signal can still saturate at high wind speeds. A disadvantage is that measurement of wind speed is still under development and the data are only available commercially, and not always for tropical cyclones.

The Coriolis/WINDSAT (Fig. 6, www.pxi.com/praxis_publicpages/WINDSAT.html) weather satellite uses a passive polarimetric microwave conical scanner that can measure surface wind vectors with less susceptibility to cloud and rain than Ku band, at 25-km resolution over a 1025-km swath. Early indications suggest a 2 m/s rms error for wind speed <20 m/s with a 25-degree error in wind direction. Recent work suggests that WINDSAT may be capable of resolving tropical cyclone winds up to 50 m/s.

AMSR-E is the Advanced Microwave Scanning Radiometer for the Earth Observing System aboard the joint NASA/Japan Space Development Agency AQUA satellite (http://sharaku.eorc.nasda.go.jp/AMSR/index_e.htm). AMSR-E has a swath of 1445 km and wind speed resolution at 21 and 38 km. Preliminary indications from AMSR-E data collected in Hurricanes Frances and Ivan of 2004 suggest that reasonable data can be collected outside major precipitation features (AMSR-E Data at NSIDC: AMSR-E Images of Hurricane Frances).

One of the big challenges for satellite microwave sensing is to develop techniques to reduce or eliminate rain and cloud contamination as well as extend the wind speed range for useful observing before the signal saturates. Recent experiments involving microwave sensors aboard research aircraft show great promise in helping to solve these problems.

5. Research and Reconnaissance Aircraft

Through the late 1990s surface wind speeds were estimated from reconnaissance and research aircraft (Fig. 7) from visual estimates of sea state, by diagnostic boundary layer models, or empirically by applying a reduction factor (Powell, 1980; Powell, 1987; Powell and Black, 1991). Sea State estimates (based on a visual assessment of the amount of foam coverage produced by breaking waves) were reasonable at low levels in high wind speeds but tend to be subjective and unavailable for intense storms since aircraft had to fly at higher levels (near 3 km) to satisfy safety requirements. Boundary layer models lose their basic assumptions (flight-level wind equivalent to a mean wind for the boundary layer) as soon as the aircraft ascends above 1.5 km, and also assume horizontal homogeneity, which is unlikely in the presence of curved flow with strong horizontal gradients (Kepert, 2001; Kepert and Wang, 2001). In the early 1990s it was recognized that the hurricane boundary layer was on the order of 500 m deep in the inner core with peak winds near the top of the boundary layer followed by a gradual decrease above due to the warm core nature of the storm (causing the

Figure 7. Left: One of two NOAA P3 hurricane research aircraft (Brad Smull). Right: GPS dropsonde.

horizontal pressure gradient to decrease with height).

In 1997, the first hurricane eyewall launches of the new GPS dropwindsonde provided the observational details to confirm boundary layer structure. The GPS sonde (Hock and Franklin, 1999) is a small cylindrical instrument package (Fig. 7) launched through a pressure tube in the aircraft skin. It falls at a speed of ~12 m/s, slowed in descent by a pyramidal shaped drogue chute. A code-less receiver collects information on GPS satellite frequencies and transmits the information back to the parent aircraft where it is processed with a full receiver into horizontal and vertical positions and wind velocities. The vertical profiles of the wind are semi-Lagrangian since the sondes move horizontally up to tens of km while falling from 1–3 km altitudes; however they certainly do not follow air parcels in a true Lagrangian sense. Based on the near surface and launch level wind speeds, empirical wind reduction factors were developed for the eyewall (90%) and outer (80%) regions of the storm (Franklin *et al.*, 2003). These empirical reduction factors were used to reclassify Hurricane Andrew as a Category 5 storm on the SS scale (Landsea *et al.*, 2004). As discussed in Powell *et al.*, 2009, vertical reduction factors based on the GPS sonde have large uncertainty and a high bias for surface winds $<70\,\mathrm{m\,s^{-1}}$. A slant reduction factor referenced to the maximum wind speed at flight level provides a more accurate estimate of the maximum surface wind speed. GPS sonde measurement uncertainties are 15% for surface values (after considering instrument, height, averaging time, and representativeness) and ~10% for values estimated from surface layer means.

NOAA aircraft have made Doppler radar measurements in hurricane since 1982 (Marks *et al.*, 1992; Gamache *et al.*, 1995; Aberson *et al.* 2009, this volume)). Aircraft Doppler radars on research aircraft can adopt sampling strategies such as perpendicular legs, fore-aft sampling, and multiple aircraft orthogonal legs to measure the three-dimensional wind field within the area comprised by the core of the hurricane (eye and eyewall). Based on comparisons of flight level observations to fore-aft scans of the airborne Doppler, uncertainty is estimated at 3 m/s but surface clutter prevents measurements below 150 m. New data assimilation and analysis techniques in development allow rapid processing of the wind field on board the aircraft which can then be sent to forecasters on the ground via satellite communication systems.

The Stepped Frequency Microwave Radiometer (SFMR) is a microwave surface wind speed sensor aboard the NOAA research aircraft (Uhlhorn and Black, 2003; Uhlhorn *et al.*, 2007). The SFMR measures the emission from the atmosphere and ocean surface at six separate frequencies from 4–7 GHz, as a brightness temperature T_b. The surface emission is from bubbles that form when waves break and inject air into the sea (Fig. 8). The accuracy of the brightness temperature measurement is ~0.5 K. The relationship between emissivity and surface wind speed is based on comparisons to GPS sondes in hurricanes. The measurement of the attenuation of the signal due to rainfall also makes it possible to remove the rain contamination effects.

Limitations of the SFMR include spurious values over sand bars, marine platforms, and land areas that are awash or any location in which the sea foam is caused by mechanisms other than wind alone (e.g. current flowing against the wind, shoaling conditions in shallow water). Independent comparisons of the SFMR to GPS sondes by Uhlhorn *et al.* (2008) found that SFMR winds were higher on average by 1.6 m s^{-1}. with an rms difference of 4 m s^{-1}.

The SFMR is able to document how the surface wind relates to the wind at flight level, and in particular, the location and magnitude of the surface wind relative to the peak wind measured at flight level. The high radial resolution of the measurements (30 s running mean of 1 s samples providing a radial resolution of ~3 km) provides an unambiguous measure of the peak surface wind that allows improved (relative to GPS sondes) estimation of flight-level wind reduction factors. For example, Fig. 8 shows the SFMR depicting peak surface wind inward of the maximum flight level wind as well as higher reduction factors on the left side than the right (relative to the storm motion). SFMR sensor technology is currently implemented on the operational C-130 reconnaissance aircraft fleet operated by the 53rd Air Force Reserve squadron at Keesler Air Force Base. Uncertainty is ~10%.

6. H*Wind Real-time Objective Wind Field Analysis

With realtime processing and a great deal of cooperation with contributing agency scientists, the various measurement platforms mentioned above are processed into a common standard framework for averaging time (maximum 1 min mean), exposure (marine or open), and height

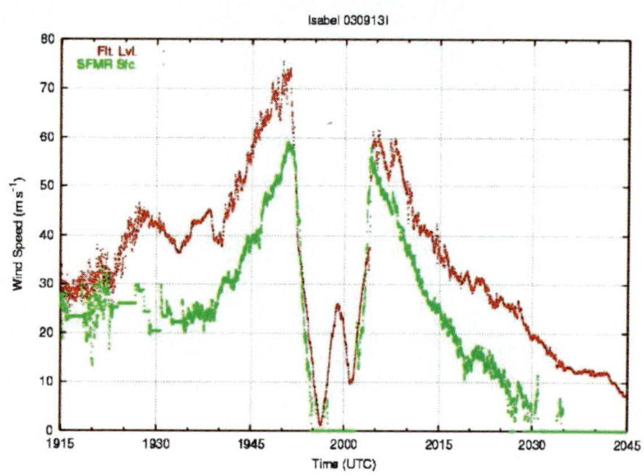

Figure 8. Left: Hurricane sea state showing foam coverage as observed from 2.2 km (Paul Chang); Right: SFMR (green) and flight level (red) radial profile of wind speed in Hurricane Isabel (2003) during CBLAST experiment.

(10 m) Powell *et al.*, 1996. A time-to-space conversion is employed to present the observations as a range and bearing from the storm center over a particular time "window" during which the storm intensity and structure is assumed to be stationary. In practice, one wants to minimize the time window length (for storms undergoing rapid intensity change) while maximizing the data coverage over the analysis domain. Typically a 4–6 h time period is required to assemble sufficient observational coverage for an analysis. This storm relative manner of displaying the observations allows one observation platform to cover a relatively large area over a period of several hours, helping to fill data voids. If multiple observing platforms are sampling the storm, scientists are faced with a large number of observations over an analysis domain, including many that may appear in the same relative patch of ocean (Fig. 9). Measurements do not always agree so it's important to conduct quality control (QC) to remove unrepresentative and erroneous observations either by subjective analyst interaction or objective statistical means (e.g. Lorenc and Hammon 1988). Subjective QC relies on the experience of the scientist who must make decisions based on how a particular observing platform is faring relative to its neighbors, given the particular storm situation and past performance. In practice, the analyst develops confidence with particular platforms in given situations, for example the GPS sonde surface wind estimated from the mean of the samples over the lowest 150 m (GPS_WL150) shows a more stable measurement than the sonde sample closest to 10 m.

In comparison to various satellite measurements nearby, there may be indications that rain contamination was affecting the satellite measurement, and that the variability in estimating a surface wind from the flight level wind may be too high to allow a confident measure of the surface wind. In this example the GPS sonde WL150 measurement would be considered the "anchor" observation and help guide which nearby platforms are representative.

The NOAA Hurricane Research Division (HRD) Real Time Hurricane Wind Analysis System, known as H*Wind, (Powell *et al.*, 1998) is an example of a graphical, interactive method for conducting QC (Figs. 9, 10) combined with an objective analysis algorithm that

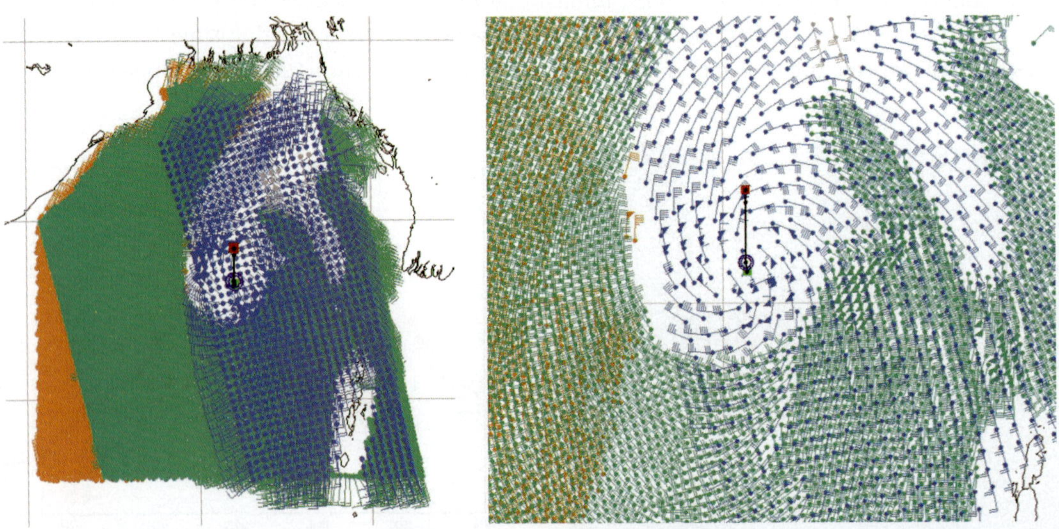

Figure 9. H*Wind QC interface data distribution for Cyclone Sidr for 1200 UTC November 2007. Left: QuikScat hi res (orange), WINDSAT (green), ASCAT (blue); Right: zoom in view of same data. Wind barbs depict direction and wind speed for color-coded observing platforms listed in table.

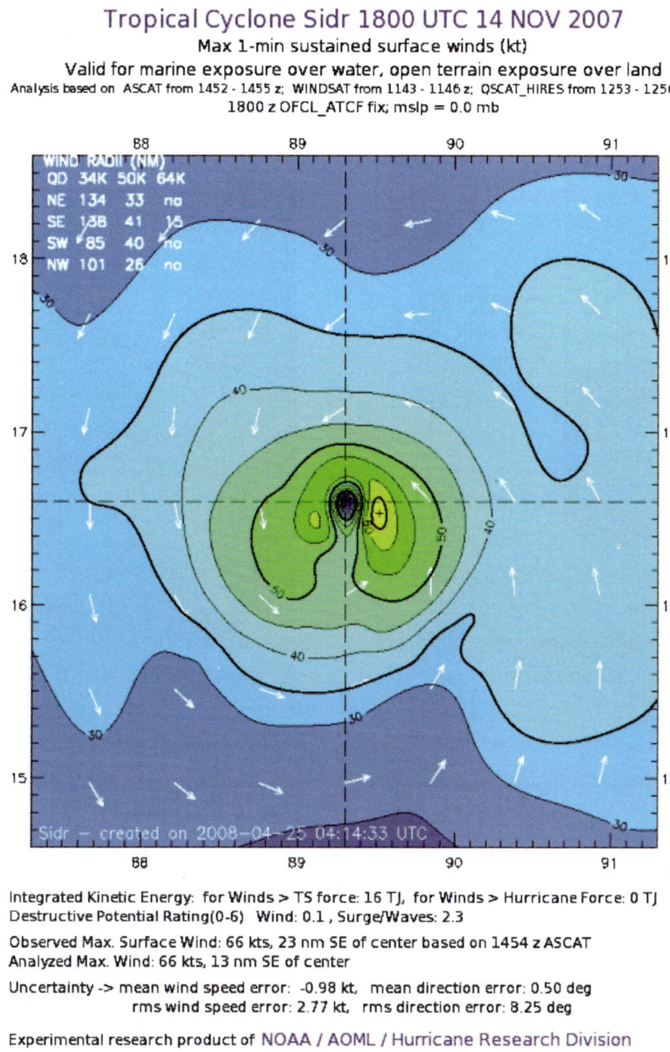

Figure 10. H*Wind analysis of Tropical Cyclone Sidr based on satellite remotely sensed ocean winds data in Fig. 9, repositioned relative to the storm center at 1800 UTC on 14 November 2007.

mechanically interpolates between observations. H*Wind uses Ooyama's (1987) "Scale Controlled Objective Analysis" technique but other analysis techniques could be applied with similar results (e.g. Kepert, 2006a,b). Once the poor quality observations are "flagged" they can be removed from further consideration and the surviving observations may be submitted for objective analysis. The objective analysis acts as a mechanical interpolator filling in regions with no data based on surrounding measurements that have survived the QC process and weighting factors that are assigned to each observing platform. The analysis is further constrained to preserve the speed and radial location of the peak observed wind by enhancing the spline fitting amplitudes so the maximum is not "smoothed out".

Real-time wind analysis "snapshot" products are designed to serve as wind field guidance on the current hurricane intensity and wind radii. An H*Wind analysis snapshot (Fig. 11) provides header information on the various platforms that contribute to the

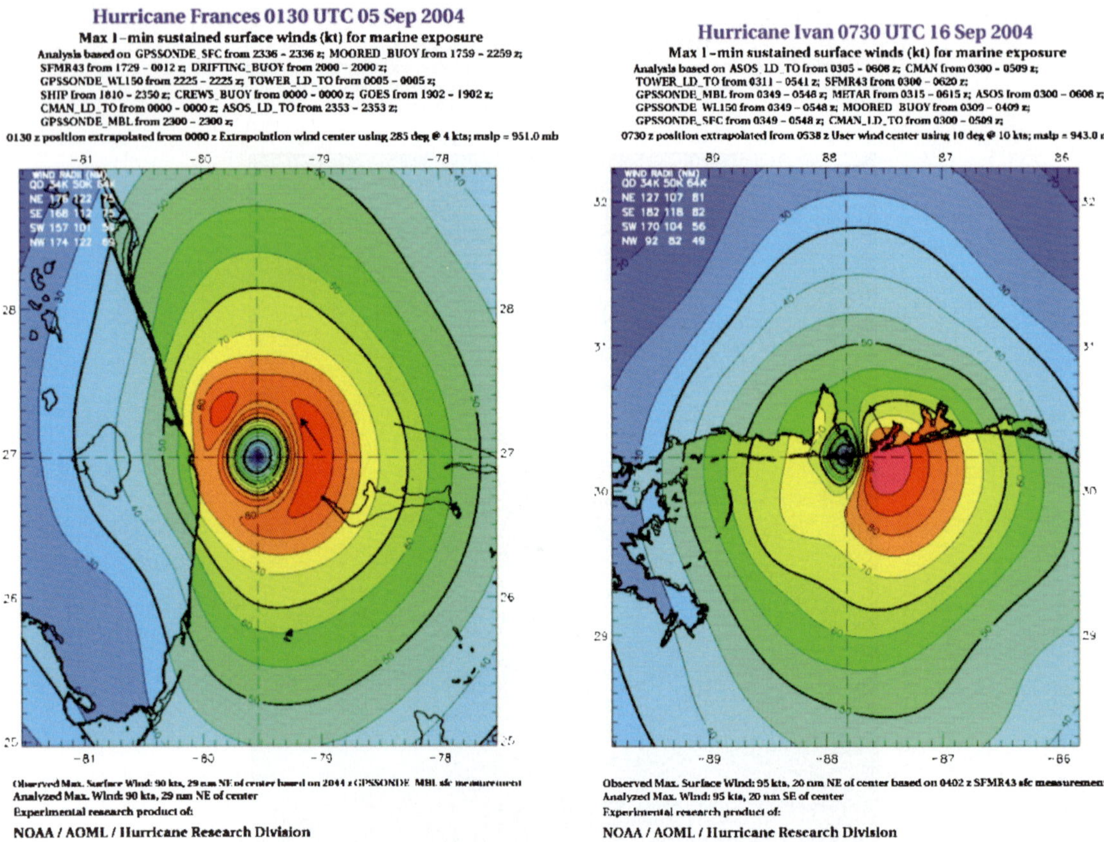

Figure 11. Realtime wind analysis conducted for left: Hurricane Frances (2004), right: Hurricane Ivan (2004).

analysis, information on the maximum surface wind and its location, the radial extent of tropical storm force, 50 kt, and hurricane force winds in each storm quadrant (NE, SE, SW, NW), and additional information on the central pressure, radius of maximum wind, and storm motion. Analysis image files, gridded files, and GIS shape files are available from the HRD web site (http://www.aoml.noaa.gov/hrd/data_sub/wind.html).

Research product real-time analyses are conducted at 6 h intervals (0130, 0730, 1330, 1930 UTC) to document the intensity, extent of damaging winds, and the changes in the wind field. During hurricane warning episodes, analyses are conducted at intervening 3 h intervals (0430, 1030, 1630, 2230 UTC). To be useful as research guidance products the delivery of the product must coincide with the forecaster's duty cycle. For example, H*Wind analyses must be available about 1.5 h before the release of official advisories by the National Hurricane Center since that time coincides with the schedule for synthesizing and evaluating guidance data needed to prepare the forecast package. H*wind analysis snapshot products are mapped to storm locations at a time near the end of the observation time window. Typically 4–6 h of observations are necessary to provide enough observational coverage to conduct an analysis. For example, a 1930 UTC wind analysis would comprise all observation available from 1530–1915 UTC, and the analysis is then mapped to an estimated 1930 UTC storm center position extrapolated from the last available reconnaissance aircraft center fix (e.g. from an 1800 UTC fix) based on the current motion of the storm. A retrospective analysis can improve on the quality of real-time analyses by including all data before and after the analysis center time (e.g. use all data from 1630–2230 UTC for a retrospective 1930 UTC analysis). Since H*Wind is a research effort and not operational, realtime analyses are not available 24–7, although we attempt to provide analyses in realtime during regular working hours and warning episodes. A recent study by Moyer *et al.* (2006) noted that H*Wind radii tend to be larger than the operational values. A likely cause of the difference is the absence of standardization (for height, averaging time, and exposure) in the operationally available data. For example, for some wind directions, coastal anemometer measurements may be affected by upstream trees or dunes. Correcting these observations to marine exposure tends to increase the winds and associated wind radii.

H*Wind is designed as a global tropical cyclone wind observing system. Locations of all active tropical cyclones are depicted and available observations are plotted. As satellite observing platforms improve their ability to measure high winds, it is conceivable that H*Wind analyses could be conducted globally within the next few years. With a suitable wind model, global automated blended analyses could begin even sooner. JAVA versions of H*Wind are under development for testing by global tropical cyclone forecast offices.

Advantages of H*Wind include an objective, observation-based measure of the magnitude and extent of damaging winds in a hurricanes using a common framework. An H*Wind analysis is only as good as the data that go into it and the fidelity of the analysis to those data. If an observing system did not sample the wind in a particular location, H*Wind will not assume that the winds could have been greater in that area. If large areas are missing observations, voids will have to be filled by background fields from earlier analyses, or, in the case of tropical cyclone basins without reconnaissance aircraft (Fig. 9 shows a data coverage that would be similar for tropical cyclone regions without reconnaissance aircraft coverage), a simple wind field model blended with observations might suffice. For the example of Sidr in Figs. 9 and 10, rain contamination would limit the accuracy of the satellite wind data near the storm center but the tropical storm force wind radii should be representative.

The H*Wind analyses are a research (rather than operational) product, but are being used to support research and a wide variety of applications including wave and storm surge modeling, disaster assessment, and analyses of record. One particular application that looks promising is inland projection of peak winds (Fig. 12) using the official forecast track and the H*Wind analysis within a few hours of landfall. After landfall the failure of ASOS stations and the lack of aircraft observations (for safety precautions, aircraft do not typically fly reconnaissance patterns over land) may leave fewer platforms available for realtime analysis so the landfall wind field can be decayed using HRD's inland wind decay model (an exponential decay with time, Kaplan and DeMaria 1995) or by empirical decay of the landfall background field based on comparisons to available post-landfall observations. The resulting wind swath map is gridded and made available on the web in shape file format. In addition, the peak winds may be evaluated at each census tract point and provided to emergency managers for use as input to their Multi HAZard U.S. (HAZUS) damage mitigation and assessment model.

Comparisons of HAZUS loss estimates compared to the Property Claims Services organization insurance claims suggest an excellent agreement when tested during the 2004 hurricane season. Another application of wind field information is for development of new metrics for estimating destructive potential. Traditionally the Saffir-Simpson Scale (SS) is used to depict the destructive potential of tropical cyclones. The SS depends only on the estimate of the maximum wind speed and lacks structural information on the wind field. Since the SS is only relevant for a very small portion of a tropical cyclone it has little relevance to damage potential of the storm as a

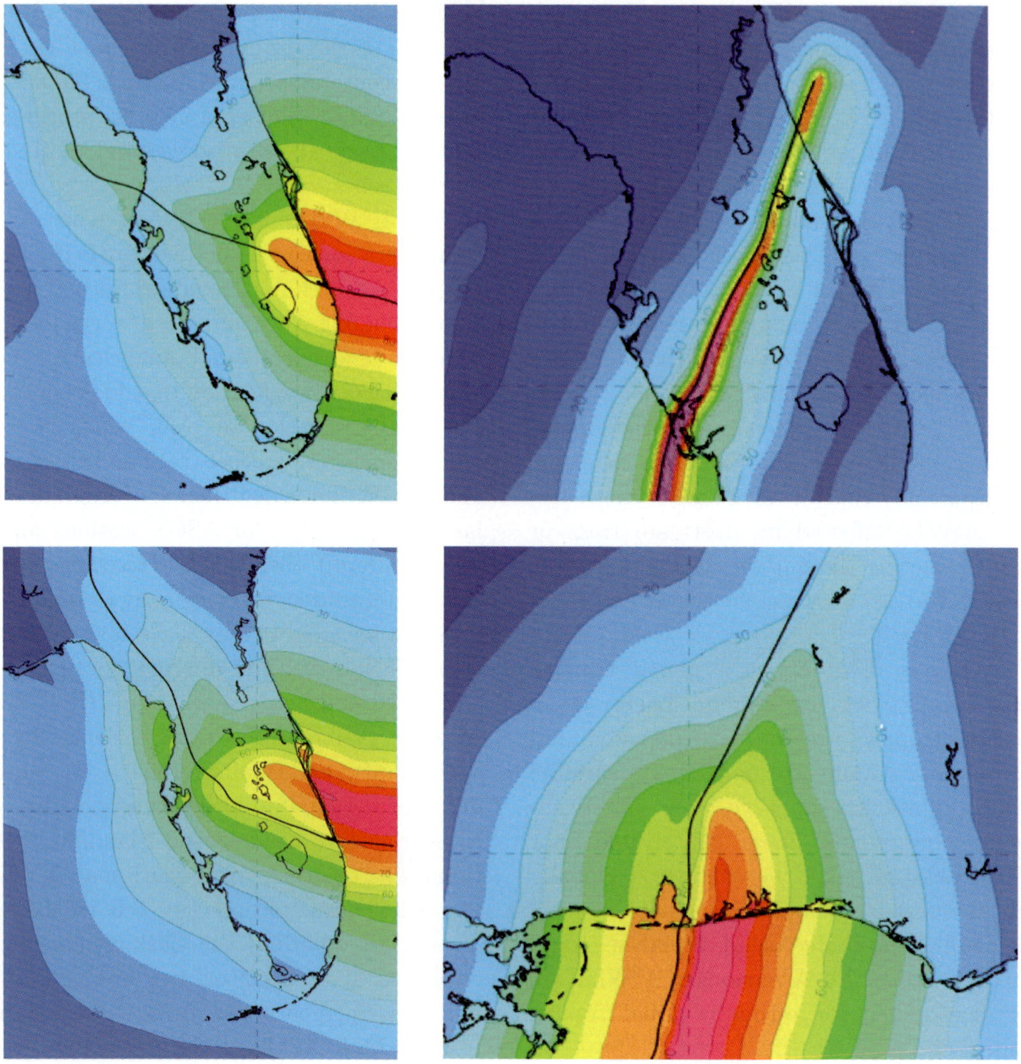

Figure 12. H*Wind swath products for 2004 Hurricanes Frances (top left), Charley (top right), Jeanne (bottom left), and Ivan (bottom right). Wind speeds are in knots.

whole. Powell and Reinhold (2007) used integrated kinetic energy to develop new scales for wind (WDP) and storm surge (SDP) destructive potential based on same numerical 1–5 range as the SS. The WDP and SDP scale take into account both the magnitude and extent of damaging winds and thus overcome the limitations of the SS. The challenge in adopting any new scale is the need to represent complex processes with a simple index. Ultimately the damage depends on local factors such as strength of building codes, building quality, roughness distribution, tidal cycle at landfall, bottom slope, and coastline orientation. With increasing coastal populations however, the lead time for evacuation necessitates actions be taken before the precise landfall location is known. Despite the lack of any information on bottom slope or coastline orientation, the SDP scale correlates well with the extent of coastal inundation at landfall (a measure very relevant for evacuation) but is no better than the SS when correlated with peak storm surge. Integrated kinetic energy also represents an improved metric for evaluation of model forecast performance. Current verification methods focus on intensity which is difficult to measure. Integrated kinetic energy evaluates both intensity and structure in a single number with less sensitivity to difficulties in measuring the maximum wind. Recent work by McClay *et al.*, 2008 suggests that integrated kinetic energy within the inner core of a tropical cyclone undergoes changes in response to storm growth due to secondary eyewall formation and replacement cycles and due to external synoptic forces like vertical wind shear and troughs.

7. Surface Wind Uncertainty

When considering use of any meteorological field representation, the user should be provided with some estimate of the product limitations in the form of an uncertainty assessment. The uncertainty of a wind field analysis depends on many factors including the measurement errors of the observing platforms contributing to the analysis, errors in how the observations are standardized (or not) to a common analysis framework, lack of sufficient measurement sampling to resolve important features of the storm circulation at the time of interest, and representativeness errors associated with how well a diverse set of measurements collected over a time period constitute the current state of the wind field. No analysis will completely represent the current state of the wind field. At best, an analysis can be considered an estimate of the field, constrained by resolution or by the the requirements of the method (e.g. to preserve the maximum observed wind speed and location). The measurements may be representative of features the analysis is unable to resolve (e.g. a GPS sonde might sample a portion of a mesovortex but the analysis cannot resolve it so the feature is effectively filtered or smoothed out.

Consider the case in which the analysis is constrained to match the peak observed wind measurement in the eyewall. Here the sampling error comes down to whether that measurement may have been exceeded in some other unsampled part of the eyewall. The only way to minimize that error is to enhance data coverage as much as possible in the eyewall region. Since hurricane wind speed gradients are much stronger in the radial than the azimuthal directions, most reconnaissance aircraft flight patterns focus on the radial resolution. Small but intense storms have relatively steep radial wind speed profiles necessitating increased radial resolution to minimize sampling error. Azimuthal resolution could increase for storms with large asymmetries associated with motion or vertical wind shear/convection to better sample the azimuth believed to contain the maximum wind. Therefore, given two storms of the same intensity, a small, fast moving storm will be more susceptible to sampling error than a slow moving large storm. If only four passes are available, the wave #1 asymmetry could be estimated from available data to provide an estimate of the maximum wind in locations that were not sampled (e.g. Willoughby, 1984).

Standardization of surface observations is an important process for removing variability from the analysis input data set. In a study of Hurricane Andrew (Powell *et al.*, 1996) random errors were estimated for adjustment of marine platforms to 10 m (3%), and adjustment for averaging time using a gust factor (6%). For land platforms the errors were 7% for height adjustment, 5% for exposure adjustment and 6% for the gust factor. Recent gust factor research (Vickery and Skerlj 2005) confirming similar gust factors for tropical cyclones and other high wind events may help reduce uncertainty. Consequently errors for not standardizing observations could be 15–30% for each process. Surface measurement platforms list 10–20% instrument accuracies in high winds. Assuming these errors are equivalent to standard deviations, they can be added to estimate the associated variance for a given wind estimate. Hence a peak wind of 50 m/s could have an error range of ±5% (for a continuously sampling 10 m platform in open terrain) to ±20% (for a platform at some other height in a non-open exposure), depending on the platform and its exposure.

Estimates for reducing the wind speed from the 3 km reconnaissance flight-level to the surface using a reduction factor or a boundary layer model have large uncertainty with reduction factor standard deviations on the order of 19% before adding additional error for averaging time/gust factor, and sampling. If the reduction factor is based on the peak flight level wind (Powell *et al.*, 2009) or if the aircraft is flying <1500 m (Dunion *et al.*, 2003), uncertainty is smaller. In most cases, the 3 km level is well above the level of maximum wind speed (500 m) and is subject to a weakening of peak wind speed with height due to a relaxation of the pressure gradient associated with the storm's warm core. If the aircraft is at 1500 m or below, there is a good correlation of the flight level wind speed with the mean boundary layer wind speed (Powell *et al.*, 2003; Dunion *et al.*, 2003). Boundary layer model (e.g. Powell, 1980) estimates of the surface wind speed compare well with GPS sonde for wind speeds <55 m/s, but show a low bias at greater wind speeds (Powell *et al.*, 1999).

Observing platforms with the smallest errors are those that already measure within the standard framework, therefore eliminating the need for height, averaging time, and exposure adjustments. Based on experiences conducting real-time and retrospective H*Wind analyses during the 2003–2008 hurricane seasons observing platforms with the least uncertainty for estimating the peak wind in a hurricane are 1) coastal 10 m towers, 2) C-MAN stations (with marine exposure, the closer to 10 m the better), 3) SFMR, 4) GPS sonde, and 5) moored buoys.

For example, the FCMP towers measure the highest 1 min and 3 s wind speeds, and turbulent intensity every 15 min at 10 m, and the TTU towers provide similar measurements, although not yet available in realtime. This information greatly reduces uncertainty, since no height adjustment or gust factor are required to estimate the maximum 1-min wind at 10 m, and upstream roughness is computed from the measured turbulent intensity. In addition, careful and repeated calibrations reduce the tower uncertainty to <5% primarily due to instrument error $< \pm 0.5$ m/s (for RM Young Wind Monitor instruments; F. Masters, J. Schroeder; personal communication). C-MAN stations or moored buoys at 10 m in marine exposure only require a gust factor to estimate the maximum 1-min wind speed (6%) and accounting for instrument error (10%) for a total uncertainty (in 50 m/s winds) of 12% (standard deviations are squared and summed to compute the combined variance and then the square root of the result is listed as the combined uncertainty).

Remote sensing platforms measure winds over a "footprint" of the ocean. H*Wind uses footprint information to compute the maximum 1 min wind speed from remote sensing platforms. To estimate a time scale for the measurement, the footprint can be divided by the measured wind speed. For Quikscat with a 25-km footprint, a 30 m/s wind speed would represent a time scale of ~800 s so a gust factor is required to compute the maximum 1 min wind over the ~14 min time period. A refined time scale estimate would need to be sensitive to differences between the streamwise and transverse scales of the turbulence relative to how the instrument scans the surface. For the SFMR, based on an aircraft motion of ~110 m/s and a height of 3 km, the footprint is ~6 km, measured at high radial resolution, resulting in a time scale of about 2 min or less in winds over 50 m/s, so the SFMR is already equivalent to a 1 min wind speed. Provided the SFMR is not over regions where mechanisms other than wind are causing sea surface foam coverage (shoaling waves, land, current opposing the wind), its fine radial resolution provides a superior estimate of the peak surface wind in the storm with uncertainty of 10% primarily due to the emissivity-wind relationship (Uhlhorn *et al.*, 2007).

The GPS dropsonde samples at a 0.5 sec rate as it falls and has a height error of ± 2 m and an instrument accuracy of ~ 2 m/s. Usually only a few GPS sonde drops are made during a reconnaissance flight. The sondes are typically launched with a strategy to attempt to target the radius of maximum wind at the surface (based on the radial flight-level wind speed profile). However, with so few sondes available on a given flight there is insufficient radial resolution to resolve the maximum wind. A sonde measurement time scale is difficult to assign because the sonde measurement is representative of whatever feature it happens to fall through. A surface measurement attributed to 10 m may be anywhere from 8–12 m and could represent a gust, lull, or mean wind speed. In addition, a forward (upward) centered differencing scheme introduces a high bias of ~ 0.5 m/s at the surface. An alternative and more stable measure of the mean surface wind is estimated from the average of the lowest 150 m layer sampled by the sonde (WL150) (Franklin *et al.*, 2003). In moderate wind speeds Houston *et al.* (2000) reported GPS sonde winds to compare to within 3.5 m/s of nearby (within 20 km) conventional buoy and C-MAN stations. In H*Wind each method for GPS sonde surface wind estimation is available to compare to neighboring observations to help determine which is most representative of the mesoscale wind field.

It is important to recognize that there is no absolute "ground truth" when evaluating various observing platforms in extreme winds. Each platform has its unique characteristics and limitations. The coastal tower might be the most accurate platform but will be difficult to place with sufficient resolution to sample the peak wind speed. The GPS dropsonde provides a high quality measurement but has insufficient radial resolution to sample the peak wind speed. The SFMR has the required radial resolution but has uncertainty in the emissivity-wind speed relationship or non-windwave sources of sea foam.

H*Wind allows combining these and other diverse observations to attempt to provide sufficient data coverage to resolve the radial and azimuthal features of the wind field. A qualitative assessment of uncertainty for various observing platforms used in H*Wind is contained in Table 1. Combined uncertainty includes instrument, exposure, averaging time, height adjustment, and representativeness errors. Platform inter-comparisons in close time and space proximity will help to identify biases and variability in a variety of atmospheric and oceanic conditions. By keeping track of these differences relative to platform, analysis, and forecast, we can assess how well an analysis or forecast reproduces the observations.

Additional methods to consider to assess analysis uncertainty include Monte Carlo simulations and bootstrap resampling. In Monte Carlo simulations the observations are perturbed with noise relevant to the variance properties of the observing platform and then objectively analyzed. After hundreds of analyses are conducted the resulting variance will provide an estimate of the uncertainty. In the bootstrap approach, observation platform errors are not required; If we have N observations, we sample N times from the observation database, with replacement, and then prepare the analysis, and repeat the process hundreds of times to compute the variance across the analyses.

8. Conclusions

Uncertainty of wind measurements in tropical cyclones has been qualitatively assessed for a variety of observing platforms. Beside instrument accuracy, additional factors affecting uncertainty include corrections for height, averaging time, and exposure, as well as sufficient resolution to provide a representative estimate of the peak wind. Combined uncertainty of 5–14% can be reached based on measurements from portable towers, SFMR, GPS sonde, C-MAN stations and moored buoys, or using revised reduction factors as discussed above.

Table 1. Uncertainties of Hurricane Surface Wind Observing Platforms.

Platform	Height (m)	Averaging time	Measurement Method	Instrument Uncertainty	Combined Uncertainty
FCMP, WEMITE Tower	5, 10	1–900 s	prop anemometer	1 m/s	5%
SFMR	10	6 km/WS	foam emissivity brightness temp.	0.5 K	2 m/s, 10% >55 m/s
C-MAN	10–40	10 min	prop anemometer	1 m/s or 10%	12%
Moored Buoy	5, 10	10 min	prop anemometer	1 m/s or 10%	12%
GPS	8–12	0.5 with 5 s filter	motion via GPS	3 m/s	15%
GPS from WL150	10	5 min	motion via GPS	3 m/s	10%
GPS from MBL	10	5 min	motion via GPS	3 m/s	10%
Max Recon (GPS based)	10	5 min	GPS/Max Recon	15%	19%
Max Recon (sfmr-based)	10	2 min	SFMR/Max Recon	10%	15%
QSCAT	10	25 km/WS	Ku Backscatter	2 m/s until hvy. rain	10% until hvy. rain
GOES Cloud Drift	10	5 min	cloud motion-> Sfc wind	2.6 m/s	15% for WS < 25 m/s
ASOS	7, 10	2 min	cup anemometer	1 m/s or 5%	10% if record survives
Ship	~20	10–30	anem or Beaufort	10%	20%
Recon 0.9	10	5 min	GPS/Recon	19%	20%

In response to the challenges listed in section 1:

1) Tropical cyclone wind observing platforms still suffer power and anemometer failures but over the last few years portable anemometer tower placement programs have helped to document wind conditions in the absence of official reporting stations. In a time with increased need for vigilance and security, international offshore, coastal, and inland observing networks must be improved to be capable of withstanding and monitoring extreme events.
2) An unprecedented number of marine wind observations are available to help monitor wind conditions in tropical cyclones approaching land. Chief among these in the Atlantic basin are the aircraft-based SFMR and GPS sonde. Elsewhere, scatterometers aboard the NASA QuikScat and ESA A-Scat satellites together with conventional moored buoys, GOES cloud drift winds, ships, and coastal automated stations may often provide sufficient observations for objective analysis. Space-based microwave remote sensing continue to improve and it is probable that these platforms could begin to measure extreme winds after a few more years. For periods of 4–6 hours, storm-relative data coverage from current (Fig. 4) aircraft reconnaissance patterns are sufficient to resolve the peak wind in typical tropical cyclones moving at average speeds of ~5 m/s. For faster moving storms, reconnaissance patterns should evolve to provide increased azimuthal resolution or analysis methods should improve to allow estimation of winds in higher order wave number fits. For small tropical cyclones, reconnaissance patterns could be enhanced to provide increased radial resolution of 0.5–1 km. These patterns might sacrifice outer wind field monitoring at the expense of more information on the structure of >25 m/s winds. The extent of tropical storm force winds is less sensitive to rain contamination of remote sensors and could be resolved with further

satellite and conventional observing platform coverage.

3) The process of assessing observation quality still depends on analyst experience in evaluating how various platforms perform relative to each other in a variety of environmental conditions. Advanced statistical methods used in data assimilation for mesoscale numerical weather prediction models may provide the promise of automated QC and objective analysis. Mesoscale models are beginning to simulate realistic hurricane features and hindcasts. The new generation of numerical hurricane prediction models (such as the H-WRF) will require detailed evaluations against observed fields to assess model performance.
4) A variety of new wind products have been developed as research products. Some of these will become operational for realtime disaster assessment and response, as well as guidance in the forecast and warning system, and input to storm surge and wave forecast models. Advances in observing network quality and remote sensing abilities point towards objective analysis as the key tool in developing the next generation of worldwide tropical cyclone wind field products. These types of products are expected to proliferate throughout the regional tropical cyclone forecast and warning centers over the next decade.

References

Barth, M. F., P. A. Miller, and A. E. MacDonald, 2002: MADIS: Providing Value-Added Observations to the Meteorological Community. FSL Forum, June, 2002.

Black, P. G. and F. D. Marks, 1991: The structure of an eyewall mesovortex in Hurricane Hugo (1989). 19th AMS Conference on Hurricanes and Tropical Meteorology, May 6–10, Miami, pp. 579–582.

Bourassa, M. A., D. M. Legler, J. J. O'Brien, and S. R. Smith, 2003: SeaWinds Validation with Research Vessels, *J. Geophys. Res.*, **108**, doi 10.1029/2001JC001081.

Cao, S., Y. Tamura, N. Kikuchi, M. Saito, I. Nakayama, and Y. Matsuzaki, 2007: Wind characteristics of a strong typhoon, *Proceedings of 12th International Conference of Wind Engineering*, July 2–6, 2007, Cairns, Australia, pp. 719–722.

Cardone, V. J., J. Greenwood, and M. Cane, 1990: On trends in historical marine wind data, *J. Climate*, **3**, 113–127.

Connor, L. N. and Paul S. Chang, 2004: Determination of ocean surface wind speeds from the TRMM Microwave Imager. In Review, *IEEE Transactions on Geoscience and Remote Sensing.*

Demuth, J. L., M. DeMaria, J. Knaff, and T. H. Vonder Harr, 2004: Evaluation of advanced microwave sounding unit tropical-cyclone intensity and size estimation algorithms, *J. Appl. Meteor.*, **43**, 282–296.

Dodge, P., R. W. Burpee, and F. D. Marks, 1991: Airborne Doppler radar analyses of the core of Hurricane Gilbert. *19th AMS Conference on Hurricanes and Tropical Meteorology*, May 6–10, Miami, pp. 551–552.

Dunion, J. P., C. W. Landsea, S. H. Houston, and M. D. Powell, 2003: A reanalysis of the surface winds for Hurricane Donna of 1960. *Mon. Wea. Rev.*, **131**, 1992–2001.

Dunion, J. P., S. H. Houston, C. S. Velden, and M. D. Powell, 2002: Application of surface-adjusted GOES low-level cloud-drift winds in the environment of Atlantic tropical cyclones. Part II: Integration into surface wind analyses. *Mon. Wea. Rev.*, **130**, 1347–1355.

Foster, R. *et al.*, 2004: Observational and theoretical evidence for roll vortices in the hurricane boundary layer. *13th AMS Conference on Interactions of the Sea and Atmosphere*, Portland, Maine.

Franklin, J. L., S. J. Lord, S. Feuer, and F. D. Marks, Jr., 1993: The kinematic structure of Hurricane Gloria (1985) determined from dropwindsonde and Doppler radar data. *Mon. Wea. Rev.*, **121**, 2433–2451.

Franklin, J. L., M. L. Black, and K. Valde, 2003: GPS dropwindsonde wind profiles in hurricanes and their operational implications. *Wea. Forecasting*, **18**, 32–44.

Gamache, J. F., F. D. Marks, and F. Roux, 1995: Comparison of three airborne Doppler sampling techniques with airborne in situ wind observations in Hurricane Gustav (1990). *J. Atmos. Ocean. Tech.*, **12**, 171–181.

Gilhousen, D. B., 1987: A field evaluation of NDBC moored buoy winds. *J. Atmos. Oceanic Technol.*, **4**, 94–104.

Goodberlet, M. A., C. T. Swift, and J. C. Wilkerson, 1989: Remote sensing of ocean surface winds with the Special Sensor Microwave/Imager, *J. Geophys. Res.*, **94**, 14,547–14,555.

Hawkins, H. F. and S. M. Imbembo, 1976: The structure of a small, intense, hurricane — Inez 1966. *Mon. Wea. Rev.*, **104**, 418–442.

Hervey, R., 1999: Are moored buoy winds too low in high seas? *Preprints, Third Conf. on Coastal Atmos. and Oceanic Prediction and Processes*, New Orleans, LA, Amer. Meteor. Soc.

Hock, T. R. and J. L. Franklin, 1999: The NCAR GPS dropwindsonde. *Bull. Amer. Meteor. Soc.*, **80**, 407–420.

Horstman, J., W. Koch, and S. Lehner, Ocean Wind Fields Retrieved from the Advanced Synthetic Aperture Radar aboard ENVISAT, *Ocean Dynamics*, in press, 2004.

Houston, S. H., P. P. Dodge, M. D. Powell, M. L. Black, G. M. Barnes, and P. S. Chu, 2000: Surface winds in hurricanes from GPS-sondes: Comparisons with observations. *Preprints, 24th Conf. on Hurricanes and Tropical Meteorology*, Ft. Lauderdale, FL, Amer. Meteor. Soc., p. 339.

Howden, S., D. Gilhousen, N. Guinasso, J. Alpert, M. Sturgeon, and L. Bender, 2008: Hurricane Katrina winds measured by a buoy-mounted sonic anemometer. *J. Atmos. Oceanic Tech.*, **25**, 607–616.

Jorgensen, D. P., 1984: Mesoscale and convective scale characteristics of mature hurricanes. II: Inner core structure of Hurricane Allen (1980), *J. Atmos. Sci.*, **41**, 1287–1311.

Kaplan, J. and M. DeMaria, 1995: A simple empirical model for predicting the decay of tropical cyclone winds after landfall. *J. App. Meteor.*, **34**.

Kepert, J. D., 2001: The dynamics of boundary layer jets within the tropical cyclone core. Part I: Linear theory. *J. Atmos. Sci.*, **58**, 2469–2484.

Kepert, J. D. and Y. Wang, 2001: The dynamics of boundary layer jets within the tropical cyclone core. Part II: Nonlinear enhancement. *J. Atmos. Sci.*, **58**, 2485–2501.

Kepert, J. D., 2002: The impact of landfall on tropical cyclone boundary layer winds. Extended abstracts, 25th Conference on Hurricanes and Tropical Meteorology, *Amer. Meteor. Soc.*, San Diego, CA, 29 April–3 May, 2002, pp. 335–336.

Kossin, J. P. and C. S. Velden, 2003: Tropical cyclone intensity estimates using geostationary satellite infrared imagery. *Proc. 57th Interdepartmental Hurricane Conference*, Miami, FL,

Landsea *et al.*, 2004: A reanalysis of Hurricane Andrew's uncertainty. *Bull. Amer. Meteor. Soc.*, **85**, 1699–1712.

Large, W. G., J. Morzel, and G. B. Crawford, 1995: Accounting for surface wave distortion of the marine wind profile in low-level ocean storms wind measurements. *J. Phys. Oceanogr.*, **25**, 2959.

Lee, W.-C., B. J.-D. Jou, P.-L. Chang, and S.-M, Deng, 1999: Tropical cyclone kinematic structure retrieved from single Doppler radar observations. Part I. Interpretation of Doppler velocity patterns and the GBVTD technique. *Mon. Wea. Rev.*, **127**, 2419–2439.

Liu, K. S. and J. C. L. Chan, 1999: Size of tropical cyclones as inferred from ERS-1 and ERS-2 data. *Mon. Wea. Rev.*, **127**, 2992–3001.

Lorenc, A. C. and O. Hammon, 1988: Objective quality control of observations using Bayesian methods. Theory, and a practical implementation. *Quart. J. Roy. Meteor. Soc.*, **114**, 515–543.

Lorsolo, S., J. L. Schroeder, P. Dodge, and F. Marks, Jr., 2008: An observational study of hurricane boundary layer small-scale coherent structures, *Mon. Wea. Rev.*, **136**, 2871–2893.

Marks, F. D., R. A. Houze, and J. F. Gamache, 1992: Dual-Aircraft investigation of the inner core of Hurricane Norbert: Part I: Kinematic structure. *J. Atmos. Sci.*, **49**, 919–942.

McClay, K. S., M. DeMaria, and T. H. Vonder Harr, 2008: Tropical cyclone inner-core kinetic energy evolution. *Mon. Wea. Rev.*, **136**, in press.

Powell, M. D., 1980: Evaluations of diagnostic marine boundary layer models applied to hurricanes. *Mon. Wea. Rev.*, **108**, 757–766.

Powell, M. D., 1987: Changes in the low-level kinematic and thermodynamic structure of Hurricane Alicia (1983) at landfall. *Mon. Wea. Rev.*, **115**(1), 75–99.

Powell, M. D. and P. G. Black, 1990: The relationship of hurricane reconnaissance flight-level measurements to winds measured by NOAA's oceanic platforms. *J. Wind Eng. Ind. Aerodyn.*, **36**, 381–392.

Powell, M. D., P. P. Dodge, and M. L. Black, 1991: The landfall of Hurricane Hugo in the Carolinas: Surface wind distribution. *Wea. Forecasting*, **6**, 379–399.

Powell, M. D., 1993: Wind measurement and archival under the automated surface observing system (ASOS): User concerns and opportunity for improvement. *Bull. Amer. Meteor. Soc.*, **74**(4), 615–623.

Powell, M. D., S. H. Houston, and T. A. Reinhold, 1996: Hurricane Andrew's landfall in South Florida. Part I: Standardizing measurements for documentation of surface wind fields. *Wea. Forecasting*, **11**, 304–328.

Powell, M. D. and S. H. Houston, 1996: Hurricane Andrew's landfall in South Florida. Part II: Surface wind fields and potential real-time applications. *Wea. Forecasting*, **11**, 329–349.

Powell, M. D., S. H. Houston, L. R. Amat, and N. Morisseau-Leroy, 1998: The HRD real-time surface wind analysis system. *J. Wind Eng. Indust. Aerodyn.*, 77–78, 53–64.

Powell, M. D., T. A. Reinhold, and R. D. Marshall, 1999: GPS sonde insights on boundary layer wind structure in hurricanes. *Wind Engineering into the 21st Century*, Larsen, Larose, and Livesey (eds.), Balkema, Rotterdam.

Powell, M. D., P. J. Vickery, and T. A. Reinhold, 2003: Reduced drag-coefficient for high wind speeds in tropical cyclones. *Nature*, **422**, 279–283.

Powell, M.D., D. Bowman, D. Gilhousen, S. Murillo, N. Carrasco, and R. St. Fleur, 2004: Tropical Cyclone Winds at Landfall: The ASOS-CMAN Wind Exposure Documentation Project. *Bull. Amer. Meteor. Soc.*, **85**, 845–851.

Powell, M. D. and T. A. Reinhold, 2007: Tropical cyclone destructive potential by integrated kinetic energy. *Bull. Amer. Meteor. Soc.*, **87**, 513–526.

Powell, M. D., E. Uhlhorn, and J. Kepert, 2009: Estimating maximum surface winds from hurricane reconnaissance aircraft. *Wea. Forecasting*, **24**, in press.

Quilfen, Y., B. Chapron, and D. Vandemark, 2001: The ERS scatterometer wind measurement accuracy: Evidence of seasonal and regional biases, *J. Atmos. and Ocean. Tech.*, **18**, 1684–1697.

Quilfen, Y. *et al.*, 2007: The potential of QuikSCAT and WindSat observations for the estimation of sea surface wind vector under severe weather conditions. *J. Geophys. Res.*, **112**, C09023.

Simpson, R. H., 1974: The hurricane disaster potential scale. *Weatherwise*, **27**, 169, 186.

Uhlhorn, E. W. and P. G. Black, 2003: Verification of remotely sensed sea surface winds in hurricanes. *J. Atmos. Oceanic Technol.*, **20**, 99–116.

Uhlhorn, E. W., P. G. Black, J. L. Franklin, M. Goodberlet, J. Carswell, and A. S. Goldstein, 2007: Hurricane surface wind measurements from an operational stepped frequency microwave radiometer. *Mon. Wea. Rev.*, **135**, 3070–3085.

Vagle, S., W.G. Large, and D. M. Farmer, 1990: An evaluation of the WOTAN technique of inferring oceanic winds from underwater ambient sound, *J. Atmos. Ocean. Tech.*, **7**, 576–595.

Vickery, P. J. and P. F. Skerlj, 2000: Elimination of exposure D along the hurricane coastline in ASCE 7. *J. Structural Engrg.*, **126**, 545–549.

Vickery, P. J. and P. F. Skerlj, 2005: Hurricane gust factors revisited. *J. Structural Engrg.*, **131**, 825–832.

Wurman and Winslow, 1998: Intense sub-kilometer-scale boundary layer rolls observed in Hurricane Fran. *Science*, **280**, 555–557.

Yueh, S. H., B. Stiles, and W. T. Liu, 2003: QuikSCAT wind retrievals for tropical cyclones, *IEEE Trans. Geoscience and Remote Sensing*, **41**, 2616–2628.

Chapter 7

Satellite Observations of Tropical Cyclones

Christopher Velden
Cooperative Institute for Meteorological Satellite Studies
University of Wisconsin
Madison, WI 53707, USA
chrisv@ssec.wisc.edu

Jeffrey Hawkins
Naval Research Laboratory
7 Grace Hopper Ave, MS#2
Monterey, CA 93943, USA
jeff.hawkins@nrlmry.navy.mil

With contributions from:
Roger Edson, Derrick Herndon, Ray Zehr, James Kossin

Meteorological satellites have virtually ended the potential for tropical cyclones to go undetected. All the major global tropical cyclone analysis and forecast centers rely heavily on satellite surveillance. In many cases, satellite imagery is the only available method for estimating location, motion and intensity. However, images are not the only useful products derived from satellites. While this chapter is not intended as an exhaustive treatise on the radiative transfer theory on which satellite remote sensing is based, it will touch on some of the quantitative ways that satellite data is being applied to the study and forecasting of tropical cyclones. After a brief historical perspective, this chapter will describe the emergence of new information contributed by satellite-borne instruments and techniques designed to exploit that information towards better observations of tropical cyclones. A concluding section provides a brief summary and outlook on the potential value of anticipated new satellite instrumentation toward improving future tropical cyclone analysis and forecasting.

1. Introduction

In 1960, the era of space-based reconnaissance began with the first successful launch of a meteorological satellite (TIROS 1). This polar-orbiting satellite was primitive, but showed promise of great results to come. Subsequent polar orbiting satellites flew more advanced sensors with visible (VIS) and Infrared (IR) footprint resolutions of 5–10 km. These satellites were the first to be used not only to locate mature storms, but also to crudely estimate maximum wind intensity from the portrayal of cloud organization in the images.

In the 1970s, a major advance in tropical cyclone detection from space came with the deployment of geostationary meteorological satellites. Able to view an area continuously with a 4–8 km ground footprint resolution, scientists at the NOAA Environmental Satellite Service were able to refine techniques to estimate tropical cyclone intensity from these improved images. In particular, NOAA scientist Vern Dvorak was developing his famous subjective image pattern recognition technique, which was first published over three decades ago (Dvorak, 1975). This method, although since refined by researchers such as NOAA scientist

Ray Zehr and operational satellite meteorologists in practice, is still in use today as the standard for estimating tropical cyclone intensity around the globe (Velden *et al.*, 2006).

Satellite applications to tropical cyclones proliferated during the 1980s and 1990s. Objective methods were developed that not only scanned images, but extracted quantitative information from them. For example, cloud-drift winds, sea surface temperatures, and rain rate algorithms all matured during this time. In addition, important new spectral information became available with the launch of the Special Sensor Microwave/Imager (SSM/I) series beginning in 1987.

The primary aims of this chapter are to briefly examine some of the more recent advances that have occurred in satellite-based remote sensing of tropical cyclones, and attempt to define the state of the art. The report is broken down into two main areas: Applications from IR-based images and microwave data. After some background in each area, the focus is on research algorithms and operational applications that have been recently developed. A final section emphasizes data fusion and information sharing between spectral frequencies as a way forward in satellite-based remote sensing of tropical cyclones.

2. IR-Based Applications

2.1. *Tropical Cyclone Intensity and Structure*

The well-known Dvorak technique (Velden *et al.*, 2006) is based on principles of cloud pattern recognition in satellite imagery. The amount of organization is directly related to hurricane intensity. Trained satellite analysts determine the cloud pattern type and relate it to storm development stages. As satellite IR sensors became more sophisticated in the 1970s, Dvorak expanded his techniques to include special enhancements (Dvorak, 1984). The technique became widely used at global tropical cyclone analysis centers, spawning regional variations (Velden *et al.*, 2006). Recently, university scientists collaborated to automate the Dvorak technique (Velden *et al.*, 1998b). The latest version of this computer-based algorithm, called the Advanced Dvorak Technique (ADT; Olander and Velden, 2007) was designed to minimize human judgment in the cloud pattern typing and rules application stages. This algorithm operates on geostationary satellite IR imagery, and has been shown to be competitive with the subjective applications. The ADT has been transitioned into NOAA/NWS operations, where it is part of the suite of satellite-based guidance tools. Other global tropical cyclone centers are also incorporating the ADT into their analysis procedures.

High spatial resolution (1-km) IR images from the AVHRR on NOAA satellites and the Operational Linescan System (OLS) on the Defense Meteorological Satellite Program (DMSP) series have been available for many years. The MODIS IR-window sensors on NASA Terra and Aqua satellites now provide 1-km resolution globally. These 1-km IR images can reveal features such as cyclonically curved thin cloud lines and transverse bands at hurricane cloud top that are not well observed by the lower resolution geostationary satellites. On occasion, this imagery can also ameliorate problems of viewing angle in the application of the Dvorak methods using geostationary imagery. In addition to the high resolution IR images, very-high resolution visible (VIS) imagery, i.e., the 250 m VIS resolution of MODIS, has provided detailed views of cloud structures in the eye that were not previously discernible from satellite imagery.

The inclusion of geostationary satellite data has also recently been shown to increase the skill in a variety of other tropical cyclone estimation and prediction algorithms. These include the estimation of surface wind structure (Mueller *et al.*, 2006; Kossin *et al.*, 2006), diagnosis of annular hurricane formation (Knaff *et al.*, 2003), probability estimation of secondary eyewall formation (Kossin and Sitkowski, 2008), diagnosis of rapid intensification events (Kaplan and DeMaria, 2006), intensity change prediction (DeMaria *et al.*, 2005), and characterizing major hurricanes (Zehr and Knaff, 2007).

Kossin *et al.* (2007a) have applied IR data to new objective methods of estimating radius of maximum wind (RMW), along with the standard operational wind radii analyses (R-35kt, R-50kt, R-64kt). Another routine was developed which generates the entire 2-dimensional wind field within a 200-km radius. These analyses provide estimates of the tropical cyclone inner-core winds, which has historically been limited to cases where aircraft reconnaissance is available. These inner core winds can be combined with wind estimates from other satellite platforms to form complete wind fields in basins where no aircraft reconnaissance is available (Knaff and DeMaria, 2006).

The recent inclusion of quantitative GOES IR-derived data into the SHIPS statistical intensity forecast model has produced small improvements in the Atlantic, and significantly improved the east Pacific forecasts by up to 7% at 12–72 h (DeMaria *et al.*, 2005). The GOES predictors are 1) the percent of the area (pixel count) from 50 to 200 km from the storm center where brightness temperature (T_b) is colder than −20°C and 2) the standard deviation of T_b (relative to the azimuthal average) averaged from 100 to 300 km. Similar GOES parameters are used as input to NHC's operational Rapid Intensity Index that provides an estimate of the probability that the maximum sustained surface wind of a tropical cyclone will increase by 25 kt or greater in the following 24 h (Kaplan and DeMaria, 2006).

2.2. *Tropical Cyclone Motion — Environmental Steering*

Tropical cyclones are steered by deep-layer environmental wind fields, which can often extend well away from the storm center. Given the scarcity of conventional tropospheric wind observations over the oceanic regions, winds derived from cloud and water vapor motions in successive geostationary satellite images have been shown to be an important input to the accurate determination of these steering analyses. Today, automated techniques have been developed that track features in multi-spectral full disk imagery from multiple geostationary platforms to provide global wind sets around the clock. These data sets have proven useful in real-time analysis (Velden *et al.*, 1998a) and in reducing hurricane track forecast error in numerical models (Goerss *et al.*, 1998; Soden *et al.*, 2001; Zapotocny *et al.*, 2008; Goerss and Hogan, 2006). Special rapid-scan schedules are now standard during tropical cyclones and can sample at a 5–15 minute frequency over targeted storms, providing very detailed wind fields over and near the center (Velden *et al.*, 2005).

2.3. *Other Important Applications of IR Satellite Data to Tropical Cyclones*

Environmental vertical shear is an important factor in tropical cyclone development and intensity change. Therefore, an accurate measure of this quantity in the near-storm environment is highly desirable. The multispectral high-density satellite-derived winds mentioned in the previous section have also been used to develop shear analyses in tropical cyclone vicinities. These analyses use the detailed wind information to quantitatively deduce the amount of vertical shear present. A study by Gallina and Velden (2002) showed that these analyses could also be used to predict short term intensity changes. The analyses are one of the most often used products from the CIMSS tropical cyclone web site by operational forecasters and researchers.

Entrainment of dry air into the circulation of a tropical cyclone is another known modulating factor on intensity. Recent work using satellite data has identified the Saharan Air Layer (SAL) as a potential influence on hurricane intensity (Dunion and Velden, 2004). The SAL is a hyper-dry air mass originating over the Saharan Desert that often traverses the tropical Atlantic Ocean, and is normally accompanied by massive dust loads. Multispectral satellite approaches have been developed to radiatively isolate the effects of the dust to denote SAL outbreaks. These westward moving air masses are often accompanying African easterly waves that can spawn tropical cyclones. While the exact mechanisms involved in the interaction between the SAL and coincident tropical cyclones is not known, Evan *et al.* (2006) did find a distinct reverse correlation between the presence of SAL dust and active tropical cyclone days over a 20+ year period in the eastern Atlantic.

Another use of IR information is the determination of sea surface temperatures (SST). SSTs have long been known to be a critical environmental influence on tropical cyclone genesis and behavior. Satellite IR radiometers are able to sense the ocean skin temperatures in the absence of clouds (i.e., Maturi *et al.*, 2008; May and Osterman, 1998). This information is particularly important as strengthening or weakening could depend on the SST gradients in the path of a hurricane. Tropical cyclone forecasters can monitor features such as the warm and cold ocean currents and associated eddies, and cool-water "wakes" left over from the heat loss and mixing associated with previous storms.

A major concern associated with landfalling hurricanes is the rainfall and flooding potential. IR imagery is used as a guidance tool to infer rainfall rates in hurricanes. Studies have been directed towards calibrating satellite IR-based rainfall estimation techniques to radar estimates at landfall. While these IR methods can provide a generally good (within a factor of two) bulk estimate of rainfall in the absence of other data, precise observations of heavy rain bands and localized accumulations are difficult from the IR alone. Improved multi-spectral satellite techniques are being developed that utilize the advantages of rapid temporal sampling from geostationary IR imagery and the cloud-penetrating properties in the microwave spectrum from polar-orbiting instruments (see section 5).

A recent topic of contention has to do with the multi-decadal trend in tropical cyclones, possibly influenced by global warming. At issue are the historical 'best track' information, and the reliability of this record in depicting true storm intensities. Reprocessed global geostationary satellite data has recently been utilized on a portion of the historic record of tropical cyclone intensity metrics (Kossin *et al.*, 2007b). This approach has the effect of homogenizing the intensity estimates for testing the fidelity of trend analyses.

In summary, IR and VIS satellite observations have become a mainstay of tropical cyclone detection and analysis. The limiting factor of passive remote sensing in these spectra is the inability to sense through clouds. The next section will address this limitation with the advent and maturing of newer satellite sensors that can penetrate clouds and provide a unique view of tropical cyclones.

3. Microwave-Based Applications

3.1. *Imagers*

Passive microwave (PMW) imagery is available from polar orbiting satellites, and applications have progressed due to the recent launch of new imagers and the extensions of existing satellite programs, i.e., DMSP Special Sensor Microwave/Imager (SSM/I). The longer wavelengths associated with microwave radiation are not sensitive to ice clouds and can, therefore, penetrate the storm's cirrus canopy to reveal important structure underneath. PMW imagers can map tropical cyclone rainband organization and eyewall development by viewing through non-raining clouds. Similar details cannot be routinely extracted in VIS/IR data due to upper-level cloud obscuration (Hawkins *et al.*, 2001; Lee *et al.*, 2002). Figure 1 illustrates a

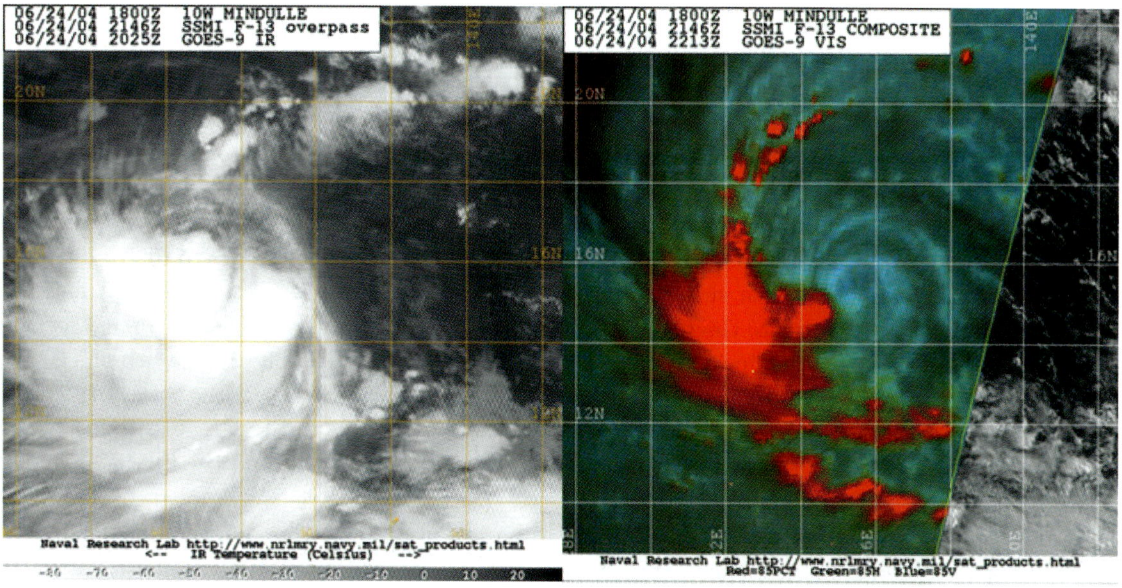

Figure 1. Typhoon Mindulle views with GOES-9 IR (left) and SSM/I 85 GHz "color" product that incorporates 85 GHz H and V polarizations and polarization corrected temperature (PCT) values (right), clearly showing low-level center well displaced from main convection (red). (Courtesy of the Naval Research Laboratory.)

classic case where clouds hide the true low-level center that can be readily identified in PMW imagery.

The U.S. Department of Defense (DoD) was the first to fly a PMW imager on a operational meteorological satellite in 1987, and this program of satellites continues today. The first SSM/I showed some of the potential of this new imaging tool, but the applications were not fully realized until multiple satellites were operating and providing more than one view per day. By the 1990s it was typical to have two to four active satellites that were capable of providing five to seven images of a storm per day. One of the SSM/I channels senses at a frequency of 85 GHz, with a spatial resolution of 13–15 km. Large, frozen hydrometeors (e.g., hail, graupel) and raindrops associated with vigorous convection dramatically scatter radiation at this frequency. Thus, the sensor can easily detect intense rain associated with tropical cyclone rainbands and the eyewall due to the lowered brightness temperatures created by the intense scattering. A time series of 85 GHz data can reveal a storm's internal convective structure and evolution by mapping the organization and vigor of the convection around the storm center (Hawkins *et al.*, 2001). It is important to note that these data have been enthusiastically used not only by the DoD sponsored Joint Typhoon Warning Center (JTWC) in Pearl Harbor, Hawaii, but by virtually all of the World Meteorological Organization (WMO) tropical cyclone warning centers over the globe.

Accurate positioning of the low level circulation center (LLCC) is a critical first step in the tropical cyclone analysis process. Position estimates are a fundamental part of every warning advisory, and are crucial to numerical weather prediction (NWP) model initialization and subsequent 1–7 day storm track forecasts. Determining the LLCC position is optimally done from information provided by multi-spectral data sets. VIS/IR data have been traditionally used to estimate LLCC positions via the use of animated geostationary imagery (typically with 30 minute refresh). This has known inadequacies when a true center is obscured by cirrus debris, and/or the true LLCC is not aligned with the apparent circulation center provided by clouds above. More recent operational availability and use of passive microwave imagers permits the detection of sub-canopy TC rainbands, eyewalls, and overall storm structure and thus enables the analyst to better estimate LLCC positions in most of the cases where VIS/IR imagery is problematic. Scatterometer ocean surface wind vectors (section 3.4) can estimate the LLCC via a combination of the vector solutions and an analyst's use of the ambiguity fields. Using all of these satellite tools, a trained satellite analyst can often efficiently locate the LLCC.

PMW data are quite valuable when trying to monitor the organization of systems undergoing rapid intensity changes. Intensifying tropical cyclones are often characterized by double (or concentric) eyewalls. Double eyewalls are rarely seen in VIS/IR imagery since upper-level clouds almost always obscure these structures. However, the PMW sensitivity to ice and rain scattering in the inner and outer eyewall typically allows this important structural signature to be observed (Fig. 2). Forecasters can observe trends in these data (a strong and constricting inner eyewall generally means intensification; a collapsing inner eyewall in favor of the outer eyewall generally implies decreasing intensity) and are beginning to learn how to use these trends as a proxy for the evolution of eyewall dynamics and as guidance in their intensity forecasts.

The first operational Special Sensor Microwave/Imager Sounder (SSMIS) was launched in October 2003 onboard DMSP F-16 (Wessel *et al.*, 2004) as the follow-on to the SSM/I. Subsequently, F-17 was launched in late 2006, F-18 in 2009 and two more are to follow over the next several years. These sensors are replacing the older SSM/I data and already aid tropical cyclone monitoring globally. The imager channels are essentially unchanged from the SSM/I (91 GHz versus 85 GHz), but with a larger scanning swath (1700 versus 1400 km), and the SSMIS includes collocated sounder channels that have multiple advantages. Further information on PMW imagery can be found at: http://www.nrlmry.navy.mil/TC.html.

Building on the success of the DoD program, NASA launched a special satellite aimed at measuring meteorological quantities over the tropics using passive and active microwave sensors. The Tropical Rainfall Measuring Mission (TRMM) satellite completed ten years of highly successful observations in November 2007. With the change in altitude from 350 km to 405 km made in August 2001, it has enough fuel to continue measurements until around 2012 if NASA continues to support it. TRMM is a low Earth-orbiting satellite with a unique orbit which oscillates about the Equator between roughly 35°N

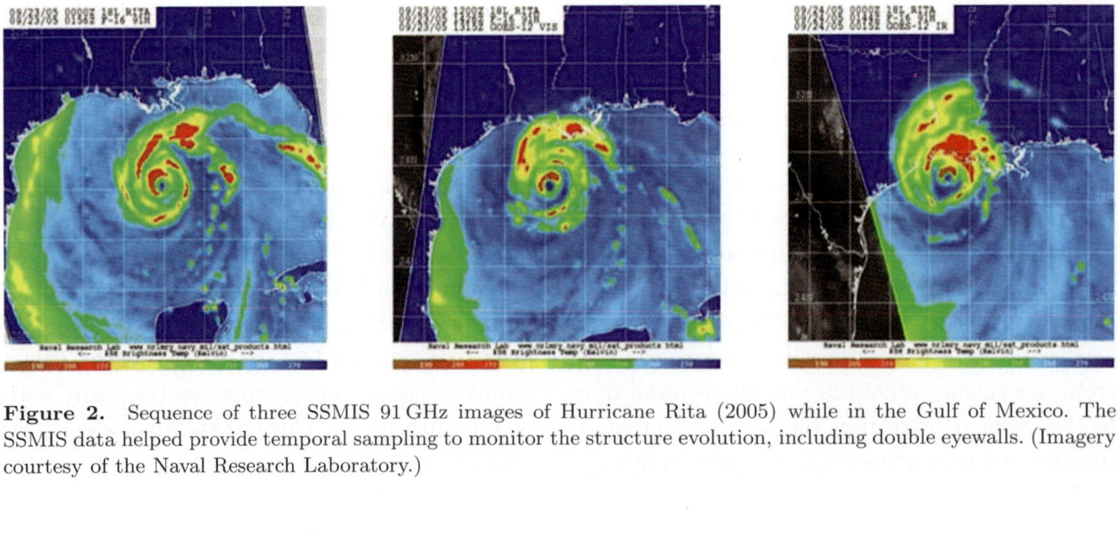

Figure 2. Sequence of three SSMIS 91 GHz images of Hurricane Rita (2005) while in the Gulf of Mexico. The SSMIS data helped provide temporal sampling to monitor the structure evolution, including double eyewalls. (Imagery courtesy of the Naval Research Laboratory.)

and 35°S. A joint mission between the U.S. and Japan, scientists recognized shortly after launch that TRMM could provide essential new information on tropical cyclones in all stages of development (Kummerow *et al.*, 2000).

The primary TRMM sensor used for monitoring tropical cyclones is the PMW TRMM Microwave Imager (TMI), which is a multichannel, dual-polarized, conically scanning instrument similar to SSM/I (Kummerow *et al.*, 1998). TRMM's physically based rain-rate observations from the combined active radar and PMW instruments allow the uncertainty in tropical rainfall to be reduced from earlier space-based estimates. Currently, these techniques are being applied to better estimate tropical cyclone rainfall. The TRMM TMI complements the SSM/I data by filling in temporal gaps and providing superior spatial resolution (the TMI 85-GHz ground footprints are 5–7 km, which is two to three times better than the SSM/I), and the data are being provided in near real time to tropical cyclone forecast centers. The great value of TRMM to the forecast and research communities are evidenced by its wide usage in operational tracking and forecasting of tropical systems globally.

Also in support of NASA's Earth Observing System (EOS), the Aqua low Earth orbiting satellite was launched in 2002. The primary goal of Aqua, as the name implies, is to gather information about the Earth's water cycle. The Advanced Microwave Scanning Radiometer-EOS (AMSR-E) is a one of the six sensors aboard Aqua. AMSR-E is a PMW radiometer that can be employed to monitor tropical cyclones in a similar manner to the SSM/I and TMI noted above. The AMSR-E's value relies on a combination of its large 1800-km swath and superb spatial resolution due to its large antenna (the 89-GHz footprint is near that of TMI (6-km) and is much improved over the viewing capabilities of SSM/I and SSMIS). The excellent resolving capabilities can capture small storm features such as compact eyewalls and eyewall cycles that often are missed by coarser resolution sensors.

Another recent addition to the PMW sensor suite is the WindSat polarimetric radiometer, a US Navy research sensor launched onboard the Coriolis spacecraft in January 2003 to test the extraction of surface wind vectors via a polarimetric radiometer (Gaiser *et al.*, 2004). The large antenna provides excellent spatial resolution brightness temperatures (T_b) at 37 GHz across the 1025-km swath and permits users to monitor tropical cyclone structure (eyewall configuration changes), similar to capabilities in existing TMI and AMSR-E data (Hawkins, *et al.*, 2006). Thus, although WindSat does not have a scattering channel (85–91 GHz), it can significantly contribute to the overall rainband mapping and is operationally used by numerous agencies to aid in center fixing. The 37-GHz channel data provides a lower-level view of structure than does the 85–91 GHz data and often yields a less ambiguous center fix (confusion can arise concerning low-level and mid-level circulation centers). The relatively poor spatial resolution of 37-GHz data from the SSM/I and SSMIS limits this channel's effectiveness from those platforms, however the high-resolution 37-GHz data from WindSat is contributing to improved satellite reconnaissance of tropical cyclones (Fig. 3).

Unfortunately, the longer PMW wavelengths require larger satellite receivers that currently limit the instrumentation to low-altitude, earth orbiting satellites. Therefore, despite multi-satellite availability, very frequent sampling (an advantage of geostationary satellites) is not yet possible. Mitigating this temporal sampling issue can be accomplished by incorporating methods to "bridge-the-gap" between two successive PMW images. Wimmers and Velden (2007) highlight a newly-developed technique to produce near real-time microwave animations that incorporate image "morphing" tools to visually illustrate how storm features evolve from one image to another, typically within 2–7 hours of one another, as

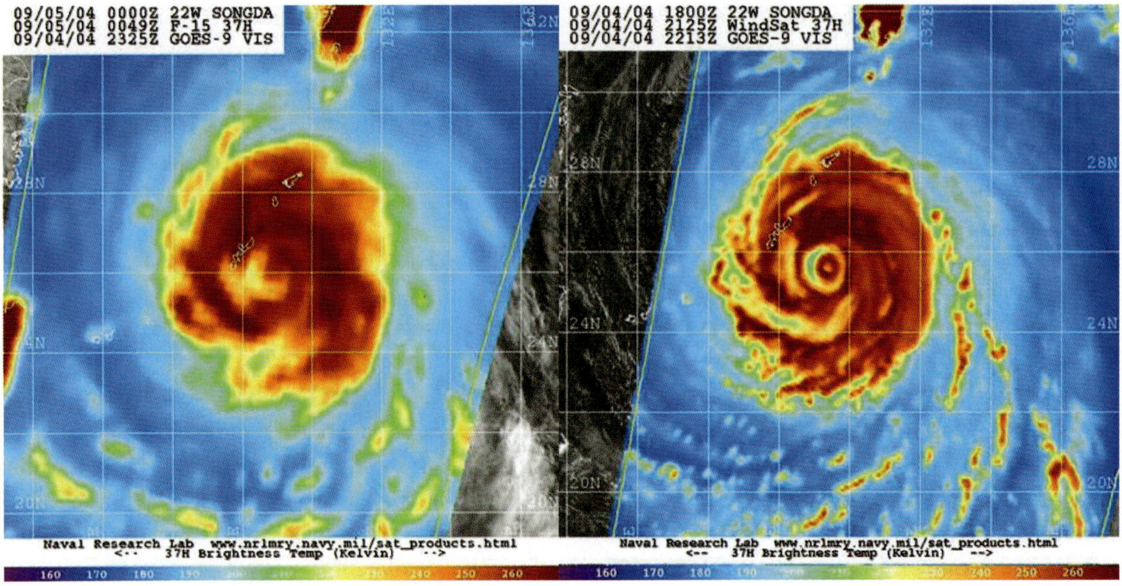

Figure 3. Comparison of 37-GHz imagery for SSM/I (left) and WindSat (right) for typhoon Songda in the western Pacific on Sept 4, 2004. Note the ability to observe inner core structure with the higher resolution WindSat. (Imagery courtesy of the Naval Research Laboratory.)

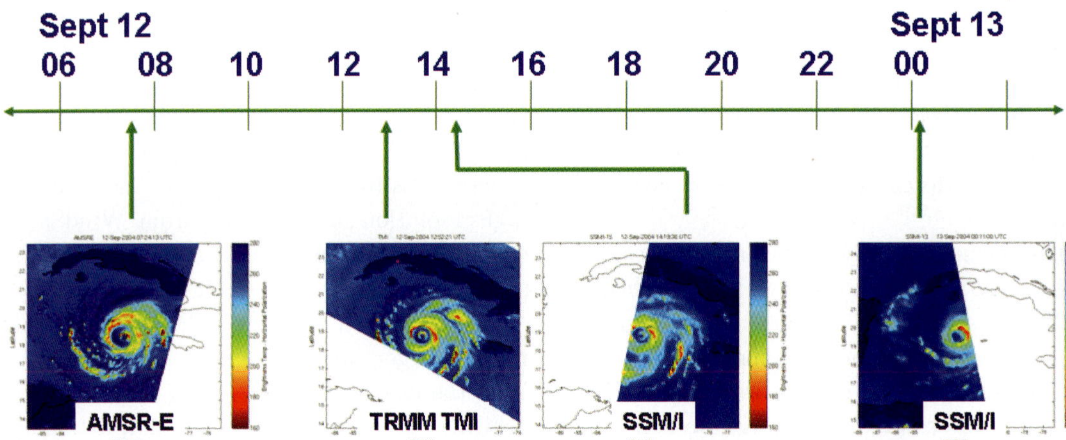

Figure 4. Example of Hurricane Ivan 85–89 GHz PMW images available to the "morphing" technique that will then attempt to fill in the gaps between successive images from the various sensors. The top scale is "hours" during Sept. 12–13, 2004. (Figure provided by Anthony Wimmers, UWisc.-CIMSS.)

noted in Fig. 4. The morphing often allows an analyst to better visualize and monitor the behavior of features such as eyewall replacement cycles that can be difficult to discern in single images. Near real-time morphing demonstrations are available from the CIMSS website in addition to more details explaining the methodology: http://cimss.ssec.wisc.edu/tropic/real-time/marti/marti.html.

In addition to the qualitative uses of the PMW imagery, attempts have been made to try and apply the data in quantitative ways. For example, Cecil and Zipser (1999) investigated the relationship between polarization corrected temperature (PCT) from the 85-GHz channel, and present and future maximum wind speed and tendency of storm intensity. More recently Hoshino and Nakazawa (2006) have proposed a new estimation method of TMI data utilizing the 10, 19, 21, 37 and 85-GHz channels to estimate tropical cyclone intensity parameters. In contrast to the previous study, Hoshino and Nakazawa (2006) found that the parameters derived from lower frequency channels of 10 or 19 GHz yield higher correlations. Clearly, more research needs to be conducted on this promising data resource toward quantitative intensity estimation applications.

Tropical cyclone eyewall replacement cycles are associated with significant, and often rapid, intensity changes, but we presently have no algorithms in place to diagnose their onset. The inception of an eyewall replacement cycle can be identified by the formation of a secondary eyewall as outer rainbands begin to encircle the existing single eyewall. Therefore, accurate intensity forecasts will depend in part on our ability to diagnose and forecast secondary eyewall formation (SEF). Ideally, SEF monitoring via geostationary data is preferred due to excellent temporal sampling, but VIS/IR data is severely challenged by obscuring cirrus residing above the strong storms undergoing SEF. Upper-level cirrus masks the convective spiral band structure and makes it difficult to observe the symmetric organization of this outer convection into a secondary eyewall. A solution to this challenge is the application of microwave satellite data. SEF signals have been identified in both infrared and microwave satellite imagery, and an objective SEF-index, based on Bayesian probability, has been created that can be applied operationally (Kossin and Sitkowski, 2008). The algorithm is being combined with a similar index that diagnoses the formation of annular hurricanes, and is being constructed by members of the NOAA/NESDIS RAMMB team (Knaff *et al.*, 2008).

3.2. *Tropical Cyclone Intensity Estimation Using Microwave Sounders*

The concept of estimating tropical cyclone structure, intensity, and intensity change using polar-orbiting satellite passive microwave sounder observations spans over three decades of active research. The microwave sounding units flown on the NOAA series of low Earth orbiters have been used to show how the characteristic warm core of tropical cyclones can be observed and, via hydrostatic approximations, related to storm intensity. Radiation sensed by oxygen emission in the 55-GHz region is used to derive upper-tropospheric temperatures in the storm core. Several factors including coarse spatial resolution and the ever-changing position of the storm within the instrument observation swath between successive observations acted to limit the accuracy of the first attempts at this approach to estimate intensity.

Within the last decade, progress has been made in understanding, treating, and reducing the undesirable effects of instrument-related warm core sampling limitations. This progress has been partly due to advanced microwave sounding instrumentation on the most recent NOAA low Earth orbiters. Starting with NOAA-15, the Advanced Microwave Sounding Unit (AMSU) has been flown. The AMSU instrument is a cross-track scanning microwave sounder containing 15 channels with frequencies ranging from 23.8 to 89 GHz. Resolution varies from 48 km at nadir to ~100 km near the limb. The primary channels used for tropical cyclone monitoring are the ~55-GHz oxygen band channels 5-8 and the moisture channels 1-4 (AMSU-A) and 16 (AMSU-B).

The qualitative relationship between the strength and vertical distribution of the AMSU-observed warm core and tropical cyclone intensity is well illustrated by a time series of NOAA-15 AMSU observations of Hurricane Floyd from 8–16 September 1999 and near-coincident aircraft reconnaissance sea-level

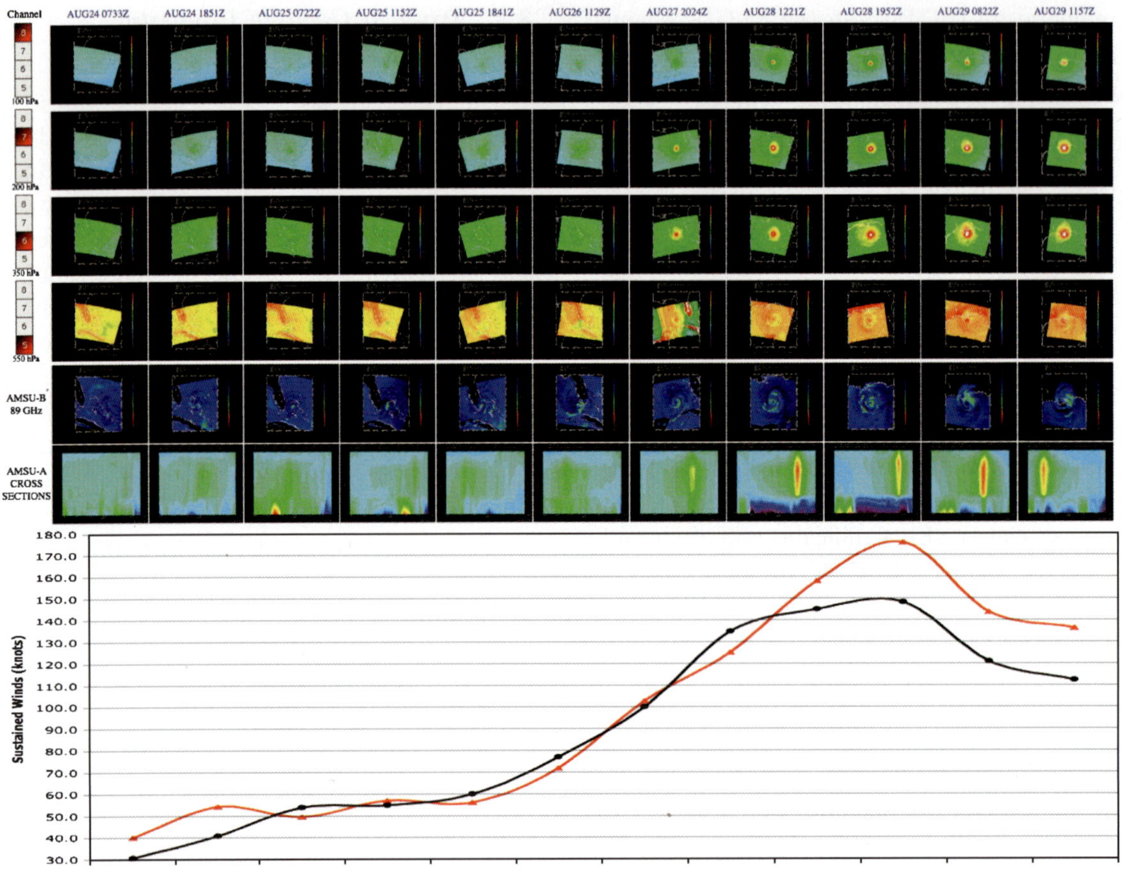

Figure 5. Image-time matrix of AMSU channels 5–8 over a period 8–16 September 1999 during Atlantic Hurricane Floyd. The coincident recon mean sea level pressure (MSLP) trace is indicated at the bottom. Floyd's strongest warm core temperature anomalies are illustrated by the yellow/red signatures in channels 6–8 late in the period.

pressure estimates (Fig. 5). Floyd's sea-level pressures are at their minimum values (<930 hPa) during periods when the warm core is well defined — particularly when upper-tropospheric temperature anomalies are at their maximum values.

Recent attempts at quantifying this relationship have led to intensity estimation algorithms that have become competitive with the IR-based Dvorak Technique. These techniques rely on the magnitude of the AMSU-observed warm core (Fig. 6). Scientists at the Cooperative Institute for Research in the Atmosphere (CIRA) have developed an algorithm that utilizes the retrieved vertical temperature structure from the multispectral AMSU radiances, and is then combined with a number of additional AMSU-based predictors to derive MSLP and maximum sustained wind (MSW) estimates. Stan Kidder and colleagues (2000) describe this method in detail. A retrieval algorithm is used to determine the temperature as a function of pressure from the AMSU radiances. The hydrostatic equation is then integrated downward from 50 hPa to the surface using an upper-level boundary condition provided from a numerical model analysis. Prior to the hydrostatic integration, an empirical correction is applied to the temperature profiles to adjust for attenuation of the AMSU radiances by liquid water and scattering by ice. Once the height field is determined as a function of pressure, the wind field is calculated assuming gradient balance. The height field is azimuthally averaged relative to the storm center for this purpose. The AMSU data does not have adequate resolution

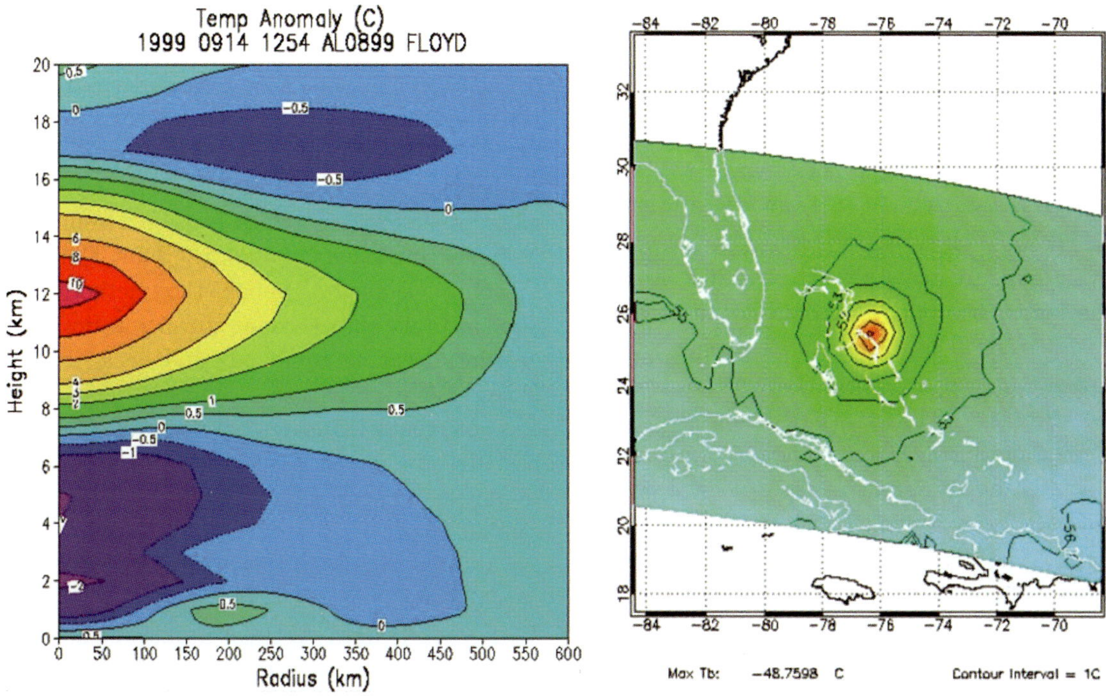

Figure 6. Representation of AMSU-A retrieved vertical temperature anomaly from CIRA (left) and horizontal raw Tb anomaly for channel 8 (~150 hPa) from CIMSS (right) for Hurricane Floyd on 14 September 1999.

to fully sample the inner core structure of intense tropical cyclones. To help correct for this problem, a statistical algorithm has been developed to estimate the maximum winds and the radii of 34 kt, 50 kt, and 64 kt winds from properties of the AMSU analyses. The inputs are analysis properties such as surface pressure drop from 600 km to the storm center, the maximum retrieved gradient wind, and the average cloud liquid water near the storm center. The output is an estimate of the observed maximum wind and wind radii (Demuth *et al.*, 2006).

Another algorithm developed at the Cooperative Institute for Meteorological Satellite Studies (CIMSS) uses the raw AMSU radiance anomalies along with corrective measures for AMSU footprint size and offsets to estimate MSLP. Limb corrections are applied to the AMSU data to compensate for decreasing brightness temperatures (T_b) as viewing angle increases from nadir. Eye size is used as a proxy for the warm core dimension, and is derived by the ADT algorithm for "clear eye" cases, or from the radius of maximum winds estimated by operational forecast centers in other cases. An initial estimate of MSLP is performed using the raw T_b anomaly. If the eye size is small compared to the AMSU FOV then a bias correction is subtracted. The T_b signal can also change slightly from one estimate to the next as a function of storm position within the scan swath. To account for this effect, a second regression predictor (distance in FOV steps from nadir) was added. The end result is a slight intensity increase for cases near the swath edge and a slight decrease for those cases near nadir. Hydrometeor contamination problems decrease effective brightness temperatures, requiring a correction using AMSU-A channels 2 and 15. A two-channel difference can determine if brightness temperatures are being depressed due to hydrometeor scattering. A correction is then applied to the raw brightness temperatures resulting in a new set of regression coefficients (Wacker and Velden, 2004). Improved performance was validated and these modifications have been adopted for real-time (Herndon and Velden, 2004).

3.3. *Applications to Tropical Cyclone Rainrate Estimation*

A recent NOAA/AOML/HRD effort has filled an important gap by formulating a tropical cyclone rainfall climatology via TMI data referred to as R-CLIPER. Lonfat *et al.* (2004), incorporated three years (1998–2000) of global TMI data to map rainfall in 260 storms. The high resolution TMI data captures much of the tropical cyclone rainfall events and is superior to coarser data sets such as SSM/I, SSMIS, and AMSU-B. TMI rain data was azimuthally averaged and binned by storm basin and intensity to permit a climatological profile of rain intensity radially outward from the storm center. The National Hurricane Center incorporates R-CLIPER and official track forecasts to create a graphical 24- and 72-h rain accumulation forecast for all landfalling storms (see Fig. 7 for an example with Hurricane Frances). The products are a starting basis forecasters can use and modify depending on specific storm characteristics (e.g., bigger than normal for Cat 2 storm).

NOAA's Tropical Rainfall Potential (TRaP) technique generates 24-hour rainfall forecasts using a combination of microwave-derived rainrates and official storm track forecasts (Ebert *et al.*, 2005). Microwave overpasses produce a snapshot of rainfall that is then advected along the forecast track output from the National Hurricane Center or the Joint Typhoon Warning Center. Although the methodology does not take into account storm rotation and growing/decaying rainbands and cells, it produces ballpark values that can assist forecasters and emergency managers. As one might expect, best results occur when the highest spatial resolution imager (TMI) serves as the base rainfall map.

A new inversion-based algorithm under development at NASA GSFC uses both TRMM TMI and PR data. Vertical profiles of precipitation, ice water, and liquid water content are derived via these merged data sets and validation statistics are quite promising (Jiang and Zipser 2006). The ability to take advantage of combined TMI/PR data and 3-D information is critical to algorithm advancement.

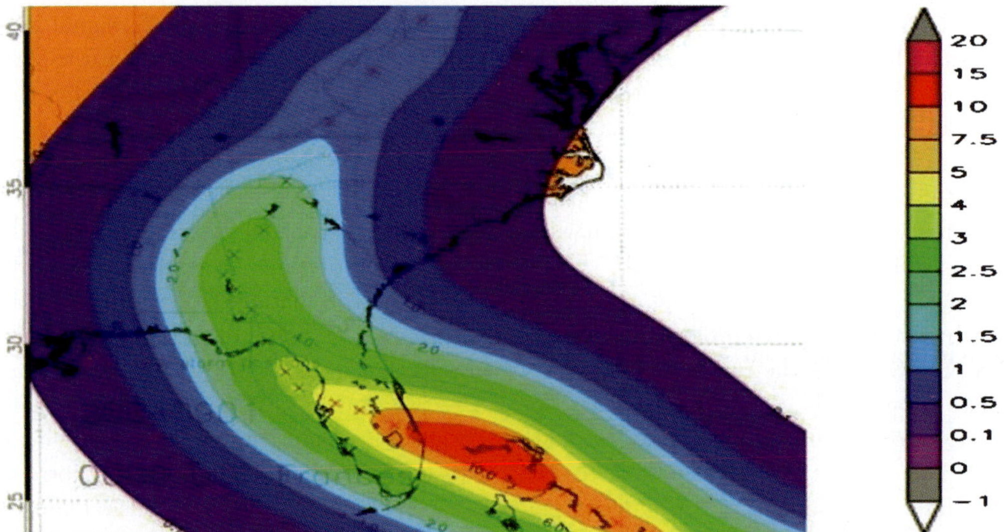

Figure 7. Example R-CLIPER rain accumulation (inches) graphic depicting forecast rainfall totals for Hurricane Jeanine during a Florida landfall in Sept. 2004. This product is available in near real-time at NHC, Miami.

3.4. *Scatterometer Data Applications*

The QuikSCAT scatterometer, launched in 1999, is an active microwave radar sensor that has proven to be a very beneficial tool in tropical cyclone analysis (Liu, 2002). The QuikSCAT is borne on a polar-orbiting, sun-synchronous satellite with an equator-ascending time of approximately 0600 local (±30 minutes). Near-surface wind vector solutions and up to four ambiguity solutions are stored in 12 km and 25 km cells over an 1800-km wide swath. Each cell also contains various flagging indicators that include possible contamination with rain, land, and ice. Since the Fifth International Workshop on Tropical Cyclones in Cairns, Australia (2002), there has been a substantial increase in the world-wide use of the QuikSCAT scatterometer to aid in almost all aspects of the tropical cyclone warning process (genesis, positioning, structure, intensity, extratropical transition/dissipation). One reason for this is that analysts have gained a better understanding of how to use the data and how to better interpret some of the known problem areas such as the range of accurate wind speeds, the ambiguity selection process and the effects of rain. This knowledge has also helped the developers of QuikSCAT improve the product as well. Additional information on the application of scatterometer data in tropical cyclone applications can be found in Edson (2002a, b). Another important development is through the sophisticated methods currently being used to obtain special high resolution data via the processing of the pulse information in the QuikSCAT source (Halterman and Long, 2006).

As mentioned above, the proper interpretation of the QuikSCAT data is often vital to its use in tropical cyclone applications. Integrating the QuikSCAT data with other imagery is a good way to determine where rain may be affecting the wind retrieval process by artificially yielding high wind speed (typically in areas where the real wind speed is less than 30 to 35 kt), and also provides a manual method to select the proper ambiguity solution. For intense systems (above ∼45–65 kt), the analyst must understand the QuikSCAT limiting factors to determine the reliability of wind speeds, such as the wind retrieval process itself, the resolution of the data, and the tendency for heavy rain regions to attenuate the signal. For weaker storms (below ∼30–40 kt), QuikSCAT wind retrievals tend to misinterpret heavy rain as a higher wind speed with an isotopic directional signal that is cross-track to the sensor's view angle.

The QuikSCAT Normalized Radar Cross-Section (NRCS) raw backscatter images (Fig. 8) have an average pixel size of 2.5 km with an effective resolution of 6 km, and offer a unique view of surface conditions in the vicinity of tropical cyclones. For intense tropical cyclones with multiple eyewalls, where the heavy rain and winds normally coincide, deducing structure from the NRCS imagery is often quite informative. The small dark (low rain and low wind region) eye provides for very precise positioning of the storm center. Other characteristics of the TC structure such as concentric eyes or an asymmetric wind field are also often evident (Fig. 8).

Determining outer wind radii (35 kt, 50 kt, or possibly 64 kt) using QuikSCAT vector data is a process that can be dependent upon the tropical cyclone intensity and extent of the rain bands. In Fig. 9, an example is shown where the PMW imagery provides guidance on the likely winds affected by hydrometeor attenuation. An experienced analyst can use the vector field along with the image data to assess the most reasonable wind radii and asymmetries. A good example of this application is for storms undergoing extratropical transition (ET). In poleward moving systems, the loss of symmetric deep convection near the storm center as viewed in PMW data, along with the expansion of the winds away from the center as indicated by the scatterometer vectors is a common signature of ET.

The recently-launched EUMETSAT Advanced Scatterometer (ASCAT) sensor flies onboard the MetOp-1 satellite platform that

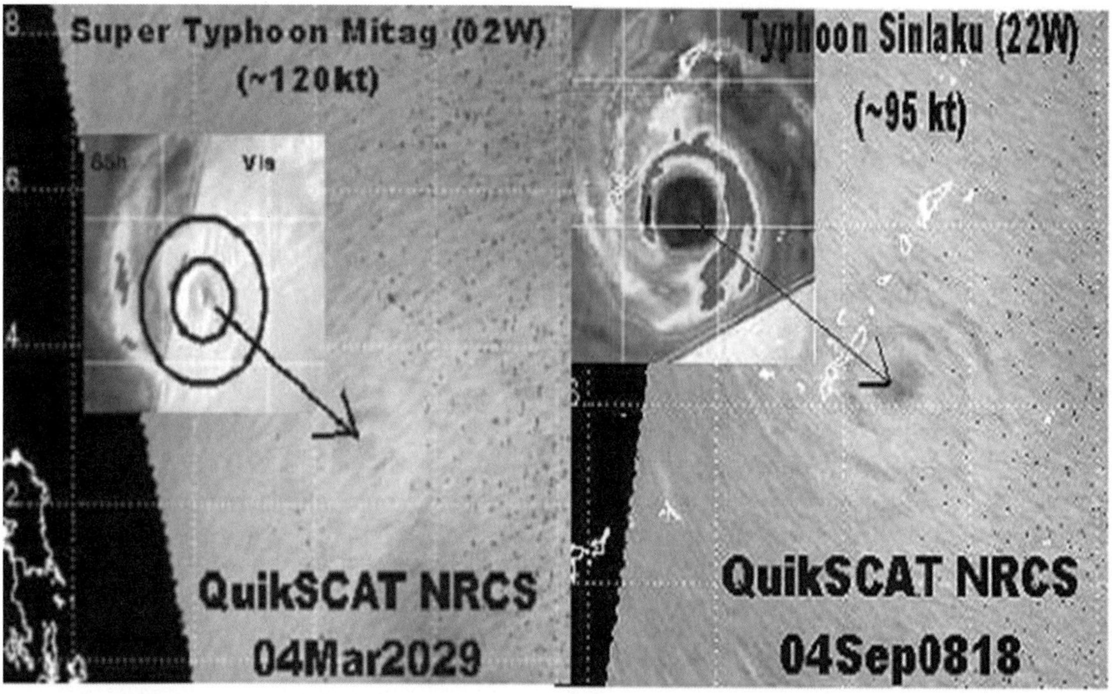

Figure 8. NRCS views of (a) Super Typhoon Mitag and (b) Typhoon Sinlaku. In each case concentric eyes are evident. Center positions are precisely indicated by the darker calm and rain-free area. (Courtesy of Roger Edson.)

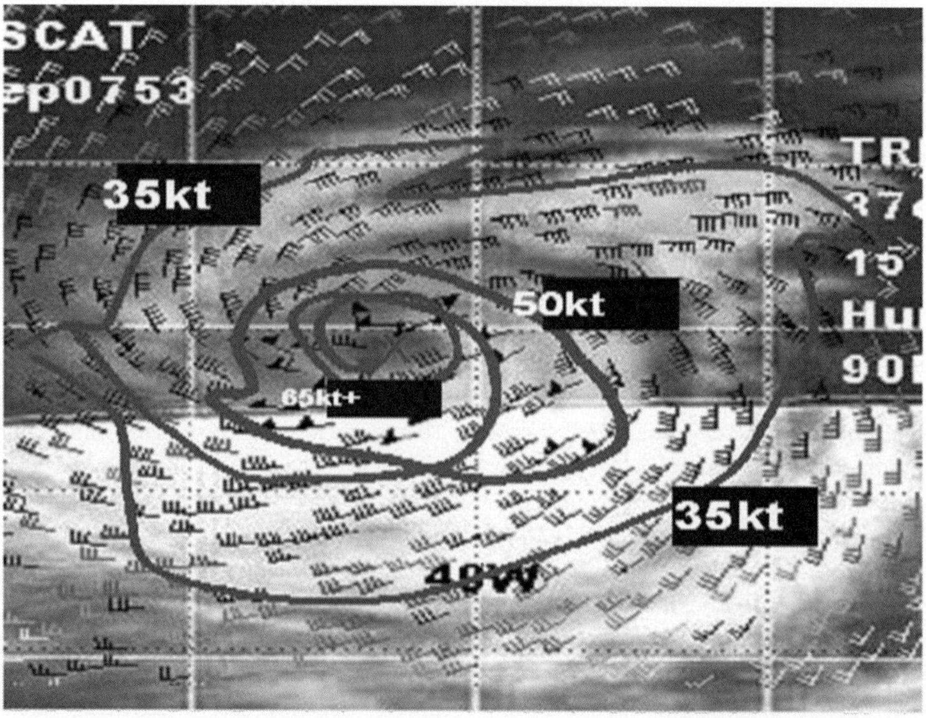

Figure 9. Wind radii determination in intense tropical cyclones using QuikSCAT and PMW imagery. (Courtesy of Roger Edson.)

is the primary "AM" or morning satellite for the joint US-EUMETSAT polar orbiter program. The ASCAT sensor is derived from the heritage scatterometer flown onboard the ERS-1/2 spacecrafts and continues a C-band active microwave sensor for wind vector retrievals. ASCAT's C-band frequency (5.25 GHz) is less impacted by rain and/or heavy moisture attenuation than QuikSCAT's Ku-band (13.4 GHz), but is fundamentally lower in spatial resolution given the same antenna size and orbital altitude. The sensor has an effective swath of 1000-km, via two 500-km swaths separated by a 768-km gap underneath the satellite. This discontinuous swath has potential issues in adequately sampling tropical cyclone wind field structure. Wind vector retrievals are nominally available at 50-km resolution, and 25-km experimental products are now available. ASCAT data is available operationally, but to date has not been validated in the full range of tropical cyclone high wind conditions. Thus, algorithm refinements are expected as these data sets and revised algorithms become available. The combination of ASCAT and QuikSCAT data is considered complimentary, since neither sensor has the temporal/spatial coverage adequately needed for continuous tropical cyclone monitoring.

As with QuikSCAT, ASCAT wind vectors are most reliable in helping to define the radius of gale force winds and to a lesser extent the radius of 50-kt winds. Due to spatial resolution issues, ASCAT will not be able to quantify the maximum sustained wind in tropical cyclones in most cases since the footprint will spread across the large wind speed gradient. Thus, ASCAT's primary value will be in providing outer wind field structure and storm size information for near real time applications.

3.5. *Use of GPS Radio Occultation Methods*

A promising new source for measurements of temperature, pressure and water vapor in the environment of tropical cyclones is the GPS radio occultation (RO) technique, which was developed by scientists at Stanford University and the Jet Propulsion Laboratory (JPL) for the remote sensing of planetary atmospheres. The Global Positioning System (GPS) is a satellite-based constellation developed for navigation and positioning. The GPS today consists of 24 low Earth orbiting satellites that transmit L-band radio signals (19 and 22 cm wavelengths) to a wide variety of users. Atmospheric soundings are obtained through the RO technique, in which Doppler-shifted frequency measurements are used to compute the bending angles of the radio waves that are a function of atmospheric refractivity.

The GPS/MET (Global Positioning System/Meteorology) Program demonstrated the relevance of this technique for accurate sounding of the Earth's atmosphere. As a result, GPS/MET technology is now recognized as a potential candidate of the global observing system in support of weather analysis and prediction, including tropical cyclones. For a detailed summary of the radio occultation sounding technique and the results from GPS/MET, the reader is referred to a special edition of Terrestrial, Atmospheric and Oceanic Sciences (TAO, 2000).

3.6. *Near Future Microwave Sensors*

The current suite of microwave imagers will be augmented/replaced with SSMIS sensors on DMSP spacecraft through at least 2015 (Fig. 10), but actual launch dates are unknown due to many variables including existing sensor health. PMW imagers were originally planned to be flown on the United States' converged National Polar-orbiting Operational Environmental Satellite System Series (NPOESS). However, the Conically scanning Microwave Imager Sounder (CMIS) has now been removed from the NPOESS sensor suite and the contract is being re-competed, with launch no sooner than the second satellite (~2016) as the Microwave Imager Sounder (MIS). A total

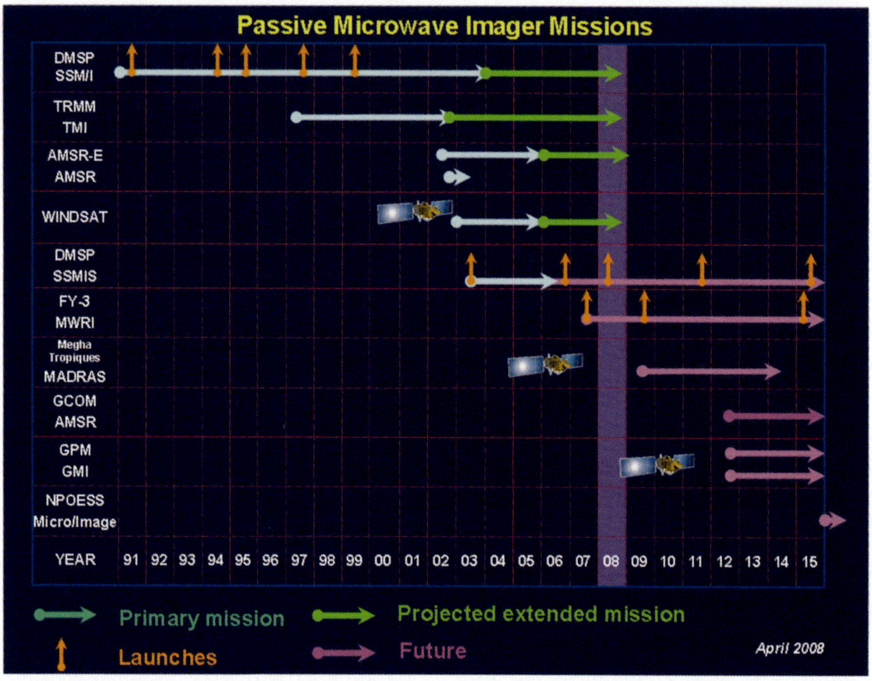

Figure 10. Recent past, present (as of 2008) and future passive microwave imagers potentially available for tropical cyclone monitoring. Note that launch dates are highly variable and subject to frequent changes and delays. (Courtesy of NRL/Ferraro/Kummerow.)

of three PMW imagers may eventually fly on NPOESS and contribute to global tropical cyclone reconnaissance.

EUMETSAT is studying the need for a microwave imager on future MetOp spacecraft and might add one for MetOp-4. If so, this would provide a welcome addition to the constellation in the mid-morning orbit. China added the MicroWave Radiometer Imager (MWRI) to their FY3 spacecraft series but sensor issues mean a look forward to the next launch. The ten-channel imager will contain 37 and 89 GHz channels that will continue to complement existing capabilities. The French/ Indian MADRAS microwave imager is scheduled to fly on the Megha-Tropiques satellite in the 2010 timeframe in a 20-degree tropical inclination. The sensor should provide a wealth of tropical cyclone views and includes the 37 and 89 GHz channels. Japan will likely fly a microwave imager similar to AMSR-E in 2012/2013.

NASA's Global Precipitation Mission (GPM) includes a microwave imager (GMI) and a dual-polarization radar (DPR) on the same platform, similar in context to the current TRMM sensor suite (Smith *et al.*, 2004 and Hou, 2005). The combined imager and radar in a 66-degree inclination orbit will provide high spatial resolution data required to capture tropical cyclone structure and sorely needed cross calibration with other microwave imagers. In addition, a second spacecraft will contain a GMI in a tropical inclination that will enhance the opportunities to fly over tropical cyclones on multiple consecutive orbits. GPM launch is tentatively slated for the 2013 timeframe.

Microwave sounders have been routinely utilized as "imagers" by creating images from their 89 GHz channels. Although the spatial resolution is poor and varies across the swath, the huge 2350-km swath and three operational sensors provide huge potential advantages. While the spatial resolution is inadequate to resolve many tropical cyclone features, the storm "context" they provide and decent resolutions at nadir often provide valuable input. Data from the Advanced Microwave Sounding Unit (AMSU-B) and the new Microwave Humidity Sounder (MHS on NOAA-18) have proven beneficial. This bodes well for continued use since both NPOESS and MetOp carry microwave sounders with 89-GHz channels. In addition, other countries are slated to contribute additional microwave sounders that should assist in temporal sampling enhancements.

3.7. *Microwave Sensor Training Efforts*

An excellent introduction to satellite-based PMW applications for forecasters and newcomers is the Cooperative Program for Operational Meteorology, Education, and Training (COMET) module available online: http://meted.ucar.edu/npoess/tc_analysis/. Taking a global approach, it introduces passive microwave imaging, discusses elementary techniques for fixing storms, explains the difference between passive and active remote sensing, covers conical vs. cross-track sensing techniques, introduces several special cases (double eye wall, cirrus covered eye, low-level detached center, etc.), and surveys the changing state of polar satellites and sensors. A primary purpose is to train forecasters at facilities like the National Hurricane Center and the Joint Typhoon Warning Center. Figure 11 shows an example graphic, highlighting a rare tropical cyclone moving into Brazil.

In addition, the Naval Research Laboratory maintains several tutorials at: http://www.nrlmry.navy.mil/training-bin/training.cgi. (Choose "Tropical Cyclones" on the left frame and then select between SSM/I and TRMM.) These tutorials contain detailed case studies to illustrate passive microwave image interpretation. The topics include polarization correction; images of 85 vs. 37 GHz; use of 4-panel composites; interpretation of rain, water vapor, and wind speed products; and the distinction between ice and liquid precipitation within storms. Special attention is given to the detection of storm centers in microwave products.

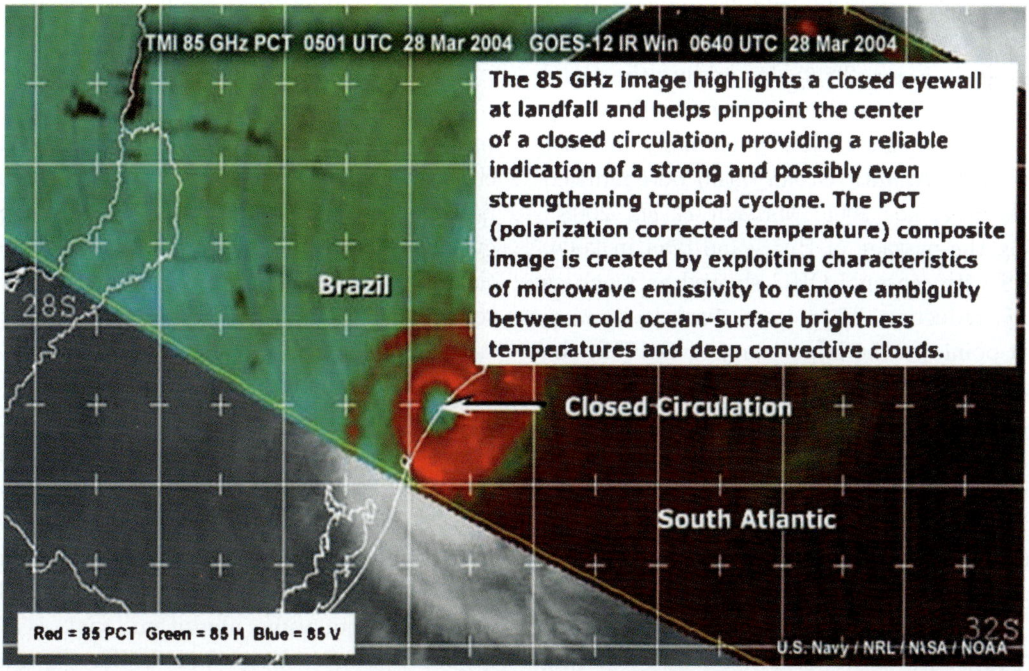

Figure 11. COMET/NRL web-based graphic illustrating the use of TMI 85 GHz polarization corrected temperature (PCT) to extract storm structure while system is landfalling into Brazil on 28 March 2004 at 0501 GMT.

4. Other Satellite Applications

Shay *et al.* (2000) and more recently Mainelli-Huber (2000) have shown that oceanic heat content variations are linked to tropical cyclone intensity change and in some cases rapid intensification. The approach is to use sea surface topography from space-based instrumentation such as TOPEX/Poseidon (T/P) and Jason-1 altimeters that measure the sea level every 9.9 days along repeat ground-track spaced 3° longitudinally at the Equator (Cheney *et al.*, 1994). The ERS-2 mission and U.S. Navy Geosat Follow-On-Missions (GFO) have repeat tracks of 35 and 17 days, respectively. When cast within a two-layer reduced model, the depth of the 26°C isotherm can be estimated from these objectively analyzed and mapped surface height anomaly fields. Using monthly, seasonal or annual climatologies, the integrated thermal structure or oceanic heat content can be estimated from that depth to the surface where the sea surface temperature is used as the surface boundary conditions. This approach provides locations of warm and cold features in front of the projected track of tropical cyclones as discussed in Shay (2008). By adding the OHC parameter to the Statistical Hurricane Intensity Prediction Scheme (SHIPS), intensity forecast errors in the Atlantic Ocean Basin were reduced (DeMaria *et al.*, 2005; Mainelli *et al.*, 2008). During the passage of Hurricane Ivan in 2004, SHIPS with seasonal OHC showed as much as a 22% reduction in forecast intensity errors, which points to the importance of OHC in the warm pool of the Caribbean Sea and the Gulf of Mexico's Loop Current and warm core rings.

CloudSat was launched in 2007 and is a NASA Earth System Pathfinder mission that utilizes a millimeter-wavelength cloud radar with the goal of advancing our understanding of cloud abundance, distribution, structure and radiative properties. CloudSat's radar is 1000 times more sensitive than existing weather radars and unlike ground-based radars that use centimeter wavelengths to detect raindrop-sized particles, CloudSat detects much smaller particles of cloud liquid water and ice that constitute cloud masses worldwide. Thus, CloudSat's 4 GHz (3 mm) nadir-viewing radar (2.5 × 1.4 km spot) has 500 m resolution (interpolated to 240 m) and provides us with a first ever capability to map the 3-dimensional profile of cloud liquid water and ice and complements the heavy to mid range precipitation sensing better accomplished by the TRMM radar. The measured backscatter signal is a function of the cloud particle number, the concentration, size and shape of the cloud droplets and is sensitive to the temperature and phase. This dependence permits us to extract basic cloud microphysical properties after extensive processing and use of auxiliary data sets. Vertical cross sections of radar reflectivity can be created. As Fig. 12 illustrates, CloudSat can provide a wealth of detail when it overpasses a tropical cyclone. CloudSat effectively maps the cloud top/base, areal extent, light and moderate rain, freezing level, strong convection and rain free zones as the eyewall is transected (Mitrescu *et al.*, 2008).

5. Integrated Multispectral Data Approaches

In addition to the methods above, several existing or promising integrated satellite-based approaches to measure/estimate tropical cyclone parameters are available (operationally or experimentally) to forecasters and researchers. Some of these utilize multiple spectral channels from the same sensor package, while others employ multi-platform information sharing approaches to achieve measurement or estimate precision beyond that of a single observing strategy.

One such example is a research effort underway at the University of Wisconsin-CIMSS to develop an integrated tropical cyclone intensity estimation algorithm that is fully automated, objective, and utilizes a multi-spectral approach. This system, called SATCON (SATellite CONsensus) builds on the latest

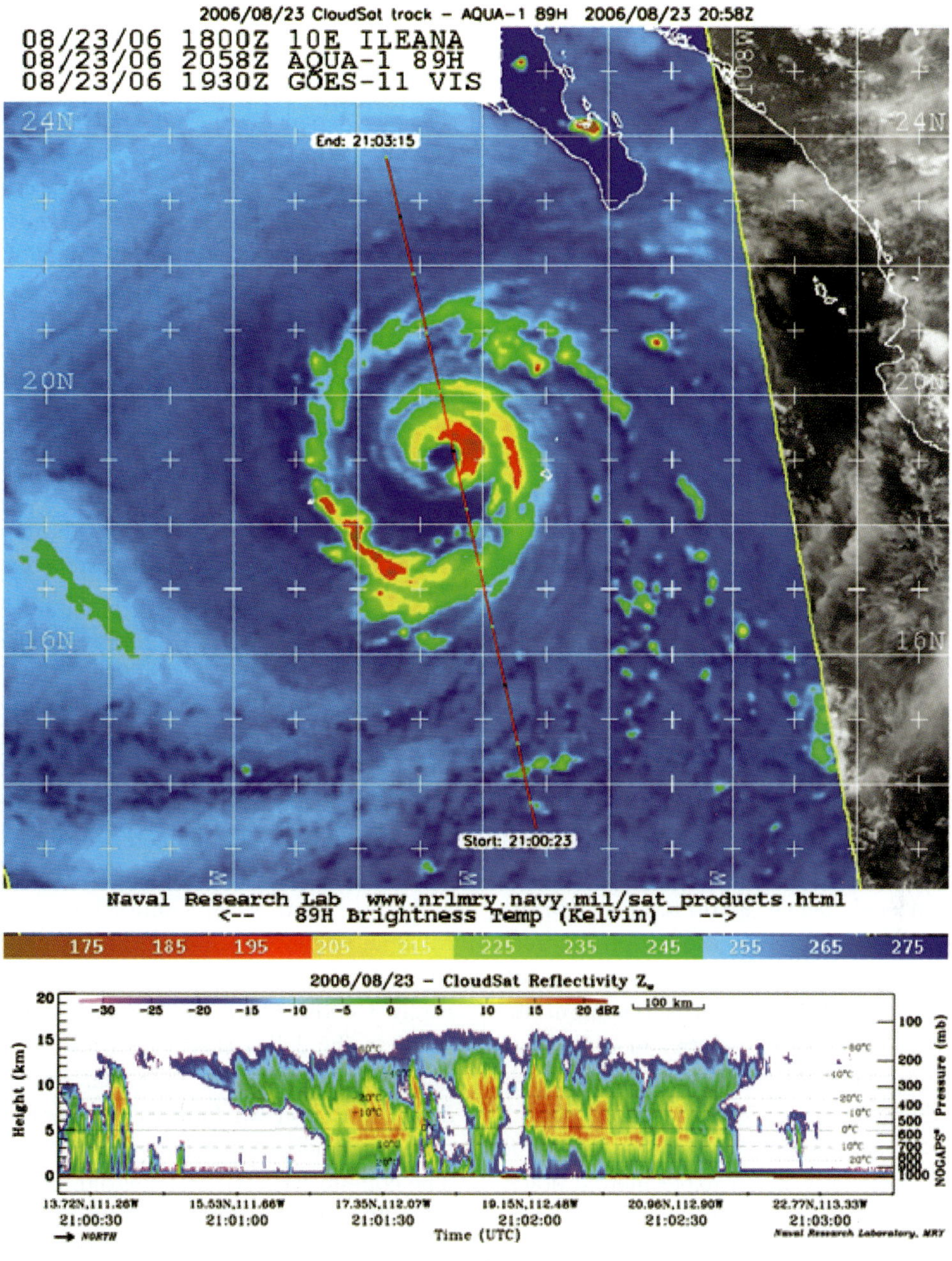

Figure 12. Vertical cross section (lower panel) of Cloudsat radar reflectivity for tropical cyclone Ileana during 2006, and the CloudSat overpass on top of coincident AMSR-E 89-GHz imagery (upper panel). Figure courtesy of Cristian Mitrescu of the Naval Research Laboratory.

science advances in existing (and emerging) methods. The components, briefly discussed in the previous sections, include the geostationary satellite IR/VIS-based Advanced Dvorak Technique, and two Advanced Microwave Sounding Unit intensity estimation methods. SATCON uses cross-technique information sharing and produces a weighted consensus of objective satellite-based estimates to determine the current intensity (Vmax and MSLP). Each of the SATCON members have documented error characteristics which are situational dependant. Each algorithm also contains unique parametric information that can be used by other members. SATCON uses this information to adjust/improve the member estimates, then considers the situational error characteristics to assign weights to each member, resulting in a final intensity estimate. Validation studies versus reconnaissance values in the Atlantic have shown SATCON to be superior in skill to the individual members or a simple average of them (Herndon and Velden, 2008). Additional information on this technique can be found at the SATCON web page: http://cimss.ssec.wisc.edu/tropic2/real-time/satcon/.

Another integrated satellite-based approach for tropical cyclone analysis has been developed at the Colorado State University — NESDIS/RAMMB/CIRA (Knaff and DeMaria, 2006). This objective method employs multi-satellite-platform information to construct a surface wind field in the tropical cyclone near-environment. Datasets for this analysis include surface wind vector estimates from Quikscat, surface wind speed estimates from SSM/I, feature track winds collected below 500 hPa, 2-d AMSU-based winds, and 700-hPa winds derived from IR imagery as described in Mueller *et al.* (2006). The resulting dataset is the input for an analysis, which is formulated in cylindrical coordinates. Each of the component datasets has its own characteristics and shortcomings, which are used to set weights in the analysis. This product is currently in an experimental state, but is expected to be transitioned to an operational product. Since it relies only on satellite data inputs (unlike the H-WIND analysis developed by AOML/HRD that employs non-satellite data as well), such an analysis could be produced for all globally occurring tropical cyclones, and could also be used for model initialization and research studies. More information on this product can be found at: http://rammb.cira.colostate.edu/products/tc_realtime/about.asp.

As briefly described above, Tropical Rainfall Potential (TRaP) forecasts provide estimates of 24-h rainfall accumulation in landfalling tropical cyclones based on the advection of a field of satellite-estimated precipitation. Ebert *et al.* (2005) are investigating a better strategy to statistically combine TRaPs based on one or more sensors and to generate an ensemble of TRaPs by adding realistic perturbations to the forecast speed and direction of the cyclone. The parameters of the microwave rain rate retrieval could also be varied to provide several initial fields. The great advantage of the ensemble approach is the ability to easily produce probability forecasts for critical rain thresholds. This should add enormous value to the TRaP product in decision-critical situations like tropical cyclones approaching landfall.

Finally, satellite estimates of SST have evolved into the multispectral domain, and are now considered highly accurate. For example, the Naval Oceanographic Office provides operational Multi-Channel Sea Surface Temperature (MCSST) retrievals from digital Advanced Very High Resolution Radiometer (AVHRR) and GOES multi-spectral imager satellite data (May and Osterman, 1998; Cummings, 2005). The MCSST retrievals are directly assimilated into the Fleet Numerical Meteorology and Oceanography Center ocean analyses. These analyses monitor the ocean thermal structure on global and regional scales in near real time and are important to ocean forecast models and atmospheric numerical weather prediction forecasts. The combined use of high spatial and temporal IR split-window approaches from geostationary satellites (i.e., Merchant *et al.*, 2008; May *et al.*, 1998) with cloud-penetrating microwave capabilities from low Earth orbiting satellites has led to algorithm development that allows a nearly continuous view of the tropical ocean surface skin temperature (Reynolds *et al.*, 2007). The newer multispectral approaches can also correct for atmospheric water vapor and aerosol attenuation. Combined with in situ ship and buoy data, tropical cyclone forecasters now rely on these derived SST fields to provide them with information on potential future intensity fluctuations.

6. Recent Satellite Data Impact Studies

There have been recent attempts to quantify the contributions to forecast accuracy made by conventional rawinsonde data and remotely sensed satellite data types. One such study that considered tropical cyclone track forecasts was by Zapotocny *et al.* (2008), who used extended-length observing system experiments (OSEs) to examine the 0–168-h forecast impact results in NCEP's Global Data Assimilation System and the Global Forecast System (GFS) during January–February and August–September 2003. In contrast to global anomaly correlations where AMSU radiances and rawinsonde data routinely provided the largest gain to forecast quality, the geostationary satellite derived wind data aided GFS tropical cyclone track forecasts in the eastern Pacific basin the most (Fig. 13). In the Atlantic basin all satellite data types provide a nearly equal contribution

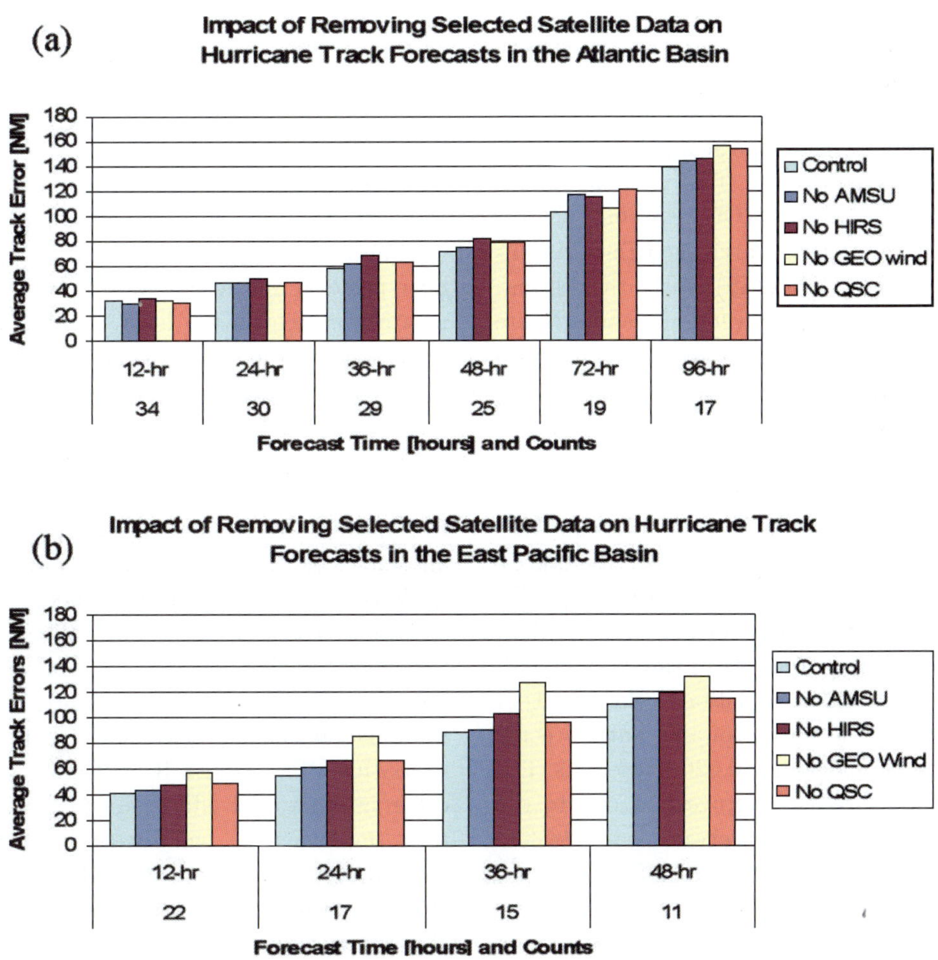

Figure 13. Average track error (n mi) by forecast hour for the control simulation and experiments where AMSU, HIRS, geostationary satellite winds (GEO), and QuikSCAT were denied. Shown are the (a) Atlantic basin and (b) eastern Pacific basin results. A small sample size in the number of hurricanes precludes presenting the 72- and 96-h results in the eastern Pacific Ocean. (From Zapotocny *et al.*, 2008.)

to the tropical cyclone forecast track quality. In general, the Atlantic basin storms are more accurately simulated by the GFS than are those in the Pacific basin, most likely due to more conventional data upstream of the storms, information from dropsondes being assimilated in the vicinity of Atlantic storms but not in the Pacific, and reconnaissance flights that are also routinely conducted in and around Atlantic storms.

In a second study, Goerss and Hogan (2006) also found that the satellite-derived winds had the greatest impact on tropical cyclone track error in the Navy Operational Global Atmospheric Prediction System (NOGAPS). The NOGAPS tropical cyclone track forecasts were evaluated for the period from 14 August to 30 September 2004. This was a particularly active period with 12 hurricanes (including Charley, Frances, Ivan, and Jeanne), 5 typhoons, and 7 tropical storms. The current operational configuration of NOGAPS consisting of the NRL Atmospheric Variational Data Assimilation System (NAVDAS) with the assimilation of all available conventional and satellite observations and a T239L30 global spectral model with Emanuel cumulus parameterization was used as the control run. It was found that the satellite observations with the most impact upon tropical cyclone track forecasts were feature-track winds from geostationary and polar-orbiting satellites, AMSU-A radiances, and SSM/I precipitable water (Fig. 14). The assimilation of the feature-track winds resulted in the largest improvement in the 24-h to 72-h forecasts while the assimilation of the AMSU-A radiances and SSM/I precipitable water become comparable in the 96-h and 120-h forecasts. The overall track forecast improvement due to the assimilation of satellite observations was 18% at 24 h, 27% at 48 h, 21% at 72 h, 17% at 96 h, and 13% at 120 h.

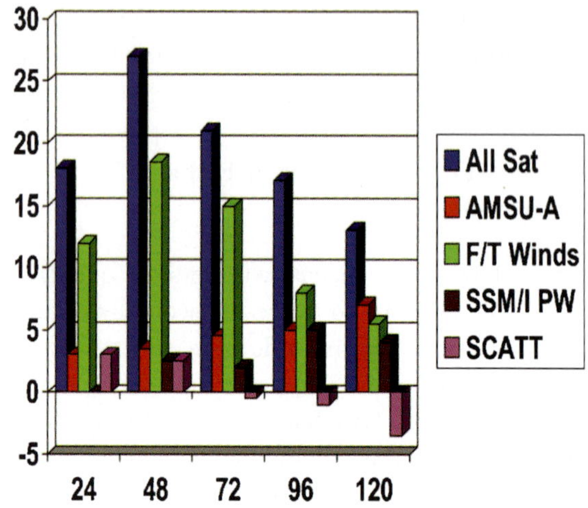

Figure 14. Percent improvement in NOGAPS 24–120-h tropical cyclone track forecast errors for 14 August to 30 September 2004 (Goerss and Hogan, 2006).

7. Summary and Outlook

From multispectral imagery to sophisticated techniques for extracting quantitative information, satellites have become the predominant platform for monitoring tropical cyclones globally. Analysts have refined and improved on methods to interpret the imagery to infer position and intensity. Observations from microwave frequencies have enabled observations beneath the storm's cloud shield and provided researchers and forecasters unique insights into the organizational structure. Improving data assimilation techniques have more successfully exploited the satellite information, leading to improved environmental numerical analyses and forecasts of tropical cyclones.

What does the future hold for our space-based observing system? Advanced multispectral imagers, sounders (infrared and microwave) and scatterometers are envisioned for launch in the next decade. Programs such as COSMIC, GPM, GOES-R, NPOESS, and many other international efforts are planned, and if fully implemented, will dramatically increase the information content available for analyzing tropical cyclones. However, there is a danger that some elements, such as existing microwave imager series, will decrease in terms of number and spatial/temporal sampling over the next 10 years. This stresses the need to foster global cooperation for gathering and freely distributing information as, for example, many countries putting satellites up in next 10 years do not have polar readout sites for near real time access.

The recent success of the Tropical Rainfall Measuring Mission (TRMM) toward improving our physical understanding of tropical weather systems points to the need to keep the satellite-based segment of the global observing system strong. It is important that the evolving space-based observational system keep one step ahead of the demands being placed by advances in numerical weather prediction. If we are successful in this endeavor, and with advancing remote sensing technologies, more sophisticated data assimilation methods, and creative multi-platform data analysis algorithms, the future appears promising for improving our analysis and prediction of tropical cyclones.

Acknowledgements

This chapter draws from the IWTC-VI special session report on the applications of satellite data to tropical cyclone intensity and structure. Further details and references can be found in this report available from WMO. The lead and contributing authors would like to thank Nick Shay (RSMAS), Jim Goerss (NRL), Jim Jung (CIMSS), John Knaff (NESDIS), Tim Olander (CIMSS) and Tetsuo Nakazawa (JMA) for their inputs to this chapter.

References

Cecil, D. J. and E. J. Zipser, 1999: Relationships between tropical cyclone intensity and satellite-based indicators of inner core convection: 85-GHz Ice-scattering signature and lightning. *Mon. Wea. Rev.*, **127**, 103–123.

Cheney, R., L. Miller, R. Agreen, N. Doyle, and J. Lillibridge, 1994: TOPEX/Poseidon the 2 cm solution. *J. Geophys. Res.*, **99**, 24,555–24,563.

Cummings, J. A., 2005: Operational multivariate ocean data assimilation, Part C. *Quart. J. Royal Met. Soc.*, **131**(613), 3583–3604.

DeMaria, M., M. Mainelli, L. K. Shay, J. A. Knaff, and J. Kaplan, 2005: Further improvement to the Statistical Hurricane Intensity Prediction Scheme (SHIPS). *Wea. Forecasting*, **20**, 531–543.

Demuth, J. L., M. DeMaria, and J. A. Knaff, 2006: Improvement of advanced microwave sounding unit tropical cyclone intensity and size estimation algorithms. *J. Appl. Meteor.*, **45**, 1573–1581.

Dunion, J. P. and C. Velden, 2004: The impact of the Saharan Air Layer (SAL) on Atlantic tropical cyclone activity. *Bull. Amer. Meteor. Soc.*, **85**, 353–365.

Dvorak, V. F., 1975: Tropical cyclone intensity analysis and forecasting from satellite imagery. *Mon. Wea. Rev.*, **103**, 420–430.

Dvorak, V. F., 1984: Tropical cyclone intensity analysis using satellite data. NOAA Technical Report NESDIS 11, NTIS, Springfield, VA 22161, 45 pp.

Ebert, E., S. Kusselson, and M. Turk, 2005: Validation of Tropical Rainfall Potential (TRaP) forecasts for Australian tropical cyclones. *Aust. Met. Mag.*, **54**, 121–135.

Edson, R. T., 2002a. Use of scatterometer data in tropical cyclone forecasting and research. Imminent uses and data assimilation, Topic 0.2. *Proceedings of the Fifth WMO International Workshop on Tropical Cyclones (IWTC-V)*, Cairns, Queensland, Australia. WMO/TD No. 1165.

Edson, R. T., 2002b. Tutorial on QUIKSCAT. Special focus topic 1.b. *Proceedings of the Fifth WMO International Workshop on Tropical Cyclones (IWTC-V)*, Cairns, Queensland, Australia. WMO/TD No. 1165.

Evan, A. T., J. Dunion, J. Foley, A. Heidinger, and C. Velden, 2006: New evidence for a relationship between Atlantic tropical cyclone activity and African dust outbreaks. *Geophys. Res. Lett.*, **33**, L19813.

Gaiser, P. G., K. M. St. Germain, E. M. Twarog, G. A. Poe, W. Purdy, D. Richardson, W. Grossman, W. L. Jones, D. Spencer, G. Golba, J. Cleveland, L. Choy, R. M. Bevilacqua, and P. S. Chang, 2004: The WindSat spaceborne polarimetric radiometer: Sensor description and early orbit performance. *IEEE Trans. Geosci. Remote Sens.*, **42**, 11, 2347–2361.

Gallina, G. M. and C. Velden, 2002: Environmental vertical wind shear and tropical cyclone intensity change utilizing enhanced satellite derived wind information. 25th AMS Conf. on Hurr. and Trop. Meteor., San Diego, CA, Amer. Meteor. Soc., pp. 172–173.

Goerss J. S. and T. F. Hogan, 2006: Impact of satellite observations and forecast model improvements on tropical cyclone track forecasts. Preprints, 27th Conf. on Hurricanes and Tropical Meteorology, Monterey, CA, Amer. Meteor. Soc., p. 5.2.

Goerss, J. S., C. S. Velden, and J. D. Hawkins, 1998: The impact of multispectral GOES-8 wind information on Atlantic tropical cyclone track forecasts in 1995. Part II: NOGAPS forecasts, *Mon. Wea. Rev.*, **126**, 1219–1227.

Halterman, R. R. and D. G. Long, 2006: A comparison of hurricane eye determine using standard and ultra-high resolution QuikSCAT winds, *Proceedings of the International Geoscience and Remote Sensing Symposium*, Denver, Colorado.

Hawkins, J. D., T. F. Lee, K. Richardson, C. Sampson, F. J. Turk, and J. E. Kent, 2001: Satellite multi-sensor tropical cyclone structure monitoring. *Bull. Amer. Meteor. Soc.*, **82**, 4, 567–578.

Hawkins, J. D., T. F. Lee, F. J. Turk, K. L. Richardson, S. Miller, C. Sampson, J. Kent, and R. Wade, 2006: NRL tropical cyclone R&D web resource augmentations, 14th AMS Conference on Satellite Meteorology and Oceanography, CD-ROM.

Herndon, D. C. and C. S. Velden, 2004: Upgrades to the UW CIMSS AMSU-based tropical cyclone intensity algorithm, 26th AMS Conf. on Hurricanes and Tropical Meteorology, CD-ROM.

Herndon, D. and C. Velden, 2008: CIMSS TC intensity Satellite Consensus (SATCON). 62nd Interdepartmental Hurricane Conference, Charleston, SC.

Hoshino, S. and T. Nakazawa, 2007: Estimation of tropical cyclone's intensity using TRMM/TMI brightness temperature data. *J. Meteor. Soc. Japan*, **85**, 437–454.

Hou, A. Y., 2005: Update on the Global Precipitation Measurement (GPM) Mission, presented to the National Academies Committee on the Future of Rainfall Measuring Missions, Washington, D.C., October 18, 2005.

Jiang, H. and E. J. Zipser, 2006: Retrieval of hydrometeor profiles in tropical cyclones and convection from combined radar and radiometer observations. *J. Appl. Meteor. Clim.*, **45**, 1096–1115.

Kaplan, J. and M. DeMaria, 2006: Estimating the likelihood of rapid intensification in the Atlantic and E. Pacific basins using SHIPS model data. *Proceedings of the 27th Conference on Hurricanes and Tropical Meteorology*, Monterey, CA.

Kidder, S., M. Goldberg, R. Zehr, M. DeMaria, J. F. W. Purdom, C. S. Velden, N. C. Grody, and S.J. Kusselson, 2000: Satellite Analysis of tropical cyclones using the Advanced Microwave Sounding Unit (AMSU). *Bull. Amer. Meteor. Soc.*, **81**, 1241–1259.

Knaff, J. A., T. A. Cram, A. B. Schumacher, J. P. Kossin, and M. DeMaria, 2008: Objective identification of annular hurricanes. *Wea. Forecasting*, **23**, 17–28.

Knaff, J. A., J. P. Kossin, and M. DeMaria, 2003: Annular hurricanes. *Wea. Forecasting*, **18**, 204–223.

Knaff, J. A. and M. DeMaria, 2006: A multi-platform satellite tropical cyclone wind analysis system. AMS 14th Conference on Satellite Meteorology and Oceanography. 29 January–3 February, Atlanta, GA.

Kossin, J., H. Berger, J. D. Hawkins, and T. Cram, 2006: Development of a secondary eyewall formation index for improvement of tropical cyclone intensity forecasting, Preprints, 60th Interdepartmental Hurricane Conference, Mobile, AL.

Kossin, J. and M. Sitkowski, 2008: An objective tool for identifying hurricane secondary eyewall formation. *Proceedings of the 62nd Interdepartmental Hurricane Conference*, Charleston, SC.

Kossin, J. P., J. A. Knaff, H. I. Berger, D. C. Herndon, T. A. Cram, C. S. Velden, R. J. Murnane, and J. D. Hawkins, 2007a: Estimating hurricane wind structure in the absence of aircraft reconnaissance. *Wea. Forecasting*, **22**, 89–101.

Kossin, J. P., K. R. Knapp, D. J. Vimont, R. J. Murnane, and B. A. Harper, 2007b: A globally consistent reanalysis of hurricane variability and trends. *Geophys. Res. Lett.*, **34**, L04815.

Kummerow, C., *et al.*, 2000: The status of the Tropical Rainfall Measuring Mission (TRMM) after two years in orbit. *J. Appl. Meteor.*, **39**, 1965–1982.

Kummerow, C., W. Barnes, T. Kozu, J. Shiue, and J. Simpson, 1998: The Tropical Rainfall Measuring Mission (TRMM) sensor package. *J. Atmos. Oceanic Technol.*, **15**, 809–817.

Lee, T. F., F. J. Turk, J. D. Hawkins, and K. A. Richardson, 2002: Interpretation of TRMM TMI images of tropical cyclones. *Earth Interactions E-Journal*, **6**, 3.

Liu, W. T., 2002: Progress in scatterometer application, *J. Oceanogr.*, **58**, 121–136.

Lonfat, M., F. D. Marks, and S. S. Chen, 2004: Precipitation distribution in tropical cyclones using the Tropical Rainfall Measuring Mission (TRMM) microwave imager: A global perspective. *Mon. Wea. Rev.*, **132**, 1645–1660.

Mainelli-Huber, M., 2000: On the role of the upper ocean in tropical cyclone intensity change. M.S. Thesis, University of Miami, Miami, FL, 73 pp.

Mainelli, M., M. DeMaria, L.K. Shay, and G. Goni, 2008: Application of oceanic heat content estimation to operational forecasting of recent category 5 hurricanes, *Wea. Forecasting*, **23**(1), 3–16.

Maturi, E., *et al.*, 2008: NOAA's Geostationary Operational Environmental Satellites — Sea Surface Temperature (GOES-SST) Products. Accepted in *Bull. Amer. Meteor. Soc.*

May, D. and W. Osterman, 1998: Satellite-derived sea surface temperatures: Evaluation of GOES-8 and GOES-9 multispectral imager retrieval accuracy. *J. Atmos. Oceanic Technol.*, **15**, 788–797.

May, D., M. Parmeter, D. Olszewski, and B. McKenzie, 1998: Operational processing of satellite sea surface temperature retrievals at the naval oceanographic office. *Bull. Amer. Meteor. Soc.*, **79**, 397–407.

Merchant, C. J., O. Embury, C. P. Old, S. N. MacCallum, E. Maturi, A. R. Harris, and J. Mittaz, 2008: Sea surface temperature estimation from the Geostationary Operational Environmental Satellite 12 (GOES 12) (manuscript submitted to *J. Atmos. Oceanic Technol.*).

Mitrescu, C., S. D. Miller, T. L'Ecuyer, J. Hawkins, J. Turk, and P. T. Partain, 2008: Near real-time applications of CloudSat data. *J. Appl. Meteor.*, **47**, No. 7, 1982–1994.

Mueller, K. J., M. DeMaria, J. A. Knaff, J. P. Kossin, and T. H. VonderHaar, 2006: Objective estimation of tropical cyclone wind structure from infrared satellite data. *Wea. Forecasting*, **21**, 990–1005.

Olander, T. and C. S. Velden, 2007: The Advanced Dvorak Technique (ADT) — Continued development of an objective scheme to estimate TC intensity using geostationary IR satellite imagery. *Wea. Forecasting*, **22**, 287–298.

Reynolds, R., *et al.*, 2007: Daily high-resolution-blended analyses for sea surface temperature. *J. Climate*, **20**, 5473–5496.

Shay, L. K., G. J. Goni, and P.G. Black, 2000: Effects of a warm oceanic feature on Hurricane Opal. *Mon. Wea. Rev.*, **125**, 1366–1383.

Shay, L. K., 2008: Air-sea interactions (Chapter 4). In *Global Perspectives of Tropical Cyclones*, ed. Johnny C. L. Chan, Elsevier Press International, Oxford, UK, 45 pp.

Smith, *et al.*, 2004: International Global Precipitation Measurement (GPM) program and mission: an overview. In *Measuring Precipitation from Space: EURAINSAT and the Future*, V. Levizzani and F. J. Turk, eds., Kluwer: Dordrecht, The Netherlands.

Soden, B. J., C. S. Velden, and R. E. Tuleya, 2001: The impact of satellite winds on experimental GFDL hurricane model forecasts. *Mon. Wea. Rev.*, **129**, 835–852.

TAO: Terrestrial, Atmospheric and Oceanic Sciences, 2000: Special Issue on Applications of the Constellation Observing System for Meteorology, Ionosphere and Climate (COSMIC), **11**(1).

Velden, C. S., T. L. Olander, and S. Wanzong, 1998a: The impact of multispectral GOES-8 wind information on Atlantic tropical cyclone track forecasts in 1995. Part I: Dataset methodology, description, and case studies. *Mon. Wea. Rev.*, **126**, 1202–1218.

Velden, C. S., T. Olander, and R. M. Zehr, 1998b: Development of an objective scheme to estimate tropical cyclone intensity from digital geostationary satellite imagery. *Wea. Forecasting*, **13**, 1720186.

Velden, C., *et al.*, 2005: Recent innovations in deriving tropospheric winds from meteorological satellites. *Bull. Amer. Meteor. Soc.*, **86**, 205–223.

Velden, C., *et al.*, 2006: The Dvorak tropical cyclone intensity estimation technique: A satellite-based method that has endured for over 30 years. *Bull. Amer. Meteor. Soc.*, **87**, 1195–1210.

Wacker, R. S. and C. S. Velden, 2004: Correcting for precipitation scattering effects in satellite-based passive microwave tropical cyclone intensity estimates, 13th AMS Conf. on Satellite Meteorology and Oceanography, CD-ROM.

Wimmers, A. J. and C. Velden, 2007: MIMIC: A new approach to visualize satellite microwave imagery of tropical cyclones. *Bull. Amer. Meteor. Soc.*, **88**, 1187–1196.

Zapotocny, T., J. A. Jung, J. F. LeMarshall, and R. E. Treadon, 2008: A two-season impact study of four satellite data types and Rawinsonde data in the NCEP global data assimilation system. *Wea. Forecasting*, **23**, 80–100.

Zehr, R. M. and J. A. Knaff, 2007: Atlantic major hurricanes, 1995–2005 — Characteristics based on best-track, aircraft, and IR images. *J. Climate*, **20**, 5865–5888.

Tropical Cyclone Satellite-Focused Web Sites

http://www.nrlmry.navy.mil/TC.html
http://cimss.ssec.wisc.edu/tropic2/
http://rammb.cira.colostate.edu/products/tc_realtime/
http://www.ssd.noaa.gov/
http://www.nasa.gov/mission_pages/hurricanes/main/index.html
http://manati.orbit.nesdis.noaa.gov/quikscat/

Chapter 8

Aircraft Observations of Tropical Cyclones

Sim D. Aberson and Joseph Cione
NOAA/AOML/Hurricane Research Division,
4301 Rickenbacker Causeway, Miami, FL 33149 USA
sim.aberson@noaa.gov
joe.cione@noaa.gov

Chun-Chieh Wu
Department of Atmospheric Sciences, National Taiwan University,
No. 1, Sec. 4, Roosevelt Rd., Taipei 10673, Taiwan
cwu@typhoon.as.ntu.edu.tw

Michael M. Bell
Naval Postgraduate School and National Center for Atmospheric Research,
Department of Meteorology, 589 Dyer Road, Root Hall,
Monterey, CA 93943-5114 USA
mmbell@nps.edu

Jeffrey Halverson
University of Maryland-Baltimore County, Department of Geography
and Environmental Systems, 211-K Sondheim Hall,
1000 Hilltop Circle, Baltimore, MD 21250 USA
jeffhalv@umbc.edu

Chris Fogarty
Environment Canada, National Laboratory for Marine and Coastal Meteorology,
45 Alderney Drive, Dartmouth, Nova Scotia B2Y 2N6, Canada
chris.fogarty@ec.gc.ca

Martin Weissmann
Deutsches Zentrum für Luft- und Raumfahrt, Institut für Physik der Atmosphäre,
Oberpfaffenhofen, Münchnerstr. 20,
D-82230 Wessling, Germany
martin.weissmann@dlr.de

Nine different types of aircraft are currently in use to observe tropical cyclones and their environments for operations and research. The following is a description of those aircraft, their instrumentation, and the field programs with which they have been involved.

1. Introduction

Tropical cyclones (TCs) have been observed by aircraft since the first flight piloted by United States Army Air Force Lt. Col. Joseph P. Duckworth on 17 July, 1943 (Sumner, 1943). The United States military conducted the first dedicated research flight into a TC (Wexler, 1945; Wood, 1945), and a subsequent flight examined the upper troposphere of a 1947 Atlantic TC (Simpson, 1954). Regular aircraft reconnaissance of the tropics by the United States military to investigate whether TCs were developing, and their locations and intensities, soon began in both the Atlantic and Western Pacific basins. With the advent of satellite data, the importance of the airborne search for developing TCs decreased, and methods to locate and estimate the intensities of TCs were derived (Dvorak, 1975; Velden *et al.*, 1998; Velden *et al.*, 2006; Olander and Velden, 2007). However, the remotely sensed satellite location and intensity data remain less accurate than in situ aircraft measurements.

In addition to operational uses, aircraft have a long history in TC research. After the disastrous Atlantic hurricane season of 1954, the United States Weather Bureau created the National Hurricane Research Project (NHRP) to advance TC science and improve forecasts. NHRP soon pioneered quantitative observations with instrumented aircraft that shaped the modern understanding of tropical cyclones (Dorst, 2007).

The following is a discussion of the operational reconnaissance, surveillance and research aircraft that have participated in TC field programs since 1997 and the instrumentation available on them. This chapter is meant to summarize the types of observations available and their uses for research and operations, with broad information available in the references.

2. NOAA P-3s

In the mid-1970s, the United States National Oceanic and Atmospheric Administration (NOAA) purchased two WP-3D (P-3) aircraft, modified United States Navy maritime reconnaissance aircraft that are versions of the civilian Lockheed Electra (Aberson *et al.*, 2006). Proposed uses of the aircraft were observations of TC structure and dynamics, participation in TC modification experiments, and monitoring of TC formation, all with the goal of improving the TC forecast and warning process to limit damage and loss of life. Since then, the NOAA Aircraft Operations Center has maintained the P-3s to be among the premier meteorological research aircraft in the world.

When the P-3s were procured, only basics of TC structure were known (Shea and Gray, 1973). The hurricane was considered to be an approximately circular, axisymmetric vortex with nearly constant wind speeds extending to at least 400 hPa, and with deviations from axisymmetry considered to be gusts or largely due to vortex motion. The eyewall was thought to be vertically aligned up to 5–10 km above the surface, and funnel-shaped above. The eye was observed to have warm, moist air below an inversion, with clear, dry air above and ice clouds streaming inward near the top. Both the eyewall and rainbands were understood to have convergent airflow in low levels and outflow above. Rainbands were seen to move inward toward the eye and to cause intensification of eyewall convection. Numerical modeling of TCs was limited to idealized 2-dimensional studies due to a lack of adequate computer power and observations of the dynamic and thermodynamic fields in three dimensions, and only primitive techniques to assimilate the data into model initial conditions (Simpson and Riehl, 1981).

Partially as a result of observations from the P-3s, the description and understanding of TC behavior and structure were revolutionized. The NOAA P-3s have conducted missions into nearly 150 TCs in the Atlantic and Eastern Pacific Oceans and near Australia. Data were obtained on the micro- to the synoptic-scale, and data analyses have led to many new insights about TC structure, dynamics, thermodynamics, and environmental interactions (Aberson *et al.*, 2006).

Sophisticated instrumentation installed on and developed for the P-3s is unique among meteorological airborne platforms (see Jorgensen, 1984 for details of the original instrument installation). At the time of delivery, both P-3s measured flight-level temperature, pressure, and moisture and had a state-of-the-art inertial navigation system allowing for flight-level wind vector calculations with 0.1–0.3 ms^{-1} accuracy; these data are transmitted in real-time at a rate of up to 0.1 Hz, and are available in research mode at 1 Hz. As electronic equipment and expendable probes became more accurate, smaller, faster and lighter, new instrumentation enhanced the potential to gather comprehensive TC research data sets. For example, P-3 instruments now include the airborne infrared radiation thermometer for remotely determining flight-level temperature in clear air (Barnes *et al.*, 1991; Eastin *et al.*, 2002), a stepped frequency microwave radiometer (SFMR) for estimating surface wind speed (Uhlhorn and Black, 2003), and Global Positioning System (GPS) navigation. Omega dropwindsondes (ODWs-Govind, 1975) were first used in hurricane research in 1982 to provide wind and thermodynamic profiles, but did not report in cloudy regions nor provide wind measurements in the boundary layer. The GPS dropwindsonde (Hock and Franklin, 1999), first used in 1996, provided reliable wind and thermodynamic profiles in the TC eyewall and down to the surface for the first time, thus revolutionizing understanding of structures in the eyewall and boundary layer. Each aircraft also carries a workstation for airborne radar and dropwindsonde processing (Griffin *et al.*, 1992).

Among the most advanced meteorological equipment on the P-3s are three digital radars. Two record the reflectivity signal: a 5.5-cm (C-band) wavelength (lower fuselage) radar extends below the fuselage and measures the horizontal distribution at all azimuth angles; a 3-cm (X-band) wavelength radar is in the tail (TA) and determines the reflectivity distribution along rays oriented either perpendicular to the aircraft track or at angles fore and aft within 25° of the aircraft heading (Marks, 1985). A third radar with a 5.2 cm (C-band) wavelength is in the nose and scans horizontally back and forth looking forward; the pilots use it to avoid turbulent weather, and the data have not been recorded since 1987. Important additions to the radar systems first occurred in 1980 when a prototype Doppler signal processing system was added to the TA radar on one P-3 (Jorgensen *et al.*, 1983) and again in 1988–1989 when refined radar data systems were installed on both aircraft (Marks, 2003). In 2006, an automated quality control and analysis system was applied to the TA radar data, and three-dimensional wind and reflectivity analyses are transmitted in near-realtime for use by operational forecasters and for testing data assimilation schemes in high-resolution regional hurricane model systems (Gamache *et al.*, 1997; Gamache, 2005).

The P-3s are also equipped with Particle Measuring System probes (a Forward Scattering Spectrometer Probe and 2-Dimensional Optical Array Probes) mounted on the wing tips to distinguish between water and ice particles and estimate their sizes (Knollenberg, 1970; Black and Hallett, 1986; Gayet *et al.*, 1993; Baumgardner and Korolev, 1997). Starting in 2008, one P-3 will have a full complement of particle imaging and non-imaging probes manufactured by Droplet Measurement Technologies (DMT). These probes include the Cloud Combination Probe, imaging particles from 0.025 mm–1.6 mm in diameter and (using forward scattering methods) measuring the diameters and number concentrations of cloud droplets in the range 0.003 mm–0.050 mm. The DMT Precipitation Imaging probe images particles in the diameter range 0.100 mm–6.4 mm. The DMT Cloud Aerosol Spectrometer is a non-imaging probe that measures particles from 0.00061 mm–0.005 mm (forward scattering) and 0.001–0.1 mm (backward scattering). These probes offer the capability to measure the size distribution and number concentrations of cloud particles over three decades of diameters with unparalleled accuracy. In addition, five field mills will allow for measurement of the three-dimensional electric field vector. This combination of instruments will provide unprecedented observations of the role of water and ice microphysics and electrification on TC convection and structure.

In addition to obtaining atmospheric data, the P-3s have the ability to collect upper ocean observations. Each aircraft has 24 chutes in its fuselage for external ejection of airborne expendable bathythermographs (Boyd, 1987; Boyd and Linzell, 1993), current profilers (Sanford, 1971), and conductivity temperature depth profilers to measure ocean temperature, current, and salinity to a depth of ~300 m. A Scanning Radar Altimeter (SRA) measures surface directional wave spectra and rain rate, and transmits these data to operational forecasters in near real-time (Walsh *et al.*, 2002; Wright *et al.*, 2001). These provide important data at and near the ocean-atmosphere interface where the TC derives its energy, allowing for greater understanding and modeling of air-sea interaction processes. All these instruments make the two P-3s the most advanced and comprehensive of all meteorological aircraft currently in use.

3. NOAA G-IV

In 1982, TC numerical modeling was in its infancy, and the most accurate track guidance available to forecasters was statistical. NOAA began to test the hypothesis that additional vertical wind and thermodynamic profiles in the TC environment (the "synoptic flow") would improve analyses and numerical model forecasts that provide operational guidance to human forecasters. The data obtained during P-3 research missions through 1996 helped to reduce errors in TC track forecasts from global and regional dynamical models significantly (Burpee *et al.*, 1996).

The success of this research program led to the development of a new generation of dropwindsonde (Hock and Franklin, 1999) and the acquisition of a high-altitude jet aircraft, the Gulfstream-IV-SP (G-IV) as an operational platform for conducting synoptic surveillance missions. These operational flights, along with concurrent research into optimal targeting and sampling strategies and data assimilation, continue to improve National Centers for Environmental Prediction (NCEP) operational global model track forecasts significantly (Aberson, 2003). A complete G-IV flight and the required dropwindsondes cost about $40,000, far less than the estimated average of $1 million needed to evacuate just six miles of coastline for landfall (Aberson *et al.*, 2006).

Because of limited flight time and dropwindsondes, optimal sampling strategies must be found. Aberson (2003) found that the most rapidly growing modes in the NCEP global model can be represented by regions of relatively large spread in the ensemble forecasting system 850–200 hPa mean wind. In testing the efficacy of this sampling strategy, the important question is whether the removal of at least one-third of the dropwindsonde data that are in non-target regions has no negative impact on subsequent forecasts. More specifically, the question is whether targeting techniques define regions where mobile platforms such as aircraft should concentrate observations and other regions in which observations will have little or no impact on subsequent forecasts. When entire regions of large ensemble spread are sampled by regularly spaced observations, a statistically significant improvement to forecast tracks above that from assimilating far more data obtained symmetrically about the storm results.

Majumdar *et al.* (2006) and Reynolds *et al.* (2006) compared this and other techniques during the 2004 hurricane season. They concluded that singular vector targets were often located within an annulus around 500 km from the storm center, and/or at long distances from the storm in locations associated with features such as mid-latitude troughs. The targets suggested by the Ensemble Transform Kalman Filter were localized around the storm in either upstream or downstream locations where wind or temperature errors were correlated with those near the verification region. Evaluation of these techniques with observing system experiments using data from operational missions is ongoing.

In addition to synoptic surveillance, by 2008, a TA Doppler radar and SFMR will be available on the G-IV to complement those instruments on the P-3s. The radar will measure reflectivity and three-dimensional wind field similarly to the P-3 TA radars, but from a much higher altitude. This may allow for three-dimensional wind measurements in the TC cirrus outflow. The SFMR will measure surface wind speed and rain rate below the aircraft in a similar manner to the instruments available on the P-3 and C-130 aircraft, but at a lower resolution and with a larger footprint due to the aircraft altitude and speed. The G-IV will have the ability to transmit these data to the Global Telecommunications System (GTS) in real time.

4. Taiwanese ASTRA SPX

At least partially due to the success of the NOAA G-IV, the Dropwindsonde Observations for Typhoon Surveillance near the Taiwan Region (DOTSTAR) began in 2003 (Wu *et al.*, 2005, 2007b). Data from dropwindsondes released from the ASTRA aircraft have substantially improved track model forecasts (Wu *et al.*, 2007b). However, the average improvement to typhoon track prediction in the Geophysical Fluid Dynamics Laboratory (GFDL) hurricane model was insignificant, likely due to the dropwindsonde signal being swamped by the synthetic vortex data used during the initialization. As a result, Chou and Wu (2008) devised an alternative technique to combine dropwindsonde and synthetic data in a mesoscale model in order to further boost the effectiveness of dropwindsonde data with the implanted storm vortex.

An additional targeting technique, the Adjoint-Derived Sensitivity Steering Vector (ADSSV), has been devised and tested during these missions (Wu *et al.*, 2007a). By appropriately defining a response function to represent the deep-layer-mean (850–300 hPa) steering flow around the TC at the verifying time, sensitive locations at the observing time are clearly shown. The ADSSV relates these sensitive areas to the steering flow at the verifying time. The ADSSV direction and magnitude at a given location indicates the change in the steering flow due to a perturbation in the initial conditions at that location and the extent of the sensitivity, respectively. This adjoint sensitivity can be used to identify important regions and dynamical features affecting the TC track and is helpful in defining target regions for surveillance. The ADSSV has been used to design missions for DOTSTAR as well as for Atlantic hurricanes.

5. United States Air Force WC130-Js

The United States Air Force Reserve 53rd Weather Reconnaissance Squadron operates a fleet of ten Lockheed-Martin C-130-J aircraft that are mostly used for operational hurricane fixing (finding of the center location and storm intensity) in the Atlantic and Eastern and Central Pacific basins when TCs threaten land. Like the NOAA P-3s, these aircraft measure and transmit flight-level wind, pressure, temperature, and moisture data, as well as surface wind speed measurements from SFMR and data from dropwindsondes. As the vast majority of operational flights are conducted by these aircraft, they are the workhorses of aircraft reconnaissance and provide the most fixes yearly to operational centers; the data are also frequently used for research. In addition to these tasks, these aircraft also participate in synoptic surveillance missions to augment coverage from the G-IV, and in research experiments by providing air deployment of large platforms such as buoys and floats necessary for ocean observations (Black *et al.*, 2007).

6. Naval Research Laboratory P-3

The Naval Research Laboratory (NRL) NP-3D "Orion" aircraft is similar in design to the NOAA P-3s, but with a different instrumentation suite. It is maintained and deployed by the NRL Military Support Division and the Scientific Development Squadron-1 (VXS-1) "Warlocks" (formerly Flight Support Detachment) for a variety of military and scientific missions. After the decommission of the National Center for Atmospheric Research (NCAR) Electra aircraft in 2001, an agreement between NRL, the National Science Foundation, and NCAR to use this aircraft as the new platform for the Electra Doppler Radar (ELDORA) was reached. An Airborne Vertical Atmospheric Profiling System (AVAPS) system was installed in 2005 to allow for dropwindsonde processing (Hock and Franklin, 1999), augmenting the radar and in situ measurements obtained from the inertial navigation system.

ELDORA is a dual beam, X-band (3.2 cm) meteorological research TA radar developed jointly by NCAR and the Centre de Recherches en Physique de L'Environnement Terrestre et Planetaire (Hildebrand *et al.*, 1994), first deployed in 1993 (Hildebrand *et al.*, 1996). It has two flat-plate, slotted waveguide antennas with oval beamwidths of $1.8° \times 2.0°$ that spin about the longitudinal axis of the aircraft. The antenna rotodome is a radome that rotates with the antennas. One antenna points slightly fore of the aircraft and the other slightly aft. As the P-3 flies, these two radars trace conical helixes through the atmosphere, providing three-dimensional precipitation and kinematic structure at resolutions finer than 400 m. This is similar to the NOAA TA radars using the Fore/Aft Scanning Technique (F/AST), but with $\sim$4 times the spatial resolution due to the dual-transmitters and a higher antenna rotation rate (up to 24 revolutions per minute). Though the high rotation rate improves the along-track resolution, it limits the number of independent samples in a radar volume necessary for accurate Doppler radial velocity measurements (i.e., the time for each beam to obtain a significant sample is shorter). The velocity statistics are improved with a complex waveform pulse within which four separate frequency sub-pulses are coded, enabling the received signals to be processed individually. A dual pulse repetition frequency technique increases the maximum unambiguous velocity to greater than 60 ms-1 while maintaining a 75 km unambiguous range, a highly desirable attribute in high-wind regimes (Wakimoto *et al.*, 1996; Wakimoto *et al.*, 2004).

The first use of the NRL P-3 for TC observations was in 2005, during the Hurricane Rainband and Intensity Change Experiment (RAINEX–Houze *et al.*, 2006). In cooperation with the three NOAA aircraft, the P-3 investigated hurricane intensity changes associated with eyewall-rainband interactions. Nine missions were flown into Hurricanes Katrina, Ophelia, and Rita, documenting many different stages of the TC life cycle and a variety of rainband configurations. One of the unique aspects of this project was the coordination among the multiple aircraft, including transmitted composite LF radar images from the NOAA P-3s to the NRL P-3 for navigation assistance. The high level of detail in TC precipitation and kinematic structure is a significant observational advancement resulting from the use of ELDORA in RAINEX.

This platform's second deployment to study TCs took place in the summer of 2008 as part of the THORPEX Pacific Area Regional Campaign (T-PARC) and Tropical Cyclone Structure 2008 (TCS-08) field programs. A Doppler wind lidar was added to the instrumentation to provide wind profiles in clear air below the aircraft, and the missions focused on the genesis, structural changes, and extratropical transition of TCs, extending the life cycle observed by the NRL P-3 and ELDORA.

Figure 1. Aircraft described in this chapter. Clockwise from top left: NOAA P-3 (top) and G-IV (bottom), AF C-130, NRL P-3, NASA DC-8, Aerosonde, DLR Falcon, Canadian Convair, NASA ER-2, and Taiwanese ASTRA.

7. NASA DC-8 and ER-2

For more than a decade, the National Aeronautics and Space Administration (NASA) Science Mission Directorate (SMD) has utilized high-altitude research aircraft to conduct scientific investigations of TC genesis, intensity change and landfall impacts. The principal airborne platforms include the McDonald Douglas DC-8 and Lockheed Martin Earth Resources-2 (ER-2).

NASA began flying sorties into TCs in Tropical Cyclone Oliver in January, 1993, in the Coral Sea during the Tropical Oceans/

Global Atmosphere Program. Since then, NASA has conducted four major hurricane research field programs. These include: the Convection and Moisture Experiments (CAMEX-3 and CAMEX-4) of 1998 and 2001 (Kakar *et al.*, 2006); the Tropical Cloud Systems and Processes experiment (TCSP) of 2005 (Halverson *et al.*, 2007); and the NASA African Monsoon Multidisciplinary Activities (NAMMA) of 2006. CAMEX was designed to intensively study TC structure and intensity change. TCSP was focused on tropical cyclogenesis, but also sampled the unusually intense, early-season Atlantic Hurricanes Dennis and Emily. NAMMA sought to investigate the metamorphosis of African easterly waves into named TCs as they emerged off the coast of western Africa.

The field experiments were jointly planned and implemented in coordination with the NOAA P-3s and G-IV aircraft. The advantages of the combined effort include the complementary nature of scientific objectives contained in both the NASA and NOAA research plans. The main advantage is that the NOAA P-3 aircraft fly low, and the NASA aircraft fly high, enabling investigation of the complete TC vertical structure from 70,000 feet down through the oceanic mixed layer. Coordinated surveys involving multiple aircraft require diligent planning, real-time mission monitoring and reliable radio contact to keep as many as five aircraft arranged in a moving, vertical stack. Flights nominally last 6–8 hours with the DC-8 sampling the storm between 30,000–40,000 feet and the ER-2 operating entirely above the storm close to 70,000 feet.

The aircraft payloads consist of a diverse suite of *in situ* and remote sensing instruments varying with each field campaign. They test new technologies, some of which may undergo transition to satellite platforms. From its very high vantage point, the ER-2 can serve as a "virtual satellite" to target observations to the particular scientific objective. The DC-8 spends most of its time within the TC core and is best suited to sample electrical fields, cloud and precipitation particles, and the three-dimensional water vapor structure. Missions may be timed to coincide with one or more NASA satellite overpasses in order to provide crucial calibration/validation information.

Several classes of instruments are common to most missions on the DC-8. Instruments designed to measure the *in situ* temperature, moisture, pressure, and winds comprise the Meteorological Measurement System, and a 4-channel AVAPS (Hock and Franklin, 1999) allows for GPS dropwindsonde processing. One or more hygrometers such as the Diode Laser Hygrometer (Heymsfield *et al.*, 1998) and Jet Propulsion Laboratory Laser Hygrometer (May, 1998) provide precise measurements of water vapor, with high accuracy at very low ambient humidity. The Lidar Atmospheric Sensing Experiment (Miloshevich *et al.*, 2004) provides high resolution, vertical profiles of water vapor mixing ratio and aerosol scatter. A series of nadir-viewing Doppler radars (Airborne Mapping Radar [Durden *et al.*, 2003]; Precipitation Radar [Im *et al.*, 2002] and the Dual-Frequency Airborne Precipitation Radar [APR-2]) provide fine-scale vertical precipitation (reflectivity) structure, particularly within stratiform rain regions, as well as radial wind measurements, and have served as prototypes of instruments flown on the Tropical Rainfall Measurement Mission (Simpson *et al.*, 1988; Kummerow *et al.*, 1998) and the planned Global Precipitation Mission (Smith *et al.*, 2002). The DC-8 also typically hosts a suite of cloud microphysics probes; these include the Cloud Particle Imager (Heymsfield and Miloshevich, 2003) to measure ice particle size, Counterflow Virtual Impactor (Twohy *et al.*, 2003) to measure cloud liquid water content, and the Cloud and Aerosol Particle Characterization (McFarquhar and Heymsfield, 1997) system of probes that measures the particle size distribution from 30 μm to above 1 mm in 30-μm increments.

Instrument payloads on the ER-2 are geared toward remote sensing and retrieval of atmospheric vertical profiles. The Advanced Microwave Precipitation Radiometer (Spencer *et al.*, 1994) uses a combination of passive microwave frequencies at high spatial and temporal resolution to map instantaneous rain rate. The ER-2 Doppler Radar (Heymsfield *et al.*, 1996) uses both nadir and forward-directed active beams to detect fine-scale precipitation structure and radial motions within convective and stratiform precipitation. Electrical field mill measurements are provided by the Lightning Imaging Package. The High Altitude Monolithic Microwave Integrated Circuit Sounding Radiometer (Lambrigtsen and Riley, 2002) obtains temperature and humidity profiles, liquid water profiles and rain rate. Vertical temperature profiles, tropopause height and high-altitude lapse rate are also retrieved using the Microwave Temperature Profiler (Denning *et al.*, 1989). The Moderate Resolution Imaging Spectroradiometer Airborne Simulator (Roskovensky *et al.*, 2004) provides multispectral visible and infrared cloudtop imagery at 50 m horizontal resolution. Recent additions to the ER-2 payload suite include the fully autonomous ER-2 High Altitude Dropsonde and the Cloud Radar System (Li *et al.*, 2004; Halverson *et al.*, 2006) that provides high sensitivity radar reflectivity, Doppler velocity and derived measurements of cloud ice.

Many of these instruments are being specifically adapted for use on the Global Hawk Unmanned Airborne System (UAS) to provide long-duration, high-altitude missions during the NASA Hurricane and Severe Storm Sentinel research initiative. The goal is to observe tropical cyclogenesis and intensification from tropical storm to hurricane during an extended time period, operating in a surveillance rather than a reconnaissance mode.

8. Canadian Convair-580

The first research aircraft mission into a hurricane (Michael) undergoing extratropical transition was arranged by the Meteorological Service of Canada using the Canadian National Research Council Convair-580 aircraft on 19 October, 2000. A comprehensive description of the aircraft specifications and of experiences during the Michael flight is provided in Wolde *et al.*, (2001). At the time, Michael was near maximum intensity and undergoing rapid transition to an extratropical storm. Since this was the first time the aircraft was being flown into a storm of that nature, the pilots chose not to fly directly through the storm center which contained vigorous convection, lightning and very strong wind in the lower levels near the boundary layer as measured by dropwindsondes and airborne conventional radar (Abraham *et al.*, 2004). The Convair-580 aircraft was equipped with cloud microphysical probes which collected information on the ice and cloud liquid water concentrations within the storm. Ice water concentrations on the western side of the storm in the stratiform precipitation region were very high, and more characteristic of concentrations seen in the center of continental cumulonimbus clouds. Data from the mission led to improved initialization techniques for numerical models (Fogarty *et al.*, 2007).

Other missions were conducted into Tropical Storms Karen (2001) and Isabel (2003), and Hurricane Juan (2003). The flight in Hurricane Juan south of Nova Scotia provided an unprecedented look at a strong hurricane making landfall in Canada with destructive effects (Fogarty *et al.*, 2006).

9. DLR Falcon

In 2008, the Falcon 20 aircraft of the Deutsches Zentrum für Luft-und Raumfahrt (DLR) in Oberpfaffenhofen, Germany was deployed to the West-Pacific for typhoon research as part of T-PARC. The main objectives were targeted observations to improve typhoon forecasts and the investigation of extra-tropical transition of tropical cyclones and downstream impact in the mid-latitudes. In coordination with the NRL

P-3, the USAF WC-130 and DOTSTAR unique typhoon observations from genesis through recurvature and extra-tropical transition, with simultaneous observations of the typhoon core and environment and sensitive areas related to the midlatitude flow or the subtropical high often highlighted by singular vector sensitivity calculations, were obtained. Typhoons Sinlaku and Jangmi were observed for 13 and 9 days, respectively. One particular focus of the data analysis will be to evaluate which observational strategies are most effective for typhoon forecasting and observations in which regions have the highest potential for forecast improvements.

Options for instrumentation including dropwindsondes, air chemistry probes, aerosol and particle probes and various lidars. For T-PARC it was configured with a dropwindsonde system, a scanning coherent wind lidar (Weissmann *et al.*, 2005) and a four-wavelength differential absorption lidar for water vapour that includes a high spectral-resolution channel to measure aerosol extinction. The individual observations are volume averages and thus seen as more useful (representative) information for assimilation in numerical models than point observations like dropwindsondes. This was pointed out by Weissmann and Cardinali (2007) using wind lidar observations during the Atlantic THORPEX Regional Campaign (A-TReC) and will be an important subject of future T-PARC research.

One restriction of lidars is that they can not penetrate optically thick clouds; if clouds are at the flight level, the aircraft must reduce its altitude to get observations. Conversely, in clear air, the wind lidar must have sufficient aerosols to determine winds or data gaps occur. The optimal mix of lidar and other observations (i.e., dropwindsondes) therefore requires further research.

10. UASs

The interaction between the ocean and the hurricane is important, complex, and not well handled in current observing systems and models. Specifically, the hurricane depends on the ocean to supply the heat necessary to form and maintain the storm. The detailed process by which a storm obtains heat from the ocean and ultimately converts it into kinetic energy (i.e., strong winds) is complex and not well understood due to the limited availability of detailed observations within the storm near the air-sea interface. To dramatically improve understanding of this rarely observed region, these low-level observations must be obtained. Unmanned aerial systems (UAS) designed to penetrate and sample the violent low-level hurricane environment can fill this critical data void.

Two groups, a NOAA/NASA team and one based in Taiwan, have used the Aerosonde (Holland *et al.*, 2001) to this end. Aerosonde missions are conducted in completely robotic mode with the aircraft monitored by a ground controller. The Aerosondes used in TC missions have measured temperature, humidity, pressure, and wind speed and direction, but they have the ability to use interchangeable instrument payloads allowing for other measurements. During a mission into hurricane Noel (2007), NOAA/NASA set endurance (17.5 h) and minimum altitude (82 m) records for a UAS in a TC environment. During the missions into Typhoon Longwang (2005), four passes through the eyewall were completed, and maximum 10-min (1-min) wind speed of 58.6 ms^{-1} ($62\,\mathrm{m\,s}^{-1}$) were measured (Lin and Lee, 2008). The aircraft is therefore able to complete long missions into severe tropical cyclones globally.

The UAS data obtained during recent missions conducted by the NOAA/NASA team (Ophelia in 2005 and Noel in 2007) have proved to be very useful. In both cases, near-surface observations were reported to the National Hurricane Center in real-time to be used during operational warning. The UAS found the highest wind speeds in both storms despite the presence of manned aircraft at the same time. Analysis of the data from these missions is proceeding, and further missions with various UAS platforms are planned. A major goal is to transmit the data operationally for assimilation into high-resolution tropical cyclone model systems and to greatly expand the amount of data available at the undersampled air-sea interface in the TC core.

11. Conclusions

Sampling of TC cores is currently performed routinely only in the Atlantic basin, though reconnaissance in the eastern and central Pacific is occasionally tasked. In all other TC-prone ocean basins, in situ information is rarely obtained. Though satellites are certainly capable of some measurements similar to those obtained by aircraft, the ability of satellites to obtain the necessary data both accurately and with the necessary spatial and temporal resolution in all TC basins remains in doubt. Aircraft remain the sole platform by which data can be obtained for operations, and are a testing ground for new instruments that may be put into orbit on future satellite systems, as well as for calibration and validation of current and future satellite observing systems, not only for operational purposes, but to improve TC monitoring for the assessment of climatic changes. These data are also required for the development of high-resolution model systems, especially next-generational data assimilation systems, that are necessary to provide accurate forecasts of TC intensity, rainfall and structure.

References

Aberson, S. D., M. L. Black, R. A. Black, R. W. Burpee, J. J. Cione, C. W. Landsea, and F. D. Marks, Jr., 2006: Thirty years of tropical cyclone research with the NOAA P-3 aircraft. *Bull. Amer. Met. Soc.*, **87**, 1039–1055.

Aberson, S. D., 2003: Targeted observations to improve operational tropical cyclone track forecast guidance. *Mon. Wea. Rev.*, **131**, 1613–1628.

Abraham, J., W. Strapp, C. Fogarty, and M. Wolde, 2004: Extratropical transition of Hurricane Michael: An aircraft investigation. *Bull. Amer. Meteor. Soc.*, **85**, 323–339.

Barnes, G. M., J. F. Gamache, M. A. LeMone, and G. J. Stossmeister, 1991: A convective cell in a hurricane rainband. *Mon. Wea. Rev.*, **119**, 776–794.

Baumgardner, D. and A. Korolev, 1997: Air speed corrections for optical array probe sample volumes. *J. Atmos. Ocean. Technol.*, **14**, 1224–1229.

Black, P. G., E. A. D'Asaro, W. M. Drennan, J. R. French, P. P. Niiler, T. B. Sanford, E. J. Terrill, E. J. Walsh, and J. A. Zhang, 2007: Air-sea exchange in hurricanes: Synthesis of observations from the coupled boundary layer air-sea transfer experiment. *Bull. Amer. Met. Soc.*, **88**, 357–374.

Black, R. A. and J. Hallett, 1986: Observations of the distribution of ice in hurricanes. *J. Atmos. Sci.*, **43**, 802–822.

Boyd, J. D., 1987: Improved depth and temperature conversion equations for Sippican AXBTs. *J. Atmos. Ocean. Technol.*, **4**, 545–551.

Boyd, J. D. and R. S. Linzell, 1993: Evaluation of the Sparton tight-tolerance AXBTs. *J. Atmos. Ocean. Technol.*, **10**, 892–899.

Burpee, R. W., J. L. Franklin, S. J. Lord, R. E. Tuleya, and S. D. Aberson, 1996: The impact of Omega dropwindsondes on operational hurricane track forecast models. *Bull. Amer. Met. Soc.*, **77**, 925–933.

Chou, K.-H. and C.-C. Wu, 2008: Development of the typhoon initialization in a mesoscale model - combination of the bogused vortex and the dropwindsonde data in DOTSTAR. *Mon. Wea. Rev.*, in press.

Denning, R. F., S. L. Guidero, G. S. Parks, and B. L. Gary, 1989: Instrument description of the airborne microwave temperature profiler. *J. Geophys. Res.*, **94**, 16757–16765.

Dorst, N. M., 2007: The National Hurricane Research Project: 50 years of research, rough rides, and name changes. *Bull. Amer. Met. Soc.*, **88**, 1566–1588.

Durden, S. L., E. Im, Z. S. Haddad, and L. Li, 2003: Comparison of TRMM precipitation radar and airborne radar data. *J. Appl. Meteor.*, **42**, 769–774.

Dvorak, V. F., 1975: Tropical cyclone intensity analysis and forecasting from satellite imagery. *Mon. Wea. Rev.*, **103**, 420–430.

Eastin, M. D., P. G. Black, and W. M. Gray, 2002: Flight-level thermodynamic instrument wetting errors in hurricanes: Part I: Observations. *Mon. Wea. Rev.*, **130**, 825–841.

Fogarty, C. T., R. J. Greatbatch, and H. Ritchie, 2007: The use of a vortex insertion technique to simulate the extratropical transition of Hurricane Michael (2000). *Wea. Forecasting*, **22**, 480–500.

Fogarty, C. T., R. J. Greatbatch, and H. Ritchie, 2006: The role of anomalously warm sea surface temperatures on the intensity of Hurricane Juan (2003) during its approach to Nova Scotia. *Mon. Wea. Rev.*, **134**, 1484–1504.

Gamache, J. F., 1997: Evaluation of a fully-three dimensional variational Doppler analysis technique. *Preprints, 28th Conference on Radar Meteorology*, Austin, TX, *Amer. Meteor. Soc.*, pp. 422–423.

Gamache, J. F., 2005: Final Report on JHT Project entitled: Real-Time Dissemination of Hurricane Wind Fields Determined from Airborne Doppler Radar Data. Available at http://www.nhc.noaa.gov/jht/2003-2005reports/DOPLRgamache_JHTfinalreport.pdf

Gayet, J. F., P. R. A. Brown, and F. Albers, 1993: A comparison of in-cloud measurements obtained from six PMS 2D-C probes. *J. Atmos. Ocean. Technol.*, **10**, 180–194.

Govind, P. K., 1975: Dropwindsonde instrumentation for weather reconnaissance aircraft. *J. Appl. Meteor.*, **14**, 1512–1520.

Griffin, J. S., R. W. Burpee, F. D. Marks, and J. L. Franklin, 1992: Real-time airborne analysis of aircraft data supporting operational hurricane forecasting. *Wea. Forecasting*, **7**, 480–490.

Halverson, J. and co-authors, 2007: NASA's tropical cloud systems and processes experiment. *Bull. Amer. Met. Soc.*, **88**, 867–882.

Halverson, J. B., J. Simpson, G. Heymsfield, H. Pierce, T. Hock, L. Ritchie, 2006: Warm core structure of Hurricane Erin diagnosed from high altitude dropsondes during CAMEX-4. *J. Atmos. Sci.*, **63**, 309–324.

Heymsfield, A. J., L. M. Miloshevich, G. Sachse, C. Twohy, and S. Oltmans, 1998: Upper tropospheric relative humidity observations and implications for cirrus ice nucleation. *Geophys. Res. Lett.*, **25**, 1343–1346.

Heymsfield, L. M. and L. M. Miloshevich, 2003: Parameterizations for the cross-sectional area and extinction of cirrus and stratiform ice cloud particles. *J. Atmos. Sci.*, **60**, 936–956.

Heymsfield, G. M. and co-authors, 1996: The EDOP radar system on the high-altitude NASA ER-2 aircraft. *J. Atmos. Ocean. Technol.*, **13**, 795–809.

Hildebrand, P. H., C. L. Frush, and C. Walther, 1994: Design of the ELDORA/ASTRAIA airborne Doppler weather radar: Goals, design, and first field test. *Proc. IEEE*, **82**, 1873–1890.

Hildebrand, P. H., W.-C. Lee, C. A. Walther, C. Frush, M. Randall, E. Loew, R. Neitzel, R. Parsons, J. Testud, F. Baudin, and A. LeCornec, 1996: The ELDORA/ASTRAIA airborne Doppler weather radar: High resolution observations from TOGA COARE. *Bull. Amer. Met. Soc.*, **77**, 213–232.

Hock, T. F. and J. L. Franklin, 1999: The NCAR GPS dropwindsonde. *Bull. Amer. Meteor. Soc.*, **80**, 407–420.

Holland, G. J., P. J. Webster, J. A. Curry, G. Tyrell, D. Gauntlett, G. Brett, J. Becker, R. Hoag, and W. Vaglienti, 2001: The Aerosonde robotic aircraft: A new paradigm for environmental observations. *Bull. Amer. Met. Soc.*, **82**, 889–901.

Houze, R. A., S. S. Chen, B. F. Smull, W.-C. Lee, and M. M. Bell, 2007: Hurricane intensity change and eyewall replacement. *Science*, **315**, 1235–1239.

Im, E. and co-authors, 2002: Second-Generation Precipitation Radar (PR-2). Final Rep. JPL D-22997, NASA Earth Science Instrument Incubator Program, Jet Propulsion Laboratory, California Institute of Technology, Pasadena, CA, 164 pp.

Jorgensen, D. P., 1984: Mesoscale and convective-scale characteristics of mature hurricanes. Part I: General observations by research aircraft. *J. Atmos. Sci.*, **41**, 1267–1285.

———, P. H. Hildebrand, and C. L. Frush, 1983: Feasibility test of an airborne pulse-Doppler meteorological radar. *J. Appl. Meteor.*, **22**, 744–757.

Kakar, R., M. Goodman, R. Hood, and A. Guillory, 2006: Overview of the Convection and Moisture Experiment (CAMEX). *J. Atmos. Sci.*, **63**, 5–18.

Knollenberg, R. G., 1970: The optical array: An alternative to scattering or extinction for airborne particle size determination. *J. Appl. Meteor.*, **9**, 86–103.

Kummerow, C., W. Barnes, T. Kozu, J. Shiue, and J. Simpson, 1998: The Tropical Rainfall Measuring Mission (TRMM) sensor package. *J. Atmos. Ocean. Technol.*, **15**, 809–817.

Lambrigtsen, B. H. and A. L. Riley, 2002: Microwave scattering observed in convective cells during CAMEX-4. *Preprints, 22d Conf. on Hurricanes and Tropical Meteorology*, San Diego, CA, Amer. Meteor. Soc., 515–516.

Li, L., G. M. Heymsfield, P. E. Racette, T. Lin, and E. Zenker, 2004: A 94-GHz cloud radar system on a NASA high-altitude ER-2 aircraft. *J. Atmos. Ocean. Technol.*, **21**, 1378–1388.

Lin, P.-H. and C.-S. Lee, 2008: *The eyewall-penetration reconnaissance observation of Typhoon Longwang (2005) with unmanned aerial vehicle*, Aerosonde. *J. Atmos. Ocean. Technol.*, **25**, 15–25.

Majumdar, S. J., S. D. Aberson, C. H. Bishop, R. Buizza, M. S. Peng, and C. A. Reynolds, 2006: A comparison of adaptive observing guidance for Atlantic tropical cyclones. *Mon. Wea. Rev.*, **134**, 2354–2372.

Marks, F. D., Jr., 1985: Evolution of the structure of precipitation in Hurricane Allen (1980). *Mon. Wea. Rev.*, **113**, 909–930.

Marks, F. D., 2003: State of the science: Radar view of tropical cyclones. *Radar and Atmospheric Science: A Collection of Essays in Honor of David Atlas*, edited by R. M Wakimoto and R. C. Srivastava, Meteorological Monographs, **52**, AMS, Boston, MA, 33–74.

May, R. D., 1998: Open-path, near-infrared tunable diode laser spectrometer for atmospheric measurements of H2O. *J. Geophys. Res.*, **103**, 19 161–19 172.

McFarquhar, G. M. and A. J. Heymsfield, 1997: Parameterization of tropical cirrus ice crystal size distributions and implications for radiative transfer: Results from CEPEX. *J. Atmos. Sci.*, **54**, 2187–2200.

Miloshevich, L. M., A. Paukkunen, H. Vömel, and S. J. Oltmans, 2004: Development and validation of a time-lag correction for Vaisala radiosonde humidity measurement. *J. Atmos. Ocean. Technol.*, **21**, 1305–1327.

Olander, T. L. and C. S. Velden, 2007: The advanced Dvorak technique: Continued development of an objective scheme to estimate tropical cyclone intensity using geostationary infrared satellite imagery. *Wea. Forecasting*, **22**, 287–298.

Reynolds, C. A., M. S. Peng, S. J. Majumdar, S. D. Aberson, C. H. Bishop, and R. Buizza, 2007: Interpretation of adaptive observing guidance for Atlantic tropical cyclones. *Mon. Wea. Rev.*, **135**, 4006–4029.

Roskovensky, J., K. N. Liou, T. J. Garrett, and D. Baumgardner, 2004: Simultaneous retrieval of aerosol and thin cirrus optical depths using MODIS airborne simulator data during CRYSTAL-FACE and CLAMS. *Geophys. Res. Lett.*, **31**, L18110.

Sanford, T. B., 1971: Motionally-induced electric and magnetic fields in the sea. *J. Geophys. Res.*, **76**, 3476–3492.

Simpson, J., R. F. Adler, and G. R. North, 1988: A proposed Tropical Rainfall Measuring Mission (TRMM) satellite. *Bull. Amer. Meteor. Soc.*, **69**, 278–295.

Simpson, R. H., 1954: Hurricanes. *Scientific American*, **109**, 22–37.

Simpson, R. H. and H. Riehl, 1981: The hurricane and its impact. Louisiana State University Press, Baton Rouge, LA, 398 pp.

Smith, E. A., A. Mehta, and J. M. Shepherd, 2002: Description of Global Precipitation Measurement (GPM) Mission. NASA/Goddard Space Flight Center, GPM Report Series 6, Tech Memo., 25 pp.

Spencer, R. W., R. E. Hood, F. J. Lafontaine, E. A. Smith, R. Platt, J. Galliano, V. L. Griffin, and E. Lobl, 1994: High resolution imaging of rain systems with the Advanced Microwave Precipitation Radiometer. *J. Atmos. Ocean. Technol.*, **11**, 849–857.

Sumner, H. C., 1943: North Atlantic hurricanes and tropical disturbances of 1943. *Mon. Wea. Rev.*, **71**, 179–183.

Twohy, C. H., J. W. Strapp, and M. Wendisch, 2003: Performance of a counterflow virtual impactor in the NASA Icing Research Tunnel. *J. Atmos. Ocean. Technol.*, **20**, 781–790.

Uhlhorn, E. W. and P. G. Black, 2003: Verification of remotely sensed sea surface winds in hurricanes. *J. Atmos. Ocean. Technol.*, **20**, 99–116.

Velden, C. S., T. L. Olander, and R. M. Zehr, 1998: Development of an objective scheme to estimate tropical cyclone intensity from digital geostationary satellite infrared imagery. *Wea. Forecasting*, **13**, 172–186.

———, and co-authors, 2006: The Dvorak tropical cyclone intensity estimation technique: A satellite-based method that has endured for over 30 years. *Bull. Amer. Met. Soc.*, **87**, 1195–1210.

Wakimoto, R. M., W.-C. Lee, H. B. Bluestein, C.-H. Liu, P. H. Hildebrand, 1996: ELDORA observations during VORTEX 95. *Bull. Amer. Met. Soc.*, **77**, 2949–2950.

Wakimoto, R. M., H. Cai, and H. V. Murphey. 2004: The Superior, Nebraska, Supercell During BAMEX. *Bull. Amer. Met. Soc.*, **85**, 1095–1106.

Walsh, E. J., C. W. Wright, D. Vandemark, W. B. Krabill, A. W. Garcia, S. H. Houston,

S. T. Murillo, M. D. Powell, P. G. Black, and F. D. Marks Jr., 2002: Hurricane directional wave spectrum spatial variation at landfall. *J. Phys. Ocean.*, **32**, 1667–1684.

Weissmann, M. and C. Cardinali, 2007: The impact of airborne Doppler lidar measurements on ECMWF forecasts. *Quart. J. Roy. Meteor. Soc.*, **133**, 107–116.

Weissmann, M., R. Busen, A. Dörnbrack, S. Rahm, and O. Reitebuch, 2005: Targeted observations with an airborne wind lidar. *J. Atmos. Ocean. Technol.*, **22**, 1706–1719.

Wexler, H., 1945: The structure of the September, 1944, hurricane when off Cape Henry, Virginia. *Bull. Amer. Meteor. Soc.*, **26**, 156–159.

Wolde, M., D. Marcotte, J. Jordan, J. Aitken, J. Abraham, and J.W. Strapp, 2001: The first Canadian experience with research flight operations in hurricane extratropical transition. *Canadian Aeronautics and Space Journal*, **47**(3), 179–189.

Wood, F. B., 1945: A flight into the September, 1944, hurricane off Cape Henry, Virginia. *Bull. Amer. Meteor. Soc.*, **26**, 153–156.

Wright, C. W., E. J. Walsh, D. Vandemark, W. B. Krabill, A. W. Garcia, S. H. Houston, M. D. Powell, P. G. Black, and F. D. Marks, 2001: Hurricane directional wave spectrum spatial variation in the open ocean. *J. Phys. Ocean.*, **31**, 2472–2488.

Wu, C.-C., J.-H. Chen, P.-H. Lin, and K.-H. Chou, 2007a: Targeted observations of tropical cyclone movement based on the adjoint-derived sensitivity steering vector. *J. Atmos. Sci.*, **64**, 2611–2626.

Wu, C.-C., K.-H. Chou, P.-H. Lin, S. D. Aberson, M. S. Peng, and T. Nakazawa, 2007b: The impact of dropwindsonde data on typhoon track forecasts in DOTSTAR. *Wea. Forecasting*, **22**, 1157–1176.

Wu, C.-C., P.-H. Lin, S. D. Aberson, T.-C. Yeh, W.-P. Huang, J.-S. Hong, G.-C. Lu, K.-C. Hsu, I.-I. Lin, K.-H. Chou, P.-L. Lin, and C.-H. Liu, 2005: Dropwindsonde Observations for Typhoon Surveillance near the Taiwan Region (DOTSTAR): An overview. *Bull. Amer. Met. Soc.*, **86**, 787–790.

PART III

Climate Variations of Tropical Cyclone Activity

Chapter 9

Tropical Cyclones and Climate Change: A Review

Thomas Knutson
NOAA/Geophysical Fluid Dynamics Laboratory,
201 Forrestal Road, Princeton, NJ 08542, USA
tom.knutson@noaa.gov

Chris Landsea
NOAA/NWS/National Hurricane Center,
11691 S.W. 17th Street, Miami, FL 33165-2149, USA
chris.landsea@noaa.gov

Kerry Emanuel
Massachusetts Institute of Technology,
77 Mass. Avenue, Cambridge, MA 02139, USA
emanuel@mit.edu

A review of the science on the relationship between climate change and tropical cyclones (TCs) is presented. Topics include changes in aspects of tropical climate that are relevant to TC activity; observed trends and low-frequency variability of TC activity; paleoclimate proxy studies; theoretical and modeling studies; future projections; roadblocks to resolution of key issues; and recommendations for making future progress.

1. Introduction

This report reviews the current state-of-the-science on possible relationships between climate variability and change and tropical cyclone (TC) activity on different time scales. The emphasis here is on multidecadal variability and longer time scale changes, including possible anthropogenic influences on tropical cyclone activity.

The present report is not a *scientific assessment* of anthropogenic influence on TCs, as we do not present likelihood statements on various key issues, as is done for example in the IPCC (Intergovernmental Panel on Climate Change) assessments. Several previous assessment reports contain specific statements on possible anthropogenic influence on TCs. One such assessment (Henderson-Sellers *et al.*, 1998) concluded: (i) there was no clear evidence for long-term trends in TC activity; (ii) the potential intensity (PI) of storms would remain the same or increase by 10–20% (in terms of central pressure fall) for a doubling of CO_2) although uncertainties remained with PI approaches; (iii) little could be said about the future distribution of intensities or about future frequencies of TCs; and (iv) the broad geographic regions of cyclogenesis and of occurrence of TCs were unlikely to change significantly. A 10–20% increase in central pressure fall would correspond to a smaller (roughly 5–10%) percentage increase in terms of maximum surface wind speeds. Other more recent assessments of this issue include WMO (2006), IPCC (2007), and CCSP (2008). WMO (2006) concluded that while there was evidence both for and against the existence of a detectable anthropogenic signal in the tropical cyclone record, no firm conclusion could be made on that point. IPCC (2007) concluded that it was "more likely than not" (meaning >50% chance) that human activities have contributed to the observed trend in intense tropical cyclone activity in some regions since 1970. CCSP (2008) concluded that "There is evidence suggesting a human contribution to recent changes in hurricane activity ... though a confident assessment will require further study." A review and discussion of the tropical cyclones and climate problem from the viewpoint of the climate change detection and attribution methodologies is given by Walsh *et al.* (2008).

The present report attempts to briefly summarize earlier published research, but with a greater emphasis on work published since the Henderson-Sellers *et al.* (1998) and Walsh (2004) reviews were completed, including more recent published work on trends in observed past TC metrics and updated TC climate projections. We do not include a focus on societal impacts of TCs in this review, while recognizing that this topic presents a further important perspective on historical records (e.g., Pielke, 2005; Pielke *et al.*, 2008) and projections.

2. Background on Tropical Climate Changes Relevant to Tropical Cyclone Activity

There is substantial evidence that the large-scale environment in which hurricanes form and evolve is changing as a result of anthropogenic emissions of greenhouse gases and aerosols. The most recent assessment of this problem is given in IPCC (2007), which concludes that most of the globally averaged temperature increase since the mid-20th century is very likely (>90% chance) due to the observed increase in human-caused greenhouse gas concentrations. A study of subsurface ocean data by Barnett *et al.* (2005) concluded that an anthropogenic warming signal is penetrating the world oceans, in broad agreement with model simulations that include the greenhouse gas forcing. Model-based attribution of early 20th century global warming (e.g., 1900–1944) to specific causes is more ambiguous, with various studies suggesting significant contributions from multiple factors, including increased greenhouse gases, solar variability, decreasing volcanic activity, and internal climate variability (e.g., Stott *et al.*, 2000; Delworth and Knutson, 2000; Meehl *et al.*, 2004; Knutson *et al.*, 2006).

2.1. *Tropical Atlantic Warming: Natural Variability vs. Anthropogenic Forcing*

Some aspects of the tropical climate of relevance to hurricane activity appear to be changing in a trend-like fashion. Based primarily on climate model assessments, there is increasing evidence that sea surface temperature increases in tropical cyclone basins, as reported in several studies (e.g., Emanuel, 2005a; Webster *et al.*, 2005), are at least partly a response to long-term increases in greenhouse gas concentrations. For example, Santer *et al.* (2006) find that observed SST increases in the Atlantic and North Pacific tropical cyclogenesis regions during the 20th century are unlikely to be due solely to unforced variability of the climate system, but are more realistically simulated in experiments using estimated historical climate forcing. Their internal climate variability assessment and external forcing results are made more robust by their use of 22 different climate models and two observed SST reconstructions. Using the models for which individual forcing experiments were available, they find that the human-induced change in greenhouse gas forcing is the main cause of the 20th century warming, and particularly of the late 20th century warming. Their results support earlier regional surface temperature trend assessments based on a more limited set (two) of models (Knutson *et al.*, 2006) or on a more limited set of forcings (Karoly and Wu, 2005) both of which found model-based support for anthropogenically forced 20th century warming trends in the tropics and other regions. In Knutson *et al.*, the simulations where anthropogenic and natural forcing agents were evaluated separately indicated significantly closer agreement with observed trends over much of the tropical oceans in the anthropogenic forcing runs than in the natural forcing or internal climate variability runs. The anthropogenic forcings in these experiments included changes in well-mixed greenhouse gases, ozone, and aerosols, as well as land use change, whereas natural forcings included solar variations and aerosols from volcanic eruptions.

For the tropical North Atlantic, the roles of naturally occurring oscillations versus radiative forcing variability and trends on tropical Atlantic SSTs have also been evaluated using statistical modeling approaches. Goldenberg *et al.* (2001) proposed that a naturally occurring oscillation of the climate system, termed the Atlantic Multidecadal Oscillation (AMO), was responsible for pronounced multidecadal variations in Atlantic major hurricane counts since the 1940s. They noted that major hurricane counts covaried with fluctuations in both an AMO index derived from detrended SST data and with vertical wind shear in the Main Development Region (MDR, a term they used to describe the tropical Atlantic and Caribbean Sea region between 10° and 20°N). Mann and Emanuel (2006) suggested an alternative interpretation by noting that late summer SSTs in the Atlantic MDR closely track, on long time scales, surface temperatures averaged over the entire Northern Hemisphere, with substantial warming over the 20th century. Using a statistical modeling approach, they suggested that most of the low-frequency (multidecadal) variation and warming trend in MDR SSTs had been produced by changing radiative forcing, as opposed to being part of a naturally occurring oscillation. They showed, using a statistical regression approach, that the evolution of summertime tropical North Atlantic sea surface temperature through the 20th century can be represented as a combination of global mean surface temperature and sulfate aerosol forcing, which is concentrated mostly in the northern hemisphere. In general, the climate forcing from aerosols, including their indirect effects via clouds and precipitation processes, is much more uncertain than the forcing due to increasing greenhouse gases (e.g., IPCC 2007).

In another statistical analysis, Trenberth and Shea (2006) show that the method of construction of AMO indices can have a significant impact on AMO anomaly values for various time periods. They proposed that the index be constructed as a residual after removal of a near-global (60 N–60 S) SST component, as opposed to residual from a linear trend (as in Goldenberg *et al.*, 2001). Using this approach, they derive a revised AMO index with smoothed anomaly values of about ± 0.2°C and a transition from negative to positive values in the mid 1990s. However, the contribution of their low-pass filtered AMO anomalies to the record summer of 2005 values is quite small (<0.1°C), and the anomalies from 1870 to 1900, are also much smaller (closer to zero) compared to those using the method of Goldenberg *et al.* In removing the global or near-global mean SST from the Atlantic SST series, both Mann and Emanuel (2006) and Trenberth and Shea (2006) include the Atlantic SST in their computation of the global or near-global mean. As can be inferred from Mann and Emanuel, this procedure could have the effect of slightly artificially damping the AMO amplitude because if there were SSTs variations in the Atlantic due to the AMO that also contribute to the global mean fluctuations, those would be partially damped as a consequence of removing the global mean variations.

Enfield and Mestas Nuñez (2000) have previously published a means of deriving an "Atlantic Multidecadal Mode" based not on linear trend removal, but on a complex empirical orthogonal function (CEOF) decomposition, in which a "Global Warming Mode" is distinguished from the Atlantic Multidecadal and Pacific interdecadal modes based on the CEOF modal decomposition. In their analysis the AMO is the third CEOF in a dataset from which an ENSO-related CEOF had previously been removed (i.e., in practical terms, the AMO is their fourth CEOF). It should be noted that different SST reconstructions have been used by various investigators, which may also contribute to differences seen in the resulting analyses (e.g., Santer *et al.*, 2006).

The existence of a robust AMO-like internal mode of the climate system is supported by some climate models, which simulate internal modes of variability that resemble the observed EOFs in several respects (Delworth and Mann, 2000; Knight *et al.*, 2005). Zhang *et al.* (2007) construct a synthetic multidecadal AMO-like change in an idealized coupled climate model and show that the Atlantic Ocean, when forced to have such a multidecadal variation, can modulate northern hemisphere mean temperature. They argue that it is plausible to interpret the long-term northern hemisphere mean temperature deviations, apart from the long-term trend, as due to either radiative forcing or to an AMO-like internal variation in ocean heat transport. Zhang (2007) presents some tentative evidence for the latter interpretation, based on the covariation of Atlantic Ocean surface and subsurface temperatures in observations and models. Model based studies also indicate that such multidecadal variations of Atlantic SSTs can have important impacts on vertical wind shear in the Atlantic MDR (Vitart and Anderson, 2001; Zhang and Delworth, 2006; Knight *et al.*, 2006), which Goldenberg *et al.* (2001) proposed can then affect Atlantic hurricane activity. Knaff (1997) described the physical mechanisms that relate vertical shear, SSTs, and sea level pressures in the Atlantic MDR. A recent review of the AMO-like variability in the Atlantic region is given in Delworth *et al.* (2007).

Enfield and Cid-Serrano (2008) also attempt to separate a global warming signal and an AMO signal in the data. Unlike the Goldenberg *et al.* (2001) approach, they first approximate the global warming signal not as a linear trend, but as a quadratic fit which better matches the increased rate of warming in recent years. This approach does not neglect AMO influences in basins outside of the Atlantic and does not dampen the AMO amplitude by not accounting for the Atlantic's contribution toward the global anomalies. Their results suggest that similar magnitudes are present for both the global warming and AMO in the Atlantic SSTs.

Santer *et al.* (2006) also provide indirect evidence that the observed low frequency variability of tropical Atlantic SST may include a substantial contribution from internal climate variability. They note that most of the 22 CMIP3 (Coupled Model Intercomparison Project 3) models they examined underestimated tropical Atlantic decadal SST variability, with average errors on the order of 50%; the Northwest Pacific was more realistically depicted. Note that some, but not all, of these models were run with indirect effects of anthropogenic aerosol forcing and/or with volcanic forcing.

Kossin and Vimont (2007) proposed that another mode of Atlantic variability, termed the Atlantic Meridional Mode (AMM), was robust in observed data and was correlated with systematic shifts in Atlantic tropical cyclogenesis regions and several measures of hurricane activity. The AMM is an apparent coupled ocean-atmosphere mode of variability that is strongly present in the tropical Atlantic and characterized by time-varying meridional SST gradients near the equator, cross-equatorial wind variability, and north-south displacements of the Intertropical Convergence Zone (ITCZ). Spatial patterns similar to those for the AMM, as derived from observed fields, also arise as normal modes of an idealized coupled dynamical system representing a tropical atmospheric boundary layer coupled to a motionless "slab" ocean. Kossin and Vimont propose that the AMO affects Atlantic hurricanes mainly though its excitation of the AMM. Wang *et al.* (2008) suggest the AMO influences Atlantic hurricanes through modulation of the Atlantic Warm Pool (Gulf of Mexico, Caribbean Sea and western tropical North Atlantic). In a study of large-scale atmospheric circulation indices, Bell and Chelliah (2006) have statistically linked multi-decadal changes in Atlantic hurricane activity to a series of large-scale circulation features, all correlated to a multi-decadal circulation signal derived from empirical orthogonal function (EOF) analysis of upper tropospheric velocity potential. The indices of Bell and Chelliah's modes and Goldenberg *et al.*'s (2001) index of Atlantic MDR vertical shear are of insufficient length to determine whether they have a cyclical or trend-like character, and the historical changes in these indices have not yet been formally attributed to internal variability or forced variability of the climate system or a combination of the two.

Wang and Lee (2008) show that from 1949 to 2006 wind shear in the MDR was positively correlated with a "global warming mode" derived from EOF analysis of global SST, although they did not attempt to attribute the warming mode to radiative forcing or internal climate variability. They argue that a weak (nonsignificant) secular decrease in U.S. landfalling hurricanes since the mid 1800s has been, at least since 1950, associated with increasing tropospheric vertical wind shear in the tropical Atlantic Main Development Region, as inferred from NCEP Reanalysis data. One source of uncertainty not explicitly examined by Wang and Lee is the possible impact of changing observing practices and capabilities on the long-term wind shear statistics (e.g., trends) in the NCEP reanalysis data. This could be an important topic area for follow-on study.

Saunders and Lea (2005) and Elsner *et al.* (2000; 2006) find statistical links between U.S. landfalling hurricane activity and large-scale circulation anomalies, although long-term climate trends in their predictors have not been firmly established. Elsner and Jagger (2008) report a significant statistical relationship between U.S. and Caribbean hurricane activity and solar activity as inferred from sunspot numbers.

Using a partial correlation statistical analysis, Elsner *et al.* (2006) examined the relationships between global temperature, tropical Atlantic SST, and an Atlantic Power Dissipation Index (PDI, see section 3.1) on high-frequency interannual time scales. They concluded that the positive influence of global temperature on PDI was limited to an indirect connection through the tropical Atlantic SSTs. After controlling for the effect of tropical Atlantic SSTs on PDI, the correlation of PDI with global temperatures was slightly negative. This result was consistent with idealized modeling studies (Shen *et al.*, 2000) and with statistical analyses of ENSO-Atlantic TC relationships (Tang and Neelin, 2004), both indicating inhibiting effects of tropospheric stabilization on TC intensity or frequency. Elsner (2006, 2007) uses Granger causality statistical analysis to demonstrate that global mean temperature can be used to predict North Atlantic SST but not the other way around. This, he argues, supports the hypothesis that greenhouse gases are the causal forcing agent for global temperatures and thus for North Atlantic SSTs and hurricanes. To eliminate nonstationarity in the data, as is required by the method, Elsner time differences both the global temperature and North Atlantic SST series before performing the causality tests. Since time differencing acts as a high-pass filter on the data, his conclusions about Granger causality strictly apply only to the higher frequency fluctuations that remain in the data and not necessarily to multidecadal time scales variability and trends.

Saunders and Lea (2008) use a multiple linear regression analysis to attempt to identify the relative roles of sea surface warming and atmospheric winds on the recent increase in Atlantic hurricane activity. While they find that both factors contribute substantially, they did not attempt to identify whether greenhouse warming in particular contributed to the hurricane activity increase.

2.2. *Tropospheric Water Vapor and Temperature Trends*

Hurricane modeling studies (Shen *et al.*, 2000) and theory (Emanuel, 1987; Holland, 1997) indicate that hurricane maximum intensities are sensitive to atmospheric temperature conditions in addition to SSTs, and modeling studies (section 6.3.1) suggest that enhanced lower tropospheric water vapor associated with climate warming can lead to enhanced rainfall rates in TCs. Therefore, in this section, we discuss evidence for secular changes in tropical atmospheric temperatures and lower tropospheric moisture and their possible causes.

Trenberth *et al.* (2005) have reported a substantial increase (1.3% ±0.3% per decade) in column-integrated atmospheric water vapor over the global oceans (1988 to 2003) as derived from the special sensor microwave imager (SSM/I) satellite data set (see also Soden *et al.*, 2005). Thus, it appears that tropical precipitable water vapor is increasing in a manner consistent with the notion of approximately constant tropical mean relative humidity, and in accord with model simulations of tropical mean relative humidity under warming conditions (e.g., Knutson and Tuleya, 2004, although Vecchi and Soden (2007b) show examples of projected non-uniform regional-scale changes in tropical 700 hPa relative humidity). Despite the relatively short available record, Santer *et al.* (2007) find preliminary evidence, from a formal detection/attribution analysis, that the signal of an anthropogenically driven increase in atmospheric moisture content is already emerging.

The vertical profile of historical tropospheric temperature trends in the tropics has been a subject of considerable debate and discussion in the climate change community. For example, CCSP (2006) reports that while tropical surface temperatures (land and ocean combined) have increased about 0.13°C per decade since 1979, two radiosonde-based and three satellite-based data sets give a range of tropospheric temperature increases of about 0.02°C to 0.19°C. Santer *et al.* (2005) examined the profile of temperature changes for the period 1979–1999 produced by a large ensemble of climate models, all incorporating a range of historical forcings including greenhouse gases and aerosols, with some of the models incorporating volcanic eruptions. The climate models generally simulate an enhanced warming of the tropical upper troposphere relative to the surface, roughly approximating the behavior of moist adiabatic lapse rates. In contrast, the observed vertical profiles of radiosonde-derived atmospheric temperature trends (and some satellite-derived trends) over this period have a distinctly different character from the model simulations, with the tropospheric warming trends that are comparable to or smaller than those at the surface. Finally, Santer *et al.* showed that both models and observations have interannual variations in upper tropospheric temperatures that are enhanced relative to the surface variations. Thus the vertical structure of interannual variations is similar to that of modeled trends (1979–1999), but is in sharp contrast with the vertical structure of observed trends (1979–1999) in tropical tropospheric temperatures. Their results are suggestive of serious remaining problems with radiosonde-derived and/or satellite-derived temperature trends — a conclusion also receiving some support from other recent studies which examined issues with radiosonde-based observations (Sherwood *et al.*, 2005) and satellite-based analyses (Fu *et al.*, 2004; Mears and Wentz, 2005). CCSP (2006) concluded that it was very likely that errors remain in adjusted radiosonde data sets in the troposphere, and that for satellite data, differences in data merging procedures were the main cause of discrepancies in tropospheric temperature trends among those data sets.

The possibility that tropospheric trend estimates from radiosonde-based observations, satellites, and reanalyses may be unreliable should be considered as a caveat when reviewing other published reports on related trend measures. Other thermodynamic measures of tropical climate relevant to hurricane formation which have been examined for possible trends include CAPE and potential intensity. For example, Gettleman *et al.* (2002) found a preponderance of upward trends in tropical CAPE since roughly the early 1960s. DeMott and Randall (2004) examined a larger number of tropical stations over a shorter period (1973–1999) and reported a more evenly divided mixture of increasing and decreasing CAPE trends. Trenberth (2005) questioned the reliability of the radiosonde data in DeMott and Randall's larger sample. Free *et al.* (2004), using a selected set of 14 tropical island radiosonde stations, found only small, statistically insignificant trends in potential intensity over the periods 1975 to 1995 and 1980 to 1995. Emanuel (2007a) reported a 10% increase in Atlantic MDR potential intensity since 1982 based on HadISST and NCEP reanalysis data.

2.3. *Other Sources of Long-Term Variability in Tropical Cyclone Activity*

Other climate variations that have been statistically linked to variations in Atlantic hurricane activity include West African monsoon activity and El Niño (e.g., Gray, 1990; Bell and Chelliah, 2006), and African dust air outbreaks (Evan *et al.*, 2006). Therefore, long-term variations in their behavior, whether anthropogenic or natural in origin, could lead to long-term variability in Atlantic hurricane activity.

Several studies (Rotstayn and Lohmann, 2002; Held *et al.*, 2005; Biasutti and Giannini, 2006) have found that 20th century trends in Sahel rainfall may have been at least partially forced by anthropogenic forcing, albeit through different physical mechanisms. In Rotstayn and Lohmann's study, anthropogenic indirect aerosol forcing, through the interaction of sulfate aerosol with cloud and precipitation processes, causes a pronounced decrease in rainfall in the Sahel. In Held *et al.* (2005) the Sahel drying from the 1950s to the 1980s is simulated by models forced by observed SST variations, and in terms of casual mechanisms, their model results suggest approximately equal contributions of internal climate variability and radiative forcing to the drying trend. The radiative forcing in the model was primarily anthropogenic "direct effect-only" aerosol forcing and increasing greenhouse gases). Biasutti and Giannini (2006) analyzed a large sample of IPCC AR4 models and found a more robust decrease in Sahel rainfall in response to late 20th century aerosol forcing, estimated at roughly one-third of the observed decline. In contrast, they found no consensus among the models on whether greenhouse warming alone contributed to late 20th century Sahel drying. Thus conceivably, human activities could have influenced Atlantic hurricane activity through their effects on Sahel rainfall.

The notion that outbreaks of dry, dust layers from the Sahara could affect Atlantic hurricane activity was hypothesized by Dunion and Velden (2004). In support of this mechanism, Evan *et al.* (2006) find a strong statistical relationship between atmospheric dust episodes over the Atlantic tropical storm regions and measures of Atlantic tropical cyclone activity during 1982–2005. This statistical connection conceivably could reflect a system in which both a drier Sahel and Atlantic TCs are influenced by a common factor: SST gradients in the Atlantic region.

Concerning the influence of El Nino, Vecchi *et al.* (2006) report evidence for a weakening trend in the Walker Circulation in the Pacific during the 20th century, similar to that which occurs during El Niño, and consistent with hindcast predictions by a number of climate models. Whether this has had any impact on Atlantic hurricane activity through remote influences remains unclear. As tentative evidence of a possible link, Wang and Lee (2008) report a statistical association between increasing vertical wind shear in the tropical Atlantic since 1949 and a global warming SST mode, along with a further statistical link to a slight decrease in U.S. landfalling hurricane activity. Vecchi *et al.* (2008) review the current status of the debate over whether greenhouse warming will lead to a more El Nino-like or more La Nina-like mean state in the tropical Pacific. The answers to this debate have important implications for future tropical cyclone activity. Merryfield (2006) reports no conclusive consensus from current (CMIP3) models on whether El Nino variability will increase, decrease, or remain essentially unchanged with future greenhouse warming.

In the Southern Hemisphere, observed trends in the Southern Annular Mode have been at least partly attributed to anthropogenic forcing (Marshall *et al.*, 2004). Pezza and Simmonds (2005), commenting on the rare atmospheric conditions associated with the first reported hurricane in the South Atlantic (Catarina, 2004), suggested that observed and predicted trends in the Southern Annular Mode could increase the probability of such conditions in the future.

The Southern Annular Mode, Pacific Walker circulation, Sahel drought, and African dust studies cited above serve as reminders that the relationship between radiative climate forcing and hurricane response may involve a variety of complex tropics-wide or even global-scale phenomena.

3. Observed Trends and Low-Frequency Variability of Tropical Cyclone Activity

Much of the recent increased attention of the hurricane research community to historical tropical cyclone records has been sparked by two observational studies of low-frequency variability and trends in several measures of tropical cyclone activity (Emanuel, 2005a; Webster *et al.*, 2005), and the follow-on studies by these and other investigators. These are reviewed according to several general topic areas that have emerged.

3.1. *Atlantic Basin Storm Intensities and Power Dissipation*

Emanuel (2005a) developed a "Power Dissipation Index" (PDI) of tropical cyclones, based on the time integrated cube of the estimated maximum sustained surface wind speeds, some of which are inferred from central pressure reports, for the Atlantic and Northwest Pacific tropical cyclone basins from the late 1940s to 2003. After adjusting for time-dependent biases due to changes in measurement and reporting practices, Emanuel reported an approximate doubling of the PDI over the period of record, with contributions from apparent increases in both intensity and mean storm duration. The two low-pass filtered PDI series in his study were significantly correlated with large-scale tropical SST indices in both basins. Landsea (2005) noted that much of the large post-2000 upswing in Emanuel's (2005) Atlantic PDI results from a problem with the treatment of endpoints from time filter used. He also noted that there is no evidence for a trend from 1949–2004 — similar to published time series of major hurricane counts or Accumulated Cyclone Energy (ACE, Bell and Chelliah 2006) — provided that wind speeds early in the record are not adjusted downward, as had been originally proposed by Landsea (1993).

More recently, Emanuel (2007a) provided additional evidence supporting the need for a downward adjustment of intensities in the early part of the record, based on available combined wind-pressure measurements, and in agreement with Landsea's (1993) earlier analysis. Emanuel's (2007a) revised analysis (Fig. 1, updated through 2007) shows that both tropical Atlantic SST and PDI increased substantially between the ~1950 and 2005, and that their low-pass-filtered variations are well-correlated, particularly from 1970 on. Landsea (2005) has argued that PDI values from the 1940s to the 1960s in the HURDAT database are likely to be substantially biased

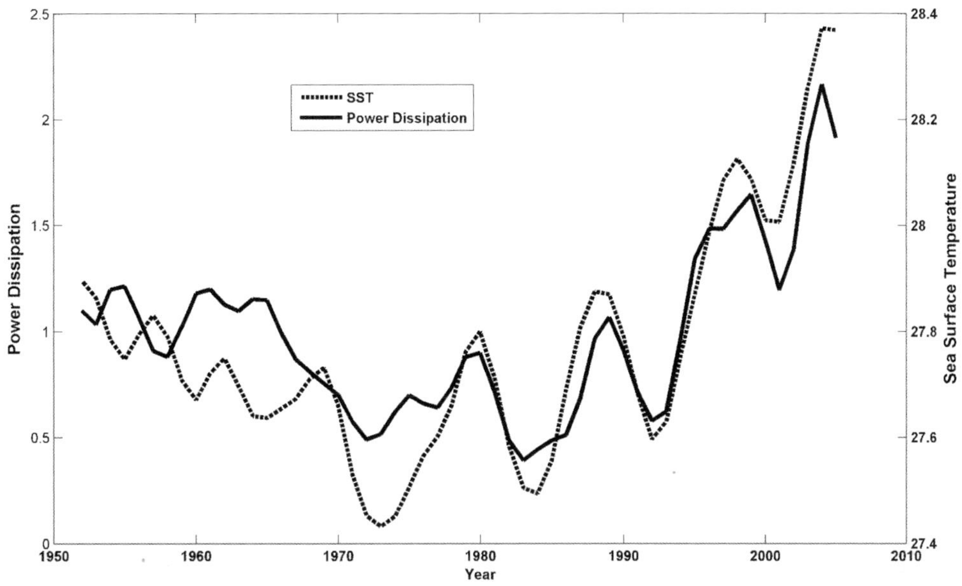

Figure 1. Low-pass filtered tropical Atlantic sea surface temperature (dashed) correlated with the Power Dissipation Index (solid) for North Atlantic hurricanes (Emanuel, 2007a; data updated through 2007). SST is averaged over 6°–18°N, 20°60°W, Aug.–Oct. The data for 1949 to 1969 are bias-adjusted to correct for a high bias in HURDAT's winds (Landsea, 1993). No adjustments are made to account for incomplete sampling of tropical cyclone frequency (Vecchi and Knutson, 2008), duration or intensity, which may adjust the earlier part of the record toward higher values of PDI.

low due to lack of routine aircraft reconnaissance far from land and no geostationary satellite monitoring of TCs. Emanuel (2007a) attributes the near doubling in Atlantic PDI during the past 25 yr to decreasing shear early in the period and increasing low-level vorticity and potential intensity over the entire period. He further attributed the 10% increase in potential intensity over the period mainly to increasing net radiative fluxes into the ocean and decreasing tropopause temperature, with SST as a cofactor, rather than a primary cause. Swanson (2008) provides an alternative perspective on statistical correlations between Atlantic PDI and SST, showing that an alternative SST index (tropical Atlantic SST minus tropical mean SST) correlates with Atlantic PDI just as well, if not better, than does local tropical Atlantic (MDR) SST. He argues that there is no straightforward statistical link between warmer SSTs in the MDR and more intense Atlantic tropical cyclone activity.

A U.S. landfalling PDI series (Landsea, 2005; updated to 2009 in Fig. 2) shows no evidence for a long-term trend since 1900 in U.S. hurricane landfalling activity. PDI for 2004 and 2005 were strong positive outliers, but a similar magnitude PDI was estimated for 1886. Emanuel (2005b) notes that a PDI series such as Landsea's (2005) based on only U.S. landfalling data, contains only about 1 percent of the data that Emanuel's (2005a) PDI contains, which is based on all storms over their entire lifetimes. Thus a trend in basin-wide PDI may not be detectable in U.S.

landfalling PDI since the former index has considerable advantage in signal to noise ratio. However, it has been previously shown that U.S. hurricane strike variability shows up robustly in both interannual forcing by El Nino-Southern Oscillation (Bove *et al.*, 1998) and AMO variability (Goldenberg *et al.*, 2001).

Using a statistical modeling approach, Briggs (2008) examined the time varying behavior of PDI of individual storms and its components. He examined three time periods

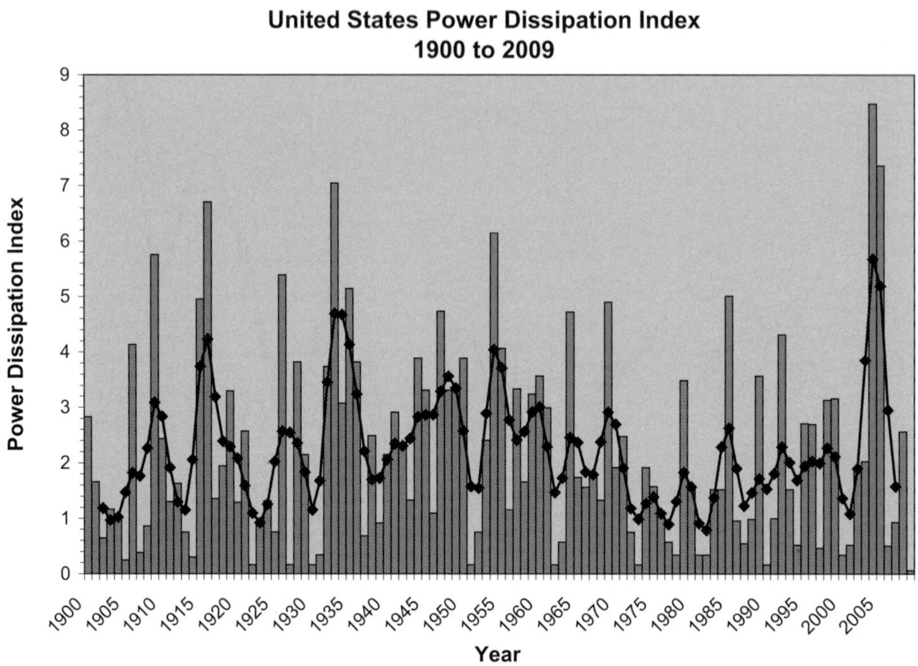

Figure 2. U.S. Power Dissipation Index at the time of impact for tropical storms, subtropical storms, and hurricanes impacting the U.S. for years 1900–2009, updated from Landsea (2005). Black line is the five-year running mean. No adjustments are made to account for incomplete sampling of tropical cyclone frequency and intensity (duration has no component for this analysis), though it is not likely that significant errors due to sub-sampling exist since about 1900 (Landsea, 2005).

(1900–2006, 1966–2006, and 1975–2006) for the Atlantic basin, with varying levels of confidence in the underlying data for these periods. Results were generally dependent on start date, but he found evidence that in recent decades the variance of both PDI and of intensity of individual storms has increased, indicating that storms have had a greater range of maximum wind speeds in more recent years, though their mean intensity has remained unchanged. Elsner *et al.* (2008) report an increase in mean Atlantic TC intensities over 1981–2007, but also find, similar to Briggs (2008), that the largest increases occurred for the more intense hurricanes. Wu *et al.* (2008) report that average Atlantic TC intensities changed little over 1975–2004, while the number and duration of storms has increased. Emanuel (2007a) discusses difficulties with attributing PDI changes to various factors, noting that there is no unique method for deconvolving PDI into different component contributions.

Elsner *et al.* (2006) and Jagger and Elsner (2006) use a data modeling approach to examine historical hurricane intensities in the U.S. coastal region, including the statistical relationship of intense hurricanes with climate phenomena such as El Nino, the Atlantic Multidecadal Oscillation (AMO), and global mean temperature. Their data model finds that during warm global mean temperature years, the maximum surface wind speeds of extreme storms are higher than in cool global mean temperature years. Their model predicts an 11% increase in the maximum surface wind speeds of a 100-yr return period storm for warm vs cool global mean temperature years. However, Jagger and Elsner (2006) note that the AMO and global mean temperature series are correlated at $r = 0.79$, and AMO is not controlled for in their regression with global temperature, so the relationship of their findings to greenhouse gas induced global warming is not presently clear. The sense of dependence they find is at least qualitatively consistent with theoretical and modeling studies of the impact of greenhouse warming on hurricane intensities (sections 6.1.2, 6.2).

Manning and Hart (2007) document how Atlantic tropical cyclones of known categories have changed over time in the ECMWF (European Centre for Medium-range Weather Forecasts) reanalysis. They find that hurricanes in the last two decades are much better resolved (more substantial warm core, deeper pressure) than the same category hurricane in the 1950s to 1970s. However, even in the last 15 years, Category 4 and 5 hurricanes in the ERA40 (ECMWF 40-Year Re-analysis) have central pressures only a few HPa lower than Category 1 hurricanes (1000 vs. 1005 HPa). Such improved representation makes finding real climate trends in TC mesoscale quantities like maximum wind/central pressure using reanalysis output very problematic, at best.

3.2. *Atlantic Basin Tropical Cyclone Counts*

Several studies have attempted to look at longer tropical cyclone records in the Atlantic by concentrating on storm counts. This requires knowing that a storm existed, and whether the storm exceeded certain intensity thresholds (e.g., tropical storm, hurricane intensity) at any point in its lifetime as a tropical cyclone. In contrast, the PDI is based on intensity information over entire lifetimes of storms, and thus studies of basin-wide PDI have generally been restricted to the aircraft reconnaissance era (~1946 and later). Mann and Emanuel (2006) pointed out that the existing record of annual counts of tropical cyclones for the Atlantic basin had a long term upward trend, and after low-pass filtering, was correlated with a similar rise in tropical Atlantic SST since 1871. Mann and Emanuel (2006) argue that detection of the existence of TCs in the years prior to the 1940s was less problematic than TC intensity estimates, since in the absence of aircraft and satellites based guidance to warn them of approaching storms, ships often encountered TCs at sea, at least peripherally.

Holland and Webster (2007) also concluded that Atlantic tropical storm counts had increased substantially, with over a 100% increasing trend in annual numbers of tropical cyclones and of hurricanes over the past 100 years, which they attributed to greenhouse warming. They interpreted the covarying behavior of tropical cyclones and SSTs during the past century in terms of three relatively stable regime periods lasting several decades each, separated by sharp transitions between regimes. While the proportion of storms becoming major hurricanes appears to have a multidecadal variation, they noted that there was no long-term trend in this proportion measure. Since they had found a long-term rising trend in tropical cyclone numbers, they concluded that major hurricanes also had an increasing trend similar to that for all tropical cyclones. In contrast to storm numbers, they report that there has been no distinct trend in the mean intensity of all Atlantic tropical storms, hurricanes, or major hurricanes.

The above studies assumed that tropical storm counts in the historical record, back to 1900 or into the late 1800s, were not systematically biased low (or high) in the earlier years. A number of other studies conclude that a low storm count bias exists in earlier parts of the record. In early work on this topic, Fernandez-Partagas and Diaz (1996) estimated an average annual storm count (June–October season) of 8.5 for the period 1851–1890 as opposed to 7.5 in the observed record. They use the fraction of tropical storms making landfall to adjust the basin-wide numbers, which assumes that all tropical storms making landfall have been accurately counted. Solow and Moore (2002) found no evidence for a trend in Atlantic hurricane activity (1900–98). Their statistical analysis assumed that the record of hurricanes making U.S. landfall is complete over that period and that the probability that a hurricane made landfall in the U.S. was constant over their period of analysis. Landsea *et al.* (2004) estimated the number of "missed" Atlantic basin tropical storms and hurricanes per year to be on the order of 0–6 for the period 1851–85 and 0–4 for the period 1886–1910. They argued that the TC record over the Atlantic should by no means be considered complete for either frequency or intensity of tropical storms and hurricanes for the years 1851 to 1910, in contrast to the more complete and accurate information available for landfalling TCs along much of the U.S. coastline.

In a more regionally focused landfalling TC study, Mock (2004) analyzed records of TC activity from 1769 to 2003 for the state of South Carolina in an effort to assess a relatively homogeneous multi-century record of tropical storm and hurricane strikes. This analysis suggests pronounced multidecadal variability, but no long-term trends. Given the well-established communities along the South Carolina coastal regions since the 18th century, it is unlikely that any significant hurricanes were not captured in this record.

In response to the Mann and Emanuel (2006) study, Landsea (2007) analyzed the fraction of Atlantic tropical storms making landfall each year since 1900 and found an apparent step-function increase in this fraction corresponding to the advent of satellite reconnaissance in 1966. (Landsea (2007) began his analysis in 1900, based upon the assumption that nearly all landfalling tropical cyclones would be reliably monitored from that year forward because of increased coastal populations by that time [Landsea *et al.*, 2004], and assumed that the fraction of storms making landfall over 1966–2006 was representative of the long-term average.) Using that fraction, he thus proposed to adjust the pre-satellite era basin-wide storm counts (1900–1965) upward by 3.2 storms per year for 1900–1965 and 1.0 storms per year for 1966 to 2002. He proposed that in the years 2003 and later, improved observational sources and analysis methods have led to the identification of about one additional storm per year since 2003. Landsea's (2007) assumption of a constant long-term fraction of landfalling tropical storms was disputed by Holland (2007) who reported that the fraction of storms making landfall was also low during periods of the 19th century, when no satellites were available. Mann *et al.* (2007) use a statistical reconstruction of storm counts based on a combination of climate indices to infer a smaller "missing storm" adjustment than Landsea (2007), specifically a range of 0.5 to 2 storms per year, most likely value of 1.2, for the years 1870–1943.

Two independent groups (Chang and Guo, 2007; Vecchi and Knutson, 2008) have attempted to estimate past numbers of missing Atlantic tropical storms using historical ship track records, since identification of storms that did not make landfall prior to aircraft and satellite reconnaissance depended on reports from ships of opportunity. The estimated number of missing storms is quite similar between these two studies, at least for periods of the 20th century that the two studies have in common (0.2 to 1.2 storms per year from the early 1900s to 1965, except the two world war periods). Vecchi and Knutson estimated larger adjustments (roughly 2 to 3 storms per year) for the two world war periods and from 1900 back to 1878. The tropical storm count series using their "base case" adjustment was estimated to contain a statistically significant positive trend over the period 1900–2006, but the higher activity levels in the late 19th century, coupled with their larger adjustment for that period, resulted in a statistically insignificant ($p = 0.05$), nominally positive trend for the period 1878–2006.

An assumption common to both the Chang and Guo (2007) and Vecchi and Knutson (2008) analyses is that any tropical cyclone of tropical storm intensity or greater in the COADS ship data would have already been included into HURDAT. Unfortunately, COADS has not yet been utilized for the reassessment of HURDAT for the period of 1851 to 1910 (Landsea *et al.*,

2004). The ship database has recently been incorporated into the reanalyses that have been completed for 1911 to 1920 (Landsea *et al.* 2008). This reanalysis using the COADS data and other sources have resulted in the addition of 13 previously unidentified tropical cyclones for the period 1911–1920. (There was also one tropical cyclone removed from HURDAT because the criteria for identifying a tropical cyclone by today's standards were not met.) Thus an additional 1.2 tropical cyclones per year in the early 20th Century should be added to the numbers estimated to be missing by Chang and Guo (2007) and 0.7 TCs per year to the estimates of Vecchi and Knutson (2008), as the latter study already includes five of the 12 (net) recently added storms in HURDAT. Chang and Guo (2007) also assumed that only one observation of gale force winds was needed for a tropical storm to be included in HURDAT. As documented by Landsea *et al.* (2008), two independent observations are required for such inclusion in practice. As shown by Vecchi and Knutson (2008), using two versus one observation, as has been done for HURDAT, results in a small, but meaningful increase in the estimate of missing tropical cyclones.

Solow and Beet (2008) expand on the Mann *et al.* (2007) statistical reconstruction mentioned previously, using a statistical model with MDR SST and the Southern Oscillation Index (SOI) as predictors of annual Atlantic tropical storm counts. In an innovation from previous work, their Poisson distribution-based model allows for a probability of observing storms which can decrease (but not increase) as one goes back in time. With this approach, they find that additional evidence that the existing Atlantic tropical storm count record is not complete for the years 1870–1966. They also infer a significant positive sensitivity of Atlantic tropical storm counts to MDR SST, with an increase of 50% (95% confidence interval of about +23% to +84%) for the roughly 0.8°C rise in MDR SST since 1870.

As noted in CCSP (2008), time series of hurricane counts (with no adjustments for possible inhomogeneities like those just discussed for tropical storms) have a significant positive trend over some analysis periods, (e.g., beginning in years 1881 through 1921 and in each case ending in 2005), whereas the trends from the earlier periods (beginning in 1851 through 1871) are not statistically significant. Elsner *et al.* (2004), Wang and Lee (2008) and Vecchi and Knutson (2008) find that there is no evidence for significant long-term trends in U.S. landfalling hurricane frequency. The unadjusted series for basin-wide major hurricane counts has significant positive trends beginning from 1901 and 1911 (CCSP, 2008). Holland and Webster (2007) find that the relative occurrence of major hurricanes and hurricanes as a proportion of the total number of Atlantic tropical storms has remained fairly stable (with no trend) going back to around 1900. They argue that this lends support to the notion of a true increase in hurricane and major hurricane counts since 1900 based on the increases in tropical storm counts since then. However, the lack of significant tropical storm trends from 1878 to recent years as reported in Vecchi and Knutson (2008) would lend support to the finding of no significant increase in hurricanes and major hurricanes since 1878. Furthermore, the impact of missing storms or underestimated intensities on the hurricane and major hurricane trend results has not yet been assessed using, for example, the methodologies applied to tropical storm counts.

Focusing on the very recent period (1983–2005) when a relatively homogeneous record of intensities is obtainable from infrared satellite imagery, Kossin *et al.* (2007) confirmed that there has been a recent increase in the number of percentage of very intense hurricanes in the Atlantic. This study is discussed in more detail in Section 3.5.

3.3. *Northwest Pacific Tropical Cyclone Analyses*

Emanuel (2005a) reported a strong increase in PDI accumulated over the Northwest Pacific and Atlantic basins, estimating that PDI had doubled over the past 30 years. The PDI increase generally accompanied the observed long-term rise in tropical and Northwest Pacific SST, similar to the behavior found for the Atlantic. However, in the Northwest Pacific, the correlation between PDI and SST on time scales of more than a few years, although still present, was weaker than found for the Atlantic. Emanuel (2007a) presented a revised version of the Northwest Pacific PDI and SST, with some further corrections for missing storms, including use of adjusted data from Kossin *et al.* (2007). Wu *et al.* (2008) attribute an increase in Northwest Pacific PDI over 1975–2004 mainly to an increase in average intensity, although Emanuel (2007a) reports some sensitivity of such results to details of the method used to deconvolve PDI into contributing factors.

Chan (2006) extended the analysis of Webster *et al.* (2005), discussed below, for the Northwest Pacific basin back to earlier years and argued that the "trend" in that basin is part of a large interdecadal variation (see also Webster *et al.*, 2006; Chan and Liu, 2004). Chan used unadjusted data from the earlier part of the record, in contrast to the adjustments for this period proposed by Emanuel (2005a) for the basin. Using observations from the U.S. Joint Typhoon Warning Center (JTWC), Chan (2007) concludes that the multi-decadal variations in intense typhoon occurrence in the Northwest Pacific for the period 1960–2005 have been driven mainly by the influence of El Nino and Pacific Decadal Oscillation (PDO) variability on large-scale atmospheric conditions. He does not find an attributable effect of global warming on the typhoon activity.

Kamahori *et al.* (2006) examine how the records of typhoon days compare between the Japanese Meteorological Agency (JMA) typhoon best tracks and those from the JTWC (which were used in Emanuel (2005a) and Webster *et al.* (2005)) from 1977 until 2004. They found a 15–30% increase in TC days with an intensity of category 2 or higher in both data sets, but there were pronounced differences between the two data sources regarding the distribution of storms within the category 2–5 range. For example, they found that the number of Category 4 and 5 typhoon days decreased from 7.2 per year in 1977–90 to 4.3 per year in 1991–2004 in the JMA database. This contrasts with the JTWC dataset, which showed for Category 4 and 5 typhoon days 9.8 per year from 1977–90 and 16.9 per year from 1991–2004.

Wu *et al.* (2006) analyzed the best track data from JMA as well as that of the Hong Kong Observatory (HKO) and found no evidence for a secular increase in western North Pacific category 4 and 5 typhoon frequency since the mid 1960s. Furthermore, neither JMA nor HKO best track data suggest an increase in western North Pacific tropical cyclone destructiveness as measured by the PDI. The discrepancies almost certainly relate to JMA/HKO vs. JTWC satellite treatment of TC intensities once aircraft reconnaissance was discontinued there in 1987. There is currently no guidance as to which, if any, of these datasets is more reliable for assessing climate trends in extreme TCs, pointing to the importance of improving future monitoring of TC activity through enhanced in situ measurement such as by aircraft reconnaissance.

3.4. *Southern Hemisphere Regional Tropical Cyclone Analyses*

Kuleshov *et al.* (2008) focused on the southern hemisphere TC basins, and examined the TC data for linear trends over 1981/82 to 2006/06, using TC data from the Australian Bureau of Meteorology, Meteo France (La Reunion), and the Meteorological Services of New Zealand and Fiji. Pre-1981 data was excluded due to concerns about data homogeneity, whereas they considered the data from 1981 TC season on as homogeneous. They reported significant positive trends in the occurrences and number of cyclone days for "severe TCs" with central pressures of 945 hPa or lower. They found no significant trends for all TCs in aggregate, nor for relatively strong TCs with central pressures of 970 hPa or lower.

Harper *et al.* (2008) reported on an analysts' review of TC intensities from 1968/69 to 2000/01 in the northwest Australian TC region. According to the review, a low bias likely existed in the earlier parts of the record (mainly pre-1980), which reduced trends over the period relative to those in the original data sets. The bias, in terms of central pressure deficit, was estimated as at least 20% in 1970, about 10% by 1980, and about 5% by 1985. To illustrate the impact of the bias on linear trend estimates (1968–2000), their calculated linear trend in PDI from the original (Neuman) data set showed a very large increase (~80%) over the period, whereas using the review-adjusted data, the estimated PDI trend was much smaller (~10% increase).

3.5. *Global Tropical Cyclone Analyses*

Webster *et al.* (2005) reported that the number of category 4 and 5 hurricanes has almost doubled globally over the past three decades. Although their analysis spans a shorter time period than Emanuel's (2007a), due to their decision to limit the analysis to the satellite era, their results indicate that a substantial increase has occurred in all six tropical storm basins that they studied. They found no trend in the numbers of tropical storms and hurricanes or in the maximum wind speed observed globally each year. While they also found an increasing trend in the duration of Atlantic tropical cyclones over this period, no significant trend in duration was identified in the other tropical basins. In a follow-on study, Hoyos *et al.* (2006) found that the increasing trends in category 4 and 5 tropical cyclones are principally correlated with SST as opposed to other environmental factors.

Using a different approach, Sriver and Huber (2006) computed power dissipation statistics from ECMWF (ERA-40) reanalysis data from 1958 to 2001 using Best Track data to identify TC tracks. Despite the coarse resolution of the reanalysis data (1.125° longitude by 1.125° latitude), their resulting global indices, normalized by their standard deviations, are well-correlated with Emanuel's (2005a) Atlantic + Western North Pacific PDI, particularly after 1978.

Sriver and Huber estimated a sensitivity of global power dissipation of roughly 60% per 0.25 deg C SST increase. In comparing these sensitivities, note that PDI depends on the cube of the wind speed and includes effects of storm duration and frequency. Nonetheless, a much higher intensity sensitivity is likely implied by Sriver and Huber's analysis than was reported for example for the Atlantic basin by Michaels *et al.* (2006), who used a statistical analog approach (see Section 5.2). As an example, if we assume based on Emanuel (2005a) that the PDI change is half due to intensity increase and half to duration increase with no frequency change, the above sensitivity estimates still differ from each other by more than a factor of 10. Maue and Hart (2007) find that most (80%) of the global power dissipation variance is described by the duration and frequency variability in the Best Track data set, and thus is not independent information from that contained in ECMWF reanalysis. Maue and Hart (2007) found that the correlation between TC winds cubed in the ECMWF (either maximum or areal averaged) versus Best Track winds cubed was only 0.3, or about 10% of the variance. Moreover, they showed that there does not exist any trend in the TC winds obtained from the ECMWF reanalyses for the years of 1958 to 2001. They conclude that Sriver and Huber's (2006) finding should not be interpreted as independent confirmation of previous findings using the Best Track data. These results also suggest that reasonable empirical estimates of PDI low-frequency variability can be obtained even in the absence of good intensity data, since the PDI low-frequency variability appears to be dominated by TC frequency and duration, with the influence of intensity being secondary.

The studies reporting strong long-term increases in multi-basin (e.g., Emanuel, 2005a, 2007a) or global tropical cyclone activity (Webster *et al.*, 2005) have been the subject of much debate in the hurricane research community, particularly with regard to the homogeneity of the tropical cyclone data over time and the required adjustments.

Landsea *et al.* (2006) propose that much — perhaps the majority — of the global increase in Category 4 and 5 TCs since 1970 may be due to data reliability issues since strong TCs are more accurately monitored in the more recent years. They documented six additional Category 4 and 5 TCs in the North Indian Ocean during the 1970s and 1980s, which were not counted as such in the Webster *et al.* (2005) study. The inclusion of these extreme TCs make the trend found in the North Indian Ocean much weaker. They argue that such systematic undercounts are endemic in the global TC records, especially in basins that rely primarily upon satellite imagery for intensity monitoring (that is, all but the Atlantic).

Over the period 1986–2005, Klotzbach (2006) finds no significant change in global net tropical cyclone activity, as described by the ACE Index, and a small trend ($\sim +10\%$) in category 4 and 5 TC frequencies. He restricted his analysis to this 20-yr period owing to data quality concerns. In particular, while he finds a large increase in TC activity in the Atlantic from 1986–2005, there is a nearly commensurate decrease in the Northeast Pacific, and the remaining global basins show negligible changes in the 20-year period.

In a major advance toward creating a more globally homogeneous tropical cyclone data set for climate studies, at least for the satellite era, Kossin *et al.* (2007) reanalyzed tropical cyclones using a more temporally consistent algorithm and data source (infrared imagery from geostationary satellites) to estimate intensities than previous studies had used. Over their period of analysis (1983–2005), they found that their global PDI decreased slightly, in contrast to the increase estimated from the original Best Track data. While their PDI trends agreed well with Best Track trends in the Atlantic and East Pacific basins, they were lower than the Best Track PDI trends for the same period in the remaining four basins, suggestive of temporal inhomogeneities in the Best Track. Emanuel (2007a) incorporated the Kossin *et al.* data into a revised analysis of PDI for the NW Pacific and reported a 35% increase since the 1970s, although the time series has much weaker correlation with low-frequency SST variability than does the PDI series for the Atlantic.

Using a modified version of the Kossin *et al.* (2007) TC intensity data set, Elsner *et al.* (2008) analyzed the lifetime maximum TC intensities globally using a quantile regression approach, exploring for possible trends in the higher intensities (upper quantiles) and possible relationships with sea surface warming. For the period 1981–2006, they reported significant upward trends for quantiles above the 70th percentile globally, with the largest increases in the North Atlantic basin. While they found some evidence for significant trends in other basins — particularly the North and South Indian Ocean basins — the evidence for significant increases in upper quantiles was much more mixed in the northwest, northeast, and southwest Pacific TC basins. Regressing maximum intensities against basin-scale seasonal-mean SST anomalies, they report a global sensitivity factor of $6.5 \pm 4.2\,\mathrm{m\,s^{-1}}$ (s.e.) per deg C SST rise, which is suggestive of a positive dependence although not statistically significant. The cause of the trends in intensity that they identified remains to be determined, as their analysis did not allow for an attribution of the trends to various causal factors, such as greenhouse warming.

Some regional low-frequency variability in prevailing typhoon tracks in the western North Pacific for the period 1965 to 2003 have been reported by Wu *et al.* (2005), although they were not able to distinguish between anthropogenic impacts or long-term natural variability. The changes in tracks were found to be consistent with expected changes based on large scale circulation (steering flow) changes.

4. Paleoclimate Proxy Studies of Past Tropical Cyclone Activity

Paleotempestology is the term for an emerging field of science that attempts to reconstruct past tropical cyclone activity using geological proxy evidence. Ancient historical documents have also been used to reconstruct past tropical cyclone activity. These techniques attempt to expand knowledge about hurricane occurrence back in time beyond the limits of conventional instrumental records, which cover roughly the last 150 years. A long-term record of hurricane activity on time scales of centuries to millennia is especially important in understanding the spatial and temporal variability of the rare but most intense landfalling hurricanes. Broader goals of paleotempestology are to help researchers explore physically based linkages between prehistoric TC activity and other aspects of past climate and to help provide independent evidence for causes of changes in hurricane climate.

Among the geologically based proxies, overwash sand layers deposited in coastal lakes and marshes appear to be quite useful (e.g., Liu and Fearn, 1993, 2000; Liu, 2004; Donnelly and Webb, 2004). The storm surge plus wave run-up during an intense hurricane can overwash a beach barrier, eroding sand and depositing a layer of the eroded sand material in a lake or marsh behind the barrier. These layers can then form a stratified record through time of intense storm overwash events. Cores of these layers can be retrieved and the layers analyzed in terms of their thickness, composition, age, and frequency of occurrence. The age of deposits is estimated through various radiometric dating techniques applied to the surrounding organic matter, supplemented by other information. In interpreting such records, confounding factors (e.g., long-term changes in sea level) must also be taken into account.

A comparison of prehistoric overwash deposits with those of well-observed storms in the modern historical record allows inferences about prehistoric storm events to be made (Liu and Fearn, 1993, 2000). Similar methods have been used to produce multi-century scale proxy records of hurricane strikes from back-barrier marshes in Rhode Island, New York, and New Jersey (Donnelly *et al.*, 2001a,b; Donnelly *et al.*, 2004; Donnelly and Webb, 2004, Scileppi and Donnelly, 2007, and from back-barrier overwash deposits in the Caribbean (Donnelly, 2005; Donnelly and Woodruff, 2007). Deposits of anomalous graded beds in the layered sediments of a near coastal lake in the Boston area, related to intense rainfall and vegetation disturbance from hurricanes, have also been used as proxy hurricane records (Besonen *et al.*, 2008). In the Australia region, the frequency of cyclone or "super-cyclone" occurrence over the past 5000 years has been inferred from chronostratigraphic series of shelly beach ridges (Nott and Hayne, 2001; Hayne and Chappell, 2001).

Stable isotope signals in tree rings (Miller *et al.*, 2005), cave deposits (Frappier *et al.*, 2007) and coral reef materials are also being actively explored for their utility in providing paleoclimate information on tropical cyclone activity. These methods attempt to exploit an oxygen isotope signal that distinguishes rain originating in hurricanes from that in low-latitude thunderstorms (Lawrence and Gedzelman, 1996). Rainwater from a hurricane is eventually incorporated into the tree-ring, cave deposit, or reef material where it may be preserved as a long-term proxy record. The above studies are beginning to show some promise as a method of inferring aspects of past tropical cyclone activity.

Historical documents apart from traditional weather service records can also be used to reconstruct some aspects of past tropical cyclone activity. For example, investigators have used sources such as newspapers, plantation diaries, various instrumental weather records, and annals in the Caribbean to reconstruct past tropical cyclone activity in the U.S., Caribbean, Gulf of Mexico, and Atlantic basin for up to several centuries before present (Ludlam, 1963; Millas, 1968; Fernandez-Partagas and Diaz, 1996; Chenoweth, 2003; Mock, 2004; Chenoweth and Divine, 2008). Spanish and British historical archives may be a useful source for further investigation (Garcia Herrera *et al.*, 2004, 2005). Even longer documentary records of tropical cyclone activity, extending back for more than 1000 years, have been found and investigated in China (Liu *et al.*, 2001; Louie and Liu, 2003; Louie and Liu, 2004).

Using these paleoclimate proxy techniques, researchers have begun to piece together a picture of hurricane variability in certain regions over the past few hundred to few thousand years. For example, Liu and Fearn (1993; 2000) have constructed a 5,000-year paleo record of inferred category 4 and 5 hurricane strikes in Alabama and northwestern Florida, showing pronounced long time scale variability in activity. Donnelly and Woodruff's (2007) proxy reconstruction of the past 5,000 years of intense hurricane activity in the western North Atlantic suggests that hurricane variability has been strongly modulated by El Nino during this time and that the past 250 years has been relatively active in the context of the past 5,000 years. Results from different paleotempestology studies do not always appear consistent with each other, suggestive of future work needed in intercomparison of results from multiple studies. For example, Besonen *et al.* (2008) report that their conclusions about activity over the last 300 yr compared with the last 1000 yr, which was based on a reconstruction from a Boston area lake, differ from conclusions about these periods from another paleo-hurricane reconstruction from a nearby region (Long Island) by Scileppi and Donnelly (2007). Nyberg *et al.* (2007) suggest that major hurricane activity in the Atlantic was anomalously low in the 1970s and 1980s relative to the past 270 years. Chenoweth and Divine (2008) reach a similar conclusion, based on a 318-yr document-based record. As with Donnelly and Woodruff, these studies measures were focused in the western part of the basin (near Puerto Rico or the Lesser Antilles), and in the Nyberg *et al.* study, hurricane activity over a broader region was inferred indirectly through statistical associations with proxies for vertical wind shear and SSTs. Neu (2008) argues that the Nyberg *et al.* (2007) reconstruction probably overestimates past missing major hurricanes, based on the much smaller estimates of missing tropical storms reported by other investigators using several methodologies (see also Nyberg *et al.* (2008)).

Paleoclimate researchers are continuing to investigate these multiple sources of information on pre-historic tropical cyclone activity, and to validate where possible, the paleoclimate proxy records against hurricane observations from the more recent, well-observed part of the historical record. These studies should provide an increasingly useful independent source of information on the tropical cyclone-climate connection, as well as a better-constrained long-term perspective on hurricane risk from rare but extreme hurricanes. Future efforts will include expansion of geographical coverage, development of new proxies, coupling of multiple proxy sources, improved calibration, and integration with modeling and advanced statistical techniques. Woodruff *et al.* (2008) provide an initial example of assessing paleohurricane data through comparisons with model simulations.

5. Use of Theory and Models to Understand Past Variations in Tropical Cyclone Activity

Theory and models of tropical cyclone activity may provide useful information both on the interpretation of past changes in activity and on possible future changes due to such factors as greenhouse gas-induced global warming or natural climate variability. The theories include potential intensity theories as well as empirical indices which attempt to relate tropical cyclone frequency to large-scale environmental conditions. The models range from global climate models to higher resolution regional models aimed at simulating hurricane structure in more detail.

In principle, extended integrations of global climate models allow for an assessment of the frequency, intensity, duration, structure, and tracks of tropical cyclone-like features in the model. In practice, simulation of realistic intensities and detailed structures of the TC's is hampered by the coarse resolution generally required of such global models, as discussed below. In addition, the fidelity of the global model's TC genesis process compared to the real world has not been well established.

In this section, the utility of the theoretical and modeling approaches is assessed based on analyses of past interdecadal variations and trends. The capabilities of these approaches for simulating present-day climatologies, seasonal cycles, and interannual (<10 yr) variability of TCs are discussed in detail in the Carmargo *et al.* chapter of this book.

5.1. *Assessment of Simulated Decadal Variability of Tropical Cyclone Activity*

In a coarse-grid global model investigation of interdecadal variability of tropical storm occurrence in the Atlantic, Vitart and Anderson (2001) were able to simulate a decrease in tropical storm frequency for the 1970s in comparison to the 1950s, similar to observations, by specifying the observed SST changes for the globe (and specifically for the tropical North Atlantic) in their model. The decreased frequency in their model was linked to increased vertical wind shear and reduced CAPE in the tropical storm formation regions. A correlation of hurricane activity with tropical vertical wind shear has also been noted in observational studies of Atlantic TC variability (e.g., Goldenberg *et al.*, 2001; Bell and Chelliah, 2006).

Using a regional model dynamical downscaling approach, Knutson *et al.* (2007) were able to reproduce the observed multidecadal increase of Atlantic hurricane activity since 1980 as well as El Nino/La Nina modulations of Atlantic storm activity. They used NCEP Reanalysis and observed SST variability to maintain the evolving large-scale solution close to the observed throughout the simulation period by an interior spectral "nudging" technique. Within this large-scale state, their 18-km grid nonhydrostatic model generated interannual variability of Atlantic hurricanes that closely approximated that observed during 1980–2005 (correlation of 0.86 in a two-member ensemble). Similarly, Emanuel *et al.* (2008) demonstrated that their statistical-deterministic downscaling framework has substantial skill at hindcasting interannual and interdecadal variations in tropical cyclone activity, when driven by information from atmospheric reanalyses. Their simulated Atlantic basin power dissipation index (1980–2006) correlated with the observed with $r = 0.7$, and for global power dissipation, the correlation was 0.47. With the exception of the eastern North Pacific, the simulated PDI shows substantial increases that are in good agreement with those derived from Best Track data, though not in agreement, outside of the Atlantic, with the reanalysis of tropical cyclone PDI undertaken by Kossin *et al.* (2007). In an even less constrained setting than the above two studies, LaRow *et al.* (2008) were able to reproduce the interannual variability of Atlantic tropical activity, using only observed SSTs in a global atmospheric model with ~100 km grid spacing. For a four-member ensemble for the years 1986–2005, they report a rank correlation of 0.78 for tropical storm counts.

5.2. *Assessment of Observed Trends in TC Activity and Related Measures*

The substantial increases in tropical cyclone Power Dissipation Indices (PDIs) reported by Emanuel (2005a) and the reported increases in the numbers and percentages of TCs attaining category 4 or 5 intensity (Webster *et al.*, 2005) raise the question of whether these changes can be reconciled with existing theory or modeling work. Some studies have demonstrated that dynamical or statistical-deterministic downscaling models can reproduce the rise in metrics such as PDI or ACE since 1980 in the Atlantic (Knutson *et al.*, 2007) or Atlantic and NW Pacific basins (Emanuel *et al.*, 2008) when constrained by observed SSTs and atmospheric reanalyses. While LaRow *et al.* (2008) demonstrate that their 100 km grid atmospheric model can reproduce the interannual variability of Atlantic ACE well over 1986–2005 using only SSTs as boundary conditions, their model's simulated trend in ACE over the period is smaller than observed. However, none of the above studies addresses the underlying cause of the changes in SSTs and reanalysis fields over the periods studied.

The idealized modeling study of Knutson and Tuleya (2004; 2005) found a peak hurricane wind speed sensitivity — to CO_2-induced climate warming — of about 3.3% per degree Celsius (or 3.7% per degree Celsius if maximum winds are inferred from central pressures following Landsea, 1993). This sensitivity is similar to that from earlier theoretical considerations (e.g., Emanuel, 1987).

Using potential intensity theory and empirical statistical analysis, Emanuel (2007a) finds that the variability of PDI since 1980 (on time scales of more than a few years) is explained by variations in potential intensity, low-level vorticity, and vertical wind shear. These three factors covary in the Atlantic on time scales of a few years or more, but not in the Pacific, where the correlation of PDI with local SST is still present though considerably weaker. Emanuel's analysis attributes the 10% increase in potential intensity in the Atlantic and 6% increase in the NW Pacific over the past 25 years to increasing net radiative flux into the ocean and decreasing tropopause temperatures in the reanalysis. However, the underlying causes of these changes (e.g., anthropogenic vs. natural variability) remains to be determined. The highly idealized studies of CO_2-induced climate change impacts imply a much smaller sensitivity of potential intensity to SST changes than has been found (Emanuel, 2007a) for recent climate change. As noted above, the recent changes in potential intensity inferred from observations have been driven by a number of factors in addition to SSTs whose relation to increases in greenhouse forcing is unclear.

Michaels *et al.* (2006) hypothesized that Atlantic tropical cyclones respond to an SST threshold such that major hurricanes are possible only for storms experiencing SSTs above the threshold value at some point in their lifetime. Using a statistical-analog approach, they infer an intensity sensitivity of about 6.3% in wind speed for a 2–3 deg C SST rise. These results were challenged by Emanuel (2007b), who pointed out that there is no physical basis for a threshold SST and that Michaels *et al.* had erred in confusing spatial variations in activity, which depend largely on strong gradients in the level of neutral buoyancy for deep convection, with temporal changes, which depend more directly on SST and tropopause temperature.

To date, only one coupled modeling study (Bengtsson *et al.* 2007) has reported on attempts to simulate the 20th century variability in tropical storm counts using historical radiative forcing. In their hindcasting experiment, they simulated little change in storm counts between 1860–1900 and the period 1900–2000, despite cooler tropical SSTs in the earlier period.

Vecchi and Soden (2007a) present potential intensity estimates for 1880 to 2006 based on climate models forced by three different historical SST reconstructions. Their results illustrate how long-term changes in potential intensity are more closely tied to patterns of SST changes than to local SST changes. A similar nonlocal SST control was reported based on an observational analysis of TC intensities by Swanson (2007, 2008), and for Atlantic hurricane activity, a similar result was also inferred by Latif *et al.* (2007) from simulations of vertical wind shear. As a result of such nonlocal effects, despite surface warming in the Atlantic, NW Pacific, and North Indian Ocean tropical cyclone regions since the late 1800s,

Vecchi and Soden (2007a) argue that potential intensity on average has declined in the NW Pacific, increased in the North Indian Ocean, and that Atlantic basin potential intensity may have peaked in the 1930s and 1950s, with recent values being close to the long-term average.

Caron and Jones (2008) use two empirical genesis parameters together with CMIP3 (IPCC AR4) model simulations of the 20th century to explore possible changes in tropical cyclone activity over that period. Their primary method uses information about convective precipitation from a climate model in place of SST threshold approach similar to that originally proposed by Royer *et al.* (1998). Using this method, little or no upward trend in global or North Atlantic TC numbers is inferred for the 20th century. As they note, caution is required when applying a genesis parameter developed for present day climate to past or future predictions as the statistical relationships may not be valid under altered climate conditions (Ryan *et al.*, 1992).

In previous sections, we noted that there remains considerable uncertainty over whether a significant long-term increase in tropical cyclone activity has been observed, particularly after consideration of uncertainties in earlier TC records is accounted for. Therefore, it is premature to judge whether current modeling efforts such as those reported above are consistent or inconsistent with the observational record, particularly for the non-landfalling storms prior to the aircraft reconnaissance and satellite eras.

6. Simulations of Future Tropical Cyclone Behavior

6.1. *Global and Regional Nested Models*

Future changes in tropical storms projected by global or regional climate models (RCMs) are subject to many sources of uncertainty, including the future climate forcing scenario, initial conditions, regional patterns and magnitudes of future climate change for various fields, model physics and dynamics, how tropical cyclones are identified and counted, and so forth. Since tropical storms are relatively rare events, large samples sizes (i.e. many years) are typically required to test the significance of any changes against natural variability, depending upon the metric being examined. The changes in frequency of storms simulated by models are often smaller than the climatological bias of the models. These errors in the tropical storm climatology add to the uncertainty of the future changes in tropical storms projected by the models.

6.1.1. *Tropical cyclone frequency*

The combined effect of all the sources of uncertainty is that there is large overall uncertainty in future changes in tropical cyclone frequency as projected by climate models forced with future greenhouse gases. The Fourth Assessment Report (IPCC 2007) concluded that a synthesis of model results showed a "less certain possibility of a decrease in the number of relatively weak tropical cyclones, increased numbers of intense tropical cyclones and a global decrease in total numbers of tropical cyclones". Results from a number of studies, including some not available for IPCC (2007) are shown in Table 1. A caveat of these results is that it is not always clear in a particular study whether reported changes are greater than the model's natural variability, or that natural variability of TC formation rates or the mechanism of the TC genesis process is properly simulated in the models.

The results in Table 1 span a range of GCMs (200 km–20 km), regional models, and idealized frameworks. The global models tend to show a signal of fewer tropical cyclones globally in a warmer climate, though this finding is still not conclusive. Also, while most global models show a global decrease in frequency (e.g., Sugi *et al.*, 2002; McDonald *et al.*, 2005; Bengtsson *et al.*, 2006; Oouchi *et al.*, 2006), there are regional variations in the sign of the changes, and the

Table 1. Summary of simulated future changes in tropical storm frequency under global warming conditions.

Reference	Model/type	Resolution/	Experiment	Ratio (%) of number of tropical storms in global warming experiment to number in control								
				Global	NH	SH	N Atl.	NW Pac.	NE Pac.	N Indian	S Indian	SW Pac.
Sugi *et al.* 2002	JMA Timeslice	T106 L21 (~120 km)	10y 1 × CO_2, 2 × CO_2	66	72	61	161	34	33	109	43	69
Tsutsui 2002	NCAR CCM2	T42 L18	10y 1 × CO_2, 2 × CO_2 from 115y +1%	102			86	111	91	116	124	99
McDonald *et al.* 2005	HadAM3 Timeslice	N144 L30 (~100 km)	15y IS95a 1979–1994 2082–2097	94	97	90	70	70	180	142	110	82
Hasegawa and Emori 2005	CCSR/NIES/ FRCGC timeslice	T106 L56 (~120 km)	5 × 20y at 1 × CO_2 7 × 20y at 2 × CO_2					96				
Yoshimura *et al.* 2006	JMA Timeslice	T106 L21 (~120 km)	10y 1 × CO_2, 2 × CO_2	85								
Bengtsson *et al.* 2006	ECHAM5-OM	T63 L31 1.5° L40	A1B 3 members 30y 20C and 21C	94								
Oouchi *et al.* 2006	MRI/JMA Timeslice	TL959 L60 (~20 km)	10y A1B 1982–1993 2080–2099	70	72	68	134	62	66	48	72	57
Chauvin *et al.* 2006	ARPEGE Climat Timeslice	~50 km	Downscale CNRM B2				118					
			Downscale Hadley A2				75					
Stowasser *et al.* 2007	IPRC Regional		Downscale NCAR CCSM2, 6 × CO_2					119				

(*Continued*)

Table 1. (*Continued*)

				Ratio (%) of number of tropical storms in global warming experiment to number in control								
Reference	Model/type	Resolution/	Experiment	Global	NH	SH	N Atl.	NW Pac.	NE Pac.	N Indian	S Indian	SW Pac.
Bengtsson *et al.* 2007	ECHAM5/MPI-OM coupled mod.	T63 (~200 km)	2071–2100		78							
Bengtsson *et al.* 2007	ECHAM5 timeslice	T213 (~60 km)	2071–2100, A1B		87		92	80	104	74		
Bengtsson *et al.* 2007	ECHAM5 timeslice	T319 (~40 km)	2071–2100, A1B		81		87	72	107	49		
Emanuel *et al.* 2008	Statistical-deterministic		Downscale 7 CMIP3 mods.: A1B, 2180–2200 Average over 7 models	93	102	87	104	106	95	93	88	85
Knutson *et al.* 2008	GFDL Zetac regional	18 km	Downscale CMIP3 ens. A1B, 2080–2100				73					
Leslie *et al.* 2007	OU-CGCM with high-res. window	Up to 50 km	2000 to 2050 control and IS92a (6 members)									~0
Gualdi *et al.* 2008	SINTEX-G coupled model	T106 (~120 km)	30 yr 1 × CO_2, 2 × CO_2	84								
Semmler *et al.* 2008	Rossby Centre regional model	28 km	16 yr control and A2, 20852100				87					

regional results vary substantially, even as to the sign of the changes, between models. The North Atlantic basin column in Table 1 illustrates the strong model dependence of results. For example, more storms are projected in the North Atlantic region in some models, despite a large reduction globally (Sugi *et al.*, 2002; Oouchi *et al.*, 2006), while fewer Atlantic TCs are projected by the N144 HadAM3 atmosphere only model (McDonald *et al.*, 2005). Chauvin *et al.* (2006) found that the sign of the changes in tropical cyclone frequency in the north Atlantic basin depended on the SST anomaly pattern in their stretched grid global model experiments (50 km over the Atlantic region). For the Australia region, Walsh *et al.* (2004) used a 30 km grid nested regional model and found little change in frequency of tropical cyclones near Australia in their $3 \times CO_2$ RCM experiments. Using a regional model downscaling approach for the NW Pacific basin, Stowasser *et al.* (2007) reported only a small statistically insignificant increase in tropical cyclone frequency in response to a very large climate change perturbation ($6 \times CO_2$).

Although Oouchi *et al.* (2006) simulated an increase in future TC frequency in the Atlantic basin, they sampled relatively short periods (20 years) from a single pair of experiments in order to examine greenhouse gas-induced changes. However, internal multidecadal variability in the Atlantic in their model, similar to the AMO mechanism proposed by Goldenberg *et al.* (2001), conceivably could produce changes in tropical-cyclone-relevant fields (such as wind shear) between two 20-year periods that are larger in terms of TC influence than the changes produced by the radiative perturbation they focused on in their study. Therefore further experimentation with their model needs to be done to place their results in the context of internal climate variability.

Building on the theme of using multiple climate model projections to assess uncertainty in future TC behavior, Vecchi and Soden (2007b) examined the changes in tropical climate characteristics believed to be important for tropical cyclone activity using outputs from a large ensemble of CMIP3 models (Meehl *et al.*, 2007) used for the IPCC AR4 assessment. An important feature of their multi-model ensemble mean projection results was a SW-NE oriented band extending across the Atlantic basin of projected less favorable conditions for tropical cyclogenesis and intensification, including enhanced vertical wind shear, reduced mid-tropospheric relative humidity, and slight decrease in potential intensity. The enhanced vertical shear feature also extends into the main cyclogenesis region of the Eastern Pacific basin; the changes in shear are relatively minor over the Atlantic Main Development Region. Physically, this projection was related to the weakening of the east-west oriented Walker Circulation in the Pacific region (Vecchi *et al.*, 2006), similar to that occurring during El Nino events. In contrast to the Atlantic, the West Pacific and Indian Oceans conditions during northern summer are projected to be more favorable for tropical cyclogenesis and intensification by these measures, with reduced vertical wind shear and increased potential intensity over most of these regions. Vecchi and Soden (2007b) also explored the consistency and range of responses of these regional climate change features among different models — an important source of uncertainty in future TC projections.

To test the impact of the late 21st century multi-model ensemble mean atmospheric changes for the Atlantic basin reported by Vecchi and Soden (2007b), Knutson *et al.* (2008) used a regional downscaling simulation of TC activity. Their control model (Knutson *et al.*, 2007) was able to reproduce the observed rise and much of the interannual variation in Atlantic hurricane counts and other metrics, when forced with observed SSTs and large-scale atmospheric conditions from reanalyses over the period 1980–2006. After modifying the large-scale climate according to the late 21st century changes projected by the ensemble of CMIP3 models, they find that Atlantic hurricane and tropical storm frequency were substantially reduced, by 27% and 18%, respectively.

Another approach to using multiple climate model projections for TC assessment was introduced by Emanuel *et al.* (2008), based on an idealized statistical deterministic model. This technique again shows considerable skill when run under present day climate conditions from atmospheric reanalyses. Run separately for late 22nd century climate warming scenarios from seven different CMIP3 models, the framework projects a mixture of increases and decreases of TC frequency across different basins, with the projected changes generally being relatively modest in magnitude (−26% to +14%) considering the substantial increases in tropical sea surface temperatures in the climate models. The frequency of tropical storms decreases in the southern hemisphere and north Indian oceans, increases in the western North Pacific, and is indeterminate in other regions.

There is no evidence from any models to date that the region of TC formation (e.g., Knutson *et al.*, 2008) or strong tropical convective activity (e.g., Dutton *et al.*, 2000) increases dramatically as the area of oceanic regions with SSTs larger that 26–27°C expands due to greenhouse warming. Rather, model results (e.g., Knutson *et al.*, 2008) suggest that the apparent threshold for TC formation (as diagnosed empirically for the present climate) is actually a climate-dependent threshold, which rises roughly along with the tropical mean SST increase in greenhouse warming experiments.

An important issue is to identify the underlying mechanisms producing changes in future TC behavior in the GCM simulations and regional downscaling experiments. Sugi *et al.* (2002) report that the simulated reduction of global TC frequency in their model was closely related to the weakening of tropical circulation, which in turn resulted from a considerable increase in the dry static stability, coupled with relatively little increase in the precipitation. Held and Soden (2006) noted that a weakening of the total upward convective mass flux in the tropics was implied as a consequence of the slower fractional rate of increase of tropical precipitation compared to lower tropospheric specific humidity in greenhouse warming simulations. However, they did not link this mechanism to smaller-scale phenomena such as tropical storms. Yoshimura and Sugi (2005) performed a series of experiments to test the relative effects of changes in SST and CO_2 upon changes in TC frequency in their model. They found that the decrease in tropical storm frequency in their model was due to the increased CO_2 (see also Sugi and Yoshimura, 2004), with the SST changes having a relatively small direct impact. This finding has not yet been confirmed by other studies. Emanuel *et al.* (2008) found that the global reduction in TC frequency in their statistical/dynamical downscaling framework was primarily due to an increase in the saturation deficit of the middle troposphere in the warm climate scenario. Regarding the regional variations in projected TC frequencies, the results of Sugi *et al.* (2002) and McDonald *et al.* (2005) and Chauvin *et al.* (2006) suggest that dynamical factors such as low level vorticity and vertical wind shear play a more important role than thermodynamical factors such as SST and moist instability.

6.1.2. *Tropical cyclone intensity*

Concerning future changes in TC intensity, there is substantial disagreement among recent global and regional modeling studies, although all of the highest resolution models (~50 km grid spacing or finer) currently available show evidence for some increase of intensity.

For global models, as discussed earlier, simulated future changes of intensity in current global models may not be reliable, since these models do not simulate the very intense TCs observed in the present climate, even in the case of the relatively high resolution (20 km grid) simulation of Oouchi *et al.* (2006). Given this caveat, Tsutsui (2002), McDonald *et al.* (2005) and Oouchi *et al.* (2006) all report evidence for intensity increases, while Sugi *et al.* (2002), Bengtsson *et al.* (2006), and Hasegawa and Emori (2005; western North Pacific only), Chauvin *et al.* (2006; North Atlantic only), and Gualdi *et al.* (2008) found either no increase or a decrease of intensity. Among these studies, Tsutsui and Bengtsson *et al.* used relatively low resolution models; McDonald *et al.*, Sugi *et al.*, Hasegawa and Emori, and Gualdi *et al.* used intermediate (~120 km) grid spacing models; while only Oouchi *et al.* used a relatively high resolution model. The Oouchi *et al.* (2006) study reports that the number of the most intense cyclones increases globally in their 20 km grid warming climate simulation, despite a large decrease in overall TC numbers. However, statistically significant intensity increases in their study were confined to only one or two individual basins. Bengtsson *et al.* (2007) presented results from a set of three global models with effective horizontal grid spacing ranging from ~200 km to ~40 km. They found a tendency for increases in tropical storm intensities for the higher resolution model, but the increase was not seen in their lower resolution model. They interpreted these results as indicating that high resolution simulations (~40 km) may be needed for a model to simulate increasing storm intensity in warmer climate.

Regional models provide another means of exploring the tropical cyclone intensity/climate issue. For example, Walsh *et al.* (2004), in a study for the Australia region with a nested regional model, found little change in overall TC frequency under $3 \times CO_2$ conditions, but a 56% increase in the number of storms with relatively high maximum winds (>30 m/sec in their model), and a 26% increase in the number of storms with central pressures less than 970 HPa. Using a global model with higher resolution focused in the Southwest Pacific region, Leslie *et al.* (2007) also found a significant (15%) increase in the number of most severe TCs simulated by the model, Knutson *et al.* (1998) simulated a significant CO_2 warming-induced increase of typhoon intensities in the NW Pacific basin, based on downscaling a sample of tropical cyclones from a high CO_2 scenario of a global climate model into a regional nested hurricane model. Stowasser *et al.* (2007) simulated a substantial increase of storm intensity in response to the climate warming from a very large ($6 \times CO_2$) perturbation using a (~50 km grid) regional downscaling approach for the NW Pacific basin. They reported that notable increases, including a 50% increase in PDI for a ~3°C tropical ocean surface warming due mainly to increased intensities, were driven in the model mainly by decreased vertical wind shear in the basin. In a regional model downscaling experiment for the tropical Atlantic using a multi-model late-21st century climate perturbation, Knutson *et al.* (2008) report a small increase (~1% per deg C) in tropical storm and hurricane wind speeds, although their control model did not simulate storms as strong as observed, nor did it simulate the observed dependence in the present climate of TC intensity on SST. Emanuel *et al.*'s (2008) statistical/deterministic downscaling approach, applied in all basins, gave a tendency for increasing intensity which, when averaged across the seven climate models they examined, ranged from about 0 to 11%, depending upon the basin considered.

In summary, among current studies, experiments with higher resolution models (<40 km grid spacing) typically find that hurricane intensities increase with greenhouse warming, although with substantial regional variation of the increases. The simulated storm intensification is much less evident in lower resolution models (>50 km grid), although such models probably give much less reliable information about hurricane intensity than higher resolution models. The intensification results from the higher resolution models in particular are also generally consistent with results from more idealized modeling and theoretical potential intensity studies which provide a mechanistic framework in which to interpret the intensity change. These alternative approaches are reviewed in section 6.2.

6.1.3. *Tropical cyclone precipitation*

A number of studies have reported increases in TC-related precipitation under warming climate conditions, using global and regional models (Knutson and Tuleya, 2004; Hasegawa and Emori, 2005; Chauvin *et al.*, 2006; Yoshimura *et al.*, 2006; Bengtsson *et al.*, 2007; Gualdi *et al.*, 2008; Knutson *et al.*, 2008). A number of investigators have interpreted the increase in hurricane-related precipitation as being due to enhanced atmospheric moisture in the warmer climate — a mechanism which has been discussed in the context of extreme precipitation in general by Trenberth (1999), Allen and Ingram (2002), and Emori and Brown (2005). Sugi *et al.* (2002) have noted that enhanced latent heating associated with increased TC precipitation does not necessarily lead to intensification of the TC, since the enhanced heating is balanced to some degree by enhanced adiabatic cooling for given amount of rising motion, due to the increased dry static stability in the simulated warmer climate.

Trenberth *et al.* (2007) used a simulation case study approach to explore precipitation in hurricanes and its dependence on environmental conditions, including idealized sea surface warming perturbation experiments. They argue that increasing atmospheric moisture content associated with human-caused global warming (Trenberth *et al.*, 2005), together with enhanced hurricane intensities based on theory and observations (e.g., Emanuel, 2005a), implies an increase in TC-related rainfall of the order of 6 to 8% since 1970.

Thus, several modeling studies support the notion of increasing hurricane-related precipitation rates due to greenhouse warming, while to our knowledge no modeling study to date provides compelling counter-evidence to this finding.

Despite the findings reported from a growing number of modeling studies on future projections, there is no confirmation of the above precipitation rate sensitivity from observed tropical cyclone precipitation studies. For example, Groisman *et al.* (2004) find no increasing trend in the total seasonal hurricane-related precipitation along the U.S. Southeast coast, despite finding that the frequency of very heavy precipitation *unrelated to TCs* has increased significantly in the same region and over the conterminous U.S. during the 20th century. They have not yet examined the behavior of hurricane-related precipitation on a per storm basis, and thus the time series they examine are influenced by changes in U.S. TC activity which has exhibited substantial multidecadal variability but no trend (Goldenberg *et al.*, 2001; Landsea, 2005). Lau and Wu (2007) report that the occurrence of heavy rain events (top 10% by rain amount in 5-day averages) has increased in the tropics during 1979–2003, based on two precipitation data sets, although these data sets have trends of opposite sign from one another in total rainfall amounts. While not focusing on tropical cyclones, their study provides more suggestive observational evidence of a changing character of rainfall toward higher rates during more extreme events, in this case over tropical regions that commonly have deep convection. Trenberth and Fasullo (2008) infer an increase in surface sensible and latent heat fluxes and total precipitation from Atlantic tropical cyclones over the period 1970–2006, accompanying the increase in tropical storm and hurricane numbers over that time.

6.2. *Theoretical or Idealized Modeling Frameworks*

Thus far, almost all theoretical or idealized modeling frameworks have focused on potential future changes in the intensities of TCs. Emanuel (1987) and later Tonkin *et al.* (1997) first presented evidence, based on potential intensity (PI) theory, that CO_2-induced climate change as simulated by several GCMs implied significant increases in future TC intensities. Limitations of their approach were discussed in Henderson-Sellers *et al.* (1998). These theory-based assessments received model-based support from Knutson and Tuleya

(1999; 2004; 2008), who used an idealized hurricane model framework to evaluate tropical climate warming scenarios from several different coupled climate models, all forced by increasing CO_2 levels. Knutson and Tuleya (2008) reported a tropical cyclone intensity increase of about 3.3% per degree Celsius SST increase (3.7% per degree C, using winds inferred from central pressures), which was roughly comparable to the increase predicted by the Holland PI theory, though somewhat larger than that predicted by the Emanuel PI for those same environments.

Vecchi and Soden (2007 a,b) show maps of the average Emanuel PI change as projected by a 22-member ensemble of CMIP3 global models for the late 21st century. For the hurricane season in the northern hemisphere (June–November), their results show a general increase in most tropical storm regions of up to about 2 m/sec per degree C global temperature increase. In the Atlantic, the projected PI increases were smaller on average than for other basins, with even a region of slight projected decrease of PI in the mid-Atlantic basin. Averaged over a region surrounding the U.S. Southeast and Gulf coast along with the Greater Antilles (18–35 N, 90–60 W), the projected average percentage change was about +1.3% per °C tropical SST change (G. Vecchi, personal communication, 2007). In contrast to the more mixed changes in the Atlantic, they found that the CMIP3 ensemble projected an increase in PI over almost the entire northern Indian and north Pacific TC basins. Southeast of Hawaii, particularly large percentage changes were projected, reaching about 7%/°C (Vecchi and Soden, 2007b).

The above methods attempt to account for changes in atmospheric temperature above the warming sea surface — an effect which acts to limit the increase of intensity for a given SST increase compared to the rate in the absence of the atmospheric temperature increases (e.g., Shen *et al.*, 2000). This influence of atmospheric temperatures is related to the work of Vecchi and Soden (2007a) and Swanson (2008). The former study connects the structure of PI change to the structure of tropical SST warming, while the latter one correlates variability of Atlantic hurricane activity to tropical Atlantic SST changes relative to changes of SST averaged over the tropics. The remote influence of non-local tropical SSTs is presumed to occur through an upper tropospheric temperature "bridge".

Concerning ocean coupling effects, Knutson *et al.* (2001) found that the CO_2-warming-induced intensification of tropical cyclones in their idealized model framework was robust to the inclusion of an ocean model (i.e., cold wake in SST) beneath the storms. This finding was contradicted by Hasegawa and Emori (2007) who used a coupled global model simulation; however, the model used for the latter study had very coarse resolution (~100 km grid spacing) for hurricane intensity simulation. As a further caveat, the addition of ocean coupling in their control climate simulation had only very weak impact on storm intensity (~2.5 Hpa) — a much smaller influence from ocean coupling than has been found using higher resolution hurricane prediction models (e.g., ~15 hPa in Bender and Ginis, 2000).

Wu and Wang (2004) performed an initial assessment of the potential impact of greenhouse gas-induced climate change on TC tracks using a trajectory modeling approach for the NW Pacific basin. Based on experiments derived from a particular climate model, they found evidence for inferred track changes, although the pattern of changes was fairly complex, as opposed to a more simply described, systematic change. The use of a parametric storm genesis approach for TC climate change was explored by Royer *et al.* (1998). Their seasonal genesis parameter was based on a measure of convective rainfall as opposed to SST or oceanic heat content. Their study found that TC frequency results for a future climate scenario depended strongly on whether the modified or unmodified genesis parameter approach was used. Using a similar approach for the CMIP3 models used in IPCC AR4, Caron and Jones (2008) find mostly small projected changes for the 21st century in TC frequency, with a small increase projected for the northwest Pacific basin.

The empirical genesis potential index (GPI) developed by Nolan *et al.* (2007) implies a positive relation between potential intensity and the likelihood of tropical cyclogenesis. Vecchi and Soden (2007b) examined the behavior of this GPI based on a multi-model ensemble of global model projections. They find for NH summer a substantial increase in genesis potential in the western and central Pacific, but with more modest changes elsewhere; in SH summer there were substantial increases in the South Indian Ocean, and modest changes elsewhere. In some regions potential intensity changes were quite small, or even negative, and vertical wind shear was enhanced in some regions, leading to reduced future genesis potential according to the index.

In further work aimed at increasing the realism of simulated TC genesis, Nolan *et al.* (2007) have undertaken a very high-resolution (4 km grid) idealized modeling approach, using the Weather Research and Forecast Model (WRF) to explore the relationship between local values of potential intensity, the Coriolis parameter, and the likelihood of tropical cyclogenesis in the absence of vertical shear. Their results show that, in radiative-convective equilibrium (RCE), the potential for TC genesis increases with increasing values of PI. The time to genesis, which decreased for increasing levels of PI, was not very sensitive to convective instability, latitude, mid-level humidity, or the GPI. They observed "spontaneous" TC genesis from random convection in RCE, suggesting that in very ideal environments, the absence of significant precursors such as easterly waves may not be a limiting factor on TC genesis.

7. The Role of Tropical Cyclones in the Climate System

The possibility that tropical cyclones play an active as opposed to essentially passive role in the climate system was proposed by Emanuel (2001). According to this hypothesis, tropical cyclones, through wind-induced mixing of tropical ocean waters and subsequent re-heating of the cold wakes, make a potentially important contribution to the meridional heat transport by the oceans. Boos *et al.* (2004) provide additional support for this idea through idealized ocean modeling experiments. If, in a warming climate, increased tropical storm activity substantially increases the poleward heat transport by the ocean through this mechanism, this process may then help explain the occurrence of distant past climates, such as the Eocene, characterized by strongly reduced equator-to-pole temperature gradients (Emanuel, 2002). With enhanced poleward oceanic heat transport, the high latitudes would warm more than otherwise, while the warming in tropical latitudes would be moderated. This, in turn, would moderate any projected increases in tropical cyclone intensity relative to those predicted on the basis of current global climate model simulations of future climates. Sriver and Huber (2007) and Sriver *et al.* (2008) provide some observational evidence of this process from examination of SSTs before and after passage of tropical cyclones. Under the assumption that all of the heat mixed down from the surface translates into poleward heat transport, they estimate that a non-negligible fraction of peak ocean heat transport may be caused by ocean vertical mixing due to tropical cyclones. However, their estimates of the magnitude of this effect are smaller than originally proposed by Emanuel (2001).

Trenberth and Fasullo (2007) used a combination of hurricane best track data and high resolution hurricane simulations to estimate the aggregate surface cooling rate due to global hurricane activity of $0.33\,\mathrm{W/m^2}$ (based on evaporation within 400 km of the storm centers) to $1.13\,\mathrm{W/m^2}$ (based on storm total precipitation). They note that existing best track data, though in need of reanalysis to improve reliability, implies a positive trend since 1970 in this global surface energy loss term. We note that on decadal and longer time scales, global surface evaporation is strongly constrained by the net surface radiative balance, so excess evaporation owing to tropical cyclones must be nearly balanced by reduced evaporation elsewhere. The surface cooling due to storm-related evaporation presumably leads to a surface condition that is cooler than the normal climate for that region, and thus a subsequent warming of the upper ocean occurs as near-surface waters are restored toward equilibrium conditions by surface fluxes and mixing, as noted by Emanuel (2001).

Pasquero and Emanuel (2008) have used idealized models to explore the interaction between tropical cyclones and upper-ocean heat content. In these idealized settings, regimes exist with intense or weak tropical cyclone activity (with deeper mixing and larger upper-ocean heat content in the intense regime). They also found that ocean feedbacks increased the sensitivity of tropical cyclone power dissipation to climate perturbations in their model.

8. Roadblocks to Further Advancements

There are substantial roadblocks both in making reliable future projections of TC activity and in determining whether a trend can be detected in historical TC data.

8.1. *Data Homogeneity in Tropical Cyclone Databases*

For the climate change detection problem, a large hurdle is the quality of the tropical cyclone historical databases. The databases were populated over time without a focus on maintaining data homogeneity, a key requirement for databases which are to be used to assess possible climate-related trends. In some cases, such as the NW Pacific basin, our ability to monitor TC intensity has diminished over time. For example, aircraft reconnaissance was conducted in the NW Pacific basin beginning in the 1940s, but was discontinued in 1987. Experience with reanalysis of the HURDAT database for the Atlantic basin (Landsea *et al.*, 2004) indicates that considerable effort and analysis is required to identify and attempt to correct, where possible, past problems with the TC databases. The possibility of such errors across all of the other ocean basins is real and problematic from both operational and climate perspectives.

Reanalysis may help provide a more uniform assessment based upon consistent use of pressure-wind relationships and flight level to surface wind analyses. Over the open oceans before that advent of satellite coverage in the 1960s, there will never be a complete, reliable TC dataset for any of the basins. Even in the Atlantic, aircraft reconnaissance typically only monitors roughly half of the region where tropical storms typically occur. However, it may be possible to have a high quality, global analysis of TC intensities and frequencies from about 1970 onward with substantial effort. One method that may be able to provide longer, reliable records is to focus upon analyses of landfalling tropical cyclones that have occurred along coastal regions with substantial populations. The tradeoff with this approach to get longer time series is that one only samples a much smaller number of tropical cyclones compared to the entire basin. Other approaches to estimating likely numbers of missing storms in the pre-satellite era have been discussed in section 3.2. The widespread concerns about problems in the TC databases reduces confidence in trends derived from those databases, and thus is an important roadblock to further advances in the topic of historical TC trends.

8.2. *Data Homogeneity Concerns with Other Tropical Cyclone-Related Climate Variables*

Improved understanding of the causes of past variations or trends in TC activity will depend on the existence of reliable climate-quality data sets for related variables, such as SST, atmospheric temperature, moisture, wind shear, etc. Although reanalysis efforts by NCEP/NCAR and ECMWF have led to important improvements in this regard, studies of upper-air data sets (e.g., Santer *et al.*, 2005; CCSP, 2006) identify likely remaining problems that could substantially affect efforts to reconcile historical TC behavior with various environmental influences. For example, if the tropical upper troposphere does not warm substantially more than the surface with greenhouse warming, as projected by climate models, the resulting destabilization would imply a larger increase in hurricane intensity, at least according to the modeling study of Shen *et al.* (2000). These data quality issues therefore also remain as an important roadblock for further advancement.

In using global reanalysis datasets such as NCAR/NCEP and ECMWF for TC-related studies (e.g., Sriver and Huber, 2006), inclusion of new observations over time complicates monitoring of trends of tropical cyclone statistics, as improved observations do lead to better identification of tropical cyclones (e.g., Manning and Hart, 2007). While the circulation of larger tropical cyclones can be identified on the synoptic scale, some systems remain smaller scale (mesoscale) and the region in which intensity is defined (the maximum sustained surface winds in the eyewall) is always on the mesoscale, which implies that these features typically cannot be well-represented in low resolution reanalysis products.

8.3. *Limitations of Climate Models*

Climate models contain hypotheses for how the climate system works in a framework which allows experiments to be performed to test various hypotheses or compare the model's historical simulations against historical observations. Nonetheless there are important uncertainties in climate models and the radiative forcings used for such experiments. For example, past aerosol forcing due to the interaction of aerosols with cloud and precipitation processes (indirect aerosol effects) remain quite uncertain. Climate change detection and attribution studies often rely in large part on climate model control runs for estimates of internal variability though paleo reconstructions also provide important information on this question. Climate models have known limitations in simulating some aspects of important internal modes of variability of the climate system (such as ENSO), although more recent models are improving in that regard (e.g., Wittenberg *et al.*, 2006; Achuta-Rao and Sperber, 2006).

The climate sensitivity to past and future radiative forcing is another important area of uncertainty, both on the global scale and with respect to important regional details, as evidenced by the wide range of likely global climate sensitivity to CO_2 doubling (2.0–4.5°C) reported in the IPCC 4th Assessment Report (IPCC, 2007). There is considerable uncertainty in projections of future warming due to uncertainties in the climate sensitivity, in the rate of future ocean heat uptake and in various climate forcing agents, including but not limited to greenhouse gas concentrations. These combined uncertainties lead to a wide range (1.1–6.4°C) in projected global warming by 2100 according to the IPCC (2007), based on a range of future emissions scenarios. For the A1B scenario, the IPCC models project a 21st century global mean warming of about 2.5°C on average (IPCC, 2007), compared with a smaller warming rate of about 2°C in the tropical Atlantic (e.g., Knutson *et al.* 2008). Although the projected warming of tropical SSTs is generally smaller than the global mean warming, the above ranges provide an indication of the relative degree of uncertainty that also applies to future projections of tropical storm basin SSTs when forcing uncertainties are considered. The forcing-related uncertainty has not yet been formally assessed in detail at the tropical storm basin scale. In addition, there is even greater uncertainty in details of the projected pattern of SST warming in the tropics than in the tropical mean warming, and these patterns could be much more important than the average warming rate in determining an impact on tropical cyclone activity (e.g., Vecchi and Soden, 2007a; Knutson *et al.*, 2008; Swanson, 2008).

The limited resolution of global climate models implies that many aspects of TC-like storms as simulated by the current models will not be very realistic, including the intensity and important smaller-scale structure such as the eye and eye-wall. This situation will gradually improve as available computing power increases (e.g., Oouchi *et al.*, 2006). Meanwhile, questions remain about the realism of the TC genesis process in the global models. For example, the fact that the number of TCs simulated in a given model depends on physical parameter settings as well as resolution (e.g., Vitart and Stockdale, 2001) raises the question of whether some parameterizations are physically justifiable. Generally, atmosphere-only models have been used for the global model-based TC-change assessments, as available computing power has been used for increasing atmospheric resolution rather than addition of ocean coupling. Eventually, this simplification will need to be relaxed, particularly in order to explore impacts of ocean coupling on TC genesis and intensity, as well as the possible role of TCs on the climate system (section 7). The important impact of model physics and physical parameterizations, even in high-resolution models, means that future progress will depend on both increased scientific understanding and increased computing power.

8.4. *Limitations of High-Resolution Idealized Models and Theory*

While high-resolution idealized models can address the problem of limited resolution to some degree, this approach has limitations and uncertainties to be addressed. For example, the nested model used for downscaling may have a substantially different climatology and climate sensitivity from the "parent" global model, even if two-way nesting is used, whereby the fine-resolution domain communicates outward and affects the solution in the low-resolution domain. This issue raises questions about the effect of such model incompatibilities (e.g., low resolution vs high resolution, physics differences, etc.) on the reliability of the overall results obtained. The simulations can also be affected by the chosen domain (e.g., Landman *et al.*, 2005). Clearly a preferred approach would be to avoid the downscaling issue altogether by using the TC statistics from the original GCM.

The potential feedback of the TC activity on the climate system (section 7) also cannot be modeled using the simple one-way nesting approaches employed to date in TC/climate studies, nor can it be reliably inferred from the present generation of global models due to resolution limitations.

In contrast to the theory of potential intensity of TCs, which is more well-established, a comparable theory of TC frequency is not well-developed at this time. (We note that even the current theories of potential TC intensity include many assumptions and, for example, do not consider any dynamical limitations on TC intensity.) The lack of theoretical underpinning of TC genesis and frequency remains an important roadblock to progress in this area, apart from global model limitations.

9. Proposals for Moving Forward

In general, hurricane-climate research is expected to progress most rapidly when a combination of theory, modeling, and observations are brought to bear on the problem.

9.1. *Improved Paleoclimate, Historical, and Future Tropical Cyclone Databases*

The need for improved climate-quality tropical cyclone databases seems clear. These will provide better information for assessing future changes, and more reliable statistical assessments of past changes in hurricane activity, including land fall, in all basins. Specific examples include the need to reanalyze historical tropical cyclone databases in all basins, and not just the Atlantic. Such efforts should be encouraged and supported. Ideally researchers should be provided with ready access to original "raw" historical observations (i.e., ship, station, buoy, radar, aircraft, and satellite data), as well as derived quantities, from all basins concerning past tropical cyclones.

Concerning future measurement systems, we advocate a comprehensive analysis of the optimal mix of observing systems in support of tropical cyclone measurement (for climate, forecasting, and other needs). Such an analysis should include consideration of both the overall costs and benefits of different mixes of observing platforms, with researchers and forecasters providing hard data on the benefits that a given mix of platforms can provide. As an example, aircraft reconnaissance was conducted in the NW Pacific basin beginning in the 1940s, but was discontinued in 1987 in favor of satellite-only intensity estimation. Is a resumption or initiation of manned or unmanned aircraft reconnaissance in various basins now justifiable in terms of costs, benefits, and alternative measurement techniques? Another alternative that may be worth pursuing is a next generation scatterometer with finer resolution and less difficulties of wind retrieval in heavy rain conditions, compared with the limited observational capabilities of today's QuikSCAT (Atlas *et al.*, 1999) in tropical cyclones.

A related issue is that future improvements in observing systems will lead, unfortunately, to more discontinuities and biases unless these are recognized and addressed. For example, in 2007 the U.S. Air Force reconnaissance aircraft has been outfitted with Stepped Frequency Microwave Radiometers to better provide continuous surface wind estimates for the first time (Uhlhorn and Black, 2003). Researchers need to be cognizant that large monitoring changes have occurred in the past and will continue to occur in the future, which can make determining actual climate-related trends problematic.

Studies of how sampling can alter monitoring of frequency, intensity and duration of tropical cyclones are one approach to investigating the data homogeneity issue. For example, what would the 2005 Atlantic hurricane season look like using only the monitoring capabilities available in 1970, 1950, or 1900? Until better quantitative estimates of how the current observational network influences the determination of numbers and intensities of tropical cyclones, climate trends may be difficult to distinguish from changes induced by monitoring improvements.

Paleotempestology research, which attempts to use information in the geological record to infer pre-historic hurricane activity, should continue to receive support from funding agencies. As the techniques themselves mature, thought should be given to a transition from technique-development research to systematic surveys designed to produce a comprehensive long-term record of tropical cyclone climatology.

9.2. *Improved Dynamical Modeling of Tropical Cyclone Activity*

Tropical cyclone/climate modeling studies will benefit from efforts to improve global climate modeling in general. Higher resolution global models are being developed and refined, and future global and regional climate models will increasingly use two-way nesting or variable resolution techniques. According to Chen *et al.* (2007) higher resolution experiments indicate that horizontal resolutions of order 1 km are needed to resolve the eye and eyewall structures in a typical hurricane. Studies which focus on simulation or downscaling of TCs could benefit from more rigorous testing of model performance at simulating a wider range of TC metrics. For example, the ability of dynamical models to simulate known interannual or interdecadal TC variability characteristics identified in various basins (e.g., Bell and Chelliah, 2006; Chan and Liu, 2004) should be further evaluated, as attempted, for example, in some recent TC downscaling studies (Knutson *et al.*, 2007; Emanuel *et al.*, 2008). Further evaluations of seasonal variability, geographical distributions of genesis and occurrence frequency, track climatologies, and storm structure are also recommended. Many of these measures should also be examined in studies using empirical approaches such as seasonal genesis parameters (see below). For dynamical model simulations, examinations of the dependence of storm characteristics on both model resolution and physics (e.g., parameterizations) are also recommended; models simulating higher intensity storms preferably would include ocean coupling (cold wake) feedbacks.

In TC climate change experiments with climate models, statistical significance testing should be emphasized to ensure that reported changes are not simply due to limited sampling. This may be particularly important in basins such as the Atlantic which feature large multidecadal variations in some observed TC metrics. By analyzing several models using a common tropical storm counting methodology, perhaps with common adjustments for resolution effects (e.g., Walsh *et al.* 2007), intercomparisons of sensitivity results between different models will be more informative. Such a procedure would help reduce differences between model results arising from differing analysis techniques alone and may give insights into the tropical simulations in the models. Analysis of individual models with perturbed physics experiments can be useful in isolating mechanisms producing model behavior. In general, there is a need for more consistency of metrics of TC activity, especially across models, and to improve understanding of the physical mechanisms producing the climate-induced changes in TC behavior in the models.

9.3. *Improved Empirical Modeling of Tropical Cyclone Activity*

Exploration of empirical approaches, such as seasonal genesis parameters, should be encouraged, including testing/evaluation and improvements aimed at reproducing characteristics of historical TC activity in different basins from both observations and climate model simulations. Based on these results, these approaches may be useful for making climate change projections of TC activity, although caution must be exercised (e.g. Ryan *et al.*, 1992). Current empirical genesis parameter approaches undoubtedly do not incorporate all processes relevant for TC formation, so further refinements would be worth pursuing.

A similar recommendation would apply to studies leading to the development of empirical approaches for tropical cyclone potential intensity. Such empirical approaches should include not only thermodynamic parameters, such as the SST and outflow temperature, but also the environmental dynamical parameters that control TC intensity, such as the vertical wind shear and translational speed (Zeng *et al.*, 2006). Current global climate models can simulate the large-scale circulation much more realistically than the individual TCs. Thus empirical approaches with environmental parameters as input to estimate TC potential intensity should be further exploited in this area.

Acknowledgments

We acknowledge the contributions of the following colleagues as members of a writing team for a preliminary version of this paper prepared for the IWTC-VI in Costa Rica, November, 2006: Seita Emori, Jenni Evans, Greg Holland, Kam-biu Liu, Ruth McDonald, David Nolan, Masato Sugi, and Yuqing Wang. Liu and McDonald contributed significant draft material for the paleoclimate and global modeling sections. A number of colleagues provided useful reviews.

References

AchutaRao, K. and K. Sperber, 2006: ENSO simulations in coupled ocean-atmosphere models: are the current models better? *Clim. Dyn.*, **27**, 1–16.

Allen, M. and W. Ingram, 2002: Constraints on future changes in climate and the hydrological cycle, *Nature*, **419**, 224–232.

Atlas, R., *et al.*, 2001: The effects of marine winds from scatterometer data on weather analysis and forecasting, *Bull. Am. Meteorol. Soc.*, **82**, 1965–1990.

Barnett, T. P., D.W. Pierce, K. M. AchutaRao, P. J. Gleckler, B. D. Santer, J. M. Gregory, and W. M. Washington, 2005: Penetration of human-induced warming into the world's oceans. *Science*, **309**, 284–287.

Bell, G. D. and M. Chelliah, 2006: Leading tropical modes associated with interannual and multidecadal fluctuations in North Atlantic hurricane activity. *J. Climate*, **19**, 590–612.

Bender, M. A. and I. Ginis, 2000: Real-case simulations of hurricane-ocean interaction using a high-resolution coupled model: effects on hurricane intensity. *Mon. Wea. Rev.*, **128**(4), 917–946.

Bengtsson, L., K. Hodges, and E. Roeckner, 2006: Storm tracks and climate change. *J. Climate*, **19**, 3518–3543.

Bengtsson, L., K. I. Hodges, M. Esch, N. Keenlyside, L. Kornblueh, J.-J. Luo, and T. Yamatata, 2007: How may tropical cyclones change in a warmer climate? *Tellus A*, **59**, 539–561.

Biasutti, M. and A. Giannini, 2006: Robust Sahel drying in response to late 20th century forcings. *Geophys. Res. Lett.*, **33**, L11706.

Boos, W. R., J. R. Scott, and K. A. Emanuel, 2004: Transient diapycnal mixing and the meridional overturning circulation. *J. Phys. Ocean.*, **34**, 334–341.

Bove, M. C., J. B. Elsner, C. W. Landsea, X. Niu, and J. J. O'Brien, 1998: Effect of El Niño on U.S. landfalling hurricanes, revisited. *Bull. Amer. Meteor. Soc.*, 79, 2477–2482.

Briggs, W. M., 2008: On the changes in number and intensity of North Atlantic tropical cyclones. *J. Climate*, **21**, 1387–1402.

Caron, L.-P. and C. G. Jones, 2008: Analysing present, past, and future tropical cyclone activity as inferred from an ensemble of coupled global climate models. *Tellus*, **60A**, 80–96.

CCSP, 2006: *Temperature Trends in the Lower Atmosphere: Steps for Understanding and Reconciling Differences.* A Report by the Climate Change Science Program and the Subcommittee on Global Change Research. [Thomas R. Karl, Susan J. Hassol, Christopher D. Miller, and William L. Murray (eds,)] Department of Commerce, NOAA's National Climatic Data Center, Washington, DC., USA, 164 pp.

CCSP, 2008: *Weather and Climate Extremes in a Changing Climate. Regions of Focus: North America, Hawaii, Caribbean, and U.S. Pacific Islands.* A Report by the U.S. Climate Change Science Program and the Subcommittee on Global Change Research. [Thomas R. Karl, Gerald A. Meehl, Christopher D. Miller, Susan J. Hassol, Anne M. Waple, and William L. Murray (eds.)]. Department of Commerce, NOAA's National Climatic Data Center, Washington, D.C., USA, 164 pp.

Chan, J. C. L., 2006: Comment on "Changes in Tropical Cyclone Number, Duration, and Intensity in a Warming Environment, *Science*, v. 311, p. 1713.

Chan, J. C. L., 2008: Decadal variations of intense typhoon occurrence in the western North Pacific. *Proc. R. Soc. A.*, **464**, 249–272.

Chan, J. C. L. and K. S. Liu, 2004: Global warming and Western North Pacific typhoon activity from an observational perspective. *J. Climate*, **17**, 4590–4602.

Chang E. K. and M. Y. Guo, 2007: Is the number of North Atlantic tropical cyclones significantly underestimated prior to the availability of satellite observations? *Geophys. Res. Lett.*, **34**, L14801.

Chauvin, F., J.-F. Royer, and M. Déqué, 2006: Response of hurricane-type vortices to global warming as simulated by ARPEGE-Climat at high resolution. *Clim. Dyn.*, **27**, 377–399

Chen, S. S., J. F. Price, Z. Wei, M. A. Donelan, and E. J. Walsh, 2007: The CBLAST-Hurricane Program and the next-generation fully coupled atmosphere–wave–ocean models for hurricane research and prediction. *Bull. Amer. Meteor. Soc.*, **88**(3), 311–317.

Chenoweth, M., 2003: The 18th century climate of Jamaica, derived from the journals of Thomas Thistlewood, 1750–1786. *Trans. Am. Philos. Soc.*, **93**(2).

Chenoweth, M. and D. Divine, 2008: A document-based 318-year record of tropical cyclones in the Lesser Antilles, 1690–2007. *Geochem., Geophys., Geosys.*, **9**, Q08013.

Delworth, T. L. and T. R. Knutson, 2000: Simulation of early 20th century global warming. *Science*, **287**(5461), 2246–2250.

Delworth, T. L. and M. E. Mann, 2000: Observed and simulated multidecadal variability in the Northern Hemisphere. *Clim. Dyn.*, **16**(9), 661–676.

Delworth, T. L., R. Zhang, and M. E. Mann, 2007: Decadal to Centennial variability of the Atlantic from observations and models. In *Ocean Circulation: Mechanisms and Impacts*, Geophys. Mono. Ser. 173, Washington, DC: American Geophysical Union; 131–148.

DeMott, C. A. and D. A. Randall, 2004: Observed variations of tropical convective available potential energy. *J. Geophys. Res.*, **109**, D02102.

Donnelly, J. P., S. S. Bryant, J. Butler, J. Dowling, L. Fan, N. Hausmann, P. Newby, B. Shuman, J. Stern, K. Westover, and T. Webb, III, 2001a: A 700 yr sedimentary record of intense hurricane landfalls in southern New England. *Geol. Soc. Amer. Bull.*, **113**, 714–727.

Donnelly, J. P., S. Roll, M. Wengren, J. Butler, R. Lederer, and T. Webb, III, 2001b: Sedimentary evidence of intense hurricane strikes from New Jersey. *Geology*, **29**, 615–618.

Donnelly, J. P. and T. Webb, III, 2004: Backbarrier sedimentary records of intense hurricane landfalls in the northeastern United States. In: *Hurricanes and Typhoons: Past, Present, and Future* (eds. R.J. Murnane, R. J. and K-b. Liu), pp. 58–95, Columbia University Press.

Donnelly, J. P., J. Butler, S. Roll, M. Wengren, and T. Webb, III, 2004: A backbarrier overwash record of intense storms from Brigantine, New Jersey. *Marine Geology*, **210**, 107–121.

Donnelly, J. P. 2005: Evidence of past intense tropical cyclones from backbarrier salt pond sediments: a case study from Isla de Culebrita, Puerto Rico, USA. *J. Coastal Res.*, **42**, 201–210.

Donnelly, J. P. and J. D. Woodruff, 2007: Intense hurricane activity over the past 5,000 years controlled by El Nino and the West African monsoon. *Nature*, **447**, 465–468.

Dunion, J. P. and C. S. Velden, 2004: The impact of the Saharan air layer on Atlantic tropical cyclone activity. *Bull. Am. Meteorol. Soc.*, **85**(3), 353–365.

Dutton, J. F., C. J. Poulsen, and J. L. Evans, 2000: The effect of global climate change on the regions of tropical convection in CSM1. *Geophys. Res. Lett.*, **27**, 3049.

Elsner, J. B., 2006: Evidence in support of the climate change-Atlantic hurricane hypothesis. *Geophys. Res. Lett.*, **33**, L16705.

Elsner, J. B., 2007: Granger causality and Atlantic hurricanes. *Tellus*, **59**, 476–485.

Elsner, J. B. and T. H. Jagger, 2008: United States and Caribbean tropical cyclone activity related to the solar cycle. *Geophys. Res. Lett.*, **35**, L18705.

Elsner, J. B., T. H. Jagger, and A. A. Tsonis, 2006: Estimated return periods for Hurricane Katrina. *Geophys. Res. Lett.*, **33**, L08704.

Elsner, J. B., J. P. Kossin, and T. H. Jagger, 2008: The increasing intensity of the strongest tropical cyclones. *Nature*, **455**, 92–95.

Elsner, J. B., K-b. Liu, and B. L. Kocher, 2000: Spatial variations in major U.S. hurricane activity: Statistics and a physical mechanism. *J. Climate*, **13**, 2293–2305.

Elsner, J. B., A. A. Tsonis, and T. H. Jagger, 2006: High-frequency variability in hurricane power dissipation and its relationship to global temperature. *Bull. Amer. Meteor. Soc.*, **87**, 763–768.

Emanuel, K. A., 1987: The dependence of hurricane intensity on climate. *Nature*, **326**, 483–485.

Emanuel, K. A., 1999: Thermodynamic control of hurricane intensity. *Nature*, **401**, 665–669.

Emanuel, K., 2000: A statistical analysis of tropical cyclone intensity. *Mon. Wea. Rev.*, **128**, 1139–1152.

Emanuel, K. A., 2001: The contribution of tropical cyclones to the oceans' meridional heat transport. *J. Geophys. Res.*, **106**, D14, 14,771–14,781.

Emanuel, K., 2002: A simple model of multiple climate regimes. *J. Geophys. Res.*, **107**(0).

Emanuel, K. A., 2005a: Increasing destructiveness of tropical cyclones over the past 30 years. *Nature*, **436**, 686–688.

Emanuel, K. A., 2005b: Emanuel replies. *Nature*, **438**.

Emanuel, K., 2007a: Environmental factors affecting tropical cyclone power dissipation. *J. Climate*, **20**, 5497–5509.

Emanuel, K., 2007b: Comment on "Sea-surface temperatures and tropical cyclones in the Atlantic basin" by Patrick J. Michaels, Paul C. Knappenberger, and Robert E. Davis. *Geophys. Res. Lett.*, **34**.

Emanuel, K., R. Sundararajan, and J. Williams, 2008: Hurricanes and global warming: results from downscaling IPCC AR4 simulations. *Bull. Amer. Meteor. Soc.*, **89**(3), 347–367.

Emori, S. and S. J. Brown, 2005: Dynamic and thermodynamic changes in mean and extreme precipitation under changed climate. *Geophys. Res. Lett.*, **32**, L17706.

Enfield, D. B. and L. Cid-Serrano, 2008: Secular and multidecadal warmings in the North Atlantic and their relationships with major hurricane activity. Accepted in *International Journal of Climatology*.

Enfield, D. B. and A. M. Mestas-Nuñez, 2000: Global modes of ENSO and non-ENSO SST variability and their associations with climate. In: *El Niño and the Southern Oscillation: Multiscale Variability and Global and Regional Impacts*, H.F. Diaz and V. Markgraf, eds., Cambridge University Press p. 89–112.

Evan, A. T., J. Dunion, J. A. Foley, A. K. Heidinger, and C. S. Velden, 2006: New evidence for a relationship between Atlantic tropical cyclone activity and African dust outbreaks, *Geophys. Res. Lett.*, **33**, L19813.

Fernandez-Partagas, J. and H. F. Diaz, 1996: Atlantic hurricanes in the second half of the nineteenth century. *Bull. Amer. Meteor. Soc.*, **77**, 2899–2906.

Frappier, A. B., D. Sahagian, S. J. Carpenter, L. A. Gonzalez, and B. R. Frappier, 2007: A stalagmite stable isotope record of recent tropical cyclone events. *Geology*, **35**, 110–114.

Free, M., M. Bister, and K. Emanuel, 2004: Potential intensity of tropical cyclones: comparison of results from radiosonde and reanalysis data. *J. Climate*, **17**, 1722–1727.

Fu, Q., C. M. Johanson, S. G. Warren, and D. J. Seidel, 2004: Contribution of stratospheric cooling to satellite-inferred tropospheric temperature trends. *Nature*, **429**, 55–58.

Garcia Herrera, R., L. Gimeno, P. Ribera, and E. Hernandez, 2005: New records of Atlantic hurricanes from Spanish documentary sources. *J. Geophys. Res.*, **110**, D03109.

Garcia Herrera, R., F. Rubio, D. Wheeler, E. Hernandez, M. R. Prieto, and L. Gimero, 2004: The use of Spanish and British documentary sources in the investigation of Atlantic hurricane incidence in historical times. In: *Hurricanes and Typhoons: Past, Present, and Future* (eds. R.J. Murnane, R. J. and K-b. Liu), pp. 149–176. Columbia University Press.

Gettelman, A. D., J. Seidel, M. C. Wheeler, and R. J. Ross, 2002: Multidecadal trends in tropical convective available potential energy. *J. Geophys. Res.*, **107**, 4606.

Goldenberg, S. B., C. W. Landsea, A. M. Mesta-Nuñez, and W. M. Gray, 2001: The recent increase in Atlantic hurricane activity: causes and implications. *Science*, **293**, 474–479.

Gray, W. M., 1990: Strong association between West African rainfall and U.S. landfall of intense hurricanes. *Science*, **249**, 1251–1256.

Groisman, P. Y., R. W. Knight, T. R. Karl, D. R. Easterling, B. Sun, and J. H. Lawrimore, 2004: Contemporary changes of the hydrological cycle over the contiguous United States: trends derived from *in situ* observations. *J. Hydormeteor.*, **5**, 64–85.

Gualdi, S., E. Scoccimarro, and A. Navarra, 2008: Changes in tropical cyclone activity due to global warming: results from a high-resolution coupled general circulation model. *J. Climate*, **21**, 5204–5228.

Harper, B. A., S. A. Stroud, M. McCormack, and S. West, 2008: A review of historical tropical cyclone intensity in northwestern Australia and implications for climate change trend analysis. *Aust. Met. Mag.*, **57**, 121–141.

Hasegawa, A. and S. Emori, 2005: Tropical cyclones and associated precipitation over the Western North Pacific: T106 atmospheric GCM simulation for present-day and doubled CO_2 climates. *SOLA*, **1**, 145148, SOI:10.2151/sola. 2005-038.

Hasegawa, A. and S. Emori, 2007: Effect of air-sea coupling in the assessment of CO_2-induced intensification of tropical cyclone activity. *Geophys. Res. Lett.*, **34**, L05701.

Hayne, M. and J. Chappell, 2001: Cyclone frequency during the last 5,000 years from Curacoa Island, Queensland. *Palaeogeography, Palaeoclimatology, Palaeoecology*, **168**, 201–219.

Held, I. M., T. L. Delworth, J. Lu, K. L. Findell, and T. R. Knutson, 2005: Simulation of Sahel drought in the 20th and 21st centuries. *Proc. Nat. Acad. Sci.*, **102**(50), 17891–17896.

Held, I. M. and B. J. Soden, 2006: Robust responses of the hydrologic cycle to global warming. *J. Climate*, **19**, 5686–5699.

Henderson-Sellers, A., H. Zhang, G. Berz, K. Emanuel, W. Gray, C. Landsea, G. Holland, J. Lighthill, S-L. Shieh, P. Webster, and K. McGuffie, 1998: Tropical cyclones and global climate change: a post-IPCC assessment. *Bull. Amer. Meteor. Soc.*, **79**, 19–38.

Holland, G. J., 1997: The maximum potential intensity of tropical cyclones. *J. Atmos. Sci.*, **54**, 2519–2541.

Holland, G. J., 2007: Misuse of landfall as a proxy for Atlantic tropical cyclone activity, *Eos Trans. AGU*, **88**, 349, 10.1029/2007EO360001.

Holland, G. J. and P. J. Webster, 2007: Heightened tropical cyclone activity in the North Atlantic: natural variability or climate trend? *Philos. Trans. Roy. Soc. A*, **365**, 2695–2716.

Hoyos, C. D., P. A. Agudelo, P. J. Webster, and J. A. Curry, 2006: Deconvolution of the factors contributing to the increase in global hurricane intensity. *Science*, 312: 94–97.

IPCC, 2007: *Climate Change 2007: The Physical Science Basis. Contribution of Working Group I to the Fourth Assessment Report of the Intergovernmental Panel on Climate Change* [Solomon, S., D. Qin, M. Manning, Z. Chen, M. Marquis, K.B. Averyt, M. Tignor and H.L. Miller (eds.)]. Cambridge University Press, Cambridge, 996 pp.

Jagger, T. H. and J. B. Elsner, 2006: Climatology models for extreme hurricane winds near the United States. *J. Climate*, **19**, 3220–3236.

Kamahori, H. N. Yamazaki, N. Mannoji, and K. Takahashi, 2006: Variability in intense tropical cyclone days in the western North Pacific. *SOLA*, **2**, 104–107.

Karoly, D. J. and Q. Wu, 2005: Detection of regional surface temperature trends. *J. Climate*, **18**, 4337–4343.

Klotzbach, P. J., 2006: Trends in global tropical cyclone activity over the past twenty years (1986-2005), *Geophys. Res. Lett.*, **33**, L10805.

Knaff, J. A., 1997: Implications of summertime sea level pressure anomalies in the tropical Atlantic region. *J. Climate*, **10**, 789–804.

Knight, J. R., R. J. Allan, C. K. Folland, M. Vellinga, and M. E. Mann, 2005: A signature of persistent natural thermohaline circulation cycles in observed climate, *Geophys. Res. Lett.*, **32**, L20708.

Knight, J. R., C. K. Folland, and A. A. Scaife, 2006: Climate impacts of the Atlantic Multidecadal Oscillation. *Geophys. Res. Lett.*, **33**, 17, L17706.

Knutson, T. R., T. L. Delworth, K. W. Dixon, I. M. Held, J. Lu, V. Ramaswamy, D. Schwarzkopf, G. Stenchikov, and R. J. Stouffer, 2006: Assessment of twentieth-century regional surface temperature trends using the GFDL CM2 coupled models. *J. Climate*, **19**(9), 1624–1651.

Knutson, T. R., J. J. Sirutis, S. T. Garner, I. M. Held, and R. E. Tuleya, 2007: Simulation of the recent multidecadal increase of Atlantic hurricane activity using an 18-km-grid regional model. *Bull. Amer. Meteor. Soc.*, **88**(10), 1549–1565.

Knutson, T. R., J. J. Sirutis, S. T. Garner, G. A. Vecchi, and I. M. Held, 2008: Simulated reduction in Atlantic hurricane frequency under 21st century warming conditions. *Nature Geoscience*, **1**, 359–364.

Knutson, T. R. and R. E. Tuleya, 1999: Increased hurricane intensities with CO_2-induced global warming as simulated using the GFDL hurricane prediction system, *Clim. Dyn.*, **15**(7), 503–519.

Knutson, T. R. and R. E. Tuleya, 2004: Impact of CO_2-induced warming on simulated hurricane intensity and precipitation: sensitivity to the choice of climate model and convective parameterization. *J. Climate*, **17**, 3477–3495.

Knutson, T. R. and R. E. Tuleya, 2005: Reply. *J. Climate*, **18**, 5183–5187.

Knutson, T. R. and R. E. Tuleya, 2008: Tropical cyclones and climate change: Revisiting recent studies at GFDL. In *Climate Extremes and Society*, R. Murnane and H. Diaz, eds., Cambridge University Press, Cambridge, 120–144.

Knutson, T. R., R. E. Tuleya, W. Shen, and I. Ginis, 2001: Impact of CO_2-induced warming on hurricane intensities as simulated in a hurricane model with ocean coupling. *J. Climate*, **14**, 2458–2468.

Knutson, T. R., R. E. Tuleya, and Y. Kurihara, 1998: Simulated increase of hurricane intensities in a CO_2 warmed climate. *Science*, **279**(5353), 1018–1020.

Kossin, J. P., K. R. Knapp, D. J. Vimont, R. J. Murnane, and B. A. Harper, 2007: A globally consistent reanalysis of hurricane variability and trends. *Geophys. Res. Lett.*, **34**, L04815.

Kossin J. P. and D. J. Vimont, 2007: A more general framework for understanding Atlantic hurricane variability and trends. *Bull. Amer. Meteor. Soc.*, **88**, 1767–1781.

Kuleshov, Y., L. Qi, R. Fawcett and D. Jones, 2008: On tropical cyclone activity in the Southern Hemisphere: trends and the ENSO connection. *Geophys. Lett.*, **35**, L14508.

Landman, W. A., A. Seth, and S. J. Camargo, 2005: The effect of regional climate model domain choice on the simulation of tropical cyclone-like vortices in the Southwestern Indian Ocean. *J. Climate*, **18**, 1253–1274.

Landsea, C. W., 1993: A climatology of intense (or major) Atlantic hurricanes. *Mon. Wea. Rev.*, **121**, 1703–1713.

Landsea, C. W., C. Anderson, N. Charles, G. Clark, J. Dunion, J. Fernandez-Partagas, P. Hungerford, C. Neumann, and M. Zimmer, 2004: The Atlantic hurricane database reanalysis project: Documentation for the 1851–1910 alterations and additions to the HURDAT database. In: *Hurricanes and Typhoons: Past, Present and Future*, R. J. Murname and K.-b. Liu, Eds., Columbia University Press, pp. 177–221.

Landsea, C. W., 2005: Hurricanes and global warming. *Nature*, **438**.

Landsea, C. W., B. A. Harper, K. Hoarau, and J. A. Knaff, 2006: Can we detect trends in extreme tropical cyclones? *Science*. **313**, 452–454.

Landsea, C. W., 2007: Counting Atlantic tropical cyclones back to 1900. *EOS*, **88**, 197 & 202.

Landsea, C. W., D. A. Glenn, W. Bredemeyer, M. Chenoweth, R. Ellis, J. Gamache, L. Hufstetler, C. Mock, R. Perez, R. Prieto, J. Sanchez-Sesma, D. Thomas, and L. Woolcock, 2008: A reanalysis of the 1911 to 1920 Atlantic hurricane database. *J. Climate*, **21**, 2138–2168.

LaRow, T. E., Y.-K., Lim, D. W. Shin, E. P. Chassignet, and S. Cocke, 2008: Atlantic basin seasonal hurricane simulations. *J. Climate*, **21**, 3191–3206.

Latif, M., N. Keenlyside, and J. Bader, 2007: Tropical sea surface temperature, vertical wind shear, and hurricane development. *Geophys. Res. Lett.*, **34**, L01710.

Lau, K.-M. and H.-T. Wu, 2007: Detecting trends in tropical rainfall characteristics, 1979-2003. *Int. J. Climatol.*, **27**, 979–988.

Lawrence, J. R. and S. D. Gedzelman, 1996: Low stable isotope ratios of tropical cyclone rains. *Geophys. Res. Lett.*, **23**, 527–530.

Leslie, L. M., D. J. Karoly, M. Leplastrier, and B. W. Buckley, 2007: Variability of tropical cyclones over the southwest Pacific Ocean using a high-resolution climate model. *Meteorol. Atmos. Phys.*, **97**, 171–180.

Liu, K.-b., 2004: Paleotempestology: Principles, Methods, and Examples from Gulf Coast Lake Sediments. In: *Hurricanes and Typhoons: Past, Present, and Future*, R.J. Murnane, R. J. and K.-b. Liu, eds., Columbia University Press, pp. 13–57.

Liu, K.-b. and M. L. Fearn, 1993: Lake-sediment record of late Holocene hurricane activities from coastal Alabama. *Geology*, **21**, 793–796.

Liu, K.-b. and M. L. Fearn, 2000: Reconstruction of prehistoric landfall frequencies of catastrophic hurricanes in northwestern Florida from lake sediment records. *Quaternary Research*, **54**, 238–245.

Liu, K.-b., C. Shen, and K. S. Louie, 2001: A 1000-year history of typhoon landfalls in Guangdong, southern China, reconstructed from Chinese historical documentary records. *Annals of the Association of American Geographers*, **91**, 453–464.

Louie, K. S. and K.-b. Liu, 2003: Earliest historical records of typhoons in China. *J. Historical Geography*, **29**, 299–316.

Louie, K. S. and K.-b. Liu, 2004: Ancient records of typhoons in Chinese historical documents. In: *Hurricanes and Typhoons: Past, Present, and Future*, R.J. Murnane, R. J. and K.-b. Liu, eds., Columbia University Press, pp. 222–248.

Ludlam, D.M., 1963: *Early American hurricanes, 1492–1870*. American Meteorological Society.

Mann, M. and K. Emanuel, 2006: Atlantic hurricane trends linked to climate change. *EOS*, **87**, 233–241.

Mann, M. E., T. A. Sabbatelli, and U. Neu, 2007: Evidence for a modest undercount bias in early historical Atlantic tropical cyclone counts. *Geophys. Res. Lett.*, **34**, L22707.

Manning, D. and R. E. Hart, 2007: Evolution of North Atlantic ERA40 tropical cyclone representation. *Geophys. Res. Lett.*, **34**, L05705.

Marshall, G. J., P. A. Stott, J. Turner, W. M. Connolley, J. C. King, T. A. Lachlan-Cope, 2004: Causes of exceptional atmospheric circulation changes in the Southern Hemisphere. *Geophys. Res. Lett.*, **31**(14), L14205.

Maue, R. N. and R. E. Hart, 2007: Comment on "Low frequency variability in globally integrated tropical cyclone power dissipation" by Ryan Sriver and Matthew Huber. *Geophys. Res. Lett.*, **34**, L11703.

McDonald, R. E., D. G. Bleaken, D. R. Cresswell, V. D. Pope, and C. A. Senior, 2005: Tropical storms: representation and diagnosis in climate models and the impacts of climate change. *Clim. Dyn.*, **25**, 19–36.

Mears, C. A. and F. J. Wentz, 2005: The effect of diurnal correction on satellite-derived lower tropospheric temperature. *Science*, **309**, 1548–1551.

Meehl, G. A., C. Covey, T. Delworth, M. Latif, B. McAvaney, J. F. B. Mitchell, R. J. Stouffer, and K. E. Taylor, 2007: The WCRP CMIP3 multimodel dataset: A new era in climate change research. *Bull. Amer. Meteor. Soc.*, **88**(9), 1383–1394.

Meehl, G. M., W. M. Washingon, C. M. Amman, J. M. Arblaster, T. M. L. Wigley, and C. Tebaldi, 2004: Combinations of natural and anthropogenic forcings in twentieth-century climate. *J. Climate*, **17**, 37213727.

Merryfield, W. J., 2006: Changes to ENSO under CO_2 doubling in a multimodel ensemble. *J. Climate*, **19**, 4009–4027.

Michaels, P. J., P. C. Knappenberger, and R. E. Davis, 2006: Sea-surface temperatures and troical cyclones in the Atlantic basin. *Geophys. Res. Lett.*, **33**, L09708.

Millas, J. C., 1968: *Hurricanes of the Caribbean and adjacent regions, 1492–1800*. Academy of the Arts and Sciences of the Americas.

Miller, D. L., C. I. Mora, H. D. Grissino-Mayer, C. J. Mock, M. E. Uhle, and Z. Sharp, 2005: Tree-ring isotope records of tropical cyclone activity. *Proc. Nat. Acad. Sci.*, **103**(39), 14294–14297.

Mock, C. J., 2004: Tropical cyclone reconstructions from documentary records: examples for South Carolina, United States. In: *Hurricanes and Typhoons: Past, Present, and Future*, R.J. Murnane, R. J. and K.-b. Liu, eds., Columbia University Press, pp. 121–148.

Neu, U., 2008: Is recent major hurricane activity normal? *Nature*, **451**, E5.

Nolan, D. S., E. D. Rappin, and K. A. Emanuel, 2007: Tropical cyclogenesis sensitivity to environmental parameters in radiative-convective equilibrium. *Q. J. R. Meteorol. Soc.*, **133**, 2085–2107.

Nott, J. and M. Hayne. 2001. High frequency of "super-cyclones" along the Great Barrier Reef over the past 5,000 years. *Nature*, **413**, 508–512.

Nyberg, J., B. A. Malmgren, A. Winter, M. R. Jury, K. Halimeda Kilbourne, and T. M. Quinn, 2007: Low Atlantic hurricane activity in the 1970s and 1980s compared to the past 270 years. *Nature*, **447**, 698–702.

Nyberg, J., B. A. Malmgren, A. Winter, M. R. Jury, K. Halimeda Kilbourne, and T. M. Quinn, 2008: Nyberg *et al.* reply. *Nature*, **451**, E6.

Oouchi, K., J. Yoshimura, H. Yoshimura, R. Mizuta, S. Kusunoki, and A. Noda, 2006: Tropical cyclone climatology in a global-warming climate as simulated in a 20km-mesh global atmospheric model: frequency and wind intensity analysis. *J. Meteorol. Soc. Japan*, **84**, 259–276.

Pasquero, C. and K. Emanuel, 2008: Tropical cyclones and transient upper-ocean warming. *J. Climate*, **21**, 149–162.

Pezza, A. B. and I. Simmonds, 2005: The first South Atlantic hurricane: unprecedented blocking, low shear, and climate change. *Geophys. Res. Lett.*, **32**, L15712.

Pielke, R. A., Jr., 2005: Are there trends in hurricane destruction? *Nature*, **438**, E11.

Pielke, R. A., J. Gratz, C. W. Landsea, D. Collins, M. A. Saunders, and R. Musulin, 2008: Normalized hurricane damage in the United States: 1900-2005. *Natural Hazards Review*, **9**, 29–42.

Rotstayn, L. and U. Lohmann, 2002: Tropical rainfall trends and the indirect aerosol effect. *J. Climate*, **15**, 2103–2116.

Royer. J.-F., F. Chauvin, B. Timbal, P. Araspin, and D. Grimal, 1998: A GCM study of the impact of greenhouse gas increase on the frequency of occurrence of tropical cyclones. *Clim. Change*, **38**, 307–343.

Ryan, B. F., I. G. Watterson, and J. L. Evans, 1992: Tropical cyclones frequencies inferred from Gray's yearly genesis parameter: validation of GCM tropical climate. *Geophys. Res. Lett.*, **19**, 1831–1834.

Santer, B. D., and Coauthors, 2005: Amplification of surface temperature trends and variability in the tropical atmosphere. *Science*, **309**, 1551–1556.

Santer, B. D., *et al.*, 2006: Forced and unforced ocean temperature changes in Atlantic and Pacific tropical cyclogenesis regions. *Proc. Nat. Acad. Sci.*, **103**, 13905–13910, 10.1073/pnas.0602861103.

Santer, B. D. and Coauthors, 2007: Identification of human-induced changes in atmospheric moisture content. *Proc. Nat. Acad. Sci.*, **104**, 15248–15253, 10.1073/pnas.0702872104.

Saunders, M. A. and A. S. Lea, 2005: Seasonal prediction of hurricane activity reaching the coast of the United States. *Nature*, **434**, 1005–1008.

Saunders, M. A. and A. S. Lea, 2008: Large contribution of sea surface warming to recent increase in Atlantic hurricane activity. *Nature*, **451**, 557–561.

Scileppi, E. and J. P. Donnelly, 2007: Sedimentary evidence of hurricane strikes in western Long Island, New York. *Geochem., Geophys., Geosys.*, **8**, Q06011.

Semmler, T., S. Varghese, R. McGrath, P. Nolan, S. Wang, P. Lynch, and C. O'Dowd, 2008: Regional climate model simulations of North Atlantic cyclones: frequency and intensity changes. *Clim. Res.*, **36**, 1–16.

Shen, W., R. E. Tuleya, and I. Ginis, 2000: A sensitivity study of the thermodynamic environment on GFDL model hurricane intensity: implications for global warming. *J. Climate*, **13**, 109–121.

Sherwood, S. C., J. R. Lanzante, and C. L. Meyer, 2005: Radiosonde daytime biases and late-20th century warming. *Science*, **309**, 1556–1559.

Soden, B. J., D. L. Jackson, V. Ramaswamy, M. D. Schwarzkopf, and X. Huang, 2005: The radiative signature of upper tropospheric moistening. *Science*, **310**(5749), 841–844.

Solow, A. R. and A. R. Beet, 2008: On the incompleteness of the historical record of North Atlantic tropical cyclones. *Geophys. Res. Lett.*, **35**, L11803.

Solow, A. R. and L. J. Moore, 2002: Testing for trend in North Atlantic Hurricane Activity, 1900-98. *J. Climate*, **15**, 3111–3114.

Sriver, R. and M. Huber, 2006: Low frequency variability in globally integrated tropical cyclone power dissipation. *Geophys. Res. Lett.*, **33**, L11705.

Sriver, R. and M. Huber, 2007: Observational evidence for an ocean heat pump induced by tropical cyclones. *Nature*, **447**, 577–580.

Sriver, R., M. Huber, and J. Nusbaumer, 2008: Investigating tropical cyclone-climate feedbacks using the TRMM Microwave Imager and the Quick Scatterometer. *Geochem., Geophys., Geosys.*, **9**, Q09V11.

Stott, P. A., S. F. B. Tett, G. S. Jones, M. R. Allen, J. F. B. Mitchell, and G. J. Jenkins, 2000: External control of 20th century temperature by natural and anthropogenic forcings. *Science*, **290**, 2133–2137.

Stowasser, M., Y. Wang, and K. Hamilton, 2007: Tropical cyclone changes in the western North Pacific in a global warming scenario. *J. Climate*, **20**, 2378–2396.

Sugi, M., A. Noda, and N. Sato, 2002: Influence of global warming on tropical cyclone climatology: an experiment with the JMA global model. *J. Meteorol. Soc. Japan*, **80**, 249–272.

Sugi, M. and J. Yoshimura, 2004: A mechanism of tropical precipitation change due to CO_2 increase. *J. Climate*, **17**, 238–243.

Swanson, K., 2007: Impact of scaling behavior on tropical cyclone intensities. *Geophys. Res. Lett.*, **34**, L18815.

Swanson, K., 2008: Nonlocality of Atlantic tropical cyclone intensities. *Geochem., Geophys., Geosys.*, **9**, Q04V01.

Tang, B. H. and J. D. Neelin, 2004: ENSO influence on Atlantic hurricanes via tropospheric warming. *Geophys. Res. Lett.*, **31**, L24204.

Tonkin, H., G. Holland, C. Landsea, and S. Li, 1997: Tropical cyclones and climate change: a preliminary assessment. In *Assessing Climate Change: Results from the Model Evaluation and Consortium for Climate Assessments*, W. Howe and A. Henderson-Sellers, eds., Taylor and Francis Group, London, pp. 327–360.

Trenberth, K., 1999: Conceptual framework for changes of extremes of the hydrological cycle with climate change. *Clim. Change*, **42**, 327–339.

Trenberth, K., 2005: Uncertainty in hurricanes and global warming. *Science*, **308**, 1753–1754.

Trenberth, K. E. and J. Fasullo, 2008: Energy budgets of Atlantic hurricanes and changes from 1970. *Geochem., Geophys., Geosys.*, **9**, Q09V08.

Trenberth, K. E., J. Fasullo, and L. Smith, 2005: Trends and variability in column-integrated atmospheric water vapor. *Clim. Dyn.*, **24**, 741–758.

Trenberth, K. E. and D. J. Shea, 2006: Atlantic hurricanes and natural variability in 2005. *Geophys. Res. Lett.*, **33**, L12704.

Trenberth, K. and J. Fasullo, 2007: Water and energy budgets of hurricanes and implications for climate change. *J. Geophys. Res.*, **112**, D23107.

Trenberth, K., C. A. Davis, and J. Fasullo, 2007: Water and energy budgets of hurricanes: Case studies of Ivan and Katrina, *J. Geophys. Res.*, **112**, D23106.

Tsutsui, J., 2002: Implications of anthropogenic climate change for tropical cyclone activity: a case study with the NCAR CCM2. *J. Meteorol. Soc. Japan*, **80**, 45–65, doi: 10.2151/jmsj.80.45.

Uhlhorn, E. W. and P. G. Black, 2003: Verification of remotely sensed sea surface winds in hurricanes. *J. Atmos. and Ocean. Tech.*, **20**, 99–116.

Vecchi, G. A., A. Clement, and B. J. Soden, 2008: Examining the tropical Pacific's response to global warming. *EOS*, **89**, 81–83.

Vecchi, G. A. and T. R. Knutson, 2008: On estimates of historical North Atlantic tropical cyclone activity. *J. Climate*, **21**, 3580–3600.

Vecchi, G. A., B. J. Soden, A. T. Wittenberg, I. M. Held, A. Leetmaa, and M. J. Harrison, 2006: Weakening of the tropical Pacific atmospheric circulation due to anthropogenic forcing. *Nature*, **441**, 73–76.

Vecchi, G. A. and B. J. Soden, 2007a: Effect of remote sea surface temperature change on tropical cyclone potential intensity. *Nature*, **450**(7172), 1066–1070.

Vecchi, G. A. and B. J. Soden, 2007b: Increased tropical Atlantic wind shear in model projections of global warming. *Geophys. Res. Lett.*, **34**, L08702.

Vitart, F. and J. L. Anderson, 2001: Sensitivity of Atlantic tropical storm frequency to ENSO and interdecadal variability of SSTs in an ensemble of AGCM integrations. *J. Climate*, **14**(4), 533–545.

Vitart, F. and T. Stockdale, 2001: Seasonal forecasting of tropical storms using coupled GCM integrations. *Mon. Wea. Rev.*, **129**, 2521–2537.

Walsh, K., 2004: Tropical cyclones and climate change: unresolved issues. *Clim. Res.*, **27**, 77–83.

Walsh, K. J. E., M. Fiorino, C. W. Landsea, and K. McInnes, 2007: Objectively-determined resolution-dependent threshold criteria for the detection of tropical cyclones in climate models and reanalyses. *J. Climate*, **20**, 2307–2314.

Walsh, K., D. Karoly, and N. Nicholls, 2008: The detection and attribution of climate change effects on tropical cyclones. To appear in: *Hurricanes and Climate* (J. Elsner, ed.), Springer.

Walsh, K. J. E., K.-C. Nguyen, and J. L. McGregor, 2004: Fine-resolution regional climate model simulations of the impact of climate change on tropical cyclones near Australia. *Clim. Dyn.*, **22**, 47–56.

Wang, C. and S.-K. Lee, 2008: Global warming and United States landfalling hurricanes. *Geophys. Res. Lett.*, **35**, L02708.

Wang, C., S.-K. Lee, and D. B. Enfield, 2008: Atlantic Warm Pool acting as a link between Atlantic Multidecadal Oscillation and Atlantic tropical cyclone activity. *Geochem., Geophys., Geosys.*, **9**, Q05V03.

Webster, P. J., G. J. Holland, J. A. Curry, and H.-R. Chang, 2005: Changes in tropical cyclone number, duration, and intensity in a warming environment. *Science*, **309**, 1844–1846.

Webster, P. J., J. A. Curry, J. Liu, and G. J. Holland, 2006: Response to comment on "Changes in tropical cyclone number, duration, and intensity in a warming environment". *Science*, **311**, 1713c.

Wittenberg, A. T., A. Rosati, N.-C. Lau, and J. J. Ploshay, 2006: GFDL's CM2 global coupled climate models Part 3: Tropical Pacific climate and ENSO. *J. Climate*, **19**(5), 698–722.

WMO (World Meteorological Organization), 2006: Atmospheric Research and Environment Programme. Statement on Tropical Cyclones and Climate Change, 13 pages, http://www.wmo.int/pages/prog/arep/tmrp/documents/iwtc_summary.pdf and Summary Statement on Tropical Cyclones and Climate Change, 1 page, http://www.wmo.int/pages/prog/arep/tmrp/documents/iwtc_statement.pdf

Woodruff, J. D., J. P. Donnelly, K. Emanuel, and P. Lane, 2008: Assessing sedimentary records of paleohurricane activity using modeled hurricane climatology. *Geochem., Geophys., Geosys.*, **9**, Q09V10.

Wu, L. and B. Wang, 2004: Assessing impacts of global warming on tropical cyclone tracks. *J. Climate*, **17**, 1686–1698.

Wu, L., B. Wang, and S. A. Braun, 2008: Implications of tropical cyclones power dissipation index. *Int. J. Climatol.*, **28**, 727–731.

Wu, L., B. Wang, and S. Geng, 2005: Growing typhoon influence on east Asia, *Geophys. Res. Lett.*, **32**, L18703.

Wu, M.-C., K.-H. Yeung, and W.-L. Chang, 2006: Trends in western North Pacific tropical cyclone intensity. *Eos. Trans.*, AGU, **87**(48), 537.

Yoshimura, J. and M. Sugi, 2005: Tropical cyclone climatology in a high-resolution AGCM-impacts of SST warming and CO_2 increase. *SOLA*, **1**, 133–136.

Yoshimura, J., M. Sugi, and A. Noda, 2006: Influence of greenhouse warming on tropical cyclone frequency. *J. Meteor. Soc. Japan*, **84**, 405–428.

Zeng, Z., Y. Wang, and C.-C. Wu, 2006: Environmental dynamical control of tropical cyclone intensity-An observational study. *Mon. Wea. Rev.*, **135**, 38–59.

Zhang, R., and T. L. Delworth, 2006: Impact of Atlantic multidecadal oscillations on India/Sahel rainfall and Atlantic hurricanes. *Geophys. Res. Lett.*, **33**, L17712.

Zhang, R., T. L. Delworth, and I. M. Held, 2007: Can the Atlantic Ocean drive the observed multidecadal variability in Northern Hemisphere mean temperature? *Geophys. Res. Lett.*, **34**, L02709.

Zhang, R., 2007: Anticorrelated multidecadal variations between surface and subsurface tropical North Atlantic. *Geophys. Res. Lett.*, **34**, L12713.

PART IV

Forecasting of Tropical Cyclones

Chapter 10

Track and Structure Forecasts of Tropical Cyclones

Julian Heming
Met Office, Exeter, UK
julian.heming@metoffice.gov.uk

Jim Goerss
Naval Research Laboratory,
Monterey, CA, USA
jim.goerss@nrlmry.navy.mil

First attempts were made at the forecasting of tropical cyclones in the Atlantic region in the 1870s and were followed by the establishment of many local and regional forecasting centres across the globe in the first half of the 20th century. During the late 1960s, the first statistical tropical cyclone forecasting techniques were developed. In this chapter a variety of tropical cyclone forecasting aids are described, starting with the statistical and climatological techniques developed in the 1960s and 1970s, through the plethora of global and regional deterministic model forecasts available today to the latest ensemble and consensus forecast aids developed in the last few years. The chapter is completed with some statistics on the performance of some of the models and forecasting aids in the prediction of tropical cyclones.

1. Introduction

Historical records exist of noteworthy tropical cyclones which occurred during the last few centuries. Details of tropical cyclones in the 16th century exist in the Atlantic and Caribbean region (Rappaport and Fernandez-Partagas, 1995) and the North Indian Ocean (Karmakar, 1998) and as far back as the 13th century in the western North Pacific (Neumann, 1975). The modern era of tropical cyclone records began in the mid-19th century. For example, Atlantic and Caribbean records began in 1851 (Blake *et al.*, 2007). However, in many locations only the strongest, most destructive storms were documented or passed to later generations through the local lore. By the early to mid 20th century records were kept of tropical cyclones in most regions of the world, although were of course constrained by the limitations of observational coverage at the time.

First attempts were made at the forecasting of tropical cyclones in the Atlantic region in the 1870s (Sheets, 1990) and were followed by the establishment of many local forecasting centres in the early 20th century. The first use of names to identify and track storms was made by Australian meteorologists in the early 20th century. This practice was taken up by United States (US) Navy meteorologists during World War II who were forecasting tropical cyclones in the Pacific. During the following decades, regional forecasting centres were established such at the US National Hurricane Center (NHC) in 1955, which became part of a global network of specialist forecast centres under the auspices of the World Meteorological Organization (WMO). During the 1960s and 1970s, the advent of geostationary satellites enabled the monitoring of tropical cyclones on a global scale and was accompanied by the development of statistical forecasting techniques in the late 1960s. In the

1970s numerical weather prediction models were first developed. By the late 1980s simple NWP models had become the fundamental tool for track forecasting and by the mid-1990's they were replaced by the more complex regional and global NWP models run by various operational centres. Improvement in numerical models' representation of physical and dynamical processes, increases in resolution, the advent of many new streams of observational data (primarily from satellites) and developments in data assimilation techniques in the last two decades have reinforced the importance of numerical model output to operational tropical cyclone forecasting centres worldwide. In recent years, the development of consensus and ensemble forecast techniques have added to the armoury of forecast aids available to the tropical cyclone forecaster.

In this chapter a variety of tropical cyclone forecasting aids are described, starting with the statistical and climatological techniques developed in the 1960s and 1970s, through the plethora of global and regional deterministic model forecasts available today to the latest ensemble and consensus forecast aids developed in the last few years. Some statistics on the performance of some of the models and forecasting aids in the prediction of tropical cyclones are also included. Since a lot of the information relating to model formulations and forecast performance will change with time a section has been included prior to the references that will include web links to the latest information available.

2. Simple Prediction Techniques

2.1. *Climatology and Persistence*

2.1.1. *HURRAN*

One of the earliest climatological models for tropical cyclone forecasting was the Hurricane Analog Technique (HURRAN) developed by Hope and Neumann (1970) and used at NHC from 1969. By identifying previous storms with characteristics in common with a current storm, HURRAN attempted to determine the most likely track of the current storm.

The technique first searched the database containing the tracks of all Atlantic tropical storms and hurricanes for analogs to the current storm. The analogs were those storms with similar location, speed and direction of motion, which occurred at about the same time of year. The tracks of the analog storms were shifted to pass through the location of the current storm. Finally, probability ellipses were calculated based on the adjusted tracks of the analog storms. The HURRAN model predicted the tracks of storms south of 25°N and before recurvature well. Since HURRAN lacked any synoptic input, it did not perform well in predicting tracks of storms just before or after their recurvature. While it provided useful information to the hurricane specialists at the time, the skill of HURRAN is so poor that the model is no longer used by NHC.

2.1.2. *Persistence*

Another of the most simple track prediction schemes is to assume the persistence of recent motion. The physical basis for this assumption is that a tropical cyclone is normally a small vortex embedded within a large-scale flow. The vortex, large-scale flow and interaction processes are often observed to change sufficiently slowly, such that the future motion approximately resembles the past motion. Hence, persistence is a reasonable predictor for short-term motion. Eventually, changes to the vortex structure and interaction of the tropical cyclone with the large-scale flow limit the skill of this technique. For more details see the chapter on movement of tropical cyclones.

2.1.3. *Climatology*

Climatological track prediction assumes that the storm will move with the average speed and direction of all past storms near that location.

To make a track prediction, the climatological vectors at the appropriate locations are multiplied by the time interval and the displacements are added to the present position. This simple application of climatology does not take into account the variability of tracks through each location. Climatology must be considered a 'baseline' forecast on minimum skill, since no meteorological information other than the present storm location is used.

2.1.4. *CLIPER*

A statistical combination of climatology and persistence (CLIPER) was developed by Neumann (1972) initially for the Atlantic region and was later extended to other regions (Leftwich and Neumann, 1977; Xu and Neumann, 1985; Leslie *et al.*, 1990). Least squares fitting of the basic predictors and various polynomial combinations are used in CLIPER to derive regression equations for future spatial displacements in increments of 12 hours. Thus, the technique makes use of the climatology of past tropical cyclone tracks and the persistence components of the present storm to generate a forecast.

The original CLIPER was able to produce a forecast out to 72 hours ahead. However, in preparation for the extension of NHC's forecasts to 120 hours, a new 5-day CLIPER was developed by Aberson (1998) based on observed tracks from the period 1956–1995. A version of the 5-day CLIPER was also developed for use in the western North Pacific Ocean (Aberson and Sampson, 2003).

Given the increasing skill of numerical models in the prediction of tropical cyclone tracks in the last 10–20 years, CLIPER is now more useful as a benchmark against which to assess the numerical forecasts rather than a vital tool for operational forecasting.

2.1.5. *SHIFOR*

As a counterpart for CLIPER, the Statistical Hurricane Intensity Forecast (SHIFOR) Model was developed for the prediction of tropical cyclone intensity in the Atlantic region (Jarvinen and Neumann, 1979). The SHIFOR model uses several climatological and persistence parameters to predict the future intensity of the storm at 12-hour periods out to 72 hours ahead. The predictor variables include the Julian day, current storm intensity, intensity change during past 12 hours, initial storm latitude and longitude and zonal (east-west) and meridional (north-south) component of the storm motion. The original SHIFOR equations were developed using data from all historic storms during the period 1900–1972 that were at least 30 nautical miles from land. In common with CLIPER, a 5-day version of SHIFOR was developed by Knaff *et al.* (2003) in preparation for the extension of NHC's forecasts to 120 hours. The new SHIFOR5 equations are based on the historic tropical cyclones from 1967–1999 that, at a minimum, intensified into tropical storms. Between five and eight predictor terms are included in each equation, most of which are second order products of the seven primary predictors listed above. The most important terms are the current intensity, the 12-hour intensity change, the Julian day and the current storm position. Unlike CLIPER, different predictors may appear in the equations for each lead time. Equations based on storms from throughout the Atlantic, Caribbean and Gulf of Mexico are used, since they out-performed those stratified by basin.

2.1.6. *Wind Radii Model*

A statistical-parametric model used to predict tropical cyclone wind structure in terms of significant wind radii (i.e., 34-, 50-, and 64-kt wind radii) was developed for use at NHC and the Joint Typhoon Warning Center (JTWC) (Knaff *et al.*, 2007). The model employs aspects of climatology and persistence to forecast tropical cyclone wind radii to 120 hours. Separate versions of the model have been created for the

Atlantic, eastern North Pacific and western North Pacific by statistically fitting a modified Rankine vortex, which is generalised to allow wavenumber-1 asymmetries, to observed values of tropical cyclone wind radii as reported by NHC and JTWC. The statistical-parametric model provides reliable operational forecasts that serve as a baseline for evaluating the skill of operational forecasts and other wind radii forecast methods in these tropical cyclone basins.

2.2. *Statistical-Dynamical Techniques*

2.2.1. *NHC98*

Techniques have been developed for forecasting of tropical cyclones which use both climatology and persistence and some form of numerically forecast environmental data. The most recent of these for track forecasting is NHC98, which is a successor to NHC90, described by Neumann and McAdie (1991). In NHC98, storms are stratified based on their latitude and their current motion, with different equations used for westward and eastward-moving storms. This stratification is used to account for the observation that storms within the easterlies tend to move to the right of the steering flow, while storms within the westerlies tend to move to the left of the steering flow. See the chapter on the motion of tropical cyclones for a further discussion of this observation. The NHC98 model produces a forecast track that is a combination of three independent track estimates. The first estimated track is that produced by CLIPER. The second is predicted using observed deep layer mean geopotential heights from the National Centers for Environmental Prediction (NCEP) Global Forecast System (GFS). Similar to CLIPER, the storm motion is separated into two components. One set of equations is used to predict the along-track movement of the storm at 12-hour intervals out to 120 hours. Another set of equations is used to predict the across-track movement of the storms. Deep-layer mean geopotential heights at two or three locations in the vicinity of the storm are used to represent the mean flow in which the storm is imbedded. The third estimated track is computed similar to the second, except forecast deep-layer mean geopotential heights from the NCEP GFS are used to produce the forecast track. The three track estimates are then combined into an optimum track forecast. NHC98 is run four times per day. The NHC98 forecasts are based on the previous runs of the NCEP GFS.

2.2.2. *SHIPS*

The SHIPS model (DeMaria and Kaplan, 1999 and DeMaria *et al.*, 2005) is a statistical-dynamical intensity prediction model. This model was developed using standard multiple regression techniques with climatological, persistence and numerical model forecasts as predictors. Estimates of future storm intensity are made for 12-hour periods out to 120 hours.

The SHIPS equations were initially developed using data from 49 storms during the period 1982–1992 that were at least 30 nautical miles from land. The collection of synoptic data began in 1989, as did the archive of operational intensity forecasts. Data for selected storms during 1982–1988 were available and included in the SHIPS developmental data set. The SHIPS equations are typically updated each year. There are 12 primary predictors — current storm intensity, day of the year, persistence, east-west component of storm motion, divergence of the wind at 200 hPa, intensification potential, vertical shear of the horizontal wind in the 850–200 hPa layer, average 200 hPa temperature, average 850 hPa vorticity, average 500–300 hPa layer relative humidity, cloud top temperature and oceanic heat content.

Sea surface temperature (SST) alone does not provide a good indication of whether a storm will intensify. However, SST does provide an upper limit to storm intensity — the Maximum Potential Intensity (MPI). Intensification potential is the difference between the current storm intensity and an estimate of the MPI. See the chapter on tropical cyclone structure for more discussion of MPI theory.

The current storm intensity predictor may be subject to error due to inherent uncertainties in estimation of storm strength when using techniques such as Dvorak (1975, 1984). However, sensitivity experiments have shown that typical ranges of initial intensity errors have a small effect on the SHIPS forecasts out to about 36 hours. Beyond 36 hours, there is almost no impact since the initial errors are dwarfed by other model deficiencies (DeMaria, personal communication).

Decay SHIPS (DSHIPS) was developed in 2000 to account for the decay of storms over land, the extension of the forecasts to 5 days in 2001, the replacement of the simple dry-adiabatic prediction model in 2001 with output from the GFS operational global model for the evaluation of the environmental predictors and the inclusion of satellite-derived parameters (specifically, cloud top temperatures and oceanic heat content) in 2004.

Unlike earlier versions, the most recent versions of SHIPS has significant skill over climatology, at least out to 3 days.

2.2.3. *LGEM*

The Logistic Growth Equation Model (LGEM) is a statistical intensity forecast model that utilises the same input as SHIPS, but in the framework of a simplified dynamical prediction system, instead of a multiple regression. The evolution of the maximum wind (i.e. intensity) is determined by a logistic growth equation which constrains the solution to lie between zero and the MPI, where the MPI is estimated from an empirical relationship with SST. The evolution of the maximum wind depends on the growth rate coefficient, which is estimated from a subset of the input to the SHIPS model. An important difference from SHIPS is that the longer range forecast depends more strongly on the environmental parameters at the later times (SST, vertical shear, etc). Most of the SHIPS predictors are averaged over the entire forecast period, while most of the LGEM predictors are averaged only over the 24 hours prior to the forecast valid time. The MPI in the LGEM prediction is the instantaneous value, rather than the forecast period average used in SHIPS. This difference makes the LGEM prediction more sensitive to environmental changes at the end of the forecast period, but also makes the prediction more sensitive to track forecast errors. Another difference from SHIPS is that because the LGEM forecast is based on a time stepping procedure, the forecast can better represent intensity changes of storms that move from water to land and then back over water.

3. Deterministic Numerical Weather Prediction

3.1. *Overview*

3.1.1. *Introduction*

The rapid increase in computer power in recent years has enabled numerical models to attain resolutions where small-scale systems such as tropical cyclones are resolved. These models are displaying real skill with motion prediction, and have the potential to handle formation. However, only a limited number of models (e.g., GFDL, Kurihara *et al.*, 1995), have attained resolutions where cyclone structure (including intensity) can be addressed.

Numerical models use a subset of the full fluid dynamics and thermodynamics equations describing the atmosphere. Subsets that are used for tropical cyclone track prediction include barotropic, non-hydrostatic, non-divergent, shallow-water equations, truncated baroclinic and full primitive-equation models. Full details of the essential elements in numerical models may be found in standard texts, such as Haltiner and Williams (1979). This section provides a brief overview. Sections 3.2 and 3.3 detail the latest configurations of many deterministic (single solution) models, which are used regularly by operational tropical cyclone warning centres in the preparation of their tropical cyclone track and intensity forecasts. Section 3.2 covers global models and 3.3 regional models. Most of these models are summarised in Table 1.

Table 1. Summary of main operational NWP models used in tropical cyclone prediction.

Numerical model	Assimilation method	Model type and resolution	Model schedule
GFS (USA)	3D-Var with TC vortex relocation	Global spectral T382L64 to 180 hours T190L64 to 384 hours	0000, 0600, 1200 and 1800 UTC to 384 hours
NOGAPS (USA)	4D-Var with synthetic TC observations	Global spectral T239L42	0000, 0600, 1200 and 1800 UTC to 180 hours
Met Office (UK)	4D-Var with synthetic TC observations	Global grid-point $0.5625° \times 0.375°$, 70 levels	0000 and 1200 UTC to 144 hours, 0600 and 1800 UTC to 48 hours
GSM (Japan)	4D-Var with synthetic TC observations	Global spectral T959L60	0000, 0600, 1800 UTC to 84 hours and 1200 UTC to 216 hours
ECMWF	4D-Var, no TC initialisation	Global spectral T799L91	0000 and 1200 UTC to 240 hours
GEM (Canada)	4D-Var, no TC initialisation	Global grid-point $0.3° \times 0.45°$, 58 levels	0000 UTC to 240 hours, 1200 UTC to 144 hours
ARPEGE (France)	4D-Var	Global spectral Variable T1291L60 to T224L60	0000, 0600, 1200 and 1800 UTC to 102, 72, 84 and 60 hours respectively
NCMRWF (India)	3D-Var	Global spectral T254L64	0000, 0600, 1200 and 1800 UTC to 168 hours
GRAPES-TCM (China)	3D-Var with synthetic TC observations	Regional spectral 17 km, 31 levels	0000, 0600, 1200 and 1800 UTC to 72 hours
GDAPS (Korea)	3D-Var with synthetic TC observations	Global spectral T426L40	0000 and 1200 UTC to 240 hours
GFDL (USA)	GFS analysis with TC vortex replacement	Regional grid-point triply nested: $0.5° \times 0.5°$, $0.167° \times 0.167°$ and $0.083° \times 0.083°$, 42 levels	0000, 0600, 1200 and 1800 UTC to 126 hours
HWRF (USA)	GFS analysis (3D-Var being developed)	Regional grid-point doubly nested $0.25° \times 0.25°$ and $0.1° \times 0.1°$, 42 levels	0000, 0600, 1200 and 1800 UTC to 126 hours
ALADIN (France)	3D-Var with synthetic TC observations	Regional spectral 10 km and 60 levels	Four times per day. 0000 and 1200 UTC to 84 hours
LBAR	GFS analysis	Regional spectral triply nested	Four times per day
QLM (India)	NCMRWF analysis with idealised TC vortex	Regional grid- point 40 km and 16 levels	0000 and 1200 UTC to 72 hours
ORSM (Hong Kong, China)	Optimal interpolation	Regional spectral doubly nested — 20 km inner mesh, 42 levels	Four times per day to 72 hours
RDAPS (Korea)	3D-Var with synthetic TC observations	Regional spectral 30 km, 10 km and 5 km and 33 levels	66 hours

3.1.2. *Model Grids*

All operational models, including spectral models, start with atmospheric fields on a regular grid array, on which the required atmospheric parameters are held. Model grids can have any shape; they do not need to be regular, or square. The grid may be distorted to provide higher resolution in a specified region or near a system of interest (e.g., ARPEGE, section 3.2). Alternatively, higher resolution grids may be successively nested within each other to provide a telescope effect with a high-resolution focus on the region of interest (e.g., GFDL, section 3.3). The grid also may be regular, with all data held at common grid-points, or it may be staggered, with data split between two overlapping grids.

3.1.3. *Model Vertical Structure*

The vertical levels, or layers, are defined by using some form of vertical co-ordinate. In the early days of modelling, the simple pressure co-ordinate was widely used. This then tended to give way to the 'sigma' co-ordinate, which has the advantage of adjusting the variable orography into a 'flat' surface with a value equal to one. Nowadays, many Numerical Weather Prediction (NWP) centres use a hybrid combination of the pressure and sigma co-ordinate systems, providing the benefits of a terrain-following system in the lower atmosphere with a pure pressure co-ordinate higher up. It is normal to vary the vertical resolution considerably, especially to provide high resolution in the atmospheric boundary layer and in the upper troposphere. In the boundary layer, high vertical resolution helps provide the best simulation of strong gradients in temperature and humidity driven by interactions with the surface. In the upper troposphere high vertical resolution helps in the modelling of the upper-level outflow region, where sharp changes in the vertical gradient occur.

3.1.4. *Data Assimilation and Bogus Observations*

Data assimilation aims to determine the best possible estimate of the current state of the atmosphere by statistically combining the available observational data with a short-term NWP forecast, optionally subject to some balance constraints. The data are obtained from a wide variety of sources and instruments, including direct temperature, moisture and wind from radiosondes, remotely sensed temperature derived from satellite sounding instruments, winds from satellite cloud drift calculations, direct radiances, etc. Typically, all of these observations are then combined together with a 'background' or 'first-guess' field into an initial analysis.

Most operational systems currently use a three-dimensional variational (3D-Var) or a four-dimensional variational (4D-Var) assimilation system each analysis cycle (e.g., Rabier *et al.*, 2000; Rawlins *et al.*, 2007). 4D-Var is considered a more advanced assimilation scheme since it gives all observations their maximum weight at their validity time rather than at the nominal start time of the model run and implicitly generates flow-dependent error covariances. This generally results in superior NWP forecast scores (Lorenc and Rawlins, 2005). Increased resolution of numerical models and the usage of high-density remotely sensed data make the issue of assimilation increasingly important (Rabier, 2005).

Satellite data coverage has increased in recent years, although limitations still exist in the use of infra-red sounding data in cloudy regions and scatterometer and microwave wind data in rainy regions. Apart from satellite data, many tropical cyclones have very few conventional observations in the vicinity and are sometimes analysed poorly, with centres ill defined and in the wrong location. Hence, attempts are made to artificially initialise cyclone vortices in many numerical models. These all utilise estimated location and intensity of cyclones from a variety of sources, together with current structural knowledge of tropical cyclones and any other available observations to generate mass and wind data representative of the system to be initialised. Without this artificial initialisation, initial errors can be large and have a major impact on the forecast of tropical cyclone tracks (Heming, 2009).

Inserting the chosen 'bogus' cyclone into the analysis is typically done using one of two methods. In the first, the original analysis is directly adjusted towards the structure of the 'bogus' cyclone. In the second the 'bogus' cyclone circulation is used to generate a set of observations that are inserted into the analysis cycle and assimilated alongside conventional observations. Details of 'bogus' schemes used by individual models are included in sections 3.2 and 3.3 below.

3.1.5. *Physical Parametrization*

All numerical models operate on a finite grid, or with a finite number of waves to describe the spectral signature. As a result, they cannot adequately resolve small-scale features, such as cumulus convection, or the interchange of energy between the surface and the ocean. Those 'sub-grid-scale' processes that have an important impact on the forecasts are included by parametrizing their effects on the larger scale.

As an example, consider cumulus convection. This is known to be related to the large-scale convergence in the lower troposphere and the degree of convective instability. One type of scheme, called convective adjustment, adjusts the model atmosphere back to a defined lapse rate whenever convective instability of a prescribed amount develops. Alternative schemes depend on both a conditionally unstable atmosphere and local convergence. Once such conditions are satisfied, convection is assumed to occur and a parametrization scheme is invoked. Such schemes can be relatively simple and based on a defined vertical heating profile with magnitude specified by the degree of moisture convergence (e.g., Kuo, 1974), or they can be very complex and based on equilibrium conditions between a population of cloud types (Arakawa and Schubert, 1974). In all cases, the aim is to provide the model with an indication of large-scale changes required as a result of the unresolved convective activity. The choice of convective parametrization scheme can greatly affect the quality of tropical cyclone forecasts.

3.1.6. *Boundary Conditions*

A very important consideration for limited area numerical forecasts is the quality of the boundary conditions (e.g., Errico and Baumhefner, 1987). This is because atmospheric waves and disturbances generated at the boundary can rapidly propagate throughout the domain and swamp the model forecast cycle. Two types of boundaries therefore need to be carefully handled: the real boundaries at the earth surface and top of the atmosphere and the artificial boundaries around the domain of limited area models. For a region of several thousand kilometres per side, the model solution will tend to be swamped by the horizontal boundary conditions after 2–3 days. For this reason, regional forecast models are always nested into global forecast systems.

3.2. *Global Models*

3.2.1. *Global Forecasting System (GFS) — USA*

The global model of the National Centers for Environmental Prediction (NCEP) is part of the Global Forecast System (GFS). It is a spectral model which operates at resolution T382L64 to 180 hours and T190L64 to 384 hours. The GFS runs four times per day, starting with analyses valid at 0000, 0600, 1200 and 1800 UTC. The model includes parametrizations of convective, radiative and boundary layer processes. The model uses a 3D-Var assimilation method. Further details on the GFS model are described by Surgi *et al.* (1997) and Campana and Caplan (2005).

A technique for initialising tropical cyclone circulations using synthetic wind observations was used during the 1990s (Lord, 1991). This was superseded by an improved technique in 2000, which involves vortex relocation. The vortex relocation procedure takes the first guess field and moves the hurricane vortex to the correct location before the new analysis is produced. This involves locating the tropical cyclone vortex centre in the first guess field, separating the model's vortex from its environmental field, moving the vortex to the position as analysed by NHC and an optional step to add a bogus vortex in the analysis if the vortex is too weak in the first guess field. Further details on the vortex relocation technique are described by Liu *et al.* (2000).

3.2.2. *US Navy Operational Global Atmospheric Prediction System (NOGAPS)*

NOGAPS is the U.S. Navy's global spectral forecast model with 42 sigma levels, a triangular truncation of 239 waves, parametrizations of convective, radiative and boundary layer and a tropical cyclone bogussing scheme. It is a global multivariate model using 4D-Var data assimilation (Xu *et al.*, 2005; Rosmond and Xu, 2006).

The initialisation of tropical cyclones is accomplished by the assimilation of synthetic observations into the global 4D-Var analysis. The synthetic observations, that represent the storm circulation, are created from the sum of an environmental flow and a symmetric vortex. The environmental wind is determined from a T20 spectral truncation of the model fields adjusted for storm motion. The symmetric vortex component is derived using a modified Rankine vortex modulated by the intensity of the tropical cyclone. The synthetic observations consist of 13 wind soundings at the mandatory levels from 1000 to 400 hPa, inclusive, plus 13 geopotential height observations at 1000 hPa. The observation locations are: one at the storm centre; four at 1.5 degree radius (north, south, east, and west of the centre), four at 3 degree radius (northeast, southeast, southwest, and northwest of the centre) and four at 4.5 degree radius (north, south, east, and west of the centre). Further details of the initialisation scheme are described by Goerss and Jeffries (1994) and Goerss *et al.* (1998).

NOGAPS runs four times per day to 180 hours, starting with analyses valid at 0000, 0600, 1200, and 1800 UTC. Track forecasts are routinely produced at 6-hourly intervals out to 144 hours using the same tracker used at NCEP (Marchok, 2002). Further details of the NOGAPS model are described by Hogan and Rosmond (1991) and Peng *et al.* (2004).

3.2.3. *Met Office Global Model — UK*

The Met Office global model is non-hydrostatic with height as the vertical co-ordinate. It has a spherical latitude-longitude grid with spacing $0.5625° \times 0.375°$ (about 40 km at mid-latitudes) and 70 levels in the vertical. This will be increased to $0.3515625° \times 0.234375°$ (about 25 km at mid-latitudes) in 2010. The model has a semi-langrangian advection scheme. Parametrizations include the Edwards-Slingo radiation scheme, convection with CAPE closure, momentum transports and convective anvils, a gravity-wave drag scheme which includes flow blocking and MOSES (Met Office Surface Exchange Scheme) surface hydrology and soil model scheme. In 2007 a new sea-surface temperature and sea ice scheme (OSTIA) was introduced (Stark *et al.*, 2007). Since 2004, the global model has been using a 4D-Var data assimilation system (Rawlins *et al.*, 2007). More detail of the model formulation can be found in Martin *et al.* (2006), Allan *et al.* (2005) and Allan *et al.* (2007).

Since 1994 a tropical cyclone initialisation scheme has been used, which utilises data from operational forecast centres around the globe. Information is extracted from these data, along with conventional observations, to derive a wind profile around the tropical cyclone at the surface and three levels in the lower troposphere. An indication of the steering flow is included by using the past 6-hour movement, and imposing this on each of the tangential wind values already derived by a simple vector addition. Bogus data are created on up to five concentric rings up to a radius of 8 degrees from the tropical cyclone centre. Full details of the formulation can be found in Heming *et al.* (1995) and latest updates to the system in Heming (2009).

The global model runs four times per day with short forecasts (to 48 hours) at 0600 and 1800 UTC and long forecasts (144 hours) at 0000 and 1200 UTC. For each of the long forecasts tropical cyclone guidance messages are produced which include, in plain text format, the forecast tracks of all active tropical cyclones and any expected to form in the next three days.

3.2.4. *Global Spectral Model (GSM) — Japan Meteorological Agency (JMA)*

The GSM has a hydrostatic approximation, uses primitive equations and a semi-Lagrangian framework. It uses a 4D-Var assimilation scheme with its own 3 to 9 hours prediction used as first guess field. It has a T959 Gaussian grid of approximately $0.1875° \times 0.1875°$(1920 $\times$ 960 points, $\sim$20 km) with 60 hybrid vertical levels above the surface. The Arakawa-Schubert (1974) cumulus parametrization is used with modifications by Moorthi and Suarez (1992), Randall and Pan (1993), and Pan and Randall (1998).

Tropical cyclones over the western North Pacific are initialised using bogus data. The axisymmetric structure based on Frank's (1977) empirical formula with parameters derived from the Dvorak method with MTSAT imagery. The asymmetric structure derived from the first guess (6-hour prediction by GSM).

The model is run out to 84 hours three times a day at 0000 UTC, 0600 UTC and 1800 UTC. The 1200 UTC run extends to 216 hours lead time. JMA NWP products are available to WMO members via its Regional Specialised Meteorological Centre Data Serving System (RSMC DSS).

3.2.5. *European Centre for Medium-range Forecasts (ECMWF) Global Model*

The ECMWF deterministic model is semi-Lagrangian with a semi-implicit time-stepping formulation. From February 2006 it was run at T799 (25 km) horizontal resolution with 91 levels (top at 0.01 hPa). No specific initialisation of tropical cyclones takes place. Research has been carried out at ECMWF to investigate the impact of using different kinds of data and of assimilation methods on tropical cyclone forecasts. Isaksen *et al.* (1998) show that the ECMWF 4D-Var data assimilation system improves the analyses of tropical cyclones compared to previous assimilation procedures. The model is run out to 10 days twice a day from 0000 UTC and 1200 UTC. A further increase in horizontal resolution to near 16 km is planned for late 2009.

3.2.6. *Global Environmental Multiscale Model (GEM) — Canadian Meteorological Centre (CMC)*

The CMC's GEM is a hydrostatic global grid point model laid on a latitude/longitude coordinate system with 0.3° latitude and 0.45° longitude (approximately 33 km at 49°N) horizontal grid spacing and 58 vertical eta levels. The eta vertical coordinate system used in the CMC's GEM depicts the bottom atmospheric layer within each grid box as a flat step. The CMC's GEM employs a 4D-Var data assimilation scheme where temporal variations in the initial data are included. No vortex specification or bogussing is performed. The convective parametrization includes the Kain-Fritsch scheme for deep moist convection and

Kuo Transient scheme for shallow convection. The Bougeault and Lacarrere (1989) mixing length is used for vertical diffusion due to atmospheric turbulence.

The CMC's GEM is run through 144 hours at 1200 UTC, 240 hours at 0000 UTC, and 360 hours on Saturdays.

3.2.7. *ARPEGE — Météo — France*

ARPEGE-IFS is a common development between Météo-France and ECMWF. ARPEGE (Action de Recherche Petite Echelle Grande Echelle) is the French name while IFS (Integrated Forecast System) is the name used by ECMWF. The model uses 4D-Var global data assimilation on a 6-hour cycle. It is a primitive equation spectral model with two time-level semi-lagrangian semi-implicit scheme. The model uses Schmidt's transformation leading to variable mesh configurations. It has a triangular truncation T538 with a stretching factor of C2.4. The resolution varies from T1291 (15 km) over France to T224 (90 km) over New Zealand. There are 60 vertical levels from screen level up to 0.1 hPa, using the hybrid co-ordinate from Simmons and Burridge (1981). The model runs four times a day with various forecast lead times. The 0000 UTC forecast runs to 102 hours, 0600 UTC forecast runs to 72 hours, 1200 UTC forecast runs to 84 hours and 1800 UTC forecast runs to 60 hours.

A version of the ARPEGE model is run to obtain more detailed forecasts in tropical regions. ARPEGE-Tropiques has a resolution of T538L60 and is initialised from the ARPEGE-IFS analysis each cycle. It is run twice per day at 0000 UTC and 1200 UTC to 72 hours.

3.2.8. *National Centre for Medium Range Weather Forecasts (NCMRWF) — India*

NCMRWF currently run a global model based on the 2006 configuration of the GFS spectral model. In the horizontal the model is T254 (approximating to a latitude-longitude grid of $0.469° \times 0.469°$). The model has 64 levels in the vertical. For further details see section on GFS model.

It is planned to replace all modelling systems at NCMRWF with systems based on the UK Met Office Unified Model from 2010 onwards.

3.2.9. *Global Data Assimilation and Prediction System (GDAPS) — Korean Meteorological Administration*

The Korean Meteorological Administration (KMA) runs its Global Data Assimilation and Prediction System (GDAPS) at 0000 UTC and 1200 UTC to 10 days. The model has a resolution of T426L40 and is hydrostatic with primitive Eulerian-form equations and has the Kuo (1974) convective parametrization. It uses 3D-Var and tropical cyclone bogus data which has an axisymmetric structure based on Fujita (1969) and an asymmetric structure derived from the first guess field from the JGSM.

KMA plans to adopt the UK Met Office Unified Model by 2010 which would supersede the GDAPS system.

3.3. *Regional Models*

3.3.1. *GFDL — USA*

The Geophysical Fluid Dynamics Laboratory (GFDL) hurricane model is a limited-area, grid-point model that was designed specifically for tropical cyclone prediction. The GFDL model consists of a triply-nested grid configuration. The outer grid spans $75° \times 75°$ at $1/2°$ horizontal resolution (approximately 56 km). The middle grid spans $11° \times 11°$ at $1/6°$ horizontal resolution (approximately 19 km). The inner grid domain size spans $5° \times 5°$ at $1/12°$ horizontal resolution (approximately 9 km). This high resolution, compared to other dynamical models, allows the GFDL model to resolve some relatively small scale features within a tropical cyclone such as the eye and eyewall, although is not able to fully resolve the highly complex structure of a tropical cyclone. The GFDL model uses a sigma vertical coordinate system with 42 vertical levels. It employs a simplified Arakawa-Schubert convective parametrization scheme and non-local boundary layer scheme. The large-scale condensation scheme contains the Ferrier micro-physics package, and the air-sea momentum flux parametrization is designed to account for strong wind conditions to give a better estimate of surface winds.

For the initialisation, the GFDL model replaces the GFS tropical cyclone vortex with an axisymmetric vortex spun up in a separate model simulation. The axisymmetric vortex model utilises tropical cyclone specifications as provided by NHC forecasters. Since the horizontal resolution of the GFDL model is sufficiently high to simulate some of the inner core tropical cyclone structure, the GFDL model is one of only a few purely dynamical models that can provide both skilful intensity and track forecasts. The GFDL model is run for up to four tropical cyclones every six hours out to 126 hours as requested by the NHC and the Central Pacific Hurricane Center (CPHC).

Another version of the GFDL model, referred to as the GFDN model, is used globally and is initialised using fields from the NOGAPS model. Further details of the GFDL model formulation can be found in Tuleya *et al.* (1995) and Kurihara *et al.* (1995).

3.3.2. *HWRF — USA*

The Hurricane Weather and Research Forecast system (HWRF) became operational at NCEP in 2007. It is a high resolution coupled air-sea-land prediction model with a movable nested grid and advanced physics for high resolution.

Over the past 15 years, the air-sea coupled GFDL model has evolved as a state of the art hurricane forecast system for track skill. Additionally, the GFDL model over the past several years has been providing increasingly skilful intensity forecasts that have now become comparable to the statistical intensity models. Therefore, using the GFDL model as the benchmark, the HWRF adapted the baseline physics of the GFDL model and the air-coupling strategy to best ensure comparable HWRF track and intensity skill forecasts with the GFDL model in both Atlantic and East Pacific basins. The HWRF has a movable two way nesting with an inner grid at 9 km and the outer grid of 27 km. over a 75×75 degree domain with a vertical resolution of 42 levels.

A simplified 4D-Var system using scale dependent background error covariances is being developed to initialise the hurricane core circulation in the HWRF using the airborne Doppler data radar from NOAA's aircraft. For storms approaching landfall, the data assimilation will also make use of the coastal 88-D high resolution data. Further plans also include the increase of the horizontal and vertical resolution to 4 km and 64 levels and the development and implementation of high resolution multi-model ensembles (Surgi *et al.*, 2008).

3.3.3. *Limited Area Sine Transform Barotropic (LBAR) model — USA*

LBAR (DeMaria *et al.*, 1992) is a dynamical track prediction model. LBAR is the NHC's implementation of the GFDL VICBAR (Vic Ooyama's Barotropic) model and runs on a 6-hour forecast cycle and produces forecasts out to 120 hours. The model has three nested grids — a fixed synoptic domain (27.5°S to 67.5°N, 10°E to 140°W), a storm environment 50 degree latitude-longitude box centred on the current storm location and a vortex domain circle of 7.4 degree (about 800 km) radius centred on the current storm location.

The synoptic scale analysis is obtained directly from the GFS model analysis. The storm environment domain analysis is produced with a two-dimensional spectral application of finite element representation, using all available observational data, with the GFS model analysis used as a low level background field. The vortex domain analysis consists of synthetic observations representing the storm circulation and current storm motion. The vortex is prescribed to be the same size and intensity in all directions (axisymmetric), with winds increasing linearly from the centre to the radius of maximum winds. Wind speeds beyond the radius of maximum winds are prescribed to decrease exponentially to the edge of the storm. In the event of multiple tropical cyclones, synthetic vortices are included for each storm.

3.3.4. *Beta and Advection Model (BAM) — USA*

The Beta and Advection Model (Marks, 1992) is a baroclinic-dynamical track prediction model. It produces a forecast track by following a trajectory in the vertically averaged horizontal wind starting at the current storm location out to 120 hours. The trajectory is corrected to account for the variation of the coriolis force with latitude, the so-called Beta effect. The conservation of absolute vorticity results in the formation of anticyclonic relative vorticity in the northeast quadrant of the storm, and the formation of cyclonic relative vorticity in the southwest quadrant of the storm. The result adds a component of motion to the northwest to the storm's trajectory.

Three versions of the BAM are run with shallow (850–700 hPa), medium (850–400 hPa), and deep (850–200 hPa) layers. All three versions of the model are run operationally four times per day.

3.3.5. *ALADIN — Météo-France*

ALADIN is a regional version of the ARPEGE model used for the prediction of tropical cyclones in the south-west Indian Ocean (the area of responsibility of RSMC La Réunion for tropical cyclone warnings). The model configuration is essentially the same with the following exceptions. Data assimilation is 3D-Var with a linear beta-plane, non linear and omega balances for upper-air observations. In the horizontal, the model has a 10 km horizontal mesh size and has 60 vertical levels. Bogus synthetic wind observations are produced around tropical cyclones from the surface to 500 hPa deduced from the Holland analytical model, plus the mean sea level pressure at the centre. The model runs on a 6-hourly assimilation cycle with 84-hour forecasts produced from data times 0000 UTC and 1200 UTC.

3.3.6. *Quasi-Lagrangian Limited Area Model (QLM) — Indian Meteorological Department*

The QLM is a multilevel primitive equation regional model using a Cartesian grid. The horizontal grid spacing is 40 km, it has 16 levels in the vertical and a $4400 \times 4400\,\text{km}^2$ domain that is centred on the initial position of the cyclone.

The initial analysis and lateral boundary conditions are generated from operational analysis and forecasts produced by the global spectral model of NCMRWF, New Delhi. The prescription of an idealised symmetric vortex is based on the storm's central pressure, the pressure of the outer most closed isobar and its distance from the centre. These parameters together with the location of the storm centre are derived from synoptic analysis and satellite imagery. A steering current, which is based on the current storm speed and direction is superimposed on the analysed fields. The steering current is computed by constructing a dipole circulation. The dipole winds and geopotential height fields (incremental heights calculated from dipole winds geostrophically) are added to the vortex fields at all levels. Full details of the model can be seen in Mathur (1991).

The model forecasts are produced for 0000 and 1200 UTC when the system attains the tropical cyclone intensity. The model provides track forecasts up to 72 hours ahead.

3.3.7. *Regional Data Assimilation and Prediction System (RDAPS) — Korean Meteorological Administration*

The KMA Regional Data Assimilation and Prediction System (RDAPS) is run with a horizontal resolution of 30 km to 66 hours, 10 km to 24 hours and 5 km to 24 hours, all with 33 vertical levels, using GDAPS as boundary conditions. KMA also runs a next generation regional model based on a version of the WRF model. However, KMA plans to adopt the UK Met Office Unified Model by 2010 which would supersede the RDAPS system.

3.3.8. *TC-LAPS — Australian Bureau of Meteorology*

The Australian Bureau of Meteorology's TC-LAPS (Davidson and Weber, 2000) has been operational over the Australian Region and northwest Pacific since 1999. It is a high-resolution prediction system with the generalised LAPS model (Puri *et al.*, 1998).

The model has a tropical cyclone vortex specification, which constructs the inner-core, imposes asymmetries consistent with the past motion, and re-locates the circulation to its observed position. High resolution objective analysis smoothly inserts the specified tropical cyclone circulation into its large-scale environment via the use of synthetic observations, using appropriate quality control checks and error covariances. The vertical motion field is then redefined to be consistent with the satellite cloud imagery, build the secondary circulation, and initialise (balance) the vortex.

The system is capable of making skilful intensity forecasts (Davidson *et al.*, 2006), but the skill is often limited by errors in initial vortex structure, errors in the large-scale environment of storms, or non-inclusion of important processes or details in the vortex structure (e.g., asymmetries). TC-LAPS is to be replaced by a regional configuration of the UK Met Office Unified Model. Following implementation of this model, important upgrades to the vortex specification, revised initialisation methods and increased resolution (~5 km, 50 levels), skilful operational intensity and structure forecasts may be achievable.

3.3.9. *Japanese Typhoon Model (JTYM) — Japan Meteorological Agency*

For many years, the storm centred JTYM was run for tropical cyclones in the western North Pacific. However, this model was retired in November 2007, co-incident with an upgrade to the Japanese Global Spectral Model and introduction of an ensemble prediction system (described in section 11.4).

3.3.10. *Operational Regional Spectral Model — Hong Kong Observatory, China*

The Hong Kong Observatory (HKO) operates a doubly nested mesoscale model, the Operational Regional Spectral Model (ORSM), with resolutions of 60 km and 20 km in the horizontal and 40 levels in the vertical. The 20 km inner mesh is run eight times each day to 42 hours, whilst the 60 km outer mesh is run four times per day to 72 hours. Outputs from JMA GSM are used as boundary conditions. Tropical cyclone bogus data based on forecasters' analysis are also employed for initializing the model.

The 60 km ORSM tropical cyclone track forecast guidance is routinely generated and disseminated via the Global Telecommunications System (GTS). The guidance is issued twice a day based on the 0000 and 1200 UTC model runs whenever a TC appears in the area of responsibility of the HKO (10–30°N, 105–125°E). The information provided in the guidance includes forecast tropical cyclone positions and intensity changes at 6-hourly intervals out to 72 hours.

The HKO plans to upgrade its ORSM with the Non-hydrostatic Model (NHM) which is based on JMA-NHM and have a horizontal resolution of 10 km in the outer mesh and 2 km in the inner mesh in 2010. Observations and TC bogus data will be assimilated into NHM with a 3D-Var data assimilation system.

3.3.11. *Global/Regional Assimilation and Prediction System (GRAPES) — China Meteorological Administration*

The Global/Regional Assimilation and Prediction System (GRAPES) is the China Meteorological Administration (CMA) next-generation unified NWP system that includes a 3D-Var data assimilation system and has a unified multi-scale dynamic core and physics for use in both regional and global configurations. GRAPES-GFS is the global configuration, which has been run at a resolution of 60 km with 36 levels out to 10 days ahead in quasi-operational mode since early 2009. GRAPES-meso is a regional configuration run at 15 km resolution with 31 levels and has been operational since 2006 (Zhang and Shen, 2008).

CMA also runs a regional typhoon prediction configuration of the model (GRAPES-TCM) at the Shanghai Regional Meteorological Centre. It became operational in 2006 and provides 3-day forecasts four times daily for tropical cyclones in the western North Pacific Ocean and the South China Sea. It has a horizontal resolution of 0.15° latitude and longitude, 31 levels and a cycling data assimilation scheme based on a model constrained 3D-Var. The bogus typhoon vortex, model predicted vortex, and the observations are assimilated simultaneously to initialize the typhoon vortex in the initial conditions. A 16-member ensemble version of GRAPES-TCM is under implementation.

3.3.12. *Philippines Atmospheric Geophysical and Astronomical Services Administration*

The Philippine Atmospheric Geophysical and Astronomical Services Administration (PAGASA) runs three different regional models for the purpose of tropical cyclone forecasting.

The High Resolution Model (HRM) is a hydrostatic limited-area model system which is being initialised by global data courtesy of the Deutscher Wetterdienst of the Federal Republic of Germany. At present, the PAGASA implementation utilises a 14 km horizontal resolution (0.125° × 0.125°) with 32 vertical layers. The model is run operationally twice a day (0000 UTC and 1200 UTC) and is fully automated from downloading input files to graphics display of outputs.

PAGASA's first limited-area non-hydrostatic model is a local implementation of MM5. The model consists of two domains; the outer and the nested domain. The outer domain has a 60 km horizontal resolution and covers an area of approximately 5500 × 5500 km^2 centred in the Philippines. The domain coverage is from 105°E–152°E and 5°S to 35°N. Nested within the outer domain is the 20 km resolution nested domain consisting of 71 by 71 grid points that covers the Philippine islands. This domain is designed to simulate small scale features and is initialised by data from the JMA GSM.

PAGASA runs a configuration of the ETA model which has a domain covering the entire Philippine archipelago and adjacent areas (101°E to 149°E and 6°S to 28°N) at a horizontal resolution of 28 km centred at 11°N latitude and 125°E longitude. It uses the output of the JMA GSM for the model's initial atmospheric conditions.

4. Ensemble Numerical Weather Prediction

4.1. *Overview*

Ensemble Prediction Systems (EPSs) are increasingly used in tropical cyclone track prediction. Single-model ensembles generally perturb initial conditions of the deterministic run that represents the best estimate of the initial state. These perturbations are done either randomly (Cheung and Chan, 1999a,b) or in a way that represents analysis uncertainty or maximises perturbation growth. These latter techniques include singular vectors (Leutbecher, 2005), bred-modes (Cheung and Chan, 1999a,b; Toth and Kalnay, 1997), or an ensemble transform (Wang and Bishop, 2003). Benefits of these techniques include state-dependent estimates of perturbation structures that are dynamically constrained and fast growing. Disadvantages include the fact that the dynamically generated perturbations are only approximations to actual analysis errors and the individual single-model ensemble runs are not fully independent of the other runs of the ensemble; this limits the ability of the ensemble to span the space of possible solutions.

In recent years, services producing ensemble forecasts have been increasing the number of ensemble members and upgrading the models. Improvements to singular vector techniques and the introduction of Ensemble Kalman Filter (EnKF) (Hamill, 2006) techniques to produce initial conditions has occurred. The emergence of techniques such as the EnKF brings with it the possibility of casting the more general problem of NWP, and specifically involving tropical cyclones, in a probabilistic framework. Since a single deterministic model forecast is only one of many realisations given the uncertainty in initial state and model physics and dynamics, it is important to focus not only on the forecast itself but also on the associated forecast error (Lewis and Tripoli, 2006). Here, the assimilation process is viewed as one of obtaining a probabilistic rather than deterministic analysis. The width of the ensemble determines the probability of the analysis (or forecast) and the mean of the ensemble determines the most likely analysis (or forecast).

EPSs are now being used to predict tropical cyclone tracks, especially to provide probabilistic track forecasts that have been useful to operational forecasters in the medium range. Also, a number of centres are introducing short-range ensembles. Some of the ensemble prediction systems used in the forecasting of tropical cyclones are described below.

4.2. *Operational Ensemble Prediction Systems*

4.2.1. *ECMWF*

The ECMWF EPS uses the same forecast model as used for the deterministic forecast (see section 11.3.2), but at lower resolution (Buizza, 1997). The ensemble comprises one control forecast (run from the operational analysis) and 50 perturbed members. Initial perturbations are generated by multi-dimensional Gaussian sampling from 50 singular vectors (SV) selected at T42 resolution in the extratropics, plus up to 30 SVs (5 per tropical cyclone) selected in the vicinity of tropical cyclones that have been reported on the GTS. The extra-tropical SVs are computed with an adiabatic version of the tangent-linear model. The SVs targeted on tropical cyclones are computed with a diabatic tangent-linear model. This is computationally more expensive, but it is important to include diabatic processes to obtain initial perturbations relevant for tropical cyclone motion (Puri *et al.*, 2001). Random perturbations of the physical tendencies (stochastic physics) are also applied to the perturbed forecasts.

In February 2006 the resolution of the forecast model was increased from T255 (80 km) to T399 (50 km) and from 40 to 62 levels. In November 2006 the range of the EPS forecast was extended from 10 days to 15 days; the last five days are run at lower horizontal resolution of T255. The EPS runs twice per day (from 0000 and 1200 UTC) (Buizza *et al.*, 2006). A further increase in horizontal resolution is planned for late 2009.

Tropical cyclones are tracked using the method in Van der Grijn *et al.* (2005) and the track forecasts are disseminated on the GTS. A range of tropical cyclone products is available to WMO members via the ECMWF website. The tropical cyclone web page includes forecast guidance products and verification for active and past tropical cyclones. Strike probability maps provide information on the likely tropical cyclone tracks. The Lagrangian EPSgram shows the forecast evolution of intensity (core pressure and 10 m wind speed) for a given tropical cyclone. Details of how to access these products are given at http://www.ecmwf.int/about/wmo_nmhs_access/index.html.

4.2.2. *National Centers for Environmental Prediction (NCEP) — USA*

The core of this model and parametrizations are the same as the GFS deterministic model. The high resolution analysis is truncated to an ensemble resolution of T126L28. Tropical cyclone initialisation and perturbation is achieved by an ensemble transform, followed by relocation of perturbed storms to the analysed position. The ensemble system has 20 members and is run four times daily at 0000, 0600, 1200 and 1800 UTC out to 16 days. Data is output 6-hourly on a 1° lat/lon grid. Output products include tracks of all members, the ensemble mean track and an experimental strike probability product based on the ensemble tracks. Note that fields are bias corrected and in future tracking will be run on the bias corrected fields. A companion short-range ensemble prediction system is also run with regional models, but no special vortex processing is done operationally. More details of the system can be found in Wei *et al.* (2008).

4.2.3. *Met Office Global and Regional Ensemble Prediction System (MOGREPS) — UK*

MOGREPS is based on the Met Office Unified Model. This system consists of two ensembles, one global and one regional using a higher resolution limited area model centred on Europe (Bowler *et al.*, 2008). The MOGREPS global ensemble is run twice a day at 0000 and 1200 UTC at N144 resolution (approximately 90 km in the mid-latitudes) with 38 vertical levels. The resolution will be increased to N216 (approximately 60 km) with 70 levels in 2010. At the Met Office it runs out to 72 hours, but it is also being run out to 15 days as part of the Met Office contribution to the THORPEX research programme.

The MOGREPS ensemble has 24 members (a control and 23 perturbed members). Perturbations to the initial conditions are calculated using the Ensemble Transform Kalman Filter, which is a computationally efficient version of the EnKF. Model uncertainties are represented in the system through a series of schemes designed to tackle the structural and sub-grid-scale sources of model error. No perturbations are specifically targeted at tropical cyclones.

The Met Office tropical cyclone identification and tracking scheme (based on mean sea-level pressure and 850hPa relative vorticity) is applied to 12-hourly output from the 15-day MOGREPS global ensemble. The system tracks existing named storms and identifies new storms out to 144 hours. An experimental tropical cyclone product includes a chart of the forecast track from each ensemble member, coloured according to lead time, a strike probability plot showing the probability of the storm passing within 75 miles and a summary of the key forecast tracks (the deterministic forecast, the ensemble control forecast, the ensemble mean of the tracks and past observations of the storm).

4.2.4. *Typhoon Ensemble Prediction System (Typhoon EPS) — Japan Meteorological Agency*

JMA run an eleven member (one non-perturbed member plus ten perturbed) typhoon EPS. It runs with a lower resolution version of the GSM–TL319 (~55 km grid) spectral discretisation in the horizontal with finite differencing on 60 p-sigma hybrid levels in the vertical. An SV method is used to make the ensemble perturbations. The linearised model and its adjoint version for this Typhoon EPS are based on the ones adopted in the 4D-Var calculus, which consists of full dynamics of Eulerian integrations and full physical processes. The model has T63 (~180 km grid) spectral discretisation in the horizontal and finite differencing on 40 p-sigma hybrid levels in the vertical. Further details are available in Yamaguchi *et al.* (2008).

Typhoon EPS is run four times a day (0000, 0600, 1200, and 1800 UTC) at most with a forecast range of 132 hours. The model run is conducted only when tropical cyclones of minimum tropical storm intensity exist in the responsibility area of RSMC Tokyo–Typhoon Center (0°N–60°N, 100°E–180°E) or are expected to move into the area within the next 24 hours.

4.2.5. *Canadian Meteorological Centre (CMC)*

The Canadian Meteorological Centre (CMC) produces 20 ensemble members twice daily at 0.9 degree 28 level resolution with GEM, with a forecast lead time of 16 days. The 20 members are integrated using different physical parametrizations and using stochastic parametrizations. The 20 initial conditions are sub-sampled from 96 different analyses that have been generated with an EnKF. The EnKF uses perturbed observations to simulate the impact of the limited accuracy of the observations. The perturbed observations are combined with the 96-member ensemble of background trajectories which are defined at 3, 4.5, 6, 7.5 and 9-hour projection. This allows the calculation of forecast to observation differences at the time of observations, which increases the accuracy and allows more satellite and aircraft data to be ingested at the beginning and the end of the 6-hour assimilation window.

A comparison of the ensemble systems operated by CMC, ECMWF and NCEP can be found at Buizza *et al.* (2005).

4.2.6. *China Meteorological Administration (CMA)*

CMA has developed an EPS for tropical cyclone tracks that has been run experimentally since 2006. It provides track forecasts and forecasts of the area of high winds and rains produced by tropical cyclones. The global model is run at T106L19 resolution to 30 days, with perturbations calculated using SVs.

4.2.7. *THORPEX (International)*

The THORPEX Asian Regional Committee is arranging the exchange of tropical cyclone track forecasts among its members. The plan is to develop a multi-model ensemble and deliver probabilistic forecasts to its members as part of a proposed THORPEX Interactive Global Grand Ensemble (TC-TIGGE) to accelerate the improvements in the accuracy of 1-day to 2-week high-impact weather forecasts for the benefit of humanity.

4.2.8. *North American Ensemble Forecast System (NAEFS)*

One major problem with ensemble forecasting systems is dissemination of large amounts of data (over 200 Gigabytes of data daily from all ensemble forecast systems globally), and the necessity of gathering these data from geographically diverse centres. To address this, the North American Ensemble Forecast System (NAEFS) is a weather modeling system run jointly by the Meteorological Service of Canada (MSC) and NCEP to provide numerical weather prediction products to weather forecasters in both countries for a forecast period that runs to two weeks. The NAEFS combines the Canadian global forecast model ensemble and the NCEP global ensemble into a joint ensemble that will create weather forecasts for all of North America. The national weather agencies in North America are participating in NAEFS–MSC, the National Meteorological Service of Mexico, and NCEP.

5. Consensus Numerical Weather Prediction

5.1. *Overview*

The meteorological community has recognised the benefits of consensus forecasting for at least three decades (Sanders, 1973; Thompson, 1977). Leslie and Fraedrich (1990) and Mundell and Rupp (1995) applied this approach to tropical cyclone track prediction and illustrated the forecast improvement that resulted from using linear combinations of forecasts from various tropical cyclone track prediction models. In 1998, Goerss (2000) first illustrated the superior tropical cyclone track forecasting performance for simple ensemble average or consensus forecasts formed using operational NWP models. Based upon this work, the first operational consensus forecast aids using global and regional NWP models exclusively were installed on the Automated Tropical Cyclone Forecasting System (Sampson and Schrader, 2000) at JTWC in 1998. Elsberry and Carr (2000) expanded upon this work and made a five-model consensus (NCON) an integral part of the expert system developed for used by the tropical cyclone forecasters at JTWC, the Systematic Approach Forecasting Aid (Peak *et al.*, 2000; Carr *et al.*, 2001).

The value of these consensus forecast aids is illustrated by Goerss *et al.* (2004), who found that consensus 72-hour tropical cyclone track forecast error is roughly the same as the 48-hour tropical cyclone track forecast error from the individual NWP models themselves. An extensive discussion of operational consensus forecasting is given by Burton *et al.* (2006). Other tropical cyclone forecast centers that have adopted consensus methods include CMA, CPHC, HKO, RSMC Fiji, Australian Bureau of Meteorology tropical cyclone Warning Centres (Perth, Darwin and Brisbane), RSMC La Réunion, Vietnam National Centre for HydroMeteorological Forecasting, and RSMC Tokyo Typhoon Centre.

Some consensus models require all of their member models to be available in order to compute the consensus, whilst others are less restrictive, requiring only two or more members to be present. The latter approach greatly increases availability, but it poses consistency issues for the forecaster, since forecasts for consecutive lead times can comprise a different set of members.

Forecast models are characterised as either 'early' or 'late', depending on whether they are available to the forecaster during the contemporary forecast cycle. For example, consider the 1200 UTC forecast cycle, which begins with the 1200 UTC synoptic time and ends with the release of an official forecast from NHC at 1500 UTC. The 1200 UTC run of the GFS model is not complete and available to the forecaster until about 1600 UTC, an hour after the forecast is released — thus the 1200 UTC GFS would be considered a 'late' model since it could not be used to prepare the 1200 UTC official forecast.

Due to their complexity, dynamical models are generally, if not always, 'late' models. Fortunately, a technique exists to take the latest available run of a late model and adjust its forecast to apply to the current synoptic time and initial conditions. In the example above, forecast data for hours 6–126 from the previous (0600 UTC) run of the GFS would be smoothed and then adjusted, or shifted, so that the 6-hour forecast (valid at 1200 UTC) would match the observed 1200 UTC position and intensity of the tropical cyclone. The adjustment process creates an 'early' version of the GFS model that becomes part of the most current available guidance for the 1200 UTC forecast cycle. The adjusted versions of the 'late' models are known, largely for historical reasons, as 'interpolated' models. Further details of this interpolation process can be found in Goerss *et al.* (2004).

The following deterministic models are used in the operational production of consensus forecasts and are all 'late' models, with the exception of DSHP and LGEM. Hence, the forecasts tracks used to produce the consensus are the 'interpolated' versions, derived using the technique described above.

DSHP — Decay SHIPS statistical-dynamical intensity model
EGRI — UK Met Office global model with manual corrections to tracker
EMXI — ECMWF global model
GFDI — GFDL regional model (initial and boundary conditions from NCEP GFS)
GFNI — GFDL regional model (initial and boundary conditions from NOGAPS)
GFSI — NCEP GFS global model (formerly known as AVNI)
GHMI — GFDI with variable intensity offset correction
HWFI — HWRF regional model (initial and boundary conditions from NCEP GFS)
JGSI — Japanese global spectral model
JTYI — Japanese typhoon model (retired in 2007)
LGEM — Logistical Growth Equation Model (statistical intensity model)
NGPI — US Navy NOGAPS global model (Atlantic version)
TCLI — Australian TC-LAPS regional model
UKMI — UK Met Office global model
WBAI — WBAR regional model

5.2. *Track Consensus Forecast Aids*

The following consensus forecasts are computed when sufficient data are available from deterministic forecasts:

5.2.1. *Florida State Superensemble (FSSE)*

The Florida State Superensemble (Krishnamurti *et al.*, 1999) makes use of a suite of operational model forecasts of tropical cyclone tracks (and intensity) and performs a collective bias reduction from the models' past performance. A least square minimisation of past errors provides a robust set of statistics that are next used for error reduction for real time storm forecasts. The success of this method requires that during the course of a hurricane season large model changes are not made to member models. It has been in operation since 1999.

5.2.2. *JTWC and NHC*

The various consensus forecast aids used by JTWC and NHC since they were first introduced in 2000 are shown in Table 2. This includes a list of members that can be included in the consensus and the minimum number required to produce the consensus.

5.3. *Intensity Consensus Forecast Aids*

More recently, consensus approaches have been applied to intensity forecasting. Some approaches are weighted averages determined

Table 2. Summary of tropical cyclone consensus forecast aids used at JTWC and NHC.

Name	Dates of operation	Used by	Members	Minimum no. members
NCON	2000–2007	JTWC	NGPI, GFNI, UKMI, JGSI, and JTYI	All
CONW	2003 onwards	JTWC	GFSI, EMXI, GFNI, JGSI, JTYI, NGPI, TCLI, UKMI, and WBAI	Two
GUNS	2000–2007	NHC	GFDI, UKMI, and NGPI	All
GUNA	2001 onwards	NHC	GFDI, UKMI/EGRI, NGPI, and GFSI	All
CONU	2004–2007	NHC	GFDI, GFNI, UKMI, NGPI, and GFSI	Two
TCON	2008 onwards	NHC	GFSI, EGRI, NGPI, GHMI, and HWFI	All
TVCN	2008 onwards	NHC	GFSI, EGRI, NGPI, GHMI, HWFI, GFNI, and EMXI	Two

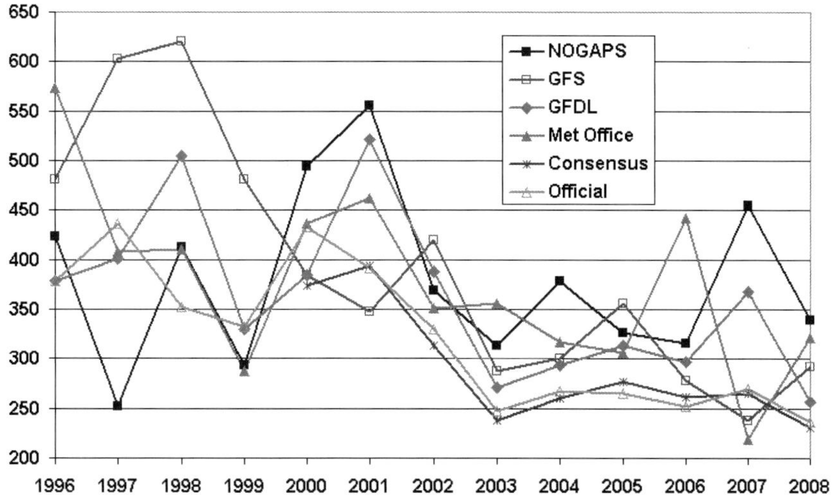

Figure 2. North Atlantic 72-hour tropical cyclone track errors (km) for selected models, their consensus and 'official' forecasts.

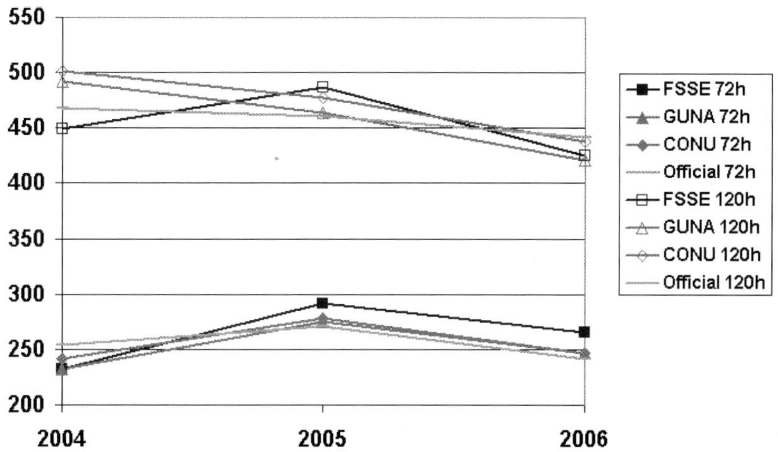

Figure 3. North Atlantic 72-hour and 120-hour tropical cyclone track errors (km) for superensemble, consensus and 'official' forecasts.

errors at longer lead times halted the downward trend in errors for a while.

No attempt is made to verify absolute values of tropical cyclone intensity in the Met Office global model since its relatively coarse resolution does not permit a realistic analysis or forecast of wind or pressure near a storm's centre. However, verification of the forecast intensity tendency skill is undertaken. This is achieved by assessing the upward or downward trend in forecast values of 850 hPa relative vorticity in each 12-hour forecast step against the observed change in maximum sustained wind speed during this period. A skill score is calculated such that values greater than zero indicate that the model forecast is better than chance. Figure 5 shows the northern hemisphere results since 2001. This indicates that the model was initially not good at predicting the strengthening of tropical cyclones. This has improved in successive seasons, but to the detriment of the weakening skill. Overall, the tendency skill

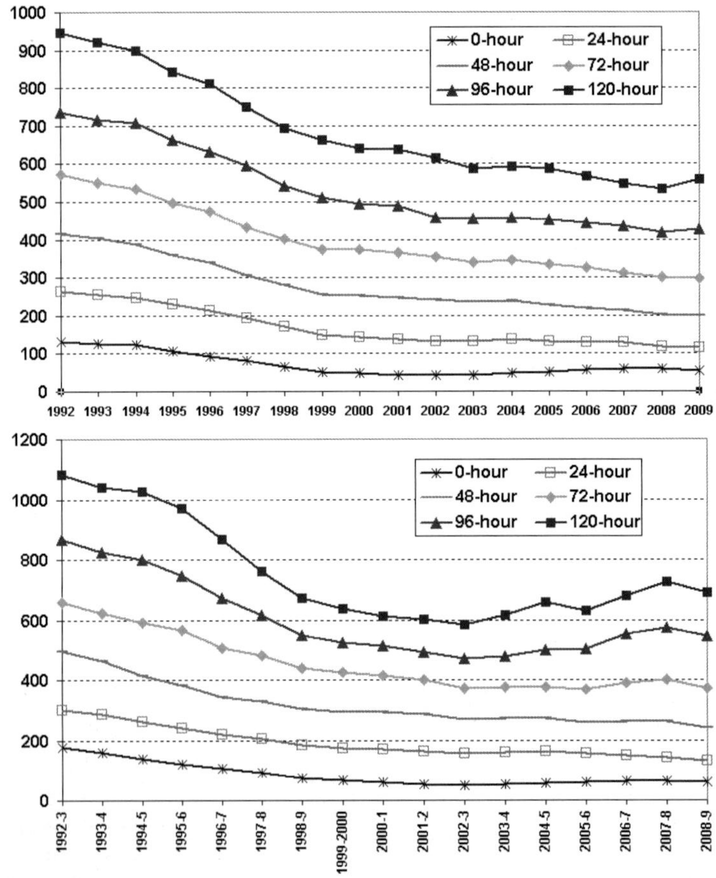

Figure 4. Met Office global model 5-year running mean of tropical cyclone track forecast errors (km) for the northern hemisphere (top) and southern hemisphere (bottom).

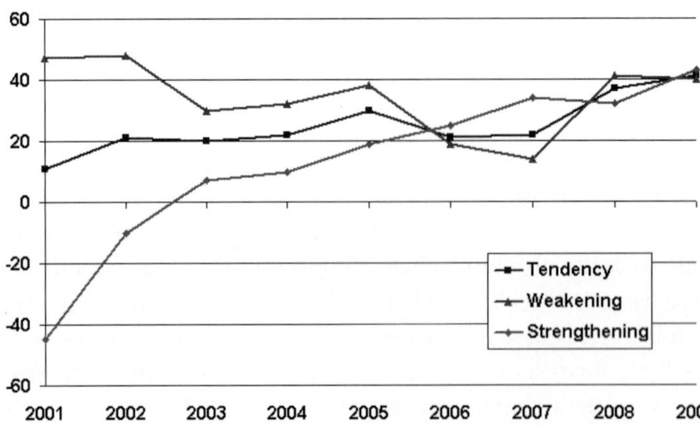

Figure 5. Met Office global model tropical cyclone intensity tendency skill scores (%) for the northern hemisphere.

(strengthening and weakening skills combined) shows a very small upward trend since verification began.

6.2.4. *Global Spectral and Typhoon Models — Japan Meteorological Agency*

A time series of the annual average position errors of the track predictions by JMA GSM and JTYM in the Western North Pacific from 1996 to 2007 are shown in Fig. 6. In 2007, the errors for 30-, 54- and 78-hour predictions were 143 km, 201 km and 252 km (GSM) and 146 km, 227 km and 301 km (JTYM) respectively.

Figure 7 represents the forecast errors of JMA GSM and JTYM's 30-hour forecasts in central pressure and maximum wind speed in 2007. It is found that JMA GSM underestimated the intensity of tropical cyclones in its predictions of both central pressures and maximum wind speeds to a much greater degree than JTYM, which has a much higher resolution and is therefore better able to represent the true pressure and wind near the centre of tropical cyclones.

6.2.5. *ECMWF*

Tropical cyclone track forecast errors for the ECMWF model are calculated at the Met Office using the same scheme used for verification of the Met Office model. Figure 8 shows the ECMWF global track forecast errors since 1994. This indicates a downward trend in errors, most notably at longer lead times where errors have reduced dramatically since 2003.

The model resolution increase in February 2006 resulted in substantial improvements in intensity, with tropical cyclones appearing deeper throughout the forecast range. The core pressure is also much better in the analysis, as well as in the forecast.

The EPS tropical cyclone forecast is presented on the ECMWF web site as a strike probability: the probability at any location that a reported tropical cyclone will pass within 120 km of that location within the next 120 hours. Verification of these probabilistic forecasts for the last five years also demonstrates improvements in performance from the high resolution system.

6.2.6. *QLM — Indian Meteorological Department*

Figure 9 shows the track forecast errors from the QLM used by IMD for forecasting of tropical cyclones in the North Indian Ocean.

6.2.7. *GDAPS — Korean Meteorological Administration (KMA)*

KMA have undertaken an intercomparison of tropical cyclone track forecasts for the KMA GDAPS model, the RSMC Tokyo official forecast and the JTWC forecast. Figure 10

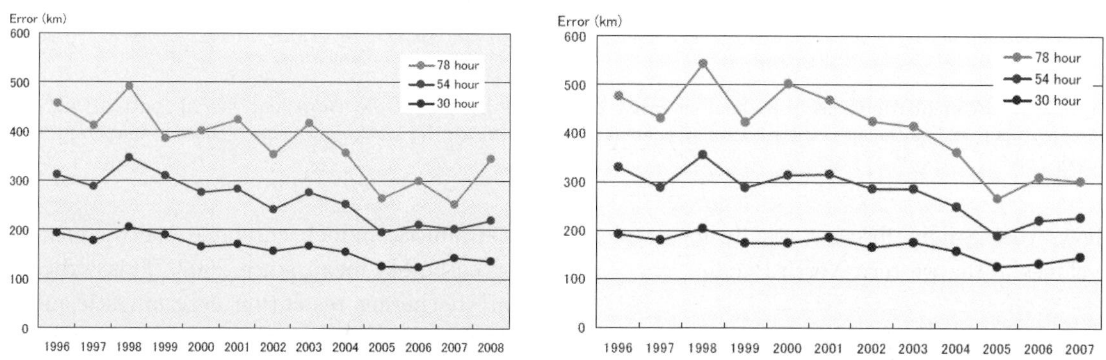

Figure 6. A time series of the annual average position errors (km) of the track predictions by JMA GSM (left) and JTYM (right) in the western North Pacific from 1996 to 2007.

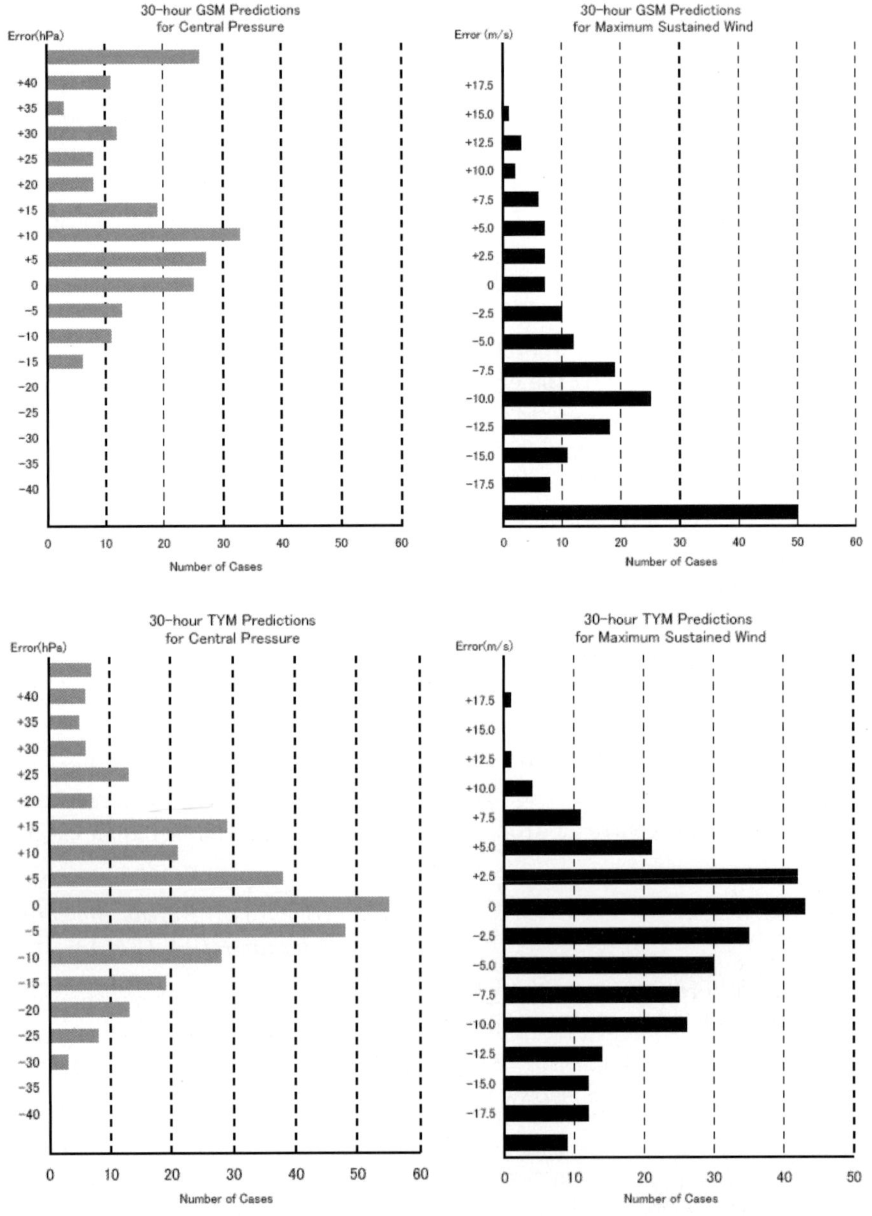

Figure 7. Error distributions of 30-hour intensity predictions by JMA GSM (top) and JTYM (bottom) in 2007. The grey figure shows the error distribution of central pressures (hPa); the black figure shows those of maximum wind speeds (ms^{-1}).

shows the 48-hour forecast results for tropical cyclones in the western North Pacific.

6.2.8. *Ensemble Means*

Figure 11 shows 48-hour and 120-hour track and pressure forecast errors from the ECMWF deterministic model, ensemble control member and ensemble mean since 2002. This indicates that the higher resolution deterministic model outperforms the ensemble mean. However, there have been periods in the past when the control member or ensemble mean 120-hour forecast was better than the deterministic model.

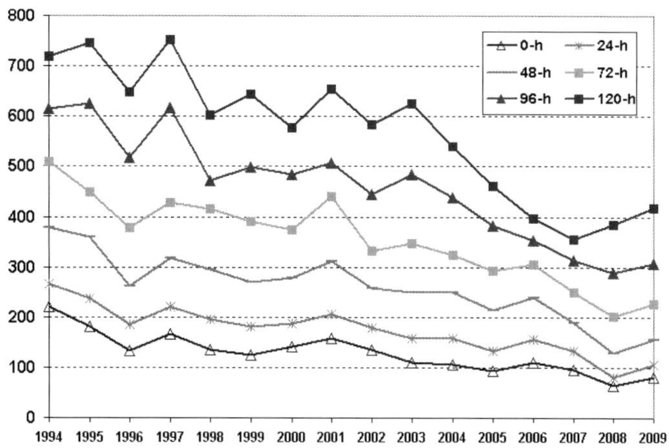

Figure 8. ECMWF deterministic global model tropical cyclone track forecast errors (km) for the whole globe.

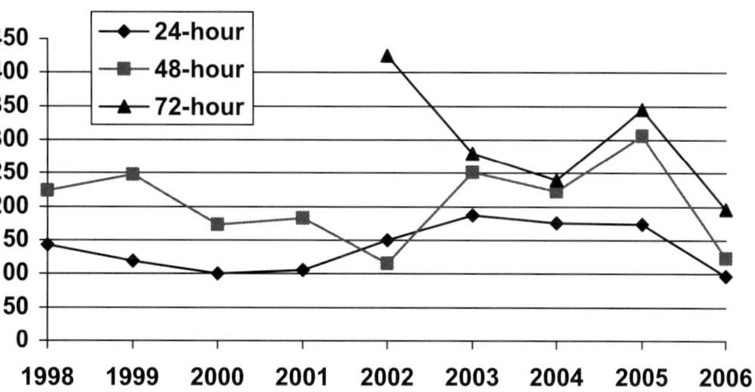

Figure 9. Track forecast errors (km) for all tropical cyclones in the North Indian Ocean from the QLM.

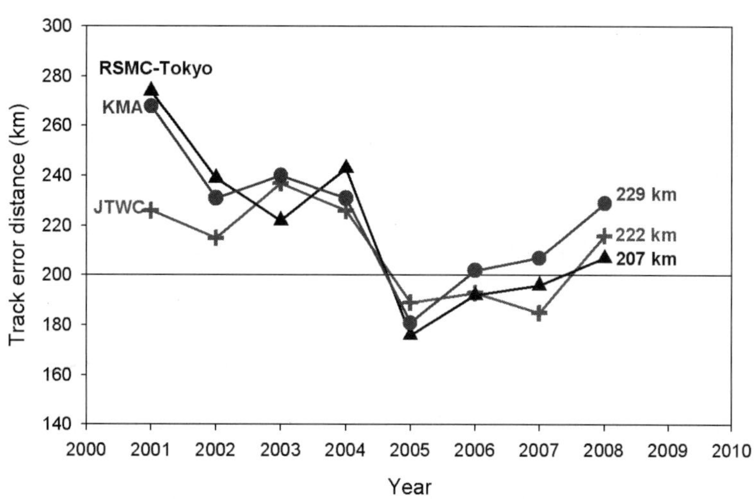

Figure 10. 48-hour forecast errors (km) in the western North Pacific for KMA, RSMC Tokyo and JTWC.

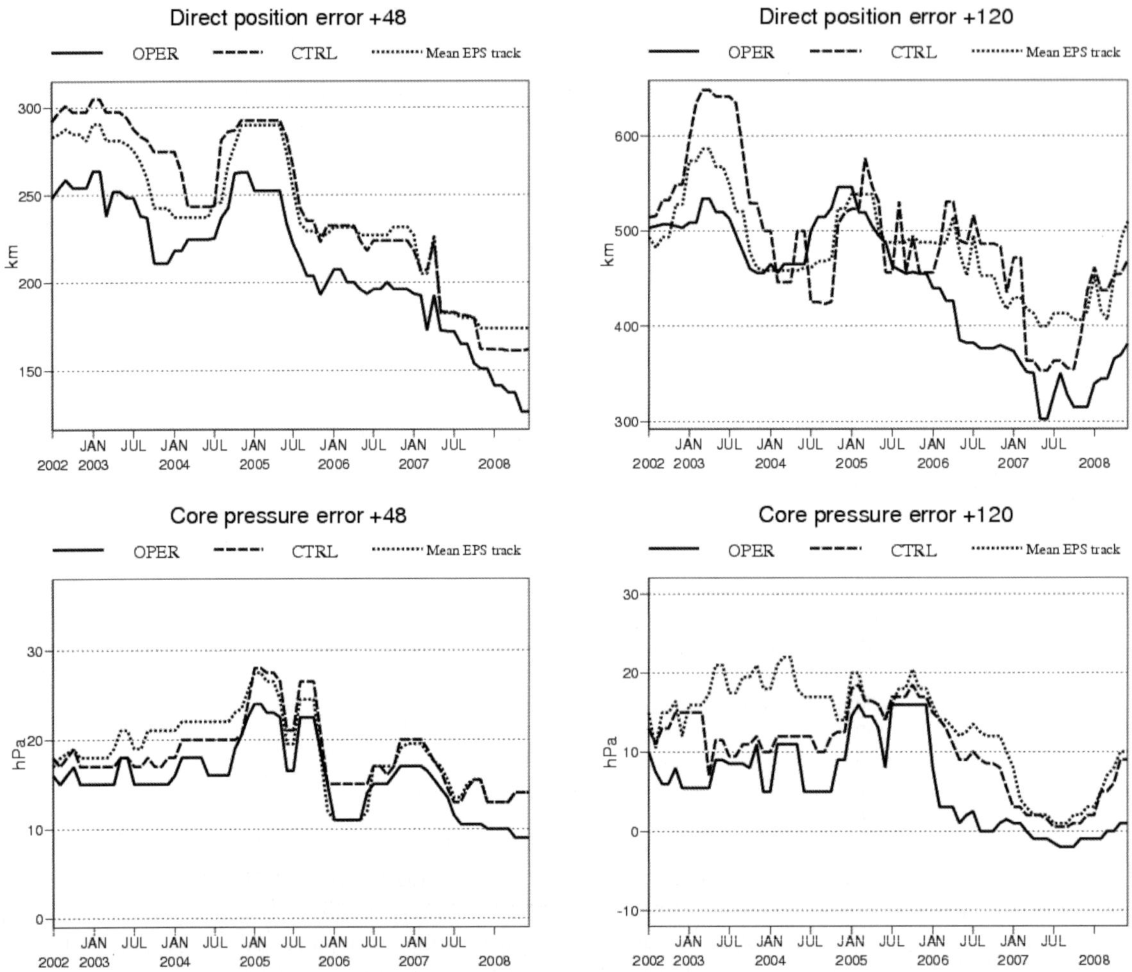

Figure 11. ECMWF 12-month running medians of the monthly-mean track error (km) and core pressure error (hPa) for the deterministic model (OPER), ensemble control member (CTRL) and the ensemble mean.

Figure 12 shows the track errors for the Met Office deterministic global model and the ensemble mean from MOGREPS from February 2007–July 2009. This indicates that the deterministic model and the ensemble mean produces very similar track forecast errors at lead times to 72 hours. At longer lead times the ensemble mean was slightly better than the deterministic model indicating that the ensemble was producing some useful information over and above that produced by the deterministic model at this range.

6.2.9. *National Hurricane Center — USA*

NHC undertakes evaluations of both the tropical cyclone forecasts it receives from external sources and its own 'official' forecast. Figure 13 shows NHC's 'official' tropical cyclone track forecast errors for the North Atlantic since 1970. A downward trend in errors up to 72 hours lead time is apparent. The figure also shows the equivalent for tropical cyclone intensity forecasts since 1990 and shows no significant trend during the period in question. Figure 14 shows the

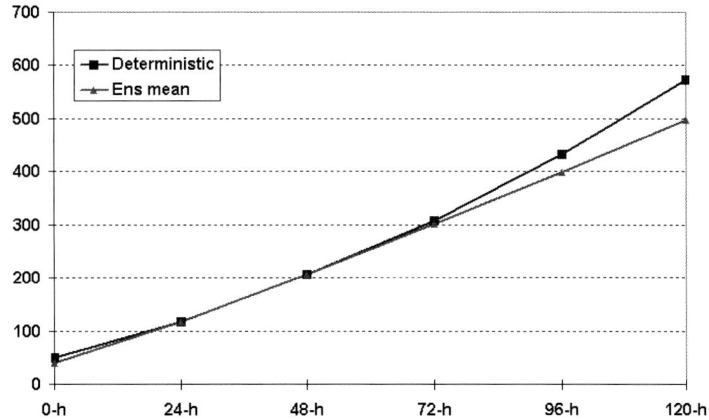

Figure 12. Met Office global model track error values (km) for the deterministic model (squares) and the ensemble mean (triangles) for February 2007–July 2009.

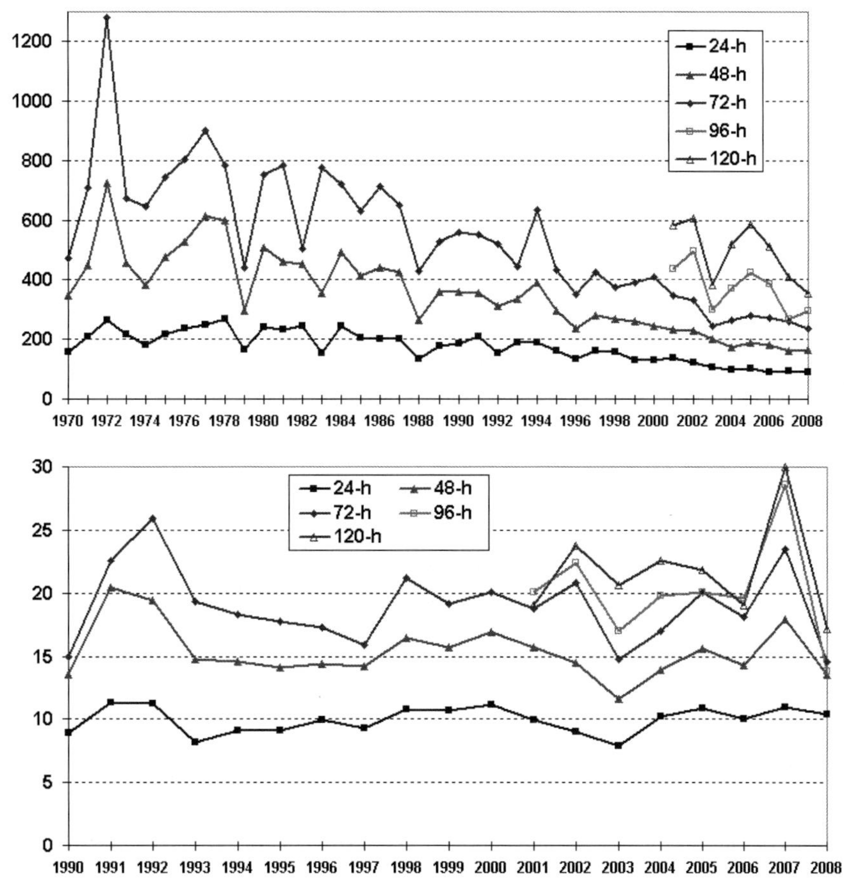

Figure 13. NHC official tropical cyclone track (top, kilometres) and intensity (bottom, knots) forecast errors for the North Atlantic.

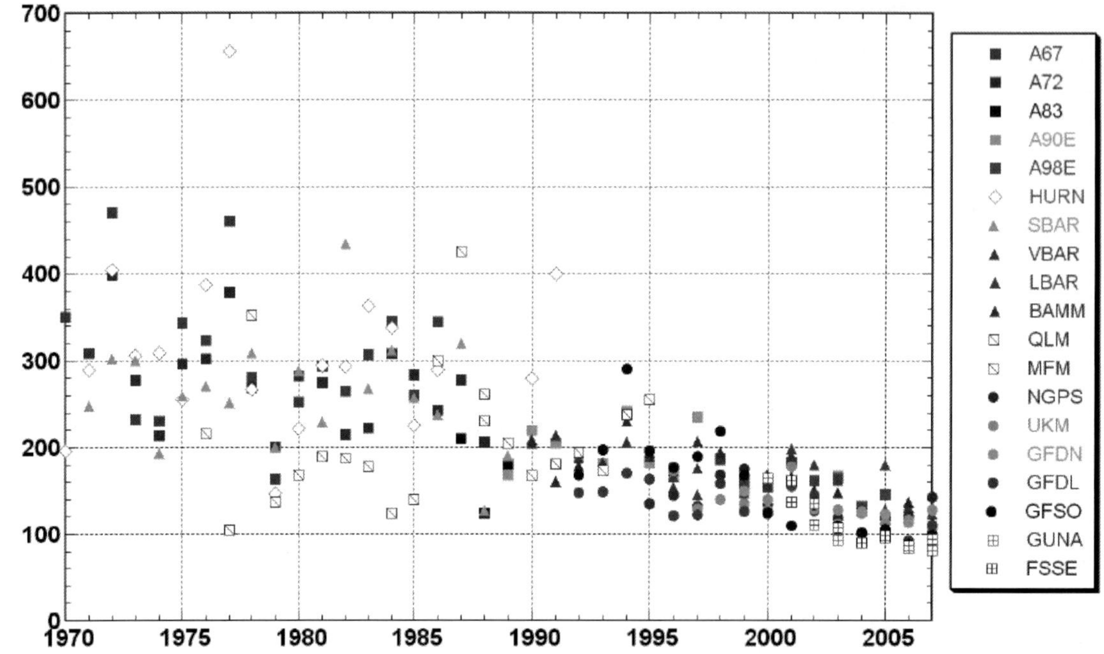

Figure 14. 48-hour tropical cyclone track forecast errors (nautical miles) for the North Atlantic for various forms of guidance available to NHC.

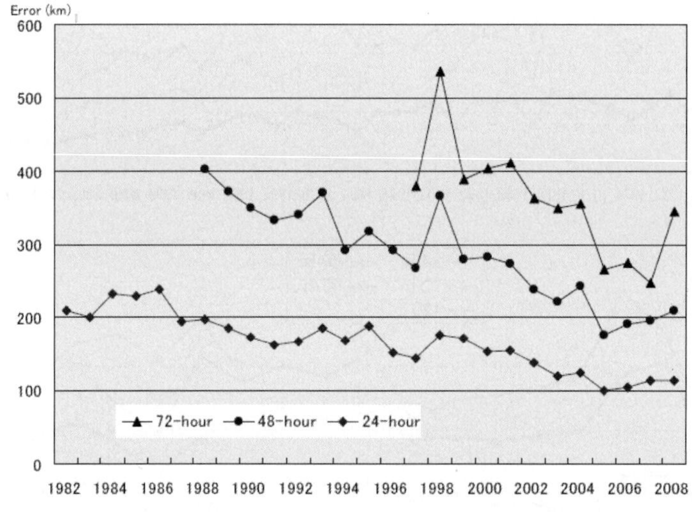

Figure 15. JMA official tropical cyclone track forecast errors (km) for the western North Pacific.

48-hour tropical cyclone track forecast errors for various objective aids used at NHC. This indicates how the forecasters have benefited from reductions in errors of all forms of forecasting aids over recent years including the introduction of consensus forecasting techniques.

6.2.10. *RSMC Tokyo Typhoon Warning Centre*

RSMC Tokyo Typhoon Warning Centre is the official tropical cyclone forecasting centre for the western North Pacific Ocean. Figure 15 shows the tropical cyclone track forecast errors for their forecasts since 1982 (for 24-hour forecasts) and 1988 and 1997 (for 48 and 72-hour forecasts respectively). In 2007, the errors for 24-, 48- and 72-hour predictions were 114 km, 196 km and 247 km. Notably, these values were lower than the equivalent errors of the JMA GSM and JTYM model runs, indicating that the forecaster was adding value.

6.2.11. *RSMC La Réunion — Météo-France*

RSMC La Réunion is the official tropical cyclone forecasting centre for the South-West Indian Ocean. Figure 16 shows the tropical cyclone track forecast errors for their forecasts since 1990–1991. A consistent downturn in track forecast errors is noted in the last three seasons shown.

6.2.12. *Australian Bureau of Meteorology*

The regional Tropical Cyclone Warning Centres at Perth, Darwin and Brisbane have responsibility for tropical cyclone forecasting in the latitude band 90°E to 160°E around Australia. Figure 17 shows the mean errors for the track forecasts issued from these centres since the

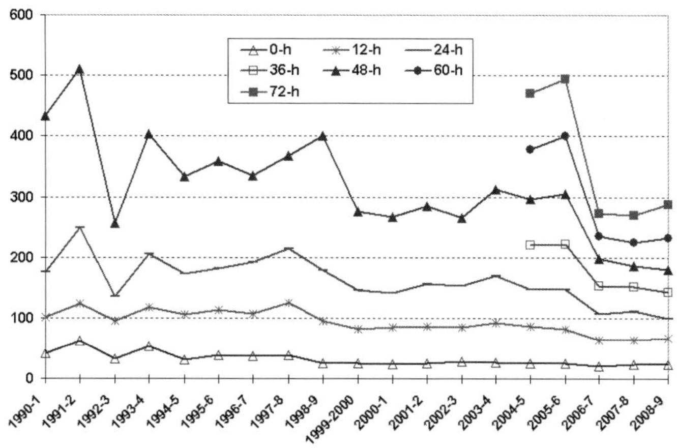

Figure 16. RSMC La Réunion annual mean tropical cyclone track forecast errors (km) for the South-West Indian Ocean.

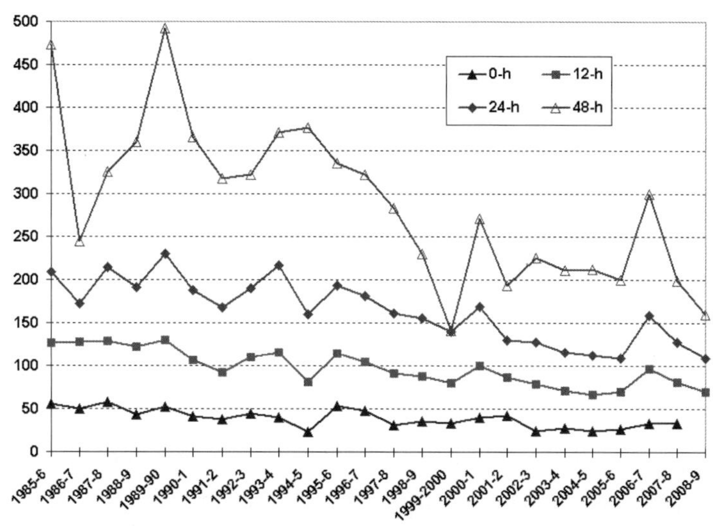

Figure 17. Australian tropical cyclone track forecast errors (km) since the mid-1980s.

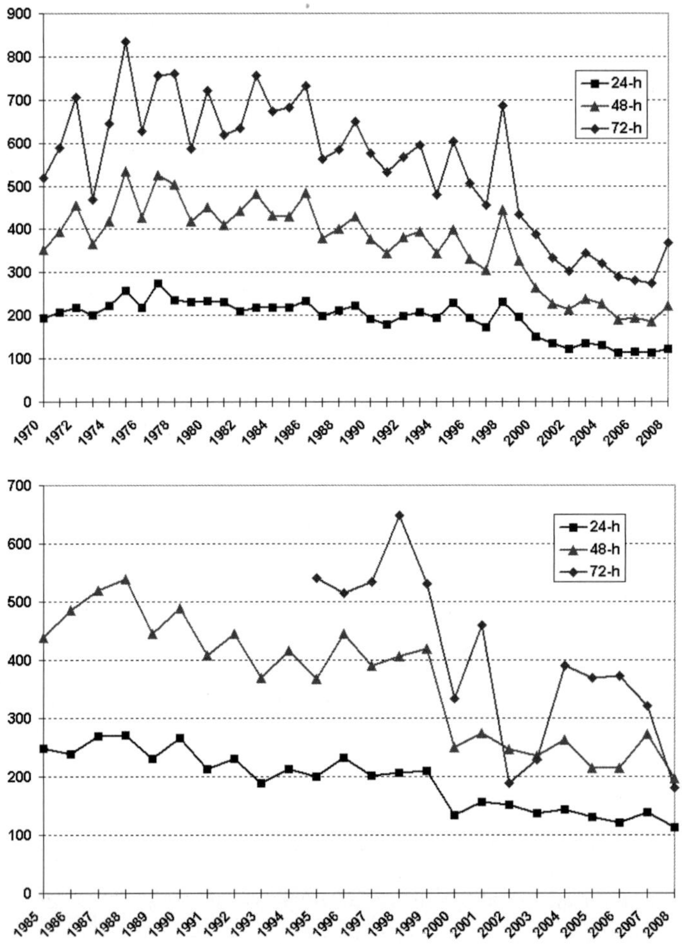

Figure 18. JTWC tropical cyclone track forecast errors (km) for the western North Pacific (top) and southern hemisphere (bottom).

mid-1980s. This indicates a long-term downward trend in short period tropical cyclone track forecast errors.

6.2.13. *Joint Typhoon Warning Center (JTWC) — USA*

JTWC produces tropical cyclone forecasts for all regions not covered by NHC and CPHC. These forecasts are subjectively produced from a variety of model guidance and consensus forecasts available to forecasters. Figure 18 shows JTWC tropical cyclone tracks forecast errors for the western North Pacific (since 1970) and the southern hemisphere (since 1985). They both indicate a long term downward trend in errors.

7. Web Links

Periodically updated information on numerical model development and forecast performance can be found in the web links in this section.

7.1. *Numerical Model Information*

Working Group on Numerical Experimentation Blue Book:
http://collaboration.cmc.ec.gc.ca/science/wgne

GFS:
http://www.emc.ncep.noaa.gov/gmb/STATS/html/model_changes.html

Met Office:
http://www.metoffice.gov.uk/research/nwp

JMA:
http://www.jma.go.jp/jma/jma-eng/jma-center/nwp/nwp-top.htm

ECMWF:
http://www.ecmwf.int/research

CMC:
http://www.msc-smc.ec.gc.ca/cmc/op_systems/global_forecast_e.html

KMA:
http://www.kma.go.kr/ema/ema03/intro_modl.html

GFDL/HWRF:
http://www.gfdl.noaa.gov/research/weather/hurricane.html

Australian Bureau of Meteorology:
http://www.bom.gov.au/nmoc/ab_nmc_op.shtml

Models used at NHC:
http://www.srh.noaa.gov/ssd/nwpmodel/html/nhcmodel.htm
http://www.nhc.noaa.gov/modelsummary.shtml

TIGGE:
http://tigge.ecmwf.int

7.2. *Forecast Verification*

Met Office, ECMWF:
http://www.metoffice.gov.uk/weather/tropicalcyclone/verification.html

JMA:
http://www.jma.go.jp/jma/jma-eng/jma-center/rsmc-hp-pub-eg/annualreport.html

NHC:
http://www.nhc.noaa.gov/verification

RSMC, La Reunion:
http://www.meteo.fr/temps/domtom/La_Reunion/Espace_boutique/BulletinClimaAnnuel/PageAccueil.html

JTWC:
http://www.usno.navy.mil/JTWC/annual-tropical-cyclone-reports

8. Concluding Remarks

This chapter has shown how techniques for tropical cyclone forecasting have developed over recent decades. Simple statistical and climatological techniques were followed by dynamical models producing a deterministic forecast solution. These models continue to improve in terms of their resolution, dynamical core, physical processes and data assimilation techniques. In more recent years, ensemble forecast techniques have helped develop a more probabilistic approach to tropical cyclone forecasting. This has been complemented by consensus forecasting aids, which are proving to be an important forecasting tool. Whilst verification of tropical cyclone forecast tracks from both numerical models and tropical cyclone warning centres has shown a steady improvement, the challenge remains to forecast changes in tropical cyclone intensity with greater skill in the future.

Acknowledgements

The authors would like to thank the following who have all provided contributions for this chapter: Sim Aberson, Michael Brennan, Kenneth Campana, Eun-Jeong Cha, S.T. Chan, Noel Davidson, Mark DeMaria, Chris Fogarty, H.R. Hatwar, Peter Houtekamer, Hiroto Kitagawa, T.N. Krishnamurti, Manon Lajoie, T.C. Lee, Adrian Lock, Bernard Meisner, Sean Milton, Chiashi Muroi, Prisco Nilo, Matthieu Plu, Alan Radford, Jamie Rhome, David Richardson, Buck Sampson, Alan Sharp, Naomi Surgi, Helen Titley, Zoltan Toth, Mozheng Wei, Samuel Westrelin, Munehiko Yamaguchi and Renhe Zhang.

References

Aberson, S. D., 1998: Five-day tropical cyclone track forecasts in the North Atlantic basin. *Wea. Forecasting*, **13**, 1005–1015.

Aberson, S. D. and C. R. Sampson, 2003: On the predictability of tropical cyclone tracks in the

Northwest Pacific Basin. *Mon. Wea. Rev.*, **131**, 1491–1497.

Allan, R. P., A. Slingo, S. F. Milton, and M. A. Brooks, 2007: Evaluation of the Met Office global forecast model using Geostationary Earth Radiation Budget (GERB) data. *Quart. J. Roy. Meteor. Soc.*, **133**, 1993–2010.

Allan, R. P., A. Slingo, S. F. Milton, and I. Culverwell, 2005: Exploitation of Geostationary Earth Radiation Budget data using simulations from a numerical weather prediction model: Methodology and data validation. *J. Geophys. Res.*, **110**, D14111.

Arakawa, A. and W. H. Schubert, 1974: Interaction of a cumulus cloud ensemble with the large-scale environment, part I. *J. Atmos. Sci.*, **31**, 674–701.

Biswas, M. K., B. P. Mackey, and T. N. Krishnamurti, 2006: Performance of the Florida State University Hurricane Superensemble during 2005. *Proc. 60th Interdepartmental Hurricane Conference*, http://www.ofcm.gov/ihc06/linking_file_ihc06.htm.

Blake, E. S., E. N. Rappaport, and C. W. Landsea, 2007: The Deadliest, Costliest and Most Intense United States Tropical Cyclones from 1851 to 2006. *NOAA Technical Memorandum NWS TPC-5.* http://www.nhc.noaa.gov/Deadliest_Costliest.shtml

Bougeault, P. and P. Lacarrere, 1989: Parameterisation of orography-induced turbulence in a mesobeta-scale model. *Mon. Wea. Rev.* **117**, 1872–1890.

Bouteloup Y., 1995: Improvement of the spectral representation of the earth topography with a variational method, *Mon. Wea. Rev.*, **123**, 1560–1573.

Bowler, N. E., A. Arribas, K. R. Mylne, K. B. Robertson, and S. E. Beare, 2008: The MOGREPS short-range ensemble prediction system. *Quart. J. Roy. Meteor. Soc.*, (in press).

Buizza, R., 1997: Potential forecast skill of ensemble prediction and spread and skill distributions of the ECMWF Ensemble Prediction System. *Mon. Wea. Rev.*, **125**, 99–119.

Buizza, R., J.-R. Bidlot, N. Wedi, M. Fuentes, M. Hamrud, G. Holt, T. Palmer, and F. Vitart, 2006: The ECMWF Variable Resolution Ensemble Prediction System (VAREPS). *ECMWF Newsletter No. 108*, 14–20. http://www.ecmwf.int/publications/newsletters/pdf/108.pdf

Buizza, R., P. L. Houtekamer, Z. Toth, G. Pellerin, M. Wei, and Y. Zhu, 2005: A comparison of the ECMWF, MSC and NCEP global ensemble prediction systems. *Mon. Wea. Rev.*, **133**, 1076–1097.

Burton, A., P. Caroff, J. Franklin, E. Fukada, T. C. Lee, C. Sampson, and T. Smith, 2006: Sharing experiences in operational consensus forecasting. Special Focus Topic 3a, WMO Sixth International Workshop on Tropical Cyclones, San Jose, Costa Rica. http://severe.worldweather.org/iwtc/document/Topic_3a_Andrew_Burton.pdf.

Campana, K. and P. Caplan, 2005: Technical Procedures Bulletin for the T382 Global Forecast System. *NCEP, Camp Springs, USA.* http://www.emc.ncep.noaa.gov/gc_wmb/Documentation/TPBoct05/T382.TPB.FINAL.htm

Carr, L. E., R. L. Elsberry, and J. E. Peak, 2001: Beta test of the systematic approach expert system prototype as a tropical cyclone track forecasting aid. *Wea. Forecasting*, **16**, 355–368.

Cheung, K. W. and J. C. L. Chan, 1999a: Ensemble forecasting of tropical cyclone motion using a barotropic model. Part I: Perturbations of the environment. *Mon. Wea. Rev.*, **127**, 1229–1243.

Cheung, K. W. and J. C. L. Chan, 1999b: Ensemble forecasting of tropical cyclone motion using a barotropic model. Part II: Perturbations of the vortex. *Mon. Wea. Rev.*, **127**, 2617–2640.

Davidson, N. E., L. J. Rikus, R. A. Dare, C. I. W. Tingwell, and H. C. Weber, 2006: Validation of TC-LAPS Structure Forecasts of Some Significant 2004–2005 US Hurricanes. Amer. Met. Soc. 27th Conference on Hurricanes and Tropical Meteorology (Monterey, USA).

Davidson, N. E. and H. C. Weber, 2000: The BMRC high resolution tropical cyclone prediction system: TC-LAPS. *Mon. Wea. Rev.*, **128**, 1245–1265.

DeMaria, M. S., S. D. Aberson, K. V. Ooyama, and S. J. Lord, 1992: A nested spectral model for hurricane track forecasting. *Mon. Wea. Rev.*, **120**, 1628–1643.

DeMaria, M. and J. Kaplan, 1999: An updated statistical hurricane intensity prediction scheme (SHIPS) for the Atlantic and Eastern North Pacific Basins. *Wea. Forecasting*, **14**, 326–337.

DeMaria, M., M. Mainelli, L. K. Shay, J. A. Knaff, and J. Kaplan, 2005: Further improvements to the Statistical Hurricane Intensity Prediction Scheme (SHIPS). *Wea. Forecasting*, **20**, 531–543.

Dvorak, V., 1975: Tropical cyclone intensity analysis and forecasting from satellite imagery. *Mon. Wea. Rev.*, **103**, 420–430.

Dvorak, V., 1984: Tropical cyclone intensity analysis using satellite data. *NOAA Tech. Report NESDIS 11*. NOAA/NESDIS, 5200 Auth Rd., Washington, DC, 20233, 47pp.

Elsberry, R. L. and L. E. Carr, 2000: Consensus of dynamical tropical cyclone track forecasts — errors versus spread. *Mon. Wea. Rev.*, **128**, 4131–4138.

Emanuel, K., 2005: Ensemble forecasting in hurricane intensity. *Proc. 59th Interdepartmental Hurricane Conference*, http://www.ofcm.gov/ihc05/linking_file_ihc05.htm

Errico, R. and D. Baumhefner, 1987: Predictability experiments using a high-resolution limited-area model. *Mon. Wea. Rev.*, **115**, 488–504.

Frank, W. M., 1977: The structure and energetics of the tropical cyclone I. Storm structure, *Mon. Wea. Rev.*, **105**, 1119–1135.

Franklin, J. L., 2006: 2005 National Hurricane Center forecast verification report. http://www.nhc.noaa.gov/verification/pdfs/Verification_2005.pdf

Fujita, T., 1952: Pressure distribution within typhoon. *Geophys. Mag.*, **23**, 437–451.

Goerss, J. S., 2000: Tropical cyclone track forecasts using an ensemble of dynamical models. *Mon. Wea. Rev.*, **128**, 1187–1193.

Goerss, J. S. and R. Jeffries, 1994: Assimilation of synthetic tropical cyclone observations into the Navy Operational Global Atmospheric Prediction System. *Wea. Forecasting,* **9**, 557–576.

Goerss, J. S., C. Velden, and J. Hawkins, 1998: The impact of multispectral GOES-8 wind information on Atlantic tropical cyclone track forecasts in 1995. Part II: NOGAPS forecasts. *Mon. Wea. Rev.*, **126**, 1219–1227.

Goerss, J. S., C. R. Sampson, and J. Gross, 2004: A history of western North Pacific tropical cyclone track forecast skill. *Wea. Forecasting*, **19**, 633–638.

Haltiner, G. J. and R. T. Williams, 1979: *Numerical Prediction and Dynamical Meteorology* (2nd Ed.). Wiley and Sons, NY, 477 pp.

Hamill, T. M., 2006: Ensemble-based atmospheric data assimilation. Chapter 6 of *Predictability of Weather and Climate*, Cambridge University Press, pp. 124–156.

Heming, J. T., 1994: Keeping an eye on the hurricane — Verification of tropical cyclone forecast tracks at the UK Meteorological Office. *NWP Gazette,* **1**, No. 2, pp. 3–8. http://www.metoffice.gov.uk/weather/tropicalcyclone/method/index.html

Heming, J. T., 2009: Evaluation of and improvements to the Met Office tropical cyclone initialisation scheme. *Meteorological Applications*, **16**, 339–351.

Heming, J. T., J. C. L. Chan, and A.M. Radford, 1995: A new scheme for the initialisation of tropical cyclones in the UK Meteorological Office global model. *Meteorological Applications*, **2**, 171–184.

Hogan, T. and T. Rosmond, 1991: The description of the Navy Operational Global Atmospheric Prediction System's spectral forecast model. *Mon. Wea. Rev.*, **119**, 1786–1815.

Hope, J. R. and C. J. Neumann, 1970: An operational technique for relating the movements of existing tropical cyclones to past tracks. *Mon. Wea. Rev.*, **98**, 925–933.

Isaksen, L., D. Le Meur, and A. Stoffelen, 1998: Impact of ERS-scatterometer wind data on ECMWF's analyses and forecasts of tropical cyclones. *European Space Agency Spec. Publ. ESA SP*, no. SP-424, 177–183.

Jarvinen, B. R. and C. J. Neumann, 1979: Statistical forecasts of tropical cyclone intensity. *NOAA Tech. Memo. NWS NHC-10*, 22 pp.

JTWC, 2008: The annual tropical cyclone reports. http://www.usno.navy.mil/JTWC/annual-tropical-cyclone-reports.

Karmakar, S., 1998: The impact of tropical cyclones on the coastal regions of SAARC countries and their influence in the region. *SMRC No.1*, SAARC Meteorological Research Centre (SMRC), Agargaon, Dhaka-1207, Bangladesh.

Knaff, J. A., M. DeMaria, C. R. Sampson, and J. M. Gross, 2003: Statistical, five-day tropical cyclone intensity forecasts derived from climatology and persistence. *Wea. Forecasting*, **18**, 80–92.

Knaff, J. A., C. R. Sampson, M. DeMaria, T. P. Marchok, J. M. Gross, and C. J. McAdie, 2007: Statistical tropical cyclone wind radii prediction using climatology and persistence. *Wea. Forecasting*, **22**, 781–791.

Krishnamurti, T. N., C. M. Kishtawal, T. LaRow, D. Bachiochi, Z. Zhang, C. E. Williford, S. Gadgil, and S. Surendran, 1999: Improved weather and seasonal climate forecasts from multimodel superensemble. *Science*, **285**, 1548–1550.

Kuo, H. L., 1974: Further studies of the parameterisation of the influence of cumulus convection on large-scale flow. *J. Atmos. Sci.*, **31**, 1232–1240.

Kurihara, Y., M. A. Bender, R. E. Tuleya and R. J. Ross, 1995: Improvements in the GFDL hurricane prediction system. *Mon. Wea. Rev.*, **123**, 2791–2801.

Leftwich, P. W. and C. J. Neumann, 1977: Statistical guidance on the prediction of eastern North Pacific tropical cyclone motion Part 2. *NOAA Technical Memo NWS WR-125*, 15 pp.

Leslie, L. M. and K. Fraedrich, 1990: Reduction of tropical cyclone position errors using an optimal combination of independent forecasts. *Wea. Forecasting*, **5**, 158–161.

Leslie, L. M., G. J. Holland, M. Glover and F. J. Woodcock, 1990: A climatological persistence (CLIPER) scheme for predicting Australian region tropical cyclone tracks. *Aust. Met. Mag.*, **38**, 87–92.

Leutbecher, M., 2005: On ensemble prediction using singular vectors started from forecasts. *Mon. Wea. Rev.*, **133**, 3038–3046.

Lewis, W. E. and G. J. Tripoli, 2006: Tropical cyclone modeling in a probabilistic framework. Amer. Met. Soc. 27th Conference on Hurricanes and Tropical Meteorology (Monterey, USA).

Liu, Q., T. P. Marchok, H.-L. Pan, M. Bender and S. Lord, 2000: Improvements in Hurricane Initialization and Forecasting at NCEP with Global and Regional (GFDL) models. National Weather Service Office of Meteorology Technical Procedures Bulletin No. 472. http://www.nws.noaa.gov/om/tpb/472.htm

Lord, S. J., 1991: A bogussing system for vortex circulations in the National Meteorological Center global forecast model. Amer. Met. Soc. 19th Conference on Hurricanes and Tropical Meteorology (Miami, USA), 328–330.

Lorenc, A. C. and F. Rawlins, 2005: Why does 4D-Var beat 3D-Var? *Quart. J. Roy. Meteor. Soc.*, **131**, 3247–3257.

Marchok, T. P., 2002: How the NCEP tropical cyclone tracker works. Amer. Met. Soc. 25th Conference on Hurricanes and Tropical Meteorology (San Diego, USA), 21–22.

Marks, D. G., 1992: The beta and advection model for hurricane track forecasting. *NOAA Tech. Memo.* NWS NMC- 70, 89 pp.

Martin, G. M., M. A. Ringer, V. D. Pope, A. Jones, C. Dearden, and T. J. Hinton, 2006: The physical properties of the atmosphere in the new Hadley Centre Global Environmental Model, (HadGEM1). Part I: Model description and global climatology. *J. Clim.*, **19**, 1274–1301.

Mathur, M. B., 1991: The National Meteorological Center's quasi-Lagrangian model for hurricane prediction. *Mon. Wea. Rev.*, **119**, 1419–1447.

Moorthi, S. and M. Suarez. 1992: Relaxed Arakawa-Schubert: A parameterization of moist convection for general circulation models. *Mon. Wea. Rev.*, **120**, 978–1002.

Mundell, D. B. and J. A. Rupp, 1995: Hybrid forecast aids at the Joint Typhoon Warning Center: Applications and results. Amer. Met. Soc. 21st Conference on Hurricanes and Tropical Meteorology (Miami, USA), 216–218.

Neumann, C. J., 1972: An alternate to the HURRAN tropical cyclone forecast system. *Mon. Wea. Rev.*, **100**, 245–255.

Neumann, C. J. and C. J. McAdie, 1991: A revised National Hurricane Center NHC83 Model (NHC90). *NOAA Tech. Memo. NWSNHC-44*, 35 pp.

Neumann, J., 1975: Great historical events that were significantly affected by the weather: I. The Mongol invasions of Japan. *Bull. Amer. Meteor. Soc.*, **56**, 1167–1171.

Pan, D.-M. and D. Randall, 1998: A cumulus parameterization with a prognostic closure. *Quart. J. Roy. Meteor. Soc.*, **124**, 949–981.

Peak, J. E., L. E. Carr, and R. L. Elsberry, 2000: Developing a systematic approach to tropical cyclone track forecasting expert system — 1: Information management, visualization and proactivity considerations. Amer. Met. Soc. 24th Conference on Hurricanes and Tropical Meteorology (Ft. Lauderdale, USA), 502–503.

Peng, M., J. Ridout, and T. Hogan, 2004: Recent modifications of the Emanuel convective scheme in the Navy Operational Global Atmospheric Prediction System. *Mon. Wea. Rev.*, **132**, 1254–1268.

Powell, M. D. and S. D. Aberson, 2001: Accuracy of United States tropical cyclone landfall forecasts in the Atlantic basin (1976–2000). *Bull. Amer. Met. Soc.*, **82**, 2749–2767.

Puri, K., G. S. Dietachmayer, G. A. Mills, N. E. Davidson, R. A. Bowen, and L. W. Logan, 1998: The new BMRC Limited Area Prediction System, LAPS. *Aust. Meteor. Mag.*, **47**, 203–223.

Puri, K., J. Barkmeijer, and T. N. Palmer, 2001: Ensemble prediction of tropical cyclones using targeted diabatic singular vectors. *Quart. J. Roy. Meteor. Soc.*, **127**, 709–731.

Rabier, F., 2005: Overview of global data assimilation developments in numerical weather-prediction centres. *Quart. J. Roy. Meteor. Soc.*, **131**, 3215–3233.

Rabier, F., H. Jarvinen, E. Klinker, J.-F. Mahfouf and A. Simmons, 2000: The ECMWF operational implementation of four-dimensional variational assimilation. Part 1: Experimental results with simplified physics. *Quart. J. Roy. Meteor. Soc.*, **126**, 1143–1170.

Randall, D. A. and D.-M. Pan, 1993: Implementation of the Arakawa-Schubert cumulus parameterization with a prognostic closure, *Monograph on Cumulus parameterization*, pp. 137–144.

Rapapport, E. N. and J. Fernandez-Partagas, 1995: The Deadliest Atlantic Tropical Cyclones, 1492–1996. *NOAA Technical Memorandum NWS NHC 47*. http://www.nhc.noaa.gov/pastdeadly.shtml

Rawlins, F., S. P. Ballard, K. J. Bovis, A. M. Clayton, D. Li, G. W. Inverarity, A. C. Lorenc and T. J. Payne, 2007: The Met Office Global 4-Dimensional Data Assimilation System. *Quart. J. Roy. Meteor. Soc.*, **133**, 347–362.

Rosmond, T. and L. Xu, 2006: Development of NAVDAS-AR: Non-linear formulation and outer loop tests. *Tellus*, **58A**, 45–58.

Sanders, F., 1973: Skill in forecasting daily temperature and precipitation: Some experimental results. *Bull. Amer. Meteor. Soc.*, **54**, 1171–1179.

Sampson, C. R. and A. J. Schrader, 2000: The Automated Tropical Cyclone Forecasting System (Version 3.2). *Bull. Amer. Meteor. Soc.*, **81**, 1131–1240.

Sampson, C. R., J. L. Franklin, J. A. Knaff, and M. DeMaria, 2008: Experiments with a simple tropical cyclone intensity consensus. *Wea. Forecasting*, **23**, 304–312.

Sheets, R. C., 1990: The National Hurricane Center — past, present and future. *Wea. Forecasting*, **5**, 185–232.

Simmons, A. J. and D. M. Burridge, 1981: An energy and angular-momentum conserving finite-difference scheme and hybrid vertical coordinates. *Mon. Wea. Rev.*, **109**, 758–766.

Stark, J. D., C. J. Donlon, M. J. Martin, and M. E. McCulloch, 2007: OSTIA: An operational, high resolution, real time, global sea surface temperature analysis system, *Oceans '07 IEEE Aberdeen, conference proceedings*. Marine challenges: coastline to deep sea. Aberdeen, Scotland. IEEE.

Surgi, N., H.-L. Pan, and S. J. Lord, 1997: Improvement of the NCEP global model over the tropics: An evaluation of model performance during the 1995 hurricane season. *Mon. Wea. Rev.*, **126**, 1287–1305.

Surgi, N., R. E. Tuleya, Q. Liu, V. Tallapragada, and Y. C. Kwon, 2008: Advancement of the HWRF for next generation hurricane prediction at NCEP's Environmental Modeling Center. Amer. Met. Soc. 28th Conference on Hurricanes and Tropical Meteorology (Orlando, USA).

Thompson, P. D., 1977: How to improve accuracy by combining independent forecasts. *Mon. Wea. Rev.*, **105**, 228–229.

Toth, Z. and E. Kalnay, 1997: Ensemble forecasting at NCEP and the breeding method. *Mon. Wea. Rev.*, **125**, 3297–3319.

Tuleya, R., M. Bender, Y. Kurihara, and S. Lord, 1995: The GFDL Multiply-Nested Moveable Mesh Hurricane System. *NWS Technical Procedures Bulletin No. 424*. NWS Office of Meteorology, Silver Spring, MD, 22 pp.

Van der Grijn, G., J. E. Paulsen, F. Lalaurette, and M. Leutbecher, 2005: Early medium-range forecasts of tropical cyclones. *ECMWF Newsletter No. 102*, 7–14. http://www.ecmwf.int/publica tions/newsletters/ pdf/102.pdf

Wang, X. and C. H. Bishop, 2003: A comparison of breeding and ensemble transform Kalman filter ensemble forecast schemes. *J. Atmos. Sci.*, **60**, 1140–1158.

Wei, M., Z. Toth, R. Wobus, and Y. Zhu, 2008: Initial perturbations based on the ensemble transform (ET) technique in the NCEP global operational forecast system. *Tellus*, **60A**, 62–79.

Xu, Y. and C. J. Neumann, 1985: A statistical model for the prediction of western North Pacific tropical cyclone motion (WPCLPR). *NOAA Tech. Memo. NWS-NHC 28*, 30 pp.

Xu, L., T. Rosmond, and R. Daley, 2005: Development of NAVDAS-AR: Formulation and initial tests of the linear problem. *Tellus*, **57A**, 546–559.

Yamaguchi, M., R. Sakai, M. Kyoda, T. Komori, and T. Kadowaki, 2008: *CAS/JSC WGNE 2008 Res. Act. in Atmos. and Ocean Modelling (submitted).*

Zhang, R. and X. Shen, 2008: On the development of the GRAPES — a new generation of the national operational NWP system in China. *Chinese Science Bulletin*, **53**(22), 3429–3432.

Chapter 11

The Influence of Natural Climate Variability on Tropical Cyclones, and Seasonal Forecasts of Tropical Cyclone Activity

Suzana J. Camargo

Lamont-Doherty Earth Observatory, Columbia University
61 Rt. 9W, Palisades NY 10960, USA
suzana@ldeo.columbia.edu

Adam H. Sobel

Department of Applied Physics and Applied Mathematics &
Department of Earth and Environmental Sciences, Columbia University
500 W. 120th Street, New York, NY 10027, USA
ahs129@columbia.edu

Anthony G. Barnston

International Research Institute for Climate and Society,
Columbia University, 61 Rt. 9W, Palisades, NY 10960, USA
tonyb@iri.columbia.edu

Philip J. Klotzbach

Department of Atmospheric Science,
Colorado State University, Fort Collins, CO 80523, USA
philk@atmos.colostate.edu

In the first part of this chapter, we give a review of the relationship of climate and tropical cyclones on various time scales, from intra-seasonal to decadal. The response of tropical cyclone activity to natural modes of variability, such as El Niño-Southern Oscillation and the Madden Julian Oscillation in various regions of the world are discussed. Genesis location, track types and intensity of tropical cyclones are influenced by these modes of variability. In the second part, a review of the state of the art of seasonal tropical cyclone forecasting is discussed. The two main techniques currently used to produce tropical cyclone seasonal forecasts (statistical and dynamical) are discussed, with a focus on operational forecasts.

1. Introduction

Natural climate variability strongly modulates the seasonal statistics of tropical cyclones (TCs). This modulation of TC activity occurs on multiple time scales (Elsberry, 1997). Here we review these influences as they have been documented in the literature.

We limit our discussion to timescales between the intraseasonal (30–60 day) and multi-decadal (20–50 year). In setting our short-timescale limit at the intraseasonal we exclude consideration of convectively coupled waves (Wheeler and Kiladis, 1999) which significantly influence TC genesis (Dickinson and Molinari, 2002; Frank and Roundy, 2006). At longer timescales, we exclude all variability outside that resolvable in the modern observational record, for which historical (e.g., Liu *et al.*, 2001; Garcia-Herrera *et al.*, 2007) or geological (e.g., Liu and Fearn, 2000; Donnelly *et al.*, 2001) and other proxy data must be used. Recent progress in paleotempestology is expanding the spatial and temporal coverage of the TC record, as well as improving the temporal resolution, dating precision, and accuracy of intensity estimates (Frappier *et al.*, 2007), but this subject is outside our scope. Some further discussion can be found in Knutson *et al.* (this volume).

Our knowledge of the relationships between various modes of climate variability and TCs comes mainly from empirical studies, in which statistical methods are used to extract these relationships. Physical insight naturally enters into such studies in a variety of ways, but ultimately our understanding of the mechanisms which determine the relationships is limited by the same factors that limit our understanding of the mechanisms of the genesis and intensification of individual TCs. For intensity, we have potential intensity theory (Emanuel, 1986, 1988) but this is not a complete theory for the actual intensity of TCs and has been recently challenged (Smith *et al.*, 2008). In any case, potential intensity theory has been relatively little used in statistical studies of TC variability, with some exceptions (Emanuel, 2000; Camargo *et al.*, 2007a; Wing *et al.*, 2007). For TC numbers, we have little theoretical guidance. While there are some theories for aspects of genesis (Bister and Emanuel, 1997; Montgomery *et al.*, 2006) and a number of interesting recent developments (e.g., Raymond *et al.*, 2008; Dunkerton *et al.*, 2008) there is no quantitative theory that goes so far as to give the probability of genesis given some large-scale conditions. We know that variables such as sea surface temperature, vertical wind shear, and midlevel humidity influence genesis (Gray, 1979), and this gives us an empirical basis for understanding how climate variations influence TC numbers. But, first-principles understanding is still very limited. Integrated indices such as accumulated cyclone energy (ACE) and power dissipation depend on intensity, number, and lifetime, and our ability to understand variations in these indices is limited by our lack of understanding of each of the individual factors.

Despite our relative lack of a theoretical foundation for understanding them, the observed relationships between large-scale climate variability and tropical cyclone statistics are strong enough that they make it possible to produce statistical forecasts of tropical cyclone activity on seasonal to interannual time scales. Dynamical forecasts using numerical models are also now feasible, and have hindcast skill that is comparable to the statistical ones.

In the first part of this chapter, the TC-climate relationship on intraseasonal to decadal time-scales will be discussed. In the following section we briefly review the different modes of climate variability whose influences on TCs have been investigated, and in section 3 we review the influences these modes have been found to have, treating each TC basin separately. In the rest of the paper, the state of the art of seasonal TC forecasting will be reviewed.

2. Modes of Variability

The El Niño–Southern Oscillation (ENSO) is a major mode of natural climate variability. ENSO is generated by coupled ocean-atmospheric dynamics in the tropical Pacific (e.g., Trenberth, 1997). ENSO is associated with sea surface temperature (SST) changes in the tropical Pacific, as well as shifts in the seasonal temperature, circulation, and precipitation patterns in many parts of the world (Bradley *et al.*, 1987; Ropelewski and Halpert, 1987). El Niño and La Niña (warm and cold) events usually recur every 3 to 7 years and tend to last for approximately a year. As the ENSO phenomenon can be predicted with some accuracy months in advance (Cane and Zebiak, 1985), ENSO forecasts are routinely used as a major component in the making of probabilistic seasonal climate forecasts (Goddard *et al.*, 2001).

ENSO affects TCs strongly in several basins, though its influence is different in each. The impact of ENSO on North Atlantic TCs (hurricanes) was first discussed by Gray (1984a), and recent reviews of the relationship of ENSO and TCs are presented in Chu (2004) and Landsea (2000). Because of this strong influence and the predictability of ENSO, it is probably the largest single factor in seasonal TC forecasts. A summary of the relationship of ENSO with TCs in all basins is given in Table 1 and is discussed in more detail below.

The North Atlantic basin is the site of several modes of variability, all of which influence TCs. Some aspects of the relationships of these Atlantic modes to each other, and especially of each to global climate change, are controversial. We will be brief on this subject here; we discuss it further to some extent in the Atlantic subsection of section 3, and Knutson *et al.* (this volume) discuss the relationship to climate change in particular in more detail.

The North Atlantic oscillation (NAO) is the dominant mode of winter climate variability in the North Atlantic region (Hurrell *et al.*, 2003); it is closely related to the global "Arctic Oscillation" pattern (Thompson and Wallace, 2000). The NAO is a large scale atmospheric seesaw between the Atlantic subtropical high and polar low. It is not a true oscillation, but rather has variation at all time scales, with a red spectrum (Stephenson *et al.*, 2000; Feldstein, 2000), and apparently owes its existence largely to the interaction of the jet streams with baroclinic eddies (Woolings *et al.*, 2008; Vallis and Gerber, 2008).

The Atlantic Multi-decadal Oscillation (AMO) is a natural mode of variability in the North Atlantic Ocean on multi-decadal time scales whose existence has been inferred from analysis of the observed sea surface temperature after removing a linear trend (Folland *et al.*, 1986; Delworth and Mann, 2000). It has also been found to exist in coupled numerical models, in which it is a manifestation of variability in the Atlantic Ocean's thermohaline circulation. Further discussion of this mode, and its relationship to other possibly related modes [such as the tropical multidecadal modes (Chelliah and Bell, 2004; Bell and Cheliah, 2006)] as well as to anthropogenically forced global climate change, can be found in Knutson *et al.* (this volume).

The Atlantic Meridional Mode (AMM; Servain *et al.*, 1999; Xie and Carton, 2004; Chiang and Vimont, 2004) is the leading mode of coupled ocean-atmosphere variability in the tropical Atlantic. The AMM is a "meridional mode" (Xie, 1999; Xie and Carton, 2004) associated with meridional displacements of the intertropical convergence zone and attendant shifts in SST and winds. Recent studies (Vimont and Kossin, 2007; Kossin and Vimont, 2007) have shown that Atlantic TC characteristics are closely related to the AMM. Vimont and Kossin (2007) provided evidence that the AMM is likely to be excited on multidecadal timescales by the AMO. While the AMM and the NAO are independent during the hurricane season (Xie *et al.*, 2005c), the AMM was shown to be excited by variations of the NAO (Xie and Tanimoto, 1998; Czaja *et al.*, 2002). The Pacific Decadal Oscillation (PDO) is a pattern of Pacific climate variability with a decadal time scale, impacts in the North Pacific and North American sectors, and secondary signatures in the tropics (Mantua, 2002). The spatial pattern of the PDO is similar to that of ENSO, and it is not entirely clear whether the PDO is a truly independent mode or simply the decadal residual of ENSO variability (e.g., Zhang *et al.*, 1997; Vimont, 2005).

The quasi-biennial oscillation (QBO) is a global-scale, zonally symmetric oscillation of the zonal winds in the equatorial stratosphere (Wallace, 1973) which owes its existence to wave-mean flow interaction (Holton and Lindzen, 1968; Plumb and McEwan, 1978). The QBO has a period of approximately 26 months and its largest amplitude occurs near 30 hPa.

Table 1. Summary of the ENSO impacts on TC activity in the different regions, including references.

Basin	Characteristic	ENSO	Change	References (e.g.)
Western North Pacific	Number of TCs	Strong EN	increase	Wang and Chan (2002)
	Number of TCs	Summer following EN	decrease	Chan (1985), Wu and Lau (1992), Chan (2000)
	Mean genesis location	EN/LN	Displacement to the Southeast/ Northwest	Chan (1985), Chen *et al.* (1998), Wang and Chan (2002), Chia and Ropelewski (2002)
	Lifetime	EN/LN	Increase/decrease	Wang and Chan (2002), Camargo and Sobel (2005)
	Number of intense typhoons and ACE	EN/LN	Increase/decrease	Camargo and Sobel (2005), Chan (2007)
	Track types	EN	Long tracks, recurve northeastward and reach more northern latitudes	Wang and Chan (2002), Wu and Wang (2004), Camargo *et al.*. (2007e)
Eastern North Pacific	Number of intense TCs	EN/LN	Increase/decrease	Gray and Sheaffer (1991), Frank and Young (2007)
	Mean genesis location	EN	Westward shift	Irwin and Davis (1999), Kimberlain (1999)
	Tracks	EN	Longer westward direction	Chu (2004), Camargo *et al.* (2008a)
Central North Pacific	Number of TCs	EN	Increase	Wu and Lau (1992), Chu and Wang (1997), Clark and Chu (2002), Camargo *et al.* (2007a)
Atlantic	Number of TCs, number of hurricanes, number of intense hurricanes, number of hurricane days, number of tropical storms, ACE	EN/LN	Decrease/Increase	Gray (1984a); Gray *et al.* (1983); Landsea *et al.* (1999); Bell *et al.* (2000)
	US landfalls	EN/LN	Decrease/increase	O'Brien *et al.* (1996); Bove *et al.* (1998); Pielke and Landsea (1998); Larson *et al.* (2005)
North Indian	Number intense TCs	EN	Decrease (Bay of Bengal)	Singh *et al.* (2000)
South Indian	Number of TCs east of 70°E	LN	Increases	Kuleshov (2003)
	Peak of the season	EN/LN	February/January	Kuleshov (2003)
	Start of the season	LN	1 month earlier	Kuleshov (2003)
	Mean genesis location	EN	Westward shift — increase (decrease) west (east) of 75°E	Ho *et al.* (2006)
	Tracks	EN	Movement further east	Ho *et al.* (2006), Kuleshov *et al.* (2008)

The Madden-Julian Oscillation (MJO; Madden and Julian, 1972, 1994; Zhang, 2005) is the strongest mode of intraseasonal variability in the tropics. The MJO has a 30–90 day period and consists of large-scale coupled patterns of deep convection and atmospheric circulation, with coherent signals in many atmospheric variables. The MJO propagates eastward across the global tropics, with signatures in deep convection primarily in the Indian and western Pacific Oceans. The MJO is stronger in boreal winter than in boreal summer (e.g., Wang and Rui, 1990) and modulates the formation of tropical cyclones in several basins.

3. Regional Variability

3.1. *Western North Pacific*

There is pronounced interannual variability in TC activity in the western North Pacific, in large part due to ENSO. Wang and Chan (2002) observed an increase in the number of TCs in the western North Pacific during strong El Niño events, though they found no significant linear relationship between TC number and indices of ENSO. A reduction in the number of TCs occurring in the summer following an El Niño event has also been found, and is related to the longitudinal shift of the Walker circulation (Chan, 1985; Wu and Lau, 1992; Chan, 2000).

ENSO has an important and well-documented impact on the mean TC genesis location, with a displacement to the southeast (northwest) in El Niño (La Niña) years, as shown in Fig. 1 (Chan, 1985; Chen *et al.*, 1998; Wang and Chan, 2002; Chia and Ropelewski, 2002). Because of this shift to the southeast, further away from the Asian continent, typhoons in El Niño years tend to last longer (Wang and Chan, 2002) and be more intense than in other years (Camargo and Sobel, 2005; Chan, 2007).

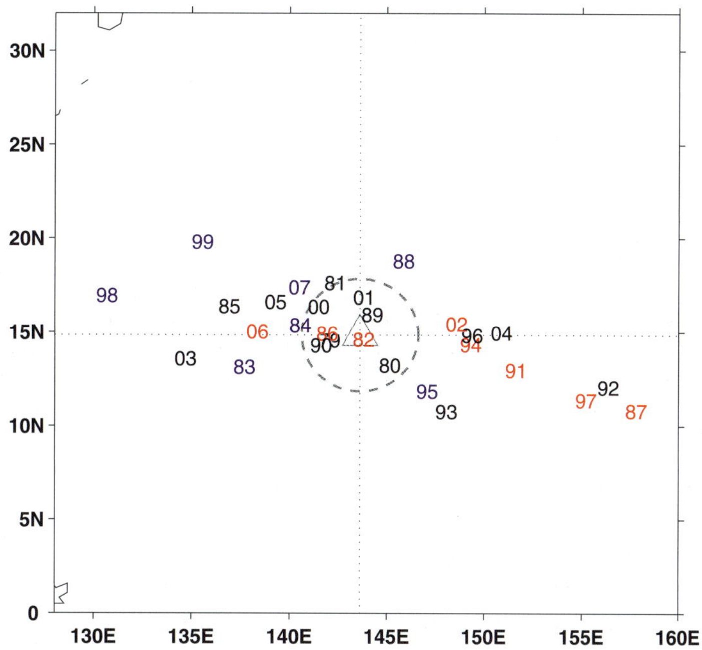

Figure 1. The seasonal mean (July to October) genesis position from 1979 to 2007, "87" is the seasonal mean from 1987, for instance. The triangle shows the long term (1979–2007) seasonal mean genesis position at the center of the circle. The radius of the circle (333 km, i.e., near 3° of latitude) is the standard deviation of the seasonal TC position. El Niño years are shown in red, La Niña years in blue. This is an updated version of a figure that appeared originally in Chia and Ropelewski (2002).

In contrast, in La Niña years, the mean genesis location is shifted to the northwest, so that the storms have shorter lifetimes and tend to be less intense (Camargo and Sobel, 2005).

The genesis location shift has been commonly attributed to the eastward extension of the monsoon trough and westerlies (associated with increased cyclonic low-level vorticity) and the reduction of vertical wind shear near the date line, both of which enhance the likelihood of genesis east of the climatological mean genesis point (Lander, 1994, 1996; Clark and Chu, 2002; Wang and Chan, 2002). The emphasis on these dynamical factors (as opposed to thermodynamic ones) in interannual variability is consistent with the emphasis placed on the monsoon trough more broadly in studies of TC genesis in the western north Pacific (e.g., Harr and Elsberry, 1995; Lander, 1996; Ritchie and Holland, 1999; Chen *et al.*, 2006). While concurring with this, Camargo *et al.* (2007e) also presented evidence that a decrease of mid-level relative humidity near the Asian continent in El Niño years may play a role in the eastward shift as well, by suppressing genesis in the western part of the basin. ENSO affects not only the starting points and lengths of TC tracks but also the shapes of the tracks. In El Niño years, the TCs have a tendency to recurve northeastward and reach more northward latitudes, due both to the shift in mean genesis location and ENSO-induced changes in the mean steering flow (Wang and Chan, 2002; Wu and Wang, 2004). Wang and Chan (2002) and Wu and Wang (2004) showed that in strong El Niño years, the deepening of the east Asian trough in the midtroposphere provides a favorable anomalous steering flow that leads to TCs recurving northward to the extratropics.

Camargo *et al.* (2007d,e) stratified western North Pacific tropical cyclone tracks into seven clusters. Two clusters which start from genesis locations southeastward of the climatological mean and are relatively intense occur more frequent in El Niño years while the largest cluster type is more typical of La Niña events. Consistent with this, storms affect the southern South China Sea more frequently during La Niña years (Zuki and Lupo, 2008), but affect the Central Pacific more frequently in El Niño years (Chu and Wang, 1997; Clark and Chu, 2002). Some Central Pacific storms reach the western North Pacific.

The ENSO's influence on western North Pacific TC tracks is reflected in the landfall rates throughout the region, with different landfall patterns according to the ENSO phase (Saunders *et al.*, 2000; Elsner and Liu, 2003). Wu *et al.* (2004) found a significant relationship between late season landfalls over China and ENSO. Fudeyasu, *et al.* (2006a) noticed an increase in landfalls in the Korean Peninsula and Japan during the early monsoon and in the Indochinese peninsula during the peak monsoon months in El Niño years. The interannual variability of typhoon landfalls in China shows a north-south anti-correlation in the historical and modern records. This is related to sea-level pressure differences between Mongolia and western China, as well as SST over the western North Pacific (Fogarty *et al.*, 2006).

While ENSO has clearly been shown to influence TCs in the western north Pacific, it has also been argued that TCs may play an active role in the dynamics of ENSO itself. El Niño years feature the relatively frequent occurrence of tropical cyclone pairs north and south of the equator over the central and western Pacific Ocean, associated with westerly wind bursts on the equator (Keen, 1982; Lander, 1990; Harrison and Giese, 1991; Ferreira *et al.*, 1996). It has been argued that the twin TCs themselves contribute to the generation of the equatorial westerlies and thus amplify El Niño events, through the standard Bjerknes' mechanism (Keen, 1982; Gau, 1988). While cyclone pairs are spectacular, they are not actually necessary to this scenario, as a single cyclone can generate equatorial westerlies if it is close to the equator (Harrison and Giese, 1991; Kindle and Phoebus, 1995). Since Pacific cyclones tend to form closer to the equator and reach greater intensities in El Niño years, there is the possibility of a positive feedback between ENSO and TCs, with single as well as twin cyclones playing a role (e.g., Sobel and Camargo 2005; see also Yu and Rienecker, 1998; Yu *et al.*, 2003; Eisenman *et al.*, 2005). On the other hand, the causality is difficult to extract from observations, because equatorial westerlies are associated with off-equatorial cyclonic vorticity and other conditions conducive to genesis. It is possible that the westerlies occur primarily for other reasons and that cyclones are largely a passive by-product, as argued by Lander (1990).

The relationship of the quasi-biennial oscillation (QBO) to TC activity in the western North Pacific was examined by Chan (1995), who found that in the westerly phase of the QBO, there are more TCs in the western North Pacific. Chan (2005) attributed this to a decrease in the upper-tropospheric vertical shear over the tropics in the westerly phase. The QBO-TC relationship will be discussed further in the Atlantic subsection, below. Associations on the interannual time scales between western north Pacific TCs and a number of other factors, besides ENSO, have been found. These factors include Tibetan plateau snow cover (Xie *et al.*, 2005b; Xie and Yan, 2007), North Pacific sea-ice cover (Fan, 2007), the large-scale circulation in the extratropical southern hemisphere (Ho *et al.*, 2005; Wang and Fan, 2007), and various other empirically-derived oscillations such as the Asian-Pacific oscillation (Zhou *et al.*, 2008), and the North Pacific oscillation (Wang *et al.*, 2007).

A few studies have examined the decadal and multi-decadal variability of tropical cyclone activity in the western North Pacific. The subject is important, but any results on this time scale must be interpreted with great caution due to the shortness of the data record. The observational record in the western north Pacific is unreliable before the 1950s, and perhaps even before the 1970s (e.g., Wu *et al.*, 2006). Matsuura *et al.* (2003) related the decadal variability in the western North Pacific to long term variations of sea surface temperature in the tropical Central Pacific and westerly winds anomalies associated with the monsoon trough. Chan (2008a) examined the decadal variability of intense typhoon occurrence. Using wavelet analysis, besides the ENSO time-scale (3–7 years), he identified another major oscillation with a multi-decadal period (16–32 years) for intense typhoons, and argued that the environmental conditions conducive to periods of intense typhoons occurrence are largely modulated by ENSO and the Pacific Decadal Oscillation. The decadal variability of TC tracks has also been largely attributed to the Pacific Decadal Oscillation (Liu and Chan, 2008). Ho *et al.* (2004) associated the inter-decadal changes of typhoon tracks with the westward expansion of the subtropical North Pacific High since 1970. The regions with the greatest inter-decadal changes identified in Ho *et al.* (2004) are the East China Sea and the Philippine Sea.

On the intraseasonal time scale, tropical cyclone activity in the western north Pacific is strongly modulated by the MJO. The MJO changes the distribution of TC tracks (Camargo *et al.*, 2007e), and genesis is favored during the active phase of enhanced deep convection (Nakazawa, 1986; Liebmann *et al.*, 1994). The environment during the active phase is favorable to genesis in multiple ways. Midlevel humidity and low-level relative vorticity are both enhanced during the active phase, and the increased occurrence of convection is more likely to generate potential seed disturbances which can undergo genesis. Sobel and Maloney (2000) argued that MJO-induced modulation of wave accumulation (Chang and Webster, 1988; Holland, 1995; Sobel and Bretherton, 1999; Kuo *et al.*, 2001), by which large-scale convergence can amplify synoptic-scale disturbances which can be precursors to TC genesis, might play a role. The idealized numerical calculations of Aiyyer and Molinari (2003) show in particular how MJO-related circulation anomalies can, through dry dynamics alone, amplify mixed

Rossby-gravity waves and change their structure to be "tropical depression type" (e.g., Takayabu and Nitta, 1993) and thus more favorable candidates for genesis.

3.2. *Central and Eastern North Pacific*

During its peak season, the eastern North Pacific is the most active tropical cyclogenesis region per unit area (Molinari *et al.*, 2000). Tropical cyclogenesis in the eastern North Pacific is influenced by wind surges, African easterly waves, topographical effects, ITCZ breakdown, upper-level potential vorticity, and the confluence between the monsoon westerlies and trade easterlies (e.g., Avila, 1991; Bosart and Bartlo, 1991; Zehnder, 1991; Bister and Emanuel, 1997; Ferreira and Schubert, 1997; Zehnder *et al.*, 1999; Molinari and Vollaro, 2000; Vincent and Fink, 2001).

There has been considerable investigation into the influence of ENSO on TC activity in the eastern north Pacific. With very few exceptions (e.g., Whitney and Hobgood, 1997), these studies have found clear relationships between ENSO and various measures of TC activity. Frank and Young (2007) obtained a strong relationship between storm counts and a multivariate ENSO index. Gray and Sheaffer (1991) found an increased number of intense hurricanes during El Niño events. Several studies have found a westward shift in the mean genesis region during El Niño events (Irwin and Davis, 1999; Kimberlain, 1999; Chu and Zhao, 2007; Wu and Chu, 2007; Camargo *et al.*, 2008a). A consequence of this westward shift is that more hurricanes propagate into the central North Pacific region (Chu, 2004).

Some studies have attempted to understand the environmental factors by which ENSO influences eastern north Pacific TCs. Using composites of a genesis potential index keyed to the ENSO state, Camargo *et al.* (2007a) concluded that wind shear is the main contributor, with potential intensity also playing a contributing role. Collins and Mason (2000) identified differing environmental parameters affecting TC activity for the regions east versus west of 116°W. They observed that in the eastern part of the region, the environmental parameters are nearly always at levels conducive to TC formation during the peak season, while in the western region, there are some years when the parameters are conducive to cyclogenesis but other years when they are less so.

Several studies have found a see-saw of TC activity between the North Atlantic and the eastern North Pacific, with enhanced activity in the eastern Pacific when the Atlantic is suppressed, and vice versa (Elsner and Kara, 1999). Stronger anti-correlations are found for stronger TCs (Frank and Young, 2007). It is not entirely clear whether this is simply a result of the fact that ENSO influences the two basins in opposite ways (which is certainly the case), or whether in fact the TC activity in one basin directly influences that in the other by another route.

On decadal time scales, Zhao and Chu (2006) examined the variability of TC activity in the eastern North Pacific using a Bayesian multiple change-point analysis. Their results indicated that hurricane activity in the eastern North Pacific had two inactive eras (1972–1981, 1999–2003) and an active era during 1982–1998.

The probability of cyclogenesis in the eastern North Pacific is strongly modulated by the MJO, as shown in Fig. 2 (Molinari *et al.*, 1997; Molinari and Vollaro, 2000; Maloney and Hartmann, 2000b, 2001; Aiyyer and Molinari, 2008; Camargo *et al.*, 2008a). In the convective MJO phase, anomalous westerlies and cyclonic vorticity occur over the eastern Pacific and Gulf of Mexico. Four times the number of tropical cyclones form in the active phase of the MJO than in the suppressed phase. Additionally, storms forming in the active MJO phase tend to do so closer to the Mexican coast, further increasing the chance of hurricane landfall compared to the suppressed phase (Maloney and Hartmann, 2000b).

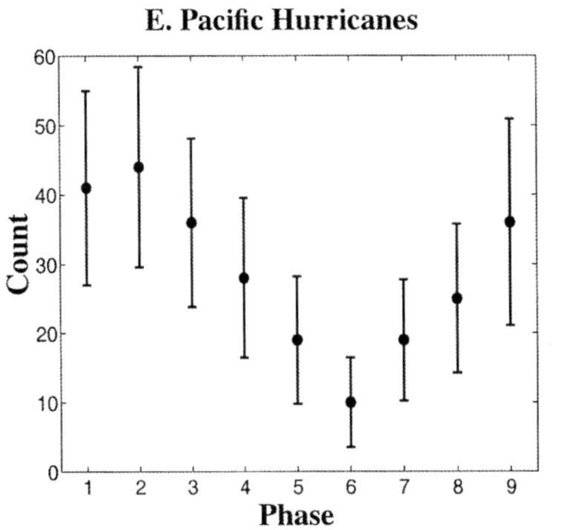

Figure 2. Number of hurricanes as a function of MJO phase for the eastern North Pacific during May — November for 1979–95. Error bars represent 95% confidence level (figure originally from Maloney and Hartmann 2000a).

Various studies have suggested that to a first order, barotropic dynamics can be used to understand the influence of the MJO on eastern north Pacific TCs. Significant barotropic energy conversion from the mean state to the eddies occurs during the convective phase of the MJO (Molinari *et al.*, 1997; Maloney and Hartmann, 2001; Hartmann and Maloney, 2001; Aiyyer and Molinari, 2008). This is consistent with, but more general than, the hypothesis of wave accumulation made in the western North Pacific by Sobel and Maloney (2000); wave accumulation involves barotropic conversion, but the conversion can also happen by other means such as barotropic instability. Aiyyer and Molinari (2008) noticed that in the active phase of the MJO, the vertical shear is relatively weak and tropical cyclones tend to form mainly within the ITCZ. In contrast, during the suppressed phase, the vertical wind shear exceeds $10\,\mathrm{m\,s^{-1}}$ over much of the region, and tropical cyclone development is shifted northward, nearer the Mexican Pacific coast. While these studies have focused on dynamics over thermodynamics, recent work (Camargo *et al.*, 2009) using a genesis potential index to compare thermodynamic and dynamic factors in a quantitative way suggests that MJO-induced variations in thermodynamic variables, particularly midlevel humidity, are important. This is consistent with results from the EPIC field experiment on the MJO's influence on generalized ITCZ convection (Raymond *et al.*, 2003).

The central North Pacific is sometimes considered as a distinct TC basin, though it can share storms (during the course of their lifetime) with either the eastern or western North Pacific. Tropical cyclone occurrence in the central North Pacific is low, with an average of four to five TCs per year. In El Niño years, the number is greater. The higher level of activity in the central North Pacific in warm ENSO events has been attributed to smaller vertical shear and greater low-level vorticity in that region (Wu and Lau, 1992; Chu and Wang, 1997; Clark and Chu, 2002; Camargo *et al.*, 2007a).

Chu and co-workers (Chu and Clark, 1999; Chu, 2002; Chu and Zhao, 2004) also examined the decadal-scale variability in the central North Pacific. They found change points in 1982 and 1995, with fewer cyclones during the 1966–1981 and 1995–2000 periods, and more during the 1982–1994 epoch. In the more active period, warmer sea surface temperatures, lower sea level pressure, stronger low-level anomalous cyclone vorticity, reduced vertical wind shear, and increased total precipitable water were present in the tropical North Pacific, compared to the reduced activity periods. The atmospheric steering flows were also different in October and November of the active period, so that eastern North Pacific cyclones had a higher chance of entering the central North Pacific. During the more active period, the storms that formed within the region were more likely to affect the Hawaiian Islands.

3.3. *North Atlantic*

North Atlantic TC activity has a strong relationship with ENSO. Gray (1984a,b) used this relationship to develop the first Atlantic TC seasonal forecasts. In El Niño years, hurricane activity in the Atlantic is reduced, while in La Niña, it is enhanced. The negative correlation of Atlantic TC activity with ENSO is evident in a wide range of TC activity indices: the number of hurricanes, number of hurricane days, number of tropical storms and ACE (accumulated cyclone energy). The influence of ENSO is very significant for the number of intense hurricanes, as in La Niña years there are usually twice as many intense hurricanes as in El Niño years. The correlations of the number of intense hurricanes with ENSO indices are higher than for weaker storms (Gray *et al.*, 1993; Landsea *et al.*, 1999).

ENSO is generally thought to influence Atlantic TC activity by altering the large-scale environment for genesis and intensification, particularly in the so-called Main Development Region (MDR), where many tropical cyclones form from African easterly waves. During an El Niño year, the vertical wind shear is larger than normal in most of the tropical Atlantic and especially in the Caribbean. This inhibits the formation of TCs. Reductions in vertical wind shear typically occur in La Niña years (e.g., Goldenberg and Shapiro, 1996; Knaff, 1997). Besides these dynamical influences of ENSO, thermodynamic effects have also been identified. Knaff (1997) argued that increased midlevel dryness during El Niño events might be responsible for some of the TC suppression during those events, while Tang and Neelin (2004) argued for the importance of increased moist stability, due to increased tropospheric temperature during El Niño (e.g., Yulaeva and Wallace, 1994; Soden, 2000; Sobel *et al.*, 2002). Camargo *et al.* (2007a) used variants on an empirical genesis index to compare the dynamic and thermodynamic effects of ENSO directly, and found the dynamic effects of wind shear to be more important than the thermodynamic effects in the North Atlantic. In addition to modulating Atlantic TC activity as a whole, ENSO also influences the probability of landfalling hurricanes in the U.S., with an increased probability of landfall in La Niña years (O'Brien *et al.*, 1996; Bove *et al.*, 1998; Pielke and Landsea, 1998; Larson *et al.*, 2005). The probability of two or more hurricane landfalls during a La Niña event is 66%, while in El Niño years it is 28% (Bove *et al.*, 1998). On a more regional basis, the differences in landfall frequencies in the cold and warm ENSO phases are particularly significant for the U.S. East coast (Georgia to Maine), and smaller for the Florida and Gulf coasts (Smith *et al.*, 2007). On a still finer scale, significant differences between specific U.S. Southeast states have been found (Xie *et al.*, 2002). U.S. hurricane losses, unsurprisingly, are also significantly related to ENSO, with much more damage occurring in La Niña events (Pielke and Landsea, 1999). An ENSO influence on landfalling storms has also been found in the Caribbean, both in the modern record (Targlione *et al.*, 2003) and in proxy records from the past 5,000 years (Donnelly and Woodruff, 2007).

A robust relationship existed between Atlantic TC activity and the QBO (Gray, 1984a; Shapiro, 1989; Gray *et al.*, 1992a,b) from approximately 1950–1994. The influence of the QBO appeared to be especially significant for intense hurricane days. Gray *et al.* (1993) found that the number of intense hurricane days which occurred in the easterly phase of the QBO was half that which occurred in the westerly phase. The largest correlation between storm activity and stratospheric winds occurred in June, three months prior to the peak of the Atlantic hurricane season (Shapiro, 1989). The relationship of QBO with the Atlantic TC activity is not statistically significant since 1995 (Klotzbach and Gray, 2004; Camargo and Sobel, 2010).

In our view the mechanism of the QBO-TC relationship is not yet satisfactorily explained. Gray (1988) and Gray *et al.* (1992a,b) argue that the influence comes about due to QBO modulation of vertical wind shear, a finding echoed by Chan (1995) in his study of the QBO influence on western north Pacific TC activity. However, the QBO modulates vertical shear substantially only in the stratosphere, with perhaps a very weak modulation in the very uppermost troposphere. If the shear variations at these high levels are responsible for the QBO-induced TC variations, it means that TCs are quite sensitive to shear even when that shear is confined to their very tops. We know of no evidence to rule this out, but on the other hand we are aware of no studies of any kind (observational, modeling, or theoretical), apart from those on the QBO influence itself, which directly examine the influence of shear at such high levels on TC formation or development, and which thus might provide support for shear as the mechanism of the QBO TC signal. There is a large modeling literature on the influence of shear on TCs, but to our knowledge all those studies examine only the role of tropospheric shear (e.g., Kepert, this volume, and references therein).

Besides the direct effect of shear on TCs, other mechanisms can be considered. Shapiro (1989) conjectured a mechanism involving the difference between the tropospheric and lower stratospheric winds and its influence on the vertical structure of pre-cyclogenesis easterly waves. The QBO also affects lower stratospheric temperature, and the temperature anomalies might be expected to influence TC intensity by the arguments of potential intensity theory. However, the phasing of the QBO-TC activity is inconsistent with this as an explanation for the observed TC variations. TC activity is enhanced during the westerly QBO phase, when the lower stratosphere is warm. This would tend to reduce the potential intensity, not increase it, and thus would be unfavorable rather than favorable for TC activity (Shapiro, 1989).

Atlantic TCs show strong variations on decadal and multi-decadal time scales in the observed record. The literature on these variations is large, and uses a variety of different concepts and nomenclature to describe and interpret them. Our understanding of the decadal and especially multi-decadal variability is somewhat related to that of anthropogenic climate change and its possible long-term influence on TC activity. These issues are discussed in detail in Knutson *et al.* (this volume) and we address this time scale only briefly here, leaving the relationship to global climate change to those authors.

At least until recently, the dominant paradigm for describing these low-frequency TC variations has been the "Atlantic Multidecadal Oscillation" (AMO). The relationship of the AMO to major hurricane activity in the Atlantic through changes in vertical wind shear was discussed by Gray, *et al.* (1997) and Goldenberg *et al.* (2001). The increase in hurricane activity in the Atlantic since 1995 was related by those authors to the modulation of the MDR by the AMO, with above normal SSTs and decreased vertical shear. More recently, a metric was created to describe the AMO based on North Atlantic SST anomalies and basin-wide North Atlantic sea level pressure anomalies, which shows a remarkable agreement with observed multidecadal variability of Atlantic TC activity and U.S. landfall frequency (Klotzbach and Gray, 2008). Normalized U.S. hurricane damage also shows a modulation following the AMO. The 1970s and 1980s saw very little damage compared with other decades, while the 1920s and 1930s, as well as the recent period from 1996–2005 saw very high levels of damage (Pielke *et al.*, 2008). Bell and Chelliah (2006) related the interannual and multidecadal variability of TC activity in the Atlantic to ENSO and two tropical multidecadal modes (TMM). Comparing periods of high activity in the Atlantic, they showed that the most recent increase in TC activity is related to one of the first tropical modes, associated with exceptionally warm SSTs in the Atlantic; while the high activity period in the 1950s and 1960s was more closely associated with the second tropical mode, associated with the West African monsoon.

Other work has related seasonal TC activity in the Atlantic on both interannual and decadal time scales to the so-called Atlantic Meridional Mode (AMM) (Xie *et al.*, 2005a,c; Vimont and Kossin, 2007; Kossin and Vimont, 2007). An attractive feature of this interpretation is that the AMM is not defined purely statistically,

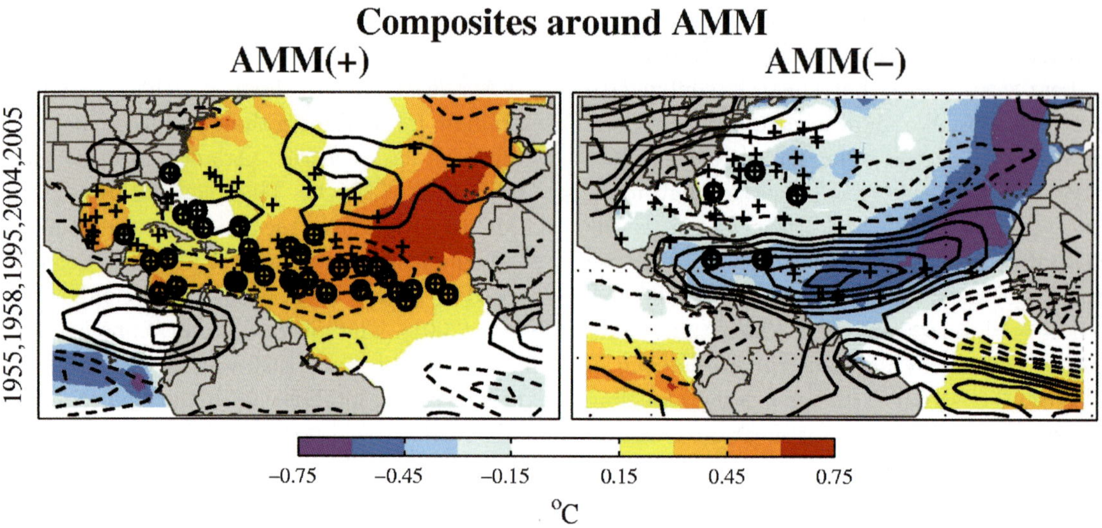

Figure 3. Tropical cyclogenesis points for the five strongest and five weakest AMM years, superimposed on composites of SST (shaded) and shear (contours) anomalies. Crosses show the genesis points for all storms that reached tropical storm strength. Storms that reached major hurricane strength also have a circle around their genesis point. Solid (dashed) shear contours denote positive (negative) values. The contour interval is 0.25 ms^{-1}, and the zero contour has been omitted. Figure originally from Kossin and Vimont (2007).

but has a dynamical interpretation as a natural mode of the climate system in the Atlantic (Xie and Carton, 2004). [Interestingly, the eastern Pacific also appears to have an analogous meridional mode, though ENSO must be removed from the data in order to see it (Chiang and Vimont, 2004). Figure 3 shows the TC genesis locations, SST and vertical shear anomalies in the different phases of the AMM. The AMM modulation comes about through its influence on a number of environmental variables that influence TC activity, such as SST and vertical wind shear. During the positive AMM phase (above normal SSTs in the North Atlantic) there is an overall increase of TC activity in the Atlantic, with the mean genesis location shifting eastward and towards the equator. Also associated with a positive AMM is an increase in storm duration and the intense hurricane frequency (Kossin and Vimont, 2007). The relationships between Atlantic TC variability and various other aspects of regional climate variability implicit in the AMM have been documented previously. For instance, the relationship of African Sahel rainfall to Atlantic hurricane activity is well known (Gray, 1990; Gray and Landsea, 1992; Landsea and Gray, 1992), as is the relationship of the AMM to Sahel rainfall (Folland *et al.*, 1986). The AMO influence on TCs can also be viewed as manifesting itself through the excitation of the AMM by the AMO (Vimont and Kossin, 2007). The value of the AMM is that it provides a unified, physically-based framework for interpreting a variety of influences on Atlantic TCs on a range of time scales.

Similarly to the Pacific, the MJO also modulates TC activity on intraseasonal time scales in the Atlantic. When the MJO is in its active phase over the eastern Pacific, four times more hurricanes form in the Gulf of Mexico and western Caribbean than in the suppressed MJO phase (Maloney and Hartmann, 2000a). During the active phase (westerly over the eastern Pacific), cyclogenesis tends to be shifted northward, to near the Mexican Pacific coast and the Gulf of Mexico. Idealized numerical calculations of the MJO show that in the active phase, easterly waves tend to propagate westward into the eastern Pacific, while during the suppressed phase, they are steered into the Gulf of Mexico (Aiyyer and Molinari, 2008).

The role of easterly waves in Atlantic TC variations on all time scales is an interesting question. In general, interannual and longer-timescale variations in Atlantic TC activity tend to be interpreted in terms of variations in the large-scale environment. This amounts to assuming a constant supply of precursor disturbances, so that variations in TC number (for example) are assigned to variations in the probability that those disturbances will undergo genesis. However, one might also ask whether TC variations might be related to variations in the supply of easterly waves impinging on the Atlantic from Africa. African easterly waves have considerable interannual variability, in number, intensity and tracks. The low-frequency (i.e., decadal) variability of the easterly waves is correlated with Atlantic TC activity, as well as with the west African monsoon and Atlantic SSTs. However, on interannual time scales the Atlantic TCs and the number of easterly waves are not significantly correlated, though a synoptic-scale measure of easterly wave activity (variance of meridional wind filtered to synoptic time scales) does show a significant positive correlation with TC activity on interannual time scales (Hopsch *et al.* 2007). Similarly, the number of easterly wave disturbances is not correlated with ENSO on interannual time scales, though the variance of filtered meridional wind is (S. Hopsch, personal communication). Since enhanced wind variance could be partly a result as well as a cause of enhanced TC activity, it seems reasonable to conclude tentatively that easterly wave variability is not an important factor responsible for the ENSO signal in Atlantic TCs. Nonetheless, the relative importance of variations in the supply of precursor disturbances vs. variations in the large-scale environment is worthy of further study in all basins, including the Atlantic.

3.4. *North Indian Ocean*

In contrast to other ocean basins, where the TC activity has a peak in the late summer or early fall of the corresponding hemisphere, North Indian Ocean cyclones primarily form in the traditional monsoon seasons — spring and autumn. One interpretation is that this is the case because it is only during these periods that the monsoon trough is located over open water (e.g., Lee *et al.*, 1989; Singh *et al.*, 2000), though it seems to us likely that the strong vertical wind shear associated with the peak monsoon may play a role in suppressing cyclones during that time.

There is a link between western North Pacific TCs and monsoon depressions over the Indian Ocean (e.g., Chen and Weng, 1999). Many monsoon depressions can be traced back to northwest Pacific TCs. Remnants of these TCs can propagate into the Bay of Bengal, where they can reintensify under the influence of warm SSTs and high moisture content that is present in the monsoon flow. Depending on the season, these monsoon depressions can then redevelop as cyclones in the Indian Ocean (Fudeyasu *et al.*, 2006b). In 1997, an El Niño year, there was a particularly close association between intense monsoon depressions and strong tropical cyclone activity in the northwest Pacific (Slingo and Annamalai, 2000). In El Niño years, during the months of July and August, there is also a higher number of monsoon depressions in the Bay of Bengal (Singh *et al.*, 2001).

There is a strong association between the variability of sea surface temperatures in the Indian and Pacific Oceans (Pan and Oort, 1983). During most El Niño events, the overall SSTs in the Indian Ocean increase, with a lag of 3–6 months after the peak SST in the central and eastern Pacific. The Indian Ocean warming is often associated with weaker surface wind speeds, though the tropospheric temperature increase throughout the global tropics, leading to increased stability to deep convection, may also play a role (Chiang and Sobel,

2002). Singh *et al.* (2000) noticed fewer intense tropical cyclones in the months of May and November in the Bay of Bengal during El Niño years.

Frank and Young (2007) observed a positive correlation between Indian Ocean TC activity and that in the Atlantic, and related this to reduced activity in the Indian Ocean during the positive phase of the NAO.

On the intraseasonal timescale, cyclone activity in the North Indian Ocean is strongly modulated by the MJO, with more activity when the MJO convective phase is over the Indian Ocean (Liebmann *et al.*, 1994).

A role for tropical cyclones in influencing large-scale climate variability in the Indian Ocean has also been suggested. Francis *et al.* (2007) argued that positive Indian Ocean dipole events (Saji *et al.*, 1999) could be triggered by the occurrence of severe cyclones over the Bay of Bengal during April–May.

3.5. *South Indian Ocean*

Several studies have examined ENSO's influence on TC activity in the South Indian Ocean basin. Jury (1993) did not find a statistically significant relationship, and attributed this to the opposing influences of increased upper-level westerly winds and enhanced convection during El Niño years. Xie *et al.* (2002) found an association between local ENSO-related SST anomalies and TC activity in the southwestern Indian Ocean. Kuleshov (2003) and Kuleshov and de Hoedt (2003) found an increase in TC numbers east of 70°E during La Niña years. Maximum TC frequency in the Southern Hemisphere occurs at the end of January during La Niña years and at the end of February to early March during El Niño years. The TC season usually starts one month earlier in the South Indian Ocean in La Niña years.

The tracks of the South Indian Ocean TCs are significantly more zonal during a La Niña event, and tend to be more frequent when local SSTs are warmer. The combination of both conditions in 2000, led to an exceptional number of TCs landfalls in Mozambique, which could be reproduced using a high-resolution coupled ocean-atmosphere model (Vitart *et al.*, 2003).

Ho *et al.* (2006) found that TC genesis is shifted westward during El Niño years, enhancing cyclogenesis west of 75°E and reducing it east of that longitude. They explained this shift through the changes in the Walker circulation associated with ENSO, with the formation of an anomalous anticyclonic low-level circulation in the eastern part of the South Indian Ocean during warm events. There are also changes in tracks between cold and warm events, with a decrease of activity southeast of Madagascar and a moderate increase in activity in the central subtropical South Indian Ocean during El Niño events compared to La Niña events. These track changes indicate that South Indian Ocean TCs move farther east during El Niño events, possibly due to anomalous southwesterly winds east of Madagascar. This shift in the TC activity was recently confirmed by Kuleshov *et al.* 2008 using an updated TC dataset for the southern hemisphere.

Jury (1993) obtained a relationship between TC frequency in the Southwest Indian Ocean and the QBO. They found that more TCs occur in the easterly phase than in the westerly phase. This is a somewhat surprising result, since it is of opposite sign to that found in the western North Pacific and North Atlantic basins, as discussed above, and the dynamical signals associated with the QBO itself are quite uniform throughout the tropics. More investigation of this relationship would be worthwhile.

Bessafi and Wheeler (2006) and Ho *et al.* (2006) analyzed the modulation of South Indian Ocean tropical cyclones by the MJO. Bessafi and Wheeler (2006) attributed the MJO modulation to perturbations in vorticity and vertical shear. Ho *et al.* (2006) noticed changes in TC tracks depending on the MJO phase.

3.6. *South Pacific and Australian Region*

Neville Nicholls was the first to relate tropical cyclone activity to ENSO in a series of papers in which he explored the environmental factors affecting TC activity in the Australian region (Nicholls, 1979, 1984, 1985, 1992). The predictability of TC activity in the Australian region was then explored, and a seasonal forecast scheme for TC activity was developed.

Years with many Australian TCs are usually preceded by below normal SST in the equatorial Pacific (La Niña event), low Darwin pressure and high North Australian SST, while years with relatively few TCs are preceded by El Niño events. As the strongest relationships occur before the start of the TC season, these variables can be used to predict TC activity in the Australian region. The relationship with ENSO is clearest for moderate intensity TCs; there is only a weak relationship with the numbers of either intense or weak TCs (Nicholls *et al.*, 1988). Various follow up studies have confirmed the relationship between ENSO and Australian TCs found in Nicholls' early studies (Dong, 1988; Hasting, 1990; Solow and Nicholls, 1990), and have shown them to be in some cases indicative of larger-scale signals for the entire basin. For example, the total number of TCs in the southwest Pacific tends to increase with El Niño (Basher and Zeng, 1995).

The region of mean TC genesis shifts with ENSO in the southwest Pacific, similarly to the western North Pacific. There is a well-documented eastward shift in the warm ENSO phase (Evans and Allan, 1992; Basher and Zeng, 1995; Kuleshov, 2003; Camargo *et al.*, 2007a; Kuleshov *et al.*, 2008) so that TCs are more likely to form away from the Australian coast and closer to the dateline in El Niño years. In some El Niño events TCs occur in French Polynesia, a location usually not affected by TCs. During La Niña events, the mean genesis location shifts westward, and TCs are more likely to make landfall in northeast Australia (Evans and Allan, 1992). The relationship of western Australian TCs with ENSO, on the other hand, is weaker than when the whole Australian region is considered (Broadbridge and Hanstrum, 1998).

In addition to the east-west shifts in TC genesis location with ENSO, there are also north-south shifts, with mean genesis shifting northward during El Niño and southward during La Niña (Revell and Goulter 1986a; Camargo *et al.*, 2007a). Among other impacts, this implies that TC landfall in Indonesia is more likely during El Niño years than in other years, a result which might be counter-intuitive since precipitation overall is suppressed there during El Niño years. Preliminary analysis of best track data (not shown) indicates that this is the case, though the statistics are poor since there are very few cyclone landfalls altogether in Indonesia.

A number of studies have used the ENSO signal to examine the influence of specific environmental factors on TC statistics. Ramsey *et al.* (2008) found larger correlations between TC activity and ENSO indices than between TC activity and local SST in the region of most frequent cyclogenesis. This is somewhat similar to findings in the western north Pacific (Chan, 2007). Correlations of TC frequency with low-level relative vorticity and vertical shear over the main genesis area are also significant.

As elsewhere, the MJO modulates TC activity in the Australian region, with more cyclones forming in the active phase of the MJO and fewer in the suppressed phase. The modulation is more pronounced to the northwest of Australia and is strengthened during El Niño events (Hall *et al.*, 2001). Low-level relative vorticity anomalies related to the MJO have been associated with changes in large-scale cyclogenesis patterns (Hall *et al.*, 2001).

4. Seasonal Forecasts of Tropical Cyclone Activity: Background

Nicholls (Nicholls, 1979, 1985, 1992) used the relationship between Australian TCs and ENSO to develop a statistical forecast scheme to forecast the number of cyclone days in a season based on the preceding July–September sea level pressure at Darwin — one of the two stations used to formulate the ENSO-related Southern Oscillation Index (SOI). Gray (Gray, 1984a) developed a statistical forecast methodology for forecasting seasonal Atlantic hurricane activity in early June and early August, which has been issued several times each year thereafter. We will discuss these forecasts in more detail below.

A discussion of the historical developments preceding the appearance of routine Atlantic seasonal hurricane forecasts appeared in Hess and Elsner (1994). They discussed the early work that uncovered the physical relationships among various factors such as sea level pressure (Brennan, 1935; Ray, 1935), easterly waves (Dunn, 1940) and mid-latitude troughs, and their influences on Atlantic hurricane frequency (Riehl, 1956). A summary of the early Atlantic forecasts as well as their verification is presented in Hastenrath (1990).

Currently, many institutions issue operational seasonal TC forecasts for various regions. In most cases, these are statistical forecasts, the seasonal typhoon activity forecasts of the City University of Hong Kong (Chan *et al.*, 1998, 2001), the Atlantic hurricane forecasts of Colorado State University (Gray, 1984a,b; Gray *et al.*, 1992c; Klotzbach and Gray, 2003, 2004; Klotzbach, 2007a,b), and Tropical Storm Risk (Saunders and Lea, 2005). The National Oceanic and Atmospheric Administration (NOAA) issues hurricane outlooks based on a blend of statistical techniques, climate model outputs, and forecasters input. More recently, dynamical forecasts have started being issued (Vitart, 2006; Vitart *et al.*, 2007; Camargo and Barnston, 2008). We will briefly discuss these forecasts, and their strengths and weaknesses. A more complete recent review of TC seasonal forecasts can be found in Camargo *et al.* (2007c).

The public and media interest in seasonal TC forecasts has increased tremendously since the time they were first produced. The disastrous impacts of the Atlantic basin hurricane seasons of 2004 and 2005 along with discussions of the possible impacts of global warming on Atlantic hurricane activity brought Atlantic seasonal hurricane forecasting to the forefront of the media.

Landfall forecasts are particularly important to users, as discussed in Elsner (2003) for the case of Florida residents. However, landfall forecast skill is quite limited. As seasonal TC forecasts improve, more attention will be able to be given to particular details such as local landfall probabilities, and the use of such specific forecasts will become more widespread and significant to decision makers and residents in coastal areas. However, the probability of a small area along the coastline being impacted by a tropical cyclone in any year is minimal, and therefore, landfall probabilities at the local level, based on seasonal forecasts, will always be quite low.

Predicting ENSO well in advance is fundamental for producing accurate seasonal TC forecasts. ENSO predictability follows a well-known seasonal cycle, in which the ENSO state for 4–6 months into the future is more accurately predicted from a starting time between July and November than between January and March. This is due to a "predictability barrier" that exists between April and June such that forecasts made just before this period are hindered by the barrier. The seasonal timing of the predictability barrier is related to the life cycle of ENSO episodes, which often emerge between April and June and endure until the following March to May. Once an episode has begun, predicting its continuation for the next 9 to 12 months is a much easier task than predicting its initial appearance. Even a strong El Niño, such as the one in 1997–98, was not well anticipated before signs of the initial onset were observed in the Northern Hemisphere spring of 1997 (Barnston *et al.*, 1999). Even after becoming apparent in the observations in late April and May 1997, the strength of this extreme El Niño event was under-predicted by most models, although a few models did correctly anticipate the rapid weakening in the spring of 1998 (Landsea and Knaff, 2000).

Due to the predictability barrier, the predictive skill for ENSO forecasts made in March is high for only 2–3 months, while for forecasts made in August the skill extends to longer lead times. Improvements in predictive skill using today's more advanced dynamical models have been small, and it remains to be seen whether or not improvements are possible (Chen and Cane, 2008) given the inherent signal-to-noise characteristics of the ocean-atmosphere system. The "slow physics" relevant to ENSO dynamics may become better predicted by both statistical and dynamical models of the future. However, better prediction of the shorter time-scale events that can also be important in triggering El Niño onset, such as the Madden-Julian Oscillation, may prove to be nearly impossible at multi-month lead times. Westerly wind bursts are often associated with MJO events and can have an impact on the magnitude and/or timing of an El-Niño event, like in 1997 (Slingo, 1998; van Oldenborgh *et al.*, 2000; Lengaigne *et al.*, 2003), although they may not be necessary for the occurrence of the event itself (Kleeman and Moore, 1997).

The ENSO predictability barrier has clear-cut implications for predictions of TC activity in the Northern Hemisphere when compared to predictions of TC activity in the Southern Hemisphere. TC activity in the Northern Hemisphere is considerably more challenging to predict because its peak TC seasons occur shortly after the ENSO predictability barrier. When an ENSO event appears somewhat later than usual, such as was the case in the late Northern Hemisphere summers of 1986 and 2006, the inhibiting effect on North Atlantic TCs is unanticipated until the peak season of August to October is already beginning. This can necessitate a sudden change as a final update to the seasonal TC prediction, and can potentially disrupt plans already being followed in accordance with an earlier seasonal TC prediction. The peak season for Southern Hemisphere TC activity occurs at least 6 months after the Northern Hemisphere spring ENSO predictability barrier, which provides a safer cushion of lead time in which to become fairly certain about the ENSO state to be expected during the peak season. Thus, last-minute surprises in seasonal TC outlooks for basins south of the equator are less likely to occur. Nonetheless, it is clear that a major hurdle in improving TC predictions for any ocean basin is the far-from-perfect quality of today's state-of-the-art ENSO forecasts. Indeed, Landsea and Knaff (2000) showed that it is still very difficult to outperform a simple statistical model that uses as predictors only the recent evolution of SST anomalies in a few tropical Pacific regions. This modest skill level for detecting El Niño onset still exists currently, as illustrated by the poor predictions of the late-starting 2006–07 El Niño. If the ENSO forecast challenge could be overcome, the skill of TC predictions could improve significantly — most notably in the Northern Hemisphere.

5. Statistical Forecasts

5.1. *Operational Statistical Forecasts*

Initial seasonal predictions for the North Atlantic basin (Gray, 1984a,b) were issued from Colorado State University in early June and early August beginning in 1984 and used statistical relationships between tropical cyclone activity and ENSO, the Quasi-Biennial Oscillation (QBO) and Caribbean basin sea-level pressures. Statistical forecast techniques for North Atlantic TCs have evolved since these early forecasts. Additional predictors have been added to the original forecast scheme, and the

QBO is not used as a predictor any more. In addition, seasonal forecasts are now issued as early as December of the previous year. Klotzbach and Gray (2004) and Klotzbach (2007a,b) explain the current forecast scheme.

Owens and Landsea (2003) examined the skill of Gray's operational Atlantic seasonal tropical cyclone forecasts relative to climatology and persistence. Their analysis indicated that for the analyzed period (1984–2001), their statistical forecasts demonstrated skill over climatology and persistence, with the adjusted forecasts being more skillful than the basic statistical forecasts. Figure 4 shows the correlation skill of the CSU seasonal forecasts for different variables.

NOAA has been issuing seasonal hurricane outlooks for the Atlantic and the eastern North Pacific regions since 1998 and 2003, respectively. These outlooks are provided to the public in both a deterministic and probabilistic format, using terciles. The deterministic prediction is given as a range — rather than a single number — and it represents about a 2/3 chance of the forecasts being in that range. The NOAA outlooks are based on the state of ENSO (Gray, 1984a) and the state of the tropical multi-decadal mode (TMM) (e.g., Chelliah and Bell, 2004), which incorporates the leading modes of tropical convective rainfall variability occurring on multi-decadal time scales. Important aspects of this signal that are related to an active Atlantic hurricane season include a strong West African monsoon, reduced vertical wind shear in the tropical Atlantic, and suppression of convection in the Amazon basin and high tropical Atlantic SSTs (Goldenberg *et al.*, 2001).

Tropical Storm Risk (TSR) issues statistical forecasts for TC activity in the Atlantic, western North Pacific and Australian regions. In a recent paper (Saunders and Lea, 2005), TSR describes their new forecast model, issued in early August,

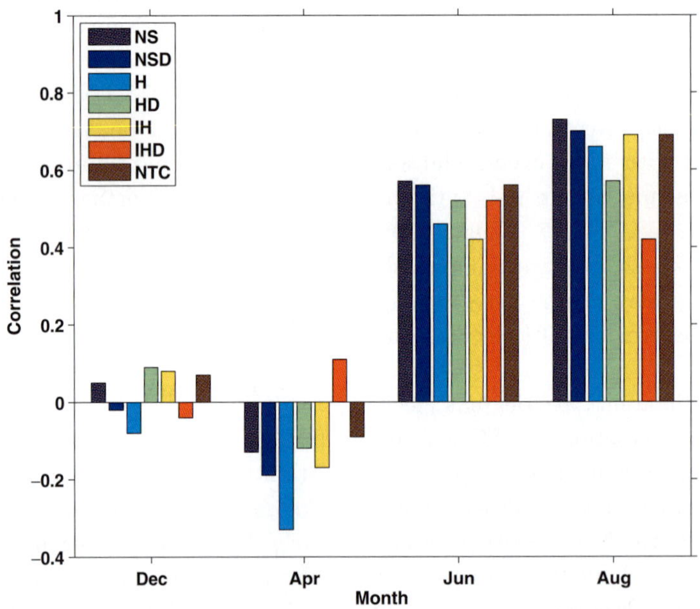

Figure 4. Correlations of the CSU seasonal forecasts for different leads: December (1992–2006), April (1995–2006), June (1984–2006 or 1990–2006) and August (1984–2006 or 1990–2006). The correlations are given for: number of named storms (NS), number of named storm days (NSD), number of hurricanes (H), number of hurricane days (HD), number of intense hurricanes (IH), number of intense hurricane days (IHD), and net tropical cyclone activity (NTC). Significant correlations at the 95% significant level are June — NS, NSD, H, HD, IH, IHD, NTC, August — NS, NSD, H, HD, IH and NTC. None of the correlations is significant for the December and April leads. Figure originally from Camargo *et al.* (2007c).

for seasonal predictions of hurricane landfall activity for the United States coastline. The model uses July wind patterns to predict the seasonal U.S. ACE index (effectively the cumulative wind energy from all U.S.-striking tropical cyclones). The July height-averaged winds in these regions are indicative of atmospheric circulation patterns that either favor or hinder hurricanes from reaching U.S. shores. The model correctly anticipates whether U.S. hurricane losses are above- or below-median in 74% of the hindcasts for the 1950–2003 period. As far as we know, this forecast is the only statistical forecasts that uses the steering flow. If this finding is corroborated, it would be a major advance for statistical foreasts. The western North Pacific seasonal prediction model uses Niño 3.75 (5°S–5°N, 180°–140°W) forecasts (Lloyd-Hughes *et al.*, 2004) to predict ACE.

Johnny Chan and colleagues at the City University of Hong Kong have issued seasonal TC forecasts for the western North Pacific basin (number of TCs and typhoons) since 1997. The statistical predictions are based on various environmental conditions in the prior year, up to the Northern Hemisphere spring of the forecast season. The most prominent atmospheric and oceanic conditions include ENSO, the extent of the Pacific subtropical ridge and the intensity of the India-Burma trough (Chan *et al.*, 1998, 2001). For a few years, forecasts of the number of TCs making landfall were also issued (Liu and Chan, 2003). Currently, the landfall forecast scheme for the South China Sea is being improved. More recently, a simple method was proposed to update seasonal prediction of the annual number of TCs in a given ocean basin, based on the cumulative number of TCs up to a given month in the early season (Chan, 2008b).

Various other agencies have been issuing forecasts for different regions of the world. The National Meteorological Services of Mexico has produced eastern North Pacific statistical seasonal forecasts since 2001. The China Meteorological Administration and the National Climate Center in China issue forecasts of typhoon activity for the western North Pacific, as well as landfalling typhoon frequency in the whole South China Sea and eastern China. The Cuban Meteorological Institute has been issuing seasonal forecasts of Atlantic hurricane activity, including hurricane landfall in Cuba, since 1996. The North Carolina State University forecast group developed a seasonal forecast methodology for Atlantic hurricane activity and landfall based on ENSO, vertical wind shear, the Atlantic dipole mode and the NAO (Xie *et al.*, 2005a,c; Keith and Xie, 2008). Forecasts for the Australian/Southwest Pacific region are presented annually in the December issue of the Experimental Long-Lead Forecast Bulletin since the 2004–5 season for the Southern Hemisphere tropical cyclone season. These forecasts are based on a Poisson regression model and use as predictors the September saturated equivalent potential temperature gradient and the SOI (McDonnell and Holbrook, 2004a,b).

5.2. *Methodology for statistical forecasts*

The first statistical seasonal forecasts for TCs in the Australian regions were developed using the values of the pressure in Darwin in the months preceding the Australian TC season to forecast the number of TCs (Nicholls, 1979). Other indices for ENSO and North Australia SST were also examined by Nicholls (1984). The performance of the forecasts using the SOI in the months prior to the season as the predictor (Solow and Nicholls, 1990) was discussed in Nicholls (1992).

James Elsner and colleagues at Florida State University (FSU) have been developing techniques for modeling seasonal hurricane activity and landfall (e.g., Elsner and Jagger, 2004). Although their forecasts are not produced operationally, their methodology is currently used to issue region-specific forecasts for various companies. The FSU group pioneered various topics in TC seasonal forecasting such as the use of a Poisson distribution for hurricane counts (Elsner and Schmertmann, 1993), the influence of the phase of the North Atlantic Oscillation (NAO) on Atlantic hurricane tracks and U.S. coastal hurricane activity (Elsner *et al.*, 2001), and the development of a skillful statistical model for seasonal forecasts of landfall probability over the southeastern United States (Lehmiller *et al.*, 1997). More recently, Elsner and Jagger (2006) built a Bayesian model for seasonal landfall over the U.S. using as predictors May–June values of the NAO, May–June values of the SOI, and May–June values of the AMO, and then extended the methodology to long-lead seasonal forecasts (Elsner *et al.*, 2006). Similar statistical techniques are used for multi-seasonal forecasts (Elsner *et al.*, 1998; Elsner *et al.*, 2008) and for building climatological models for extreme hurricane winds near the U.S. (Jagger and Elsner, 2006). It should be noted though, that only a statistical link between May–June NAO and subsequent Atlantic hurricane tracks was found. There has been no corroboration of the physical connection between the NAO and the tropospheric steering flow, either concurrent or predictive.

Two different methods, a binary classification scheme and a Poisson regression, are used to forecast annual Atlantic TC counts. Modeling the annual counts as a state-dependent Poisson process using a binary classification approach, only two factors are necessary to explain a large portion of the variance: ENSO and MDR SST. If a Poisson regression is used, the most skillful statistical model also considers the NAO as a predictor (Sabatelli and Mann, 2007).

Chu and Zhao (2007) developed a seasonal forecast for the Central North Pacific in the peak season (July to September), using a linear regression model in a Bayesian framework. Five large-scale environmental variables in the antecedent May and June months are used as predictors and the cross-validated procedure applied to the period 1966–2003 produced satisfactory results.

A statistical multivariate prediction model for the number of tropical cyclone days in the Southwest Indian Ocean one season in advance was developed by Jury *et al.* (1999). The chosen predictors are SST anomalies, OLR anomalies, upper level winds and surface winds in different regions, depending on the forecast lead time.

6. Dynamical Forecasts

6.1. *TCs in climate models*

Extended integrations of global climate models in principle allow for an assessment of the frequency, intensity, duration, structure, and tracks of tropical cyclone-like features in the model. In practice, simulation of realistic intensities and detailed structures of TCs is hampered by the coarse resolution generally required of such global models, as discussed below. In addition, the fidelity of the global model's TC genesis process compared to that of the real world has not been well established.

For the global models, TCs are located and tracked in model data using objective techniques that are usually based on a local maximum of cyclonic relative vorticity at 850 hPa and often involving other criteria such as: system lifetime, evidence of a warm core, maximum winds above a threshold, and a local minimum in sea level pressure (e.g., Tsutsui and Kasahara, 1996; Vitart *et al.*, 1997; Camargo and Zebiak, 2002; Sugi *et al.*, 2002; McDonald *et al.*, 2005). The location and tracking method tends to be unique to each study which makes it difficult to compare the results of the different studies directly. Walsh *et al.* (2006) provide recommendations for providing homogeneous comparisons of various resolution models for determining tropical cyclone frequencies. They base this upon an analysis of how minimal tropical storms (with maximum winds at 17.5 m/s) would be depicted under various resolutions. Use of such resolution-based criteria for determining tropical cyclone occurrence should allow for more rigorous quantitative comparisons of global (and regional) climate model output of tropical cyclones frequencies. For seasonal forecasting, however, simulated interannual variability of TC activity is more important than evaluation of the models' absolute performance in TC simulation. Model-specific detection thresholds which adjust for model biases (Camargo and Zebiak, 2002) may be more appropriate for this application.

The global climate models used for tropical cyclone analysis have tended to be of low (300 km) horizontal resolution (e.g., Vitart *et al.*, 1997; Tsutsui, 2002; Bengtsson *et al.*, 2006) or of medium (120 km) resolution (e.g., Sugi *et al.*, 2002; McDonald *et al.*, 2005; Hasegawa and Emori, 2005; Yoshimura and Sugi, 2005; Bengtsson *et al.*, 2007; LaRow *et al.*, 2008). The grid-scale of the low and medium resolution models is larger than the typical scale of tropical cyclones. This can lead to a poor simulation of tropical cyclones (e.g., Vitart *et al.*, 1997; McDonald *et al.*, 2005). The simulated cyclones tend to have a larger horizontal scale, and although they have warm cores, the intense inner core is not well-simulated. Thus, the simulated cyclones have lower wind speeds than occurs in observed tropical cyclones (Vitart *et al.*, 1997). Minimum central pressures tend to be better simulated than the maximum surface wind speeds. Recent studies have used higher resolutions of 50 km (Chauvin *et al.*, 2006) and 20 km (Oouchi *et al.*, 2006), but models of this resolution are too expensive for most modeling centers to use for long climate change experiments, though they may soon become practical for seasonal forecasting. An alternative approach is to use a global model with a stretched grid (i.e., higher resolution) over the region of interest (e.g., Chauvin *et al.*, 2006), although this limits the study to the region where the resolution is high. Even at that resolution, the highest simulated TC intensity reported by Oouchi *et al.* was about 932 hPa, compared with the observed record of 870 hPa, indicating the limitation of their global model in simulating very intense TCs.

Even though smaller-scale features of the individual cyclones are typically not well simulated in the global models, these models are able to reproduce some aspects of the observed climatology and inter-annual variability of tropical cyclones (Tsutsui and Kasahara, 1996; Sugi *et al.*, 2002; Camargo *et al.*, 2005; McDonald *et al.*, 2005; Bengtsson *et al.*, 2007). Most models are able to simulate tropical cyclone-like disturbances in roughly the correct location and at the correct time of the year, although all models exhibit some biases. Several models simulate tropical cyclones in the South Atlantic (Vitart *et al.*, 1997; Sugi *et al.*, 2002; McDonald *et al.*, 2005; Oouchi *et al.*, 2006) where they are rarely observed (Pezza and Simmonds, 2005). It should also be pointed out that not all models simulate storms in the South Atlantic (Camargo *et al.*, 2005).

The global models' simulated TC tracks are sometimes shorter than observed (Tsutsui and Kasahara, 1996; Sugi *et al.*, 2002; Camargo *et al.*, 2005). High resolution models such as that of Oouchi *et al.* (2006) are able to better simulate the length of TC tracks. Low resolution models such as ECHAM3 and ECHAM4 exhibit TC tracks that are too long (Camargo *et al.*, 2005). Some of these differences in length may be due to the objective techniques used to identify and track the cyclones in the model data.

The simulated global annual frequency of TCs in global models varies, with some models simulating too many TCs (e.g., McDonald *et al.*, 2005) while others simulate too few (Camargo *et al.*, 2005). Both from comparing results among different models (e.g., Camargo *et al.*, 2005) and from sensitivity experiments with a given model (Vitart *et al.*, 2001; Emori *et al.*, 2005), it is evident that both model resolution and model physics can play important roles in determining the frequency of TC occurrence in the global models. The dynamics of the genesis process in a coarse resolution model was explored by Camargo and Sobel (2004).

The genesis process in that model was found to be qualitatively similar to that in observations (though with major quantitative differences), featuring a convectively coupled vortex, which intensifies, partly due to enhanced surface fluxes, after an initial period of constant intensity or slight weakening.

While models in many existing studies have demonstrated an ability to simulate many aspects of the seasonal variability of tropical cyclone frequency in each basin, all of the models have some errors in both frequency and timings. These errors are basin-, season- and model-dependent. Increasing the horizontal resolution of global models typically improves the simulation of individual cyclones (Bengtsson *et al.*, 1995) but may not improve the tropical cyclone climatology and interannual variability. One reason for that may be biases of the variability in large-scale climate itself. On the other hand, a comparison of several models suggests that differences in the simulation of the large-scale environment for TC genesis do not translate in a simple way to differences in simulated TC statistics (Camargo *et al.*, 2007f).

More realistic maximum TC intensities have been simulated by downscaling individual storm cases from a coarse-grid global model into an operational regional high-resolution hurricane prediction system (Knutson *et al.*, 1998) or into a regional climate model (Walsh and Ryan, 2000). Another promising approach involves embedding a regional climate model within a global model or atmospheric reanalysis in order to simulate the seasonal evolution of more realistic TC formation, evolution, and intensities (e.g., Walsh *et al.*, 2004; Knutson *et al.*, 2007; Stowasser *et al.*, 2007; Feser and Von Storch, 2008; Knutson *et al.*, 2008). However, the TC simulations using a regional climate model have uncertainties due to model domain choices, parameterizations, and other issues (Landman *et al.*, 2005; Camargo *et al.*, 2007b).

Climate models also must simulate realistic ENSO and decadal variability under present day and future climate conditions as a necessary condition for providing reliable future projections of TC activity in these regions (e.g., Nguyen and Walsh, 2001). The interannual variability of TC occurrence in global models can be tested by comparing cyclones simulated in models forced with observed SSTs to tropical cyclone observations from the same period (Wu and Lau, 1992; Tsutsui and Kasahara, 1996; Vitart *et al.*, 1997; Sugi *et al.*, 2002; McDonald *et al.*, 2005; Camargo *et al.*, 2005; Bengtsson *et al.*, 2007; LaRow *et al.*, 2008). The nine-member ensemble of Vitart *et al.* (1997) and the 40-yr experiments used by Camargo *et al.* (2005) are better suited for analysis of the interannual variability than are the shorter experiments used by Tsutsui and Kasahara (1996), Sugi *et al.* (2002) and McDonald *et al.* (2005) because of the larger sample sizes. The correlation of the global annual number of tropical cyclones with the observed number varies from 0.15 in the JMA model (Sugi *et al.*, 2002) to 0.41 in the GFDL model (Vitart *et al.*, 1997). The correlations are better in some seasons, basins and models than in others. The correlations tend to be highest in the western North Pacific and North Atlantic basins (Vitart *et al.*, 1997; Camargo *et al.*, 2005), possibly because of the importance of ENSO in those regions. In the case of the Atlantic, La Row *et al.* (2008) obtained correlations of 0.78 for the interannual variability of number of tropical cyclones using a T126 horizontal resolution model, four-ensemble member, and 20 years of integration. The interannual variability performance of the 20 km grid global model of Oouchi *et al.* (2006) has not yet been assessed.

The longer time-scale variability of TCs in global models was explored in a few studies. Vitart and Doblas-Reyes (2007) analyzed the inter-decadal variability of tropical storm frequency for the period from 1958–2001 in various regions. The inter-decadal variability of TCs in the model is more realistic for varying greenhouse gas concentrations than for fixed concentrations. However, the natural inter-decadal variability plays a more important role in this period than does anthropogenic forcing. Matsuura *et al.* (2003) and Yumoto *et al.* (2003) showed that a coupled ocean-atmosphere global model could reproduce the inter-decadal TC frequency variability in the western North Pacific, which was related to the decadal variability of the monsoon westerlies in the model and in the reanalysis.

Nguyen and Walsh (2001) analyzed the interannual and decadal variability of TCs in the South Pacific using a regional climate model. A higher incidence of TCs on the Australian east coast occurred in La Niña conditions, while in El Niño conditions, the formation and occurrence shifted towards the central Pacific, as in observations. The TC activity in the model also exhibited coherent decadal variability similar to observed patterns on these time-scales. Knutson *et al.* (2007) used a high-resolution regional model over the North Atlantic and were able to reproduce the observed ENSO and multidecadal variability in Atlantic TC activity over the period from 1980–2006, with a 0.87 correlation between observed and model-derived hurricane frequency.

An alternative approach to explicit global model simulation is to use an empirical "seasonal genesis parameter" (e.g., Ryan *et al.*, 1992; Watterson *et al.*, 1995; Thorncroft and Pytharoulis, 2001) to infer a genesis frequency from climate model data (Tsutsui and Kasahara, 1996; Royer *et al.*, 1998; McDonald *et al.*, 2005; Chauvin *et al.*, 2006; Camargo *et al.*, 2007f). In this approach, the focus is on using the models' predictions of seasonal variations in the large-scale environment for TC genesis and intensification, rather than their predictions of the TCs themselves. The strength of this approach is based on the fact that the ability of climate models to simulate large-scale climate is clearly superior to their ability to simulate TCs. Great caution is required when applying a parameter developed for present day climate to future predictions, as the statistical relationships may not be valid under altered climate conditions (Ryan *et al.*, 1992). Royer *et al.* (1998) and Emanuel and Nolan (2004) (see also Nolan *et al.*, 2007; Caron and Jones, 2008) have proposed refined versions of Gray's (1979) genesis index that avoid the use of factors such as threshold SSTs that themselves may well vary in an altered climate (e.g., Henderson-Sellers *et al.*, 1998). These methods typically produce plausible maps and seasonal cycles of TC genesis. Camargo *et al.* (2007a) showed that the index developed by Emanuel and Nolan (2004) was able to reproduce the interannual variations of the observed frequency and location of genesis in several different basins.

6.2. *Current operational dynamical forecasts*

While low-resolution simulations are not adequate for forecasting individual cyclones' tracks and intensities, some climate models have skill in forecasting levels of seasonal TC activity. These climate models are able to reproduce typical ENSO influences on seasonal TC activity (e.g., Vitart *et al.*, 1997).

Currently, three institutions produce seasonal forecasts of tropical cyclone activity by dynamical methods. The European Centre for Medium-range Weather Forecasts (ECMWF) issues experimental seasonal forecasts of tropical storm frequency based on dynamical models for various regions since 2001 and are only available to institutions affiliated with the ECMWF. The ECMWF forecasts are based on coupled ocean-atmosphere models (Vitart and Stockdale, 2001; Vitart, 2006), with tropical cyclone-like vortices being identified and tracked in the atmospheric model output. The European multi-model (EUROSIP) dynamical forecasts of TC frequency skillfully distinguished the very-active Atlantic hurricane season in 2005 from the below-average season in 2006, as shown in Fig. 5 (Vitart *et al.*, 2007). The EUROSIP forecasts have been produced in real time since 2005, but are not currently available to the public. The United Kingdom Met Office started issuing dynamical seasonal forecasts for the

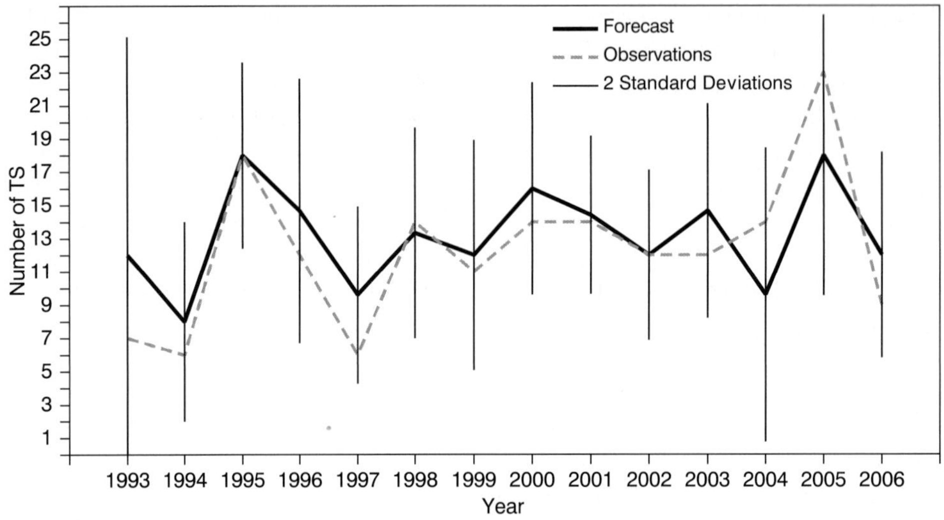

Figure 5. Number of tropical storms from July to November predicted by the EUROSIP (median) starting on 1 June (black thick solid line) for the period 1993–2006. Hindcasts were used for the period 1993–2004, and real-time forecasts in 2005–2006. The observations are given in the dashed line and vertical lines represent two standard deviations within the multi-model ensemble distribution. (Reproduced from Vitart *et al.*, 2007, by permission of the American Geophysical Union.)

Atlantic basin operationally in June 2007, also using a coupled ocean-atmosphere model (one of the EUROSIP models, discussed in Vitart *et al.*, 2007) and a methodology similar to the ECMWF forecasts.

The International Research Institute (IRI) for Climate and Society also issues experimental seasonal forecasts based on dynamical models for tropical storm frequency and accumulated cyclone energy (ACE; Bell *et al.*, 2000) in Northern Hemisphere regions since 2003. The experimental IRI forecasts are obtained using a two-tier procedure. First, various possible scenarios for SSTs are predicted using statistical and dynamical models. Then, atmospheric models are forced with those predicted SSTs. Similar to the ECMWF procedure, the tropical cyclone-like vortices are then identified and tracked (Camargo and Zebiak, 2002). The IRI forecasts are probabilistic by tercile category (above normal, normal, below normal).

The skill of some of the best performing dynamical models in predicting the frequency of tropical storms is comparable to the skill of statistical models in some ocean basins. Over the North and South Indian Oceans, dynamical models usually perform poorly (Camargo *et al.*, 2005; Bengtsson *et al.*, 2007). It is not clear to what extent this is due to model errors or due to a lack of predictability. Similar to the experience with seasonal climate forecasts, combining different model forecasts (multi-model ensemble forecasts) appears to produce overall better forecasts than individual model ensemble forecasts (Vitart, 2006). The hindcast skill of various dynamical climate models in predicting seasonal TC activity is discussed in Camargo *et al.* (2005) and Vitart (2006), and the real time performance of the IRI forecasts is analyzed in Camargo and Barnston (2009).

Seasonal prediction of tropical cyclone landfall represents a major challenge for dynamical models. Tropical cyclones take an unrealistically poleward track in some of the models used in seasonal forecasting systems due partly to the coarse horizontal model resolution, which leads to larger vortices than observed ones. These larger vortices would likely be more influenced by the beta effect (Rhines, 1975) leading to stronger "beta drift", (e.g., Chan, 2005) and thus more poleward tracks. Finer resolution climate models are able to reproduce landfall differences related to ENSO impacts, such as in Mozambique (Vitart *et al.*, 2003).

7. Summary

In this chapter, we reviewed the influence of various climates modes on tropical cyclone activity in intraseasonal to decadal time scales. Globally, ENSO and the MJO, and to some extent the QBO modulate TC activity. How this modulation occurs depends on the region considered. On a regional scale, other climate modes also affect TC activity, modifying the timing, genesis location, frequency, tracks and landfall frequency.

Using the relationships between TC activity and these climate modes on intraseasonal and interannual time scales, statistical forecasts of TC activity can be constructed. Dynamical TC seasonal forecasts based on TCs in climate models are now also issued by some centers.

In most cases, the modulation of TC activity by the climate has been focused on specific time scales and modes of variability. It would be interesting in the future to have a better understanding of the interactions and relationships of these modes. This would lead to a better framework for understanding TC variability.

Reflecting the literature, our discussion has focused on the influence of climate on TCs. The influence of TCs themselves on the large-scale climate has been mentioned only briefly, because relative few studies have tried to assess that influence. Besides a possible role in ENSO dynamics (Keen, 1982; Sobel and Camargo, 2005) or the Indian Ocean dipole (Francis *et al.*, 2007), roles for TCs in controlling the global thermohaline circulation (Emanuel, 2001; Korty *et al.*, 2008) and in creating a multi-year persistence in TC activity (Pasquero and Emanuel, 2008) have been hypothesized. A better understanding of the role of TCs in the global climate is desirable and more work in this area would be worthwhile.

Acknowledgments

The second part of this chapter (seasonal forecasts) is partly based on the paper Camargo *et al.* (2007c), and we thank Chris Landsea for his contribution to that paper. We also thank Ruth McDonald and Tom Knutson for their contribution, which was originally in the Knutson *et al.* chapter in this book. We also thank Ed Gerber for advice on NAO references and Susanna Hopsch for discussions on easterly waves.

References

Aiyyer, A. and J. Molinari, 2003: Evolution of mixed Rossby-gravity waves in MJO type environments. *J. Atmos. Sci.*, **60**, 2837–2855.

Aiyyer, A. and J. Molinari, 2008: MJO and tropical cyclogenesis in the Gulf of Mexico and eastern Pacific: Case study and idealized numerical modeling. *J. Atmos. Sci.*, **65**, 2691–2704.

Avila, L. A., 1991: Atlantic tropical systems of 1990. *Mon. Wea. Rev.*, **119**, 2027–2033.

Barnston, A. G., M. H. Glantz, and Y. He, 1999: Predictive skill of statistical and dynamical climate models in SST forecasts during the 1997–98 El Niño episode and the 1998 La Niña onset. *Bull. Amer. Meteor. Soc.*, **80**, 217–243.

Basher, R. and X. Zheng, 1995: Tropical cyclones in the southwest Pacific: Spatial patterns and relationships to the Southern Oscillation and sea surface temperature. *J. Climate*, **8**, 1249–1260.

Bell, G. D. and M. Chelliah, 2006: Leading tropical modes associated with interannual and multidecadal fluctuations in North Atlantic hurricane activity. *J. Climate*, **19**, 590–612.

Bengtsson, L., H. Böttger, and M. Kanamitsu, 1982: Simulation of hurricane-type vortices in a general circulation model. *Tellus*, **34**, 440–457.

Bengtsson, L., M. Botzet, and M. Esch, 1995: Hurricane-type vortices in a general circulation model. *Tellus* **47A**, 175–196.

Bengtsson, L., K. I. Hodges, and E. Roeckner, 2006: Storm tracks and climate change. *J. Climate*, **19**, 3518–3543.

Bengtsson, L., K. I. Hodges, and M. Esch, 2007: Tropical cyclones in a T159 resolution global climate model: Comparison with observations and re-analysis. *Tellus*, **59A**, 396–416.

Bessafi, M. and M. C. Wheeler, 2006: Modulation of South Indian Ocean tropical cyclones by the Madden-Julian oscillation and convectively coupled equatorial waves. *Mon. Wea. Rev.*, **134**, 638–656.

Bister, M. and K. A. Emanuel, 1997: The genesis of hurricane Guillermo: TEXMEX analyses and a modeling study. *Mon. Wea. Rev.*, **125**, 2662–2682.

Bosart, L. F. and J. A. Bartlo, 1991: Tropical storm formation in a baroclinic environment. *Mon. Wea. Rev.*, **119**, 1979–2013.

Bove, M. C., J. J. O'Brien, J. B. Elsner, C. W. Landsea, and X. Niu, 1998: Effect of El Niño on U.S. landfalling hurricanes, revisited. *Bull. Amer. Meteor. Soc.*, **79**, 2477–2482.

Bradley, R. S., H. F. Diaz, G. N. Kiladis, and E. I. Eisched, 1987: ENSO signal in continental temperatures and precipitation records, *Nature*, **327**, 497–501.

Brennan, J. F., 1935: Relation of May-June weather conditions in Jamaica to the Caribbean tropical disturbances of the following season. *Mon. Wea. Rev.*, **63**, 13–14.

Broadbridge, L. W. and B. N. Hanstrum, 1998: The relationship between tropical cyclones near western Australia and the Southern Oscillation index. *Aust. Meteor. Mag.*, **47**, 183–190.

Camargo, S. J. and A. G. Barnston, 2009: Experimental seasonal dynamical forecasts of tropical cyclone activity at IRI. *Wea. Forecasting*, **24**, 472–491.

Camargo, S. J. and A. H. Sobel, 2004: Formation of tropical storms in an atmospheric general circulation model. *Tellus*, **56A**, 56–67.

Camargo, S. J. and A. H. Sobel, 2005: Western North Pacific tropical cyclone intensity and ENSO. *J. Climate*, **18**, 2996–3006.

Camargo, S. J. and A. H. Sobel, 2010: Revisiting the influence of the quasi-biennial oscillation on tropical cyclones, *J. Climate*, submitted.

Camargo, S. J. and S. E. Zebiak, 2002: Improving the detection and tracking of tropical cyclones in atmospheric general circulation models. *Wea. Forecasting*, **17**, 1152–1162.

Camargo, S. J., A. G. Barnston, and S. E. Zebiak, 2005: A statistical assessment of tropical cyclone activity in atmospheric general circulation models. *Tellus*, **57A**, 589–604.

Camargo, S. J., K. A. Emanuel, and A. H. Sobel, 2007a: Use of a genesis potential index to diagnose ENSO effects on tropical cyclone genesis. *J. Climate*, **20**, 4819–4834.

Camargo, S. J., H. Li, L. Sun, 2007b: Feasibility study for downscaling seasonal tropical cyclone activity using the NCEP regional spectral model. *Int. J. Climate*, **27**, 311–325.

Camargo, S. J., A. G. Barnston, P. J. Klotzbach, and C. W. Landsea, 2007c: Seasonal tropical cyclone forecasts. *WMO Bull.*, **56**, 297–309.

Camargo, S. J., A. W. Robertson, S. J. Gaffney, P. Smyth, and M. Ghil, 2007d: Cluster analysis of typhoon tracks: Part I: General Properties. *J. Climate*, **20**, 3635–3653.

Camargo, S. J., A. W. Robertson, S. J. Gaffney, P. Smyth, and M. Ghil, 2007e: Cluster analysis of typhoon tracks: Part II: Large-scale circulation and ENSO. *J. Climate*, **20**, 3654–3676.

Camargo, S. J., A. W. Robertson, A. G. Barnston, and M. Ghil, 2008a: Clustering of eastern North Pacific tropical cyclone tracks: ENSO and MJO effects. *Geochem. Geophys. Geosys.*, **9**, Q06V05, doi: 10.1029/2007GC001861.

Camargo, S. J., A. H. Sobel, A. G. Barnston, and K. A. Emanuel, 2007f: Tropical cyclone genesis potential index in climate models. *Tellus*, **59A**, 428–443.

Camargo, S. J., M. C. Wheeler, and A. H. Sobel, 2009: Diagnosis of the MJO modulation of tropical cyclogenesis using an emperical index. *J. Atmos. Sci.*, **66**, 3061–3074.

Cane, M. A. and S. E. Zebiak, 1985: A theory for El Niño and the Southern Oscillation. *Science*, **228**, 1085–1087.

Caron, L. P. and C. G. Jones, 2008: Analysing present, past and future tropical cyclone activity as inferred from an ensemble of coupled global climate models. *Tellus*, **60A**, 80–96.

Chan, J. C. L., 1985: Tropical cyclone activity in the Northwest Pacific in relation to the El Niño/Southern Oscillation phenomenon. *Mon. Wea. Rev.*, **113**, 599–606.

Chan, J. C. L., 1995: Tropical cyclone activity in the western North Pacific in relation to the stratospheric quasi-biennial oscillation. *Mon. Wea. Rev.*, **123**, 2567–2571.

Chan, J. C. L., 2000: Tropical cyclone activity over the western North Pacific associated with El Niño and La Niña events. *J. Climate*, **13**, 2960–2972.

Chan, J. C. L., 2005: The physics of tropical cyclone montion. *Ann. Rev. Fluid Mech.*, **37**, 99–128.

Chan, J. C. L., 2007: Interannual variations of intense typhoon activity. *Tellus*, **59A**, 455–460.

Chan, J. C. L., 2008a: Decadal variations of intense typhoon occurrence in the western North Pacific. *Proc. R. Soc. A.*, **464**, 249–272.

Chan, J. C. L., 2008b: A simple seasonal forecast update of tropical cyclone activity. *Wea. Forecasting*, **23**, 1016–1021.

Chan, J. C. L., J. Shi, and C. M. Lam, 1998: Seasonal forecasting of tropical cyclone activity over the western North Pacific and the South China Sea. *Wea. Forecasting*, **13**, 997–1004.

Chan, J. C. L., J. E. Shi, and C. M. Lam, 2001. Improvements in the seasonal forecasting of tropical cyclone activity over the western North Pacific, *Wea. Forecasting*, **16**, 997–1004.

Chauvin, F., J.-F. Royer, and M. Déqué, 2006: Response of hurricane-type vortices to global warming as simulated by ARPEGE-Climat at high resolution. *Clim. Dyn.*, **27**, 377–399.

Chelliah M. and G. D. Bell, 2004: Tropical multidecadal and interannual climate variability in the NCEP/NCAR reanalysis. *J. Climate*, **17**, 1777–1803.

Chen, T.-C. and S.-P. Weng, 1999: Interannual and intraseasonal variations in monsoon depressions and their westward-propagating predecessors. *Mon. Wea. Rev.*, **127**, 1005–1020.

Chen, T.-C., S.-P. Weng, N. Yamazaki, and S. Kiehne, 1998: Interannual variation in the tropical cyclone activity over the western North Pacific. *Mon. Wea. Rev.*, **126**, 1080–1090.

Chen, T.-C., S. Y. Wang, and M. C. Yen, 2006: Interannual variation of tropical cyclone activity over the western North Pacific. *J. Climate*, **19**, 5709–5720.

Chia, H. H. and C. F. Ropelewski, 2002: The interannual variability in the genesis location of tropical cyclones in the northwest Pacific. *J. Climate*, **15**, 2934–2944.

Chiang J. C. H. and D. J. Vimont, 2004: Analogous Pacific and Atlantic meridional modes of tropical atmosphere-ocean variability. *J. Climate*, **17**, 4143–4158.

Chu, P.-S., 2002: Large-scale circulation features associated with decadal variations of tropical cyclone activity over the Central North Pacific. *J. Climate*, **15**, 2678–2689.

Chu, P.-S., 2004: ENSO and tropical cyclone activity, in *Hurricanes and Typhoons, Past, Present and Future*, R. J. Murnane and K.-B. Liu, eds., New York, pp. 297–332, Columbia University Press.

Chu, P.-S. and J. D. Clark, 1999: Decadal variations of tropical cyclone activity over the Central North Pacific. *Bull. Amer. Meteor. Soc.*, **80**, 1875–1881.

Chu, P.-S. and J. Wang, 1997: Tropical cyclone occurrences in the vicinity of Hawaii: Are the differences between El Niño and non- El Niño years significant? *J. Climate*, **10**, 2683–2689.

Chu, P.-S. and X. Zhao, 2004: Bayesian changepoint analysis of tropical cyclone activity: The Central North Pacific case. *J. Climate*, **17**, 4893–4901.

Chu, P.-S. and X. Zhao, 2007: A Bayesian regression approach for predicting seasonal tropical cyclone activity over the Central North Pacific. *J. Climate*, **20**, 4002–4013.

Clark, J. D. and P.-S. Chu, 2002: Interannual variation of tropical cyclone activity over the Central North Pacific. *J. Meteor. Soc. Japan*, **80**, 403–418.

Collins, J. M. and I. M. Mason, 2000: Local environment conditions related to seasonal tropical cyclone activity in the Northeast Pacific basin. *Geophys. Res. Lett.*, **27**, 3881–3884.

Czaja, A., P. van der Vaart, and J. Marshall, 2002: A diagnostic study of the role of remote forcing in tropical Atlantic variability. *J. Climate*, 15, 3280–3290.

Davidson, N. E. and H. H. Hendon, 1989: Downstream development in the southern hemisphere monsoon during FGGE/WMONEX. *Mon. Wea. Rev.*, **117**, 1458–1470.

Delworth, T. L. and M. E. Mann, 2000: Observed and simulated multidecadal variability in the Northern Hemisphere. *Clim. Dyn.*, **16**, 661–676.

Dickinson, M. and J. Molinari, 2002: Mixed Rossby-gravity waves and western Pacific tropical cyclogenesis. Part I: Synoptic evolution. *J. Atmos. Sci.*, **59**, 2183–2196.

Dong, K., 1988: El Niño and tropical cyclone frequency in the Australian region and the northwest Pacific. *Aust. Met. Mag.*, **36**, 219–225.

Donnelly, J. P. and J. D. Woodruff, 2007: Intense hurricane activity over the past 5,000 years controlled by El Niño and the West African monsoon. *Nature*, **447**, 465–468.

Donnelly, J. P., S. S. Bryant, J. Butler, J. Dowling, L. Fan, N. Hausmann, P. Newby, B. Shuman, J. Stern, K. Westover, and T. Webb III, 2001: A 700-year sedimentary record of intense hurricane landfalls in southern New England. *Geol. Soc. Am. Bull.*, **113**, 714–727.

Dunkerton, T. J., M. T. Montgomery, and Z. Wang, 2008: Tropical cyclogenesis in a tropical wave critical layer: Easterly waves. *Atmos. Chem. Phys. Discuss.*, **8**, 11149–11292.

Dunn, G. E., 1940: Cyclogenesis in the tropical Atlantic. *Bull. Amer. Meteor. Soc.*, **21**, 215–229.

Elsberry, R. L., 1987: Tropical cyclone motion. *A Global view of tropical cyclones*, R. L. Elsberry, Ed., Office of Naval Research, 189, pp.

Elsner, J. B., 2003: What seasonal hurricane forecasts mean to residents of Florida. *Bull. Amer. Meteor. Soc.*, **84**, 556–558.

Elsner, J. B. and T. H. Jagger, 2004: A hierarchical Bayesian approach to seasonal hurricane modeling. *J. Climate*, **17**, 2813–2827.

Elsner, J. B. and T. H. Jagger, 2006: Prediction models for annual U.S. hurricane counts. *J. Climate*, **19**, 2935–2952.

Elsner, J. B. and A. B. Kara, 1999: *Hurricanes of the North Atlantic: Climate and society.* Oxford University Press, New York.

Elsner, J. B. and B. Kocher, 2000: Global tropical cyclone activity: A link to the North Atlantic Oscillation. *Geophys. Res. Lett.*, **27**, 129–132.

Elsner, J. B. and K. B. Liu, 2003: Examining the ENSO-typhoon hypothesis. *Climate Res.*, **25**, 43–54.

Elsner, J. B. and C. P. Schmertmann, 1993: Improving extended-range seasonal predictions of intense Atlantic hurricane activity. *Wea. Forecasting*, **8**, 345–351.

Elsner, J. B., B. H. Bossak, and X.-F. Niu, 2001: Secular changes to the ENSO – U.S. hurricane relationship. *Geophys. Res. Lett.*, **28**, 4123–4126.

Elsner, J. B., A. B. Kara, and M. A. Owens, 1999: Fluctuations in North Atlantic hurricane frequency. *J. Climate*, **12**, 427–437.

Elsner, J. B., R. J. Murnane, and T. H. Jagger, 2006. Forecasting U.S. hurricanes 6 months in advance. *Geophys. Res. Lett.*, **33**, L10704, doi:10.1029/2006GL025693.

Elsner, J. B., X. Niu, and A. A. Tsonis, 1998: Multi-year prediction model of North Atlantic hurricane activity. *Meteor. Atmos. Phys.*, **68**, 43–51.

Elsner, J. B., T. H. Jagger, M. Dickinson, and D. Rowe, 2008: Improving multiseason forecasts of North Atlantic hurricane activity. *J. Climate*, **21**, 1209–1219.

Emanuel, K., 2005: Increasing destructiveness of tropical cyclones over the past 30 years. *Nature*, **436**, 686–688.

Emanuel, K., 2007: Environmental factors affecting tropical cyclone power dissipation. *J. Climate*, **20**, 5497–5509.

Emanuel, K. and D. S. Nolan, 2004: Tropical cyclone activity and the global climate system. Proc. 26th AMS Conf. on Hurr. and Trop. Meteor., #10A.2, pp. 240–241.

Emanuel, K. A., 1986: An air-sea interaction theory for tropical cyclones. Part I: Steady-state maintenance. *J. Atmos. Sci.*, **43**, 585–604.

Emanuel, K. A., 1988: The maximum intensity of hurricanes. *J. Atmos. Sci.*, **45**, 1143–1155.

Emanuel, K. A., 2000: A statistical analysis of tropical cyclone intensity. *Mon. Wea. Rev.*, **128**, 1139–1152.

Emanuel, K. A., 2001: Contribution of tropical cyclones to meridional heat transport by the oceans. *J. Geophys. Res.*, **106**, 14771–14781.

Emori, S., A. Hasegawa, T. Suzuki, and K. Dairaku, 2005: Validation, parameterization dependence, and future projection of daily precipitation simulated with a high-resolution atmospheric GCM. *Geophys. Res. Lett.*, **32**, L06708.

Evans, J. L. and R. J. Allan, 1992: El Niño/Southern Oscillation modification to the structure of the monsoon and tropical cyclone activity in the Australasian region. *Int. J. Climatol.*, **12**, 611–623.

Fan, K., 2007: North Pacific sea ice cover, a predictor for the western North Pacific typhoon frequency? *Science in China Series D-Earth Sciences*, **50**, 1251–1257.

Fedelstein, S. B., 2000: The timescale, power spectra and climate noise properties of teleconnection patterns. *J. Climate*, **13**, 4430–4440.

Ferreira, R. N. and W. H. Schubert, 1997: Barotropic aspects of ITCZ breakdown. *J. Atmos. Sci.*, **54**, 261–285.

Feser, F. and H. Von Storch, 2008: A dynamical downscaling study for typhoons in southeast Asia using a regional climate model. *Mon. Wea. Rev.*, **5**, 1806–1815.

Fogarty, E. A., J. B. Elsner, T. H. Jagger, K.-b. Liu, and K.-s. Louie, 2006: Variations in typhoons landfalls over China. *Adv. Atmos. Sci.*, **23**, 665–677.

Folland, C. K., T. N. Palmer, and D. E. Parker, 1986: Sahel rainfall and worldwide sea temperatures. *Nature*, **320**, 602–607.

Francis, P. A., S. Gadgil, and P. N. Vinayachandran, 2007: Triggering of the positive Indian Ocean dipole events by severe cyclones over the Bay of Bengal. *Tellus*, **59A**, 461–475.

Frank, W. M. and P. E. Roundy, 2006: The role of tropical waves in tropical cyclogenesis. *Mon. Wea. Rev.*, **134**, 2397–2417.

Frank, W. M. and G. S. Young, 2007: The interannual variability of tropical cyclones. *Mon. Wea. Rev.*, **135**, 3587–3598.

Frappier, A., T. Knutson, K.-B. Liu, K. Emanuel, 2007: Perspective: coordinating paleoclimate research on tropical cyclones with hurricane-climate theory and modelling. *Tellus*, **59A**, 529–537.

Fudeyasu, H., S. Iizuka, T. Matsuura, 2006a: Impact of ENSO on landfall characteristics of tropical cyclones over the western North Pacific Turing the summer monsoon season. *Geophys. Res. Lett.*, **33**, L21815.

Fudeyasu, H., S. Iizuka, T. Matsuura, 2006b: Seasonality of westward-propagating disturbances over Southeast and South Asia originated from typhoons. *Geophys. Res. Lett.*, **33**, L10809.

Garcia-Herrera, R., P. Ribera, E. Hernandez, and L. Gimeno, 2007: Northwest Pacific typhoons documented by the Philippine Jesuits, 1566–1900. *J. Geophys. Res.*, **112**, D06108.

Goddard, L., S. J. Mason, S. E. Zebiak, C. F. Ropelewski, R. E. Basher, and M. A. Cane, 2001: Current approaches to seasonal to interannual climate predictions. *Int. J. Climatol.*, **21**, 1111–1152.

Goldenberg, S. B. and L. J. Shapiro, 1996: Physical mechanisms for the association of El Niño and West African rainfall with Atlantic major hurricane activity. *J. Climate*, **9**, 1169–1187.

Goldenberg, S. B., C. W. Landsea, A. M. Mestas-Nuñez, and W. M. Gray, 2001: The recent increase in Atlantic hurricane activity: Causes and implications. *Science*, **293**, 474–479.

Gray, W. M., 1979: Hurricanes: their formation, structure and likely role in the tropical circulation. Meteorology over the tropical oceans. Royal Meteorological Society, Bracknall, England.

Gray, W. M., 1984a: Atlantic seasonal hurricane frequency. Part I: El Niño and 30 mb quasi-biennial oscillation influences. *Mon. Wea. Rev.*, **112**, 1649–1668.

Gray, W. M., 1984b: Atlantic seasonal hurricane frequency. Part II: Forecasting its variability. *Mon. Wea. Rev.*, **112**, 1669–1683.

Gray, W. M., 1990: Strong association between West African rainfall and U. S. landfall of intense hurricanes. *Science*, **249**, 1251–1256.

Gray, W. M., 1992: African rainfall as a precursor of hurricane-related destruction on the U.S. East coast. *Bull. Amer. Meteor. Soc.*, **73**, 1352–1364.

Gray, W. M. and J. D. Sheaffer, 1991: El Niño and QBO influences on tropical cyclone activity, in *Teleconnections linking worldwide anomalies*. M. H. Glanz, R. W. Katz, and N. Nicholls, eds., Cambridge University Press, New York, pp. 257–284.

Gray, W. M., J. D. Sheaffer, and J. A. Knaff, 1992a: Hypothesized mechanism for stratospheric QBO influence on ENSO variability. *Geophys. Res. Lett.*, **19**, 107–110.

Gray, W. M., J. D. Sheaffer, and J. A. Knaff, 1992b: Influence of the stratospheric QBO on ENSO variability. *J. Meteor. Soc. Japan*, **70**, 975–995.

Gray, W. M., J. D. Sheaffer, and C. W. Landsea, 1997: Climate trends associated wit multidecadal variability of Atlantic hurricane activity. *Hurricanes: climate and socioeconomic impacts*. H. F. Diaz and R. S. Pulwarty, eds., Springer-Verlag, Berlin.

Gray, W. M., C. W. Landsea, P. W. Mielke Jr., and K. J. Berry, 1992c: Predicting Atlantic seasonal hurricane activity 6–11 months in advance. *Wea. Forecasting*, **7**, 440–455.

Gray, W. M., C. W. Landsea, P. W. Mielke Jr., and K. J. Berry, 1993: Predicting Atlantic basin seasonal tropical cyclone activity by 1 August. *Wea. Forecasting*, **8**, 73–86.

Hall, J. D., A. J. Matthews, and D. J. Karoly, 2001: The modulation of tropical cyclone activity in the Australian region by the Madden-Julian oscillation. *Mon. Wea. Rev.*, **129**, 2970–2982.

Harr, P. A. and R. L. Elsberry, 1991: Tropical cyclone track characteristics as a function of large-scale anomalies. *Mon. Wea. Rev.*, **119**, 1448–1468.

Harr, P. A. and R. L. Elsberry, 1995: Large-scale circulation variability over the tropical western North Pacific. Part I: Spatial patterns and tropical cyclone characteristics. *Mon. Wea. Rev.*, **123**, 1225–1246.

Hartmann, D. L. and E. D. Maloney, 2001: The Madden-Julian oscillation, barotropic dynamics, and North Pacific tropical cyclone formation. Part II: Stochastic barotropic modeling. *J. Atmos. Sci.*, **58**, 2559–2570.

Hasegawa, A. and S. Emori, 2005: Tropical cyclones and associated precipitation over the western North Pacific: T106 atmospheric GCM simulation for present-day and doubled CO2 climates. *SOLA*, **1**, 145–148, SOI:10.2151/sola. 2005–038.

Hastenrath, S., 1990: Tropical climate prediction: A progress report, 1985–1990. *Bull. Amer. Meteor. Soc.*, **71**, 819–825.

Hastings, P. A., 1990: Southern Oscillation influences on tropical cyclone activity in the

Australian/Southwest Pacific region. *Int. J. Climatol.*, **10**, 291–298.

Henderson-Sellers, A., H. Zhang, G. Berz, K. Emanuel, W. Gray, C. Landsea, G. Holland, J. Lighthill, S.-L. Shieh, P. Webster, and K. McGuffie, 1998: Tropical cyclones and global climate change: a post-IPCC assessment. *Bull. Amer. Meteor. Soc.*, **79**, 19–38.

Hess, J. C. and J. B. Elsner, 1994: Historical developments leading to current forecast models of annual Atlantic hurricane activity. *Bull. Amer. Meteor. Soc.*, **75**, 1611–1621.

Ho, C.-H., J.-J. Baik, J.-H. Kim, D.-Y. Gong, and C.-H. Sui, 2004: Interdecadal changes in summertime typhoon tracks. *J. Climate*, **17**, 1767–1776.

Ho, C.-H., J.-H. Kim, H.-S. Kim, C.-H. Sui, and D.-Y. Gong, 2005: Possible influence of the Antarctic oscillation on tropical cyclone activity in the western North Pacific. *J. Geophys. Res.*, **110**, D19104.

Ho, C.-H., J-H. Kim, J.-H. Jeong, and H.-S. Kim, 2006: Variation of tropical cyclone activity in the South Indian Ocean: El Niño-Southern Oscillation and Madden-Julian Oscillation effects. *J. Geophys. Res.*, **111**, D22191, doi:10.1029/2006JD007289.

Holland, G. J., 1995: Scale interaction in the western North Pacific monsoon. *Meteorol. Atmos. Phys.*, **1–2**, 57–79.

Hopsch, S. B., C. D. Thorncroft, K. Hodges, A. Aiyyer, 2007: West African storm tracks and their relationship to Atlantic tropical cyclones. *J. Climate*, **20**, 2468–2483.

Hsu, H.-H., C.-H. Hung, A. K. Lo, and C.-W. Hung, 2008: Influence of tropical cyclone on the estimation of climate variability in the tropical western North Pacific, *J. Climate*, **21**, 2960–2975.

Hurrell, J. W., Y. Kushnir, G. Ottersen, and M. Visbeck (eds.), 2003: The North Atlantic oscillation climate significance and impacts. *Geophysical Monograph Series*, **134**.

Irwin, R. P. and R. Davis, 1999: The relationship between the southern oscillation index and tropical cyclone tracks in the eastern North Pacific. *Geophys. Res. Lett.*, **26**, 2251–2254.

Jagger, T. H. and J. B. Elsner, 2006. Climatology models for extreme hurricane winds near the United States. *J. Climate*, **19**, 3220–3236.

Joseph, P. V., J. K. Eischeid, and R. J. Pyle, 1994: Interannual variability of the onset of the Indian summer monsoon and its assocation with atmospheric features, El Niño, and sea surface temperature anomalies. *J. Climate*, **7**, 81–105.

Jury, M. R., 1993: A preliminary study of climatological associations and characteristics of tropical cyclones in the SW Indian Ocean. *Meteorol. Atmos. Phys.*, **51**, 101–115.

Jury, M. R., B. Pathack, B. Parker, 1999: Climatic determinants and statistical prediction of tropical cyclone days in the Southwest Indian Ocean. *J. Climate*, **12**, 1738–1746.

Keen, R. A., 1982: The role of cross-equatorial tropical cyclone pairs in the Southern Oscillation. *Mon. Wea. Rev.*, **110**, 1405–1416.

Keith, E. and Xie, L., 2009: Predicting Atlantic tropical cyclone seasonal activity in April. *Wea. Forecasting*, **24**, 436–455.

Kemball-Cook, S. and B. Wang, 2001: Equatorial waves and air-sea interaction in the boreal summer intraseasonal oscillation. *J. Climate*, **14**, 2923–2942.

Kim, J.-H., C.-H. Ho, C-H. Sui, and S.-K. Park, 2005: Dipole structure of interannual variation of summertime tropical cyclone activity over East Asia. *J. Climate*, **12**, 1738–1746.

Kimberlain, T. B., 1999: The effects of ENSO on North Pacific and North Atlantic tropical cyclone activity, in *Proc. 23rd Conference on Hurricanes and Tropical Meteorology*, Dallas, TX, AMS, Boston.

Kleeman, R. and A. M. Moore, 1997: A theory for the limitation of ENSO predictability due to stochastic atmospheric transients. *J. Atmos. Sci.*, **54**, 737–767.

Klotzbach, P. J., 2007a: Recent developments in statistical prediction of seasonal Atlantic basin tropical cyclone activity. *Tellus*, **59A**, 511–518.

Klotzbach, P. J., 2007b: Revised prediction of seasonal Atlantic basin tropical cyclone activity from 1 August. *Wea. Forecasting*, **22**, 937–949.

Klotzbach, P. J. and W. M. Gray, 2003: Forecasting September Atlantic basin tropical cyclone activity. *Wea. Forecasting*, **18**, 1109–1128.

Klotzbach, P. J. and W. M. Gray, 2004: Updated 6–11 month prediction of Atlantic basin seasonal hurricane activity. *Wea. Forecasting*, **19**, 917–934.

Klotzbach, P. J. and W. M. Gray, 2008: Multidecadal variability in North Atlantic tropical cyclone activity. *J. Climate*, **21**, 3929–3935.

Knaff, J. A., 1997: Implications of summertime sea level pressure anomalies in the tropical Atlantic region. *J. Climate*, **10**, 789–790.

Knutson, T. R., R. E. Tuleya, and Y. Kurihara, 1998: Simulated increase of hurricane intensities in a CO2-warmed climate. *Science*, **279**(5353), 1018–1020.

Knutson, T. R., J. J. Sirutis, S. T. Garner, I. M. Held, and R. E. Tuleya, 2007: Simulation of the recent multidecadal increase of Atlantic hurricane activity using an 18-km-Grid Regional Model. *Bull. Amer. Meteor. Soc.*, **88**, 1549–1565.

Knutson, T. R., J. J. Sirutis, S. T. Garner, G. A. Vecchi, and I. M. Held, 2008: Simulated reduction in Atlantic hurricane frequency under twenty-first-century warming conditions. *Nature Geoscience*, **1**, 359–364.

Korty, R. L., K. A. Emanuel, and J. R. Scott, 2008: Tropical cyclone-induced upper-ocean mixing and climate: Application to equable climates. *J. Climate*, **21**, 638–654.

Kossin, J. P. and D. J. Vimont, 2007: A more general framework for understanding Atlantic hurricane variability and trends. *Bull. Amer. Meteor. Soc.*, **88**, 1767–1781.

Kuleshov, Y., 2003: Tropical cyclones in the southern hemisphere: Influence of the El Niño-Southern Oscillation phenomenon. *Proc. 7th Int. Conf. Southern Hemisphere Meteorology and Oceanography*, Wellington, New Zealand, 24–28 March 2003, American Meteorological Society, Boston, MA, #13.3.

Kuleshov, Y. and G. de Hoedt, 2003: Tropical cyclone activity in the southern hemisphere. *Bull. Austr. Meteorol. Soc.*, **16**, 135–137.

Kuleshov, Y., L. Qi, R. Fawcett, and D. Jones, 2008: On tropical cyclone activity in the southern hemisphere: Trends and the ENSO connection. *Geophys. Res. Lett.*, **35**, L14S08.

Kumar, V. and R. Krishan, 2005: On the association between the Indian summer monsoon and the tropical cyclone activity over northwest Pacific. *Current science*, **88**, 602–612.

Lander, M. A., 1994: An exploratory analysis of the relationship between tropical storm formation in the western North Pacific and ENSO. *Mon. Wea. Rev.*, **122**, 636–651.

Lander, M. A., 1996: Specific tropical cyclone tracks and unusual tropical cyclone motions associated with a reverse-oriented monsoon trough in the western North Pacific. *Wea. Forecasting*, **11**, 170–186.

Landman, W. A., A. Seth, and S. J. Camargo, 2005: The effect of regional climate model domain choice on the simulation of tropical cyclone-like vortices in the Southwestern Indian Ocean. *J. Climate*, **18**, 1253–1274.

Landsea, C. W., 2000: El Niño–Southern Oscillation and the seasonal predictability of tropical cyclones, in *El Niño: Impacts of multiscale variability on natural ecosystems and society*, H. F. Díaz and V. Markgraf, eds., pp. 149–181. Cambridge University Press, UK.

Landsea, C. W. and W. M. Gray, 1992: The strong association between western Sahelian monsoon rainfall and intense U.S. landfalling hurricanes. *J. Climate*, **5**, 1528–1534.

Landsea, C. W. and J. A. Knaff, 2000: How much skill was there in forecasting the very strong 1997–1998 El Niño? *Bull. Amer. Meteor. Soc.*, **81**, 2107–2119.

Landsea, C. W., R. A. Pielke, Jr., A. M. Mesta-Nuñez, J. A. Knaff, 1999: Atlantic basin hurricanes: Indices of climatic changes. *Climatic Change*, **42**, 89–129.

Larson, J., Y. Zhou, and R. W. Higgins, 2005: Characteristics of landfalling tropical cyclones in the United States and Mexico: Climatology and interannual variability. *J. Climate*, **18**, 1247–1262.

LaRow, T. E., Y.-K. Lim, D. W. Shin, E. P. Chassignet, and S. Cocke, 2008: Atlantic basin seasonal hurricane simulations, *J. Climate*, **21**, 3191–3206.

Lee, C.-S., R. Edson, and W. M. Gray, 1989: Some large-scale characteristics associated with tropical cyclone development in the North Indian Ocean during FGGE. *Mon. Wea. Rev.*, **117**, 407–426.

Lehmiller, G. S., T. B. Kimberlain, and J. B. Elsner, 1997: Seasonal prediction models for the North Atlantic basin hurricane location. *Mon. Wea. Rev.*, **125**, 1780–1791.

Lengaigne, M., J.-P. Boulanger, C. Menkes, G. Madec, P. Delecluse, E. Guilyardi, and J. Slingo, 2003: The March 1997 westerly wind event and the onset of the 1997/98 El Niño: Understanding the role of the atmospheric response. *J. Climate*, **20**, 3330–3343.

Leroy, A. and M. C. Wheeler, 2008: Statistical prediction of weekly tropical cyclone activity in the southern hemisphere. *Mon. Wea. Rev.*, **10**, 2637–2654.

Liebmann, B., H. H. Hendon, and J. D. Glick, 1994: The relationship between tropical cyclones of the western Pacific and Indian oceans and the Madden-Julian oscillation. *J. Meteor. Soc. Japan*, **72**, 401–411.

Liu, K. B. and M. L. Fearn, 2000: Reconstruction of prehistoric landfall frequencies of catastrophic hurricanes in northwestern Florida from lake sediment records. *Quaternary Res.*, **54**, 238–245.

Liu, K. B., C. M. Shen, and K. S. Louie, 2001: A 1,000 year history of typhoon landfalls in Guangdong, southern China, reconstructed from

Chinese historical documentary records. *Ann. Assoc. Am. Geogr.*, **91**, 453–464.

Liu, K. S. and J. C. L. Chan, 2003: Climatological characteristics and seasonal forecasting of tropical cyclones making landfall along the South China coast. *Mon. Wea. Rev.*, **131**, 1650–1662.

Liu, K. S. and J. C. L. Chan, 2008: Interdecadal variability of western North Pacific tropical cyclone tracks. *J. Climate*, **21**, 4464–4476.

Lloyd-Hughes, B., M. A. Saunders, and P. Rockett, 2004: A consolidated CLIPER model for improved August-September ENSO prediction skill. *Wea. Forecasting*, **19**, 1089–1105.

Madden, R. A. and P. R. Julian, 1972: Description of global circulation cells in the tropics with a 40–45 day period. *J. Atmos. Sci.*, **29**, 1109–1123.

Madden, R. A. and P. R. Julian, 1994: Observations of the 40–50-day tropical oscillation — A review. *Mon. Wea. Rev.*, **122**, 814–837.

Maloney, E. D. and D. L. Hartmann, 2000a: Modulation of eastern Pacific hurricanes by the Madden-Julian oscillation. *J. Climate*, **13**, 1451–1460.

Maloney, E. D. and D. L. Hartmann, 2000b: Modulation of hurricane activity in the Gulf of Mexico by the Madden-Julian oscillation. *Science*, **287**, 2002–2004.

Maloney, E. D. and D. L. Hartmann, 2001: The Madden-Julian oscillation, barotropic dynamics and North Pacific tropical cyclone formation. Part I: Observations. *J. Atmos. Sci.*, **58**, 2545–2558.

Mantua, N., 2002. The Pacific decadal oscillation. *J. Oceanogr.*, **58**, 35–44.

Matsuura, T., M. Yumoto, and S. Tizuka, 2003: A mechanism of interdecadal variability of tropical cyclone activity over the western North Pacific. *Clim. Dyn.*, **21**, 105–117.

McDonald, R. E., D. G. Bleaken, D. R. Cresswell, V. D. Pope, and C. A. Senior, 2005: Tropical storms: Representation and diagnosis in climate models and the impacts of climate change. *Clim. Dyn.*, **25**, 19–36.

McDonnell, K. A. and N. J. Holbrook, 2004a: A Poisson regression model of tropical cyclogenesis for the Australian-southwest Pacific Ocean region. *Wea. Forecasting*, **19**, 440–455.

McDonnell, K. A. and N. J. Holbrook, 2004b: A Poisson regression model approach to predicting tropical cyclogenesis in the Australian/southwest Pacific Ocean region using the SOI and saturated equivalent potential temperature gradient as predictors. *Geophys. Res. Lett.*, **31**, L20110.

Molinari, J. and D. Vollaro, 2000: Planetary- and synoptic-scale influences on eastern Pacific cyclogenesis. *Mon. Wea. Rev.*, **128**, 3296–3307.

Molinari, J., D. Knight, M. Dickinson, D. Vollaro, and S. Skubis, 1997: Potential vorticity, easterly waves, and eastern Pacific tropical cyclogenesis. *Mon. Wea. Rev.*, **125**, 2699–2708.

Molinari, J., D. Vollaro, S. Skubis, and M. Dickinson, 2000: Origins and mechanisms of eastern Pacific cyclogenesis: A case study. *Mon. Wea. Rev.*, **128**, 125–139.

Montgomery, M. T., M. E. Nicholls, T. A. Cram, and A. B. Saunders, 2006. A vortical hot tower route to tropical cyclogenesis. *J. Atmos. Sci.*, **63**, 355–386.

Nakazawa, T., 1988: Tropical super clusters within intraseasonal variations over the western Pacific. *J. Meteor. Soc. Japan*, **64**, 17–34.

Nguyen, K. C. and K. J. E. Walsh, 2001: Interannual, decadal, and transient greenhouse simulation of tropical cyclone-like vortices in a regional climate model of the South Pacific. *J. Climate*, **14**, 3043–3054.

Nicholls, N., 1979: A possible method for predicting seasonal tropical cyclone activity in the Australian region. *Mon. Wea. Rev.*, **107**, 1221–1224.

Nicholls, N., 1984: The southern oscillation, sea-surface temperature, and interannual fluctuations in Australian tropical cyclone activity. *J. Climatol.*, **4**, 661–670.

Nicholls, N., 1985: Predictability of interannual variations of Australian seasonal tropical cyclone activity, *Mon. Wea. Rev.*, **113**, 1144–1149.

Nicholls, N., 1992: Recent performance of a method for forecasting seasonal tropical cyclone activity. *Aust. Meteor. Mag.*, **40**, 105–110.

Nicholls, N., C. Landsea, and J. Gill, 1988: Recent trends in the Australian region tropical cyclone activity. *Meteor. Atmos. Phys.*, **65**, 197–205.

Nolan, D. S., E. D. Rappin, and K. A. Emanuel, 2007: Tropical cyclogenesis sensitivity to environmental parameters in radiative-convective equilibrium. *Q. J. R. Meteorol. Soc.*, **133**, 2085–2107.

Oouchi, K., J. Yoshimura, H. Yoshimura, R. Mizuta, S. Kusunoki, and A. Noda, 2006: Tropical cyclone climatology in a global-warming climate as simulated in a 20km-mesh global atmospheric model: frequency and wind intensity analysis. *J. Meteorol. Soc. Japan*, **84**, 259–276.

Owens, B. F. and C. W. Landsea, 2003: Assessing the skill of operational Atlantic seasonal tropical cyclone forecasts, *Wea. Forecasting*, **18**, 45–54.

Pan, Y. H. and A. H. Oort, 1983: Global climate variations connected with sea surface temperature anomalies in the eastern equatorial Pacific for the 1958–1973 period. *Mon. Wea. Rev.*, **111**, 1244–1258.

Pasquero, C. and K. Emanuel, 2008: Tropical cyclones and transient upper-ocean warming. *J. Climate*, **21**, 638–657.

Pezza, A. B. and I. Simmonds, 2005: The first South Atlantic hurricane: unprecedented blocking, low shear, and climate change. *Geophys. Res. Lett.*, **32**, L15712, doi:10.1029/2005GL023390.

Pielke, Jr. R. A. and C. W. Landsea, 1999: La Niña, El Niño and Atlantic hurricane damages in the United States. *Bull. Amer. Meteor. Soc.*, **80**, 2027–2033.

Pielke, Jr. R. A., J. Gratz, C. W. Landsea, D. Collins, M. A. Saunders, R. Musulin, 2008: Normalized hurricane damage in the United States: 1900–2005. *Nat. Hazards Rev.*, **9**, 29–42.

Plumb, R. A. and A. D. McEwan, 1978: The instability of a forced standing wave in a viscous stratified fluid – Laboratory analog of quasi-biennial oscillation. *J. Atmos. Sci.* **35**, 1827–1839.

Ramsay, H. A., L. M. Leslie, P. J. Lamb, M. B. Richman, and M. Leplastrier, 2008: Interannual variability of tropical cyclones in the Australian region: Role of large-scale environment. *J. Climate*, **21**, 1083–1103.

Ray, C. L., 1935: Relation of tropical cyclone frequency to summer pressures and ocean surface water temperatures. *Mon. Wea. Rev.*, **63**, 10–12.

Raymond, D. J., G. B. Raga, C. S. Bretherton, J. Molinari, C. Lopez-Carillo, and Z. Fuchs, 2003: Convective forcing in the intertropical convergence zone of the eastern Pacific. *J. Atmos. Sci.*, **60**, 2064–2082.

Raymond, D. J. S. Sessions, and Z. Fuchs, 2008: A theory for the spinup of tropical depressions. *Q. J. R. Meteor. Soc.*, **133**, 1743–1754.

Reed, R. J. and E. E. Recker, 1971: Structure and properties of synoptic-scale wave disturbances in the equatorial western Pacific. *J. Atmos. Sci.*, **28**, 1117–1133.

Revell, C. G. and S. W. Goulter, 1986a: South Pacific tropical cyclones and the Southern Oscillation. *Mon. Wea. Rev.*, **114**, 1138–1145.

Revell, C. G. and S. W. Goulter, 1986b: Lagged relationships between the Southern Oscillation and the numbers of tropical cyclones in the southern Pacific region. *Mon. Wea. Rev.*, **114**, 2669–2670.

Rhines, P., 1975: Waves and turbulence on a beta-plane. *J. Fluid Mech.*, **69**, 417–433.

Ritchie, E. R. and G. J. Holland, 1999: Large-scale patterns associated with tropical cyclogenesis in the western North Pacific. *Mon. Wea. Rev.*, **127**, 2027–2043.

Riehl, H., 1956: Sea surface temperature anomalies and hurricanes. *Bull. Amer. Meteor. Soc.*, **37**, 413–417.

Ropelewski, C. F. and M. S. Halpern, 1987: Global and regional scale precipitation patterns associated with the El Niño/Southern Oscillation. *Mon. Wea. Rev.*, **115**, 1606–1626.

Royer, J.-F., F. Chauvin, B. Timbal, P. Araspin, and D. Grimal, 1998: A GCM study of the impact of greenhouse gas increase on the frequency of occurrence of tropical cyclones. *Clim. Change*, **38**, 307–343.

Ryan, B. F., I. G. Watterson, and J. L. Evans, 1992: Tropical cyclones frequencies inferred from Gray's yearly genesis parameter: validation of GCM tropical climate. *Geophys. Res. Lett.*, **19**, 1831–1834.

Sabbatelli, T. A. and M. E. Mann, 2007: The influence of climate state variables on Atlantic tropical cyclone occurrence rates. *J. Geophys. Res.*, **112**, D17114.

Saji, N. H., B. N. Goswani, P. N. Vinayachandran, T. Yagamata, 1999: A dipole mode in the tropical Indian Ocean. *Nature*, **401**, 360–363.

Saunders, M. A. and A. S. Lea, 2005: Seasonal prediction of hurricane activity reaching the coast of the United States. *Nature*, **434**, 1005–1008.

Saunders, M. A., R. E. Chandler, C. J. Merchant, and F. P. Roberts, 2000: Atlantic hurricanes and NW Pacific typhoons: ENSO spatial impacts on occurrence and landfall. *Geophys. Res. Lett.*, **27**, 1147–1150.

Servain, J. I., I. Wainer, J. P. McCreary, and A. Dessier, 1999: Relationship between the equatorial and meridional modes of climatic variability in the tropical Atlantic. *Geophys. Res. Lett.*, **26**, 485–488.

Shapiro, L. J., 1989: The relationship of the quasi-biennial oscillation to Atlantic tropical storm activity. *Mon. Wea. Rev.*, **117**, 1545–1551.

Singh, O. P., T. M. A. Khan, and M. S. Rahman, 2000: Changes in the frequency of tropical cyclones over the North Indian Ocean. *Meteor. Atmos. Phys.*, **75**, 11–20.

Singh, O. P., T. M. A. Khan, and M. S. Rahman, 2001: Probable reasons for enhanced cyclogenesis in the Bay of Bengal during July-August of ENSO years. *Glob. Planet. Change*, **29**, 135–147.

Slingo, J. M., 1998: Extratropical forcing of tropical convection in a northern winter simulation with the UGAMP GCM. *Quart. J. Royal Meteor. Soc.*, **124**, 27–51.

Slingo, J. M. and H. Annamalai, 2000, 1997: The El Niño of the century and the response of the Indian summer monsoon. *Mon. Wea. Rev.*, **128**, 1778–1797.

Smith, R. K., M. T. Montgomery, and S. Vogl, 2008: A critique of Emanuel's hurricane model and potential intensity theory. *Quart. J. Roy. Met. Soc.*, **134**, 551–561.

Smith, S. R., J. Brolley, J. J. O'Brien, and C. A. Tartaglione, 2007: ENSO's impact on regional U.S. hurricane activity. *J. Climate*, **20**, 1404–1414.

Sobel, A. H. and C. S. Bretherton, 1999: Development of synoptic-scale disturbances over the summertime tropical northwest Pacific. *J. Atmos. Sci.*, **56**, 3106–3127.

Sobel, A. H. and S. J. Camargo, 2005: Influence of western North Pacific tropical cyclones on their environment. *J. Atmos. Sci.*, **62**, 3396–3407.

Sobel, A. H. and E. D. Maloney, 2000: Effect of ENSO and the MJO on the western North Pacific tropical cyclones. *Geophys. Res. Lett.*, **27**, 1739–1742.

Sobel, A. H., I. M. Held, and C. S. Bretherton, 2002: The ENSO signal in tropical tropospheric temperature. *J. Climate*, **15**, 2702–2706.

Soden, B. J., 2000: The sensitivity of the tropical hydrological cycle to ENSO. *J. Climate*, **13**, 538–549.

Solow, A. and N. Nichols, 1990: The relationship between the Southern Oscillation and tropical cyclone frequency in the Australian region. *J. Climate*, **3**, 1097–1101.

Stephenson, D. B., V. Pavan, and R. Bojariu, 2000: Is the North Atlantic Oscillation a random walk? *Int. J. Climatol.*, **20**, 1–18.

Stowasser, M., Y. Wang, and K. Hamilton, 2007: Tropical cyclone changes in the western North Pacific in a global warming scenario. *J. Climate*, **20**, 2378–2396.

Sugi, M., A. Noda, and N. Sato, 2002: Influence of global warming on tropical cyclone climatology: An experiment with the JMA global model. *J. Meteor. Soc. Japan*, **80**, 249–272.

Takayabu, Y. N. and T. Nitta, 1993: 3–5 day period disturbances coupled with convection over the tropical Pacific Ocean. *J. Meteor. Soc. Japan*, **71**, 221–246.

Tang, B. H. and J. D. Neelin, 2004: ENSO influence on Atlantic hurricanes via tropospheric warming. *Geophys. Res. Lett.*, **31**, L24204.

Targlione, C. A., S. R. Smith, and J. J. O'Brien, 2003: ENSO impact on hurricane landfall probabilities for the Caribbean. *J. Climate*, **16**, 2925–2931.

Thompson, D. W. J. and J. M. Wallace, 2000: Annular modes in the extratropical circulation. Part I: Month-to-month variability. *J. Climate*, **13**, 1000–1016.

Thorncroft, C. and I. Pytharoulis, 2001: A dynamical approach to seasonal prediction of Atlantic tropical cyclone activity. *Wea. Forecasting*, **16**, 725–734.

Trenberth, K. E., 1997: The definition of El Niño. *Bull. Amer. Meteor. Soc.*, **78**, 2771–2777.

Tsutsui, J., 2002: Implications of anthropogenic climate change for tropical cyclone activity: a case study with the NCAR CCM2. *J. Meteorol. Soc. Japan*, **80**, 45–65.

Tsutsui, J. I. and A. Kasahara, 1996: Simulated tropical cyclones using the National Center for Atmospheric Research community climate model. *J. Geophys. Res.*, **101**, 15013–15032.

Vallis, G. K. and E. P. Gerber, 2008: Local and hemispheric dynamics of the North Atlantic Oscillation, annular patterns and the zonal index. *Dyn. Atmos. Oceans*, **44**, 184–212.

Van Oldenborgh, G. J., G. Burges, and A. K. Tank, 2000: On the El Niño teleconnection to spring precipitation in Europe. *Int. J. Climatol.*,**20**, 565–574.

Vecchi, G. A. and T. R. Knutson, 2008: On estimates of historical North Atlantic tropical cyclone activity. J. Climate, **21**, 3580–3600.

Vimont, D. J, 2005: The contribution of the interannual ENSO cycle to the spatial pattern of decadal ENSO-like variability. *J. Climate*, **18**, 2080–2092.

Vimont, D. J. and J. P. Kossin, 2007: The Atlantic meridional mode and hurricane activity. *Geophys. Res. Lett.*, **34**, L07709.

Vincent, D. G., and A. H. Fink, 2001: Tropical cyclone environments over the northeastern and northwestern Pacific based on ERA-15 analyses. *Mon Wea. Rev.*, **129**, 1928–1948.

Vitart F., 2006: Seasonal forecasting of tropical storm frequency using a multi-model ensemble. *Q. J. R. Meteorol. Soc.*, **132**, 647–666.

Vitart, F., F. Doblas-Reyes, 2007: Impact of greenhouse gas concentrations on tropical storms in coupled seasonal forecasts. *Tellus*, **59A**, 417–427.

Vitart, F., J. L. Anderson, and W. F. Stern, 1997: Simulation of interannual variability of tropical storm frequency in an ensemble of GCM integrations. *J. Climate*, **10**, 745–760.

Vitart, F., M. R. Huddleston, M. Déqué, D. Peake, T. N. Palmer, T. N. Stockdale, M. K. Davey, S. Inenson, A. Weisheimer, 2007: Dynamically-based seasonal forecasts of Atlantic tropical storm activity issued in June by EUROSIP, *Geophys. Res. Lett.*, **34**, L16815.

Vitart F. D. and T. N. Stockdale, 2001: Seasonal forecasting of tropical storms using coupled GCM integrations. *Mon. Wea. Rev.*, **129**, 2521–2537.

Vitart F., D. Anderson, and T. Stockdale, 2003: Seasonal forecasting of tropical cyclone landfall over Mozambique. *J. Climate*, **16**, 3932–3945.

Wallace, J., 1973: General circulation of the tropical lower stratosphere. *Rev. Geophys.*, **11**, 191–222.

Walsh, K. J. E. and B. F. Ryan, 2000: Tropical cyclone intensity increase near Australia as a result of climate change. *J. Climate*, **13**, 3029–3036.

Walsh, K. J. E., K.-C. Nguyen, and J. L. McGregor, 2004: Fine-resolution regional climate model simulations of the impact of climate change on tropical cyclones near Australia. *Clim. Dyn.*, **22**, 47–56, DOI: 10.1007/s00382–003–0362–0

Walsh, K. J., M. Fiorino, C. W. Landsea, and K. L. McInnes, 2007: Objectively determined resolution-dependent threshold criteria for the detection of tropical cyclones in climate model and reanalyses. *J. Climate*, **20**, 2307–2314.

Wang, B. and J. C. L. Chan, 2002: How strong ENSO events affect tropical storm activity over the western North Pacific. *J. Climate*, **15**, 1643–1658.

Wang, B. and H. Rui, 1990: Synoptic climatology of transient tropical intraseasonal convection anomalies: 1975–1985. *Meteor. Atmos. Phys.*, **44**, 53–61.

Wang, H. J. and K. Fan, 2007. Relationship between the Antarctic Oscillation in the western North Pacific typhoon frequency. *Chinese Sci. Bull.*, **52**, 561–565.

Wang, H. J., J. Q. Sun, and K. Fan, 2007: Relationships between the North Pacific oscillation and the typhoon/hurricane frequencies. *Science in China Series D- Earth Sciences*, **50**, 1409–1416.

Watterson, I. G., J. L. Evans, and B. F. Ryan, 1995: Seasonal and inter-annual variability of tropical cyclogenesis: diagnostics from large-scale fields. *J. Climate*, **8**, 3052–3066.

Wheeler, M. C. and H. H. Hendon, 2004: An all-season real-time multivariate MJO index: Development of an index for monitoring and prediction. *Mon. Wea. Rev.*, **132**, 1917–1932.

Wheeler, M. C. and G. N. Kiladis, 1999. Convectively coupled equatorial waves: Analysis of clouds and temperature in the wavenumber-frequency domain. *J. Atmos. Sci.*, **56**, 374–399.

Whitney, L. D. and J. Hobgood, 1997: The relationship between sea surface temperature and maximum intensities of tropical cyclones in the eastern North Pacific. *J. Climate*, **10**, 2921–2930.

Wing, A. A., A. H. Sobel, and S. J. Camargo, 2007: The relationship between the potential and actual intensities of tropical cyclones on interannual time scales. *Geophys. Res. Lett.*, **34**, L08810.

Woolings, T., B. Hoskins, M. Blackburn, and P. Berrisford, 2008: A new Rossby wave-breaking interpretation of the North Atlantic Oscillation. *J. Atmos. Sci.*, **65**, 609–626.

Wu, G. and N.-C. Lau, 2002: A GCM simulation of the relationship between tropical-storm formation and ENSO. *Mon. Wea. Rev.*, **120**, 958–977.

Wu, L. and B. Wang, 2004: Assessing impacts of global warming on tropical cyclone tracks. *J. Climate*, **17**, 1686–1698.

Wu, M. C., W. L. Chang, and W. M. Leung, 2004: Impacts of El Niño-Southern Oscillation events on tropical cyclone landfalling activity in the western North Pacific. *J. Climate*, **17**, 1419–1428.

Wu, M.-C., K.-H. Yeung, and W.-L. Chang, 2006: Trends in western North Pacific tropical cyclone intensity,*EOS Trans. AGU*, **87**, 537–548.

Wu, P. and P.-S. Chu, 2007: Characteristics of tropical cyclone activity over the eastern North Pacific: The extremely active 1992 and the inactive 1977. *Tellus*, **59A**, 444–454.

Xie, L. and T. Yan, 2007: Western North Pacific typhoon track patterns and their potential connection to Tibetan Plateau snow cover. *Nat. Hazards*, **42**, 317–333.

Xie, L., L. J. Pietrafesa, and K. Wu, 2002: Interannual and decadal variability of landfalling tropical cyclones in the southeast coastal states of the United States. *Adv. Atmos. Sci.*, **19**, 677–686.

Xie, L., T. Yan, and L. J. Pietrafesa, 2005a: The effect of Atlantic sea surface temperature dipole mode on hurricanes: Implications for the 2004 Atlantic hurricane season. *Geophys. Res. Lett.*, 32, L03701.

Xie, L., T. Yan, L. J. Pietrafesa, T. Karl, and X. Xu, 2005b: Relationship between western North Pacific typhoon activity and Tibetan Plateau winter and spring snow cover. *Geophys. Res. Lett.*, **32**, L16703.

Xie, L., T. Yan, L. J. Pietrafesa, J. M. Morrison, and T. Karl, 2005c: Climatology and interannual variability of North Atlantic hurricane tracks. *J. Climate*, **18**, 5370–5381.

Xie, S.-P., 1999: A dynamic ocean-atmosphere model of the tropical Atlantic decadal variability. *J. Climate*, **12**, 64–70.

Xie, S.-P. and J. A. Carton, 2004: Tropical Atlantic variability: Patterns, mechanisms and impacts. *Earth's Climate: The ocean-atmosphere interaction.* C. Wang, S.-P. Xie, and J. A. Carton, Eds., AGU Press, 121–142.

Xie, S.-P., H. Annamalai, F. A. Schott, and J. P. McCreary, 2002: Structure and mechanisms of South Indian Ocean climate variability. *J. Climate*, **15**, 864–878.

Xie, S.-P. and Y. Tanimoto, 1998: A pan-Atlantic decadal climate oscillation. *Geophys. Res. Lett.*, **25**, 2185–2188.

Yoshimura, J. and M. Sugi, 2005: Tropical cyclone climatology in a high-resolution AGCM - impacts of SST warming and CO2 increase. *SOLA*, **1**, 133–136, DOI: 10.2151/sola.2005–035.

Yulaeva, E. and J. M. Wallace, 1994: The signature of ENSO in global temperature and precipitation fields derived from the microwave sounding unit. *J. Climate*, **7**, 1719–1736.

Yumoto, M., T. Matsuura, and S. Tizuka, 2003: Interdecadal variability of tropical cyclone frequency over the western North Pacific in a high-resolution atmosphere-ocean coupled GCM. *J. Meteor. Soc. Japan*, **81**, 1069–1086.

Zehnder, J. A., 1991: The interaction of planetary-scale tropical easterly waves with topography: A mechanism for the initiation of tropical cyclones. *J. Atmos. Sci.*, **48**, 1217–1230.

Zehnder, J. A., D. M. Powell, D. L. Ropp, 1999: The interaction of easterly waves, orography, and the intertropical convergence zone in the genesis of eastern Pacific tropical cyclones. *Mon. Wea. Rev.*, **127**, 1566–1585.

Zhang, C. 2005: Madden-Julian Oscillation, *Rev. Geophys.*, **43**, RG2003, doi:10.1029/2004RG000158.

Zhang, Y., J. M. Wallace, and D. S. Battisti, 1997: ENSO-like interdecadal variability: 1900–93. *J. Climate*, **10**, 1004–1020.

Zhao, X. and P.-S. Chu, 2006: Bayesian multiple changepoint analysis of hurricane activity in the eastern North Pacific: A Markov chain Monte Carlo approach. *J. Climate*, **19**, 564–578.

Zhou, B. T., X. Cui, and P. Zhao, 2008: Relationship between the Asian-Pacific oscillation and the tropical cyclone frequency in the western North Pacific. *Science in China, Series D- Earth Sciences*, **51**, 380–385.

Zuki, Z. M. and A. R. Lupo, 2008: Interannual variability of tropical cyclone activity in the southern South China Sea. *J. Geophys. Res.*, **113**, D06106.

PART V

Hydrological Aspects of Tropical Cyclones

Chapter 12

Storm Surge Modeling and Applications in Coastal Areas

Shishir K. Dube

Centre for Atmospheric Sciences,
Indian Institute of Technology, New Delhi, India
skdube@cas.iitd.ac.in

Tad S. Murty

Department of Civil Engineering, University of Ottawa, Ottawa, Canada

Jesse C. Feyen

Office of Coast Survey Development Laboratory, National Ocean Service,
National Oceanic and Atmospheric Administration,
Silver Spring, Maryland, USA

Reggina Cabrera

Eastern Region Headquarters, National Weather Service,
National Oceanic and Atmospheric Administration, Bohemia, New York, USA

Bruce A. Harper

Systems Engineering Australia Pty Ltd, Brisbane, Australia

Jerad D. Bales

North Carolina Water Science Center, U.S. Geological Survey,
Raleigh, North Carolina, USA

Saud Amer

International Water Resources Branch, U.S. Geological Survey,
Reston, Virginia, USA

This chapter introduces the reader to a wide spectrum of storm surge modeling systems used to assess the impact of tropical cyclones, covering a range of numerical methods, model domains, forcing and boundary conditions, and purposes. New technologies to obtain data such as deployment of temporary sensors and remote sensing practices to support modeling are also presented. Extensive storm surge modeling applications have been made with existing modeling systems and some of them are described in this chapter.

The authors recognize the importance of evaluating river-ocean interactions in coastal environments during tropical cyclones. Therefore, the coupling of hydraulic (riverine) and storm surge models is discussed. In addition, results from studies performed in the coast of India are shown which generated maps to help emergency managers and reduce risk due to coastal inundation.

1. Introduction

Storm surges resulting from tropical cyclones expose the population in coastal areas to flooding and destruction. Accurate and understandable forecasts and flood information products play an important role in preparing for and mitigating these events, including flood protection and evacuation. The purpose of this chapter is to discuss theory for modeling tropical cyclone storm surge and the application of river flow and storm surge data for modeling flooding conditions in coastal areas.

Although numerous models are available for forecasting storm surge water levels in coastal areas, many of the models still remain in the field of research. One of the primary distinctions between research and operational models is that operational models must perform under all conditions, which constitutes an ongoing testament to their robustness. In the operational framework[a], there needs to be assurance that forecast products will be available at all times for use by emergency managers. Conversely, research applications are able to explore areas of technological advancement that benefits both future operations (operational models can be continuously improved by research) as well as other significant needs. These include but are not limited to calculations of long term risk, emergency planning, and coastal management and engineering.

Even though the authors recognize the importance of precipitation forecast due to tropical cyclones, this topic is not discussed in this section. This section will address only storm surge models and their coupling with hydraulic models for the application of coastal inundation.

2. Storm Surge Simulation

As a tropical cyclone approaches land, its impact on the sea leads to a rise in water level at the coast that can be extremely severe, inundating vast low-lying areas. This *storm surge* is defined as the rise in water level caused by the surface wind stress, atmospheric pressure deficit, and momentum transfer from breaking surface wind waves (wave setup). The term *storm tide* is commonly used to describe the water level change caused by the combined effects of the *storm surge* and the astronomical tide (Hicks *et al.*, 2000, p. 25). Both the *storm surge* and *storm tide* refer to a long gravity wave (so called because its wavelength is much greater than depth) that is of similar size to the storm itself. Wind-generated short waves (with wavelengths smaller than depth), while possibly devastating, are considered a separate phenomenon because of their short time and spatial scale. Figure 1 illustrates the relevant contributions to the total water level during an extreme storm surge event, and further description follows.

In deep water, storm surge is minimized because the effects of the wind stress are distributed over depth and counter currents can be generated in deep water. The primary water surface elevation response in deep water is the atmospheric pressure deficit which causes the so-called "inverted barometer" effect; the rise is approximately 1 cm for each hPa drop in pressure (see, e.g., Pugh, 2004). However, as the storm passes over the shelf, reduced bathymetric depths lead to a set-up of the water surface due to the currents caused by the surface wind stress. A large shallow shelf can therefore enhance surge generation. Local dynamics become increasingly important as the surge approaches the coast. The bathymetry, coastline, topography, and inundation barriers such as levees and dunes are capable of modifying the surge. Surges of several meters are commonly generated by land-falling tropical cyclones, and extreme surges approaching 10 meters can occur under certain conditions. The storm's strength, size, track/angle of attack and forward speed effectively determines the surge at any one location.

Other processes can affect the total water level and resulting damage caused by land-falling tropical cyclones. First, wind-generated waves can be severe and significant. They

[a]The term "operational" in the context used in this chapter, means executed in real time.

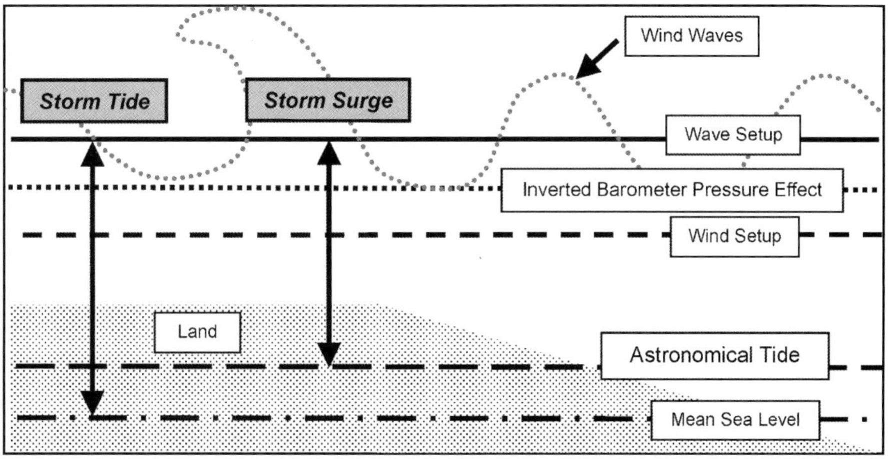

Figure 1. Contributions to the total water level during a severe storm surge event as it reaches land (gray shading). The individual effects vary significantly, depending on a variety of factors and this illustration cannot be construed as specific to any one case.

can contribute to the surge by a transfer of momentum, mainly when breaking, which is called wave set-up. They also are extremely destructive when impacting structures inundated by the surge. Second, flooding from rainfall can occur in conjunction with a storm surge. While there is generally a delay in the flooding caused by heavy tropical cyclone precipitation of a few days after landfall, saturated conditions and flooding caused by previous events can exacerbate inundation. Third, the total water level is also affected by seasonal and annual variations such as persistent (e.g., trade) winds and solar heating of the large ocean basins, which can cause a variation in water levels compared to the long-term mean.

Storm surge and the superimposed wind waves cause most of the damage from tropical cyclones at landfall. Extensive low-lying coastal areas (e.g., Louisiana, Bangladesh, and Myanmar) can be flooded, destroying communities, damaging property and infrastructure, and resulting in casualties. Furthermore, coastal development is rapidly increasing and population centers are growing (Crossett *et al.*, 2004). These factors are combining to raise the economic and social risk caused by severe storm surge. A clear example is the widespread impact that Hurricane Katrina had on coastal Louisiana and Mississippi, and on New Orleans in particular (Knabb *et al.*, 2006).

Considering the severe impact of storm surge, it is important to develop effective mitigation systems for supporting resilient coastal communities. This requires an analysis of the vulnerability due to storm surge activity at the coast. However, the frequency of severe storm activity at any particular location may be limited, hampering statistical analysis because of limited historical observations and precluding adequate in situ data acquisition (and thus effective data assimilation analysis strategies). Storm surge is so dependent upon the individual characteristics of each storm and geographical location that surge heights from a handful of historical hurricanes may not reliably predict future inundation risk. Predictive modeling capability is needed to study the potential threat due to storm surge as a function of meteorological forcing and geographic conditions.

A model capable of accurate storm surge simulation is important for minimizing risks to coastal communities. These simulations are needed to forecast potential flooding from an incoming hurricane, to develop and implement effective evacuation plans, to design flood protection and coastal management projects, and to assess surge risk for flood insurance programs. There are two distinct types of users of storm surge simulations: real-time responders and long-term planners. These users include emergency managers; meteorologists and forecasters; land-use, coastal zone, floodplain, and natural resource managers; engineers and public infrastructure managers; insurance companies; and commercial companies and non-governmental organizations. Their needs lead to a wide range of models and applications in use, and the most common are discussed below.

2.1. *Theory and Methodology*

The need for modeling storm surge and developing surge forecasting systems has led to a range of numerical modeling approaches utilizing an array of techniques. The predominant approach is to solve the two-dimensional (2-D), depth integrated form of the shallow water equations, which disregard baroclinic effects and variations in the velocity field over the vertical direction. The numerical techniques employed primarily include finite difference and finite element algorithms but other approaches have been used, such as finite volume schemes. The domains chosen tend to be regional, shelf based models, although some models use nested or large scale unstructured grids in order to apply far field boundary conditions (albeit with limited, if any, dynamic two-way coupling between domains). Bode and Hardy (1997) provides a review of developments in the field of storm surge modeling that identifies many approaches.

2.1.1. *Governing Equations*

Coastal ocean models are generally based on the shallow water equations (SWE). These governing equations are appropriate since they describe the dynamics in most free surface geophysical flow problems. The SWE are based on the principles of conservation of mass for an incompressible fluid, and the conservation of momentum as described by the Navier-Stokes equations subject to Reynold's averaging to describe the mean flow. Simplifications are made in the formulation of the SWE which lead to a more manageable set of equations: the hydrostatic approximation makes the vertical acceleration and associated viscous terms in the vertical momentum equation unimportant, the Boussinesq approximation allows gradients of fluid density to be ignored, and the shallow water assumption leads to a depth-integrated set of 2-D equations. The diffusion and dispersion of momentum due to unresolved lateral scales of motion as well as the effects of depth-averaging are accounted for by a turbulence closure model. The resulting SWE are applied to simulate ocean and estuarine circulation. When these equations are applied on a scale appropriate for geophysical flow problems, the Coriolis term is included to account for the rotational acceleration of the earth, and they can be solved in a spherical coordinate system (with horizontal dimensions of longitude and latitude) in order to more accurately account for the earth's curvature. For further detail on the derivation of the SWE see, for example, Pedlosky (1987), Cushman-Roisin (1994), or Weiyan (1992).

The SWE describe 2-D, depth-integrated circulation, which is a valid approach for modeling storm surge since the response due to hurricane forcing is of a relatively short time scale. This limits the generation of three-dimensional (3-D) circulation (namely, a return flow opposite to the surface gradient). Rather, storm surge propagates as a shallow water wave with height on the order of 10 meters and length on the scale of hundreds of kilometers. This satisfies the condition describing shallow water conditions: horizontal scales of flow are much larger than depth. When computations of vertical variations in currents are required a 3-D model can be used. However, 2-D models are much more economical and describe the dominant characteristics for storm surge simulations.

In order to improve model efficiency, nonlinear terms are omitted within some surge models. In particular, the nonlinear advective terms can pose a constraint on computational efficiency (namely the numerical time step size) of a surge model algorithm. (The nonlinear bottom friction and water surface elevation terms are more generally included.) It is in principle preferable to implement a fully nonlinear model, but linearizing the equations can be an acceptable choice if the compromise in accuracy is minor compared to the improvement in simulation cost. In practice the uncertainties inherent in the forcing, parameterizations, and grid design (including elevation specification) limit many gains in accuracy possible through use of a fully nonlinear model.

2.1.2. *Numerical Methods*

Computational models of the SWE have been in existence for over three decades, employing various numerical methods for simulating storm surge. The earliest numerical techniques employed finite difference (FD) schemes. The modeling community's experience with FD modeling of the Navier-Stokes equations led to its early application to the SWE. The resulting computational codes have proven to be efficient and robust and are widely applied. However, a disadvantage of FD methods comes from their use of structured grids. This hinders their ability to refine the complex coastal geometries and localized flow patterns near the coast and over floodplains. Because these features significantly affect surge height, accuracy can be affected by the grid resolution and orientation.

The application of the finite element (FE) method to the SWE has allowed use of unstructured meshes able to map complex geometry quite accurately. This produces domains that minimize computational effort while maximizing accuracy by varying nodal spacing over several orders of magnitude (Blain *et al.*, 1998; Westerink *et al.*, 1994). This allows for high resolution of coastal features such as inlets and barrier islands while using much larger node spacing further away in what may be very large grids. The method that proved to be most successful and widely used for FE modeling of the SWE is the wave equation formulation, due to its suppression of spurious oscillations (Lynch and Gray, 1979; Kinnmark, 1986). However, it has been established that for shallow, nonlinearly dominated flows there can be mass balance error, as the continuity equation is not being satisfied in the wave equation, but rather its time derivative (Walters and Carey, 1984; Kolar *et al.*, 1994). Additionally, this FE approach has been found to require very small time steps below the Courant condition, leading to costly simulations. When combined with the highly resolved grids that FE models generally apply, the computational resources required for FE storm surge simulations can be significantly more than FD models, limiting their use in time-constrained applications.

Recent research has been focused on combining the strengths of FD (mass conserving) and the strengths of FE (high resolution unstructured grid) methods. This has led to various forms of finite volume (FV) techniques, but use of these models has not yet been widespread although results have been reliable. The same cost issues regarding the size of highly resolved grids in FE modeling also apply to FV models as well.

Considering the cost of storm surge applications, which can rely on large, highly refined grids and time-consuming numerical schemes, state-of-the-art computational techniques are required to make these simulations feasible. For example, parallel processing techniques are being used to run models on distributed memory computers. With these approaches computational platforms among the most advanced available can be utilized, and wall clock times are reduced by a factor close to the number of processors.

2.1.3. *Grid Design*

Storm surge elevations can be impacted by complex local geometries, as flooding is controlled by geographic features. Therefore, simulations capturing these features improve storm surge prediction, and the corresponding model requires fine resolution. However, the size and resolution of the model domain must be balanced by consideration of the computational requirements. For example, forecasts are constrained by timeliness requirements, but pre- and post-storm investigations for engineering and risk determination purposes can benefit from more complex approaches that require greater computational resources.

FD storm surge models apply structured computational grids with regularly spaced discretization points. Curvilinear FD models provide fine resolution near the coastline and coarser resolution as the grid extends away from the coast. These grids can be cost efficient but if the boundary location is not extended far enough, these models may not correctly capture the dynamics of the basin-shelf interaction. However, nested modeling techniques can be employed which couple high resolution coastal grids with coarser larger scale domains. These larger basin effects can produce a set-up on the shelf which is not known *a priori* and therefore cannot be specified as a boundary condition. Additionally, structured grids can have difficulty sufficiently resolving irregular coastal features or may poorly represent them because of the regular structure of the grid. Flood amplitude and propagation are controlled by details such as the shoreline, inlets, and levees, so it is important to ensure the coast is resolved as accurately as possible. Therefore, some models link to subgrid-scale one-dimensional (1-D) models to represent flooding barriers, conveyance, or storage at a resolution higher than permitted by the native grid. Figure 2 shows an example of a structured orthogonal curvilinear mesh for the Northwest Florida Panhandle, including a close-up of the model resolution around Pensacola Bay. This model domain contains approximately 63,500 cells.

A storm surge model that uses an unstructured grid methodology can apply fine resolution of the coastal region along with far-field boundary conditions in the deep ocean. The paradigm recommended is a large domain/unstructured grid strategy that couples the ocean, shelf, and coastal floodplain while resolving local features. However, this methodology typically leads to costly model grids and it requires sufficient bathymetric and topographic detail to support its high resolution. Figure 3 shows a finite element mesh from a storm surge model of coastal Alabama and the Northwest Florida Panhandle. Detail of the model grid is shown for the inlet to Pensacola Bay, Florida, along with the view of the larger model domain and the increased resolution for the inundation region. This model grid has resolution of approximately 100 m at the coast but expands out to tens of kilometers in the deep basins, where the open ocean boundary conditions are applied. It has a total of approximately 450,000 nodes.

Key to any numerical storm surge simulation is the representation of the bathymetry, topography, and coastal features. The configuration of bathymetry (sea floor depth), its resolution, and its age/correctness is a critical factor in surge generation. Similarly, the topography (land elevation) of a coastal region is essential to correctly simulating the overland surge. This is because flooding depths over land are defined in part by the shape of topographic contours and features impacting inundation (e.g., levees, barrier island dunes, raised roadways). Waterways can propagate surge well inland, while concave coasts focus and amplify surge. Bathymetry and topography can change quickly due to both natural and anthropogenic causes, so it needs to be regularly updated. Considering the accuracy (errors reaching less than one-third of a meter) and resolution (reaching less than 100 meters) of widely used storm surge models (Jarvinen and Lawrence, 1985; Westerink *et al.*, 2008), data

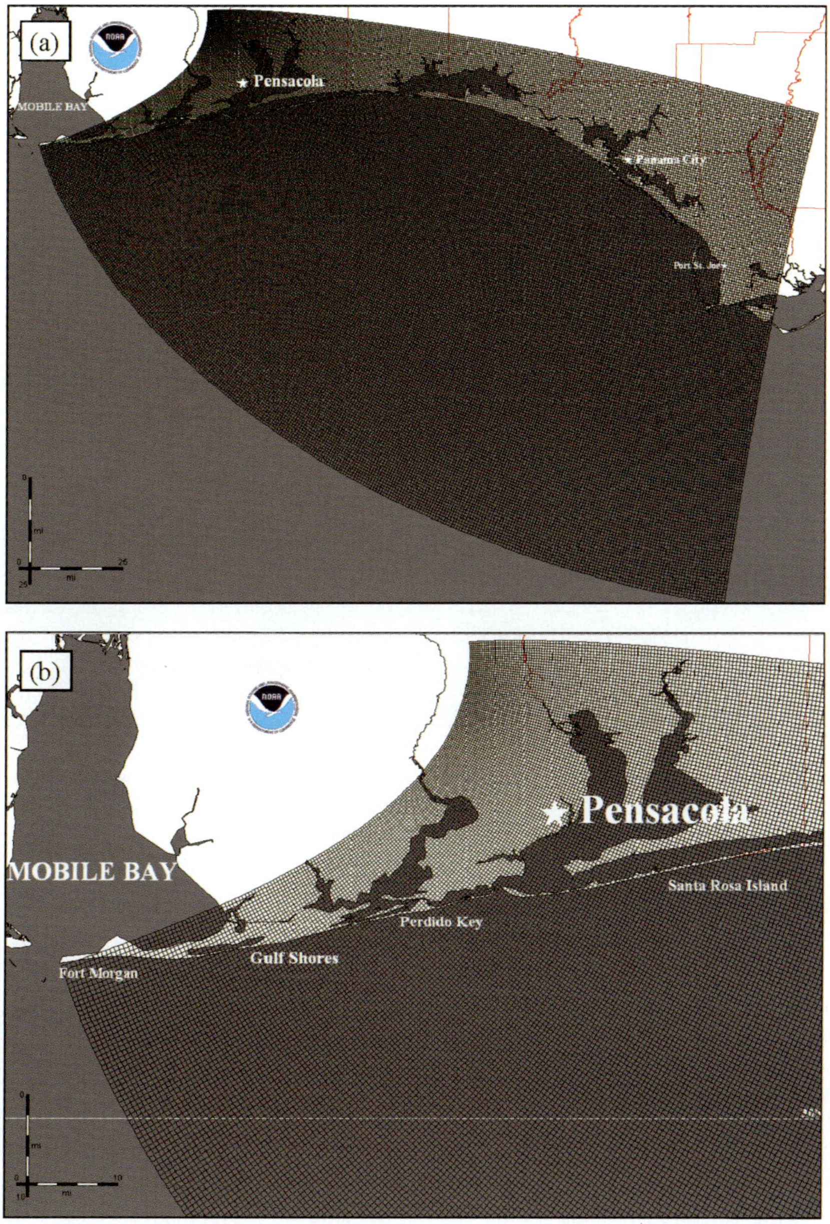

Figure 2. Finite difference storm surge model grid: (a) orthogonal curvilinear grid domain and (b) grid detail around Pensacola Bay. Figures provided by the US National Oceanic and Atmospheric Administration National Weather Service.

collection and processing need to meet or exceed these constraints. Many new remote sensing technologies are being employed for bathymetric and topographic data collection (e.g., sidescan and multibeam sonar, Light Detection and Ranging (LiDAR), interferometric radar). It is important that expensive data collection efforts include: necessary ties to vertical datums; accurate post-processing to produce bare-earth data sets; and appropriate collation into a useful Digital Elevation Model (DEM) or similar product.

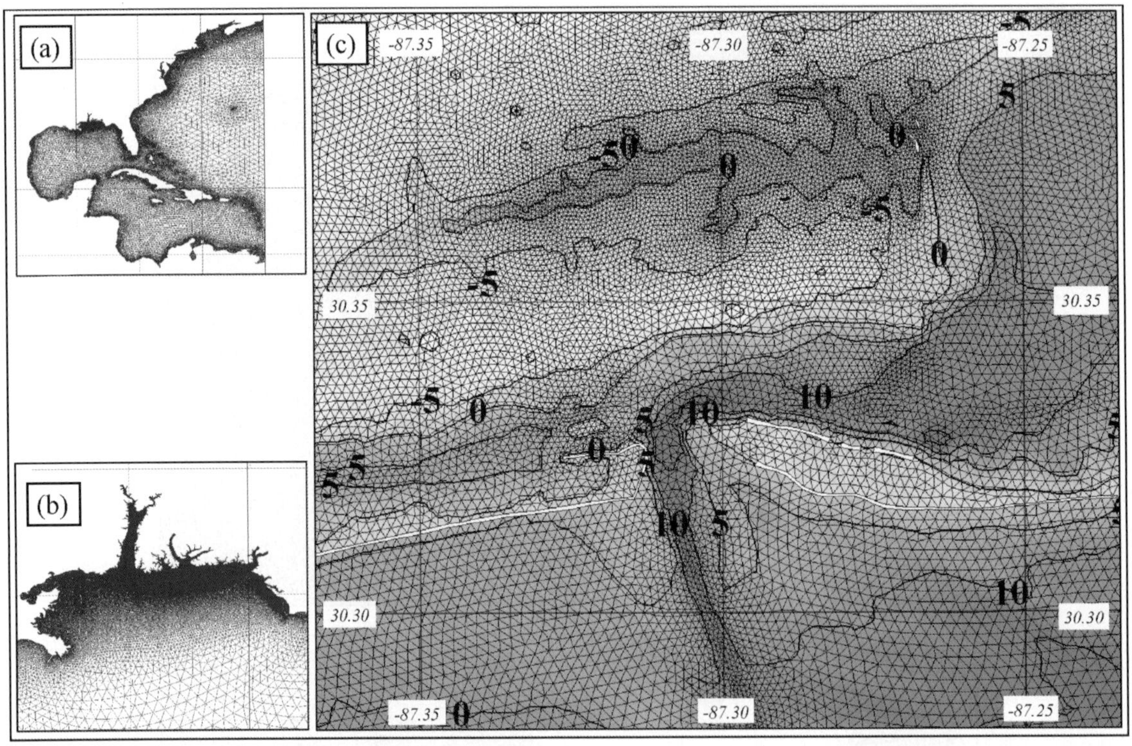

Figure 3. Finite element storm surge model grid: (a) unstructured grid domain, (b) high resolution grid in project region, and (c) model grid and bathymetry for the inlet to Pensacola Bay (m NAVD 88). Figure provided by the US National Oceanic and Atmospheric Administration National Ocean Service.

There are often a number of vertical datums used as a reference for elevation measurements. Bathymetry is generally referenced to a tidal datum, which is determined by computing the mean of a stage of an observed tide. For example, in the US, bathymetry is generally provided relative to Mean Lower Low Water (MLLW). Topography is usually referenced to a land-based geodetic datum, which is determined by surveying techniques from a national set of established benchmarks. In the US this geodetic datum standard is the North American Vertical Datum of 1988 (NAVD 88). While NAVD 88 is constrained to Mean Sea Level (MSL) at one location along the US/Canadian border, the difference between NAVD 88 and MSL varies along the US coast. Water level observations are generally collected relative to a tidal datum but peak surge values can be referenced to another datum. Finally, vertical datums also include 3-D ellipsoidal datums used by global navigation satellite systems. Storm surge simulations require consistent referencing of all data at the coast, where tidal, orthometric, and ellipsoidal datums are all used. Therefore, adjusting all datasets (bathymetry, topography, and water level observations) to a common vertical datum is required when building a surge model which describes propagation of water level from offshore to the coast and inland. This can be done by careful examination of data at locations where ties to multiple datums are provided, commonly at tidal gauges which have geodetic benchmarks. For example, the US National Oceanic and Atmospheric Administration (NOAA) is developing a vertical datum transformation tool called VDatum (Myers 2005) based upon these datasets to provide these adjustments for all of the US coastline (see http://vdatum.noaa.gov).

2.1.4. *Boundary Condition Specification*

The boundaries to a storm surge model have a significant influence on model response. For example, forcing at lateral boundaries (e.g., astronomic tides, river inflow, and inverted barometer pressure signal) requires accurate boundary conditions in order to generate the proper surge. Many models also employ specialized boundary conditions to characterize geographic features that can not be adequately modeled at the mesh resolution (e.g., levees, streams). Finally, storm surge simulation has the special case of a moving boundary at the wet/dry interface which must be handled in a way that correctly models inundation and recession.

Initial conditions specify the water level before the storm arrival. Models generally apply an elevation-specified boundary condition at the open ocean to define water level variations occurring outside the domain. Tidal constituent data from global ocean models or satellite altimetry can be applied if tidal forcing is desired. However, care must be taken to ensure that the constituents provided at the open ocean boundary are sufficient and accurate, since global databases are often not accurate in shallow water and many areas of the world lack adequate shallow water observations. Storm-induced set-up in the water level surface can be applied as well, if known. Also, the inverted barometer effect of the hurricane's low pressure system can be readily applied if known.

A model can define river inflows or a no-normal flow condition for land boundaries. However, a simple specified flux at river inflow boundaries can reflect surface gravity waves back into the domain, so use can be made of a wave radiation boundary condition that allows such waves to pass out of the domain.

Modeling inundation requires simulation of wetting and drying fronts. However, this is a challenge because the size of grid resolution (hundreds to thousands of meters) is coarse compared to features which locally impact flooding. Generally, the inundation front is described by allowing a region to flood if a sufficient volume of water can enter it from neighboring grid cells.

During hurricane flooding conditions, storm surge may overtop levees, dunes, road systems and similar features which are local barriers to flood propagation. However, these barriers generally fall below grid scale and represent a non-hydrostatic flow scenario. Therefore, it may be effective to treat these structures as subgrid scale objects within the domain, generally using boundary conditions which apply weir formulae. Similar techniques have also been applied for subgrid scale hydraulic features such as small streams and channels.

2.1.5. *Surface and Bottom Boundary Specification*

The specification of boundary conditions defines the forcing entering a storm surge model domain. However, the simulation also requires the stresses at its surface and bottom boundaries that guide the circulation. Due to the significance of the storm as a forcing mechanism, meteorological conditions provide the most significant contribution to model accuracy. Tropical cyclone wind and pressure fields input to the hydrodynamic model essentially control the generation of storm surge. A hurricane's track, strength, and size are critical factors which determine storm surge height. A shift in track location to either side of a river, for example, will significantly alter the impact that the storm has on that waterway.

The dominant forcing mechanism for storm surge is the surface wind stress (the inverted barometer and wave set-up effects generally contribute less than a meter of what may be a several meter surge). Therefore, accurate definition of the wind field (including its strength, size, and shape) as well as parameterization of the wind stress at the water surface critically affect accuracy. While an analysis of closely spaced observations can produce an accurate, objectively analyzed wind field (Powell *et al.*, 1998), this approach is not viable for forecast or hypothetical storm simulations, but only for hindcast scenarios. Therefore, meteorological tropical cyclone models are applied for these scenarios. Considering the importance of the wind field, the quality of the tropical cyclone model is very important to the surge prediction. Therefore, validation of the wind fields should be undertaken when possible, and probabilistic approaches may be employed during forecast scenarios to account for uncertainty in tropical cyclone predictions.

There are two primary classes of atmospheric models used to generate tropical cyclone wind and pressure fields. The first class solves the 3-D mathematical equations describing the dynamics of the atmosphere, and they are therefore called dynamical models. Dynamical models benefit from their fully 3-D solution of the governing equations but are costly to run and require extensive initial and boundary conditions. This type of model is used to guide forecasts of a tropical cyclone's track (and also intensity from some models); the wind and pressure fields computed during these predictions can also be used to drive a storm surge model. Three predominant dynamical models used by US National Weather Service (NWS) are the Global Forecast System (GFS; Global Climate and Weather Modeling Branch 2003; see http://wwwt.emc.ncep.noaa.gov/gmb/STATS/html/model_changes.html for the latest version), the Geophysical Fluid Dynamics Laboratory hurricane model (GFDL; Bender *et al.*, 2007), and the Hurricane Weather Research and Forecasting model (HWRF; see http://www.emc.ncep.noaa.gov/HWRF/index.html). The GFS is a suite of global model products produced for numerical weather prediction at about 35 km resolution. It has an incomplete description of the tropical cyclone and therefore is only relied upon for track predictions. The GFDL and HWRF models both use high resolution nested grids (smaller than 10 km) to represent the storm's dynamics and can predict the storm's track and intensity variation.

The second class of atmospheric tropical cyclone models develops an idealized representation of the storm based on a few key input parameters to characterize its location, size, and intensity; these are referred to as parametric models. This type of model is effective when estimated storm conditions are desired despite a lack of specific data about the event. This makes them good candidates for simulating hypothetical storms or when dealing with forecast uncertainty (e.g., ensembles of predictions). Parametric wind models include the one incorporated with the Sea, Lake, and Overland Surges for Hurricanes (SLOSH) storm surge model (Jelesnianski *et al.*, 1992), the tropical cyclone model of the planetary boundary layer from Thompson and Cardone (1996), and the Holland model of the gradient winds in a storm (Holland, 1980), among others. Typical input parameters are the track (to define location and translational speed), peak wind speed or minimum central pressure (to characterize intensity), and the radius to maximum winds (to define size). While each parametric model approaches the calculation of the tropical cyclone wind and pressure fields from a different basis, the resulting wind and pressure fields tend to be simplified, fairly uniform approximations that do not account for the variations in structure that occur in individual storms. Parametric models also lack meteorological conditions outside the storm. However, they solve simplified equations and therefore are very efficient and are often used to run large numbers of hypothetical storms (e.g., for planning purposes or as ensembles of possible land-falling tracks).

Air-sea interaction governs storm surge generation caused by the wind field. The surface shear stress parameterization is particularly important since this is the predominant forcing mechanism for surge generation. Wind shear stress at the water surface causes momentum transfer from the hurricane wind field to the water column, causing a gradient in the elevation of sea surface (set-up). Use of a wind drag law dictates that shear stress is governed by the wind velocity and the wind drag coefficient. The drag coefficient is defined by empirical data collected to define the relationship between wind velocity and surface roughness. Examples of drag coefficient relationships include Garratt (1977) and Large and Pond (1981), although recent research (Powell *et al.*, 2003; Makin, 2004) indicates that the generally linear correlation these laws provide for drag as a function of wind speed may not apply as winds reach hurricane strength, where drag may level off. Finally, overland regions have increased roughness in comparison to marine conditions, reducing the wind speed and the surface shear stress. Overland roughness can vary from relatively low values for open terrain to high values in heavily vegetated and urban regions. However, drag coefficient relationships are developed based upon data collected under open marine conditions. Therefore, surge models which are applying parametric-type tropical cyclone models which assume open marine conditions often reduce wind speeds according to land type classification (Jelesnianski *et al.*, 1992; Westerink *et al.*, 2008).

In addition to wind shear stress, the action of wind-generated short gravity waves can affect the surge through several mechanisms and, in consequence, the total water level. First, waves directly impart moment to the water column. This action can be described by radiation stresses which are applied at the water surface, leading to a localized wave-generated set-up in areas where waves are breaking sharply. However, when wave breaking occurs gradually and energy is dissipated by bottom friction it does not produce a significant increase in water elevation. Second, the sea state is characterized by the condition of the wind waves (e.g., height and steepness), which depends on their age. Changes in sea state therefore affect the roughness of the ocean surface and thus the drag coefficient. Modeling efforts which vary the drag coefficient dependent upon wave conditions improve momentum transfer from wind to water (e.g., Mastenbroek *et al.*, 1993). Third, waves affect local bottom friction effects, which impacts hydrodynamic circulation. Therefore, provided that good quality nearshore land and seabed elevation data is available, surge simulations might be improved when combined with wave models that describe the wave field.

Accurate representation of bottom shear stress is also important in governing the propagation of the surge inland. In 2D models the bottom shear stress is generally defined as a quadratic function of the depth-averaged current and a coefficient of friction. Bottom shear stress parameterization is especially significant during flooding processes due to shallow water depths and complex variations in topographic roughness (e.g., from sandy beaches and parking lots to grassy fields and wooded areas). While there can be a lack of empirical data for characterizing bottom friction under this range of conditions, many models do allow variable friction coefficients spatially to account for land cover type, such as by applying varying Manning's n values (e.g., see Chow, 1959). One advantage of 3D models is their ability to describe the bottom boundary layer according to local roughness heights, which can be more accurate than parameterizations used in the 2D equations.

2.1.6. *Calibration and Validation*

Hindcasts of recorded historical tropical cyclones are used to validate and calibrate surge model implementations. Ideally large datasets from multiple storms are used so that calibration and validation can be conducted independently. Considering the model grid's important role in representing coastal geometry, this process needs to be done separately for each model domain.

In order to validate storm surge models, observations of water level variations caused by tropical cyclones are required. Most valuable are time series observations of water level, generally from a preexisting water level gauge. It is important that the datums for these gauges be accurately tied into a common, documented datum such as a national network of benchmarks so that adjustments can be made to match the model's datum. The design and location of a water level gauge is normally intended to filter out short gravity wind waves, leaving the remaining storm tide signal. Unfortunately, water level time series observations are not widely available due to the cost of installing and maintaining gauge sites. Additionally, many gauges subject to the strong storms are destroyed, eliminating data from severe events.

In order to expand the suite of data available to characterize the impact of the storm, many times a post-event survey of peak storm tide levels is undertaken. These surveys collect High Water Marks (HWMs) which are generally water or debris lines on structures or similar features that represent the peak storm tide. However, these observations often include not only the storm tide but also wave run-up, flooding from precipitation, and other local variability. Therefore confidence is increased by collecting a HWM from a building interior and by ensuring consistency with marks in close proximity. Still, their values may vary by 20 percent or more against nearby marks (Jelesnianski *et al.*, 1984; Harper *et al.*, 2001b), and one study has found that only 15 percent of marks were high quality indicators of peak storm tide (Ebersole *et al.*, 2007).

Once a storm surge model has been developed for a particular geographical region and a hindcast has been made of a historical storm which has good observations of the storm tide, a skill assessment of the model can be performed. This is necessary to validate the model for use. The correlation between the modeled and observed peak surge can be calculated to evaluate model skill. State-of-the-art storm surge model simulations have been found to produce mean error measures of one half of a meter and less (Westerink *et al.*, 2008).

3. Storm Surge Systems

The storm surge community continues to develop and apply systems for a variety of purposes. These include providing local high resolution predictions, improving modeling skill, examining location-specific environmental and management issues, and more. Descriptions of some prominent storm surge modeling systems are provided in section 3.1 below to illustrate the types of efforts underway.

3.1. *Models in Use*

There is a wide range of storm surge modeling systems in use for predicting the impact of tropical cyclones, covering a range of numerical methods, model domains, forcing and boundary conditions, and purposes. However, in the US there are two primary surge models: the Sea, Lake, and Overland Surges for Hurricanes (SLOSH) model and the Advanced Circulation (ADCIRC) model. While sometimes used for similar tasks, they are quite different in approach, have different strengths and weaknesses, and each is likely more well-suited for a particular application. Of course, there are many other models in use, and a survey of some notable approaches is provided.

3.1.1. *SLOSH*

The SLOSH storm surge model was developed by the US NWS (Jarvinen and Lawrence, 1985; Jelesnianski *et al.*, 1992). It is based on a linearized form of the governing SWE; the advective terms in the equations are discarded. The SLOSH model is a FD model that employs an orthogonal curvilinear grid. SLOSH grids for all vulnerable areas of the US coast have been constructed by the NWS for use in emergency management (http://www.weather.gov/tdl/marine/Basin.htm). The structure of a SLOSH grid results in fine resolution near the pole of the grid (which is placed near the area of interest) and larger elements at its outer boundary. These grids are limited in domain size and have their open ocean boundaries located on the shelf. SLOSH has the capability to simulate wetting and drying as well as parameterize sub-grid scale features such as 1-D channel flow with contractions and expansions, vertical obstructions to flow with overtopping (levees, roads, and banks that include cuts) and increased friction drag in heavily vegetated areas. However, it is not able to account for inflow, rainfall, or tides.

The SLOSH model contains an internal parametric tropical cyclone wind model used for forcing both forecast and hypothetical storm simulations. The tropical cyclone track, central pressure deficit, and radius of maximum winds are used as input to create wind and pressure fields which are interpolated onto the grid. A wind drag coefficient that is constant with respect to wind speed is applied, although it changes according to vegetation over land.

SLOSH has an extensive history of use by the U.S. National Weather Service (NWS) over nearly four decades. An extensive skill assessment found peak surge errors to be less than 0.6 m for nearly 80 percent of evaluations (Jarvinen and Lawrence, 1985). However, while it is a capable model, it does have limitations. First, at present it cannot provide astronomical tide forcing or river inflow. Second, it is a linearized model subject to error in locations where advection is important, such as in the vicinity of tidal inlets and similar constrictions. Finally, a SLOSH grid is generally limited in size to the coastal shelf surrounding the study area. Two issues arise from this limited domain strategy. First, the use of a structured grid limits the capability to provide localized refinement. While SLOSH grids do have increased refinement at their center, it is not always possible to resolve additional features that may be elsewhere (e.g., an inlet along the coastline that is away from the center of the grid). Conversely, the semi-annular structure of the grid also leads to over-resolution in regions outside of the area of interest (e.g., inland beyond the floodplain). Second, the shelf-based, regional nature of a SLOSH domain limits accurate specification of boundary conditions during storm surge events because of lack of knowledge of set-up at the open boundary, and prevents dynamic coupling to larger basins.

In the US, flood-prone areas are usually determined by using the SLOSH model and input parameters from thousands of hypothetical hurricanes. These model runs are used to create atlases of potential surge, which guides emergency managers in creating evacuation plans (Jarvinen and Lawrence, 1985). The US NWS's National Hurricane Center (NHC) also runs SLOSH for prediction of storm surge from potentially land-falling tropical cyclones. The storm's parameters (track, size, and intensity) are provided by the official hurricane forecast. These forecasts are distributed in real time to the forecast and emergency management community as Geographic Information System (GIS) products. They begin when a storm is within 24 hours of landfall due to the limitations of storm forecast accuracy; therefore they are only used to augment the pre-computed atlases of surge potential, which is the basis for emergency management activities. A probabilistic component based on the official NHC track forecast has been developed which forecasts the likelihood of surge as a function of the historical range of track error; see http://www.weather.gov/mdl/psurge/.

3.1.2. *ADCIRC*

The ADCIRC model was originally developed for the US Army Corps of Engineers for high resolution coastal ocean modeling (Luettich *et al.*, 1992). It has since been used by federal agencies, academic researchers, and private companies for a wide range of modeling applications, include storm surge, basin-scale tidal modeling, and tidal inlet circulation studies. ADCIRC solves the SWE discretized in space using the FE method, which allows for highly flexible, unstructured grids. All non-linear terms have been retained in these equations. ADCIRC can be run either as a 2-D depth-integrated (2DDI) model or as a 3-D model. In either case, elevation is obtained from the solution of the depth-integrated continuity equation in Generalized Wave-Continuity Equation (GWCE) form. The GWCE is implemented to prevent generation of spurious oscillations; however, it replaces the continuity equation with its time derivative, meaning continuity is not satisfied and mass balance errors exist. Velocity is obtained from the solution of either the 2DDI or 3D momentum equations. ADCIRC requires input of wind and pressure fields to consider the effect of storm surge. These wind and pressure fields are developed by meteorological models independent of ADCIRC. For further description of the model implementation see the ADCIRC theory report (Luettich and Westerink, 2004).

In order to accurately simulate storm surge, several features have been included within ADCIRC. It can model the wetting and drying of inundated areas, it can represent subgrid scale obstructions to flow as weirs, and it can apply the transfer of momentum from breaking wind waves as surface radiation stresses. Most importantly, its unstructured grid methodology allows for very high refinement of coastal regions (successfully modeled at scales less than 50 meters). These highly refined coastal regions are built into basin scale domains by smoothly varying element size. With the high resolution of inundation regions and the small time steps required by the semi-explicit solution scheme comes high computational cost, however. The model can be run on a single processor, but most often is run in parallel on high performance computing systems with hundreds of processors using the Message Passing Interface (MPI).

ADCIRC is presently in wide use in the US by federal agencies for storm surge modeling. Following 2005's Hurricane Katrina, the US Army Corps of Engineers partnered with other federal agencies, academic, and private institutions to extend an existing ADCIRC storm surge application for Southern Louisiana (Westerink *et al.*, 2008). The extension of the model built for studying the hurricane protection system for Louisiana was to conduct an extremely large and highly refined forensic study of Katrina's disastrous storm surge. As part of the Interagency Performance Evaluation Task Force (https://ipet.wes.army.mil/), this project applies very high resolutions of 50 to 100 m across coastal Louisiana and Mississippi, resulting in grids exceeding one million nodes. This model was adopted by FEMA to produce flood insurance rate maps for the Gulf Coast. Furthermore, both NOAA and the US Navy have been applying ADCIRC as a storm surge model for specific research and development projects.

3.1.3. *Survey of Other Approaches*

In addition to the two predominant storm surge models in the US (SLOSH and ADCIRC), a number of other modeling systems have been developed and applied worldwide. A few of the predominant systems are presented here as an overview in order to illustrate how storm surge simulations are being improved and the range of issues being addressed.

One of the predominant FD coastal ocean models is the Princeton Ocean Model (POM; Blumberg and Mellor, 1987). POM is a 3-D, fully nonlinear hydrodynamic model that has been in development and use by a wide range of user over the past three decades. It has been applied for use in tropical storm surge modeling in North Carolina (Xie *et al.*, 2003; Peng *et al.*, 2004, and Pietrefesa *et al.*, 2007). In this application it has been coupled with meteorological (wind, pressure, and precipitation), wave, and river discharge models in order to provide a comprehensive description of a storm's impacts.

Another well-developed FD model that has been applied to storm surge is the Curvilinear-grid Hydrodynamics model in 3D (CH3D; Sheng, 1986; Sheng and Alymov, 2002). It is a parallelized model capable of running large grids using MPI on high performance computing systems. CH3D uses a structured curvilinear grid and has been applied on domains with resolution reaching less than 20 meters. It is a fully nonlinear model that has been coupled with wave models and has been coupled with larger scale circulation models. Various wind models have been used to provide input.

In Australia, the FD MMUSURGE model developed at James Cook University has been widely applied along the north-eastern coastline and especially optimized for the complex Great Barrier Reef environment using subgrid boundary controls (e.g., Bode and Mason, 1994; Bode *et al.*, 1997; Harper *et al.*, 2001b (Appendix D), Harper *et al.*, 2007). Also, the commercially available TUFLOW hydrodynamic model (www.tuflow.com) is well suited to linked 2D and 1D applications and has been successfully used for many storm surge modeling studies.

Another FD SWE model has been developed in Australia for tropical storm surge modeling. The Global Environmental Modeling Systems 2D Coastal Ocean Model (GCOM2D; McInnes *et al.*, 2002) solves for depth-averaged flow and includes overland flooding and runoff as well as wave setup effects through coupling to atmospheric and wave models. In applications, the researchers have employed a coarse (1 km) far-field model and coupled that to a finely resolved (100 m) local model using a nested grid strategy.

The Chesapeake Inundation Prediction System (CIPS) focuses on the forecast needs of emergency managers and first responders (Stamey *et al.*, 2007). CIPS is a combined surge-tide-river discharge flood prediction system for the Potomac River near Washington, DC. It applies the mass-conserving unstructured model ELCIRC (Eulerian-Lagrangian Circulation; Zhang *et al.*, 2004), which is based upon a low order finite volume technique. While the domain is of limited extent (the Potomac River), it is highly resolved and includes inland regions at scales of 50 to 100 meters. CIPS combines the storm surge model with a high resolution (4 km) atmospheric prediction system and river discharge forecasts from NOAA in order to provide a comprehensive characterization of the flooding threat for a coastal riverine system. Model output is provided in GIS formats that allow for mapping and visualization applications in conjunction with other datasets.

Two additional FV coastal ocean models have been used for tropical cyclone surge modeling applications. The fully nonlinear 3-D model FVCOM has been applied to storm surge simulations in Florida (Weisberg and Zheng 2006). This modeling application demonstrates a mass-conserving 3-D simulation on an unstructured grid extending past the continental shelf.

Several projects have focused on approaches that leverage models available in the community. The Southeastern Universities Research Association's Coastal Ocean Observing and Prediction (SCOOP) project is a community approach employing Grid computing techniques that leverage resources over a wide range of locales (Bogden *et al.*, 2007). By combining a suite of models available in the community (including ADCIRC, ELCIRC, CH3D, and wave models) the system is not tied to one model but rather is an infrastructure for enabling model predictions. It does this by providing streamlined access to model inputs and outputs over a high speed network in standardized formats. Similarly, a project funded by the National Oceanographic Partnership Program (NOPP) combines readily available models for surge (ADCIRC), waves (the Wave Analysis Model (WAM)), and tropical cyclones (NOAA's real-time analysis system H*WIND and Oceanweather's Interactive Objective Kinetic Analysis (IOKA)) for simulations of events (Graber *et al.*, 2006). By constructing an ensemble of forecast tracks this networked system of models is used as a probabilistic forecast tool.

Finally, commercial hydrodynamic modeling systems by the Danish Hydraulic Institute (DHI) and Deltares (Delft Hydraulics) are widely used in storm surge risk assessment studies around the world. For example, DHI's MIKE 21 model has a FV version that has been used to simulate storm surges for the coast of Bangladesh (Paudyal, 2002).

3.2. *Storm Surge Measurements*

3.2.1. *Storm Surge Measurement on Land*

Storm surge height typically is based on high water marks (see section 2.1.6) or eye-witness accounts. This type of information can be useful, and combined with information such as debris piles, and structural and vegetation damage can provide estimates of the inland extent of the storm surge. These somewhat qualitative records, however, do not provide information about the timing of the surge and can be subject to major errors. In 2005, the US Geological Survey (USGS) initiated a program to collect more detailed storm surge information. The program consists of deploying a network of unvented pressure transducers in advance of cyclone landfall along the predicted path of the storm. Some transducers are positioned well above the expected storm surge height so that barometric pressure can be measured and used to correct the measurements from the unvented transducers used to measure water-surface elevation. These transducers then record the timing, spatial extent, and magnitude of storm surge over the affected region. The first deployment was along the coasts of southwest Louisiana and southeast Texas, USA (Fig. 4) in anticipation of the landfall of Hurricane Rita in September 2005 (Fig. 5). Subsequent deployments were made along the east coast of Hurricane Eduardo, but storm surge was minimal during that event.

A total of 34 storm surge sites were deployed for Hurricane Rita. Sensor locations ranged from a few meters from the coast to about 50 km inland. An additional 13 sensors were deployed to measure barometric pressure. Data were recorded at 30-second intervals and are available from McGee *et al.* (2006b). This unique network provides some of the most comprehensive storm surge data yet available. Time-series of storm surge elevations are available for the almost 2-day period when the coast was inundated, and maps showing inundation at 3-hour intervals have been prepared

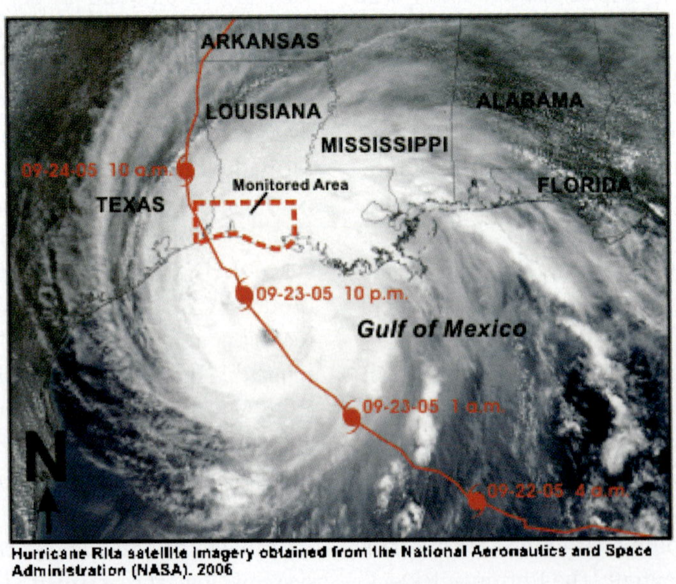

Figure 4. Path of Hurricane Rita, September 22–24, 2005; satellite imagery obtained from National Aeronautics and Space Administration (2006); from McGee *et al.*, 2006a.

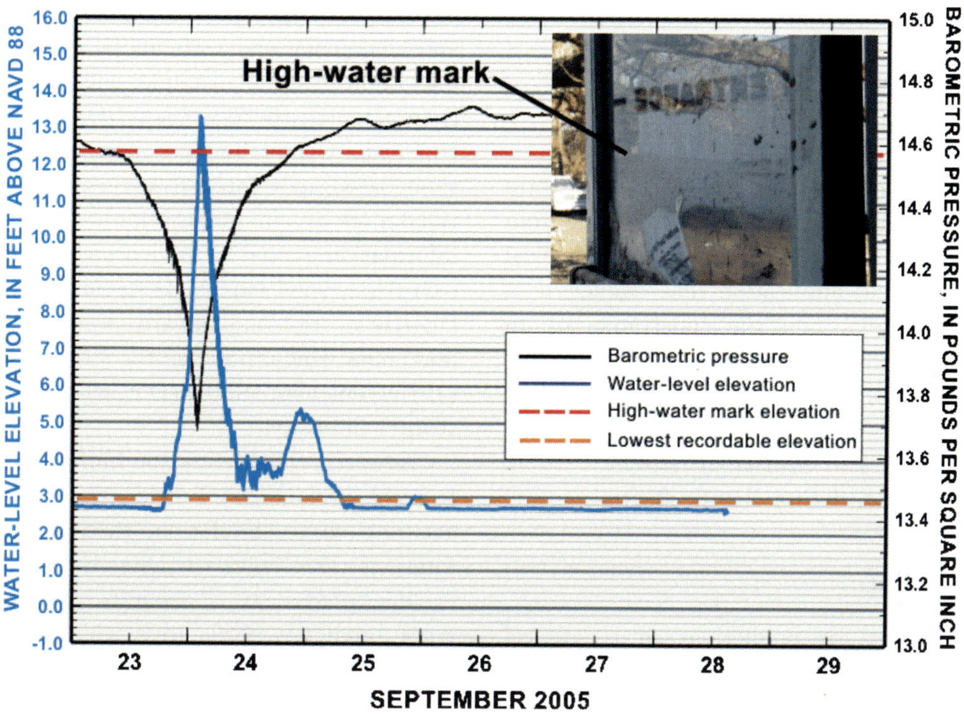

Figure 5. Example of Hurricane Rita storm surge data, along with a well-established high-water mark (from McGee *et al.*, 2006b).

(Berenbrock and Mason, 2008). One key finding from the data network was that high-water marks may underestimate storm surge heights by as much as 0.5 m, perhaps because the high-water marks are established as water is retreating to the ocean rather than during the initial surge.

3.2.2. *Remote Sensing Techniques*

For more than a decade, radar technology on board orbital satellites have been utilized to track, measure and monitor the ocean surface, including the wind, waves, and eyes of damaging hurricanes. Satellite altimeters provide indirect measures of ocean surface currents through monitoring sea level anomalies. The currents are then interpreted to generate a topographic map of the sea surface, which relates to wave height and direction. Ocean surface wave measurements were made from airborne Scanning Radar Altimeter (SRA) in tropical cyclones in 1998. At this time, wave heights, wave variations and direction of motion around the tropical cyclones were measured and provided near real-time estimates of wave height and storm surge at landfall (http://www.aoml.noaa.gov/hrd/hrd_sub/milestones.html).

Researchers from both the hydrodynamic and the remote sensing communities should work closely to have remotely sensed derived data not only for wave characteristics as an input into coupled model systems but also in support of determination for storm surge. The remotely sensed data can be used to fill gaps in the absence of sufficient in-situ data while hydrodynamic models outputs could be used to validate remotely sensed derived storm surge products. The new satellites missions, such as the Advanced Land Observing Satellite (ALOS), Sentinels, and the RADARSAT, in addition to their primary missions, are designed to address ocean tidal characteristics based on some of the concerns from researches (Johannessen *et al.*, 2006).

4. Storm Surge System Applications

A properly implemented, well-validated storm surge model can be used to simulate surge from historical, predicted, or hypothetical tropical cyclones. Each application has value in a different context, and the constraints of an individual project can define which modeling approach will be the most appropriate.

4.1. *Review and Status of Storm Surge Modeling Applications in the US*

4.1.1. *Forecasts*

It is common for predictions of a tropical cyclone to be made in order to provide information needed to implement emergency management measures such as evacuations. These predictions can be combined with hydrodynamic models to forecast storm surge. However, tropical cyclone storm surge is very sensitive to the bathymetry and topography of the ultimate landfall location. Equally important are the internal characteristics of the storms themselves — especially intensity and structure. Regardless of the model(s) chosen for calculating storm surge, their operational use presently is limited by the quality of the meteorological input. Better meteorological forecasts will result in better storm surge calculations. For example, cyclone landfall accuracy better than 20 miles at 36 to 48 hours in advance, along with intensity forecasts to within 10 mb central pressure deficit are needed to provide accurately useful storm surge calculations for evacuation purposes. Therefore, the most promising avenue for operational storm surge calculations useful for evacuations lies in improvements in meteorology.

Many national meteorological service agencies have implemented operational surge prediction systems for vulnerable regions. These model implementations provide guidance to forecasters and emergency managers about possible flooding conditions. For example, when a hurricane is threatening the US coast, NOAA's National Hurricane Center initiates storm surge predictions based upon forecast advisories it provides (Glahn *et al.*, in preparation). The following information is used to drive a parametric tropical cyclone wind model: (1) storm track to characterize location and forward speed, (2) radius of maximum winds to define the storm's size, and (3) the central pressure deficit, which defines the storm's intensity.

However, considering the importance of the tropical cyclone wind field in producing an accurate surge simulation, these storm surge forecasts are only valuable if the tropical cyclone prediction is accurate. The error in track prediction leads to a wide variation in distribution of wind speeds and thus possible surge predictions depending on where a storm makes landfall (Ginis, 1995). Similarly a tropical cyclone forecast of landfall 24 hours in advance has an error in track location that exceeds 113 km two-thirds of the time (National Hurricane Center, 2007). Considering a typical storm forward speed of 25 km/h, the time of landfall has an uncertainty of 4.5 h. Astronomical tidal conditions cannot be accurately predicted for landfall because of this uncertainty in timing. This makes it difficult to sequence surge simulations with astronomical tide forcing (Jelesnianski *et al.*, 1992). Therefore, it is difficult to combine surge and astronomical tide predictions in a forecast simulation. Instead, the tide level can be added linearly to the peak surge prediction once more confidence in land-falling time is gained; this approach is taken by US National Hurricane Center forecasts (Jelesniaski *et al.*, 1992). Finally, predictions of tropical cyclone intensity can be less reliable than track predictions, also enhancing surge forecast errors.

Emergency managers and first-responders still find forecasts of storm surge valuable, despite the uncertainty. One strategy under development for dealing with the inherent errors in tropical cyclone predictions is to implement a probabilistic forecast of surge based upon an ensemble of simulations (e.g., Graber *et al.*, 2006). A suite of possible tracks can be constructed based upon knowledge of the range of tropical cyclone prediction errors. These hypothetical tracks can vary the location, speed, size, and intensity of the predicted storm. This suite of possible storm conditions can then be fed into a storm surge model which produces the corresponding suite of results. From this output the distribution of risk can be inferred along the coast. One difficulty in this approach is that now instead of needing sufficient resources to complete one surge simulation; computational requirements increase several-fold so that all runs can be finished in a timely manner. However, an Australian forecasting system has implemented a hybrid approach that uses dynamic models to inform parametric probabilistic models that allow for variation of a wide variety of variables, such as track, size, and intensity (Systems Engineering Australia, 2005; Harper *et al.*, 2007). This method provides rapid computation and is well suited to regions where more sophisticated dynamical modeling is either unaffordable, impractical or the lack of accurate data (atmospheric, bathymetric or topographic) provides little improvement. Another challenge is communicating information about the resulting distribution of risk in a manner that can be easily understood and acted upon by emergency managers.

4.1.2. *Planning and Management*

Although forecasts of potential surge heights may be valuable in real time, emergency management plans such as evacuation routes must be in place well before the storm approaches in order to be effective. Therefore, characterization of the vulnerability from future storms is needed to guide these types of planning and management activities.

By generating a suite of possible tropical cyclone parameters that could affect a region, an atlas of possible storm surges can be produced from a storm surge model to guide planning. By varying important parameters such as track, size, and intensity, a complete set of hypothetical storms can be created. The output from simulations of these storms can then be compiled into useful tools that describe the distribution of risk across a region. In the US, NOAA produces maps of maximum surges generated by a specific class of tropical cyclones (e.g., Category 3 storms approaching from the south) for a specific impact zone. This map is called the Maximum Envelope of Waters (MEOW). A set of maps can then be referred to when developing evacuation plans or when a storm of similar characteristics is approaching land. For more detailed discussion of the implementation of this methodology, see Jelesnianski *et al.*, (1992).

While an atlas of possible storm surges is a useful planning tool for emergency management, it does not provide information about the probability of surge for each region. By combining a suite of potential storm conditions with statistical information about the likelihood of occurrence for these events, an understanding of flooding risk can be determined for coastal regions. This information can then be used to define insurance rates and building codes. In the US, the Federal Emergency Management Agency (FEMA) applies storm surge models to construct flood insurance rate maps according to the recurrence interval of potential tropical cyclones (Massey *et al.*, 2007).

4.1.3. *Coastal Engineering*

Storm surge simulations are needed for engineering design projects to predict the impacts of future storms. One significant example is in the design of hurricane protection systems, which need to simulate not only the potential surge height but also how it will be changed by proposed design projects. Along the US Gulf Coast many design projects are underway to restore barrier islands, dunes, and beaches as well as improve levee systems in order to reduce storm surge impacts in highly populated regions (e.g., see http://lacpr.usace.army.mil/). Considering the detail involved in designing these public works, it is common for very high resolution models to be utilized. Similarly, building standards are generally defined by the risk that exists from natural hazards within local areas. By computing the surge risk from a stochastically chosen set of storms, the potential likelihood of surge occurrence can guide implementation of building standards.

4.2. *Review of Recent Developments in Predicting the Storm Surges in the Bay of Bengal and Arabian Sea*

Storm surges are extremely serious hazards along the east coast of India, Bangladesh, Myanmar, and Sri Lanka. Although Sri Lanka is affected only occasionally by storm surge, tropical cyclones of November 1964, November 1978 and November 1992 caused extensive loss of life and property damage in the region. Storm surges affecting Myanmar appear to be much less intense than those occurring in India and Bangladesh. Notable storm surges, which have affected Myanmar, occurred during May 1967, 1968, 1970, 1975, 1982, 1992, 1994, and 2008. Of these, Nargis 2008 was the worst cyclone. Nargis generated storm surge in excess of 4 m near Ayeyarwady deltaic region. The entire deltaic coast of Myanmar was flooded with surges ranging from 1.5–4.5 m.

Of all the countries surrounding the Bay of Bengal, Bangladesh suffers most from storm surges. The main factors contributing to disastrous surges in Bangladesh are: (a) shallow coastal water, (b) geographical convergence of the bay, (c) large astronomical tides, (d) densely-populated lowlying islands, (e) favorable cyclone track, and (f) innumerable number of inlets including those in the world's largest river system (Ganga-Brahmaputra-Meghna). Detailed reviews of the problem of storm surges in the Bay of Bengal are given by Ali (1979), Rao (1982), Roy (1984), Murty (1984), Murty *et al.* (1986), Das (1994a,b), Dube *et al.* (1997, 1999, 2000a, 2006, 2009), Rao *et al.* (1997), Chittibabu (1999), and Gönnert *et al.* (2001),

Although storm surges are less frequent in the Arabian Sea than in the Bay of Bengal, major destructive surges occasionally have occurred along the Gujarat Coast of India and Pakistan. Storm surge only infrequently affects the Gulf of Oman, but there have been some cases when storm surge caused destruction along the coasts of Oman. The most recent example was the super cyclonic storm Gonu that struck Oman and caused about $4 billion in damages and 49 deaths (JTWC, 2007).

4.2.1. *Real Time Storm Surge Predictions System for the Bay of Bengal and the Arabian Sea*

In India, the study of numerical storm surge prediction was pioneered by Das (1972). Subsequently several workers attempted the prediction of storm surges in the Bay of Bengal (Das *et al.*, 1974; Ghosh, 1977; Johna and Ali, 1980; Johns *et al.*, 1981; Murty and Henry, 1983; Dube *et al.*, 1985a; etc.). Dube *et al.* (1994), Dube and Gaur (1995) and Chittibabu *et al.* (2000) developed a real-time storm surge prediction system for the coastal regions of India. Real-time storm surge prediction systems have also been developed for Bangladesh, Myanmar, Pakistan, Sri Lanka, and Oman (Dube *et al.*, 2004; Jain *et al.*, 2006 a, b; Rao *et al.*, 1994; and Chittibabu *et al.*, 2002).

The national meteorological and national hydrological services of many countries surrounding the North Indian Ocean have achieved some success in providing storm surge warnings and implementing improved models through cooperative and coordinated sharing of responsibilities. Much of this work is done within the framework and overall guidance and supervision of the Tropical Cyclone Programme (TCP) of the World Meteorological Organization (WMO). The TCP of WMO supported technology transfer to run and operationalize storm surge models for Bangladesh, Myanmar, Oman, Pakistan, and Sri Lanka from the Indian Institute of Technology (IIT)-Delhi/Kharagpur.

The forecasting system developed at IIT Delhi is based on the vertically-integrated numerical storm surge models that were developed by the group (Johns *et al.*, 1981, 1983; Das *et al.*, 1983; Dube *et al.*, 1985a,b). The model is fully nonlinear and is forced by wind stress and uses a quadratic bottom friction formulation. The nonlinear advection terms in the model were found to have a significant effect on the simulations, especially in the shallow coastal waters of the head Bay of Bengal. Therefore, although the numerical solution of these terms affects the computational speed of the model, the nonlinear terms must be included in the operational model in order to provide the best forecasts. The details of the model and the numerical solution procedure are described in Das *et al.* (1983) and Dube *et al.* (1985b).

Surface winds associated with a tropical cyclone are derived from a dynamic storm model of Jelesnianski and Taylor (1973). The only meteorological inputs required for the model are the positions of the cyclone, pressure drop and radii of maximum winds at any fixed interval of time, and the model can be run in a few minutes on a PC in an operational office. The system is operated via a terminal menu and the output consists of the peak sea surface elevations and currents. One of the significant features of this storm surge prediction system is that the meteorological input needed for surge prediction can be periodically updated with the latest observations and forecasts of national weather services. The model has extensively been tested with severe cyclonic storms.

4.2.2. *Location-Specific Models*

The evolution of storm surges near the coast is known to be very sensitive to the coastal geometry and offshore bathymetry at the location of the landfall of the cyclone. As a result, operational models should include these factors as accurately as possible. This means that in addition to the large-scale storm surge prediction models, operational offices should also use high-resolution location-specific models for accurate prediction of the surges. Based on this need for location-specific models, Chittibabu (1999), Chittibabu *et al.* (2000, 2002); Dube *et al.* (2000b, 2000c; 2004) and Jain *et al.* (2006a, b) have developed location specific high-resolution models for the Andhra, Orissa, Tamil Nadu, and Gujarat coasts of India, and for Bangladesh, Pakistan, Myanmar, Oman and Sri Lanka, following an approach similar to that of Dube *et al.* (1994). An important feature of each of these location-specific models is that they use accurate and highly-detailed bathymetry for the offshore waters. A simple wetting -drying scheme also has been included in the models in order to avoid the exposure of land near the coast to strong negative surges. The reliability of these models has been tested using data from several severe cyclones, which struck the coastal regions of the countries in the Bay of Bengal and the Arabian Sea. The following sections describe results of selected numerical experiments carried out using location-specific models to simulate the surges.

4.2.2.1. Myanmar: Mala cyclone (May 2006)

Peak storm surge of about 4 m was reported for the May 2006 Mala cyclone on the Myanmar deltaic coast near Gwa by the Department of Meteorology and Hydrology, Yangon. The simulated surge contours (Fig. 6) show a maximum surge of 3.6 m to the right of the landfall point near Gwa, and the peak surge at Pathein is about 2 m. The coast from Ye to Moulmein also was affected by surges of 1 to 2 m.

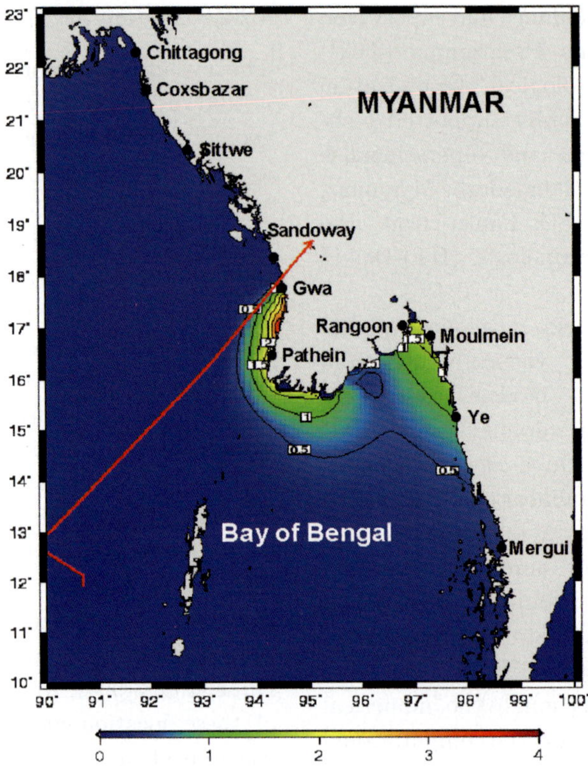

Figure 6. Simulated peak surge envelope for the May 2006 Mala cyclone (after Dube *et al.*, 2009).

4.2.2.2. Sri Lanka: Rameswaram cyclone (December 1964)

The December 1964 Rameswaram cyclone was one of the most severe storms to have affected Sri Lanka and the southern Indian peninsula. The simulated maximum surge of about 5.4 m occurred near Tondi (Fig. 7). Heavy inland flooding in the region, with an estimated peak elevation of about 5 m was reported by Rao and Mazumdar (1966), which is in good agreement with the simulations. It can also be seen from the figure that the coastal region to the north of Mannar Island along the west coast of Sri Lanka was flooded with a surge of more than 4 m.

4.2.2.3. Oman: Gonu cyclone (June 2007)

Simulations for the June 2007 Gonu cyclone were made using a pressure drop of 98 hPa and radius of maximum wind of 40 km (Fig. 8). A maximum surge of 2.6 m was simulated to have occurred near the eastern tip of Oman, which is where the storm was nearest the shore. Peak surges of about 2.6 m and 2.5 m were simulated at Sur and Nazwa Fins, respectively. The peak surge value at Muscat was around 2 m. About 100 km of the coast in northeastern Oman was flooded with surges of about 1 m. The computed maximum surge value of 2.6 m is comparable with the estimated range of surge heights.

4.3. *Review of Recent Developments in Predicting Storm Surge in Australia*

Australia has an extensive tropical coastline that is regularly impacted by severe tropical cyclones (TCs) and, in spite of its sparsely settled population, has recorded significant storm tide impacts dating from the late 1800s.

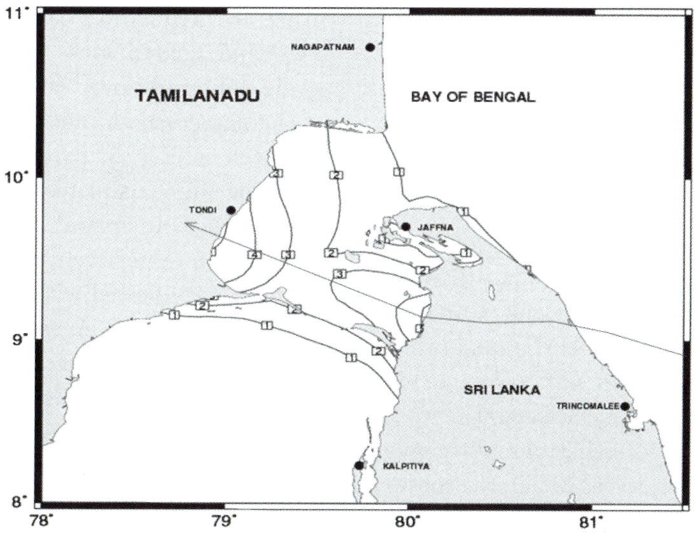

Figure 7. Simulated peak surge envelope associated with the December 1964 Rameswaram cyclone (after Dube *et al.*, 2009).

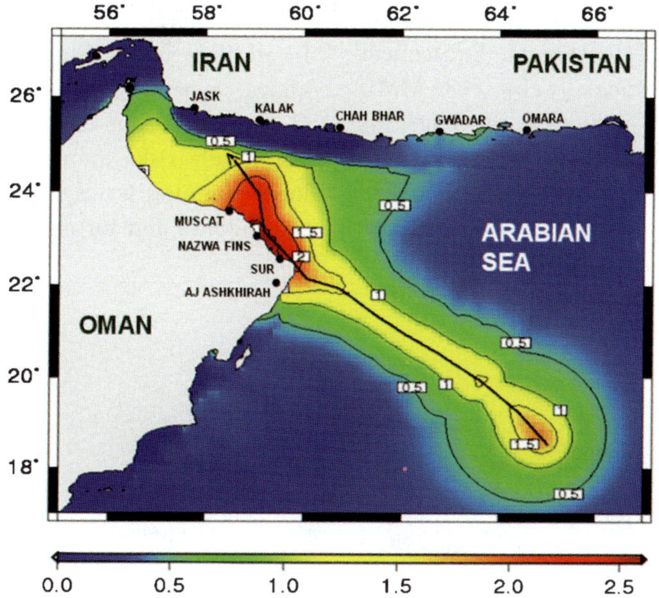

Figure 8. Simulated peak surge envelope associated with June 2007 Gonu Cyclone (after Dube *et al.*, 2009).

Accordingly there has been a history of development of storm surge modeling, risk assessment and associated atmospheric modeling capabilities since the availability of significant computational power in the 1970s. This work has focused on the east coast of Queensland (Harper, 1999) where TCs occur along a settled and rapidly developing coastal margin, much of which is bounded by the Great Barrier Reef (GBR). The most intense TCs in Australia are observed in the northwest of Western Australia, although the population there is relatively sparse and storm tide studies have been typically limited to specific towns or industrial facilities. This brief summary follows Harper *et al.* (2007).

4.3.1. *Queensland Climate Change Study (QCCS)*

This refers to a number of investigations undertaken over the period 2000 to 2004 that were funded primarily through Government enhanced-greenhouse research allocations. The initial work (Harper, 2001b) was conducted by Systems Engineering Australia (SEA) in association with the James Cook University Marine Modeling Unit (JCU-MMU) and others. This initial reference sets out recommended methodologies for storm tide studies that would also be capable of addressing climate change (enhanced-greenhouse) issues within the context of tropical cyclones. Importantly, a holistic approach was advocated for identifying the physical forcing mechanisms, ocean responses, vulnerabilities and impacts that would lead to informed decision making and the long term mitigation of storm tide threats. Subsequent studies conducted principally by JCU-MMU provided updated TC-induced surge plus tide updates for selected east coast sites, all with and without allowance for predicted enhanced-greenhouse effects by 2050. This work was done using the JCU-MMU storm surge, wave, and TC wind field models that had been extensively calibrated over many years. The extent of the numerical 'A' and 'B' modeling domains is shown in Fig. 9, covering some 3500 km of coastline and completely enclosing the Great Barrier Reef at spatial resolutions of 13.9 km and 2.78 km respectively. A further 20 'C' model domains were nested within the 'B' domains at 0.55 km resolution. A series of demonstration hindcasts of significant historical storms indicated that peak surge levels were likely reproducible within 5% of measured values provided that adequate meteorological and bathymetric data was available. An example of one such event, Severe Tropical Cyclone *Althea* in 1971, is shown in Fig. 10.

4.3.2. *Local Coastal Inundation Studies*

In addition to the State-wide QCCS described above, coastal Local Government regions in Queensland have been active in assessing storm tide risks for the purposes of updating local planning levels in response to enhanced-Greenhouse and for emergency response. Local

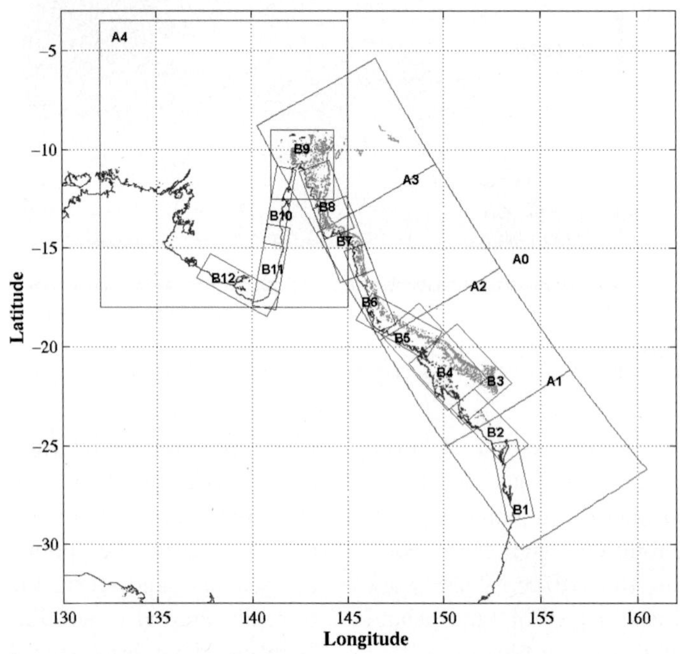

Figure 9. Queensland climate change 'A' and 'B' numerical domains (after Harper *et al.*, 2001b).

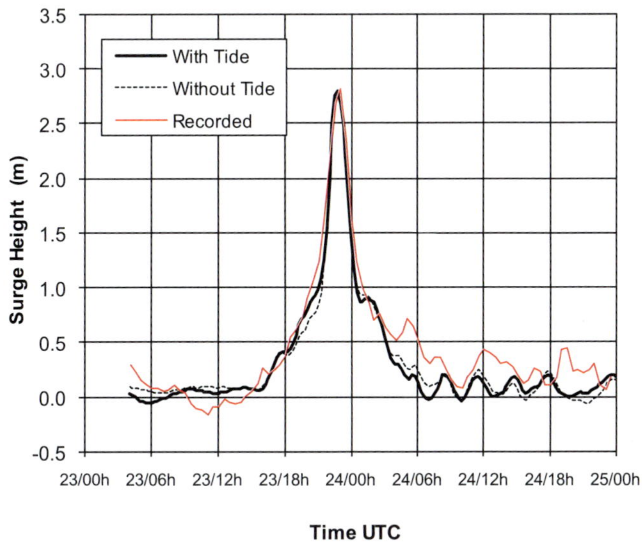

Figure 10. Measured and modeled surge component at Townsville Harbour for *Althea* (1971) (after Harper *et al.*, 2001b).

inundation studies are typically conducted at spatial resolutions of 50 m or finer, depending on the available topographic data and the need for representation (e.g., Fryar *et al.*, 2004). Many such studies have adopted a methodology similar to that described as the "hybrid" approach recommended in Harper (2001b). To achieve this, a parametric storm tide model (surge and wave) is developed based on comprehensive hydrodynamic model parameter response mapping of the study region. The parametric model then provides an economical kernel for Monte Carlo simulation and a flexible test bed for sensitivity testing of a range of model assumptions, El Niño Southern Oscillation (ENSO) or climate change influences. Water level persistence and joint probabilities of surge, tide, wave height and period are also available from these types of studies.

4.3.3. *Coral Atoll and Reef Environments*

Coral atoll and fringing reef environments can present with significantly different vulnerabilities from the typical continental margin environments described previously and require consideration of concurrent non-linear tide, surge and wave interactions. The principal threat of inundation in such low lying environments is due to extreme water levels on the outer reefs due to the combined effects of the pressure deficit component of a storm surge acting coincidentally with a high tidal level and high wave setup, caused by wave breaking on the outer reefs and reef entrances. Breaking wave setup can be expected to be the dominant water level controller with magnitudes as high as several meters. Wind-stress induced storm surge setup is small because of the surrounding deep ocean environment but potentially more significant in some parts of shallow lagoons.

Because of the complex non-linear interactions of storm surge, tide and breaking wave setup, a statistical simulation methodology is again advisable and a combination of parametric surge, wave and wave setup models have been used successfully within a Monte Carlo driven TC climatology (e.g., Harper *et al.*, 2001a). The method has been validated against historical events and delivers sequences of time-varying water level components that would be generated during a severe tropical cyclone event, with breaking wave setup on the outer reef flats and local wind stress influencing the shallow lagoon regions. Both these effects are critically modulated in a non-linear fashion by the stage of the tide.

4.3.4. *Storm Tide Warning Systems*

One of the aims of the QCCS was to develop self-consistent methodologies for both mitigation/planning and forecast/warning applications that could be efficiently developed and deployed to provide coverage over large sections of coastline. The proposed "hybrid" approach, a combination of detailed deterministic hydrodynamic modeling and derived parametric models, satisfied these criteria by providing a self-consistent framework for long term climatology modeling and short term warning needs. This method differs from the static Maximum Envelope of Waters (MEOW) approach in that it provides a rapid response tool that can incorporate tidal variation and storm parameter uncertainty.

Based on these recommendations, the Bureau of Meteorology (BoM) in Queensland subsequently adopted this approach in developing a parametric surge model for the east coast that would augment their existing access to real time hydrodynamic modeling. This provided a significant benefit in operational flexibility and also in training of forecasters, who could quickly examine and test a range of forecast scenarios. Importantly, this approach recognizes that the greatest uncertainty in the storm tide forecast relates to the variability of the forecast meteorological parameters. This is especially relevant to the vast majority of tropical cyclone forecast centers around the world that do not have aerial reconnaissance capability, where the Dvorak satellite image interpretation method is the primary intensity forecast tool. Even with the rapid development of advanced atmospheric models this situation is unlikely to change significantly within the next few decades.

The Northern Territory BoM engaged SEA in 2005 to provide a turnkey parametric storm tide forecasting system that incorporates tidal prediction, storm surge, waves and wave setup for the entire northern Australian coast from the Gulf of Carpentaria west to the Kimberley region in Western Australia. Figure 11 shows the area that was comprehensively modeled using seven along-coast domains, each of about 400 km extent, to provide the basis for the necessary

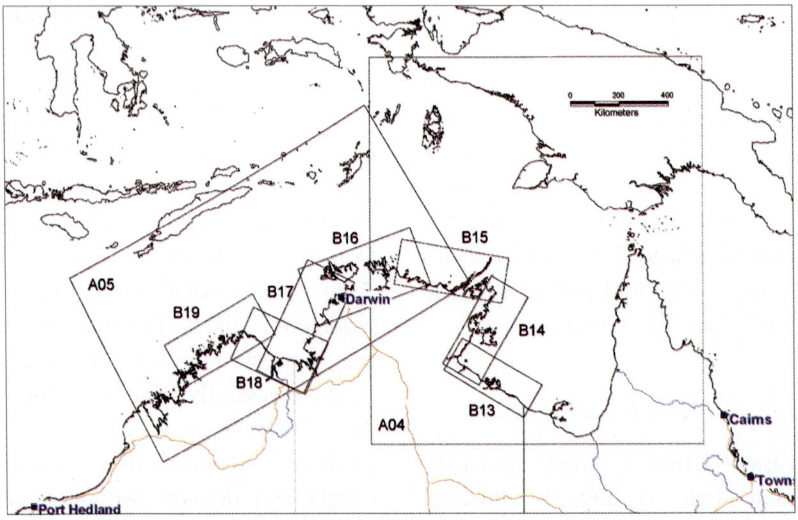

Figure 11. Extent of the Northern Territory storm tide warning modeling system (after Harper *et al.*, 2007).

parametric models. Approximately 25,000 individual TCs were modeled as part of this development and storm tide predictions are available for over 1,000 named localities. However, the resulting parametric prediction model SEAtide can generate more than 100 probabilistic storm tide forecasts within one minute on a typical desktop computer.

5. Future Development

Extensive storm surge modeling applications have been made with existing modeling systems. These efforts have been extremely valuable in reducing the damages caused by extreme tropical cyclones by provided emergency managers with reliable planning tools. However, there are many issues where future development is needed to further decrease the potential for loss. A brief discussion follows, including some of the main issues that are outstanding or under development by the research community.

5.1. *Improved Hurricane Forecasts*

The most important factor affecting the accuracy of a storm surge model is the forcing; that is, the ability of a tropical cyclone model to construct realistic wind and pressure fields. Since the most significant improvement in storm surge forecasting depends on improvements in predicting tropical cyclones, particularly for track and intensity, surge simulations are critically dependent on the accuracy of these storm characteristics. Without confidence in tropical cyclone forecasts, storm surge models have limited use for prediction.

5.2. *High Quality Bathymetry and Topography*

The most significant factor affecting model skill besides meteorological forcing is the accurate description of geographic features in the model grid. These errors can be minimized by careful grid construction and inclusion of sub-grid scale features in the model, but this depends upon the quality of the geographic data. Coordinated collection of high-resolution topographic and bathymetric data, especially LiDAR datasets, is needed to accurately model coastal inundation. Furthermore, vertical datum transformations are essential for combining coastal geospatial data. Elevations that are referenced to inconsistent vertical datums can cause artificial discontinuities in model grids, especially at the coast. Therefore, the capability to transform coastal elevations to a common datum is crucial for developing the geospatial infrastructure for the coastal zone. Finally, the creation of seamless elevation datasets (such as Digital Elevation Models) for storm surge model development would enable accurate grid construction.

5.3. *Air-Sea Interaction*

The interaction between the wind, sea surface and water column is not well understood, particularly at hurricane wind speeds. What is required is an accurate description of the physics of momentum transfer from wind to waves to water column. However, data used to parameterize wind drag coefficient relationships is difficult to collect and is nearly non-existent from tropical cyclones. Research indicates that the assumption of a linear increase in surface stress may not apply at hurricane strength winds (Powell *et al.*, 2003). Furthermore, wind drag laws are developed under the assumption of open marine conditions, but increased roughness is likely to exceed these conditions in near-shore and overland conditions. Surface waves also affect air-sea interaction; this is especially true as waves break and inundation occurs, since roughness increases and becomes more localized. Further research is needed to characterize air-sea interaction under tropical cyclone conditions.

5.4. *Observations of Surge*

There is a shortage of comprehensive data available to evaluate storm surge simulations.

Observations of water levels during hurricanes are not extensive. Gauges are not widespread and are often destroyed during severe events. Post-storm HWM surveys are unreliable because of the difficulty in collecting accurate data and are not done for all storms. Therefore, there is an important need for expanded and coordinated collection of water level data from tropical cyclone systems. A network of gauges that are hardened to survive strong storms and collect data throughout is needed. Furthermore, this permanent network should be augmented by a suite of temporary pressure gauges installed in the path of a storm, as done in the US for Hurricane Rita in 2005 (McGee *et al.*, 2006b). Third, post-storm collection of HWMs should be systematically completed and archived. Fourth, all water level observations need to be accurately tied to a geodetic network so that a common vertical datum can be used for both model output and observations.

5.5. *Communication of Uncertainty*

There is inherent uncertainty in tropical cyclone forecasting and corresponding predictions of surge. However, users of such products need estimates of this uncertainty in order to guide decision-making processes. Development of probability distributions of storm forecasts will be increasingly required, created by an ensemble of simulations whose results need to be combined in a fashion that communicates risk appropriately.

5.6. *Improved Model Accuracy and Efficiency*

Although there is a large number of storm surge models available in the community, there are still outstanding needs to improve model accuracy by including additional physics and increasing resolution. For example, surge water levels are affected not only by wind and pressure fields, but also by tides, waves, river inflow, precipitation, and erosion, factors that few if any models comprehensively consider. Conversely, it is important that the high-resolution models being implemented run quickly in order to satisfy the large suites of runs required for probabilistic forecasts, creation of storm surge atlases, and calculation of historical flooding probability.

5.7. *Wave Modeling*

One of the predominant effects of a tropical cyclone is the large wind waves generated by the storm, which affect the storm surge in a number of ways. While research and development is ongoing into coupling surge and wave models (see section 2.3.3), many systems are lacking a full description of the physics that occurs. A wind wave model will provide wave radiation stresses which drive wave-current interaction and increase storm surge height. Furthermore, wave-current interaction modifies the bottom friction stresses in shallow areas, increasing the bottom friction as the wave height increases. Finally, the wave model describes the conditions of the sea state as the hurricane travels from offshore to land, enabling improved description of the drag coefficient based upon sea state.

5.8. *River Forecasting*

There are important interactions between rivers and the coastal environment during tropical cyclones. While the flooding caused by rainfall generally occurs after the surge conditions have passed, the occurrence of multiple storms in succession requires the accurate knowledge of river inflow and water levels to accurately model surge throughout an estuarine system. Inclusion of river inflow and precipitation can address these issues.

5.9. *Coastal Geomorphology*

Future focus is needed on study of geomorphological changes that can occur during surge events, as these can be important in defining the volume of water that passes over an eroding barrier island, for example. The present state of sediment transport models is generally not sufficiently development to consider these conditions. Sediment transport models are needed that accurately predict the time rate of change of the coast, and especially dune systems, under the effects of waves and surge from a tropical cyclone.

5.10. *Decision-Support Tools*

Increased utilization of surge model output can be made by providing it in GIS formats and combining it with other data layers. This leads to easily utilized maps and visualizations, and enables incorporation of socioeconomic and demographic data with model output. This enables decision makers to evaluate risks in a more thorough manner.

5.11. *Community Approach*

Improvements in storm surge modeling benefit both the public and private sector. When there are advancements in the knowledge and prediction of storm surge impacts, appropriate mitigation steps can be taken to minimize losses. A community approach to model development that shares the funding burden as well as the benefits of research and development advances all interests. A community approach is one where both resources and developments are shared across all model users, whether government, academic, or private. It allows all members of the community to benefit from the advancements that are made so that the state of the technology is continually moving forward. This strategy enhances both research and applications by allowing researchers to focus on making scientific improvements without having to recreate the larger set of technology needed to support such work, and facilitating operational application of developments which improve skill and further the use and value of a model. Therefore, a community approach streamlines the research to operations pathway by sharing common technology. It also promotes uniformity of techniques which enhances reliability, and it provides the best leverage of limited funding. Open sharing of data, tools, models, and approaches increases the value and reliability of the system as a whole. Leadership from the public sector is necessary to support a community approach as they are the major provider of storm surge mitigation strategies and they develop and implement relevant technical standards.

One major benefit of a community approach is a standardization of terminology that increases understanding of storm surge threat in the user community. It avoids confusion caused when federal agencies provide information described in a multitude of ways; uncertainty about differences in technical language has confused the public in the past.

A community approach does not mean that a single model is selected for use by the community. Different models will likely continue to be advantageous for specific applications. Rather, it establishes an infrastructure that enables developments to be shared across models. Tools such as standards and documentation will enable technology advancements to benefit the community at large.

Finally, an open, coordinated forum for a community approach to modeling can provide standards and regional test beds can be used for model comparisons which will guide future applications. This will enable the community to select an appropriate model implementation for specific applications.

6. Integrated Modeling in Coastal Areas

Generally, it takes a long time for riverine flooding to reach the coast, whereas storm surges are roughly coincident with landfall. However, high river flows occurring simultaneously with storm surge likely will have adverse impacts on coastal inundation. The discussion below will refer to coupling of river flow and coastal hydrodynamic models to address potential coastal inundation.

6.1. *Hydraulic and Storm Surge Model Coupling*

Most of the studies reviewed address models that forecast either storm surge or river flow, but few studies have described the coupling of both inland flood and storm surge models for forecasting purposes. Even though in most cases, surge and river flooding occur separately, with storm surge coming at time of land-fall and river flooding occurring later. There are situations when both effects might co-exist and setup a severe flooding situation. This was experienced in 1999, when Hurricane Floyd affected the East Coast of the United States. Not only Floyd's storm surge was affecting coastal areas of North Carolina but also torrential rainfall occurred in an area hit by Hurricane Dennis just weeks earlier. Rivers were already in flood by the time Floyd made land-fall. In the USA, coupled systems using 1-D models are operational for several coastal rivers and some of them were utilized in pilot projects for generation of inundation mapping in real-time. Reed and Stucky (2005) described the application of the NWS Dynamic Wave Operational Model (DWOPER) on the lower Ohio/Mississippi Rivers to the Gulf of Mexico. This model uses the Sea, Lake, and Overland Surges for Hurricanes (SLOSH) water level forecasts as the downstream boundary. Brown *et al.* (2007) recently described "the first attempt to model storm surge flooding of an urban area with 2-D hydraulic model and to explore the uncertainties associated with model predictions" for the Canvey Island, UK.

In order to accurately reproduce the passage of a flood wave through a river reach, the responses (water level, flow, velocity) are simulated using unsteady-state hydraulic models. Many hydraulic models are available to simulate water levels and river flow, but the local site characteristics might require specific model capabilities. Low-lying coastal estuaries are complex hydrodynamic systems that often require 2-D models for accurate simulation, and the 1-D approach may not be appropriate. Issues of computational grid resolution must sometimes be addressed during the coupling process. For example, if the downstream boundary of the 1-D hydraulic model includes multiple surge model computational cells, then averaging of surge elevation across the hydraulic model boundary may be required. Likewise, in the case of a 2-D depth-average hydraulic river model, the downstream boundary will likely include multiple river cells which must be linked with one or more surge model computational cells, again requiring some averaging or smoothing across the boundary. The Tar River example discussed below illustrates this issue.

The Tar River, North Carolina, USA, was modeled using the 1-D NWS hydraulic model called Flood Wave (FLDWAV) and coupled with a tidal model (Riverside Technology, Inc., 2006). The lower river reach of the Tar River is described as follows (p. 8):

> "The Tar River drains into the Pamlico Sound, which also receives flow from several other rivers in North Carolina. The Sound is protected by a series of barrier islands with a number of inlets through which tidal influences are transmitted and hydrologic inflows drain to the Atlantic Ocean. Tidal influences predominate in the determination of water levels in the Pamlico Sound and have a strong influence on stages in the Tar River as far upstream as Greenville under normal flow conditions and as far as Rock Springs under very low flow conditions."

This is an example in which the use of a 1-D hydraulic model might be inadequate for simulating conditions near the Sound; a 2-D model might be more appropriate for simulating flood and storm surge in the estuary. However, because of operational constraints it was modeled using a 1-D model.

Following selection of the hydraulic and tidal models, the general approach for simulating the coupled surge-inland flooding response of the river system includes: (i) selection of river-reach boundaries, type and location; (ii) collection of cross-section geometry to build the channel and flood plain geometry for the hydraulic model, and (iii) determination of local flows using data or previously-developed hydrologic models. The upstream boundary of the model reach is usually defined by a time-series of flow; the water-level time series for the downstream boundary will be provided by the surge model. This rather simplistic approach might become cumbersome when considering operational implementation. Data demand for research and model development is usually confined to historical data for calibration. Real-time operations require either observed or forecast information on rainfall, a forecast inflow hydrograph, and forecast surge elevations derived from a surge model driven by meteorological forecast.

Issues to be considered for the hydraulic/tidal coupling include:

1. The calibration period should be selected to ensure there is enough data for all the time series required by the model system. In addition, the period selected should include a wide range of river flows (low, medium and high flows) and minor, medium and major surge effects.
2. The forecast window for individual models (including the hydrologic and meteorological models) must be examined to define the window for the coupling system and to ensure that data from each model is available as needed by the dependent models.
3. The time interval for the system coupling must be selected based on the minimum allowed to describe the process while maintaining feasibility of model execution for forecasting purposes. For example, in the USA most of the operational models used to forecast river flow are performed for incremental periods between 1-hour and 6-hours. Storm surge models can produce output for time increments on the order of a few minutes. However, because the forecast system architecture might be too rigid and because of operational limitations for execution of the hydraulic models, the coupling is performed in hourly increments.
4. Consistency of reference datums is essential for the accuracy of results. This includes vertical and horizontal datums for cross sections, water levels, and tide gages.
5. Consistency of boundary conditions must be maintained for situations in which multiple river computational cells (2-D hydraulic model) coincide with a single surge model cell. Multiple iterations between the hydraulic and estuarine models might be required.
6. Continuous execution of the coupled system ensures that the models are initialized correctly, but this might not be possible for a situation in which a surge model is not run continuously and no downstream real-time tidal data are available.

A coupled system for St. Johns River, Florida, USA has been in operation since 2001 (Fig. 12). The US NWS Southeast River Forecast Center (SERFC) is currently forecasting water levels in the lower end of the St. Johns River near Jacksonville using the NWS hydraulic model FLDWAV. The model is executed for a 300-mile reach; the upstream boundary (flow hydrograph) as well as the local inflows from tributaries is derived from the hydrological model Sacramento Soil Moisture Accounting. The downstream boundary selected is the tide gage located at Mayport at the mouth of the river near Jacksonville, Florida. A forecast window of 5 days is used.

The time series for the water level at the downstream boundary are composed of the observed water level up to the forecast time. For the forecast period, the astronomical tide forecast provided by the National Ocean Service (NOS) is combined with the output from the

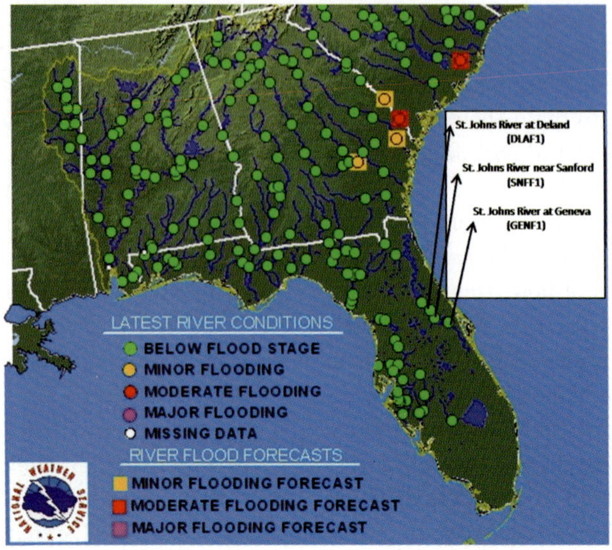

Figure 12. St. Johns River, Florida, USA. The river flows from South to North.

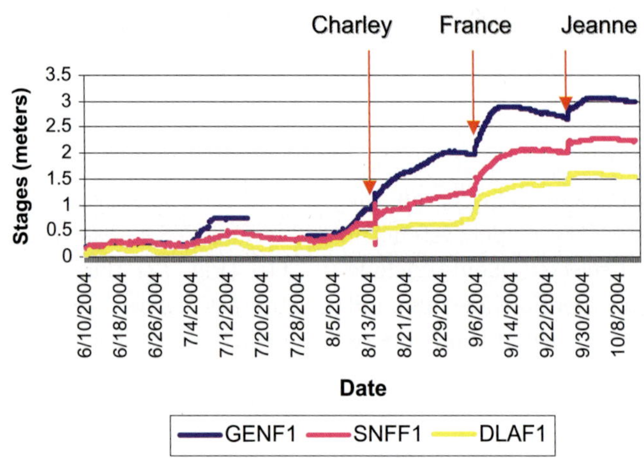

Figure 13. Water levels in St. Johns River during 2004 hurricane season (Charley in August, Frances and Jeanne in September).

Sea, Lake and Overland Surges from Hurricanes (SLOSH) model. The NWS developed blending techniques to correct the downstream water level once observed data becomes available

The 2004 hurricane season provided insight on the performance of the hydraulic model. Three hurricanes affected this area: Charley, Frances, and Jeanne (Fig. 13). Rainfall produced by these systems affected the water levels in the St. Johns River. Water levels were low at the beginning of August, and the rainfall associated with Hurricane Charley produced a significant rise, but it wasn't until the passage of Hurricane Frances that the rises reached flood stage and remained above flood stage until the beginning of November 2004.

During the operational performance of the system, the wind acting on the river reach was observed to greatly influence the propagation of the surge and forecasts of water levels. The jump in water levels due to wind effect (Fig. 13) was more noticeable at the station of Deland

(DLAF1) when Charley affected the area and the river was at low levels. The hydraulic model, used by the NWS at that time, (FLDWAV) had the capability to incorporate wind direction and magnitude as a single value for the whole reach, but this is not practical for the entire 300-mile reach. As a result, wind effects on the river are not included in the coupled system. The NWS is in the process of incorporating this variable into the hydraulic modeling.

6.2. *Inundation Mapping*

Several countries are working on projects involving potential inundation mapping. In China, the production of flood maps has been in practice since 1986. Today all available Flood Hazard Mapping are being posted on the internet (WMO Draft Report, Typhoon Committee, Macao, Republic of China, 2006). A pilot project in the Philippines produced a preliminary flood hazard map of flood-prone areas in the San Juan River Basin. Flood maps were produced for Kuala Lumpur in Malaysia, as well as maps showing minimum, moderate, and severe flooding for the Gomback basin. Most of the mapping projects mentioned above involved steady-state models for the generation of static-map libraries of inundated areas at different water levels. Maps from these libraries are then called up for application based on water-level forecasts.

6.2.1. *Development in the USA*

Inundation maps are based on predictions of water levels along the river reach and the corresponding terrain information. The whole process to determine inundation mapping in coastal areas due to topical cyclones involves meteorological forecasting (storm tracking, intensity, and precipitation totals), oceanographic, estuarine, and riverine hydrodynamic modeling (including wave effects), watershed modeling of storm runoff, and spatial mapping of inundation (Fig. 14).

In addition to forecasts of surge timing and height, flood inundation maps are increasingly being requested by emergency managers and decision makers. Maps are needed not only of coastal storm surge, but also inland flooding resulting from high rainfall associated with cyclones. In the USA, since the NWS instituted a program to model tropical cyclone

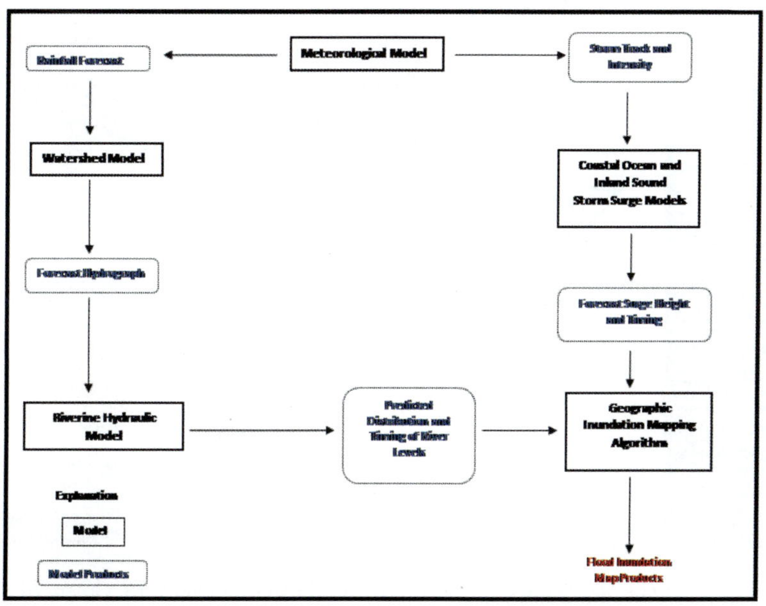

Figure 14. Suite of models required to simulate inundation from storm tides and upland flooding.

storm surge, more fatalities occur from inland flooding resulting from tropical cyclones than from storm surge (not including statistics from Hurricane Katrina), although this is not the case in all areas of the world. Flood inundation maps are useful in advance planning, as well as for response during events and for post-assessment after events.

In the USA, after Hurricane Floyd (1999) the State of North Carolina began a project to generate maps depicting inundated areas. This was the first NWS project involving the generation of inundation map libraries for a river reach (Fig. 15). The graph below is a representation of flood inundation for NWS flood categories. These maps are based on steady-state

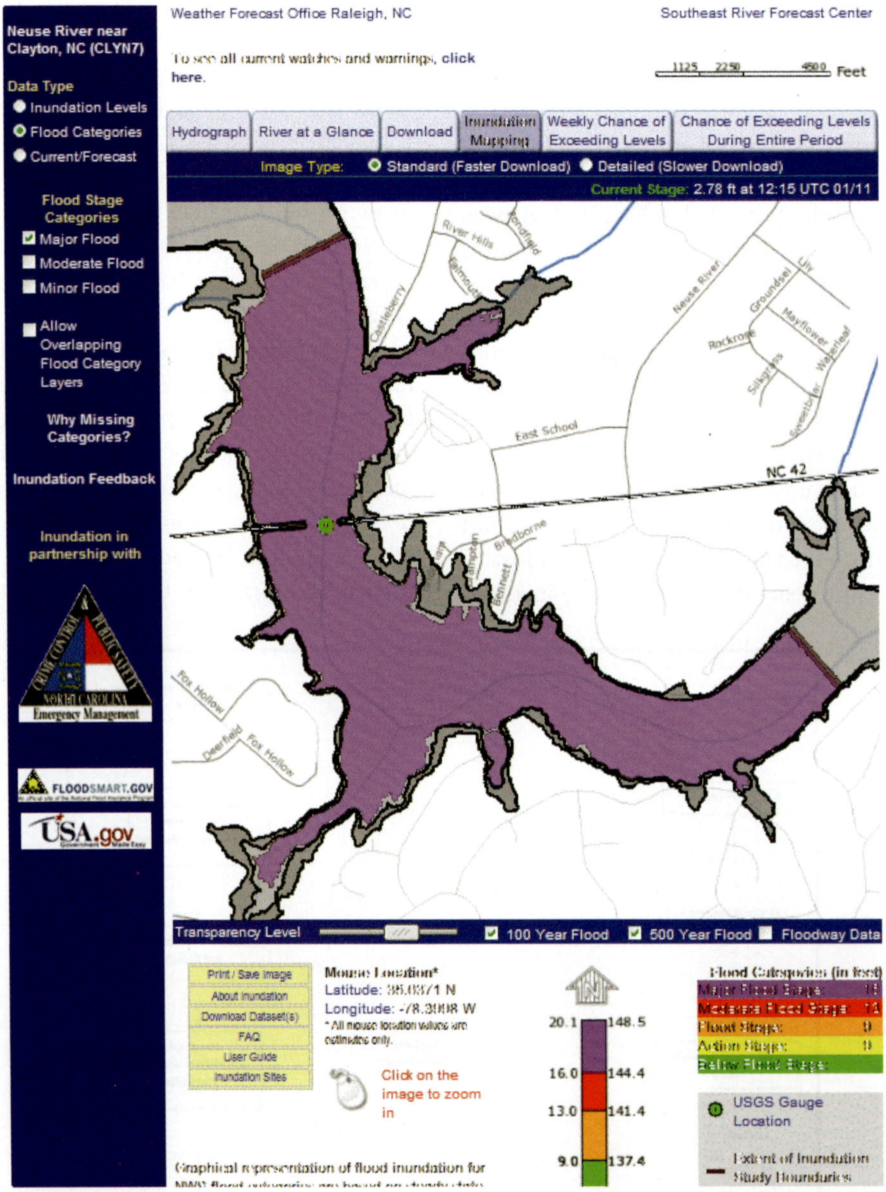

Figure 15. Inundation map for the Neuse River near Clayton, North Carolina, USA. (http://newweb.erh.noaa.gov/ahps2/inundation/inundation.php?wfo=rah&gage=clyn7)

hydraulic modeling of water surface elevations for incremented discharges.

The alternative to the use of the steady-flow assumption and development of inundation map libraries is to estimate inundated areas in real time during, or immediately prior to, a flood so that the particular characteristics of the rainfall and flood hydrograph are well represented in the hydraulic modeling. The limitation of this approach is that the models must be executed in real time for each event and that results must be distributed quickly to emergency management officials and all other interested parties. Moreover, there will be some uncertainty in the forecast flows for which the inundation modeling is to be conducted. A recent pilot project in the USA to test and evaluate dynamic mapping concluded that ongoing efforts in static mapping should be fully tested and evaluated before embracing implementation of operational, real-time (dynamic) inundation mapping (NWS, Office of Hydrologic Development, R-Time Report, 2007, in review).

Several approaches have been taken to develop inundation maps and include: (i) assuming that a storm surge of a given height will inundate or impact up to that height of the land contour, (ii) pre-generating a library of maps for a range of water levels (static maps), and (iii) generating inundation maps from real-time forecasts of water level (dynamic maps) based on unique features of a given event. Although dynamic mapping might seem the best option, implementation can be difficult and costly.

An example of static maps is presented by Bales *et al.*, 2007. He generated a set of water-surface profiles at 0.305 m (1 ft) increments for a reach of the Tar River. Based on the water-surface profile, a water-surface elevation was assigned to each cross section in the reach; the water surface was assumed to be level across the cross section, which is consistent with the 1-D modeling approach. Water-surface elevations between cross sections were estimated using a spline interpolation. Inundated areas were identified by subtracting the water-surface elevation in each grid cell from the land-surface elevation in the cell. An automated procedure was developed to identify all inundated cells that were hydraulically connected to the cell at the downstream-most gauge in the model domain. This process resulted in a set of inundation map libraries for each modeled reach. Inundation polygons were merged with a variety of other geospatial data to provide information for flood mitigation and emergency response.

At this time, the NWS does not use 2-D hydraulic models for operational purposes, so tests and pilot projects developed for dynamic maps have mainly focused on areas for which the 1-D approach is valid or can be approximated. However, in one of the pilot projects, St. Johns River, Florida there was an opportunity to test the coupling by using the 1-D hydraulic model to generate flow outputs which in turn were used as inputs for a 2-D estuarine model. The estuarine model was also used to forecast salinity and temperature.

Currently, the USA's strategy is to develop static maps for flood-prone areas and gradually develop hydrodynamic models for estuaries to provide real-time flood maps. The increased demand of probabilistic inundation maps by emergency managers is being recognized, but the need for developing operational procedures remains, including priorities for addressing mapping uncertainty.

6.2.2. *Case Study: Andhra Pradesh Coast of India*

Rao (1968) classified the Indian coastline into three categories based on combined storm surges and wind waves. According to this classification, the Andhra Coast of India from 14° N to 16.5° N falls in the B-category (2 to 5 m surge) with a short C-type belt ($>$ 5 m surge) near Nizampatnam bay. According to an analysis of historical records by Jayanthi (1999), the Andhra coast is high-risk prone with a small very high-risk prone zone near Nizampatnam Bay. The storm surges occurred that during 1977 and 1990 near Machilipatnam further support the vulnerability of Andhra coast for disastrous surges. In recent years, there has been considerable concern regarding the vulnerability of coasts due to cyclones and associated surges in view of projected global warming and sea level rise. In this section, we have undertaken as a case study the development of Disaster Management Plan (DMP) for cyclones and associated storm surges for mitigation in the nine coastal districts of the State of Andhra Pradesh (AP), India.

Based on historical cyclone data, through a simple statistical analysis, Delta P (atmospheric pressure deficit) was determined for cyclones making landfall on the AP coast, for return periods of 2, 5, 10, 25 and 50 years. The Storm Surge Model developed by IIT-Delhi was applied with the 50-year Delta P value for a set of synthetic tracks, which were developed by compositing actual tracks, ensuring that each coastal district was covered. The results of the computer simulations, calibrated with observed surge data for each region of the coast, provided maximum probable surge amplitudes at the mandal level, which is the geographical unit immediately below the district level, and is made up of several villages and maybe towns.

A generally accepted procedure in determining the extent of land inundation by a storm surge is to assume that, a water level of 5-m at the coastline would have an impact up to the 5-m land contour, a 10-m water level would impact up to the 10-m contour, and so on. This is a standard approach when very detailed orographic information is not available and might somewhat over-estimate the extent of inundation, but is an acceptable approach for coastal zone storm mitigation planning purposes.

In summary, the approach for determination of the Physical Vulnerability (PV) is as follows:

(a) A database of Tropical Cyclone (TC) generated Storm Surges (SS) impacting the AP Coast was drawn from the India Meteorological Department (IMD) and from several other national and international sources.
(b) Because of climate change, projections into the future were limited to 50 years all the available cyclone tracks for AP were synthesized into composite tracks to cover each of the coastal districts of AP.
(c) Making use of the projected pressure drop, the IIT-Delhi Storm Surge Model was applied using the synthetic tracks to determine the maximum possible storm surge amplitude (during a 50 year period) at various locations along the AP Coast.
(d) The Total Water Level Envelope (TWLE) was determined by superimposing the tidal amplitudes and wind wave setup on the surge amplitudes.
(e) These water levels were then projected onto the coastal land using onshore topography data to demarcate the horizontal extent of inundation.
(f) This conservative approach may slightly over-estimate the extent of inundation, but is desirable for Hazard Mitigation and for Coastal Zone Management, and is widely used around the world.
(g) Maps of regions subjected to possible wind damage from cyclones also were prepared.

6.2.2.1. Physical Vulnerability (PV) Maps for the Coastal Districts

Inundation by storm surge and regions subjected to wind damage were mapped for the districts of Prakasham (Fig. 16) and Guntur (Fig. 17) of coastal AP. The PV maps were prepared for four scenarios: (a) frequent (10-percent

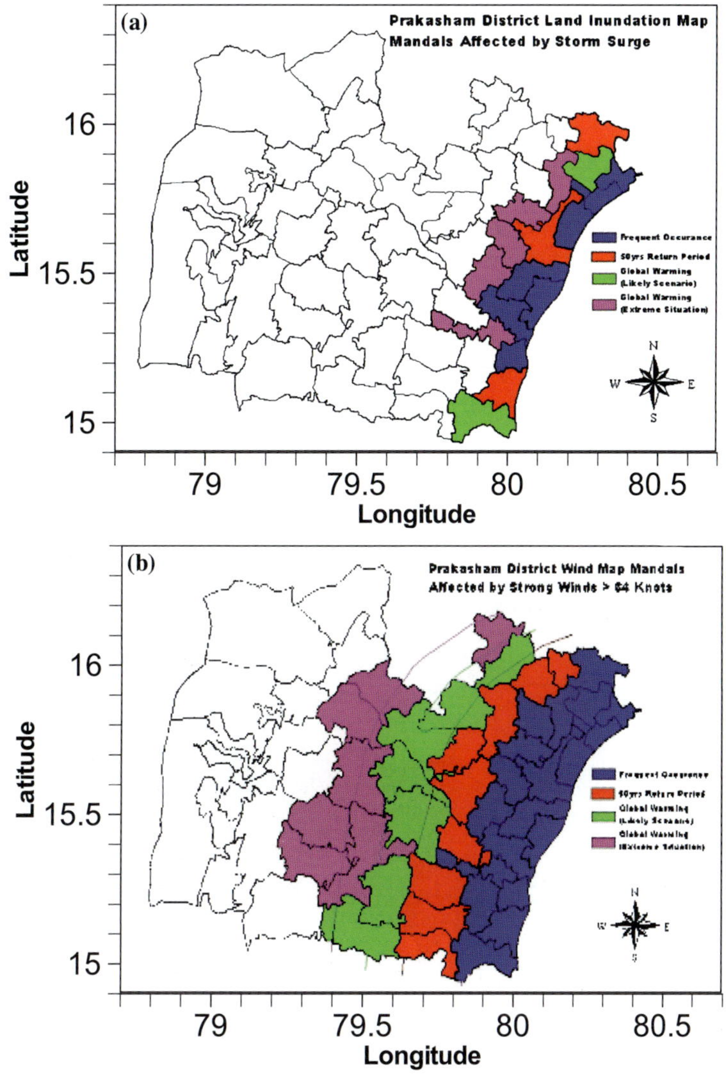

Figure 16. (a) Land inundation map affected by storm surge, (b) regions affected by cyclonic winds for Prakasham District, Andhra Pradesh, India (after Dube *et al.*, 2009).

annual recurrence interval), (b) infrequent (2-percent annual recurrence interval), (c) a future climate scenario resulting in an intensification of the pressure field by 5 percent, and (d) a more extreme case of intensification of 7 percent.

The three large rivers in AP, Godavari, Krishna and Pennar, are subject to storm surge penetration. The storm surge penetration into these rivers was determined by projecting the surge water levels into the rivers. It was assumed that for a river with many meanders, the

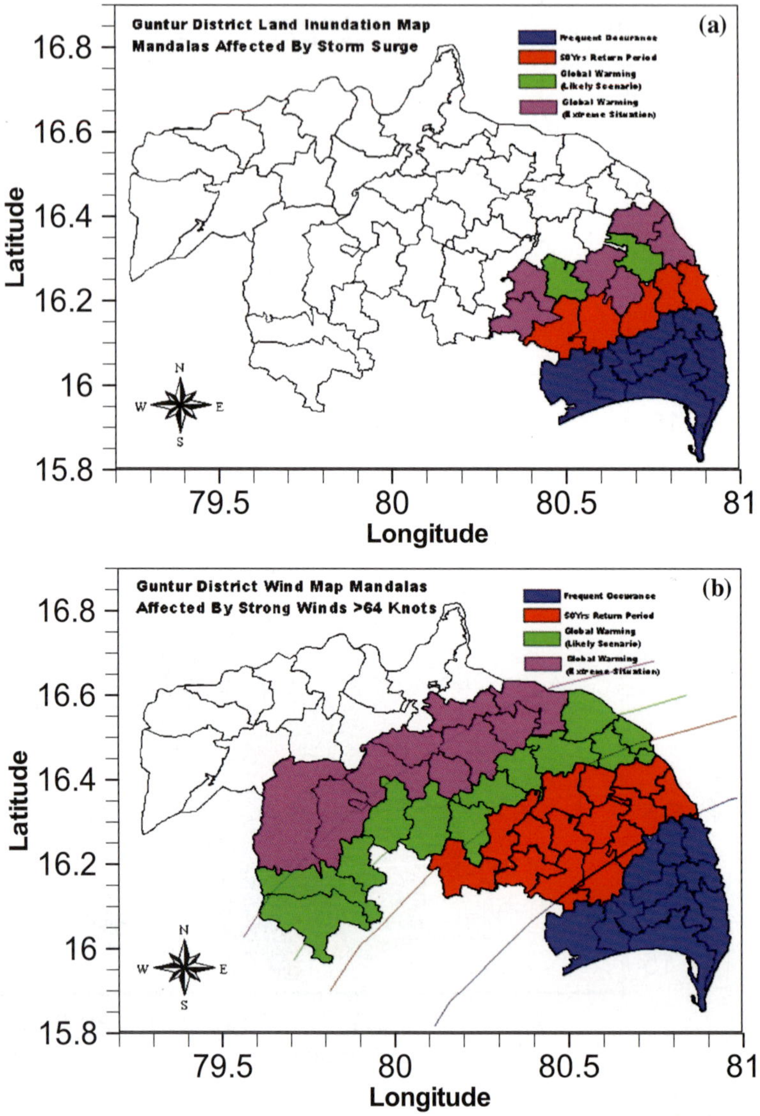

Figure 17. (a) Land inundation map affected by storm surge, (b) regions affected by cyclonic winds for Guntur District, Andhra Pradesh, India (after Dube *et al.*, 2009).

storm surge would penetrate 10 percent farther than on land. If the river had few meanders, the increased penetration was 15 percent. The 10–15 % numbers are arrived at based upon actual observations of storm surge penetration through these rivers (Murty, 1984). PV maps for storm surge penetration up the rivers were then prepared (Fig. 18).

6.2.2.2. Social Vulnerability (SV)

Social Vulnerability was developed for physically vulnerable mandals. By using available population and other data, along with the PV maps, overall cyclone vulnerability index maps were developed. Figure 19 shows the map for one of the districts of coastal Andhra Pradesh.

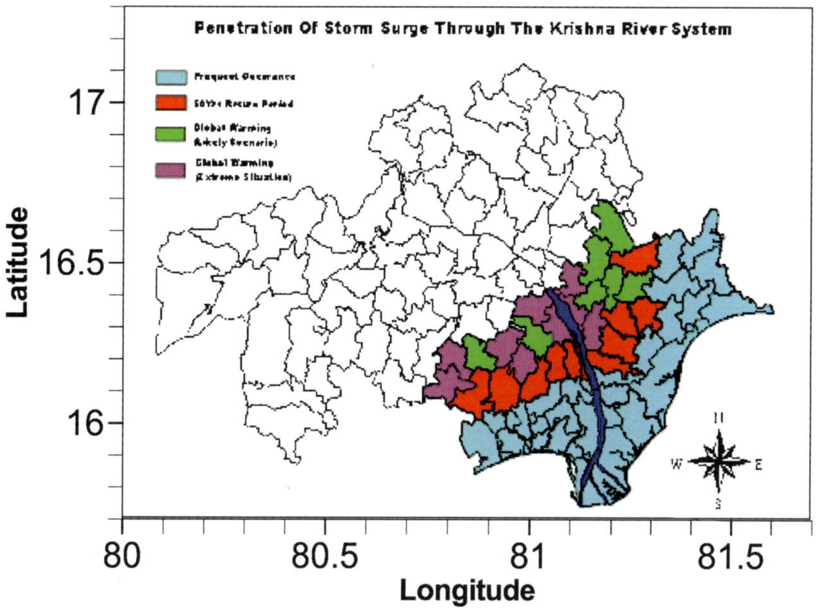

Figure 18. Storm surge penetration through the Krishna River System (after Dube *et al.*, 2009).

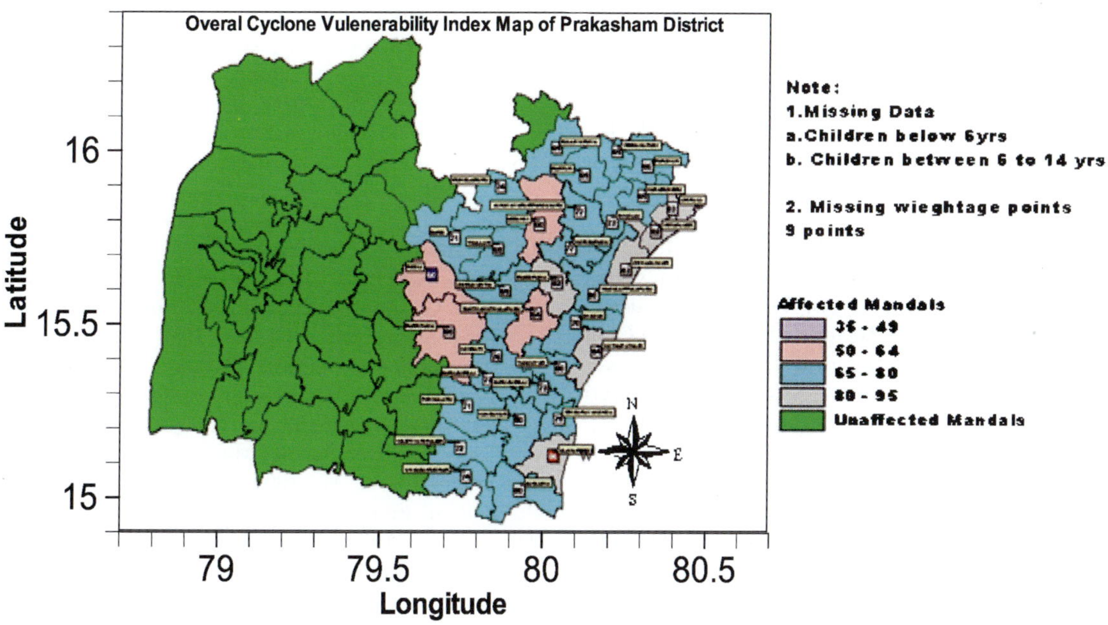

Figure 19. Overall Cyclone Vulnerability map for Prakasham District, Andhra Pradesh, India (after Dube *et al.*, 2009).

References

Aijun, Z., E. Myers, and F. Aikman III, 2006: The St. Johns River Operational Forecast System (SJROFS) and Its Skill Assessment. Coastal Survey Development Laboratory, National Ocean Service, NOAA, Silver Spring, Maryland 20910. February 10, 2006.

Ali, A., 1979: Storm surges in the Bay of Bengal and some related problems, Ph.D. Thesis, University of Reading, England, 227 pp.

Atlantic Oceanographic and meteorological laboratory, Hurricane Research Division, (http://www.aoml.noaa.gov/hrd/hrd_sub/milestones.html

Bales, J. D., C. R. Wagner, K. C. Tighe, and S. Terziotti 2007: LiDAR-derived flood-inundation maps for real-time flood-mapping applications, Tar River basin, North Carolina: US Geological Survey Scientific Investigations Report 2007–5032m, 42 p.

Bender, M. A., I. Ginis, R. Tuleya, B. Thomas, and T. Marchok, 2007: The operational GFDL coupled hurricane-ocean prediction system and a summary of its performance. *Mon. Wea. Rev.*, **135**(12), 3965–3989.

Berenbrock, C. and R. R. Mason, 2008: Mapping Hurricane Rita inland storm-tide: Toronto, *Proceedings of the 4th International Symposium on Flood Defence*, Institute for Catastrophic Loss Reduction, 8 p.

Blain, C. A., J. J. Westerink, and R. A. Luettich, Jr., 1998: Grid convergence studies for the prediction of hurricane storm surge. *Int. J. Numer. Methods in Fluids*, **26**, 369–401.

Blumberg, A. F. and G. L. Mellor, 1987: A description of a three-dimensional coastal ocean circulation model, in *Three-Dimensional Coastal Ocean Models*, N. S. Heaps, (ed.), American Geophysical Union, Washington, DC, pp. 1–16.

Bode, L. and T. Hardy, 1997: Progress and recent developments in storm surge modeling. *J. Hydraulic Engr.*, **123**(4), 315–331.

Bode, L., and L. B. Mason, 1994: Application of an implicit hydrodynamic model over a range of spatial scales, *Proceedings of 6th Computational Techniques and Applications Conference* (CTAC93), D. Stewart, H. Gardner and D. Singleton, (eds.), World Scientific, Singapore pp. 112–121.

Bode, L., L. B. Mason and J. H. Middleton, 1997: Reef parameterisation schemes for long wave models, *Progress in Oceanography*, **40**, 285–324.

Bogden, P., T. Gale, G. Allen, J. MacLaren, G. Almes, G. Creager, J. Bintz, L. D. Wright, H. Graber, N. Williams, S. Graves, H. Conover, K. Galluppi, R. Luettich, W. Perrie, B. Toulany, Y. P. Sheng, J. R. Davis, H. Wang, and D. Forrest, 2007: Architecture of a community infrastructure for predicting and analyzing coastal inundation, *Marine Technology Society Journal*, **41**(1), 53–61.

Brown, J. D., T. Spencer, and I. Moeller, 2007: Modeling storm surge flooding of an urban area with particular reference to modeling uncertainties: A case study of Canvey Island, UK: *Water Resources Research*, **43**(6):W06402, 22 p.

Cheung, K. F., A. C. Phadke, Y. Wei, R. Rojas, Y. J.-M. Douyere, N. Dodd, S. Liao, and E. Nakazaki, 2003: Modeling of storm-induced coastal flooding for emergency management. *Ocean Engineering* **30**, 1353–1386. www.sciencedirect.com

Chittibabu, P. 1999: Development of storm surge prediction models for the Bay of Bengal and the Arabian Sea, Ph.D. Thesis, IIT Delhi, India, 262 pp.

Chittibabu, P., S. K. Dube, A. D. Rao, P. C. Sinha, and T. S. Murty, 2000: Numerical simulation of extreme sea levels using location specific high resolution model for Gujarat coast of India, *Marine Geodesy* **23**, 133–142.

Chittibabu, P., S. K. Dube, P. C. Sinha, A. D. Rao, and T. S. Murty, 2002: Numerical simulation of extreme sea levels for the Tamil Nadu (India) and Sri Lanka coasts, *Marine Geodesy* **25**, 235–244.

Chow, V. T., 1959: *Open Channel Hydraulics*, McGraw-Hill, New York.

Crossett, K. M., T. J. Culliton, P. C. Wiley, and T. R. Goodspeed, 2004: Population trends along the coastal United States: 1980–2008. Coastal Trends Report Series, National Oceanic and Atmospheric Administration/National Ocean Service/Management and Budget Office/Special Projects, Silver Spring, Maryland.

Cushman-Roisin, B. 1994: *Introduction to Geophysical Fluid Dynamics*. Prentice-Hall, 320 p.

Das, P. K., 1972: A prediction model for storm surges in the Bay of Bengal, *Nature* **239**, 211–213.

Das, P. K., 1994a: Prediction of storm surges in the Bay of Bengal, *Proc. Indian. Natn. Sci. Acad.* **60**, 513–533.

Das, P. K., 1994b: On the prediction of storm surges, *Sadhana*, **19**, 583–595.

Das, P. K., M. C. Sinha, and V. Balasubramanyam, 1974: Storm surges in the Bay of Bengal, *Quart. J. Roy. Meteor. Soc.* **100**, 437–449.

Das, P. K., S. K. Dube, U. C. Mohanty, P. C. Sinha, and A. D. Rao, 1983: Numerical simulation of the surge generated by the June 1982 Orissa cyclone, *Mausam* **34**, 359–366.

Dube, S. K. and V. K. Gaur, 1995: Real time storm surge prediction system for the Bay of Bengal, *Current Science* **68**, 103–113.

Dube, S. K., P. C. Sinha, and G. D. Roy, The numerical simulation of storm surges along the Bangladesh coast, *Dyn. Atmos. Oceans* **9**, 121–133.

Dube, S. K., P. C. Sinha, A. D. Rao, and G. S. Rao, 1985b: Numerical modeling of storm surges in the Arabian Sea, *Appl. Math. Modeling* **9**, 289–294.

Dube, S. K., P. C. Sinha, and G. D. Roy, 1986: The effect of continuously deforming coastline on the numerical simulation of storm surges in Bangladesh, *Math. Comput. Simul.* **28**, 41–56.

Dube, S. K., A. D. Rao, P. C. Sinha, and P. Chittibabu, 1994: A real time storm surge prediction system: An application to East coast of India, *Proc. Indian Natn. Sci. Acad.* **60**, 157–170.

Dube, S. K., A. D. Rao, P. C. Sinha, T. S. Murty, and N. Bahulayan, 1997: Storm surge in the Bay of Bengal and Arabian Sea: The problem and its prediction, *Mausam* **48**, 283–304.

Dube, S. K., A. D. Rao, P. C. Sinha, and P. Chittibabu, 2000a: Storm surge modeling and prediction, in *Mathematical Analysis and Applications* (ed. Dwivedi), Narosa Publishing House, New Delhi, India, pp. 109–124.

Dube, S. K., P. Chittibabu, A. D. Rao, P. C. Sinha, and T. S. Murty, 2000b: Extreme sea levels associated with severe tropical cyclones hitting Orissa coast of India, *Marine Geodesy* **23**, 75–90.

Dube, S. K., P. Chittibabu, A. D. Rao, P. C. Sinha, and T. S. Murty, 2000c: Sea levels and coastal inundation due to tropical cyclones in Indian coastal regions of Andhra and Orissa, *Marine Geodesy* **23**, 65–73.

Dube, S. K., P. Chittibabu, P. C. Sinha, A. D. Rao, and T. S. Murty, 2004: Numerical modeling of storm surges in the head Bay of Bengal using location specific model, *Natural Hazards* **31**, 437–453.

Dube, S. K., P. C. Sinha, A. D. Rao, I. Jain and N. Agnihotri, 2005: Effect of Mahanadi river on the development of storm surge along the Orissa coast of India: A numerical study, *Pure Appl. Geophys.* **162**, 1673–1688.

Dube, S. K., I. Jain, and A. D. Rao, 2006: Numerical storm surge prediction model for the North Indian Ocean and the South China Sea, *Disaster and Development* **1**, 47–63.

Dube, S. K., I. Jain, A. D. Rao and T. S. Murty, 2009: Storm surge modeling for the Bay of Bengal and Arabian Sea, *Natural Hazards*, **51**, 3–27.

Ebersole, B. A., D. T. Resio, and J. J. Westerink, 2007: A community approach to improved prediction and characterization of coastal storm hazards, *Marine Technology Society Journal*, **40**(4), 56–68.

Fread, D. L., NWS Operational Dynamic Wave Model, verification of mathematical and physical models, in *Hydraulic Engineering, Proceedings of 26th Annual Hydraulic Division Specialty Conference*; 04/16/2002 II.4-DWOPER-13 rfs: 24dwoper.wpd, College Park, MD, August 1978, pp. 455–464.

Fryar, R. M, B.A. Harper and I. Botev, 2004: Storm tide modeling of the Whitsunday coast and resort islands, Proc Conference Hydraulics, Engineers Australia, July, 2004.

Ghosh, S. K., 1977: Prediction of storm surges on the east coast of India. *Ind. J. Meteo. Geophys*, **28**, 157–168.

Glahn, B., A. Taylor, N. Kurkowski, and W. A. Shaffer. In preparation 2003: SLOSH's role in National Weather Service Storm Surge Forecasting.

Global Climate and Weather Modeling Branch, 2003: The GFS Atmospheric Model, NCEP Office Note 442, National Oceanic and Atmospheric Administration/National Weather Service/National Centers for Environmental Prediction/Environmental Modeling Center, Camp Springs, Maryland, 14 p.

Gönnert, G., S. K. Dube, T. Murty, and W. Siefert, 2001: *Global Storm Surges: Theory, Observations and Applications*, Die Kueste, 623 pp.

Graber, H. C., V. J. Cardone, R. E. Jensen, D. N. Slinn, S. C. Hagen, A. T. Cox, M. D. Powell, and C. Grassl, 2006: Coastal forecasts and storm surge predictions for tropical cyclones, *Oceanography*, **19**(1), 130–141.

Harper, B. A., Storm tide threat in Queensland: History, prediction and relative risks, Conservation Technical Report No. 10, Queensland Government, Dept. of Environment and Heritage, January 1999.

Harper, B.A, M.R. Gourlay and C.M. Jones, 2001a: Storm tide risk assessment for the Cocos

(Keeling) Islands, *Proc. 15th Australasian Conference on Coastal and Ocean Engineering*, Coasts and Ports 2001, Engineers Australia, Gold Coast, September, pp. 75–80.

Harper, B.A. (ed.), T. Hardy, L. Mason, L. Bode, I. Young, and P. Nielson, 2001b: *Queensland Climate Change and Community Vulnerability to Tropical Cyclones: Ocean Hazards Assessment — Stage 1*, State of Queensland Department of Natural Resources and Mines, 383 pp.

Harper, B., T. Hardy, L. Mason, and R. Fryar, 2007: Developments in storm tide modeling and risk assessment in the Australian region, *Proceedings of JCOMM Scientific and Technical Symposium on Storm Surges*, Seoul, Korea, 2–6 October 2007.

Heaps, N. S., 1973: Three-dimensional numerical model for the Irish Sea, *Geophys., J. R. Astron. Soc.* **35**, 99–120.

Hicks, S. D., R. L. Sillcox, C. R. Nichols, B. Via, and E. C. McCray, *Tide and Current Glossary*. National Oceanic and Atmospheric Administration/National Ocean Service/Center for Operational Oceanographic Products and Services, 2000, Silver Spring, Maryland, http://co-ops.nos.noaa.gov/publications/glossary2.pdf (accessed 22 April 2008).

Holland, G. J. 1980: An analytic model of the wind and pressure profiles in hurricanes, *Mon. Wea. Rev.*, **108**, 1212–1218.

IMD Report, Report on cyclonic disturbances over North Indian Ocean during 1998, Abridged report for circulation during the meeting of WMO/ESCAP panel on Tropical cyclones. RSMC-Tropical cyclones. New Delhi, 1999, 71 pp.

IOC-UNESCO, Project proposal on storm surges disaster for the northern part of the Indian Ocean, Project no. IOC/INF-1131, Paris, 1999.

Jain, Indu, P. Chittibabu, Neetu Agnihotri, S. K. Dube, P. C. Sinha, and A. D. Rao, 2006a: Simulation of storm surges along Myanmar coast using a location specific numerical model, *Natural Hazards* **39**, 71–82.

Jain, Indu, P. Chittibabu, Neetu Agnihotri, S. K. Dube, P. C. Sinha, and A. D. Rao 2006b: Numerical storm surge prediction model for Gujarat coast of India and adjoining Pakistan coast, *Natural Hazards* **42**, 67–73.

Jarvinen, B. and M. Lawrence, 1985: An evaluation of the SLOSH storm-surge model. *Bulletin of the American Meteorological Society*, **66**(11), 1408–1411.

Jayanthi, N., Storm surges and its risk assessments over the coastal areas of Bay of Bengal and Arabian Sea. Presented at National Conference on Tropical Meteorology (Tropmet 99), Chennai, India. 16–19, February 1999.

Jelesnianski, C., J. Chen and W. Shaffer, SLOSH: Sea, lake, and overland surges from hurricanes. Technical Report NWS 48, National Oceanic and Atmospheric Administration, Silver Spring, Maryland (1992). At URL http://www.csc.noaa.gov/hes/images/pdf/SLOSH_TR48.pdf

Jelesnianski, C., J. Chen, W. A. Shaffer, and A. J. Gilad, 1984: SLOSH — a hurricane storm surge forecasting model. *Preprints, Oceans 84*, Washington, D. C., Marine Technology Society and IEEE/Oceanic Engineering Society, pp. 314–317.

Jelesnianski, C. P. and A. D. Taylor, 1973: *NOAA Technical Memorandum*, ERL, WMPO-3, 33 pp.

Johannessen, J., Lofti, A., and Harald J, 2006: ESA, Benefiting our Economy, Observing the Earth, http://www.esa.int/esaEO/SEMKZSMVGJE_economy_0.html

Johns, B. and A. Ali, 1980: The numerical modelling of storm surges in the Bay of Bengal, *Quart. J. Roy. Meteor. Soc.* **106**, 1–8.

Johns, B., S. K. Dube, U. C. Mohanty, and P. C. Sinha, 1981: Numerical simulation of the surge generated by the 1977 Andhra Cyclone, *Quart. J. Roy. Meteor. Soc.* **107**, 915–934.

Johns, B., P. C. Sinha, S. K. Dube, U. C. Mohanty, and A. D. Rao, 1983: Simulation of storm surges using a three-dimensional numerical model: an application to the 1977 Andhra Cyclone, *Quart. J. Roy. Meteor. Soc.* **109**, 211–224.

Johns, B., A. D. Rao, S. K. Dube, and P. C. Sinha, 1985: Numerical Modeling of the tide surge interaction in the Bay of Bengal, *Phil. Trans. Roy Soc. London* **A 313**, 507–535.

Joint Typhoon Warning Center (JTWC), *Northern Indian Ocean Tropical Cyclone Best Track Data*. Retrieved on 2007-06-04, 2007.

Kinnmark, I., *The Shallow Water Wave Equations: Formulation, Analysis, and Application*. Lecture Notes in Engineering No. 15, Springer-Verlag, 1986.

Knabb, R. D., J. R. Rhome, and D. P. Brown, Tropical Cyclone Report, Hurricane Katrina, 23–30 August 2005, National Oceanic and Atmospheric Administration/National Weather Service/National Hurricane Center, 2006, http://www.nhc.noaa.gov/pdf/TCR-AL122005_Katrina.pdf (accessed 23 April 2008).

Kolar, R., J. J. Westerink, M. Cantekin, and C. A. Blain, 1994: Aspects of nonlinear simulations using shallow models based on the wave continuity equation. *Computers and Fluids*, **23**(3), 523–538.

Luettich, R. and J. J. Westerink, 2004: Formulation and Numerical Implementation of the 2D/3D ADCIRC Finite Element Model Version 44.XX, http://adcric.org/adcirc_theory_2004_12_08.pdf (access 8 May 2008).

Luettich, Jr., R., J. J. Westerink and N. Scheffner, ADCIRC: An advanced three-dimensional circulation model for shelves, coasts, and estuaries. Technical Report DRP-92-6, US Army Corps of Engineers, Washington, D.C. (1992).

Lwin, T., Review of methods of storm surge prediction currently used in Burma, Report to the WMO Workshop on Storm Surges, Nov. 10–15, Rangoon, Burma, 1980, 8p.

Lynch, D. and W. Gray, 1979: A wave equation model for finite element tidal computations. *Computers and Fluids*, **7**(3), 207–228.

Massey, W. G., J. W. Gangai, E. Drei-Horgan, and K. J. Slover, 2007: History of Coastal Inundation Models, *Marine Technology Society Journal*, **41**(1), 7–17.

Mastenbroek, C., G. Burgers, and P. A. E. M. Janssen, 1993: The dynamical coupling of a wave model and a storm surge model through the atmospheric boundary layer. *J.Phys.Oceanogr.*, **23**, 1856–1867.

McGee, B. D., R. W. Tollett, and R. R. Mason, 2006a: Monitoring inland storm surge and flooding from Hurricane Rita: US Geological Survey Fact Sheet 2006–3136, 4 p.

McGee, B. D., B. B. Goree, R. W. Tollett, B. K. Woodward, and W. H. Kress, 2006b: Hurricane Rita Surge Data, Southwestern Louisiana and Southeastern Texas, September to November 2005. US Geological Survey Data Series 220. http://pubs.usgs.gov/ds/2006/220/

McInnes, K., G. Hubbert, D. Abbs and S. Oliver, 2002: A numerical modelling study of coastal flooding. *Meteorology and Atmospheric Physics*, **80**: 217–233.

Murty, T. S., *Storm Surges: Meteorological Ocean Tides*, Department of Fisheries and Oceans, Ottawa, Canada, 1984.

Murty, T. S. and R. F. Henry, 1983: *Tides in the Bay of Bengal. Journal of Geophysical Research*, **88**(c-10), 6069–6076.

Murty, T. S., R. A. Flather, and R. F. Henry, 1986: *The storm surge problem in the Bay of Bengal, Prog. Oceanog.* **16**, 195–233.

National Aeronautics and Space Administration, 2006: Life on Earth — Looking at Earth: accessed May 16, 2006, at URL http://www.nasa.gov/vision/earth/lookingatearth/h2005_rita.html

National Hurricane Center 2007: Average NHC 5-Year Average Forecast Errors (2002–2006), All Tropical Cyclones. http://www.nhc.noaa.gov/verification/pdfs/OFCL_5-yr_averages.pdf, accessed June 3, 2008.

Office of Hydrologic Development (OHD), Real-Time (Dynamic) Inundation Mapping Evaluation (R-Time) Team Report, December 2007 (in review).

Office of Hydrologic Development (OHD), Evaluation of Hydraulic Models, June 2007.

Paudyal, G. N. 2002: Forecasting and warning of water-related disasters in a complex hydraulic setting—the case of Bangladesh. *Hydrological Sciences*, **47**(S), S5–S18.

Pedlosky, J. 1987: *Geophysical Fluid Dynamics*, 2nd edition, Springer, New York, 728 pp.

Peng, M., L. Xie and L. Pietrafesa, 2004: A numerical study of storm surge and inundation in the Croatan-Albemarle-Pamlico estuary system." *Estuarine, Coastal and Shelf Science*, 59, 121–137.

Pietrafesa, L. J., K. Kelleher, T. Karl, M. Davidson, M. Peng, S. Bao, D. Dickey, L. Xie, H. Liu, and M. Xia, 2007: A new architecture for coastal inundation and flood warning prediction, *Marine Technology Society Journal*, **40**(4), 71–77.

Powell, M., S. Houston, L. Amat and N. Morrisseau-Leroy, 1998: The HRD real-time hurricane wind analysis system. *Journal of Wind Engineering and Industrial Aerodynamics*, **77–78**, 53–64.

Powell, M., P. Vickery and T. Reinhold, 2003: Reduced drag coefficient for high wind speeds in tropical cyclones. *Nature*, **422**: 279–283.

Pugh, D. T., 2004: *Changing Sea Levels: Effects of Tides, Weather, and Climate.* Cambridge University Press, Cambridge, 280 p.

Rao, A. D., 1982: *Numerical storm surge prediction in India*, Ph.D. thesis, IIT Delhi, New Delhi, 211 pp.

Rao, A. D., S. K. Dube, and P. Chittibabu, 1994: Finite difference techniques applied to the simulation of surges and currents around Sri Lanka and Southern Indian Peninsula, *Computational Fluid Dynamics* **3**, 71–77.

Rao, N. S. B., 1968: On some aspects of local and tropical storms in the Indian area. Ph.D. Thesis, University of Jadavapur, India, 262 pp.

Rao, N. S. B. and S. Mazumdar, 1966: A technique for forecasting storm waves, *Indian Journal of Meteorology and Geophysics* **17**, 333–346.

Rao, Y. R., P. Chittibabu, S. K. Dube , A. D. Rao, and P. C. Sinha, 1997: Storm surge prediction and frequency analysis for Andhra coast of India, Mausam **48**, 555–566.

Reed, D. B. and B. E. Stucky, 2005: Forecasting Hurricane Storm Surge on the Mississippi River. *Proceedings of ASCE Conference "Solutions to Coastal Disasters 2005"*, Charleston, SC, May 8–11.

Riverside Technology, Inc., Tar River Calibration, 2006.

Riverside Technology, Inc., FLDVIEW Evaluation, 2007.

Roy, G. D., 1984: *Numerical storm surge prediction in Bangladesh.* Ph.D. Thesis, Indian Institute of Technology, Delhi, 188 pp.

Roy, G. D., A. B. M. Humayuan Kabir, M. M. Mandal, and M .Z. Haque, 1999: Polar coordinates shallow water storm surge model for the coast of Bangladesh, *Dyn. Atmos. Oceans* **29**, 397–413.

Schuler, D. L. and Lee, J. S, 2004: Measurement of Ocean Wave Spectra and Surface Slopes by Polarimetric SAR, Remote Sensing Division, http://www.nrl.navy.mil/content.php?P=04REVIEW186

Systems Engineering Australia Pty Ltd, 2005: *Darwin TCWC Northern Region Storm Tide Prediction System — System Development Technical Report*, Bureau of Meteorology Darwin, 208 pp.

Sheng, Y. P., 1986: *A Three-Dimensional Mathematical Model of Coastal, Estuarine and Lake Currents Using Boundary-Fitted Grid.* Technical Report No. 585, Aeronautical Research Associates of Princeton, Princeton, New Jersey.

Sheng, Y. P. and V. Alymov, 2002: Coastal Flooding Analysis of Pinellas County Using ALSM Data: A Comparison between UF's 2-D Method and Results vs. FEMA's Method and Results. Report, Department of Civil and Coastal Engineering, University of Florida.

Sielecki, A., 1968: An energy-conserving difference scheme for the storm surge equations, *Mon. Weather Rev.* **96**, 150–156.

Stamey, B., H. Wang, and M. Koterba, 2007: Predicting the next storm surge flood, *Sea Technology*, **48**(8).

Thompson, E. and V. Cardone, 1996: Practical modeling of hurricane surface wind fields. *Journal of Waterway, Port, Coastal, and Ocean Engineering*, **122**(4), 195–205.

Unisys, http://weather.unisys.com/hurricane/indian_oc/index.html.

Walters, R. and G. Carey, 1984: Numerical noise in ocean and estuarine models. *Advances in Water Resources*, **7**, 15–20.

Weisberg, R. and L. Zheng, 2006: Hurricane storm surge simulations for Tampa Bay. *Estuaries and Coasts*, **29**(6A), 899–913.

Weiyan, T., 1992: *Shallow Water Hydrodynamics.* Elsevier Oceanography Series, V. 55, Amsterdam, 434 p.

Westerink, J. J., R. A. Luettich, J. C. Feyen, J. H. Atkinson, C. N. Dawson, M. D. Powell, J. P. Dunion, H. J. Roberts, E. J. Kubatko, and H. Pourtaheri, 2008: A Basin to Channel Scale Unstructured Grid Hurricane Storm Surge Model Applied to Southern Louisiana. *Mon. Wea. Rev.*, 136, 833–864.

Westerink, J. J, R. A. Luettich Jr., and J. A. Muccino, 1994: Modeling tides in the western North Atlantic using unstructured graded grids. *Tellus*, **46A**, 178–199. .

WMO Draft Report. Reports of the Workshop on Integrated Activities of the Hydrology, Meteorology and DPP Components of the Typhoon Committee into Related International Frameworks for Disaster Risk Management for Better Impacts and Visibility: Hydrology Component. Economical and Social Commission for Asia and the Pacific and World Meteorological Organization, Macao, Republic of China 4–9 September 2006.

Wright, C.W., E. J. Walsh, W. B. Krabill, D. Vandemark, A. W. Garcia, P. G. Black, F. D. Marks, Jr., R. A. Luettich, Jr., 2002: Real-time storm surge measurement with a scanning radar altimeter. Geoscience and Remote Sensing Symposium, 2002. IGARSS apos; 02. 2002 IEEE International, **3**, pp. 1492–1495.

Xie, L., L. Pietrafesa and K. Wu, 2003: A numerical study of wave-current interaction through surface and bottom stresses: Coastal ocean response to Hurricane Fran of 1996. *Journal of Geophysical Research*, **108**(C2).

Zhang, Y.-L., A. M. Baptista, and E. P. Myers, 2004: A cross-scale model for 3D baroclinic circulation in estuary-plume-shelf systems: I. Formulation and skill assessment. *Continental Shelf Research*, **24**, 2187–2214.

PART VI

Societal Impacts of Tropical Cyclones

Chapter 13

Disaster Mitigation and Societal Impacts

David King
Centre for Disaster Studies, James Cook University,
Townsville, Qld 4811, Australia
David.King@jcu.edu.au

Jim Davidson
(Australian) Bureau of Meteorology
GPO Box 413, Brisbane, Qld 4001, Australia
j.davidson@bom.gov.au

Linda Anderson-Berry
(Australian) Bureau of Meteorology
GPO Box 1289, Melbourne, Vic 3001, Australia
l.anderson-berry@bom.gov.au

Throughout the world societies are experiencing rapid social change that is making them more vulnerable to the impacts of natural hazards. One of the aspects of social change is human impact upon the environment that has contributed to climate change, which may be contributing to a likely increase in the severity and possible increase in frequency of tropical cyclones (IPCC, 2007). Alongside increased vulnerability, societies exhibit strengths and resilience that are targeted at enhancing social capacity that in turn will mitigate the impacts of tropical cyclones.

1. Introduction

Disaster mitigation and societal impact are separate, interrelated concepts. Within each of these we can identify many subgroups and connecting factors. In order to understand the types and range of factors, Figure 1 is used to deconstruct the key elements of disaster mitigation and societal impact in the tropical cyclone context.

These elements have been grouped into four main areas that form the basis of this chapter — tropical cyclone impacts, vulnerability, resilience and mitigation (preparation and prevention).

1.1. *Global Impact of Tropical Cyclones*

It is the gradual onset of tropical cyclones that enable people and institutions to make preparations to reduce the extent of the impact, although a few storms intensify rapidly close to landfall. A further predictability of tropical cyclones is their seasonality and regularity of occurrence. These characteristics make it incumbent upon people, their communities, organisations and governments to take actions that will mitigate the impact and reduce the level of the disaster that will inevitably accompany landfall of tropical cyclones.

Tropical Cyclone Impacts	Phenomena (intensity, frequency, incidence) — structural and societal (property, environment, infrastructure, lifelines, industry) — intangible social losses
Vulnerability	Understanding and classifying vulnerability — risk equation — societal — community — household — individual — influence of climate change
Resilience	Indicators of resilience (social capital) — role of organisations (direct, indirect, informal) — best practice strategies for resilient communities
Mitigation	Total risk — physical measures — education and awareness — tropical cyclone warnings — capacity building (partnerships, empowerment, good governance)

Figure 1. Key elements of disaster mitigation and societal impact.

The format of this review begins by identifying the impacts of tropical cyclones on society and its institutions, structures and infrastructure, and moves on to examine social vulnerability and the positive aspects of social resilience, thereby providing a basis for an examination of mitigation through prevention and preparedness.

Globally disasters consume enormous quantities of resources. The following figure and tables summarise global impacts from the beginning of the 20th century, since the International Decade for Natural Disaster Reduction (IDNDR) commenced in 1990, and from the turn of the 21st century. Data are derived from the Emergency Events Database (EM-DAT) of the Centre for Research on the Epidemiology of Disasters (CRED). The disasters recorded are identified on the basis of each event meeting at least one of four criteria: 10 or more people were reported to have been killed; 100 or more people were reported to have been affected; the affected government made a call for international assistance; or there was a declaration of a state of emergency.

Figure 2 displays the natural disasters that have occurred worldwide since 1900. However, to understand the impact of tropical cyclones it is important to include the context of all types of disasters, as they involve many of the

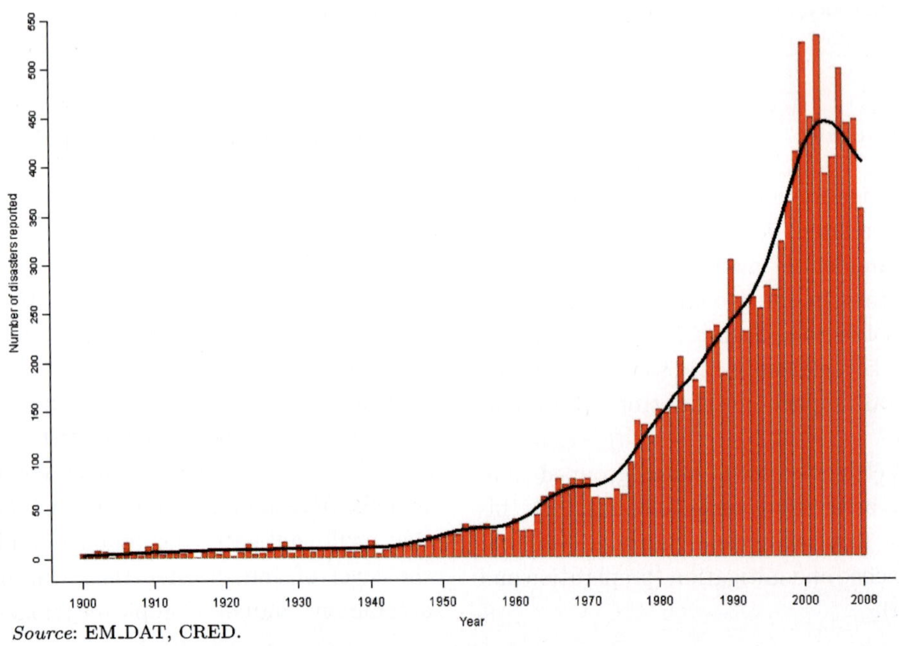

Source: EM_DAT, CRED.

Figure 2. Number of natural disasters worldwide 1900 to 2008.

same emergency managers and institutions that respond to both natural and human induced disasters. There is also a fuzzy boundary in the case of complex disasters, when the impact of natural hazards is enhanced as a consequence of war or the breakdown of governance.

Tropical cyclones fall into the category of wind storms, which also include tornadoes and non-tropical storms, but this category is dominated by tropical cyclones, which are also responsible for some of the wave and surge events, even though impacts in this category have been dominated by tsunamis. Additionally tropical cyclones are directly responsible for some of the flood events and indirectly responsible for many more, as a consequence of ex-cyclonic weather systems continuing to inundate without destructive winds, or others that transform after interacting with frontal systems in the middle latitudes.

As seen in Table 1, tropical cyclones and floods are the primary natural disasters, with wind storms causing an average of 97 disasters a year and 125 floods annually during this and the previous decade: a regularity and predictability that demands strong mitigation efforts from all governments. This is further exemplified by the number of deaths shown in Table 2, when in the last eight years almost 28,000 people or 3,500 a year have perished in wind storms. Deaths in floods were almost 20,000 higher. Under "wave and surge", the latter was a minor killer with the exception of the 2004 tsunami.

The trend of the last two decades is the recent sharp increase in the number of deaths, possibly as a consequence of the numbers and severity of tropical cyclones having recently increased. The trend in the 20th century had been a decrease in deaths, with the costs of disasters continuously escalating (ISDR, 2008) particularly in the more developed parts of the world. Developing countries had experienced rapid population growth and continued to experience much higher death rates. By the mid-1990s there were encouraging signs that death rates were reducing. However, the late 1990s and the present decade have been at a time of increasing numbers of natural hazards as

Table 1. Numbers of natural disasters by type 1990 to 2007 worldwide.

	Drought	Earthquake	Epidemic	Extreme Temp.	Flood	Insect Infestation	Slides	Volcano	Wave/ Surge	Wildfire	Wind Storm	Total
1990	7	43	19	13	57	0	7	2	0	4	136	288
1991	12	29	43	7	75	1	12	10	1	8	66	264
1992	18	25	27	7	61	1	10	5	0	8	76	238
1993	11	20	4	5	85	3	21	6	1	2	108	266
1994	12	25	9	10	88	0	7	6	1	13	81	252
1995	11	26	24	13	95	4	15	5	0	7	80	280
1996	7	12	40	5	92	0	24	5	2	5	77	269
1997	14	22	57	13	93	1	13	4	3	16	80	316
1998	22	29	78	12	94	1	22	4	1	18	88	369
1999	14	33	75	8	119	0	18	5	2	22	107	403
2000	22	31	112	31	153	2	29	5	2	30	103	520
2001	29	25	65	23	158	2	21	6	0	13	108	450
2002	28	36	87	14	172	0	19	7	1	23	118	505
2003	15	40	58	23	154	0	20	2	0	14	86	412
2004	14	30	35	19	134	12	17	5	11	8	126	411
2005	16	25	50	22	189	0	12	8	0	12	130	464
2006	18	23	58	32	233	0	19	12	1	10	77	483
2007	11	19	47	22	203	0	10	6	3	18	103	442
Total	281	493	888	279	2255	27	296	103	103	231	1750	6632

Source: EM-DAT, CRED.

Table 2. Deaths worldwide 2000 to 2007 by disaster type.

	2000	2001	2002	2003	2004	2005	2006	2007	Total	%
Complex Crisis	0	0	610,000	0	0	0	0	134	610,134	49.3
Drought	31	261	722	152	80	14	147	0	1,407	0.1
Earthquake	216	21,348	1,634	29,617	882	76,241	5,890	654	136,482	11.0
Epidemic	6,973	6,672	10,772	3,945	2,897	3,921	3,555	8,826	47,561	3.8
Extreme Temp	941	1,787	2,669	74,797	856	536	5,464	926	87,976	7.1
Flood	6,014	5,060	4,272	3,276	7,345	5,618	5,994	8,349	45,928	3.7
Industrial	1,807	1,279	1,112	1,444	1,797	2,281	1,822	1,544	13,086	1.1
Misc. Accident	1,341	1,341	2,013	1,438	2,115	2,669	1,120	793	12,830	1.0
Slides	1,023	725	1,145	451	612	646	1,647	264	6,513	0.5
Transport	8,168	5,913	9,303	7,868	6,386	6,669	7,021	4,122	55,450	4.5
Volcano	0	0	200	0	2	3	5	11	221	0
Wave/surge	1	0	0	0	190,934	0	802	64	191,801	15.5
Wild Fire	47	33	6	47	14	44	18	151	360	0
Wind Storm	1,361	1,914	1,358	1,031	6,556	5,238	4,428	5,970	27,856	2.3
Total	27,923	46,333	645,206	124,066	220,476	103,880	37,913	31,808	1,237,605	100

Source: EM-DAT, CRED.

Table 3. Damage in million US$ worldwide 2000 to 2007 by disaster type.

	2000	2001	2002	2003	2004	2005	2006	2007	Total	%
Drought	2,010	5,131	8,972	938	3	3,450	3,135	0	23,640	3.6
Earthquake	413	7,363	2,068	8,253	28,614	6,705	3,376	16,312	73,104	11.2
Epidemic	0	0	0	0	0	0	0	0	0	0
Extreme Temperature	370,159	200	0	12,520	0	400	1,000	0	14,490	2.2
Flood	25,801	4,753	27,006	20,533	10,687	17,598	8,145	23,588	138,109	21.0
Infestation	120	0	0	0	0	0	0	0	120	0.02
Slides	462	71	13	45	11	55	40	0	696	0.1
Volcano	2	17	9	0	0	0	150	0	178	0.03
Wave/Surge	0	0	0	0	8,204	0	55	0	8,259	1.3
Wild Fires	2,559	0	452	6,095	3	3,747	939	4,597	18,392	2.8
Wind Storm	12,462	14,520	14,756	21,334	84,189	184,760	17,736	29,888	379,645	57.8
Total	44,199	32,054	53,275	69,717	131,711	216,715	34,577	74,385	656,634	100

Note: No values were recorded for complex disasters and epidemics.
Source: EM-DAT, CRED.

well as an increase in the number of deaths. 2008 saw a marked increase in the number of deaths and economic losses compared to the 2000–2007 yearly average.

The financial impact, as shown in Table 3, shows that the impact of tropical cyclones is dominant, with 58% of the costs of all natural disasters during this decade. A great deal of this cost has been the remorseless impact of hurricanes on densely populated highly developed areas of the USA. Floods also made a significant financial impact. During 2006 natural hazards cost 0.5% of the world's GDP and in 2007 1.0% (Wikipedia, 2008; World Bank, 2009). An impact of around 1% of global GDP is both significant and economically and socially unsustainable (Sathaye, 2007) when it is put against the growth rate of GDP of 7.3% (based on World Bank Purchasing Power Parity adjusted figures). This cost of more than 1% of GDP is consumed by disaster recovery and rebuilding as well as absolute loss. The grand mean of all GDP growth rates of the countries of the world, reduces the global rate of growth to only 5.1%. This is a more realistic growth rate for the unequal impact that often falls on less developed countries. Many small and poor countries experienced disaster losses greater than their GDP or more frequently greater than the total government budget. The approximately 1% of global GDP that is lost in disaster impact represents resources that would otherwise contribute to development and increase in standards of living.

Furthermore, on top of the direct and tangible costs are the greater intangible losses of lack of new developments, and the economic constraints of loss of both existing and planned infrastructure. The trend of increasing costs of tropical cyclones is an unsustainable burden that has to be the target of strengthened mitigation efforts.

Deaths and financial losses are at the sharp end of the impact of tropical cyclones. Additionally there are many more people who are injured, and even greater numbers made homeless and an extensive population that is affected by the impact. Table 4 summarises the impact of wind storms since 1900. Over the last 108 years over 796 million people were affected by wind storms, and over 3 billion impacted by floods, in both cases primarily in Asia, with the North American impact much less but still very significant. The global damage bill from wind storms over this period was US$690 billion, averaging US$223 million and 258,000 people affected by each event. In comparison, each flood affected more people but resulted in less damage.

Pielke *et al.* (2006) warn that because CRED's database is compiled from a range of reports and estimates, the data should be regarded as indicative rather than absolute, and have consequently recommended the use of cost-benefit analyses to measure the impact of mitigation strategies. Pielke also examined contrasting scenarios of global warming in relation to economic impacts and concluded that "there is evidence that changing patterns of extreme events are drivers for recent global losses." (Pielke *et al.*, 2006, p. 555) The impacts of both wind and flood events are enormous and have consumed significant resources throughout history, while global climate change may contribute to increases in these impacts.

1.2. *Definitions of Relevance*

This chapter is grounded within the structures and thus the terminology of emergency management, as it is this sector that has the primary responsibility for the safety and preparation of communities that are at risk from the impact of tropical cyclones. The formal terminology of emergency management also has an advantage in being relatively accessible and understood between countries. The primary terms that are used in this chapter are as they are defined in the glossary produced by Emergency Management Australia (Emergency Management Australia 1998).

Tropical cyclone is used throughout this chapter to include hurricanes and typhoons as it is a broader categorisation, but specific events are referred to by their local name such as Hurricane Katrina, Cyclone Zoe or Typhoon Tokage.

Societal risk is formally defined as "*the risk of a number of fatalities occurring.*" (Emergency Management Australia, 1998). This can

Table 4. Summarised table of wind storms from 1900 to 2007.

	Number of events	Killed	Injured	Homeless	Affected	Total Affected	Damage $US Millions
Global	3,086	1,213,742	1,262,193	50,641,659	744,313,701	796,217,553	689,627
Mean per event		393	409	16,410	241,190	258,009	223.469

Note: Affected means people who required immediate assistance during the period of an emergency; it can also include displaced or evacuated people.
Source: EM-DAT, CRED.

be related directly to the risk equation — risk = hazard × vulnerability — or — risk = hazard × vulnerability × elements at risk. Society's risk is identified with vulnerability, although the elements at risk may seriously impact upon social losses and recovery.

Impact is "*a sudden occurrence without prior warning*", and the impact area is defined as "*any area which is likely to bear, is bearing, or has borne the full impact of any disaster and in which major lifesaving operations are necessary*" (Emergency Management Australia, 1998).

Disaster is formally defined as "*a serious disruption to community life which threatens or causes death or injury in that community and/or damage to property which is beyond the day to day capacity of the prescribed statutory authorities and which requires special mobilisation and organisation of resources other than those normally available to those authorities*" (Emergency Management Australia, 1998).

Thus **mitigation** which is "*measures taken in advance of a disaster aimed at decreasing or eliminating its impact on society and environment*" and **prevention** which is "*regulatory and physical measures to ensure that emergencies are prevented, or their effects mitigated — measures to eliminate or reduce the incidence or severity of emergencies*" (Emergency Management Australia, 1998) are partially overlapping.

Vulnerability is seen as a function of susceptibility to loss and the capacity to recover while the capacity to recover from a disastrous impact is termed **resilience**.

The concepts of **prevention** or **reduction** in the Reduction Readiness Response and Recovery (**RRRR**) or Preparedness, Prevention, Response and Recovery (**PPRR**) models of emergency management refer to physical structures such as levees, barriers, building codes etc. that could lessen the damage caused by the hazard.

Readiness or **preparation** concerns those actions related to education and awareness, that individuals or households could undertake to reduce their personal risk.

Over the past decade all of these emergency management concepts of reduction, prevention, readiness and preparation have been grouped under the broader umbrella of **mitigation**. When emergency managers and institutions refer to mitigation it can be assumed that this includes both the physical actions as well as the awareness raising and community activities.

The **Tropical Cyclone Category Scale** referenced later in the chapter is the one used operationally in Australia and the Southwest Pacific. This choice is merely one of convenience. In simple terms, Category 1 systems have gale force winds associated with them, Category 2 have storm force winds, while Categories 3 to 5 have hurricane force winds of increasing strength.

2. Tropical Cyclone Impacts

Having considered the global impact of hazards, the issue of mitigation requires a focus at the local or community level. Mitigating tropical cyclone disaster means that measures have to be in place that both reduce the destruction and damage and prepare people to endure the event as safely as possible.

Impacts can be separated into physical characteristics of tropical cyclones, and their impact on human structures, and the effects of that damage on people and communities. Furthermore there are non-structural social impacts that result from isolation, displacement and refugee migration. Some of these impacts are temporary but many can be long term or permanent. So what are the impacts of tropical cyclones?

2.1. *Phenomena Impacts*

Phenomena impacts of tropical cyclones are the consequence of the characteristics of the storms. Wind damage is the primary destructive force, but storm surge, wave set up and extensive flooding can wreak the most devastating effect. The greatest likelihood of fatalities results from inundation and consequent death by drowning. Severe cyclones in Bangladesh, Orissa in India (Chittibabu *et al.*, 2004) and New Orleans USA (Congleton, 2006) provide examples of high death rates from surge and flooding. Historically many ships and fishing fleets were lost in tropical cyclones. Australia's single greatest loss of life in a disaster remains the destruction of a pearling fleet by Cyclone Mahina in 1899 (Holthouse, 1971). As communications have improved and ships have become larger, loss of life at sea diminished throughout the 20th century, but low technology fishing boats in less developed countries and recreational sailing craft remain vulnerable and continue to be lost at sea. A further impact of surge is the permanent damage caused to buildings, infrastructure such as roads and drains, and to farmland and coastal freshwater sources by saltwater inundation.

Wind damage is influenced by terrain and vegetation. While no landscape features can be relied on to provide protection, exposed locations and landforms that enhance funnelling effects can increase the impacts of the wind. Some types of vegetation can provide protection and withstand cyclonic winds, but vegetation can become missiles that increase damage to structures. Wind damage is complicated by the reversal or change in direction that accompanies the passage of the eye of the cyclone. Damage debris is picked up from a new direction to attack previously protected or undamaged sides of buildings. These impacts are further mitigated or enhanced by the speed with which the tropical cyclone crosses a community (as well as the size of the storm). Prolonged battering that successively adds to the flying debris, multiplies the scale of damage as well as adding to the trauma of the experience for the sheltering inhabitants (Henderson and Harper, 2003).

The phenomena impact of a tropical cyclone is therefore a combination of a number of factors that create a multiplier of damage and stress. We can qualify these in terms of intensity, frequency and onset.

2.1.1. *Influence of Tropical Cyclone Intensity*

There is an obvious correlation between structural damage and the intensity of a tropical cyclone, as predicted by the standard warnings of types of damage associated with each level of intensity. As far as social impact is concerned the levels of impact are not simply incremental. At Categories 1 to 3 most structural damage is relatively superficial, or at least within the realms of human understanding. Building codes may ensure that residential dwellings are built to standards that may withstand winds at Category 3 intensity. At this level vegetation damage is especially severe with flimsy buildings, out-houses and fences suffering greatest damage. In most societies, even relatively poor communities have the capacity to clear up and repair damage wrought by Categories 1 to 3 tropical cyclones. Intensities of Categories 4 and 5 go significantly beyond the general community resilience and capacity to recover. Damage levels and losses are multiplied at higher intensities with Category 5 cyclones wreaking extreme damage and heavy losses. Exacerbating structural damage at higher category levels are vortices within tropical cyclones that function like tornadoes, resulting in random and localised severe destruction.

2.1.2. *Influence of Tropical Cyclone Frequency*

Societal recovery from a single catastrophic tropical cyclone is a long and traumatic experience for impacted communities. Where tropical cyclones return in a short period, or bring destruction year after year, communities may find themselves unable to recover. This is especially the case in developing countries where poor households lack the resources to return to a pre-tropical cyclone state. The social impact of frequent catastrophic tropical cyclones (Pielke and Pielke, 1997) is a long-term loss of economy and population. Younger people migrate away from the disaster zone to seek opportunities elsewhere (Hunter, 2005), while the elderly, having lost a lifetime's possessions, become particularly vulnerable to sickness and depression and many are often described as having died from heartbreak.

Climate change scientists are predicting increasing intensity and possible increase in the frequency of tropical cyclones (ISDR, 2007; IPCC, 2007). With increasing frequencies we may see the development of long term disaster zones with dislocated economies and declining populations, while with increasing intensity there is a likelihood of higher death rates (Terry, 2007).

2.1.3. *Influence of Tropical Cyclone Incidence*

Improved technology is providing enhanced knowledge of the genesis, tracks and regions of impact of tropical cyclones. Effective warnings enable communities to make preparations to reduce physical impacts by clearing loose items and securing property before taking effective shelter. However, tropical cyclones are highly variable in their development, their onset and landfalling track. They also occur within the context of earlier weather conditions, such that a region may already be flooded, or saturated before the cyclone arrives. This would increase the physical impact as support services are unable to access the region and vegetation is more easily whipped from the ground, becoming the windborne debris that will exacerbate damage levels.

These factors are aspects of the physical impact that may increase hazard risk. Actions to mitigate such physical impacts are the responsibility of households and communities, which require more detailed knowledge of physical impacts in order to make decisions about structures and family safety. In many contexts, such as isolated regions and poorly resourced developing countries, this information may not be available and communities are left to make decisions with limited knowledge and warnings.

2.2. *Structural and Societal Impacts*

Physical impacts of tropical cyclones translate into specific structural and social losses. These losses can be summarised as being the result of damage to structures, the environment, infrastructure, lifelines, commerce, industry, and the economy. Social losses are often intangible.

2.2.1. *Impacts on Buildings, Houses and Property*

Damage to houses ranges from minor repairs to complete destruction and homelessness. Less substantial structures in developing countries fare the worst, and although they may be cheaply replaced from local materials, tropical cyclones often strip away the palm fronds and vegetation that are essential for reconstruction. Homelessness may contribute to extended periods of relocation and dependency. While wind damage may be the principal damage agent to people's homes, it is accompanied by heavy rain that adds water damage to all furniture and belongings, including personal and irreplaceable items. The loss of household goods and memorabilia increases the trauma and loss for people and may contribute to inability to recover, including stress related and mental illness (Henderson and Harper, 2003).

Water damage alone may make a structure unrepairable. Storm surge is even more destructive, but even where buildings survive flooding by seawater, its corrosive damage may often result in a building having to be condemned especially if significant levels of seawater have entered the building. Apart from residential dwellings, households lose or sustain damage to outbuildings, sheds, farm buildings, vehicles, fences, landscaping, access roads and power lines. In many cases these represent more than an additional cost, as the loss of such external structures and facilities may compromise a household's business or livelihood.

2.2.2. *Impacts on the Environment*

Vegetation provides society with meaning, amenity, ecology and livelihood. Meaning attaches to trees and plants of iconic, spiritual or historical significance. These meanings may be highly culturally specific or may simply relate to individual families. The loss of trees or groves of cultural significance intensifies the sense of loss, and contributes negatively to recovery. Related to this is the ecological significance of forests, woodlands and fragile ecosystems (Everham and Brokaw, 1996), where cyclone damage that kills or displaces species of birds and animals not only alters the ecological balance but may contribute further loss to the community. Amenity is the attractiveness and usefulness of vegetation, providing diversity and softening of the landscape, shade, meeting places, buffers and privacy. These intangible social characteristics are lost or transformed after the storm, especially during the short term of the recovery period.

Farmed plants, crops and trees are especially vulnerable to wind and flood damage. Economy, livelihoods and food supplements are impacted or lost in tropical cyclones, with recovery constrained, especially in poorer societies. Cyclone Zoe that struck Tikopia in the Solomon Islands in 2002, caused extensive damage to taro gardens, destroyed food bearing trees, and storm surge removed topsoil and freshwater sources (Anderson-Berry *et al.*, 2006).

2.2.3. *Impacts on Infrastructure and Lifelines*

Wind and water damage to roads, drains, fences, powerlines, community buildings, bridges, harbours and commercial and retail premises bring about layers of secondary crises that prevent or slow relief efforts and may require the rebuilding of social infrastructure before recovery can start. Frequently community services and infrastructure are concentrated in more vulnerable locations, such as town centres adjacent to ports or rivers. Lifelines are the critical infrastructure necessary to service the recovery of an impacted community. Often school buildings are designated as recovery centres, but may be unusable. Additionally the loss or closure of a school also holds back the recovery of children. International emphasis has been placed on the particular vulnerability of critical infrastructure (Buckle, 2005; Queste, 2005; Shinozuka, 2005).

2.2.4. *Impacts on Commerce, Industry and the Economy*

Public service and government organisations may continue to function after a disaster, albeit under altered circumstances and priorities. People employed in the public sector continue to be on the payroll and are directly involved in recovery activities. However, many farms, small businesses, shops and manufacturing facilities are likely to be put out of business in the aftermath of a tropical cyclone, either through direct damage to premises, crops and machinery, or through disruption of supply and communications. Even where insurance covers business losses the disruption to the market may be unsustainable, especially if the population is reduced through evacuation and homelessness. Some builders and tradespeople may benefit from additional work, but for many private enterprises a crisis level is reached through the loss of economic activity and consequent unemployment, loss of income and an inability to rebuild and recover (Pielke, 2006; Wong, 2006).

2.2.5. *Intangible Social Impacts*

The physical impacts of wind and water damage create layers of crisis and disruption that contribute to the complexity of both the impact and recovery. Most of the areas of impact outlined above are direct and measurable physical impacts, although some of these contribute to intangible impacts. The loss of a farmer's crop can be measured, but the loss of a sacred tree or roadside and riverside forest with their associated ecosystems cannot be measured, but have profound impacts on people and communities.

Other intangible impacts of tropical cyclone disaster concern damage to the social fabric through isolation, loss of services, displacement of families and evacuation. These contribute to the trauma of the event although reactions are very individual. Some people may even find themselves strengthened by a crisis, especially where they provide leadership, but displacement, evacuation and homelessness throw people into a temporary nowhere place of dependency and loss.

With any tropical cyclone there are also a number of related secondary health impacts. Sewerage systems are compromised and may leak into the environment or become unusable, resulting in gastric and bacterial illnesses and infections. This may contribute directly to pollution of a region's water supply, but even without sewerage problems water supplies are often cut off by infrastructure damage and power supply disruption, and polluted from general run off. High humidity levels that accompany cyclone impact, result in mould and additional bacterial infections, as well as an increase in mosquito borne diseases. In industrial areas toxic spills or releases are a further health and environmental hazard that can result from both wind and flood damage.

These experiences are made much more severe by repeated impacts of tropical cyclones. While community preparedness and awareness is dramatically enhanced by the direct experience of a tropical cyclone impact, households that endure repeated impacts will ultimately be pushed to the limits of their endurance. Following any disaster some people may leave the community, but after repeated disasters both economic necessity and the traumatic stress will force significant outmigration and a long-term reduction in the resilience of the community.

3. Vulnerability

3.1. *Understanding and Classifying Vulnerability*

The impact of tropical cyclones on society varies in relation to the physical factors that have already been outlined, and by the vulnerability and resilience of individuals, households and communities. Individual characteristics of vulnerability and resilience to hazards are effectively contained within the household which, with its dependency on the dwelling is the most basic component of the larger community. When we refer to the societal or social impact, it is generally subdivided to the community or communities. While this term is increasingly used as a substitute word for "the public" it has specific meanings that provide the basis of understanding hazard vulnerability, resilience and mitigation.

The community is a group of people with common attributes who have a sense of connection and belonging. Emergency managers consider a community to be a geographical locality and they assume connections through neighbourliness and proximity. The problem with such a view of community is that in urban areas people form complex, spatially extensive communities, based on networks and associations that transcend the geographical neighbourhood (Ilf, 1995). Additionally communities are of different ages and exhibit a great deal of variability in their strengths. On top of such variations, people are also simultaneously members of multiple communities — workplace, school, religious, recreational etc as well as place of residence. These networks of communities contribute to social capital which is a powerful component of resilience. Some of the social networks are well-suited to respond to a disaster, while other communities may reinforce their structural social disadvantages (Mileti, 1999). These variations in vulnerability and resilience strongly influence the strength of disaster impact on the capacity to recover.

Vulnerability and resilience are not opposite sets of characteristics, although some individual variables may exist at opposite ends of the scale, such as the vulnerability of poverty and some resilience afforded by wealth (Passerini, 2000). Generally characteristics of society or community may be classified as increasing vulnerability or increasing resilience, but often on parallel scales. A community may be both highly vulnerable and extremely resilient at the same time — for example a poor coastal village that has strong social cohesion and well developed leadership.

The risk equation — Total Risk = Hazard × Vulnerability × Elements at Risk — (as modified by Granger *et al.*, 1999) establishes vulnerability as a major contributor to hazard risk. It is a human construct made up of individual, household and community characteristics of demography, society, knowledge, experience, resources and attitudes. Very broadly it consists of those characteristics that increase susceptibility to harm or loss (Blaikie *et al.*, 1994). At-risk groups in the developed world have been recognised in North American, European and Australian studies as low income households, women, the very young, the elderly, unemployed, the disabled, large families, single-parent families, newcomers and migrants, ethnic minorities and female headed households (Peacock *et al.*, 1997; Blaikie *et al.*, 1994; Mileti, 1999; Bolin, 1994; Fothergill *et al.*, 1999; Granger, 1997).

A difficulty in measuring vulnerability to tropical cyclones is deciding which characteristics of people and communities to incorporate into a list of vulnerability characteristics and then how to balance and weight specific characteristics. Many individual, household and community characteristics can be extracted from censuses, which are easily available to all levels of government and allow quantification at small geographic locations. However there are some dangers in quantifying these characteristics. One problem is that of assigning a relevant weighting to a population characteristic. Then there are all the characteristics of people and their households that are not recorded in a census, including social involvement, knowledge and awareness and experience of cyclones. Finally there are the issues of type of community, membership of communities, networks and locations of other communities to which people belong. At the very least, the extended family and a network of close friends is usually of far greater significance than immediate neighbours.

At a fundamental level, the census and local government inventories of facilities and services enable the identification and quantification of specific vulnerable groups, such as the elderly, single parent families, or families in locations where evacuation may be necessary. In the cyclone warning phase, knowledge of actual numbers of people in households that may require assistance is a necessary part of logistical planning, especially in protecting lives. In the same way vulnerable lifelines and infrastructure may be identified and quantified in terms of protection and first response, but these are only indicators of specific vulnerabilities which do not explain the bigger picture of vulnerability that is necessary for effective mitigation and recovery.

Anderson-Berry's research of vulnerability to tropical cyclones in North Queensland Australia, identified a wide range of characteristics, variables and physical elements that together help us to identify vulnerability and contribute to its understanding (Anderson-Berry and King, 2005). These indicators are illustrated in Fig. 3 as a model of vulnerability and resilience.

Characteristics of individuals, communities and the social setting are expressed in the model as indicators. In the figure the indicators are grouped into categories, such as income for example, where a specific indicator might be the percentage or number of people, or households, living in poverty, as it is defined in a particular country. A corresponding indicator of capacity or lower vulnerability might be the percentage or number of people or households with high incomes, where the levels of income are defined by individual countries. Initially the model contains both the vulnerability and

Figure 3. A model of vulnerability and resilience (Anderson-Berry and King, 2005).

resilience indicators. In some cases they may be at opposite ends of a scale, such as income levels, but for most indicators the vulnerability measure is a different statement to that of resilience. For example, the very complex area of gender generates many indicators of both vulnerability and resilience according to male/female, age, roles in mitigation and shelter and recovery, all qualified by culture.

3.2. *Societal Vulnerability*

Anderson-Berry's model groups indicators according to individual, community and social infrastructure characteristics. Implicit in this model is the extreme variability between communities in terms of access to all or some of the physical elements, which represent lifelines, making communities with very few services and facilities more vulnerable to tropical cyclone impact. On the other hand, highly dependent urban societies may suffer far greater dislocation from the loss of these lifelines in comparison to self-sufficient, simpler communities that make up for a shortage of infrastructure through stronger social capital.

People shelter from a cyclone in a designated safe shelter or in their own homes, but in many places safe shelters do not exist or are very limited in the number of places they have available. Where storm surge or severe destruction threaten, evacuation to other locations is recommended or enforced. If safe evacuation centres do not exist, people have to stay with friends or family or relocate out of the impact area at their own expense, utilising their own transportation. Individual capacity can make households more vulnerable if they lack transport or friends to shelter with. Vulnerability is increased through dependency on institutional or public transport, and a transport network that is not designed for a massive one-way movement, often in poor weather conditions and probable flooding.

Power supply, communications, water supply and sewerage are all likely to be disrupted or cease functioning for a period of days following tropical cyclone impact. The loss of these facilities increases vulnerability as fuel and ATMs will no longer be available, and non-functioning sewerage systems contribute to secondary health crises. The vulnerability of these lifelines combines with the impact on services such as hospitals, administration and retail facilities to increase the vulnerability of those households that were initially the most vulnerable to begin with. It is the poorest and least aware and least prepared households that are most likely to run out of food at a time when shops may have ceased to function. In complex inter-dependent urban communities the vulnerability of lifelines and infrastructure will multiply the overall vulnerability. In cities poorer households are much more likely to be concentrated and isolated in poorly serviced suburbs. Furthermore most urban services and infrastructure are not under the control of government and emergency management, but are provided and operated by private enterprise. Thus it is the very complexity of cities that increases their physical vulnerability.

3.3. *Community Vulnerability*

Communities that are in economic and population decline are especially vulnerable, although places of rapid growth and in-migration may also have services that are extremely stretched as well as a population that may be new to the tropical cyclone risk. Some communities are dysfunctional for very diverse reasons; wealthy or suburban communities where there is no local interaction; ghettoes, slums and squatter settlements that contain divisive ethnic diversity (on the resilience side some economically disadvantaged communities may have very strong social capital and interaction); conservative introspective communities that exhibit no interest in external institutions or in changing realities. For similar reasons some local governments and institutions that lack effective governance may fail to mitigate the hazard and prove ineffective in recovery. There are failed governments, where the dysfunctionality derives from a culture of greed or corruption or lack of care that overwhelms the good work and intentions of many hard-working individuals in the system. Thus community vulnerability can be an intangible concept. It is easier to identify community resilience as this is visible in the social capital of networks, organisations, volunteerism, civic pride and responsibility, sense of place, trust in authorities and unity of purpose and belonging. It is much easier to measure the existence of these indicators of community resilience, than it is to measure their absence.

Isolation increases community vulnerability. However strong a remote community may be, the very fact of isolation through distance or physical barriers increases the likelihood that lifelines will be minimal and response and recovery constrained by distance. Isolation may also be extended to communications, where warnings, information and advice are limited or constrained. Villagers in developing countries are particularly vulnerable to isolation from communications. Community isolation may also occur in the form of barriers such as language, culture, social and economic differences that exacerbate disadvantage, and political separateness and differences that may diminish trust and contact with the mainstream institutions.

The increasing trend of retirement communities, especially coastal relocation, may also contribute to vulnerability. In one sense the concentration of the elderly in well serviced communities may increase resilience, but simultaneously there is a danger that such communities may become isolated from the mainstream, developing a dependency on the service sector within a community that is not set up to deal with a major disaster, while at the same time community members have lost contact with the family and communities in which they established routines and social capital during their more active years. Retirement communities, in being separated from more physically active members of the wider community, make themselves more vulnerable to cyclone impact.

3.4. *Household and Individual Vulnerability*

Throughout most of the tropical cyclone prone regions of the world, the household is the basic level of human organisation. The characteristics of the household are a composite of its individual members with each individual adding to both its vulnerability and resilience. Household size and composition vary between cultures and between less-developed countries and developed countries, but as international censuses attest, there is a fundamentally similar definition of household and a basic core of a nuclear family, common to the vast majority of households.

Tropical cyclone knowledge within the family and household is pooled and reinforces vulnerability and resilience. Awareness in preparation for tropical cyclone impact comes from both adults and children (King, 2006; Izadkhah and Hosseini, 2005) and is compounded by experience of past events. Households communicate knowledge and actions within the broader community, but the vulnerability of individual households is more variable than that of the community and is measurable from the characteristics and knowledge of its individual members.

Anderson-Berry's model identifies individual indicators and groups in three areas of socio-demographic characteristics, hazard awareness and knowledge, and psycho-social indicators, within the context of social norms and values. Socio-demographic indicators are easily available from census data and in most countries may be accessed at community level. These are the basic indicators that have been identified in the literature (Peacock *et al.*, 1997; Blaikie *et al.*, 1994; Mileti, 1999; Bolin, 1994; Fothergill *et al.*, 1999; Granger, 1997).

Many indicators correlate strongly and may be grouped into composite socio-economic or socio-demographic indices, using principal components analysis or similar methods. This enables identification of at-risk communities (but not households as they may not be identified individually in censuses). Specific indicators may be grouped to identify evacuation needs, such as households without transport, single parents with small children and the very elderly etc. Absolute numbers of such people or households can be used in targeted planning for intervention or evacuation. Disabled individuals and people with critical medical needs are a further category of vulnerability that is frequently not quantified or spatially identified. These indicators are guides to intervention rather than to crudely quantify, label or categorise communities.

Research in North Queensland Australia (Anderson-Berry and King, 2005) has shown that there is little relationship between social and demographic indicators and the groups of indicators that define awareness and knowledge. A high level of education, for example, does not necessarily equate with a high level of tropical cyclone awareness. There are too many other variables in the population to make a neat connection or even an overall ranking of vulnerability. People may be very vulnerable demographically, such as being old, but may possess enormous knowledge and personal resilience.

After cyclone Larry devastated Innisfail in North Queensland Australia in 2006, an eighty-year old informant explained that she spent the morning before the cyclone (which crossed the coast at dawn on the following day) clearing up rubbish and potential missiles and securing her house before heading off to play bowls for the rest of the afternoon because as she stated "life must go on" (King *et al.* 2006). In terms of age and household structure she was vulnerable, but her high level of tropical cyclone awareness and previous experience reduced her vulnerability, while her optimistic attitude and perception of control made her extremely resilient. People are both vulnerable and resilient simultaneously, but their personal attributes are further reinforced or diminished by the ethos of their cultures and social values.

A further control on the vulnerability and resilience of individuals is added by their worldview, psychological makeup and personality. Some survivors of tropical cyclones found the experience terrifying, while others found it exciting. For some individuals recovery is a challenge, for others it is a time of depression and misery. The psychology of individuals is not directly related to either awareness and knowledge or socio-demographic characteristics. Thus in post disaster analyses psychological assessments have been conducted of the post-traumatic impacts on individuals, often in longitudinal studies (King, 2002).

The vulnerability to tropical cyclone impact is as multifaceted and complex as the people and communities that experience the hazard. Indicators can predict areas of risk and communities that may need intervention, but as an overall concept vulnerability is intangible or at the very least often unquantifiable. Prediction of climate change suggests greater intensities of tropical cyclones in some regions of the world, thereby increasing the overall risk from the physical hazard perspective, but additionally negating some of the accumulated community experience and knowledge. A disaster overwhelms peoples' prior knowledge and experience (King, 2000). Many communities are entering a future that will not replicate the immediate past. Vulnerability is further increased by the uncertainty of climate change, although significant variability and frequencies of tropical cyclones have occurred in past decades worldwide.

4. Resilience

While vulnerability is socially relative, multifaceted and, in an overall sense difficult to quantify, the idea of resilience provides much more tangible and applicable characteristics that may be incorporated into mitigation strategies. Vulnerability stresses negatives in society or throws up constraints that cannot be tackled by emergency managers. Only in the area of tropical cyclone awareness and knowledge can emergency managers directly intervene in the vulnerability of a lack of knowledge, by providing information and education, while most of the vulnerability characteristics are out of the control of hazard mitigators. Such basic indicators as low income or poverty are structural disadvantages that might be targeted by government or institutional strategies and policies for social or ideological reasons, but not for hazard risk.

Most indicators of vulnerability are about disadvantage or uncontrollable characteristics of people, communities and infrastructure, while resilience indicators identify strengths and capacities that may be built upon to enhance mitigation, to respond effectively to a disaster, and to tackle recovery. Effective mitigation builds on resilience rather than trying to change aspects of vulnerability that are outside the scope of emergency management, or local government. High income is not in itself an indicator of resilience. Similarly while the very young and very old are clearly vulnerable to hazard risk, there is nothing that can be done to change that characteristic, and it does not therefore follow that adults are resilient. Rather they are less vulnerable. Resilience indicators are a different set of characteristics from vulnerability; a parallel scale, whereby individuals, households and communities may simultaneously be vulnerable and resilient. Recognition of characteristics of resilience enables an entry for effective mitigation.

4.1. *Indicators of Resilience*

The model of vulnerability and resilience presented in Figure 3 lists general categories of indicators of each. Resilience is enhanced by the provision of good infrastructure, as long as it functions, or can rapidly be brought back to functionality after the impact of a tropical cyclone. The construction of protective features such as barriers and levees does not increase community resilience. It may even have the reverse effect as people develop a false security and as Katrina's destruction of the New Orleans levees illustrated, protective or preventative structures may increase vulnerability and risk. Generally a well-connected and functioning infrastructure gives communities and individual households' flexibility in preparing for tropical cyclone impact.

In the area of individual attributes of resilience, hazard knowledge, awareness and previous experience all contribute highly, with knowledge and awareness being open to direct targeting of educational campaigns. The targeting of education and information is most effective when it is presented to specific groups and communities. Tropical cyclone awareness education for children, men, women, young adults, the elderly, cultural and language groups and different socio-economic groups all need to be crafted differently (United Nations, 2005) and must be specific to the needs of each group. A vulnerability indicator may be built upon to generate greater resilience.

The other individual indicators, which are defined as well-being, sense of control, access to media, and trusted authorities (Fig. 3) assume good governance. Even though these are individual indicators, governance is a significant contributor to resilience in providing the structures of a fully functioning society. The community indicators of resilience build on and contribute to good governance — all of these are summarised in the concept of strong social capital. In defining social capital it is these community indicators that are the main components. Each of these factors is a positive and constructive facet of society that may be developed or enhanced through community programs, settlement design and land use planning, community engagement and empowerment, and participatory planning. These may not all be under the control of emergency management, but local government hazard planning that engages with both the communities and the service providers that support them, can direct hazard resilience by building and encouraging social capital.

4.2. *Role of Organisations*

National Meteorological Services (NMSs) are the recognised experts on the meteorological aspects of severe weather hazards, and are therefore a key stakeholder in the development of a country's natural disaster risk management plans. Early warnings are the main contribution by NMSs, underscored by their role in public education and outreach programs and support given to hazard impact and risk assessments conducted by emergency management and local government organisations. (Davidson and Wong, 2005; Falls, 2002; Elsberry, 1995).

The reality of a disaster is that the local emergency services and the civil functions of local government organisations are unable to cope. Emergency Management is a civil government activity, coordinated from the local level through state or provincial governments, up to a central government response depending upon the severity of the disaster (Douglas, 1999). In developing nations a lack of resources, as well as traditions of military involvement in government, may result in a stronger reliance on military intervention in disasters, although there are many developing nations that have a small or even no defence force, especially in the Pacific where only PNG and Fiji have armies.

Emergency management is structured through laws that apportion responsibility for preparedness and prevention, and for response and recovery. Where legislation puts responsibility on local levels of government to prepare for disaster, the capability to respond adequately is limited. Apart from local councils, disaster preparation is usually a requirement for all other government departments that provide social and welfare services to the community. Emergency Management coordinates the responses of all these government organisations, but they are separate entities answerable to different political bosses. Each government department has its own priorities, agendas and needs, even when cooperating. While many of these organisations have clearly defined, exclusive roles, organisations responsible for specific community members, such as the aged, or specific ethnic groups, may end up providing similar services.

Government is only one of many types of organisations that become involved in disasters. A trend towards less government and more privatisation has meant that former government roles in welfare and service provision are now provided by private enterprise. There is a blurred line between companies that are providing services to the public, and the traditional Non Government Organisations that began life as charities, whose members are driven principally by beliefs, altruism and often a strong political agenda. At the business end of this continuum of organisations are companies that are in business to make profits for their shareholders, but which possess plant, machinery and expertise or infrastructure that may contribute to community recovery. Some companies are devastated or go out of business in a disaster, but for many others there is an opportunity for expansion, although individuals working for these companies may be no less motivated by altruism than members of government and relief organisations.

Directly involved organisations as in Table 5 are those that have a primary responsibility for emergency management and disaster mitigation. Government organisations include all levels of government and the NGOs include the traditional relief organisations and charities, but there are also NGOs that are indirectly involved in hazard response and recovery, such as the formal organisations of religions, and various types of youth groups that provide volunteers. These are separate from organisations which rely on volunteers, but are sponsored directly by government. Grass Roots Organisations are primarily local community groups. Within the developing world, many GROs were formed as progress associations for the sake of advancing development, but adapted to relief and recovery in the event of a disaster. In the developed world GROs have often been single-issue, but otherwise quite loose organisations. Some of these have responded directly to disasters or the threat of a hazard, as victim support or lobby groups, but there are far more GROs that are indirectly involved in disaster response and recovery (Fahey, 2003). They may not have been formed with any thought of hazard or disaster, but once formed they create and maintain community links and networks which may rapidly be used in a crisis.

The indirect group of organisations are those which exist primarily for a purpose other than hazard mitigation, but which contribute resources, personnel, networks and knowledge during and after a disaster. Included here is the media, in all of its forms and roles. The media is primarily a private enterprise industry, but it is crucial to government organisations and emergency managers who depend upon media response to get information, warnings and messages to the community. The media are often the earliest responders to a disaster, providing initial information and immediate impressions. When Category 5 cyclone Zoe hit the island of Tikopia in 2002, a remote and extremely isolated

Table 5. Organisations involved in disaster response and recovery.

Direct	Indirect	Informal
International	Businesses	Culture
Government Organisations	Economic Organisations	Community Networks
Non Government Organisations	Recreational Organisations	Internet
Privatised Specialists	Religious Organisations	Residual Leadership
Grass Roots Organisations	Cultural Groups	Volunteers
Community Organisations	Interest Groups	Tradespeople
	Political Groups	Family and Household
	Media	Individuals and Visitors

outlier of the Solomon Islands, in the Pacific Ocean, the first outsider to provide any news of the islanders' survival was a freelance journalist. His initial report of devastation and survival described people fishing. Thus although the media was the first outsider, the Tikopians had got on with their own process of survival and recovery.

The media is also extremely powerful in advertising a crisis and prompting political and citizen response. The corollary of this power is the invisibility of unreported, or poorly reported disasters. Disasters are, in part, *constructed* by the media. They often exaggerate human interest elements, they look for scandals and government ineptitude, assign blame and instigate generous relief responses. They self censor some images of suffering, but are also guilty of misusing distressing images to increase the impact of their stories. Emergency managers have no control over the media and need their positive support at all phases of emergency management. All organisations must therefore devote some of their resources to media liaison and must continually work at these networks. NGOs especially work hard at developing their media image and ensuring media visibility as citizen generosity is a key component of their income.

There is also a level of informal organisation comprising residual and spontaneous "organisations." The spontaneous response and resulting "organisation," or community, or network, occurs directly as a consequence of the disaster event. The residual "organisations," are those elements of civil society that underpin all that we do, the ways in which we define and construct our society, the traditional, and the informal. Residual "organisations," are culturally constructed, grounded in community and civil society, and are representative of the ways we organise and network.

The size and numbers of organisations that participate in any disaster vary with the spatial extent, severity and population involved. The more organisations that are involved, the more complex is the task of emergency management. These organisations bring to the community a wide range of skills and approaches. It is inevitable in any disaster that problems of organisation and leadership will emerge. The effectiveness with which these issues are resolved will have a powerful influence on the effectiveness of response and recovery. Organisations will pursue their own agendas generating both contradiction and cooperation, while communities will pursue their own, often parallel, routes to response and recovery and mitigation (Loosemore and Hughes, 2001).

4.3. *Best Practice Strategies for Resilient Communities*

The International Federation of Red Cross and Red Crescent Societies summarises strategies for enhancing resilience at the community level. On the basis of their experience in many aid projects and disaster relief they provide six basic conclusions.

"1. *Systematic assessment of what enables people to cope with, recover from and adapt to various risks and adversities — at household and community level — is badly needed.*" More knowledge of how communities function and what constitutes resilience is required by emergency managers in order to help them engage effectively with communities.

"2. *Strengthening social capital should be the key objective of all disaster interventions, whether in relief, recovery or risk reduction — rather than a by-product.*" The top-down approach of emergency managers tells people what to do without paying attention to their social capital.

"3. *People-centred approaches to development provide models that can improve the humanitarian aid and disaster risk management.*" As many NGOs work in both development and disaster relief, experience in poverty reduction and development are as important in reducing risk.

"4. *New institutional strategies and cross sectoral coalitions are required to boost the resilience of local livelihoods in the face of multi-dimensional risks.*" Communities have many greater priorities than risk reduction, such that natural hazards are relatively low down as a priority. Complex and multilayered communities require many partners and organisations.

"5. *Good governance is essential to create the environment in which more resilient communities can thrive.*" Governance extends across all societies, countries and institutions. Good governance is a continuum from the household to the international community.

"6. *Scaling up strategies based on the aspirations and capacities of people at risk remains the greatest challenge.*" This is the bottom up, grass roots basis of community resilience driving disaster mitigation that is based on resilience and community capacity. (International Federation of Red Cross and Red Crescent Societies, 2004, pages 32 to 33, Jaquemet, 2002)

5. Tropical Cyclone Mitigation

The basic risk equation starts with a relationship between hazard and vulnerability. As risk assessment leads to strategies to reduce risk, the risk equation can be modified as follows:

$$\text{Total Risk} = \frac{\text{Hazard} \times \text{Vulnerability} \times \text{Elements at Risk}}{\text{Resilience} \times \text{Mitigation}} \quad (1)$$

The integrated emergency management framework of prevention or reduction, preparedness or readiness, response and recovery presents a linear model of stages of emergency management and areas of responsibility. In recent years the terms prevention, reduction, preparation and readiness have become subsumed under the umbrella term of mitigation. Internationally emergency managers have increased emphasis on mitigation even though they have a primary responsibility for response and the coordination of recovery. It is clear from the experiences of tropical cyclone disasters that response and recovery occur differentially, with some communities moving ahead while others remain isolated or recovering very slowly. At the same time individuals, organisations and communities learn from the recent impact of the disaster as soon as it happened and incorporate that experience into their mitigation knowledge, such that they will rebuild and reconstruct to reduce the impact of future events. Mitigation does not stop at the moment of impact. It is ongoing and iterative.

Figure 4 illustrates both the impact of tropical cyclones and the multi-dimensional nature of mitigation and recovery. Tropical cyclone impact is distributed by sector and geography, with the hazard threat developing over a period of time, occurring for a shorter period of time, and leaving a legacy of destruction over an extended period. Mitigation must consequently be three-dimensional, or even in a sense four dimensional, as places and communities are in a two-dimensional space, overlain with sectors of the community, which are simplified in Fig. 4 as the physical/environment, economy/infrastructure, social/community, and household/individual. On the timescale of planning for and mitigating the impact of a tropical cyclone there are

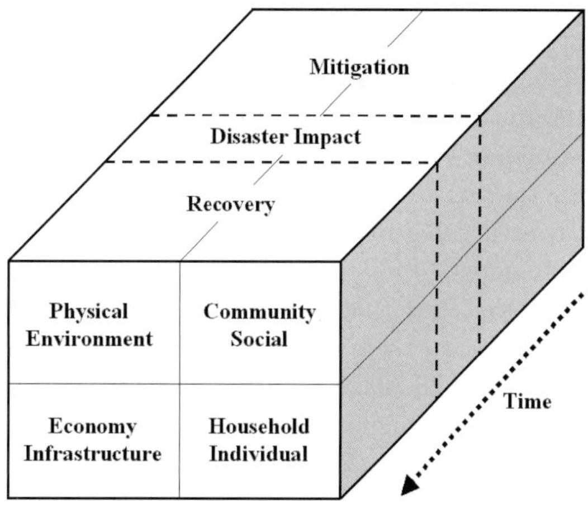

Figure 4. The multi-dimensional nature of mitigation, impact and recovery.

distinctive time periods when different activities predominate.

1. **Long-term mitigation** — way in advance of a likely tropical cyclone, prevention and reduction operate through land use planning, building codes, drainage, barriers and coastal defences etc. Alongside physical measures there is the process of educating and informing the community to be prepared and to protect itself. National Meteorological Services (NMSs) work to improve tropical cyclone forecasting skills.

2. **Short-term mitigation** — when a tropical cyclone threatens, the primary mitigation tool is NMS warnings, which usually contain recommended protective actions. In this phase people and organisations must carry out these protective actions while emergency management agencies will ready their equipment, personnel and supplies.

3. **Impact phase** — during the passage of the tropical cyclone the NMSs remain active in monitoring and warning. People continue to receive information and take a series of actions within their homes or shelters. Emergency management agencies commence response actions.

4. **Post impact phase** — the immediate post impact period is one of response and the commencement of recovery. People begin to observe and absorb the details of the impact and carry information into long-term mitigation strategies.

5. **Recovery phase** — mitigation activities during recovery are particularly directed to rebuilding to withstand future events, as well as the sharing of experience and knowledge, and converting that experience into tropical cyclone awareness.

In the context of vulnerability and resilience it is clear that these different types of mitigation must all take place within the framework of societal engagement and civic responsibility and good governance. The following review of mitigation is structured in four broad areas of (1) physical mitigation measures, (2) education and awareness, (3) tropical cyclone warnings and (4) capacity building: partnerships, empowerment and good governance.

5.1. *Physical Mitigation Measures*

Short term physical mitigation measures are practised in many countries prior to landfall of a tropical cyclone. These include measures such as shutting down electricity grids and gas supplies to prevent damage and fires, turning off and securing hazardous facilities, and securing transportation centres. The construction of coastal and river defences such as walls, surge barriers, levees and related drainage channels probably provide more psychological reassurance than real long-term prevention of storm surge or flooding. Apart from construction costs, ongoing maintenance of artificial structures can lead to neglect or skimping on proper repairs. This can result in the extremely dangerous scenario of a complacently secure population unaware of the danger of their compromised protection. The danger to New Orleans was well-known long before Katrina's impact, yet the situation of the city's physical vulnerability was not adequately addressed. Apart from the catastrophe of a sudden collapse of walls and barriers, once water has entered low-lying areas it is ponded by the protective walls thereby exacerbating the crisis and severely constraining response and recovery. Inevitably some barriers will be constructed around dense urban areas or around "safe" islands to which people are evacuated, but generally, protective structures are not sustainable, especially for low density settlements or residential areas.

On the other hand natural features in tropical areas are extremely effective in controlling storm surge and flooding. Coral reefs, absorb some of the power of tropical cyclone wind generated waves and surge offshore (Kench *et al.*, 2006). Mangroves are even more effective in protecting coastal areas (Williams *et al.*, 2006). Behind coastal dune ridges and mangrove forests, lagoonal wetlands absorb inundation. These three common features of tropical coasts are physical mitigators of tropical cyclone impact. Unfortunately throughout the tropical and subtropical regions all three of these are under threat from human encroachment and modification. Mangroves and coastal wetlands have suffered the greatest impact, as coastal zones have been cleared, settled and built over. Access to the sea, especially as a recreational resource, and extensive flat land has resulted in coastal wetlands being drained and mangroves cleared. It is obviously foolhardy development, but population pressures and greed place people and their dwellings in harm's way, both in developed and developing countries.

The mitigation response to settlement in extremely hazard-prone environments can be achieved through responsible land use planning. Piecemeal planning has tackled the coastal development issue in an ineffective way. Much damage has been done with extensive and relatively recent coastal development already in place. There is an urgent need for land use planning to pay greater attention to risk avoidance, even to the point of sequestering flood and surge prone land, and planners must accept responsibility to ensure risk reduction (King, 2006).

The building codes have existed for a long period of time and here the cost is passed directly to the residents rather than being borne by the community or local government. People in poorer societies are not able to meet rigid and expensive building code requirements, but many effective programs in villages and informal settlements have encouraged stronger building methods that use local and bush materials. Additionally, residential dwellings are made safer by raising structures above surge or flash flood levels on compacted pads or stilts and stumps.

While building codes establish minimum standards, which vary considerably from one country to another, architecture can pursue best practices. Most residential dwellings are not architect designed, but public and commercial buildings usually are, and are able to implement a greater range of cyclone safe designs. Public and commercial large buildings are often much stronger than people's houses, with many structures delegated a shelter function during a tropical cyclone. Some countries and regions have much greater sheltering strategies than others, but problems of litigation and responsibility are constraints to the opening up of buildings for shelter and many schools and religious buildings that may function as community centres are not structurally safe during a tropical cyclone. However, if storm surge or severe flooding threaten, evacuation to a safe shelter becomes essential. This may be a move to a public shelter, but is probably more often to other family or friends, or even to a hotel in the safe zone.

People in surge and flood prone areas must have their own evacuation plan. This is part of the education and awareness mitigation, but physical mitigation measures must be in place in order to enable rapid and safe evacuation. At the very least this necessitates the signposting of evacuation routes, but a greater responsibility of local government is the provision of safe roads that will not be blocked by flood or fallen vegetation. As many people have no transport of their own an additional physical requirement may be the provision of public or chartered transport.

5.2. *Education and Awareness*

Physical structures that reduce tropical cyclone impact have a role in the community, although legislation in the form of building codes and land use planning are broadly more effective. None of these measures contributes to mitigation if the people do not know the hazard risk, if they do not have household cyclone plans or if they are unaware of evacuation routes, sheltering strategies, and the appropriate response to warnings. Tropical cyclone education strategies that enhance awareness are the core of mitigation.

People who have experienced a tropical cyclone impact can be expected to have a raised awareness of the hazard and of their risk level, but experience alone is not a substitute for education and information. Many people who may have been through a Category 1 or 2 cyclone have no awareness from that experience of what a Category 3 or 4 will do. Even if someone has experienced a severe cyclone, the uniqueness of each event will not necessarily prepare them for the next one, especially if their previous experience was with the less severe side of the storm track. Similarly someone who has moved to a different house or location cannot rely on previous experience as a predictor of a future event. Those people who have gone through many tropical cyclones in the same place, may have the breadth of experience to contribute to awareness and knowledge, but do not necessarily have the communication skills to share it with the community, although media attention during a storm may help to focus and interpret the experience.

Many people in tropical cyclone prone regions have not had any direct experience and must first be made aware of the cyclone risk before they can be educated to mitigate that risk. Awareness, like warnings, has to happen before people will listen and absorb the messages. Second to direct experience in terms of education and information is word-of-mouth through contact with family, friends and neighbours. Unfortunately community knowledge is frequently distorted or is even incorrect. We also extract from what we see and hear, so that particles of information and misinformation are woven into the fabric of community knowledge. The emergency manager does not pour hazard education and information into an empty community vessel. There is a pre-existing awareness and knowledge that may contradict the "official" message, in which case the official message may be rejected, or distorted.

A fundamental challenge to community hazard education and awareness is overcoming existing prejudice, misinformation and lack of interest. Additional to this is the gap between knowledge and action. Some people may possess highly developed tropical cyclone knowledge but make no direct connection to action. An optimistic view of the world, and competing priorities, enable hazard aware people to continue to live on the beach, or to fail to clear up and repair property or to go out picking up debris during the onslaught of the storm. Tropical cyclone awareness raising and education has to combat pre-existing attitudes and risk-taking behaviour.

Neither communities nor members of the public are homogenous. There must be a different focus, language and approach for children, the elderly, men and women, well-educated and less educated people, and cultural and ethnic minorities, as well as the variability in world-views and psychology (O'Neil, 2004). Messages must be targeted to the diversity of society.

One of the most effective groups to target educationally is schoolchildren. Material aimed at primary and high school students will vary but all school children are in a learning environment within a context of discipline and structure. Hazard education can be incorporated into social science, environmental science and science modules, and support materials such as games, scenarios and stories can be made available through the Internet and publications. Learning is actively reinforced through projects and assignments and especially in the case of younger children it is frequently shared at home with family and neighbours. Once learned this knowledge remains with the individual for life. A necessary part of raising awareness for children lies in first raising the awareness of teachers, principals and education departments (Izadkhah and Hosseini, 2005; ISDR, 2007). Institutions beyond the school also participate in the education of children, reinforcing the school experience. Hong Kong has been innovative in its use of media and education to enhance mitigation (Anderson-Berry *et al.*, 2006).

Adults receive more passive education messages and materials without the compulsion for them to pay attention. Being a mainly public sector activity emergency management funds are usually limited for the more expensive advertising media, such as television, so the media in all its forms must be engaged and encouraged to report accurately, to convey stories and historical accounts, and to give time to emergency managers to explain mitigation behaviour. Most mitigation education is conveyed through print, radio, internet and television media. Increased funding for hazard education is essential in all countries, but a close working relationship with the media will always be a priority. Knowledge empowers, and empowered communities practise effective cyclone mitigation.

A number of NMSs conduct education and awareness campaigns prior to the cyclone season. These campaigns are normally undertaken in partnership with emergency management agencies. Interactive demonstrations and workshops are featured during the campaigns and meetings held with local government authorities. Interviews are also conducted with the local media. The principal aims of these campaigns are (1) to establish a high level of awareness of natural hazards — and how to prepare for and deal with them, (2) to strengthen partnerships between stakeholder organisations, and (3) to provide information on the role of the various organisations — what products and services are available — and how best to access them.

5.3. *Tropical Cyclone Warnings*

In the transmission of tropical cyclone warnings the media plays a central role, and is the conduit between the experts and the community. Warnings are generated by specialist meteorologists who know the areas of uncertainty and inaccuracy in both their data and their forecasts and need to balance that uncertainty against the clarity of the warning message. Governments and institutions play a part in transmitting, modifying or explaining the message. Transmission of tropical cyclone warnings varies between countries according to political regimes and communications. Government institutions play a greater role in developing countries than in the developed world, with media under significant government control in some countries. However, it is the print, radio, television and internet media that ultimately convey the warning message on behalf of a top-down hierarchy.

The process of communicating a tropical cyclone warning ideally assumes that the experts have full knowledge, the media and government organisations are responsible and truthful and that the public — diverse communities — absorbs the warning message and responds rationally. These assumptions are seldom correct. Figure 5 illustrates the tropical cyclone communication process. There are two possible routes which allow for the differences in transmission of warnings between countries that function in significantly different ways. In some societies the warning may not be transmitted to the public without going through stages of government approval (**dashed lines**), while in most cases the NMSs communicate directly to the media who in turn convey the message to communities, households and individuals (**full lines**).

Figure 5 therefore presents the idea of warning messages coming from quite different sources. At each breakpoint in transmission there may be issues and constraints that can

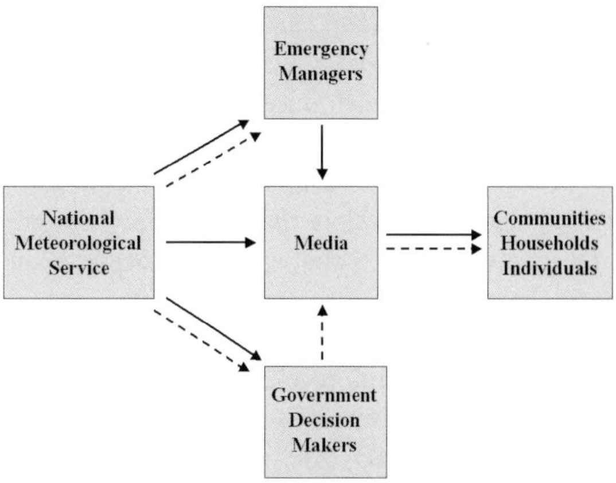

Figure 5. The warning communication process.

distort or slow the warning message. The first issue of hazard recognition involves the expert making a series of decisions in interpreting data to estimate the size, intensity, likely track and speed of the tropical cyclone. At this forecast stage data and information can change rapidly, with decisions like repositioning the centre or track of a tropical cyclone on the basis of later information being interpreted by the media and the public as uncertainty or unreliability of the forecast.

Where government and emergency management organisations make decisions about when and whom to inform and roles in preparing for a potential disaster, political agendas and misunderstanding can slow transmission or even break the warning chain The relationship between government and the media is determined by levels of governance, but in an open society with a free press government trusts NMSs to communicate directly with the community through the media, thereby circumventing some of the barriers. However, the media can also distort the message through its timing, exaggeration or even lack of interest in the story.

At the household and community levels the warning message relies upon knowledge and understanding of the tropical cyclone hazard in order to be effective in facilitating mitigation response and behaviour. Firstly there must be information systems and networks. Geographical, social and cultural isolation reduce communication, although strong social capital can counteract a lack of communications through active social networks that transmit information locally.

All stakeholders, communicators and individuals need education that explains, engages and involves them in the warning process. Education must address the wording of warnings and responsive behaviour, the timing and frequency of transmission and receipt of warnings, and appropriate media messages targeted to diverse communities (WMO, 2005). Messages must be diverse, but significantly standardised, and they must engage and empower the community to make them actors in receiving and actioning the message (Mileti quoted in WMO, 2005).

This is not a strategy that is only for the community. It must involve private enterprise — organisations and companies that all too frequently are excluded, or exclude themselves from hazard mitigation. In confronting tropical cyclone threats individuals often neglect their role as employers or employees, instead placing emphasis on their own family and household. This is an inevitable reaction, but the consequence is of businesses remaining open, employees being pressured to remain at work, and people clearing up their own backyards while neglecting works yards, building sites etc.

Finally, warnings and their associated education campaigns must be monitored and evaluated. Evaluation of education campaigns can be carried out by household surveys, but evaluation of warnings at the time of tropical cyclone impact has to be done in the form of post disaster studies. The first type of evaluation, following an educational campaign, tests knowledge of terminology and wording as well as receipt of educational information. The second type of evaluation, after a cyclone has passed, concentrates on the source and timing of warning information as well as its clarity and appropriateness. This type of survey will usually evaluate the media's role and effectiveness in transmitting a message.

5.4. *Capacity Building: Partnerships, Empowerment and Good Governance*

The warning communication process illustrates the interconnected dependency of many stakeholders, organisations, communities and individuals. Both tropical cyclone hazard education and warning messages are dependent upon an extensive network of connections and relationships. From the old command and control paradigm, a hierarchy of connections takes education and warnings as mitigation actions from the top, down to the community at the bottom.

Many emergency managers think within this structure. At the grassroots community level, there is some awareness of the existence of a hierarchy of information, but communities and individuals exist primarily in a lattice type of network, which can best be compared to the Web of the internet. People are members of many communities and information comes from all directions, both within each type of community and as external connections between communities, and from formal as well as informal organisations.

Increasingly the population of tropical cyclone prone coastal zones is concentrated in large dense urban and peri-urban settlements. These places depend upon the complex lattice of networks; a society that resembles the complexity and interdependence of the internet. Within these complex urban communities are billions of connections, millions of households and all of the organisations, infrastructure and institutions that sustain society.

Building capacity, enhancing resilience, utilising social capital and empowerment are tools to mitigate tropical cyclone risk that must be employed at all levels and in multiple partnerships. Partnerships are both formal and informal, primarily involving areas of common concern and responsibility where strategies, knowledge, needs and information are shared towards the achievement of a common purpose. Anderson-Berry *et al.* (2006) have documented examples of partnerships and capacity-building in the Pacific, where the knowledge that is gained from post disaster analyses contributes to a better understanding of endemic community capacity. Partnerships would include "*i) other government agencies with missions involving the protection of life and property, (such as National meteorological agencies) national regional or local emergency management agencies, first responders, infrastructure managers; ii) the media; iii) non-government organisations; iv) emergency relief organisations, such as the Red Cross and Red Crescent Society; v) academic institutions and schools; vi) trained volunteers associated with the meteorological agencies, such as co-operative observers, storms spotters, and amateur radio operators; vii) meteorological societies and other professional associations in risk management disciplines; viii) private sector weather companies, and ix) utility services, telecommunication operators and other operation critical or weather sensitive businesses.*" (WMO, 2005, p. 8).

The advantages of partnerships in enhancing cooperation and exchange of information includes; expertise from a range of disciplines such as engineering, planning, social science; achieving outcomes beyond the capacity of a single organisation; targeting common and shared budgetary goals; resources for research, awareness and education; sharing knowledge and experience; standardising messages and extending the distribution and communication of tropical cyclone awareness (WMO, 2005; WMO, 2007). The core necessity in mitigation is partnerships between diverse and multiple organisations in a complex and dynamic society. Effective partnerships must transcend sectors, regions and nations and will bring together people and organisations in the common purpose of tropical cyclone mitigation.

6. Conclusion

A small number of significant factors may be restated in conclusion. The global impact of tropical cyclones is highly significant socially and economically, and reduces sustainability and development in many societies. The vulnerability of society to tropical cyclones is increasing as a consequence of migration, urban growth, rapid social change, increasing standards of living with associated investment in the built environment, and population growth. On the other hand resilience and especially social capital are central to effective mitigation and targeting of education and awareness campaigns. Climate change is altering the tropical cyclone hazard risk and requires new emphases in mitigation and adaptation. Achievement of effective and sustained mitigation of the risk also requires extensive partnerships and inter-sectoral cooperation within a complex, inter-dependent social system. The key areas for implementation of societal mitigation strategies that were outlined in section 5 involve new attitudes to land use planning, education and awareness building throughout the community, effective communication of tropical cyclone warnings and capacity building through partnerships, empowerment and good governance.

References

Anderson-Berry, L., D. King, H. Lam, T. Hart, and M. Bergin, 2006: Topic 5.3 Mitigation strategies and community capacity building for disaster reduction. *Sixth International Workshop on Tropical Cyclones*, WMO, Geneva.

Anderson-Berry, L. and D. King, 2005: Mitigation of the Impact of Tropical Cyclones in Northern Australia through Community Capacity Enhancement. Special issue of *Mitigation and Adaptation Strategies for Global Change* **10**, 367–392.

Blaikie, P., T. Cannon, I. Davis, and B. Wisner, 1994: *At Risk — Natural Hazards, People's Vulnerability, and Disasters.* Routledge, London.

Bolin, R., 1994: Post-disaster shelter and housing: social processes in response and recovery, in R. Dynes, and K. Tierney, (eds.) *Disaster, Collective Behaviour and Social Organisation*, University of Delaware Press, Newark, pp. 115–127.

Buckle, I., 2005: Enhancing the resilience of highway networks against extreme events. Session 4.8. Vulnerability of modern societies towards natural disasters — the impact on critical infrastructures. Kobe Report of thematic segment. WCDR. http://www.unisdr.org/wcdr/ accessed on 13/2/08.

Congleton, R., 2006. The story of Katrina: New Orleans and the political economy of catastrophe. *Public Choice*, **127**(1–2).

Chittibabu, P., S. K. Dube, J. B. Macnabb, T. S. Murty, A. D. Rao, U. C. Mohanty, and P. C. Sinha, 2004: Mitigation of flooding and cyclone hazard in Orissa, India. *Natural Hazards*, **31**, (2).

CRED, 2008: Emergency Events Database (EM-DAT). Centre for Research on the Epidemiology of Disasters. Accessed on 30/1/08 at http://www.emdat.be/Database/DisasterProfile/natural-table-emdat_disasters.php?dis_type=Flood&Submit=Display+Disaster+Profile

Davidson, J. and M. C. Wong, 2005: Guidelines on Integrating Severe Weather Warnings into Disaster Risk Management. WMO/TD No. 1292. WMO, Geneva.

Douglas, R., 1999: Administrative law and response to emergencies. *Australian Journal of Emergency Management*, **14**(2).

Elsberry, R., (ed.) 1995: Tropical Cyclone Disaster Mitigation. Global Perspectives on Tropical Cyclones. WMO/TD-No. 693, Tropical Cyclone Programme Report No. TCP-38. Secretariat of the World Meteorological Organization, Geneva.

Everham, E. M. and N. V. L. Brokaw, 1996: Forest damage and recovery from catastrophic wind. *The Botanical Review*, **62**(2).

Emergency Management Australia, 1998: Australian Emergency Management Glossary. Australian Emergency Manuals Series, Part I: The Fundamentals, Manual 3. Commonwealth Government of Australia.

Fahey, C., 2003: Working with communities to build social capital — reflecting on old and new thinking about volunteers. *Australian Journal of Emergency Management*, **18**(4).

Falls, R., 2002: Tropical Cyclone Impacts. *Proceedings of the Fifth WMO International Workshop on Tropical Cyclones (IWTC-V)*, Cairns, Australia, WMO Report/TD No. XXX, Geneva.

Fothergill, A., E. Maestas, and J. Darlington, 1999: Race, ethnicity and disasters in the United States: A review of the literature. *Disasters*, **23**(2), 156–173.

Granger, K., T. Jones, M. Leiba, and G. Scott, 1999: *Community Risk in Cairns: A Multi Hazard Risk Assessment*, Australian Geological Survey Organisation, Canberra.

Granger, K. J., 1997: Lifelines and the AGSO cities project, *The Australian Journal of Emergency Management*, **12**(1), pp. 16–18.

Henderson, D. and B. Harper, 2003: *Climate Change and Tropical Cyclone Impact on Coastal Communities' Vulnerability.* Project with Queensland State Government's Department of Natural Resource and Mines and Department of Emergency Services. Cyclone Testing Station, James Cook University and Systems Engineering Australia Pty Ltd.

Holthouse, H., 1971: *Cyclone.* Rigby, Adelaide.

Hunter, L., 2005. Migration and environmental hazards. *Population and Environment*, **26**(4).

Ilf, J., 1995: *Community Development. Creating Community Alternatives — Vision Analysis and Practice*, Longman, South Melbourne, pp. 1–3, 14–17, 89–94.

International Federation of Red Cross and Red Crescent Societies, 2004: *World Disasters Report 2004: Focus on community resilience.* International Federation of Red Cross and Red Crescent Societies, Geneva.

IPCC, 2007: Climate Change 2007: Synthesis Report. Contribution of Working Groups I, II and III to the Fourth Assessment Report of the Intergovernmental Panel on Climate Change (Core Writing Team, Pachauri, R.K and Reisinger, A. (eds.)). IPCC, Geneva, Switzerland, 104 pp.

ISDR, 2007: *Acting with Common Purpose: Proceedings of the first session of the Global Platform for Disaster Risk Reduction*, Geneva, June 2007. International Strategy for Disaster Reduction, UN, Geneva.

ISDR, 2008: International Strategy for Disaster Reduction: Disaster statistics 1991–2005. http://www.unisdr.org/disaster-statistics/introduction.htm accessed 30/1/08.

Izadkhah Y. O. and M. Hosseini, 2005: Towards resilient communities in developing countries through education of children for disaster preparedness. *International Journal of Emergency Management (IJEM)* **2**(3), 138–148.

Jaquemet, I., 2002: Assessing vulnerabilities and capacities. *World Disasters Report 2002.* International Federation of Red Cross and Red Crescent Societies. Accessed on 31/1/8 at http://www.ifrc.org/publicat/wdr2002/chapter6.asp

Kench, P., R. McLean, R. Brander, S. Nichol, S. Smithers, M. Ford, K. Parnell, and M. Aslam, 2006: Geological effects of tsunami on mid-ocean atoll islands: The Maldives before and after the Sumatran tsunami. *Geology*, **34**(3), 177–180.

King, D., 2000: You're on your own: Community vulnerability and the need for awareness and education for predictable natural disasters, in *Proceedings of the International Emergency Management Society*, 6th Conference, Delft, Netherlands. *Journal of Contingencies and Crisis Management*, **8**(4), 223–228.

King, D., 2002: Post disaster surveys: Experience and methodology. *Australian Journal of Emergency Management*, **17**(3).

King, D., 2006: Planning for hazard resilient communities. Chapter in Paton, D. & Johnston, D. (eds.), *Disaster Resilience: An Integrated Approach.* Charles C. Thomas, Springfield, Illinois.

King, D., D. Goudie, and D. Dominey-Howes, 2006: Cyclone knowledge and household preparation — some insights from Cyclone Larry. *Australian Journal of Emergency Management*, **21**(3).

Loosemore, M. and W. P. Hughes, 2001: Confronting social defence mechanisms: Avoiding disorganisation during crises. *Journal of Contingencies and Crisis Management*, **9**(2).

Mileti, D., 1999: *Disasters by Design: A Reassessment of Natural Hazards in the United States.* Joseph Henry Press, Washington, DC.

O'Neil, P., 2004: Why don't they listen — developing a risk communication model to promote community safety behaviour, in *The International Emergency Management Society: 11th Annual Conference Proceedings*, May 18–21, 2004. Melbourne, Victoria, Australia.

Passerini, E., 2000: Disasters as agents of social change in recovery and reconstruction, *Natural Hazards Review*, **1**(2), 67–72.

Peacock, W., B. Morrow, and H. Gladwin, 1997: *Hurricane Andrew: Ethnicity, Gender and the Sociology of Disaster*, Routledge, London.

Pielke, R. A. and R. Pielke, 1997: *Hurricanes: Their Nature and Impacts on Society.* John Wiley and Sons, U.K.

Pielke, R. A., Jr., R. Crompton, E. Faust, J. Gratz, M. Lonfat, Y. Qian, and S. Raghavan, 2006: Topic 5.2: Factors Contributing to Human and Economic Losses. *Sixth International Workshop on Tropical Cyclones*, WMO, Geneva.

Queste, A., 2005: Vulnerability of modern societies towards natural disasters — the impact on critical infrastructures. Session 4.8. Vulnerability of modern societies towards natural disasters — the impact on critical infrastructures. Kobe Report of thematic segment. WCDR. http://www.unisdr.org/wcdr/

Sathaye, J., A. Najam, C. Cocklin, T. Heller, F. Lecocq, J. Llanes-Regueiro, J. Pan, G. Petschel-Held, S. Rayner, J. Robinson, R. Schaeffer, Y. Sokona, R. Swart, and H. Winkler, 2007: Sustainable development and mitigation, in *Climate Change 2007: Mitigation. Contribution of Working Group III to the Fourth Assessment Report of the Intergovernmental Panel on Climate Change* [B. Metz, O. R. Davidson, P. R. Bosch, R. Dave and L. A. Meyer (eds.)], Cambridge University Press, Cambridge.

Shinozuka, M., 2005: Enhancing the resilience of power grids against extreme events. Session 4.8.

Vulnerability of modern societies towards natural disasters — the impact on critical infrastructures. Kobe Report of thematic segment. WCDR. http://www.unisdr.org/wcdr/

Terry, J., 2007. *Tropical Cyclones Climatology and Impacts in the South Pacific*. Springer, New York.

United Nations, 2005: Draft Hyogo Declaration. World Conference on Disaster Reduction, January 2005, Kobe. http://www.unisdr.org/wcdr/

Wikipedia, 2008: List of countries by GDP: World Bank 2006, accessed on 31/1/8 at http://en.wikipedia.org/wiki/List_of_countries_by_GDP_%28nominal%29

World Bank, 2009: Data and Statistics. Website accessed on 23/3/09 at http://web.worldbank.org/WBSITE/EXTERNAL/DATASTATISTICS/0,,contentMDK:20535285~menuPK:1390200~pagePK:64133150~piPK:64133175~theSitePK:239419,00.html

Williams, M., R. Coles, and J. Primavera, 2006: A lesson from cyclone Larry: An untold story of the success of good coastal planning. *Estuarine, Coastal and Shelf Science*, no. 71, pp. 364–367.

WMO, 2005: Climate Information for Adaptation and Development Needs. WMO-No.1025. World Climate Programme and the World Climate Research Programme, WMO, Geneva.

WMO, 2007: Madrid Conference Statement and Action Plan. WMO, Geneva.

Wong, M. C., 2006: Topic 5: Disaster Mitigation, Warning Systems and Societal Impact. *Sixth International Workshop on Tropical Cyclones*, WMO, Geneva.